Conference Proceedings of the Society for Experimental Mechanics Series

Series Editor

Kristin B. Zimmerman, Ph.D.
Society for Experimental Mechanics, Inc.,
Bethel, CT, USA

The Conference Proceedings of the Society for Experimental Mechanics Series presents early findings and case studies from a wide range of fundamental and applied work across the broad range of fields that comprise Experimental Mechanics. Series volumes follow the principle tracks or focus topics featured in each of the Society's two annual conferences: IMAC, A Conference and Exposition on Structural Dynamics, and the Society's Annual Conference & Exposition and will address critical areas of interest to researchers and design engineers working in all areas of Structural Dynamics, Solid Mechanics and Materials Research

More information about this series at http://www.springer.com/series/8922

Brandon Dilworth • Michael Mains
Editors

Topics in Modal Analysis & Testing, Volume 8

Proceedings of the 38th IMAC, A Conference and Exposition on Structural Dynamics 2020

Springer

Editors
Brandon Dilworth
Massachusetts Institute of Technology
Lincoln Laboratory
Lexington, MA, USA

Michael Mains
University of Cincinnati
Cincinnati, Ohio, USA

ISSN 2191-5644 ISSN 2191-5652 (electronic)
Conference Proceedings of the Society for Experimental Mechanics Series
ISBN 978-3-030-47716-5 ISBN 978-3-030-47717-2 (eBook)
https://doi.org/10.1007/978-3-030-47717-2

This Springer imprint is published by the registered company Springer Nature Switzerland AG.
The registered company address is: Gewerbestrasse 11, 6330 Cham, Switzerland

Preface

Topics in Modal Analysis & Testing represents one of eight volumes of technical papers presented at the 38th IMAC, A Conference and Exposition on Structural Dynamics, organized by the Society for Experimental Mechanics, and held in Houston, Texas, February 10–13, 2020. The full proceedings also include volumes on *Nonlinear Structures and Systems*; *Dynamics of Civil Structures*; *Model Validation and Uncertainty Quantification*; *Dynamic Substructures*; *Special Topics in Structural Dynamics & Experimental Techniques*; *Rotating Machinery, Optical Methods & Scanning LDV Methods*; and *Sensors and Instrumentation, Aircraft/Aerospace, Energy Harvesting & Dynamic Environments Testing*.

Each collection presents early findings from experimental and computational investigations on an important area within Structural Dynamics. Topics in Modal Analysis represents papers on enabling technologies for Modal Analysis measurements and applications of Modal Analysis in specific application areas.

The organizers would like to thank the authors, presenters, session organizers, and session chairs for their participation in this track.

Lexington, MA, USA

Brandon Dilworth

Contents

Chapter 1
Modal Analysis on a Wind Turbine Blade Based on Wind Tunnel Experiments

L. G. Trujillo-Franco, H. F. Abundis-Fong, R. Campos-Amezcua, and R. Gomez-Martinez

Abstract This paper describes the evaluation of a time domain algebraic modal parameters identification methodology. This methodology is applied on a wind turbine blade. The natural frequencies and modal damping factors associated to the blade are estimated from measurements of velocities. A comparison with usual modal identification techniques is performed in order to evaluate and establish the main contributions of the proposed approach. The modal parameter identification algorithms are implemented to run (but not limited to) on a Matlab platform running in a PC using measurements obtained from a laser vibrometer and the corresponding data acquisition system. The results show the performance and parametric estimation of the proposed algebraic identification approach.

Keywords Experimental modal analysis · Operational modal analysis · Wind tunnel experiments

1.1 Introduction

The field of structural dynamics has a set of applications that is in a constant expansion due to the advances in mechanical design procedures and testing protocols and structural health monitoring schemes applied to engineering structures involved in the considerably diverse universe of mechatronic systems, for example the specific case of wind turbines that involve supporting structures and blades that are subjected to harmonic excitation product of their natural interaction with the air in normal or nominal operating conditions. In this context, modal analysis and modal testing are powerful technological tools with a solid theoretical and experimental background [1–4] widely applied to the validation of the mathematical models of the dynamic response of the mechanical systems. In both of the two main presentations of modal analysis: experimental modal analysis (EMA) and operational modal analysis (OMA) the mechanical design engineer have a reliable source of information about the dynamic nature of the system or constituting part of it, like the case of the blades, a vital part of a wind turbine for an energy generation system.

In this work, we present experimental results of a modal analysis procedure performed on a wind turbine blade designed for small power applications. The test was performed in the two common formats: experimental modal analysis at laboratory conditions and operational modal analysis in real life-like operational conditions, the latter were simulated by wind tunnel experiments with turbulence generated by using a fixed pattern grid. We use velocity measurements in the analysis to determine the natural frequencies and modal damping factors.

L. G. Trujillo-Franco (✉)
Licenciatura en ingeniería mecánica automotriz, Universidad Politécnica de Pachuca, Zempoala, Hidalgo, México
e-mail: luis.trujillo@upp.edu.mx

H. F. Abundis-Fong
División de Estudios de Posgrado e Investigación, Tecnológico Nacional de México/I.T. Pachuca, Pachuca de Soto, Hidalgo, Mexico
e-mail: hugo.af@pachuca.tecnm.mx

R. Campos-Amezcua
Tecnológico Nacional de México/Centro Nacional de Investigación y Desarrollo Tecnológico, Interior Internado Palmira S/N, Col. Palmira, Cuernavaca 62490, Mexico
e-mail: rafael.ca@cenidet.tecnm.mx

R. Gomez-Martinez
Instituto de Ingeniería, Universidad Nacional Autónoma de México, Mexico City, CDMX, Mexico
e-mail: RGomezM@iingen.unam.mx

B. Dilworth (ed.), *Topics in Modal Analysis & Testing, Volume 8*, Conference Proceedings of the Society for Experimental Mechanics Series, https://doi.org/10.1007/978-3-030-47717-2_1

1.2 Modal Analysis and Operational Calculus

It is well known that the mathematical model for most of the engineering structures considered as flexible mechanical systems of n degrees of freedom (DOF) under free vibration condition is given by the so called and continuously cited in the literature ordinary differential equation, in matrix form:

$$\mathbf{M}\ddot{\mathbf{x}} + \mathbf{C}\dot{\mathbf{x}} + \mathbf{K}\mathbf{x} = \mathbf{0}, \qquad \mathbf{x} \in \mathbf{R}^n \tag{1.1}$$

here, $\mathbf{x}$ is the vector of displacements, and $\mathbf{M}$, $\mathbf{C}$ and $\mathbf{K}$ are symmetric inertia, damping and stiffness $n \times n$ matrices, respectively. Those matrices do not obey or follow a known or fully established pattern, therefore it is common to make assumptions in order to guarantee the stability and physical coherence of the model. In general one can assume that (1.1) represents a real system when $\mathbf{K}$ is positive definite and $\mathbf{C} \equiv \mathbf{0}$, and asymptotically stable when $\mathbf{C}$ is positive definite (see, e.g., Inman [12]). On the other hand it is also well known that the coupled differential Eq. (1.1) can be transformed to modal coordinates q_i, $i = 1, 2, \cdots, n$, as follows:

$$\ddot{q}_i + 2\zeta_i\omega_i\dot{q}_i + \omega_i^2 q_i = 0 \tag{1.2}$$

$$\mathbf{x}(t) = \boldsymbol{\Psi}\mathbf{q}(t) \tag{1.3}$$

where ω_i and ζ_i are the natural frequencies and damping ratios associated to the i-th vibration mode, respectively, and $\boldsymbol{\Psi}$ is the so-called $n \times n$ modal matrix In notation of Mikusiński's operational calculus [1, 5], the modal analysis representation or modal model (1.2) is described as:

$$\left(s^2 + 2\zeta_i\omega_i s + \omega_i^2\right) q_i(s) = p_{0,i} + p_{1,i}s \tag{1.4}$$

where $p_{o,i}$ are constants depending on the system initial conditions at the time $t_0 \geq 0$. Using Eqs. (1.3) and (1.4), we obtain the expression for the physical displacements x_i, in the form of a summation of independent single degree of freedom responses known as vibration modes:

$$x_i(s) = \sum_{j=1}^{n} \frac{\psi_{ij}\left(p_{0,j} + p_{1,j}s\right)}{s^2 + 2\zeta_j\omega_j s + \omega_j^2} \tag{1.5}$$

It is easy to prove that for each physical displacement$x_i$we have:

$$p_c(s)\, x_i(s) = r_{0,i} + r_{1,i}s + \cdots + r_{2n-2,i}s^{2n-2} + r_{2n-1,i}s^{2n-1} \tag{1.6}$$

where $p_c(s)$ is the characteristic polynomial of the mechanical system and $r_{i,j}$ are constants which can be easily calculated by using the values of the system initial conditions as well as the modal matrix components ψ_{ij}, and has the general polynomial form given by:

$$p_c(s) = s^{2n} + a_{2n-1}s^{2n-1} + \cdots + a_1 s + a_0 \tag{1.7}$$

When using velocity measurements, instead of displacement measurements, one could simply multiply Eq. (1.5) by s and then describe the velocity output as:

$$y_i(s) = \sum_{j=1}^{n} \frac{\psi_{ij}\left(p_{0,j}s + p_{1,j}s^2\right)}{s^2 + 2\zeta_j\omega_j s + \omega_j^2} \tag{1.8}$$

The resulting output $y_i(s) = sx_i(s)$ is a velocity output variable; described in a complex s domain. It is widely known that the roots of the characteristic polynomial (1.7) provide the damping factors and the damped natural frequencies, and hence the natural frequencies and damping ratios of the flexible structure. Here, we use a modal parameters identification approach to estimate the modal parameters of the mechanical system through the estimation of the positive coefficients a_k

of the system's characteristic polynomial using only velocity measurements of some output variable, obtained from a wind tunnel experiment.

1.3 Output Only Modal Parameters Estimation

We perform an algebraic identification approach to estimate the modal parameters of the mechanical system through the real-time estimation of the coefficients a_k of the system's characteristic polynomial as reported in [8, 9] using only velocity measurements of a single point of the blade. The application of the online algebraic identification scheme is performed using cumulative trapezoidal numerical integration with fixed sampling period of 4.16 μs. The specific algebraic identification scheme applied here is described on detail in [7] where is shown that by solving the algebraic Eq. (1.9) also detailed in [7–9] one obtains the parameter vector $\boldsymbol{\theta}$ as:

$$\boldsymbol{\theta} = \mathbf{A}^{-1}\mathbf{B} = \frac{1}{\Delta}\begin{bmatrix} \Delta_1 \\ \Delta_2 \\ \vdots \\ \Delta_{n-1} \\ \Delta_n \end{bmatrix} \tag{1.9}$$

Then, the algebraic identifiers to estimate the coefficients a_k of the characteristic polynomial, avoiding singularities when the determinant $\Delta = \det(\mathbf{A}(t))$ crosses by zero, are obtained with:

$$\hat{a}_k = \frac{\int |\Delta_{k-1}|}{\int |\Delta|}, \qquad k = 1,\ 2, \cdots, 2n-1 \tag{1.10}$$

Thus, one could implement the algebraic identifiers (1.10) using only any available acceleration, velocity or position measurements of any specific floor or degree of freedom. From the estimated coefficients$\hat{a}_k$, one can obtain the roots of the characteristic polynomial. Hence, the estimates of the natural frequencies $\hat{\omega}_{ni}$ and damping ratios $\hat{\xi}_i$ are given by:

$$\hat{\omega}_{ni} = \sqrt{\hat{\sigma}_i^2 + \hat{\omega}_{di}^2}, \qquad \hat{\xi}_i = -\frac{\hat{\sigma}_i}{\sqrt{\hat{\sigma}_i^2 + \hat{\omega}_{di}^2}} \tag{1.11}$$

where $\hat{\sigma}_i$ and $\hat{\omega}_{di}$ are respectively estimates of the damping factors and damped natural frequencies of the mechanical system. The proposed algebraic identification scheme is shown in Fig. 1.1, where the block diagram shows the general structure of the proposed approach. Notice that the data acquisition system samples the velocity at only one specific point or test location of the wind turbine blade (in the horizontal axis direction at a precise fixed sample rate of 2.4Khz), and then, those samples are sent to a standard PC running under *Windows 10*® to finally perform the time domain identification scheme.

1.4 Wind Turbine Blade

In the present work, we report the results of a vibrations analysis of a wind turbine blade made of a composite material (glass fiber). First, we performed a traditional modal testing based on impact hammer response analysis assuming free clamped boundary conditions as shown in Fig. 1.2a. The impact hammer testing was performed by acquiring velocity measurements using a Polytec® portable laser vibrometer model PVD-100 at 2.4ksps at one fixed reference point and 9 different locations of excitation according to Maxwell's reciprocity principle [2, 3].

The complete set of measurements, corresponding to 9 different points of excitation is shown in Fig. 1.3, where Fig. 1.3a shows the free decay response, in the time domain, whereas Fig. 1.3b shows the resulting frequency response functions (FRF) of each point. Finally, the estimation, by applying the classic peak picking technique, of the first 7 natural frequencies and modal damping factors are reported in Table 1.1.

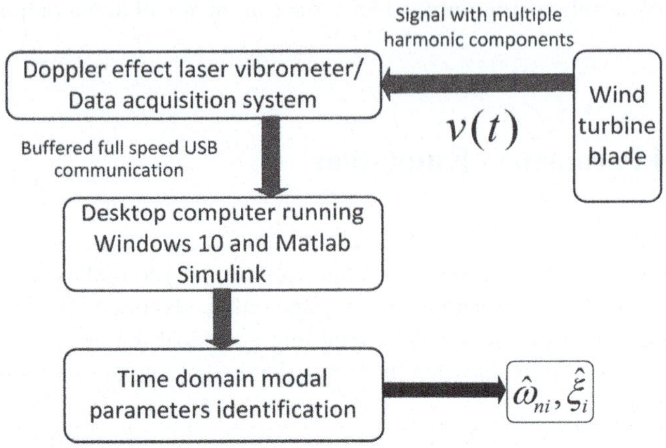

Fig. 1.1 Flowchart of the proposed modal parameters estimation scheme

Fig. 1.2 Wind tunnel set up for modal testing of the wind turbine blade: (**a**) Free clamped boundary conditions and (**b**) Power-distance mount and focus stabilization of the portable laser vibrometer

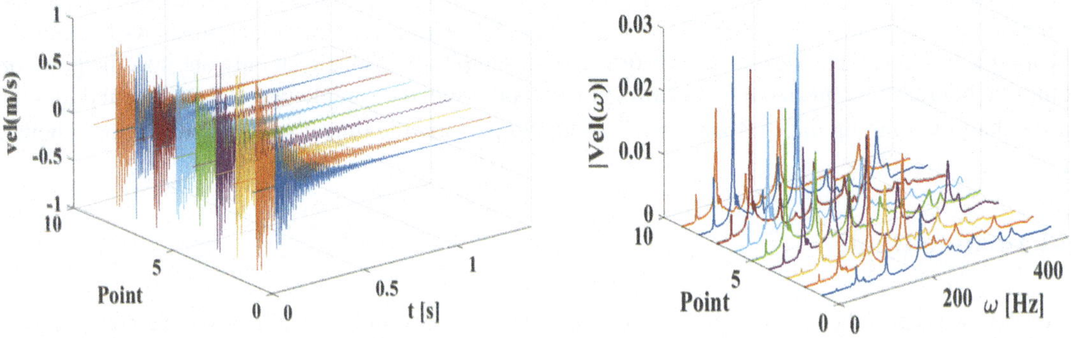

Fig. 1.3 Impact response of the blade; (**a**) Free vibration decay in the time domain (**b**) FRF at different excitation points

The detailed FRF corresponding to the point number 7 is shown in Fig. 1.4 where the first six resonances are marked in red.

By a brief examination of the experimental FRF, we can assume the dynamic behavior of the wind turbine blade in the specific bandwidth of 300 Hz is dominantly linear and lightly damped considering the velocity range of 0.6 m/s. Even though

Table 1.1 Impact hammer test modal parameters identification

Mode	Natural frequency [Hz]	Modal damping %
1	9.15	0.05
2	32.6	0.19
3	75.8	0.09
4	141.0	0.26
5	175.8	0.06
6	214.24	0.05
7	275.03	0.30

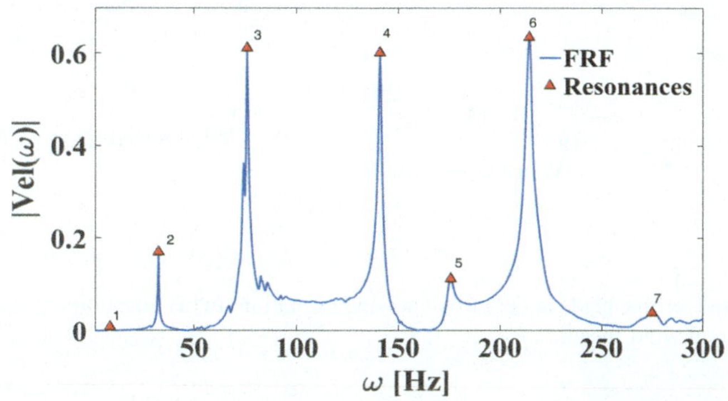

Fig. 1.4 Experimental FRF of point 7 corresponding to excitation point 2

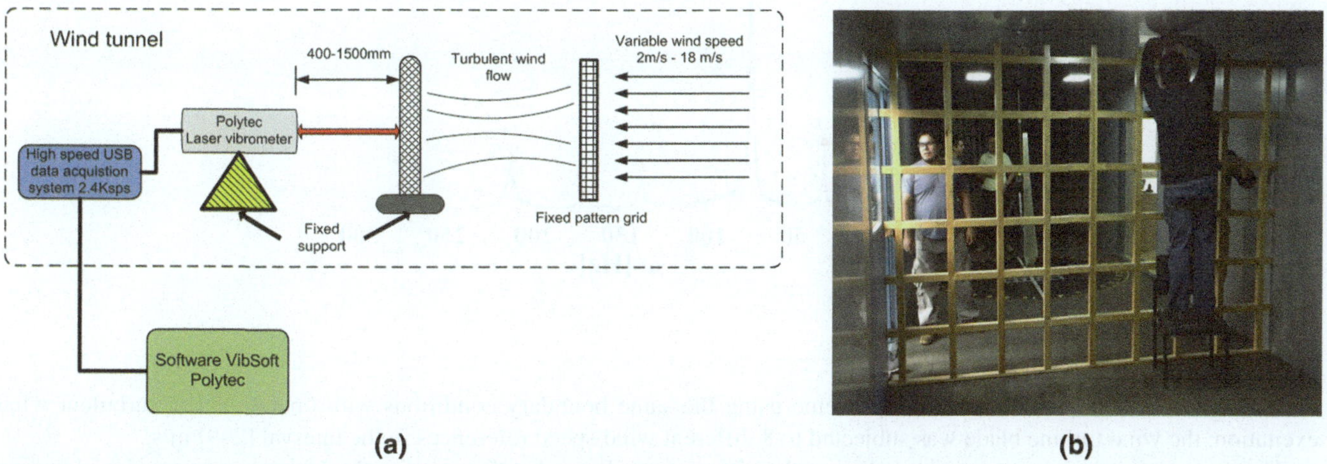

Fig. 1.5 Wind turbine experiments set up; (**a**) flowchart and (**b**) fixed pattern (passive) turbulence generator grid

the construction material of the blade is a composite material (glass fiber) the performed modal testing does not evidence the presence of non-linearity or distortion in the experimental FRF.

1.5 Wind Tunnel Experiment

It is well known that in normal operation conditions, the blades, as the main eolic energy conversion elements of the wind turbines (including small power systems), are subjected to harmonic excitations. In the ideal case of laminar flow (not common in real environmental conditions) the main stresses are static; nevertheless, the fatigue issues are considerably common in real life applications. In order to perform a realistic operational conditions test, we introduce a fixed pattern (square panel) grid in the wind tunnel (Fig. 1.5), with the purpose of generating turbulence in the laminar-like flux wind produced by the turbine of the wind tunnel that produces the controlled and variable speed wind excitation as shown in Fig. 1.5.

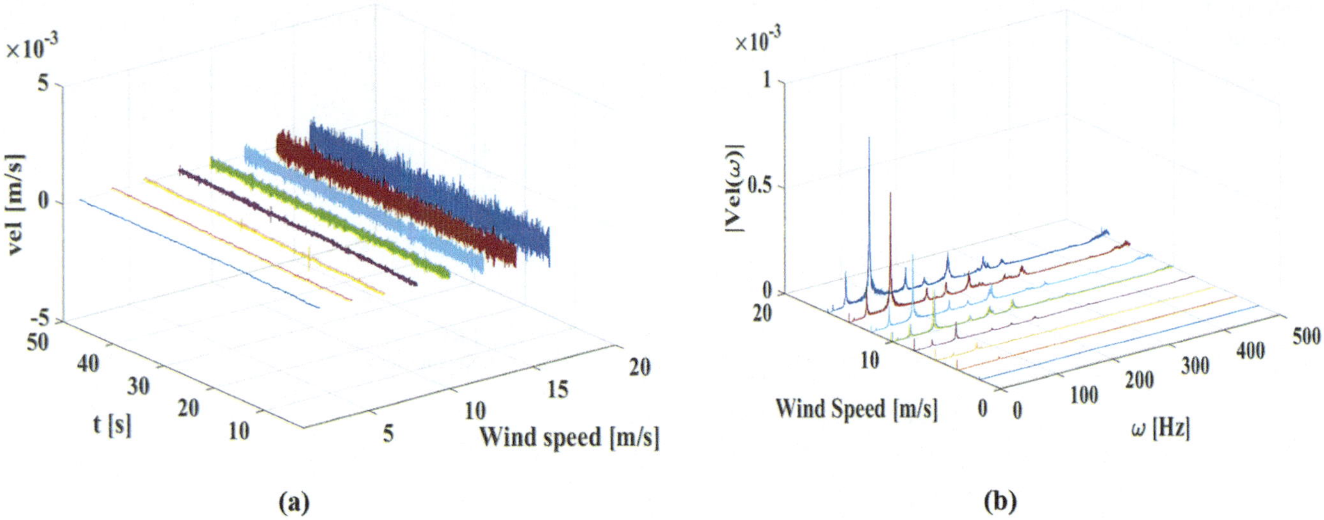

Fig. 1.6 Dynamic response of the wind turbine blade under turbulent wind excitation. In (**a**) time domain velocity signals at different wind speeds, and (**b**) FRF a corresponding to different wind speeds

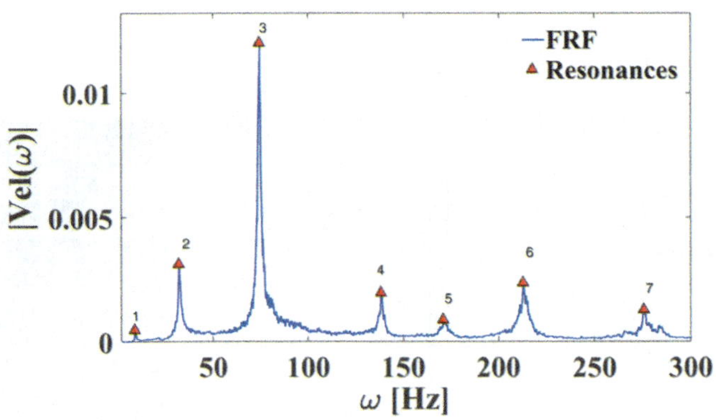

Fig. 1.7 FRF at wind speed 18 m/s

In order to evaluate the identification scheme using the same boundary conditions with operational or turbulent wind excitation, the wind turbine blade was subjected to 8 different wind speed references in the interval [2–9] m/s.

The time domain responses corresponding to the different wind speeds referenced to the 15th point or response location are those shown in Fig. 1.6a. Naturally, the amplitude of the velocity signal increases according to the wind speed, however, the frequency content of the signal is (in terms of harmonic content) the same and it shows the inherent dynamic behavior of the blade [1] as it is depicted in Fig. 1.6b. The FRF corresponding to the wind speed of 18 m/s is reported with detail in Fig. 1.7, where the resonances are marked with red and only the first seven bending modes are analyzed.

1.6 Application of the Time Domain Identification Scheme

The proposed time domain modal parameters estimation scheme has been reported and detailed in several previous works [6, 7–12] where the online methodology of this approach has been evaluated. Here, we use the same methodology in its off-line configuration. The same algebraic problem expressed on (1.9) is solved in a post-processing context. Roughly speaking, we take a buffer of 2400 samples as shown in the flowchart of Fig. 1.1, and then, we apply the algebraic identification scheme (1.11) to the time domain array of experimental values. The results of the application of the time domain identification scheme are reported in Table 1.2. where a comparison with those obtained with the impact hammer testing is shown.

Table 1.2 Wind tunnel test modal parameters identification

Mode	Natural frequency [Hz]		Modal damping %	
	Peak picking	Time domain algebraic	Peak picking	Time domain algebraic
1	9.04	8.82	0.027	0.032
2	32.02	32.0	0.17	0.23
3	74.45	75.2	0.08	0.07
4	138.21	137.8	0.24	0.26
5	170.82	169.6	0.06	0.08
6	212.66	211.56	0.03	0.04
7	275.56	273.2	0.3	0.37

1.7 Conclusion

An algebraic and time domain identification approach for the estimation of the natural frequencies and damping ratios of a lumped parameters vibrating mechanical system is presented, specifically, a wind turbine blade, subjected to turbulent wind excitation in a wind tunnel environment. The values of the coefficients of the characteristic polynomial of the mechanical system are firstly estimated from a data buffer, and then the modal parameters are obtained. In the design process, we have considered that measurements of only one velocity output, or measurement point is available for the implementation. It is also important to consider that one could easily extend the results for situations where acceleration or position measurements are available. The algebraic modal parameter identification was tested for a lumped parameters N-DOF mechanical system excited by a turbulent wind in wind tunnel conditions. In general, the experimental results show a satisfactory performance of the proposed identification.

Acknowledgements We appreciate the support of the wind tunnel facility's technical staff of Instituto de Ingeniería of Universidad Nacional Autónoma de México. We express our gratitude to R. Sanchez, I.M. Arenas, O. Rosales and M.A. Mendoza for their special attention and disposition to collaborate in this project.

References

1. Brincker, R., Ventura, C.E.: Introduction to operational modal analysis. Wiley (2015)
2. Heylen, W., Lammens, S., Sas, P.: Modal analysis, theory and testing. Katholieke Universiteit Leuven, Belgium (2003)
3. Soderstrom, T., Stoica, P.: System identification. Prentice-Hall, New York (1989)
4. Chopra, A.K.: Dynamics of structures theory and applications to earthquake engineering. Prentice-Hall (1995)
5. Mikusiński, J.: Operational calculus, vol. 1, 2nd edn. PWN & Pergamon, Warsaw (1983)
6. Fliess, M., Sira-Ramírez, H.: An algebraic framework for linear identification. ESAIM: Control, Optimization and Calculus of Variations. **9**, 151–168 (1996)
7. Beltran-Carbajal, F., Silva-Navarro, G., Trujillo-Franco, L.G.: Evaluation of on-line algebraic modal parameter identification methods. In: Proceedings of the 32nd international modal analysis conference (IMAC XXXII), vol. 8, pp. 145–152 (2014)
8. Beltran-Carbajal, F., Silva-Navarro, G., Chávez-Conde, E.: Design of active vibration absorbers using on line estimation of parameters and signals. In: Beltrán-Carbajal, F. (ed.) Vibration analysis and control. New trends and developments. InTech, Croatia (2011)
9. Beltran-Carbajal, F., Silva-Navarro, G.: Algebraic parameter identification of multi-degree-of-freedom vibrating mechanical systems. In: Proceedings of the 20th international congress on sound and vibration (ICSV20), Bangkok, Thailand, pp. 1–8 (2013)
10. Beltran-Carbajal, F., Silva-Navarro, G.: Adaptive-like vibration control in mechanical systems with unknown parameters and signals. Asian J. Control. (2013). https://doi.org/10.1002/asjc.727, to appear, 2013
11. Beltran-Carbajal, F., Silva-Navarro, G., Arias-Montiel, M.: Active unbalance control of rotor systems using on-line algebraic identification methods. Asian J. Control. (2013). https://doi.org/10.1002/asjc.744, to appear, 2013
12. Inman, D.J.: Vibration with control. Wiley, New York (2006)

Chapter 2
An Investigation of Vibrational Characteristics of Lap Joints Using Experimental and Analytical Methods

Thomas Roberts and Phillip J. Cornwell

Abstract Many structures are assembled with components that are joined together with connections such as lap joints, making it important to understand how to effectively model and identify damage, such as loosening bolts, in these connections. The structures examined in this work were two bars joined by simple lap joints – a solid structure, a welded structure, and a bolted structure. Experimental modal analysis and finite element models were used to determine the natural frequencies, damping ratios, and mode shapes for each of the different structure configurations. The first goal of this work was to determine if changes in the natural frequencies and damping ratios were large enough to distinguish between different types of structures and between experimental and analytical models. Although differences were present, results showed that natural frequencies and damping ratios are not extremely reliable metrics for determining the differences in these structures. Damage in the bolted structure was investigated by loosening or removing a bolt. Bonded contact regions were implemented within the bolted structure FE models to simulate the effects of a loosened connection while maintaining linearity for modal analysis. The end-goal in this aspect of the research was to ascertain whether a fractional strain energy method via mode shape curvature could be used to determine the location and intensity of damage in a structure. For convenience, a MATLAB GUI was developed to implement this technique. The strain energy method was unsuccessful in identifying differences between structures or damage within the bolted structure, for the differences in mode shape curvatures was not significant enough. Results from the finite element model, however, exhibited significant enough differences to distinguish the bolted structure from the solid and welded structures as well as to detect several different simulated forms of damage.

Keywords Bolted joints · Welded joints · Structural health monitoring · Strain energy · Modal analysis

2.1 Introduction

This work involved experimental modal analysis, finite element models, and the structural health monitoring technique called the fractional strain energy method. The structures examined in this study have two different types of lap joints – welded and bolted. These types of joints are very common in real structures, and it is important to understand how the joints change the dynamic characteristics of the structure. McCarthy et al. emphasized the importance of analyzing mechanical joints, as they are likely to be the weakest points within a structure [1]. Mechanical joints introduce factors such as bolt bending, bolt pre-loading, and non-linear stress/strain relationships, which are very hard to model analytically.

Previous research on the vibrational characteristics of lap joints, particularly bolted ones, points out that it can be difficult to distinguish different levels of bolt pretension based on natural frequencies alone [2, 3]. For that reason, the experimental aspects of this work are focused mostly on detecting more severe cases of damage, such as a missing bolt. Sun et al. found that there is some correlation between a tighter bolted lap joint and a lower resulting damping ratio [2]. Other comparisons between different types of lap joint structures, such as welded and bolted, are verified in this work. Zaman et al. studied the differences between a welded lap joint structure and bolted lap joint structure. The results from that work show that even when mass properties are similar, welded lap joints tend to be stiffer than bolted lap joints [4]. Another common method for analyzing the vibrational characteristics of lap joints is utilizing guided waves. This work aims to take a simpler approach to the problem, but Du et al. and Kedra et al. describe the guided wave analysis process in detail [5, 6].

T. Roberts (✉) · P. J. Cornwell
Department of Mechanical Engineering, Rose-Hulman Institute of Technology, Terre Haute, IN, USA
e-mail: Thomas.Roberts@Utah.edu; robertt1@rose-hulman.edu; cornwell@rose-hulman.edu

B. Dilworth (ed.), *Topics in Modal Analysis & Testing, Volume 8*, Conference Proceedings of the Society for Experimental Mechanics Series, https://doi.org/10.1007/978-3-030-47717-2_2

Previous research has shown multiple different ways to represent structures with joints in a finite element model. Many studies have attempted to analyze these factors using a two-dimensional model [1], as the computation cost of a three-dimensional model is much greater. If a three-dimensional model is used, Kim et al. offers multiple suggestions for accurately modeling a bolted joint. Pretension can be modeled rather simply as a thermal deformation or an external strain in the bolt. The most simplistic representation for the fastener is a solid model as opposed to an assembly of bolts, washers, and nuts [7]. Using a solid representation of the fasteners allows for more simple external geometries and thus a smoother, more accurate mesh using brick elements [7]. More complex fastener models, such as assemblies of nuts, bolts, and washers with contact conditions, can be used to produce a model that more closely resembles the real world. The main disadvantages of more complex models are the uncertainty in contact parameters and the increased simulation time. However, the increased complexities of this type of model may allow for a more accurate representation of energy dissipation within the bolted joint. Unfortunately, modeling techniques that use pre-stress of any kind or contact parameters that are anything but bonded cannot be used in modal analysis. Modal analysis studies require that all properties of the model be linear. If more complex modeling techniques are desired, then time-dependent solvers can be used to determine the dynamic response from the FE model. If the computational capability is available, some researchers prefer to use nonlinear FE solvers to analyze problems like this. For example, Sun et al. used penalty stiffness contact in the entirety of a bolted lap joint model and solved for the frequency response with a nonlinear solver [2]. Kim et al. proposed a method for simulating bolt pretension that requires only boned contact regions in the model. A cylindrical-shaped region around the bolt hole was used for applying bonded contact constraints between the two plates in the joint [7]. Additionally, the solid bolt model was bonded to those same contact regions.

Many methods have been proposed to model welded joints using finite elements. The most practical method of those researched breaks up the model into three zones: base metal, weld metal, and heat-affected metal. Each of these zones is a solid model, and the zones are assembled together and connected with contact parameters within the finite element program [8]. This method allows for the inside of the joint (where the weld has not penetrated) to be modeled without contact, similar to the actual specimen. Additionally, pre-stresses due to the intense heat of the welding process can be applied to specific regions of the model; since the heat-affected zone is a separate body, it can be isolated in order to apply the heat pre-stress. Lastly, the amount of filler material left as a product of the welding process can be controlled very easily using this method of assembling the model.

Another objective of this work was to utilize structural health monitoring techniques to compare structures with and without damage. The chosen method was a fractional strain energy method, proposed by Stubbs for beam-like structures [9] and extended by Cornwell [10] for plate-like structures. The fractional strain energy method uses the curvature of mode shapes to calculate a damage index. Experimental mode shapes and FE mode shapes were used in the method for detecting differences between the solid, welded and bolted structures. Additionally, the fractional strain energy method was used to detect differences between an undamaged bolted structure and a damaged one with a missing bolt. This comparison was again made with both experimental and analytical mode shape information.

2.2 Experimental Methods

Roving hammer tests were performed on three structures – one solid, one welded, and one bolted, as shown in Fig. 2.1. Each structure was constructed from 6061 Aluminum bar stock. The solid structure was machined from a solid piece of ½" × 2″ stock. The bolted and welded structures were constructed from two pieces of ¼" × 2″ stock. Welded joints contain a filler material that was assumed to be 4043 Aluminum, and the bolts in the bolted structure were medium carbon alloy steel.

The first method used for characterizing the modal properties of the lap joint structures was a roving hammer modal test. Each structure was tested at 45 different points in order to create a grid with nodes spaced one-inch by one-inch apart. Mode shapes, natural frequencies, and damping ratios were computed for each of the three lap joint structures. In addition, the bolted structure was tested with one of the four bolts removed from the assembly. In the context of structural health monitoring (SHM), a structure with a missing bolt would be considered damaged when being compared to a structure with all four bolts adequately tightened to 25 inch-lbs. The bolted structure was only tested with all bolts fully tightened or one bolt completely removed.

The experimental modal tests were mapped to a planar geometry with one-inch by one-inch node spacing. The geometry used for the modal test had 45 nodes, and 72 tracelines. For the fractional strain energy method, a total of 28 four-node shells and three separate fifteen-node beams were defined. Figure 2.2 shows the geometry used for all experimental modal tests. The planar geometry of the structures was set to the x- and y-planes and the response degree-of-freedom was in the positive

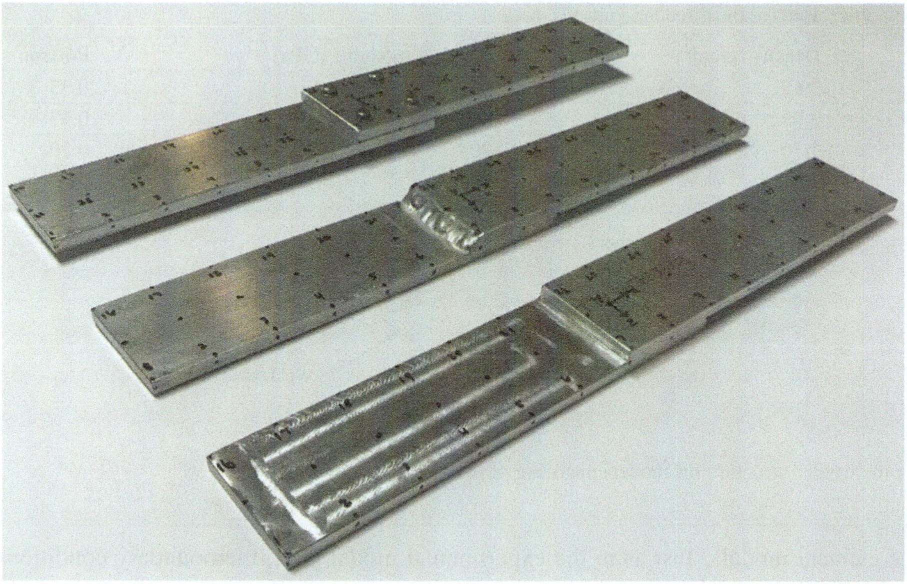

Fig. 2.1 Solid, welded, and bolted lap joint structures (front to rear, respectively) used for experimental analysis

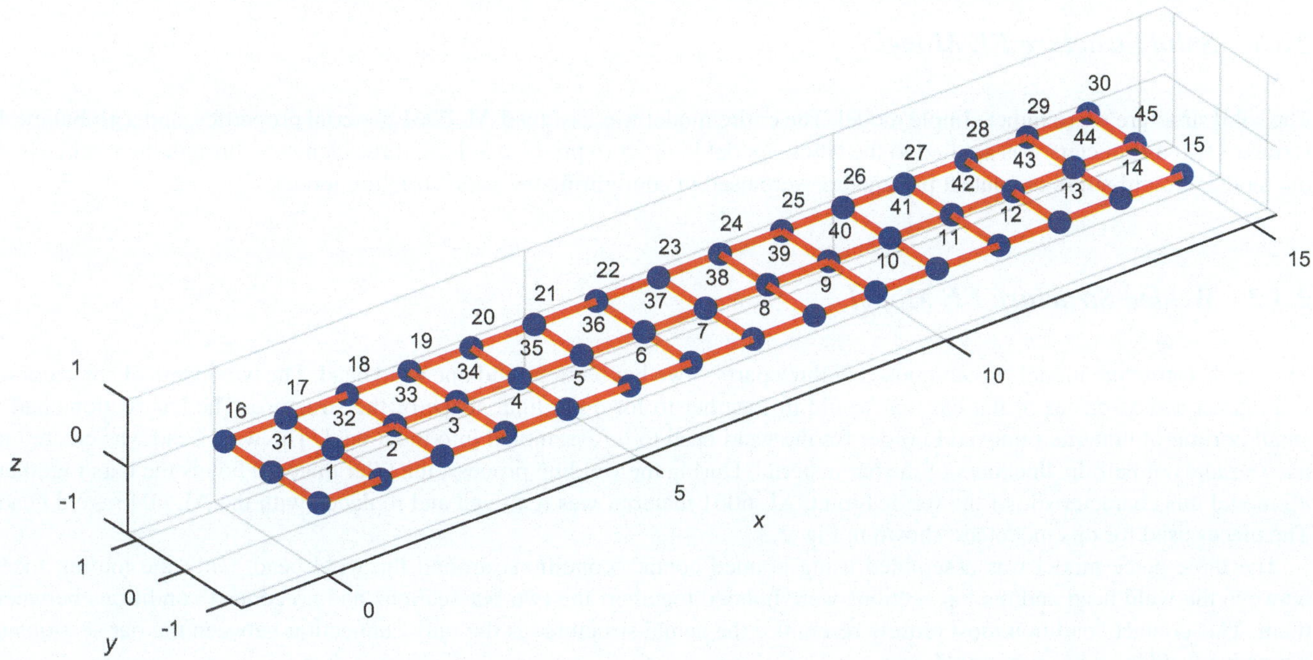

Fig. 2.2 Geometry used for mapping experimental modal data to the mode shapes of the structure

z-axis for detecting in-plane (I.P) modes. In order to observe out-of-plane modes (O.P.), the response degree of freedom was changed to the negative y-axis. Each structure was tested three times in a randomized order.

2.3 Analytical Methods

Finite element models of each of the three lap joint structures were used for modal analysis studies, and the results from the modal studies were compared to the results from the experimental modal tests. Each model was configured with the same meshing parameters and appropriate material properties. Table 2.1 summarizes the material properties used to configure the

Table 2.1 Material properties used for the three lap joint FE models

Material	Density [g/cm^3]	Young's modulus [GPa]	Poisson's ratio
AL 6061	2.65	66.0	0.33
AL 4043	2.68	75.0	0.33
Steel alloy	7.70	205	0.29

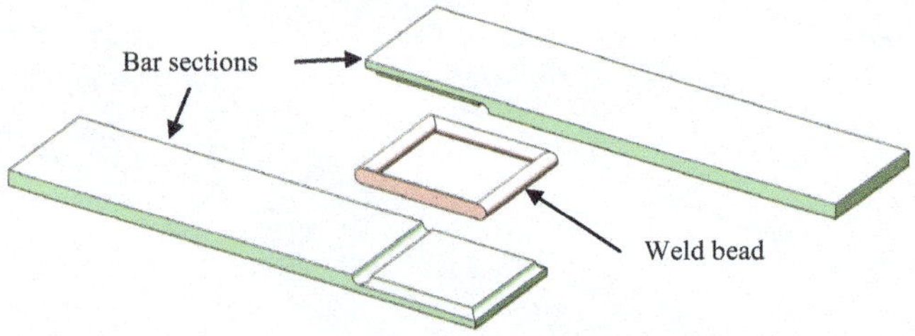

Fig. 2.3 Geometry for the welded structure FE model consisting of two bar sections and a weld bead

three different finite element models. Just as in the experimental modal tests, the boundary conditions for each FE model were assumed to be free-free.

2.3.1 *Solid Structure FE Model*

The solid structure was a rather simple model. The entire model was assigned AL 6061 material properties, and a global mesh sizing of two millimeters was applied to the whole model in order to produce at least three elements through the thickness of the model. No pre-stress or contact parameters were used in configuring the solid structure model.

2.3.2 *Welded Structure FE Model*

The welded structure model was composed of three parts – two bar sections and one weld bead. The two identical bar sections were generated as pieces of flat bar that would fit together to form the final shape of the structure. The bar sections had a small portion of material removed in order for the weld bead to be assembled into the model. The weld bead was created to have a radius of half the thickness of the bar material. During the welding process, the filler material bonds the bars together. To model this characteristic of the welded joint, AL 6061 material was removed and replaced with the AL 4043 weld bead. The pieces used for this model are shown in Fig. 2.3.

The three-piece model was assembled using bonded contact conditions around the weld bead. Only the mating faces between the weld bead and the bar sections were bonded together; the two bar sections had no contact conditions between them. This contact condition most closely resembles the actual structure, as the only connection between the bar sections is where the weld bead has penetrated.

2.3.3 *Bolted Structure FE Model*

The bolted structure model was constructed of two bar sections with matching holes and a bolt for each set of holes (four bolts in all). The bar sections and bolts were assigned AL 6061 and Steel Alloy material properties, respectively. The geometry used for the bolted structure model is shown in Fig. 2.4.

The bolts were modeled as an assembly of a bolt, nut, and two washers. Creating a solid geometry of the bolt assembly allowed for a much simpler model as the contact between components and the interactions of bolt/nut threads were not a concern for this analysis. Since the FE models were to be used in a modal study, all contact parameters were bonded, and no pretension was included in the model.

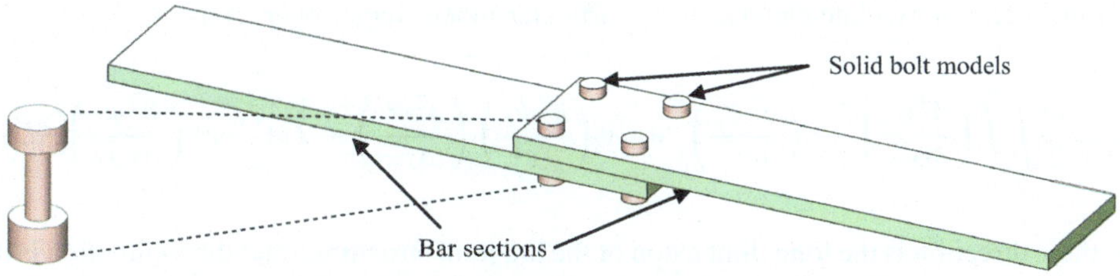

Fig. 2.4 Geometry for the bolted structure FE model consisting of two bar sections and four solid bolt models

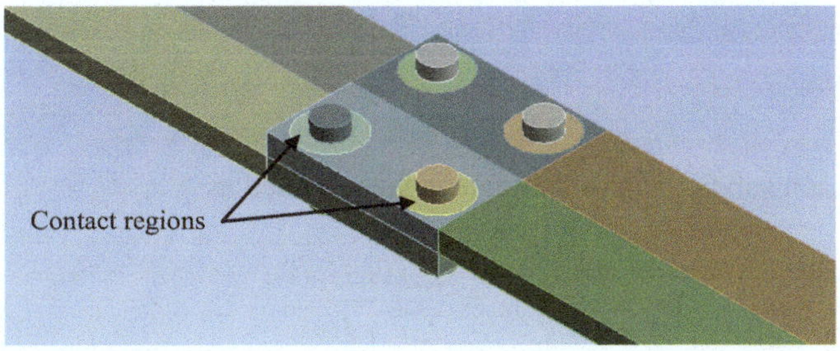

Fig. 2.5 Contact regions used to simulate pretension in the bolted structure FE model

In order to allow for the approximation of a loosened connection in the modal study, the bars were bonded to the bolts using a circular region around each bolt hole, as shown in Fig. 2.5. A similar method was successfully employed by Liao et al. [11]. These circular regions had shared topology with their respective bar sections, and contact was assigned between the bars and between the bars and the bolt head/nuts. Different levels of pretension were approximated by changing the size of the circular contact regions around the bolt heads.

Multiple trials of the bolted structure FE model were solved. First, the sizing of the contact region around each bolt was adjusted to model the effects of a loosened connection in the model. In general, a smaller contact region around each bolt resulted in a model that was less stiff, thus simulating the effects of loosening a bolt in the model while mass properties remain constant. Next, one bolt was completely removed from the model along with its contact parameters. In this type of model, the mass properties and the component contact are different from the undamaged structure. SHM methods were also used to compare the differences between complete (healthy) structures and structures with missing components or differing size contact regions (damaged).

2.3.4 *Fractional Strain Energy Method*

Changes in natural frequencies and damping ratios do not often provide sufficient information to determine the presence of damage in a structure. Even if natural frequencies and damping ratios are sensitive to damage, they cannot be used to locate the damage. In this work, a fractional strain energy method was implemented to compare experimental data and analytical results.

The fractional strain energy method (SEM) was implemented for generic planar structures, such as a beam or plate, that have only a single degree-of-freedom of modal response. The fractional strain energy for a beam element for a particular mode shape, $\psi_i(x)$, is

$$U_i = \frac{1}{2}\int_0^l EI\left(\frac{\partial^2 \psi_i}{\partial x^2}\right)^2 dx. \tag{2.1}$$

and the fractional strain energy for a plate element for a particular mode shape, $\psi_i(x, y)$, is

$$U_i = \frac{D}{2}\int_0^b\int_0^a \left(\frac{\partial^2\psi_i}{\partial x^2}\right)^2 + \left(\frac{\partial^2\psi_i}{\partial y^2}\right)^2 + 2\nu\left(\frac{\partial^2\psi_i}{\partial x^2}\right)\left(\frac{\partial^2\psi_i}{\partial y^2}\right) + 2(1-\nu)\left(\frac{\partial^2\psi_i}{\partial x\partial y}\right)^2 dxdy. \tag{2.2}$$

For this analysis, the x-direction is the long dimension of the lap joint structures, and the y-direction is the short dimension. Refer to Fig. 2.2 for a visualization. The plates or beams are then subdivided and the strain energy for each sub-region within the model, U_{ijk}, can be calculated by integrating over the area of each sub-region. The fractional strain energy at location jk is defined to be

$$f_{ijk} = \frac{U_{ijk}}{U_i} \tag{2.3}$$

The damage index, β_{jk}, for each subdivision of the plate is then defined as

$$\beta_{jk} = \frac{\sum_{i-1}^{m} f_{ijk}^{*}}{\sum_{i-1}^{m} f_{ijk}}, \tag{2.4}$$

where f_{ijk}^{*} is the fractional strain energy for the damaged structure, and m is the number of measured modes. The fractional strain energy damage index is lastly normalized by its mean, $\overline{\beta}_k$, and standard deviation, σ_k, to create the final metric used to compare healthy and damaged structures:

$$Z_k = \frac{\beta_k - \overline{\beta}_k}{\sigma_k}. \tag{2.5}$$

Further detail about the development of the fractional strain energy method can be found in [10].

2.4 User Interface Development

The algorithm for the fractional strain energy method described in the previous section was originally part of the MATLAB program *DIAMOMD* [6, 12]. *DIAMOND* (Damage Identification and Modal Analysis of Data) is a program created at Los Alamos National Laboratory under the auspices of the US Department of Energy. The strain energy algorithms for both beam and shell element computations were extracted from *DIAMOND* and modified to run independently using simpler forms of data input. The splash page of the new strain energy user interface (UI) is shown in Fig. 2.6.

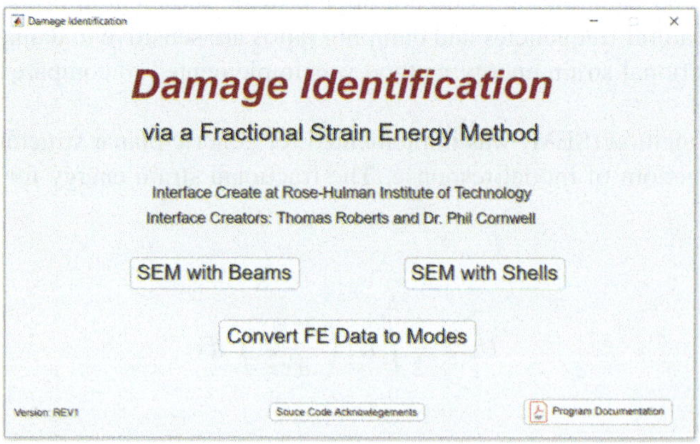

Fig. 2.6 Main window of the strain energy method user interface. From here, the user can access the UI's three main functions

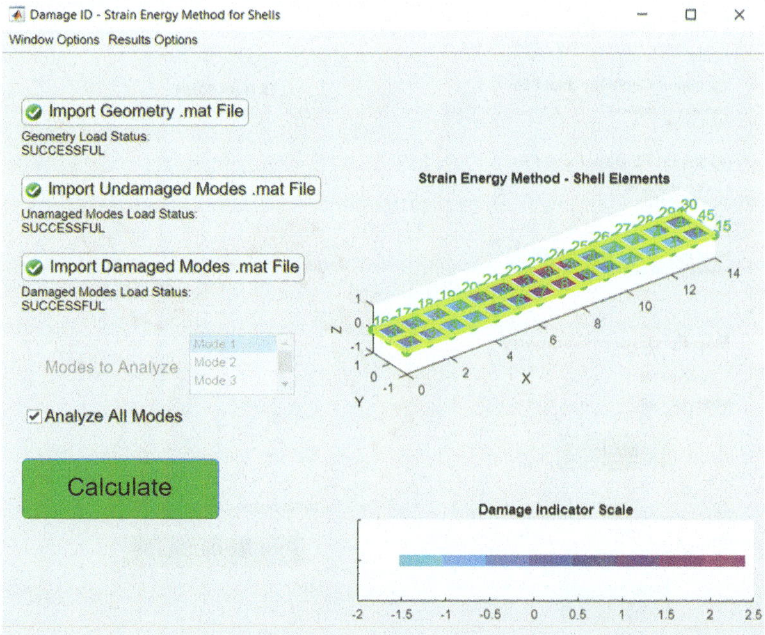

Fig. 2.7 SEM with Shells window. Here the user inputs geometry, undamaged and damaged modes, and selects which modes to use in the computation

The window shown in Fig. 2.6 is where the user accesses the UI's three main functions – SEM with Beams, SEM with Shells, and Convert FE Data to Modes. The two strain energy methods work in similar ways and require the same inputs. Once given the correct input, each function executes its respective method and produces damage identification results. Figure 2.7 shows the window for executing the SEM with Shells function (the window for the SEM with Beams functions is visually identical with the exception of the figure's title and resulting plot format).

To utilize the SEM functions within the UI, the user must input a valid geometry file, an undamaged modes file, and a damaged modes file. The UI requires these files to be in MATLAB data format. Geometry is specified by Nodes, Tracelines, Beams, Quads, and Shells. Files for undamaged and damaged modes can be either direct modal analysis output from *DIAMOND*, or the user can generate their own mode shape files from other modal analysis software packages. The SEM UI only requires a MATLAB Structure data type of the nodal displacements that pertain to the geometry and a confirmation that the response degree-of-freedom is in the positive z-axis. Some notable assumptions within the SEM UI are that it requires planar geometries with the response degree-of-freedom being in the positive z-axis. The SEM with Beams function requires that the beams used in the computation be parallel in either the x- or y-axis, but beams need not be the same length in this function.

The third function within the SEM UI is mapping FE mode shapes to geometry that can be used within the strain energy methods. The Convert FE Data to Modes window, shown in Fig. 2.8, has two capabilities – mapping FE mode shapes to an existing geometry and mapping FE mode shapes to a new geometry. The user inputs a reference geometry (the geometry in Fig. 2.8 is the same as shown in Fig. 2.2) MATLAB file and a Microsoft Excel file containing the FE mode shapes that are defined by paths in the geometry. The format required for the FE Displacement Excel File can be found in the open-source example data that accompanies the UI. The UI then configures itself to map the Current Path to the nodes that the user inputs in the Nodes for Path dialog box (node numbers separated by commas). This window in the UI relies on similar geometry assumptions as previously stated with some extra constraints. The Convert FE Data to Modes functions require that paths be parallel in either the x- or y-axis and the same length, and it is advantageous to list the paths in ascending order with the most negative x- or y-direction path first.

ANSYS Workbench 19.1 was used in this work to create FE models and do the modal analysis. ANSYS allows the user to create paths along edges in a geometry, and this functionality allows the user to extract nodal displacements to be used in the SEM UI. Figure 2.9 shows a screen shot of a path and its corresponding data from within an ANSYS modal analysis.

The second functionality of the Convert FE Data to Modes window (which is still in development) is mapping FE displacements to a new geometry. For this function, the user only inputs the FE Displacements Excel File and the number of nodes to discretize each path into. In this function it is critical that all paths be parallel, the same length, and listed in the correct order. Once given the proper information, the UI generates the mode shape data and the geometry file to be used

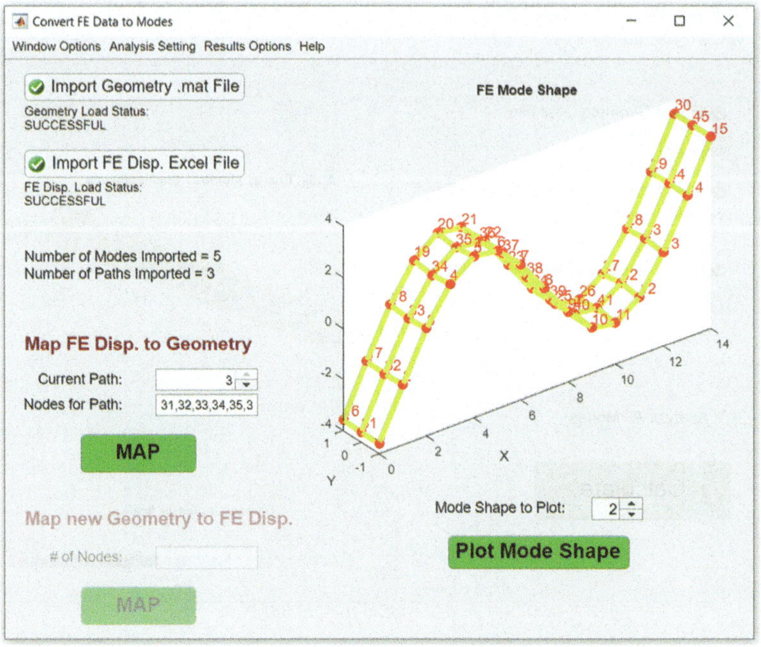

Fig. 2.8 Convert FE Data to Modes window. Here the user can perform two functions – mapping FE mode shapes to a current geometry and mapping FE mode shapes to a new geometry

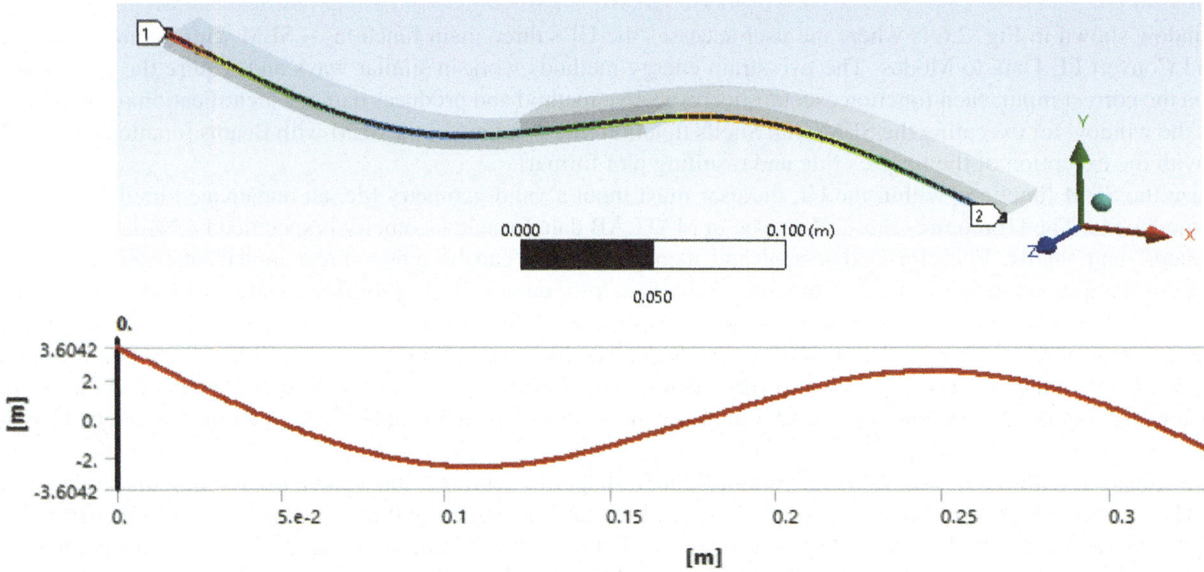

Fig. 2.9 Path and tabular data from within an ANSYS modal analysis of the solid FE model

within the SEM windows. Once the FE displacements are mapped to existing or new geometry, the UI allows the user to plot the mode shapes to ensure that information was mapped correctly. Finally, the user can save the information from this window to be used later with the strain energy methods.

2.5 Results and Discussion

In this work, modal properties of different structures and models were compared using several different methods – natural frequencies, damping ratios, and strain energy. Natural frequencies, damping ratios, and mode shapes were extracted from the experimental modal tests of the three structures. Only natural frequencies and mode shapes were extracted from the FE models. The three previously mentioned methods were used to compare different structures and like structures with and without damage.

2.5.1 Experimental Modal Comparisons

Each structure was tested three times. In order to extract natural frequencies, damping ratios, and mode shapes, a rational-polynomial curve fitting method was used with the complex mode indicator function (CMIF) for each separate test. Figure 2.10 shows the average CMIF for each of the structures.

Natural frequencies and damping ratios for each of the three structures were used as a first level of comparison. Differences in the natural frequencies for each of these structures were expected based on differences in mass and material properties. The masses of the solid, welded and bolted structures were 348 grams, 356 grams, and 369 grams, respectively. Based on this property alone, it was expected that the solid structure would have the highest natural frequencies, followed by the welded and then bolted structures. However, the average experimental modal frequency results summarized in Table 2.2 show that the welded structure actually had the highest natural frequencies. The welded structure contained a significant amount of weld filler material, AL 4043, that has higher stiffness than the rest of the structure. The material stiffness as well as any heat-induced stress from the welding and quenching process appear to have had a significant effect on the overall stiffness of the welded structure, thus causing some natural frequencies to be relatively higher than expected.

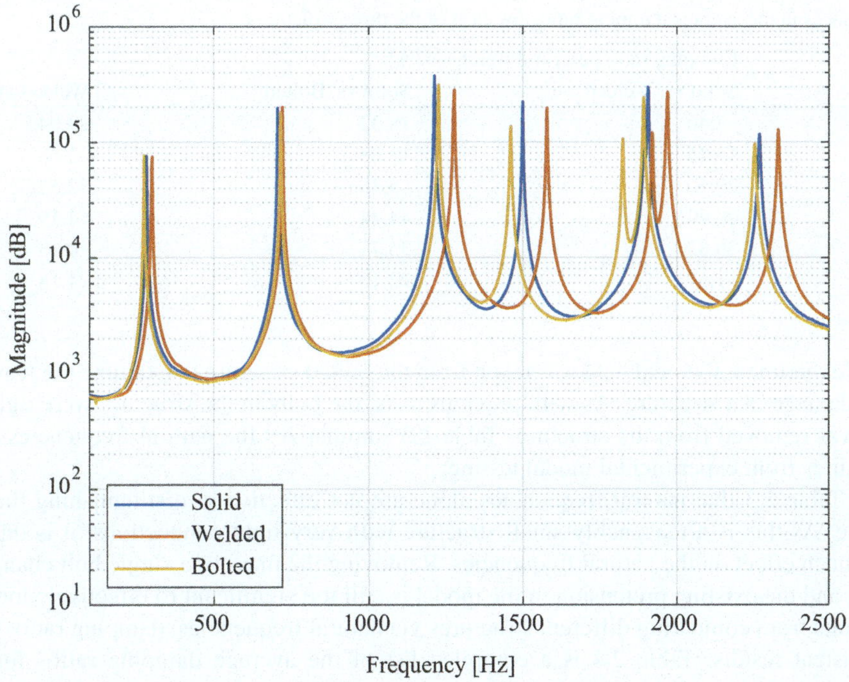

Fig. 2.10 Average CMIF used with a rational polynomial curve-fitting method for determining each structure's modal properties

Table 2.2 Average natural frequencies for each structure and percent differences for comparison

	Natural frequencies [Hz]			Percent differences [%]		
Mode	Solid	Welded	Bolted	Solid vs. Welded	Solid vs. Bolted	Welded vs. Bolted
1st bending	284.18	301.27	275.88	6.01	2.92	8.43
2nd bending	704.59	719.24	713.38	2.08	1.25	0.81
1st torsion	1212.8	1275.8	1225.5	5.19	1.05	3.94
3rd bending	1499.5	1577.1	1459.9	5.18	2.64	7.43
1st O.P. bending	1893.5	1919.4	1823.7	1.37	3.69	4.99
2nd torsion	1906.2	1967.7	1891.1	3.23	0.79	3.90

Table 2.3 Average natural frequencies and respective percent differences for the undamaged and damaged bolted structures. The O.P. direction was not tested for the damaged structure

	Natural frequencies [Hz]		Percent differences [%]
Mode	Bolted (Undamaged)	Bolted (Damaged)	Damaged vs. Undamaged
1st bending	275.88	273.17	0.97
2nd bending	713.38	713.38	0.001
1st torsion	1225.5	1217.6	0.65
3rd bending	1459.9	1445.4	0.99
1st O.P. bending	N/A	N/A	N/A
2nd torsion	1891.1	1904.6	0.71

Table 2.4 Average damping ratios and standard deviations for each structure

	Solid damping ratios		Bolted damping ratios		Welded damping ratios	
Mode	Avg.	Std. Dev	Avg.	Std. Dev	Avg.	Std. Dev
1st bending	0.00366	0.00093	0.00390	0.00112	0.00299	0.00100
2nd bending	0.00119	0.00053	0.00101	0.00013	0.00135	0.00047
1st torsion	0.00108	0.00017	0.00086	0.00035	0.00113	0.00025
3rd bending	0.00071	0.00019	0.00090	0.00024	0.00103	0.00007
1st O.P. bending	0.00172	0.00011	0.00108	0.00021	0.00109	0.00015
2nd torsion	0.00086	0.00035	0.00085	0.00039	0.00105	0.00024

Table 2.5 Percent differences in the average damping ratios for each of the three structures

	Damping ratio percent differences [%]		
Mode	Solid vs. Welded	Solid vs. Bolted	Welded vs. Bolted
1st bending	7.40	18.32	23.94
2nd bending	15.45	13.03	33.68
1st torsion	20.50	5.35	32.52
3rd bending	26.49	44.44	14.19
1st O.P. bending	36.82	36.73	0.13
2nd torsion	1.67	22.10	24.18

A successive set of experimental modal data was taken from the bolted structure to examine the feasibility of using natural frequencies to detect damage in a structure. As with previous tests, the bolts in the structure were tightened to 25 in-lbs, but in this test, one bolt was removed from the structure. Table 2.3 summarizes the natural frequencies of the undamaged and damaged bolted structures from experimental modal testing.

It is obvious from Table 2.3 that natural frequencies alone are not effective in distinguishing the undamaged structure from the damaged one. As this is a reasonably small structure with very tight connections, it is unlikely that removing a single bolt will have much effect on the natural frequencies. Removing the mass of a single bolt changes the overall mass of the structure by 1.6%, and the existing pretension in the model is still too significant to effectively loosen the joint.

Contrary to the results from comparing different structures via natural frequencies, damping ratios for the three structures did not provide consistent results. Table 2.4 is a compiled list of the average damping ratios for each structure along with respective standard deviations. Table 2.5 contains the percent differences between the average damping ratios of the structures.

On average, the standard deviations of the damping ratios for all structures was 26% of the average damping ratio value. In some cases, such as the 2nd Torsion mode for the welded structure, the standard deviation in the damping ratios was over 45% of the damping ratio average value. These statistics suggest that there is too much variability between damping ratios for each individual structure to allow for a meaningful comparison of results between structures. To reinforce this observation, Fig. 2.11 shows the individual damping ratio results from each test of each structure.

Consider the results of the 1st Bending mode. There is more variability in the individual damping ratios for the bolted structure than there is between the averages of all three structures. Figure 2.11 shows that the variability between structures is not significant enough to consider damping a proper metric for distinguishing between different structures or damaged/undamaged structures.

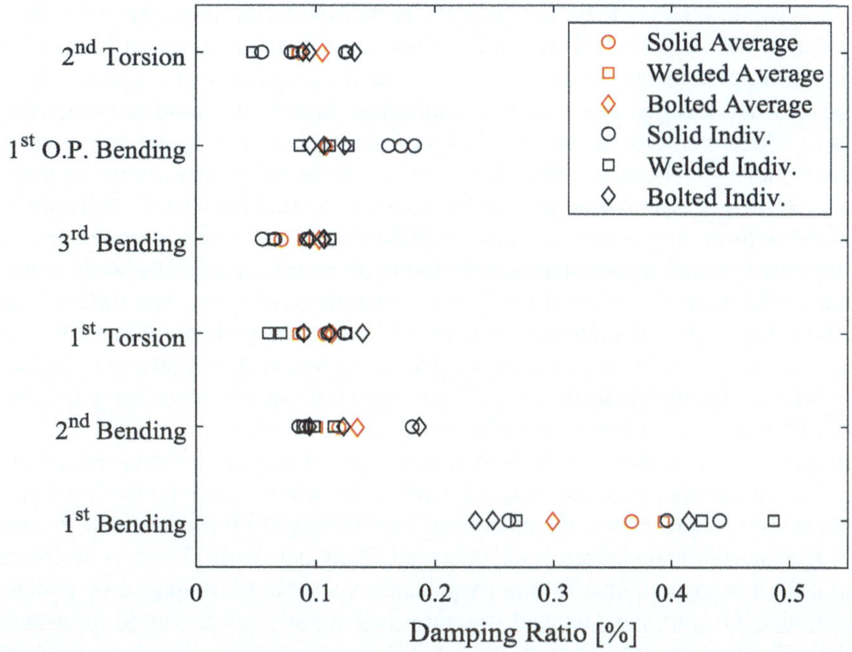

Fig. 2.11 Damping ratio results from each of the three experimental modal tests done on each structure

Table 2.6 Natural frequencies and percent differences for the solid structure and the solid structure finite element model

	Natural frequencies [Hz]		Percent differences [%]
Mode	Solid	Solid FE	Solid vs. Solid FE
1st bending	284.18	284.52	0.12
2nd bending	704.59	707.49	0.41
1st torsion	1212.8	1214.5	0.13
3rd bending	1499.5	1505.8	0.41
1st O.P. bending	1893.5	1889.8	0.19
2nd torsion	1906.2	1903.4	0.14

Table 2.7 Natural frequencies and percent differences for the welded structure and the welded structure finite element model

	Natural frequencies [Hz]		Percent differences [%]
Mode	Welded	Welded FE	Welded vs. Welded FE
1st bending	301.27	295.05	2.06
2nd bending	719.24	718.46	0.10
1st torsion	1275.8	1250.1	2.02
3rd bending	1577.1	1558.8	1.16
1st O.P. bending	1919.4	1938.4	0.98
2nd torsion	1967.7	1933.8	1.72

2.5.2 FE Model Comparisons

The FE models for the solid and welded structures needed very little refinement in order to produce results with reasonable percent error when compared to the experimental results. The only parameters varied in these models were the densities and Young's modulus of the two aluminum alloys. The material model for AL 6061 was modified so that mass properties and natural frequencies matched for the solid model. This AL 6061 material model was then shared with the remaining FE models. The material model for AL 4043 was modified in a similar fashion until experimental and FE results for the welded structure matched. Tables 2.6 and 2.7 summarize the natural frequencies from the experimental modal analysis and the FE modal analysis for the solid and welded structures, respectively.

An inherent source of uncertainty for the bolted structure is the pretension within each bolt. Depending on the torque applied to each bolt when assembling the joint, natural frequencies (and damping ratios) are subject to change with the pretension. Intuitively, one would imagine that the higher the pretension in the joint, the stiffer the system will be and the higher the natural frequencies will become. This possibility poses an interesting question from the perspective of structural health monitoring. If changing pretension has an effect on the modal properties of a system, then those properties could be used to detect the severity and location of a loosened connection. In order to simulate loosened connections, the size of the contact zone (i.e. area where elements are bonded together) was varied and modal properties were extracted for each different size contact zone. Table 2.8 shows the natural frequencies of the bolted structure from experimental data and from two different FE models – Bolted FE with 10-mm contact regions and Bolted FE with 17-mm contact regions.

As expected, the model with the larger contact region resulted in higher natural frequencies. However, the FE model results still vary from the experimental results significantly (up to about 6%), especially when comparing higher frequency modes. Since the FE natural frequencies are consistently lower than those from the experimental modal tests, it is possible that the material properties of the bolts should be adjusted. By increasing the stiffness of the bolts, the natural frequencies of the models would likely increase, but these adjustments may lead to a model that is falsely accurate. The bolted connection creates much more variability within the model than material properties do, so it is more likely that the connection parameters are modeled incorrectly rather than the material model being the source of error.

The last of the FE studies were two models with one bolt missing – 10-mm and 17-mm contact regions with a bolt and its contact region missing. The mass properties of the model as well as the stiffness from pretension will be affected by removing a bolt. Table 2.9 lists the natural frequencies of the undamaged and damaged FE models with 10-mm contact regions. Table 2.10 lists the natural frequencies of the undamaged and damaged FE models with 17-mm contact regions.

From Tables 2.9 and 2.10, it is apparent that natural frequencies change in the damaged vs. undamaged models; however, the results are counterintuitive. Overall, the damaged structures had slightly lower natural frequencies than the undamaged structures, especially in the 1st Torsion, 3rd Bending, and 1st O.P. Bending modes. The damaged model with 17-mm contact

Table 2.8 Natural frequencies and percent differences for the bolted structure, bolted structure FE model with 10-mm contact regions, and the bolted structure FE model with 17-mm contact regions

	Undamaged natural frequencies [Hz]			Percent difference [%]	
Mode	Bolted	Bolted FE 10 mm	Bolted FE 17 mm	Bolted vs. FE 10 mm	Bolted vs. FE 17 mm
1st bending	275.88	260.13	269.85	5.70	2.18
2nd bending	713.38	693.10	697.27	2.84	2.25
1st torsion	1225.5	1167.0	1195.7	4.77	2.43
3rd bending	1459.9	1388.2	1438.6	4.91	1.46
1st O.P. bending	1823.7	1754.3	1812.2	3.80	0.63
2nd torsion	1891.1	1777.9	1790.9	5.98	5.29

Table 2.9 Natural frequencies and percent differences for the undamaged and damaged bolted structure FE model with 10-mm contact regions

	Bolted FE model 10-mm contact region		
	Natural frequencies [Hz]		Percent differences [%]
Mode	Undamaged	Damaged	Damaged vs. undamaged
1st bending	260.13	254.20	2.14
2nd bending	693.10	693.39	0.04
1st torsion	1167.0	1121.1	3.74
3rd bending	1388.2	1348.4	2.72
1st O.P. bending	1754.3	1690.7	3.48
2nd torsion	1777.9	1775.6	0.12

Table 2.10 Natural frequencies and percent differences for the undamaged and damaged bolted structure FE model with 17-mm contact regions

	Bolted FE model 17-mm contact region		
	Natural frequencies [Hz]		Percent differences [%]
Mode	Undamaged	Damaged	Damaged vs. undamaged
1st bending	269.85	248.29	7.81
2nd bending	697.27	695.21	0.28
1st torsion	1195.7	1096.4	8.10
3rd bending	1438.6	1314.1	8.52
1st O.P. bending	1812.2	1644.6	9.18
2nd torsion	1790.9	1785.3	0.29

regions, which would be expected to have higher natural frequencies than the damaged model with 10-mm contact regions, consistently had the lowest natural frequencies of all the models. These results suggest that the size of the contact region has a profound effect on the overall stiffness of the model, as removing one 17-mm contact region had a much greater effect on the natural frequencies than removing one 10-mm contact region. Additionally, these results show that the FE models were much less sensitive to changes in material properties and overall mass properties – removing a bolt decreases mass and would be expected to increase natural frequencies, which was not the case in either model.

2.5.3 Strain-Energy Method Comparisons

The following comparisons utilize the strain energy method with shell elements. The SEM comparison of the experimental mode shapes produced similar results to those seen in the natural frequency comparison. The differences between structures was not extreme enough to be detected. Both the comparisons of the solid to the welded structure and the solid to the bolted structure resulted in a damage index mapping that did not allow the structures to be distinguished from each other, as shown in Figs. 2.12a, b.

A successive set of experimental modal tests was done on the bolted structure using the geometry shown in Fig. 2.2 for finer discretization. Finer geometries allow for more resolution within each shell. Additionally, geometries with more than two points in the y-axis, as shown in Fig. 2.13, allow for curvature to be computed between nodes in both the x and y-directions. Figure 2.13 shows the SEM comparison (from experimental modal data) of the undamaged bolted structure and the bolted structure with a bolt missing near Node 7. This comparison also uses a finer geometry than the previous SEM comparisons.

From Fig. 2.13 it is clear that it was not possible to distinguish a difference between the undamaged and damaged bolted structures. The failure of the strain energy method in this context is likely due to the same reasons as the natural frequency method's deficiencies. The mass properties of the structures are not different enough, and the pretension in the model is high enough, that with only one bolt missing, the joint remains tight.

The second application for the strain energy method UI was to compare the FE mode shape results. FE models of the solid, welded, and multiple bolted structures were created using ANSYS Mechanical. These models were discretized to the same geometry as seen in Fig. 2.13. First, the FE results from the solid and welded models and the solid and bolted models were compared using the SEM with shell elements, as shown in Fig. 2.14a, b, respectively.

From Fig. 2.14, it is clear that the SEM was not able to identify any differences between the models of the solid structure and the welded structure, but it did identify a difference in the region around the bolts between the model of the solid structure and the model of the bolted structure. In the first case, the masses of these two models are closer to each other than the masses of the solid and bolted models. Second, the welded model has bonded contact conditions at all points on the perimeter of the joint seam, whereas the bolted model is only bonded around each bolt. Since the bolted model's joint seam is allowed

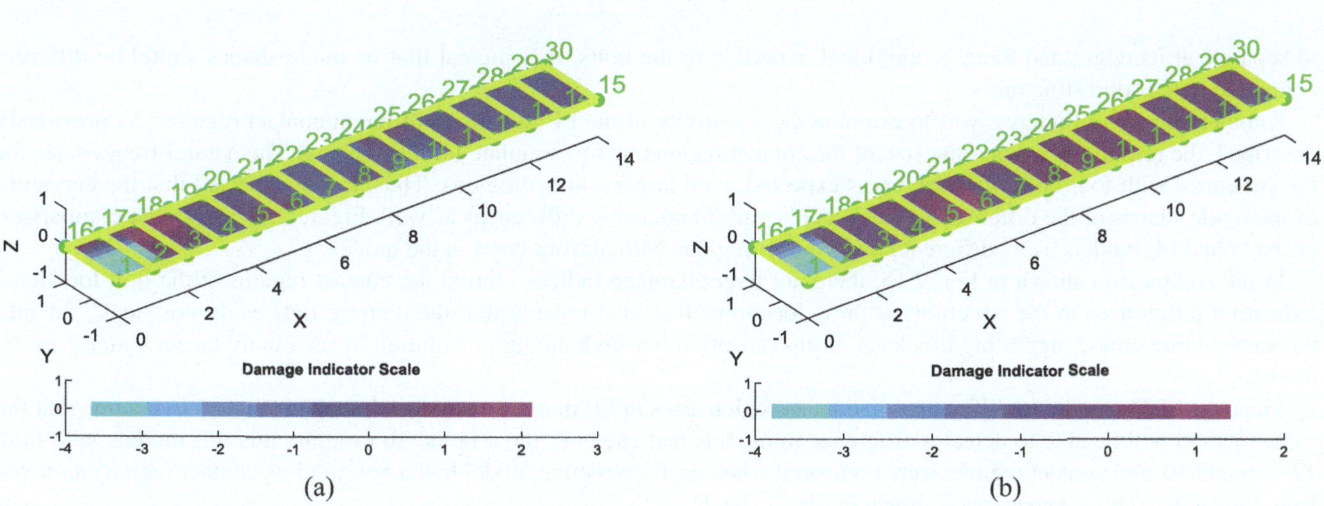

Fig. 2.12 SEM with shell elements comparing (**a**) the solid and welded structure and (**b**) the solid and bolted structure

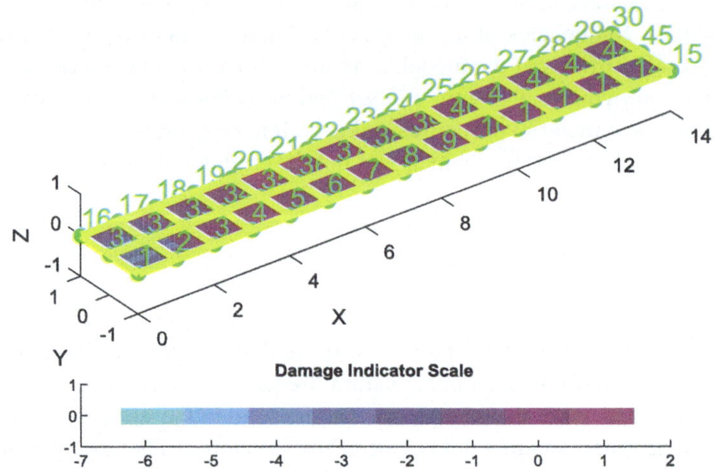

Fig. 2.13 SEM with shell elements comparing the undamaged and damaged (bolt near Node 7 missing) structures

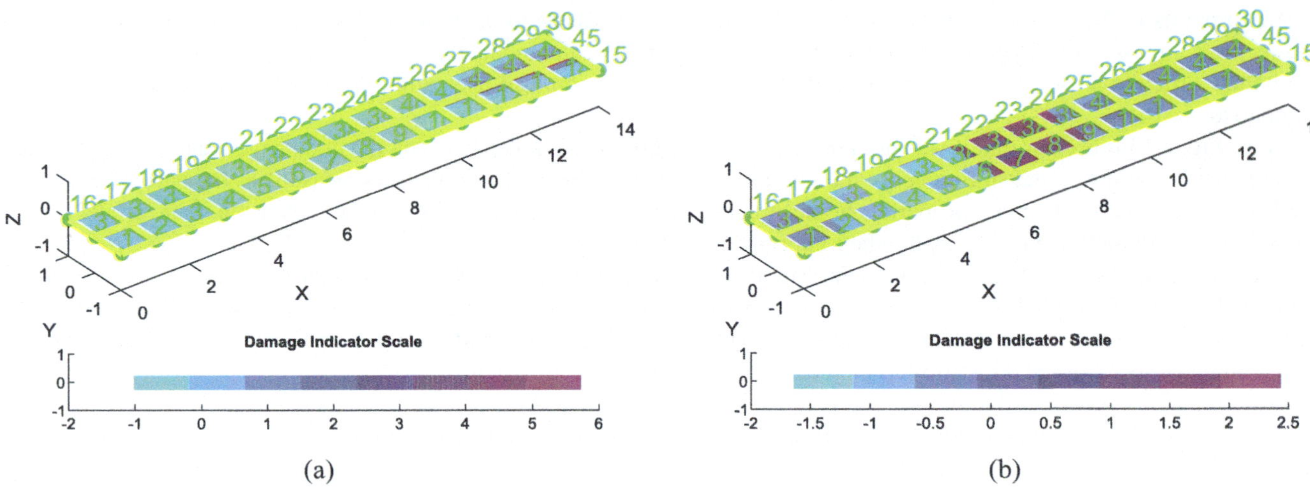

Fig. 2.14 SEM comparing (**a**) the solid and welded FE models and (**b**) the solid and bolted FE models. The bolted model used 17-mm contact regions with four bolts

to separate at its edges and there is additional mass due to the bolts, it is logical that its mode shapes would be different compared to the solid structure's.

Another question of interest was to examine the sensitivity of the FE models to different contact regions. As previously described, the reason for varying the size of the contact regions was to simulate joint loosening. The natural frequencies for the structures with four bolts were larger, as expected, with larger contact regions. Thus, it was expected that the curvature of the mode shapes of the different models would exhibit noticeable differences as well. Figure 2.15 shows the comparison of the bolted FE models using different sized contact regions with all four bolts in the model.

In the comparison shown in Fig. 2.15, there are larger damage indices around the contact regions at the bolt locations, indicating differences in the structures at these locations. It is also notable that the damage indices do not "leak" far into the surrounding areas, suggesting this level of discretization has high enough resolution to accurately locate damage in the model.

Since the SEM can detect differences in contact region sizes in FE models using four bolts, it is logical to assume that the same method will be able to detect a difference in models that use only three bolts. To examine this, FE models with both 17-mm and 10-mm contact regions were compared when each respective model had a bolt (and its contact region) removed from the model. This comparison is shown in Fig. 2.16a, b.

Figure 2.16a shows that the SEM is capable of detecting damage in a structure when a single bolt and its contact region are removed from the model. The damage is expected near Node 7, and the method correctly detects the general area of the

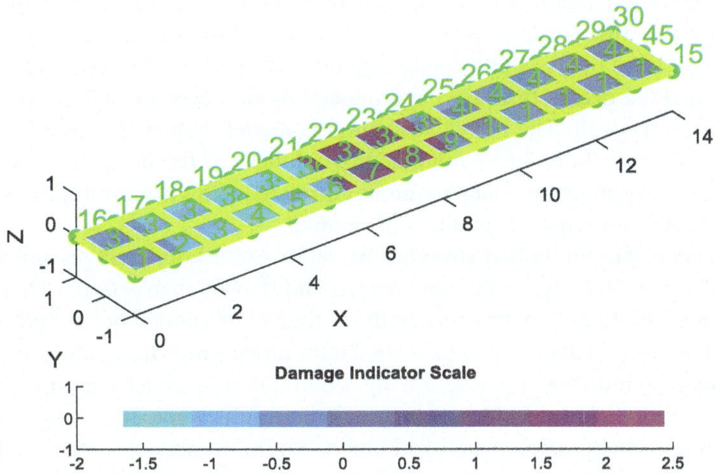

Fig. 2.15 SEM with shell elements comparing two FE models of the bolted structure. One model used 17-mm contact regions and the other used 10-mm contact regions

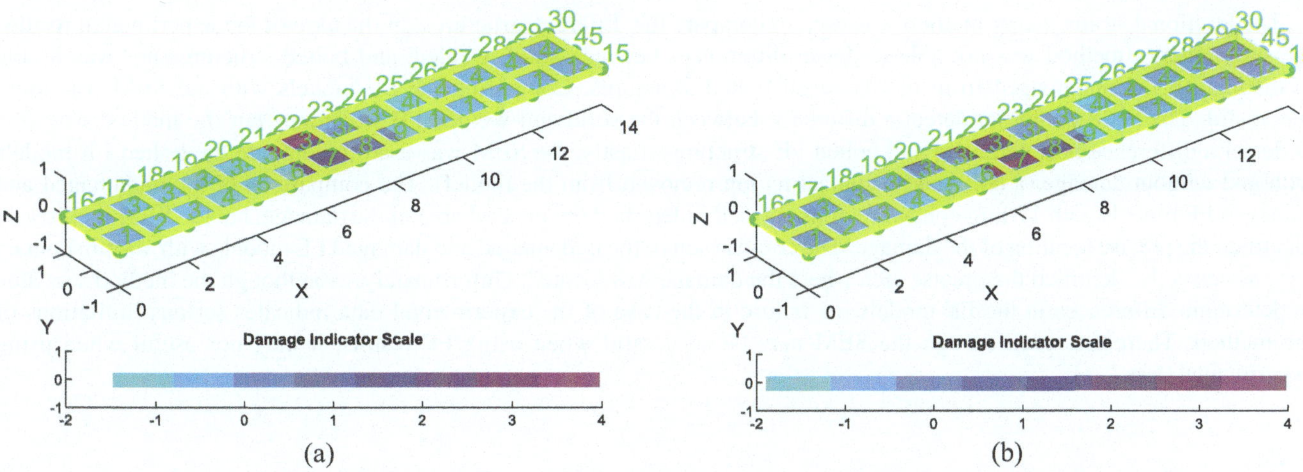

Fig. 2.16 SEM with shell elements comparing undamaged and damaged bolted FE models. (**a**) model used 17-mm contact regions, and the damaged structure was missing a bolt near Node 7. (**b**) model used 10-mm contact regions, and the damaged structure was missing a bolt near Node 7

damage. However, the SEM was not capable of precisely locating the damage in this model, as the damage indices are higher on the opposite side of the structure (near Node 21) where there was no damage. Figure 2.16b, on the other hand, correctly identified the location of the removed bolt. Thus, the SEM was capable of detecting the general area of the damage, but not always the exact location.

2.6 Conclusions

In this work, the differences between a solid structure, a welded structure, and a bolted structure were examined. Even though it was sometimes possible to distinguish one structure from another using modal properties, some properties had significant uncertainty and variability, making it difficult to draw conclusions. Damping ratios extracted from the experimental modal tests proved to be an ineffective property for comparing the three structures. The variability between different tests of the same structure was sometimes greater than the variability in damping ratios between structures. The experimental natural frequencies, however, provided distinct enough results to distinguish each structure. It was expected that the solid structure would have the highest natural frequencies, followed by the welded and bolted structures, based on their masses; however, the welded structure had consistently higher natural frequencies than the other two structures. Although the welded structure was heavier than the solid structure, the stiffer material properties of the weld-filler material resulted in higher natural frequencies. The mass properties were found to be of less significance than other properties of the structures. The FE models of the solid and welded structures confirmed the results from the experimental modal tests; the natural frequencies from the FE models and the experiments compared very well. The bolted structure FE model, however, did not match the experimental results quite as well as the other two models. In order to simulate the tightness of the bolted joint, different sized contact regions were used in the FE models. In general, larger contact regions correlated to higher natural frequencies, and the larger contact regions tended to compare better to the experimental modal results.

When damage was introduced into the bolted structure by removing a bolt, changes in experimental modal properties proved to be inadequate to distinguish between the undamaged and damaged structures. There was virtually no difference in the natural frequencies between the bolted structure with all four bolts (undamaged) and that with one bolt completely missing (damaged). However, natural frequency results from the FE models did show some differences between the damaged and undamaged structures. One would expect that removing a bolt and its contact region from the model would lower the natural frequencies; this expectation held true, although some of the results were still counter-intuitive. Before removing the bolt from the model, the 17-mm contact region model had higher natural frequencies than the 10-mm contact region model. When a bolt was removed from the model, however, the 10-mm contact region model had higher natural frequencies than the 17-mm contact region model. These results, though not expected, confirm that the models are very sensitive to the contact parameters used to represent the bolted joint and that other properties, such as mass and material parameters, have a smaller effect on the model.

The fractional strain energy method was used to compare the different structures. In the case of the experimental results, the strain energy method was not able to detect differences between solid, welded, and bolted structures nor was it able to distinguish the undamaged from the damaged bolted structures. Comparing the FE models with the SEM was more successful. The SEM could not detect a difference between the solid and welded FE structures, but the method was able to detect a difference between solid and bolted FE structures. Lastly, the SEM was used to compare the bolted FE models with and without damage (a bolt and its contact region removed from the model). The comparison of the undamaged and damaged FE models with 17-mm contact regions correctly identified the general area of the damage, but the SEM incorrectly identified the precise location of the damage. The comparison of the undamaged and damaged FE models with 10-mm contact regions correctly identified the precise area where the damage was located. Unfortunately, even though the method was able to determine differences in the FE models, its failure in the case of the experimental data indicates serious limitations of the method. Therefore, even though the SEM may be successful when using FE results, it may not useful when using experimental data.

References

1. McCarthy, M., McCarthy, C., Lawlor, V., Stanley, W.: Three-dimensional finite element analysis of single-bolt, single-lap composite bolted structures. Compos. Struct. **71**(2), 140–158 (2005)
2. Sun, D., Liao, R.: Damping prediction technique of the bolted structure considering pretension force. Open Civil Eng. J. **9**, 622–626 (2015)
3. Xu, W.: Effect of Bolted Joint Preload on Structural Damping. University of South Florida (2013)
4. Zaman, I., Khalid, A., Araby, S., Ghazali, M.: The effects of bolted joints on dynamic responses of structures. In: IOP Conferene Series: Materials Science and Engineering 50 (2013)
5. Kedra, R., Rucka, M.: Damage detection in a bolted lap joint using guided waves. In: *International Conference on Structural Dynamics* (2017)
6. Du, F., Xu, C., Ren, H., Yan, C.: Structural health monitoring of bolted joints using guided waves: a review. Struct. Health Monit.Sens. Process. **8**, 163–180 (2018)
7. Kim, J., Yoon, J., Kang, B.: Finite element analysis and modeling of a structure with bolted joints. Appl. Math. Model. **31**(5), 895–911 (2007)
8. Mohammed, R.: Finite Element Analysis of Fillet Welded Joint. University of Southern Queensland, Toowoomba (2015)
9. Stubbs, N., Kim, J.-T., Farrar, C.R.: Verification of a nondestructive damage localization and sensitivity estimatory algorithm. In: *13th International Modal Analysis Conference* (1995)
10. Cornwell, P., Doebling, S., Farrar, C.: Application of the strain energy damage detection method to plate-like structures. J. Sound Vib. **224**(2), 359–374 (1999)
11. Liao, X., Zhang, J., Xu, X.: Analytical model of bolted joint structure and its nonlinear dynamic characteristics in transient excitation. Shock Vib. **2016**, 11 (2016)
12. Doebling, S.W., Cornwell, P.J., Farrar, C.R.: DIAMOND: a graphical interface toolbox for comparative modal analysis and damage identification. In: Sixth International Conference on Recent Advances in Structural Dynamics, Southampton (1997)

Chapter 3
Using Strain Gages as References to Calculate Free-Free Frequency Response Functions

Kevin Napolitano and David Cloutier

Abstract The objective of many modal tests to experimentally measure free-free modes is achieved by suspending the test article on soft supports such that the resulting rigid body modes are adequately decoupled in frequency from the flexible body. The soft supports can also be placed at the nodal locations of the primary free-free modes to provide further decoupling.

Oftentimes, however, the rigid body modes cannot be separated from the flexible body modes for a variety of reasons, such as the flexible modes being very low in frequency, the test article being so large that it is not cost effective to build a suspension system, or a combination of both. In these cases, the rigid body and flexible modes are intermingled, which means that effort must be spent updating the model of the suspension system instead of the test article itself. Sometimes, one may also want to use the experimentally measured free-free modes directly in an analysis without building an analysis model.

This paper proposes a method that uses strain gage measurements at the structure's boundary as references for calculating frequency response functions. The resulting frequency response functions, associated with a free-free structure, can then be used to estimate modes.

Keywords Modal testing · Vibrations · Constraint shapes · Boundary condition correction · Strain gage

3.1 Introduction

Recently implemented methods by ATA Engineering have used acceleration measurements on the boundary of a test article as references to calculate frequency response functions (FRFs) that result in fixed base modes [1–5]. The key to implementing these methods is to have at least one independent input on the boundary for every independent acceleration degree of freedom (DOF) on the boundary. In the case where a set of accelerations on the boundary are highly correlated, a reduced number of constraint shapes can be used as references.

It would be advantageous if a similar method could be developed to measure FRFs associated with a free-free structure mounted on a flexible boundary. Conceptually, this can be achieved if all interface forces' patterns can be measured and separated into independent references when FRFs are calculated.

Unfortunately, it is often difficult or impossible to mount interface load cells between a structure and its boundary. However, it is straightforward to mount strain gages at the structure interface; in this way, while forces are not measured directly, the measured forces are proportional to the forces flowing between the structure and boundary.

This paper proposes a method that uses strain gage measurements at the structure's boundary as references for calculating FRFs. The resulting FRFs, associated with a free-free structure, are then used to estimate modal parameters.

3.1.1 Calculation of Free-Free FRFs Using Interface Forces as References

For this calculation, assume that the test article is mounted to a boundary such as a shake table through a series of interface load cells, and also assume that the system is excited on the test article at some locations as well as on the boundary with

K. Napolitano (✉) · D. Cloutier
ATA Engineering, Inc., San Diego, CA, USA
e-mail: kevin.napolitano@ata-e.com; dcloutier@ata-e.com

B. Dilworth (ed.), *Topics in Modal Analysis & Testing, Volume 8*, Conference Proceedings of the Society for Experimental Mechanics Series, https://doi.org/10.1007/978-3-030-47717-2_3

as many shakers as there are independent force patterns associated with the interface load cells. The baseline FRFs can be defined as

$$\{R\} = [H]\{F\}, \tag{3.1}$$

where {R} is all of the measured responses including accelerations and interface load measurements, [H] is the FRF matrix, and {F} is all the applied forces. To simplify this derivation, assume that the responses are accelerations on the test article {a}, and interface forces $\{f_r\}$, such that

$$\{R\} = \begin{Bmatrix} a \\ f_r \end{Bmatrix}. \tag{3.2}$$

Also assume that the applied forces {F} can be separated into forces applied directly to the test article $\{f_i\}$, and forces applied to the boundary $\{f_b\}$, such that

$$\{F\} = \begin{Bmatrix} f_i \\ f_b \end{Bmatrix}. \tag{3.3}$$

The FRF matrix [H] is partitioned to obtain

$$\begin{Bmatrix} a \\ f_r \end{Bmatrix} = \begin{bmatrix} H_{ai} & H_{ab} \\ H_{ri} & H_{rb} \end{bmatrix} \begin{Bmatrix} f_i \\ f_b \end{Bmatrix}. \tag{3.4}$$

In the case where the number of independent reaction force patterns is less than the number of interface load cell measurements, the number of reaction force DOFs can be reduced using force pattern constraint shapes such that

$$\{f_r\} = [\Psi]\{f_E\}, \tag{3.5}$$

where $[\Psi]$ is a matrix of independent force patterns and $\{f_E\}$ is the force pattern DOFs. Inserting Eq. (3.5) into Eq. (3.4) yields

$$\begin{Bmatrix} a \\ f_E \end{Bmatrix} = \begin{bmatrix} H_{ai} & H_{ab} \\ H_{Ei} & H_{Eb} \end{bmatrix} \begin{Bmatrix} f_i \\ f_b \end{Bmatrix}, \tag{3.6}$$

where, since $\{f_E\} = [\Psi]^+\{f_E\}$, $[H_{Ei}]=[\Psi]^+[H_{ri}]$ and $[H_{Eb}]=[\Psi]^+[H_{rb}]$. Note that the term "+" denotes the pseudo-inverse of a matrix.

Assuming that an equal number of forces are applied to the boundary as there are independent force patterns, one can perform a partial inversion of the FRF matrix to move the force patterns to the right-hand side and the boundary forces to the left-hand side to obtain

$$\begin{Bmatrix} a \\ f_b \end{Bmatrix} = \begin{bmatrix} \overline{H}_{ai} & \overline{H}_{ab} \\ \overline{H}_{bi} & \overline{H}_{bE} \end{bmatrix} \begin{Bmatrix} f_i \\ f_E \end{Bmatrix}, \tag{3.7}$$

where $\overline{H}_{ai} = H_{ai} - H_{ab}{H_{Eb}}^{-1}H_{Ei}$, $\overline{H}_{ab} = H_{ab}{H_{Eb}}^{-1}$, $\overline{H}_{bi} = -{H_{Eb}}^{-1}H_{Ei}$, and $\overline{H}_{bE} = {H_{Eb}}^{-1}$.

For any linear relationship in the form of $\{x\} = [C]\{y\}$, the matrix element C_{ij} is equal to the value of $x_i$due to a unit input at y_j, holding all other elements $\{y\}$ in equal to zero. Applying this property to Eq. (3.7) means that the submatrix $\overline{H}_{ai}$ is the FRF matrix associated with all interface force patterns equal to zero—the definition of a free-free boundary condition.

Most load cells are based on strain gages that are calibrated to give an estimate of force. However, instead of measuring forces directly, assume one measured a series of strain gages of which a linear combination captured all the interface force patterns such that

$$\{f_r\} = [\mathrm{D}]\{\varepsilon\}, \tag{3.8}$$

where $\{\varepsilon\}$ is a vector of strain gages and [D] is a calibration matrix used to calculate forces. With no loss in generality, one can measure strains directly and then use them to calculate a series of force patterns, which can then be used to measure free-free FRFs. Combine Eq. (3.8) with Eq. (3.5) to obtain $\{f_r\} = [\mathrm{D}]\{\varepsilon\} = [\Psi]\{f_E\}$, and therefore $\{\varepsilon\} = \left\lceil \Psi \right\rceil \{f_E\}$, where $\left\lceil \Psi \right\rceil = [\mathrm{D}]^+ [\Psi]$ and $\left\lceil \Psi \right\rceil$is a matrix of independent strain patterns.

In practice, one would not calculate the calibration matrix and would instead calculate the matrix $\left\lceil \Psi \right\rceil$ directly from the strain measurements.

3.1.2 Numerical Example

This example is a simplified version of a structure mounted on a rigid six DOFs shake table with four three-axis load cells (twelve load cell measurements in total) mounted between the structure and its rigid base.

A picture of the simplified model is presented in Fig. 3.1. Nodes 1, 11, and 21 are coincident. Node 1 is not connected to nodes 11 and 21, but nodes 11 and 21 are connected through a six-DOF spring. Nodes 2 through 5 are coincident with nodes 12 through 15 and are connected to each other through three-DOF springs. Nodes 2 through 5 are rigidly connected to node 1, nodes 12 through 15 are rigidly connected to node 11, and node 31 is connected rigidly to node 21. Mass elements are applied to the independent nodes 1, 11, and 31.

The properties of the simplified model are listed in Table 3.1. Spring and mass element values were varied arbitrarily to introduce some nonsymmetry and to demonstrate that one only needs to measure relative displacement, and not interface force, for the method to work.

There are 18 independent DOFs. The baseline full system modal parameters including six rigid body modes are listed in Table 3.2.

FRFs were calculated by applying forces to all six DOFs at node 1 and three translational DOFs at node 31 and measuring acceleration responses at all nodes. Damping of 0% was used for the six rigid body modes, and 1% was used for the twelve flexible modes. Relative displacement (similar to strain) was measured across nodes 2 through 5 and nodes 12 through 15, respectively. A total of twelve relative displacement FRFs were calculated.

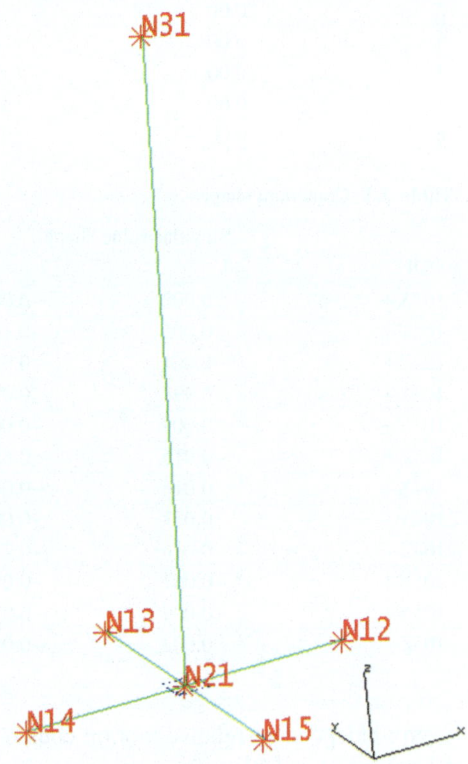

Fig. 3.1 Display model of structure. Nodes 1, 11, and 21 are coincident. Nodes 2 through 5 are coincident with nodes 12 through 15, respectively

Table 3.1 Structure properties and node dependencies

NODES LOCATIONS

Node Number	X	Y	Z	Master Node
1	0.0	0.0	0.0	1
2	1.0	0.0	0.0	1
3	0.0	1.0	0.0	1
4	-1.0	0.0	0.0	1
5	0.0	-1.0	0.0	1
11	0.0	0.0	0.0	11
12	1.0	0.0	0.0	11
13	0.0	1.0	0.0	11
14	-1.0	0.0	0.0	11
15	0.0	-1.0	0.0	11
21	0.0	0.0	0.0	21
31	0.1	-0.2	4.0	21

MASS PROPERTIES

Node Number	Mass	Ixx	Iyy	Izz
1	1	1	1	1
11	1	1	2	3
31	1	1	1	1

SPRING ELEMENTS

Node 1	Dir 1	Node 2	Dir 2	Spring Value
Load Cell Springs				
2	1	12	1	10000
2	2	12	2	100000
2	3	12	3	10000
3	1	13	1	1000000
3	2	13	2	10000
3	3	13	3	100000
4	1	14	1	100000
4	2	14	2	10000
4	3	14	3	100000
5	1	15	1	100000
5	2	15	2	10000
5	3	15	3	10000
Structure Base Springs				
11	1	21	1	10000
11	2	21	2	10000
11	3	21	3	10000
11	4	21	4	8000
11	5	21	5	9000
11	6	21	6	10000

Table 3.2 Baseline full system modes

Mode No.	Frequency Hz	Mode No.	Frequency Hz	Mode No.	Frequency Hz
1	0.00	7	9.45	13	67.90
2	0.00	8	10.47	14	70.43
3	0.00	9	17.78	15	76.42
4	0.00	10	19.08	16	117.47
5	0.00	11	35.83	17	122.33
6	0.00	12	64.86	18	299.73

Table 3.3 Constraint shapes

	Singular value shapes					
DOF	1	2	3	4	5	6
102X+	0.002	−0.001	−0.036	0.002	−0.026	−0.001
102Y+	0.005	−0.101	−0.584	−0.039	0.786	0.017
102Z+	−0.001	0.081	−0.009	−0.102	0.014	−0.698
103X+	0.997	0.022	−0.041	0.004	−0.037	0.000
103Y+	0.009	−0.009	−0.023	−0.006	0.104	0.002
103Z+	−0.016	0.659	−0.043	0.735	0.092	−0.099
104X+	0.016	−0.010	−0.362	0.020	−0.259	−0.005
104Y+	0.017	−0.008	0.013	−0.008	0.130	0.003
104Z+	0.016	−0.738	0.067	0.659	−0.010	−0.112
105X+	−0.067	−0.023	−0.720	0.040	−0.515	−0.010
105Y+	0.009	−0.009	−0.023	−0.006	0.104	0.002
105Z+	0.002	−0.059	0.002	−0.110	0.004	−0.700

Six independent relative motion constraint shapes were calculated by performing a singular value decomposition of the relative motion FRFs. The constraint shapes are shown in Table 3.3. DOFs 102 through 105 correspond to the relative displacements across nodes 2 through 5 and nodes 12 through 15, respectively.

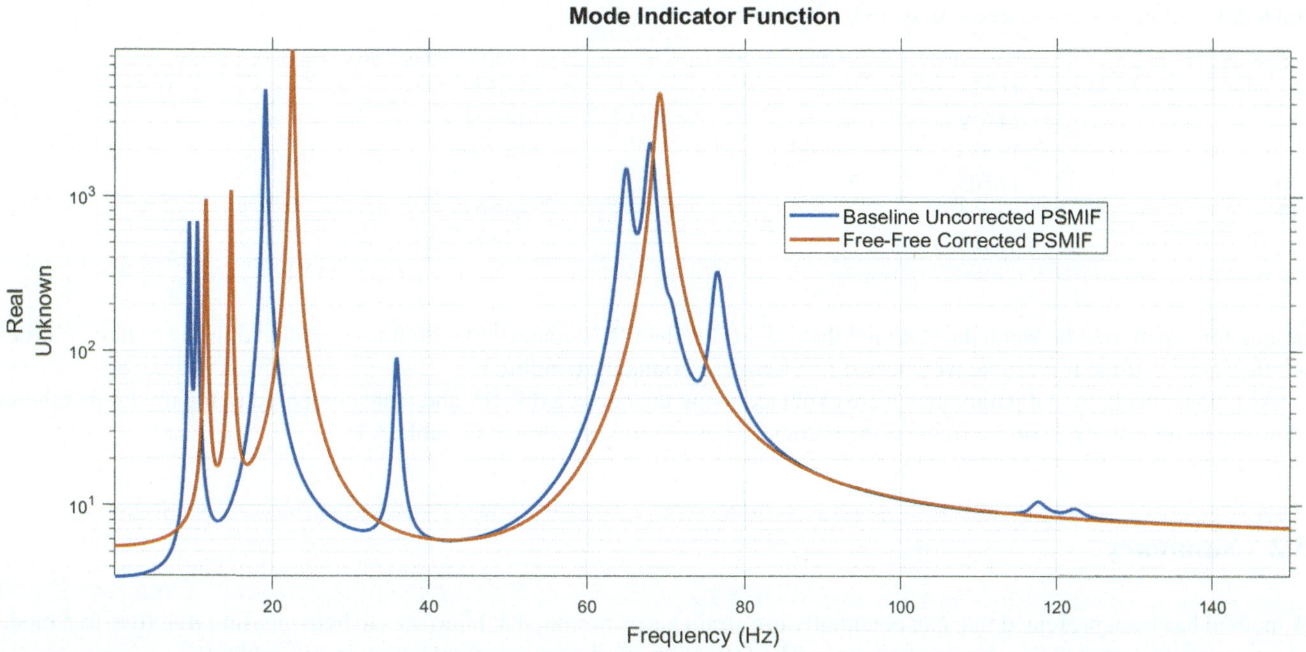

Fig. 3.2 PSMIF for baseline uncorrected FRF (blue) and free-free corrected FRF (red). The red curve is equivalent to FRF from a test article in a free-free boundary condition

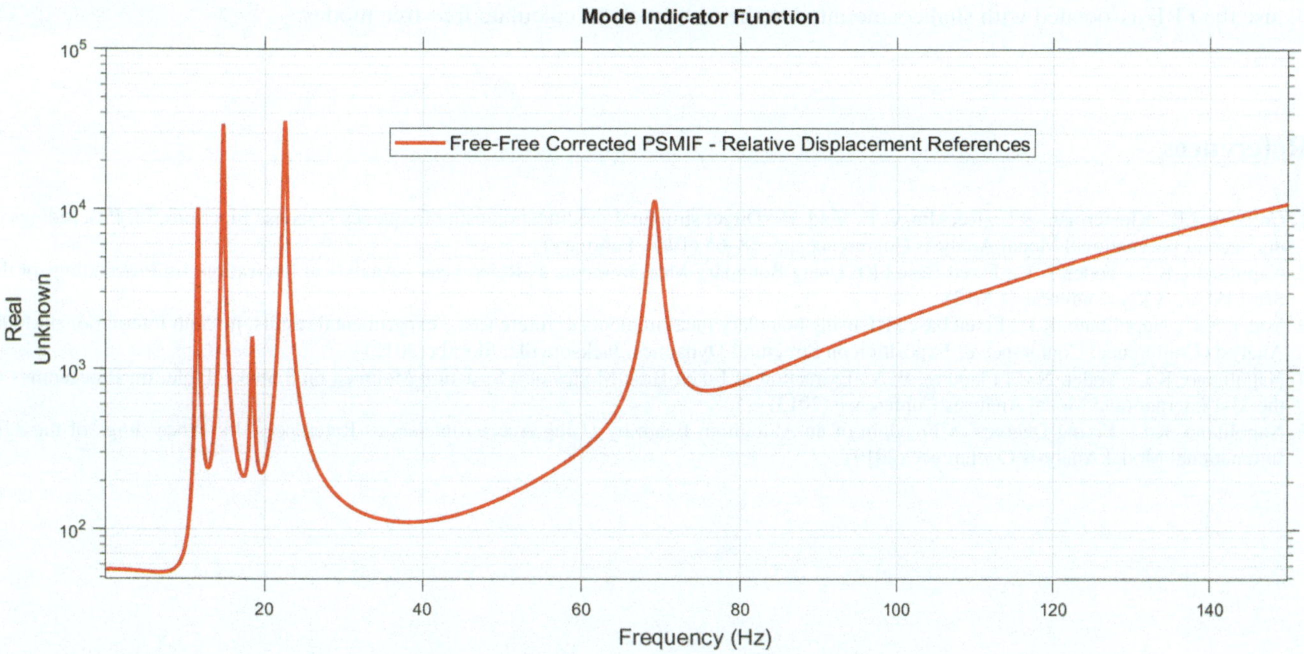

Fig. 3.3 PSMIF for free-free corrected FRF (red) associated with constraint shape references

These constraint shapes were used to calculate FRFs from the nine applied loads through the coordinate transformation $\{f_E\} = [\Psi]^+ \{\varepsilon\}$, and then a partial matrix inversion was performed using Eq. (3.7) to generate FRFs with the constraint shapes and the forces at node 31 as references. A power spectrum mode indicator function (PSMIF) for all acceleration DOFs on the test article—using the three DOFs at 31 as references for both the baseline and the corrected FRF—are shown in Fig. 3.2. Figure 3.3 shows the PSMIF of the corrected FRF of the test article using the relative displacement constraint shapes. Note that the PSMIF continues to increase as frequency increases. This is because the damping forces across the interface are not captured. However, since the constraint shapes capture all the unique relative motion patterns (and therefore all the

Table 3.4 Test article free-free modal parameters

Mode No.	Baseline Free-Free (Analysis) Hz	Corrected Free-Free (Fit FRF Data) Hz
7	11.537	11.537
8	14.699	14.699
9	18.380	18.380
10	22.508	22.508
11	69.002	69.002
12	69.145	69.145

unique force patterns) between the base and the test article, the FRFs associated with forces applied directly to the test article are the same as if the test article were tested in a free-free boundary condition.

As a final check, modal parameters were extracted using the corrected FRFs and were compared to the modal parameters of an eigenvalue analysis of the free-free test article. These results are shown in Table 3.4.

3.2 Summary

A method has been presented that can potentially use strain gages mounted at interfaces to help measure true free-free modes of a test article mounted to a flexible boundary. The following are the keys to implementing the method:

1. ensure that the strain gages, or constraint shapes of a combination of strain gages, capture all the interface forces,
2. apply at least one independent excitation to the base for each unique strain gage (or constraint shape) reference, and
3. use the FRF associated with shakers mounted to the test article to calculate free-free modes.

References

1. Crowley, J.R., Klosterman, A.L., Rocklin, G.T., Vold, H.: Direct structural modification using frequency response functions. In: Proceedings of the Second International Modal Analysis Conference, pp. 58–65 (1984, February)
2. Napolitano, K.L., Yoder, N.C.: Fixed Base FRF Using Boundary Measurements as References—Analytical Derivation. In: Proceedings of the SEM IMAC XXX Conference (2012)
3. Yoder, N.C., Napolitano, K.L.: Fixed base FRF using boundary measurements as references—experimental results. In: 30th International Modal Analysis Conference: Conference & Exposition on Structural Dynamics, Jacksonville, Florida (2012)
4. Napolitano, K.L., Yoder, N.C., Fladung, W.A.: Extraction of Fixed-Base Modes of a Structure Mounted on a Shake Table. In: Proceedings of the 31st International Modal Analysis Conference (2013)
5. Napolitano, K.L.: Fixing Degrees of Freedom of an Aluminum Beam by Using Accelerometers as References. In: Proceedings of the 37th International Modal Analysis Conference (2019)

Chapter 4
A Principle for Obtaining Pragmatic Uncertainty Bounds on Modal Parameters

Jonas G. Kjeld and Anders Brandt

Abstract In the field of structural dynamics, determination of damping contributions of civil structures is a subject influenced with considerable uncertainties. Many different modal parameter estimation techniques are available to describe the dynamic behavior using only the modal response, both in frequency and time domain, but common for all is the struggle to obtain stable damping values. This paper deals with the extraction of modal parameters on simulated data through Operational Modal Analysis with focus on establishing uncertainty bounds for the damping estimates. We investigate a pragmatic principle to bound the errors, including both random and bias errors. Modal parameters are extracted by applying a low order and a high order parameter estimation method which makes it possible to establish uncertainty bounds of the total damping. These uncertainty bounds represent a lower and a higher boundary for the damping estimates corresponding to each of the identified natural modes of the test structure. Attaching a statistical confidence interval to the estimates allows for a better understanding of the uncertainties related to the damping values obtained through Operational Modal Analysis.

Keywords Operational modal analysis · Damping quantification · Uncertainty bounds · Signal processing · Offshore structures

4.1 Introduction

In Operational Modal Analysis (OMA) of large civil structures, it is usually possible to identify consistent modal parameters in terms of natural frequencies and mode shapes but when it comes to damping of the structure, it appears to be much more difficult to quantify consistent estimates. If we consider an offshore wind turbine (OWT), the structure is exposed to a number of different operational conditions in terms of climatic changes, scour depth, the active system of the OWT itself and many other parameters. The variations in operational conditions influence the dynamic behavior of the structure and this is potentially the reason why it is so difficult to obtain stable damping estimates. Some of these challenges are described in [1, 2].

Today, a variety of modal parameter estimation (MPE) methods are available and while they all are capable of identifying and extract modal parameters of a given system, the results turn out to be slightly different. It is possible to establish a general formulation for many of the MPE methods based on either the Frequency Response Function (FRF) or Impulse Response Function (IRF) which makes it more intuitive to grasp the difference in modal parameters each method computes. A unified formulation for modal identification is presented in [3] which makes it possible to compare the most commonly used MPE methods.

The motivation for this paper comes from the experience that it is very difficult to obtain reliable damping estimates. If we include multiple MPE methods and investigate their characteristic equations, we expect that a confidence interval can be defined which bounds the errors, including both random and bias errors. In this paper, the pragmatic uncertainty bounds are defined by studying the modal parameters obtained by using the Mulitple-reference Ibrahim Time Domain (MITD) method

J. G. Kjeld (✉)
Department of Technology and Innovation, University of Southern Denmark, Odense, Denmark

Civil and Structure, Vattenfall Vindkraft A/S, Kolding, Denmark
e-mail: jonas.kjeld@vattenfall.com; jokje@iti.sdu.dk

A. Brandt
Department of Technology and Innovation, University of Southern Denmark, Odense, Denmark
e-mail: abra@iti.sdu.dk

B. Dilworth (ed.), *Topics in Modal Analysis & Testing, Volume 8*, Conference Proceedings of the Society for Experimental Mechanics Series, https://doi.org/10.1007/978-3-030-47717-2_4

and the Modified Multiple-reference Ibrahim Time Domain method. Among other MPE methods, the MITD and MMITD method are presented in [4].

In [5] and [6], numerical methods for calculating the uncertainty on the modal parameters have been presented with focus on simulated and measured vibration data.

4.2 Theory

The FRF matrix of a system, $[H]$, can be described as the inverse of the system impedance matrix, $[Z]$ which can be obtained by applying the Laplace transformation of Newton's equation based on physical properties including the mass, $[M]$, stiffness, $[K]$, and damping matrices, $[C]$. For large systems, such as most civil structures, we rarely know the damping matrix and extraction of modal parameters based on the entire modal matrices is inefficient in terms of computation. Instead, the Laplace transformation of Newton's equation can be rewritten using the mode shape matrix, $[\Psi]$, and pole matrix, $[S]$

$$[H(f)] = [\Psi][S^{-1}][\Psi^T] \tag{4.1}$$

In time domain, this can be written as

$$[h(t)] = [\Psi]\lceil e^{s_r t} \rfloor [L]^T \tag{4.2}$$

where Ψ is the eigenvector matrix, $\lceil e^{s_r t} \rfloor$ is the diagonal pole matrix and L is the modal participation factor (MPF) matrix.

4.2.1 Multiple-Reference Ibrahim Time Domain (MITD) Method

The MITD method is an extension of the Ibrahim time domain (ITD) method presented in [7]. The MITD method allows for multiple references to be included in the parameter estimation and is applicable for Operational Modal Analysis (OMA) as it works on free decay measurements and thus correlation functions.

The common equation given in Eq. (4.2) is also valid for the MITD method. Next step is to repeat this equation at different times, a total of m times in row direction and n times in column direction

$$[H_{mn}(t)] = \begin{bmatrix} [h(t)] & [h(t+\Delta t)] \cdots & [h(t+(n-1)\Delta t)] \\ [h(t+\Delta t)] & [h(t+2\Delta t)] \cdots & [h(t+n\Delta t)] \\ \vdots & \vdots \quad \cdots & \vdots \\ [h(t+(m-1)\Delta t)] & [h(t+n\Delta t)] \ldots & [h(t+(m+n-2)\Delta t)] \end{bmatrix} \tag{4.3}$$

This equation is known as the block Hankel matrix (referred to as the Hankel matrix). By using the Hankel matrix, Eq. (4.2) can be expanded into

$$[H_{mn}(t)] = [\tilde{\Psi}]\lceil e^{s_r t} \rfloor [\tilde{L}]^T \tag{4.4}$$

The expanded mode shapes are given as

$$\left[\tilde{\Psi}\right] = \begin{bmatrix} [\Psi] \\ [\Psi]\lceil e^{s_r \Delta t} \rfloor \\ \vdots \\ [\Psi]\lceil e^{s_r (n-1)\Delta t} \rfloor \end{bmatrix} \tag{4.5}$$

and the extended MPF matrix

$$\left[\tilde{L}\right] = \left[[L]^T \; \lceil e^{s_r \Delta t} \rfloor [L]^T \; \cdots \; \lceil e^{s_r (n-1)\Delta t} \rfloor [L]^T \right] \tag{4.6}$$

It is now possible to rewrite Eq. (4.4) by post multiplying with the pseudo-inverse of the expanded mode shape matrix $\left[\tilde{\Psi}\right]$

$$[\tilde{\Psi}]^{+}[H_{mn}(t)] = \lceil e^{s_r t} \rfloor [\tilde{L}]^{T} \tag{4.7}$$

If a new Hankel matrix is defined, shifted Δt in time, the diagonal pole matrix becomes $\lceil e^{s_r(t+\Delta t)} \rfloor$ which can be rewritten to $\lceil e^{s_r t} \rfloor \lceil e^{s_r \Delta t} \rfloor$. Using Eq. (4.7) with the time shifted Hankel matrix, we get

$$[H_{mn}(t+\Delta t)] = [\tilde{\Psi}]\lceil e^{s_r \Delta t} \rfloor [\tilde{\Psi}]^{+}[H_{mn}(t)] \tag{4.8}$$

Defining a system matrix $[A] = [\tilde{\Psi}]\lceil e^{s_r \Delta t} \rfloor \tilde{\Psi}]^{+}$ results in

$$[A][H_{mn}(t)] = [H_{mn}(t+\Delta t)]. \tag{4.9}$$

The system matrix can be calculated as

$$\begin{aligned} &[A][H_{mn}(t)] = [H_{mn}(t+\Delta t)] \implies [A][H_{mn}(t)][H_{mn}(t)]^{T} = [H_{mn}(t+\Delta t)][H_{mn}(t)]^{T} \\ &\implies [A_1] = [H_{mn}(t+\Delta t)][H_{mn}(t)]^{T}([H_{mn}(t)][H_{mn}(t)]^{T})^{-1} \end{aligned} \tag{4.10}$$

From Equation (4.8) and (4.10) an eigenvalue problem for each column in $\left[\tilde{\Psi}\right]$ can be established which needs to be solved in order to obtain the modal parameters

$$[A]\left[\tilde{\Psi}\right] = \left[\tilde{\Psi}\right]\lceil e^{s_r \Delta t} \rfloor \tag{4.11}$$

Here, λ_r are the eigenvalues and correspond to $e^{s_r \Delta t}$. The poles of the system, s_r can now be calculated as $s_r = f_s \ln(\lambda_r)$.

4.2.2 *Modified Multiple-Reference Ibrahim Time Domain (MMITD) Method*

For the MMITD method, the transposed Hankel matrix at time t is required

$$[H_{mn}(t)]^{T} = \left[\tilde{L}\right]\lceil e^{s_r t} \rfloor \left[\tilde{\psi}\right]^{T} \tag{4.12}$$

The time shifted Hankel matrix then becomes

$$[H_{mn}(t+\Delta t)]^{T} = \left[\tilde{L}\right]\lceil e^{s_r \Delta t} \lceil e^{s_r t} \rfloor \left[\tilde{\psi}\right]^{T} \tag{4.13}$$

Similar to Equation (4.7), we get

$$\lceil e^{s_r t} \rfloor \left[\tilde{\psi}\right]^{T} = \left[\tilde{L}\right]^{+}[H_{mn}(t)]^{T} \tag{4.14}$$

Inserting this into Eq. (4.13), the equation becomes

$$[H_{mn}(t+\Delta t)]^{T} = \left[\tilde{L}\right]\lceil e^{s_r \Delta t} \rfloor \left[\tilde{L}\right]^{+}[H_{mn}(t)]^{T} \tag{4.15}$$

Post multiplication of $[H_{mn}(t)]$ into Eq. (4.15) gives

$$[H_{mn}(t+\Delta t)]^{T}[H_{mn}(t)] = \left[\tilde{L}\right]\lceil e^{s_r \Delta t} \rfloor \left[\tilde{L}\right]^{+}[H_{mn}(t)]^{T}[H_{mn}(t)] \tag{4.16}$$

A system matrix is defined as

$$[A] = \left([H_{mn}(t+\Delta t)]^T [H_{mn}(t)]\right)\left([H_{mn}(t)]^T [H_{mn}(t)]\right)^{-1} \tag{4.17}$$

Finally, in order to obtain the eigenvalue problem, post multiplying Eq. (4.16) with $\left([H_{mn}(t)]^T [H_{mn}(t)]\right)^{-1}$ and $\left[\tilde{L}\right]$ is needed

$$[A]\left[\tilde{L}\right] = \left[\tilde{L}\right]\lceil e^{s_r \Delta t}\rfloor \tag{4.18}$$

It is important to note that the MMITD method gives us the modal participation factors ($[L]$) and not directly the mode shapes, in contrary to the MITD method. The mode shapes can be estimated in a second step, by using the least squares time domain method [4].

Relevant for both MITD and MMITD is data compression. In this paper, Singular Value Decomposition (SVD) is applied which condenses the frequency response matrix so that the resulting matrix contains the same information as the original.

When working with MPE methods, a number of decisions have to be made on how to solve the system equation for the modal parameter estimation. Some of the decisions are described below.

4.2.3 Low Order Versus High Order Method

Instead of using a physically-based mathematical model to describe the various MPE methods, it is also possible to consider a matrix coefficient polynomial model where the common characteristics can be more intuitively identified. By considering a particular response point p and reference point q, Eq. (4.2) can be written as

$$h_{pq}(t_i) = \frac{u_p(t_i)}{f_q(t_i)} = \frac{\beta_n(z)^n + \beta_{n-1}(z)^{n-1} + \cdots + \beta_0(z)^0}{\alpha_m(z)^m + \alpha_{m-1}(z)^{m-1} + \cdots + \alpha_0(z)^0} \tag{4.19}$$

Where $z = e^{s\Delta t}$ and in this case, t_i denotes an arbitrary time – this can be seen as the measured time. By collecting the terms in Eq. (4.19), the polynomial can be written as

$$h_{pq}(t_i) = \frac{u_p(\omega_i)}{f_q(\omega_i)} = \frac{\sum\limits_{l=0}^{n} \beta_l(z)^l}{\sum\limits_{k=0}^{m} \alpha_k(z)^k} \tag{4.20}$$

If the MPE method considered is based on the original FRF or IRF, the polynomial order describing the denominator (α_k) in Eq. (4.20) will in most cases be low (one or two) and the MPE method will be described as a low order method, accordingly. The resulting dimensions of α_k is $[N_S x N_S]$ and β_k is $[N_L x N_L]$. Here, S denotes the *short* dimension which refers to the number of references (inputs) and L denotes the *long* dimension which refers to the number of responses (outputs). If the MPE method is based on the transposed FRF or IRF, instead, the dimensions of α_k and β_k are reversed and thus the polynomial order of the denominator (that is α_k) will become higher. The method is now defined as a high order method.

Equation (4.2) directly shows that the MITD method is a low order method since it is built on the original FRF whereas the MMITD method is built on the transposed FRF seen from Eq. (4.12) making it a high order method.

4.2.4 Normalization to the Lowest or Highest Coefficient of the Matrix Polynomials

The roots of the matrix coefficients can be determined by solving for the eigenvalues of the companion matrix. This requires normalization of either the lowest order coefficient ($\alpha_0 = 1$) or the highest order coefficient ($\alpha(m) = 1$). It can be worth to consider the influence of the normalization since it may influence the results to some extent. A study comparing low order and high order normalization has been carried out in [8] which, however, generally showed an insignificant difference in the estimated modal parameters.

4.2.5 Calculating the System Matrix with Respect to Either the Original Hankel Matrix or the Time Shifted Hankel Matrix

If we consider Eq. (4.10), the system matrix [A] can also be calculated using the time shifted Hankel matrix ($[H_{mn}(t+\Delta t)]$). If this is the case, $[A]$ will become

$$[A_2] = [H_{mn}(t+\Delta t)][H_{mn}(t+\Delta t)]^T([H_{mn}(t)][H_{mn}(t+\Delta t)]^T)^{-1} \tag{4.21}$$

It turns out, that the resulting modal parameters change slightly depending on which Hankel matrix is used for calculating the system matrix. According to [9], it is good practice to estimate the modal parameters by considering both cases and average the extracted modal parameters.

4.3 Methodology

The modal parameters are extracted from simulated acceleration time data based on an FE-model representing an OWT with 9 references located evenly spaced along the structure. In the simulation, only the first 4 non-symmetrical bending modes are included and the damping of each mode has been set to 1%. In Table 4.1, the settings for simulation of time data are listed. The natural frequencies of the first 4 non-symmetrical bending modes are 0.309, 1.197, 2.006 and 4.300 Hz, respectively.

For most MPE methods, it is necessary to determine the optimal model order which yields better results in terms of modal parameters. Here, it is very popular to utilize the so-called stabilization diagram which visualizes the calculated poles as a function of increasing model order. In order to select poles that represent the actual natural modes of the system, certain engineering skills and judgment is required. An algorithm for automated operational modal analysis (AOMA) is presented in [10] that successfully has estimated modal parameters from different data sets including experimental measurements from a bridge test. The algorithm utilizes a statistical representation of modal parameters and is complemented by a number of decision rules based on the modal assurance criterion. The estimated modal parameters are only dependent on a few specified tolerances which reduces any bias that may have been introduced by the user. A similar approach is carried out in this paper to estimate the modal parameters of the simulated data.

In this paper, a low order and a high order method are considered which refers to MITD and MMITD, respectively, but the coefficient matrices of the system are only normalized to the highest order. Both cases of either using the first Hankel matrix or the time shifted Hankel matrix are also taken into account. In order to truly investigate the outcome of these considerations, it has been important to ensure that as many input parameters to the MPE method as possible remain unchanged. One of the more important input parameters in the MPE methods, is the first time lag and the total number of time lags considered in the correlation functions. Since the correlation function of the noise (typically broadband noise) dies out fast, the first few time lag values should be skipped – in this case, the first 10 were skipped. The total number of time lags is fixed but for the modal parameter estimation, the start value will vary for each computation.

In Table 4.2, the input parameters for the estimation of modal parameters are given.

Table 4.1 Settings for simulation of acceleration time data

Data simulation			
Sampling frequency f_s [Hz]	Measurement time T [s]	Number of references [–]	Damping of each mode [z] [%]
10	9000	9	1

Table 4.2 Input parameters for modal parameter estimation

Modal parameter estimation				
Time lags N_{lines} [–]	Start lag N_s [–]	Block size N [–]	Model order $\text{Max}_{\text{modes}}$ [–]	MAC tolerance MAC_{tol} [–]
100	10:10:100	2048	60	0.99

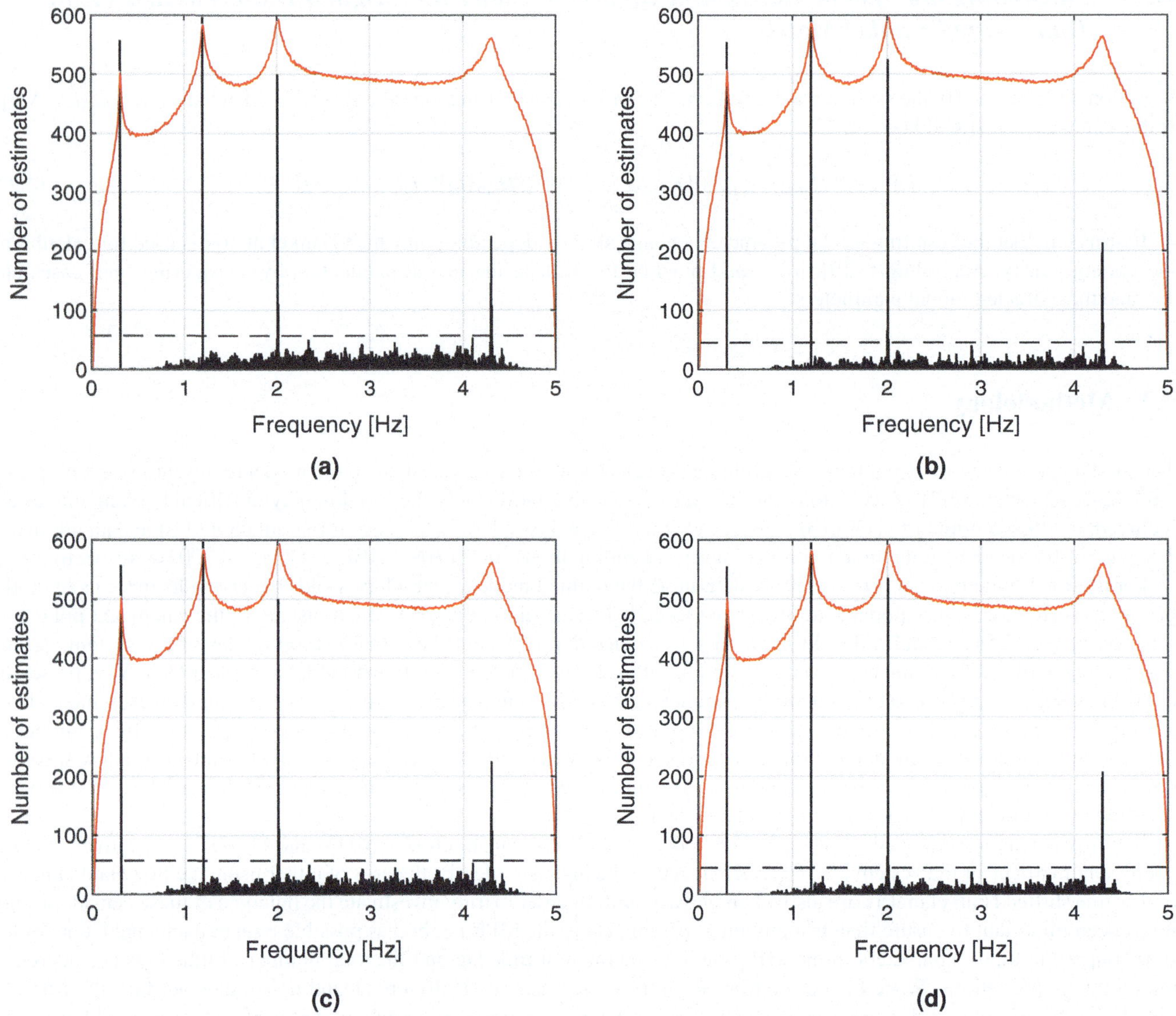

Fig. 4.1 Histograms of calculated poles. (——) shows the MIF overlay. (▬) shows the bins of the histogram and (- - -) indicates the minimum number of estimates required in a bin to be considered in the further analysis. (**a**) MITD and $[A_1]$. (**b**) MITD and $[A_2]$. (**c**) MMITD and $[A_1]$. (**d**) MMITD and $[A_2]$

4.4 Results and Discussion

During modal parameter estimation, a total of 4 separate calculations are carried out. Results of the first two are visualized in Fig. 4.1a and b which are based on the MITD method with changes to the system matrix $[A]$ as to how it is calculated – that is either $[A_1]$ from Eq. (4.10) or $[A_2]$ from Eq. (4.21). The results of the last two are visualized in Fig. 4.1c and d which are based on the MMITD method, similarly with changes to the system matrix. The Mode Indicator Function (MIF) overlay shown in the figures, is an average of the Power Spectral Density using Welch method with a 50% overlap Hanning window for all output channels. From the figures, it is very difficult to tell any difference between the MITD method and MMITD

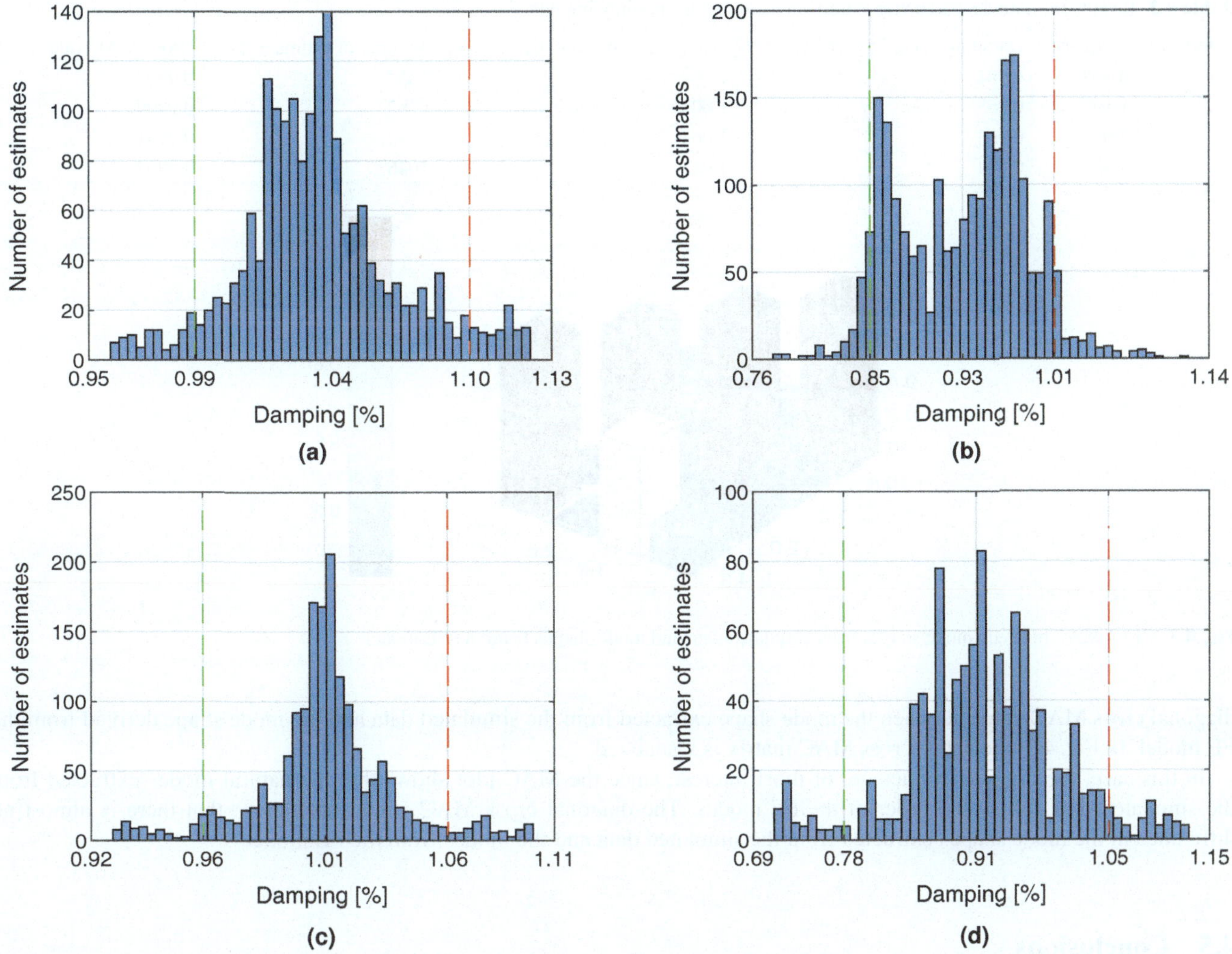

Fig. 4.2 Histograms of damping estimates from identified modes. (- - -) indicates the lower boundary (5%-fractile) and (- - -) indicates the higher boundary (95%-fractile). (**a**) Mode 1 at 0.308 Hz. (**b**) Mode 2 at 1.197 Hz. (**c**) Mode 3 at 2.005 Hz. (**d**) Mode 4 at 4.307 Hz

method whereas the largest difference can be seen when switching between $[A_1]$ and $[A_2]$. Nevertheless, the estimates for all 4 calculations are very similar.

Estimates for all 4 calculations are collected into one set of parameters for each mode which has been identified as one of the FE modes. In order to ensure that each mode from all calculations have similar modal properties, a MAC criterion of 0.99 is applied. In Fig. 4.2, histograms of the collected damping estimates for the 4 modes (which match the FE model) are shown together with the 5%-fractile and 95%-fractile (90% confidence interval).

From the plots in Fig. 4.2a–d, there is no apparent distribution that fits them all. This is somehow unexpected since many of the studies on uncertainty bounds of modal parameters assume that the distribution of measured or simulated data is Gaussian. Mode 3 comes closest to a Gaussian distribution but the remainder seems far from – especially mode 2 and 4. For the first 3 modes, the majority of damping estimates are within a small range and the target damping of 1% lies within the 90% confidence interval. The 1% target damping is also within the boundaries for mode 4 but based on the lower boundary (5%-fractile), the damping is generally estimated lower than expected. It is also evident to see from the histograms in Fig. 4.1 that the number of damping estimates within one bin for the fourth mode is much lower than the number of estimates for the other modes. The fourth mode also has multiple bins very close in frequency with a significant number of estimates whereas the estimates lie in a single bin for each of the first 3 modes. This is possibly a result of a small bin width which is based on the first frequency, in fact, the frequency of the first mode divided by 50 (0.006 Hz). In Table 4.3, the 90% confidence interval is listed for all 4 modes. The total number of estimates for each mode is also listed and here, this number also tells that there are significantly fewer estimates of the fourth mode compared to the rest. The last column of the table displays the

Table 4.3 Results from modal parameter extraction including uncertainty bounds

Mode no.	Frequency	Damping value 95%-fractile [%]	Damping value 95%-fractile [%]	Number of estimates [–]	Cross-MAC value [–]
1	0.308	0.991	1.099	1854	0.9997
2	1.197	0.855	1.008	2282	0.9999
3	2.005	0.963	1.064	1742	0.9991
4	4.307	0.784	1.045	1046	0.9999

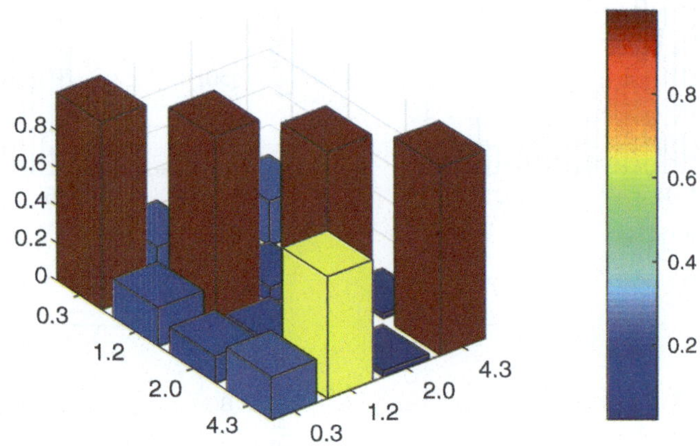

Fig. 4.3 Cross MAC between mode shapes from simulated data and mode shapes from the FE model

diagonal cross MAC value between the mode shape extracted from the simulated data and the mode shape derived from the FE model. In Fig. 4.3, the entire cross MAC matrix is visualized.

In this case, the diagonal values are of most interest, since the MAC plot shows that the natural modes extracted from the simulated data follow the order of the FE modes. The diagonal cross MAC values also show, that there is almost no difference in the mode shapes extracted from the simulated data and the modes from the FE model.

4.5 Conclusions

This paper presented pragmatic uncertainty bounds for modal parameters obtained from simulated data in terms of a statistical confidence interval. In the first step, acceleration time data were simulated based on an FE model which represents an offshore wind turbine. In the second step, modal parameters of the system were extracted using the Multiple-reference Ibrahim Time Domain method and Modified Multiple-reference Ibrahim Time Domain method and finally, the results of both methods were compared. A statistical representation in terms of histograms showed that the distribution of the damping estimates is not Gaussian. It was shown that a 90% confidence interval bounded the true damping for all modes.

Acknowledgments The research presented in this paper is funded by Vattenfall Vindkraft A/S and the Innovation Fund Denmark.

References

1. Carswell, W., Johansson, J., Løvholt, F., Arwade, S., Madshus, C., DeGroot, D., Myers, A.: Foundation damping and the dynamics of offshore wind turbine monopiles. Renew. Energy **80**, 724–736 (2015)
2. Shirzadeh, R., Weijtjens, W., Guillaume, P., Devriendt, C.: The dynamics of an offshore wind turbine in parked conditions: a comparison between simulations and measurements. Wind Energy **18**(10), 1685–1702 (2014)
3. Allemang, R.J., Brown, D.: A unified matrix polynomial approach to modal identification. J. Sound Vib. **211**(3), 301–322 (1998)
4. Allemang, R.J., Brown, D.L.: Experimental modal analysis and dynamic component synthesis. Volume 3: Modal parameter estimation. Technical report, Department of Mechanical and Industrial Engineering, Cincinnati University Ohio (1987)
5. Pintelon, R., Guillaume, P., Schoukens, J.: Uncertainty calculation in (operational) modal analysis. Mech. Syst. Signal Process. **21**(6), 2359–2373 (2007)

6. Reynders, E., Pintelon, R., De Roeck, G.: Uncertainty bounds on modal parameters obtained from stochastic subspace identification. Mech. Syst. Signal Process. **22**(4), 948–969 (2008)
7. Ibrahim, S. R., Mikulcik, E. C.: A Method for the Direct Identification of Vibration Parameters from the Free Response. Shock Vib. Bull. **47**(47), 183–198 (1977)
8. Christensen, S.S., Brandt, A.: Time domain modal parameter estimation methods used in OMA: a comparison. In: 8th International Operational Modal Analysis Conference, IOMAC 2019, pp. 507–509 (2019)
9. Brincker, R., Ventura, C.: Introduction to operational modal analysis. John Wiley & Sons Ltd, Chichester, United Kingdom (2015)
10. Christensen, S.S., Brandt, A.: Automatic operational modal analysis using statistical modelling of pole locations. In: 28th International Conference on Noise and Vibration Engineering, pp. 2927–2938 (2018)

Jonas G. Kjeld is MSc in Structural Engineering from the University of Southern Denmark. He is currently engaged in an industrial PhD which is a collaboration together with Vattenfall and the University of Denmark. Jonas works with the dynamics of offshore wind turbine foundations and how to quantify the damping contribution of such structures.

Chapter 5
On Partitioning of an SHM Problem and Parallels with Transfer Learning

George P. Tsialiamanis, D. J. Wagg, Paul A. Gardner, N. Dervilis, and K. Worden

Abstract In the current work, a problem-splitting approach and a scheme motivated by transfer learning is applied to a structural health monitoring problem. The specific problem in this case is that of localising damage on an aircraft wing. The original experiment is described, together with the initial approach, in which a neural network was trained to localise damage. The results were not ideal, partly because of a scarcity of training data, and partly because of the difficulty in resolving two of the damage cases. In the current paper, the problem is split into two sub-problems and an increase in classification accuracy is obtained. The sub-problems are obtained by separating out the most difficult-to-classify damage cases. A second approach to the problem is considered by adopting ideas from transfer learning (usually applied in much deeper) networks to see if a network trained on the simpler damage cases can help with feature extraction in the more difficult cases. The transfer of a fixed trained batch of layers between the networks is found to improve classification by making the classes more separable in the feature space and to speed up convergence.

Keywords Structural health monitoring (SHM) · Machine learning · Classification · Problem splitting · Transfer learning

5.1 Introduction

Structural health monitoring (SHM) refers to the process of implementing a damage detection strategy for aerospace, civil or mechanical engineering infrastructure [1]. Here, damage is defined as changes introduced into a system/structure, either intentionally or unintentionally, that affect current or future performance of the system. Detecting damage is becoming more and more important in modern societies, where everyday activities depend increasingly on engineering systems and structures. One the one hand, safety has to be assured, both for users and for equipment or machinery existing within these structures. On the other hand, infrastructure is often designed for a predefined lifetime and damage occurrence may reduce the expected lifetime and have a huge economic impact as a result of necessary repairs or even rebuilding or decommison. Damage can be visible on or in structures, but more often it is not, and has to be inferred from signals measured by sensors placed on them.

An increasingly useful tool in SHM is *machine learning* (ML) [1]. In many current applications large sets of data are gathered by sensors or generated by models and these can be exploited to gain insight into structural dynamics and materials engineering. Machine learning is employed because of its efficiency in classification, function interpolation and prediction using data. Data-driven models are built and used to serve SHM purposes. These models can also be used to further understand how structures react to different conditions and explain their physics. However, one of the main drawbacks of such methods is the need for large datasets. ML models may have many parameters which are established during *training* on data which may need to span all the health conditions of interest for the given structure or system. Larger datasets assist in better tuning of the models as far as accuracy and generalisation are concerned. However, even if large datasets are available, sometimes there are very few observations on damaged states, which are important in SHM. In the current paper, increased accuracy of a data-driven SHM classifier will be discussed in terms of two strategies: splitting the problem into two sub-

G. P. Tsialiamanis (✉) · D. J. Wagg · P. A. Gardner · N. Dervilis · K. Worden
Dynamics Research Group, Department of Mechanical Engineering, University of Sheffield, Sheffield, UK
e-mail: g.tsialiamanis@sheffield.ac.uk; david.wagg@sheffield.ac.uk; p.gardner@sheffield.ac.uk; n.dervilis@sheffield.ac.uk; k.worden@sheffield.ac.uk

B. Dilworth (ed.), *Topics in Modal Analysis & Testing, Volume 8*, Conference Proceedings of the Society for Experimental Mechanics Series, https://doi.org/10.1007/978-3-030-47717-2_5

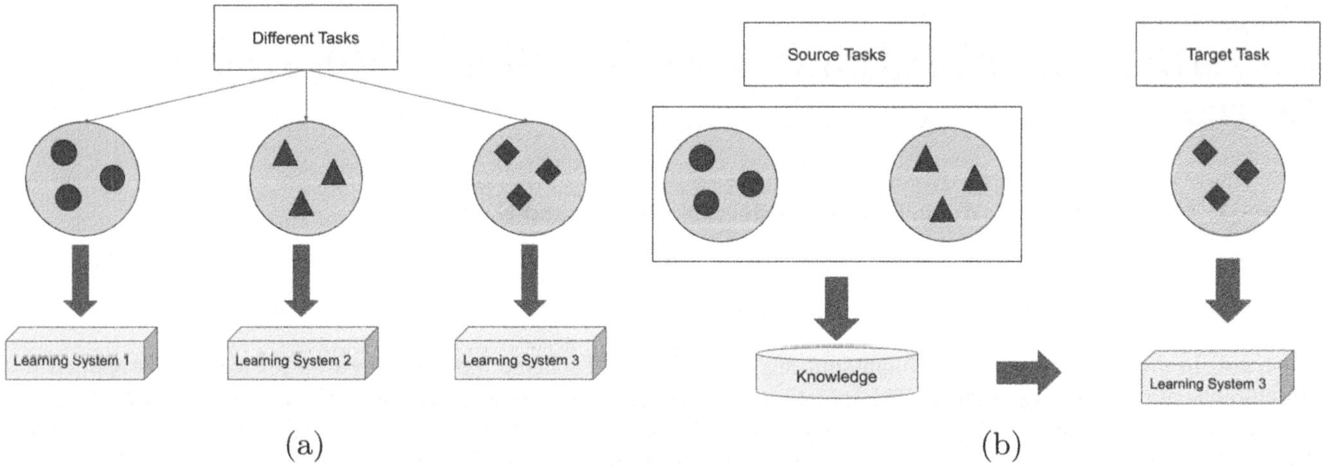

Fig. 5.1 Traditional (**a**) and transfer (**b**) learning schemes (following [2])

problems and attempting transfer of information between the two sub-problems in a manner motivated by transfer learning [2].

Transfer learning is the procedure of taking knowledge from a source domain and task and applying it to a different domain and task to help improve performance on the second task [2]. Transfer learning is useful because a model trained on a dataset can not naturally be applied on another due to difference in data distribution, but can be further tuned to also apply on the second dataset. An accurate representation of the difference between traditional and transfer learning schemes can be seen in Fig. 5.1. The SHM problem herein will be addressed using neural networks [3], for which transfer learning has been proven quite efficient (although usually in deeper learning architectures [7, 8]). Due to the layered structure of the networks, after having created a model for a task, transferring a part of it (e.g. some subset of the layers) is easy. The method is used in many disciplines, such as computer vision [4, 5]. The most commonly-used learners are Convolutional Neural Networks (CNNs), which can be very slow to train and may need a lot of data, which in many cases can be hard to obtain (e.g. labelled images). These problems can be dealt with by using the fixed initial layers of pre-trained models to extract features of images, and then train only the last layers to classify in the new context. In this way, both the number of trainable parameters and the need for huge datasets and computation time are reduced. Another topic that transfer learning has been used in is natural language processing (NLP) [6], where the same issues of lack of labelled data and large amounts of training time are dealt with by transferring of pre-trained models into new tasks. Further examples of the benefits of transfer learning can be found in web document classification [7, 8]; in these cases, in newly-created web sites, lack of labelled data occurs. To address this problem, even though the new web sites belong to a different domain than the training domain of the existing sites, the same models can be used to help classify documents in the new websites.

In the context of the current work, transfer learning is considered in transferring knowledge from one sub-problem to the other by introducing pre-trained layers into new classifiers. The classification problem that will be presented is related to damage class/location. A model trained to predict a subset of the damage classes (source task) with data corresponding of that subset (source domain), will be used to boost performance of a second classifier trained to identify a different subset of damage states.

5.2 Problem Description

Similar to the aforementioned applications, in SHM machine learning is also used for classification and regression. In data driven SHM one tries to identify features that will reveal whether a structure is damaged or what type of damage is present and so, labelled data are necessity. Therefore, in SHM applications lack of labelled data about damage location or severity is a drawback. SHM problems can be categorised in many ways but are often broken down according to the hierarchical structure proposed by Rytter [9]:

1. Is there damage in the system (*existence*)?
2. Where is the damage in the system (*location*)?

3. What kind of damage is present (*type/classification*)?
4. How severe is the damage (*extent/severity*)?
5. How much useful (safe) life remains (*prognosis*)?

A common approach to the first level is to observe the structure in its normal condition and try to find changes in features extracted from measured signals that are sensitive to damage. This approach is called *novelty detection* [10, 11], and it has some advantages and disadvantages. The main advantage is that it is usually an *unsupervised* method, that is only trained on data that are considered to be from the undamaged condition of the structure, without a specific target class label. These methods are thus trained to detect any *changes* in the behaviour of the elements under consideration, which can be a disadvantage, since structures can change their behaviour for benign reasons, like changes in their environmental or operational conditions; such benign changes or *confounding influences* can raise false alarms.

In this work a problem of damage localisation is considered (at Level 2 in Rytter's hierarchy [9]); the structure of interest being a wing of a Gnat trainer aircraft. The problem is one of supervised-learning, as the data for all damage cases were collected and a classification model was trained accordingly. Subsequently, the classifier was used to predict the damage class of newly-presented data. The features used as inputs to the classifier were novelty indices calculated between frequency intervals of the transmissibilities of the normal condition of the structure (undamaged state) and the testing states. The transmissibility between two points of a structure is given by equation (5.1), and this represents the ratio of two response spectra. This feature is useful because it describes the response of the structure in the frequency domain, without requiring any knowledge of the frequency content of the excitation. The transmissibility is defined as,

$$T_{ij} = \frac{FRF_i}{FRF_j} = \frac{\frac{\mathcal{F}_i}{\mathcal{F}_{excitation}}}{\frac{\mathcal{F}_j}{\mathcal{F}_{excitation}}} = \frac{\mathcal{F}_i}{\mathcal{F}_j} \tag{5.1}$$

where, $\mathcal{F}_i$ is the Fourier Transform of the signal given by the ith sensor and FRF_i is the *Frequency Response Function* (FRF) at the ith point.

The experiment was set up as described in [12]. The wing of the aircraft was excited with a Gaussian white noise using an electrodynamic shaker attached on the bottom surface of the wing. The configuration of the sensors placed on the wing can be seen in Fig. 5.2. Responses were measured with accelerometers on the upper surface of the wing, and the transmissibilities

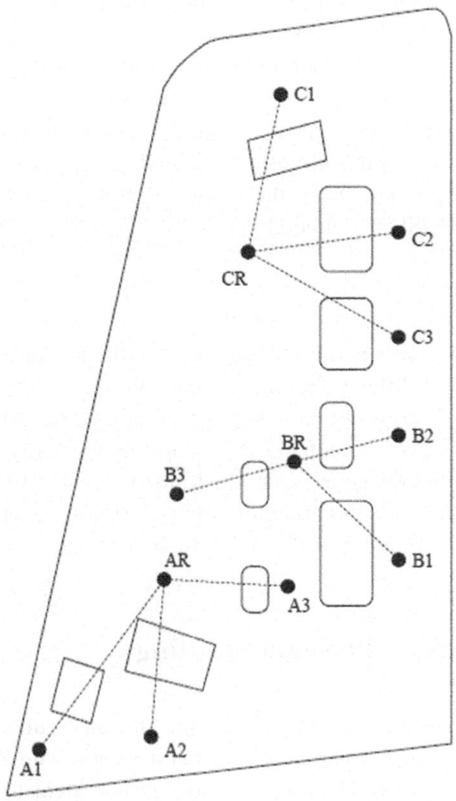

Fig. 5.2 Configuration of sensors on the Gnat aircraft wing [13]

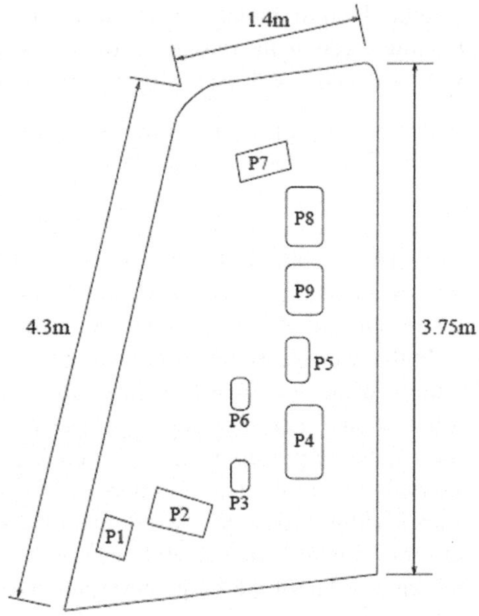

Fig. 5.3 Schematic showing wing panels removed to simulate the nine damage cases [13]

between each sensor and the corresponding reference sensor were calculated. The transmissibilities were recorded in the 1–2 kHz range, as this interval was found to be sensitive to the damage that was going to be introduced to the structure. Each transmissibility contained 2048 spectral lines.

Initially, the structure was excited in its normal condition, i.e. with no introduced damage. The transmissibilities of this state were recorded and subsequently, to simulate damage, several panels were removed from the wing, one at a time. In each panel removal, the wing was excited again with white Gaussian noise and the transmissibilities were recorded. The panels that were removed are shown in Fig. 5.3. Each panel has a different size, varying from 0.008 to 0.08 m^2 and so the localisation of smaller panels becomes more difficult, since their removal affects the transmissibilities less than the bigger panels. The measurements were repeated 200 times for each damage case, ultimately leading to 1800 data points belonging to nine different damage cases/classes. The data were separated into training, validation and testing sub sets, each having 66 points per damage case.

For the purposes of damage localisation, features had to be selected which would be sensitive to the panel removals; this was initially done manually [13], selecting by visual 'engineering judgement' the intervals of the transmissibilities that appeared to be more sensitive to damage and calculating the novelty indices of each state by comparison with the transmissibilities of the undamaged state. The novelty indices were computed using the Mahalanobis squared-distance (MSD) D_ζ^2 of the feature vectors $\mathbf{x}_\zeta$, which in this case contained the magnitudes of transmissibility spectral lines. The MSD is defined by,

$$D_\zeta^2 = (\mathbf{x}_\zeta - \overline{\mathbf{x}})^T S^{-1} (\mathbf{x}_\zeta - \overline{\mathbf{x}}) \tag{5.2}$$

were $\overline{\mathbf{x}}$ is the sample mean on the normal condition feature data, and S is the sample covariance matrix.

After selecting 'by eye' the most important features for damage detection [13], a genetic algorithm was used [14] to choose the most sensitive features, in order to localise/classify the damage. Finally, nine features were chosen as the most sensitive and an MLP neural network [3] with nine nodes in the input layer, ten nodes in the hidden layer and nine nodes in the decision layer was trained. The confusion matrix of the resulting classifier is shown in the Table 5.1. It can be seen that the misclassification rate is very low and that the damage cases that are most confused are the ones where the missing panel is Panel 3 or Panel 6, which were the smallest ones.

5.3 Problem Splitting

As mentioned in [15], the rule-of-thumb for a network that generalises well is that it should be trained with at least ten samples per weight of the network. The aforementioned network had 180 trainable weights (and another 19 bias terms) so the 596 training samples are not ideal for the neural network. As a solution, a splitting of the original problem into two

Table 5.1 Confusion Matrix of neural network classifier, test set, total accuracy: 98.14% [14]

Predicted panel	1	2	3	4	5	6	7	8	9
Missing panel 1	65	0	0	0	0	0	0	0	1
Missing panel 2	0	65	0	1	0	0	0	0	0
Missing panel 3	1	0	62	0	0	1	0	1	1
Missing panel 4	0	0	0	66	0	0	0	0	0
Missing panel 5	0	0	0	0	66	0	0	0	0
Missing panel 6	0	3	0	0	0	62	0	1	0
Missing panel 7	0	0	0	0	0	0	66	0	0
Missing panel 8	1	0	0	0	0	0	0	65	0
Missing panel 9	0	0	0	0	0	0	0	0	66

sub-problems is considered here to try and reduce the misclassification rate on the testing data even further. The dataset is split into two parts, one containing all the damage cases except Panels 3 & 6 and the second containing the rest of the data. Subsequently, two neural network classifiers were trained separately on the new datasets. This was thought to be a good practice, since the panels are the smallest, and their removal affects the novelty indices less than the rest of the panel removals. The impact is that the points appear closer to each other in the feature space, and are swamped by points belonging to other classes, so the initial classifier cannot separate them efficiently. By assigning the tasks to different classifiers, an increase in performance is expected, especially in the case of separating the two smallest panel classes.

To illustrate the data feature space, a visualisation is attempted here. Since the data belong to a nine-dimensional feature space, principal component analysis (PCA) was performed on the data and three of the principal components, explaining 71% of total variance, are plotted in scatter plots shown in Fig. 5.4a. Points referring to data corresponding to the missing panels 3 and 6 (grey and magenta points respectively) are entangled with other class points causing most of the misclassification rate shown above.

Random initialisation was followed for the neural networks. Initial values of the weights and biases of the networks were sampled from a normal zero-mean distribution. The two networks were initialised several times and trained for different sizes of the hidden layer to find the ones with optimal structure for the newly-defined problems. After randomly initialising and training multiple neural networks for both cases and keeping the ones with the minimum loss function value the best architectures were found to be networks with nine nodes in the hidden layer for both cases and seven output nodes for the first dataset and two for the second. The loss function used in training was the categorical cross-entropy function given by,

$$L(y, \hat{y}) = -\frac{1}{N}\sum_{i=1}^{N}\sum_{j=1}^{n_{cl}}[y_{i,j} log \hat{y}_{i,j} + (1 - y_{i,j}) log(1 - \hat{y}_{i,j})] \tag{5.3}$$

In Equation (5.3), N is the number of samples during training, n_{cl} is the number of possible classes, $\hat{y}_{i,j}$ the estimated probability that the ith point belongs to the jth class and $y_{i,j}$ is 1 if the ith sample belongs to the jth class, otherwise it is 0.

Confusion matrices on the test sets for the classifiers are shown in Tables 5.2 and 5.3. By splitting the dataset into two subsets the total accuracy is slightly increased from 98.14% to 98.82%. This is best considered in terms of classification error, which has been reduced from 1.86% to 1.18%, and this is an important reduction in SHM terms. Reduction of the number of trainable parameters has certainly contributed to this improvement, since the amount of training data is small. Performance on the task of separating only the two smallest panel classes was also increased because it is an easier task for the classifier than trying to discriminate them among the panel removals with greater impact on the novelty indices. This fact is also clear in Fig. 5.4c, where the principal components of samples belonging to the classes of missing Panels 3 and 6 are clearly separable.

5.4 Knowledge Transfer Between the Two Problems

Having split the problem into two sub-problems, a scheme motivated by transfer learning in deeper learners was examined. The idea being to establish if the features extracted at the hidden layer in one problem, could be used for the other. In transfer learning termology, the seven-class problem specifies the source domain and task, while the two-class problem gives the target domain and task. The transfer is carried out by using the fixed input and hidden layers from the classifier in the source

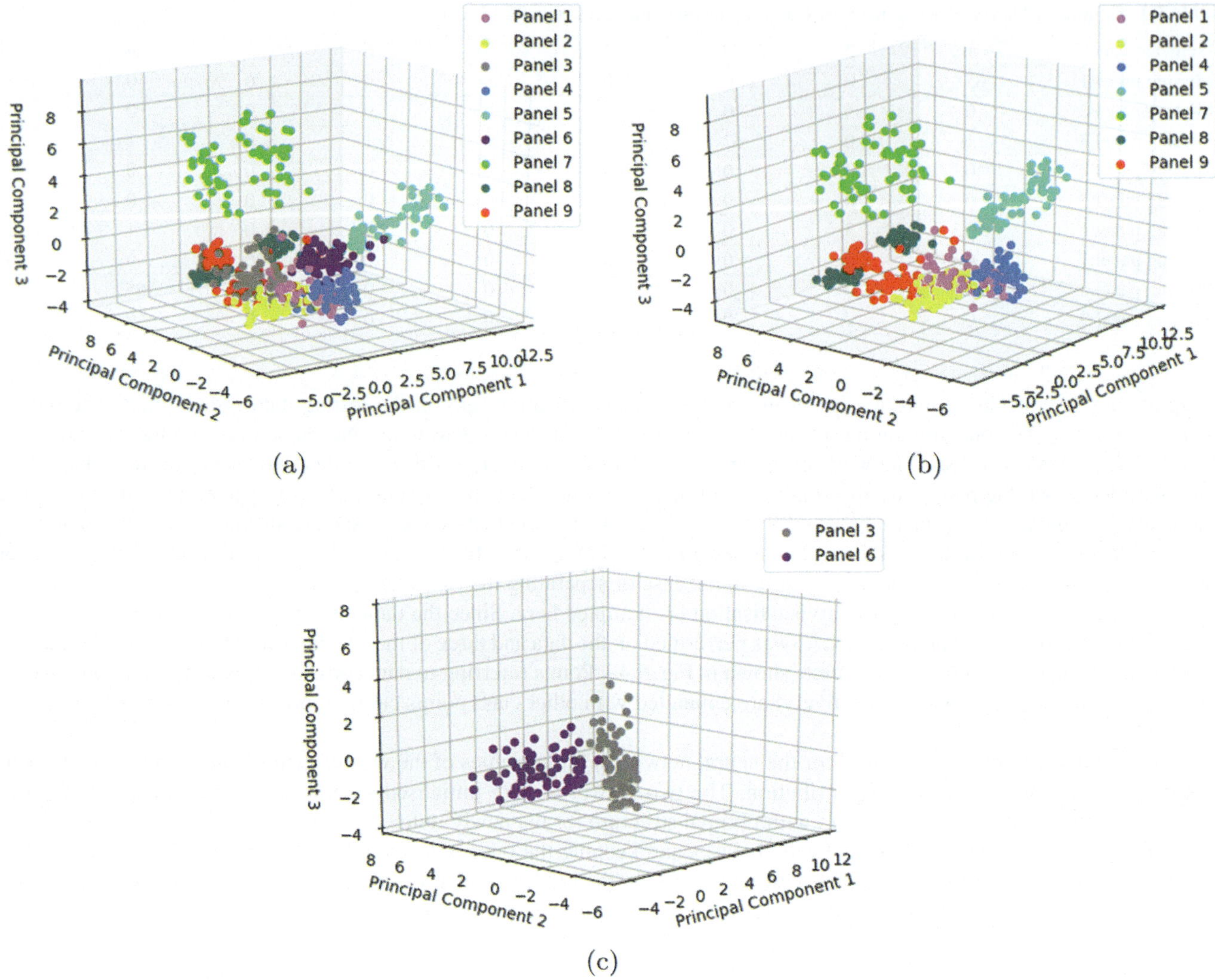

Fig. 5.4 Principal components of all samples (**a**), samples excepting panels 3 and 6 (**b**) and samples of panels 3 and 6 (**c**)

Table 5.2 Confusion Matrix of neural network classifier trained on the first dataset, test set, total accuracy: 98.48%

Predicted panel	1	2	4	5	7	8	9
Missing panel 1	65	1	0	0	0	0	0
Missing panel 2	0	63	1	0	0	0	2
Missing panel 4	1	0	65	0	0	0	0
Missing panel 5	0	0	0	66	0	0	0
Missing panel 7	0	0	0	0	66	0	0
Missing panel 8	1	0	0	0	0	65	0
Missing panel 9	1	0	0	0	0	0	65

task, as the input and hidden layers of the target task; this means that only the weights between the hidden and output layers remain to be trained for the target task. This strategy reduces the number of parameters considerably. The functional form of the network for the source task is given by,

$$\mathbf{y} = f_0 W_2(f_1(W_1\mathbf{x} + b_1)) + b_2 \tag{5.4}$$

where f_0 and f_1 are the non-linear activation functions of the output layer and the hidden layer respectively, $W_{1,2}$ are the weight matrices of the transformations between the layers, $b_{1,2}$ are the bias vectors of the layers, $\mathbf{x}$ is the input vector and $\mathbf{y}$ the output vector. The *softmax* function is chosen to be the activation function of the decision layer, as this is appropriate

Table 5.3 Confusion Matrix of neural network classifier trained on the first dataset, test set, total accuracy: 100%

Predicted panel	3	6
Missing panel 3	66	0
Missing panel 6	0	66

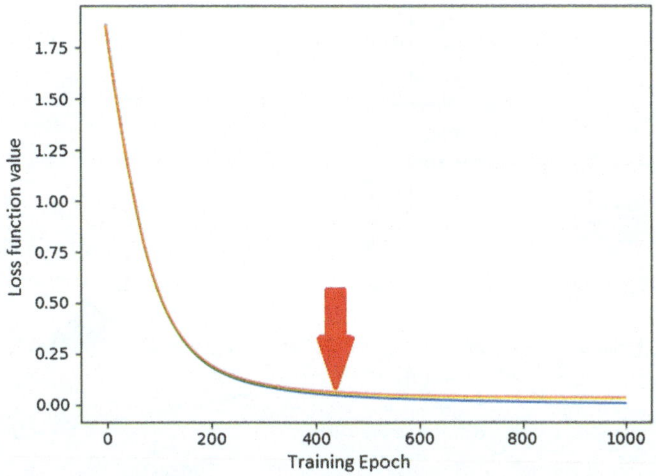

Fig. 5.5 Training and validation loss histories and the point of early stopping (red arrow)

Table 5.4 Confusion Matrix of neural network classifier trained on the original data of the second dataset, test set, total accuracy: 97.72%

Predicted panel	3	6
Missing panel 3	65	1
Missing panel 6	2	64

to a classification problem. The prediction of the network, concerning which damage class the sample belongs to, is the index that maximises the output vector **y**; the outputs are interpreted as the a posteriori probabilities of class membership, so this leads to a Bayesian decision rule. Loosely speaking, one can think of the transformation between the hidden and output layers as the actual classifier, and the transformation between the input layer into the hidden layer as a map to latent states in which the classes are more easily separable. In the context of deep networks, the hope is that the earlier layers carry out an automated feature extraction which facilitates an eventual classifier. In the deep context, transfer between problems is carried out by simply copying the 'feature extraction' layers directly into the new network, and only training the later classification layers. The simple idea explored here, is whether that strategy helps in the much more shallow learner considered in this study. The transfer is accomplished by copying the weights W_1 and biases b_1 from sub-problem one directly into the network for sub-problem two, and only training the weights W_2 and biases b_2.

As before, multiple neural networks were trained on the first dataset. In a transfer learning scheme, it is even more important that models should not be overtrained, since that will make the model too case-specific and it would be unlikely for it to carry knowledge to other problems. To achieve this for the current problem, an early stopping strategy was followed. Models were trained until a point were the value of the loss function decreases less than a percentage of the current value. An example of this can be seen in Fig. 5.5 where instead of training the neural network for 1000 epochs, training stops at the point indicated with the red arrow.

After multiple networks were trained following the early stopping scheme above, the network with the lowest value on validation loss was determined and the transfer learning scheme was applied to the second problem. The nonlinear transformation given by the transition from the input layer to the hidden layer was applied on the data of the second dataset. Consequently, another neural network was trained on the transformed data, having only one input layer and one output/decision layer. To comment on the effect of the transformation, another two-layer network was trained on the original second dataset and the results were compared.

The confusion matrices of the two neural networks on the testing data are given in Tables 5.4 and 5.5; the misclassification rates are very similar. However, it is interesting to also look at the effect of the transfer on the convergence rate of the network trained on the transferred data and also to illustrate the feature transformation on the first and the second datasets.

Table 5.5 Confusion Matrix of neural network classifier trained on the transformed data of the second dataset, test set, total accuracy: 96.96%

Predicted panel	3	6
Missing panel 3	65	1
Missing panel 6	3	63

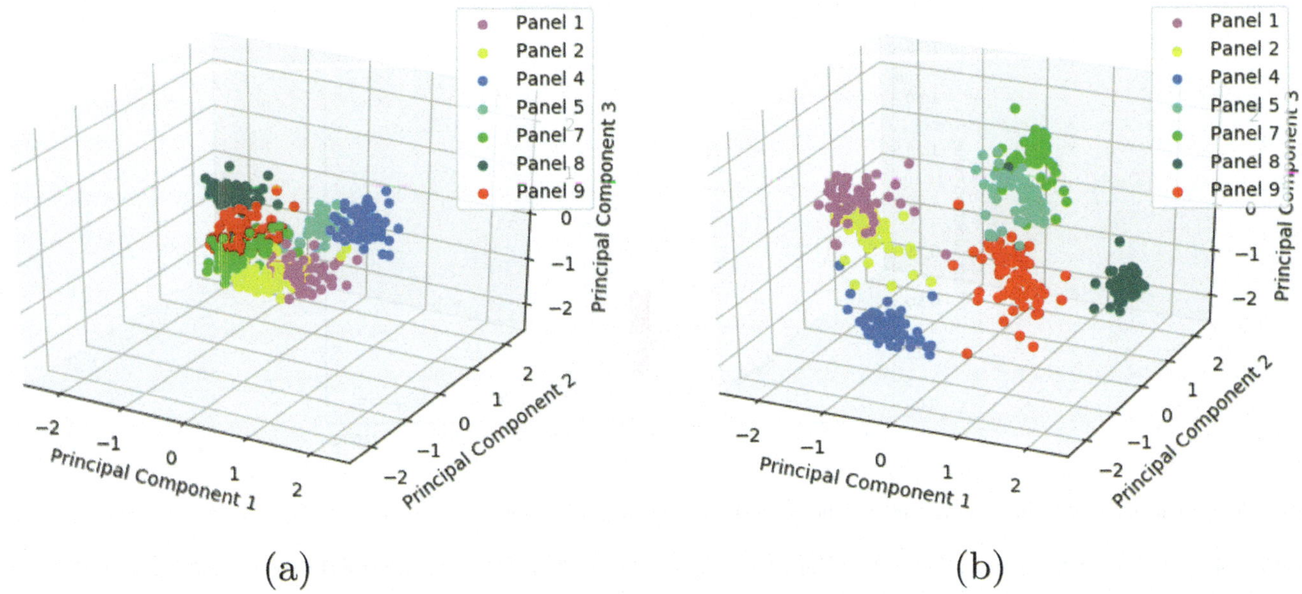

Fig. 5.6 Principal components of original features of the first dataset (**a**) and transformed features (**b**)

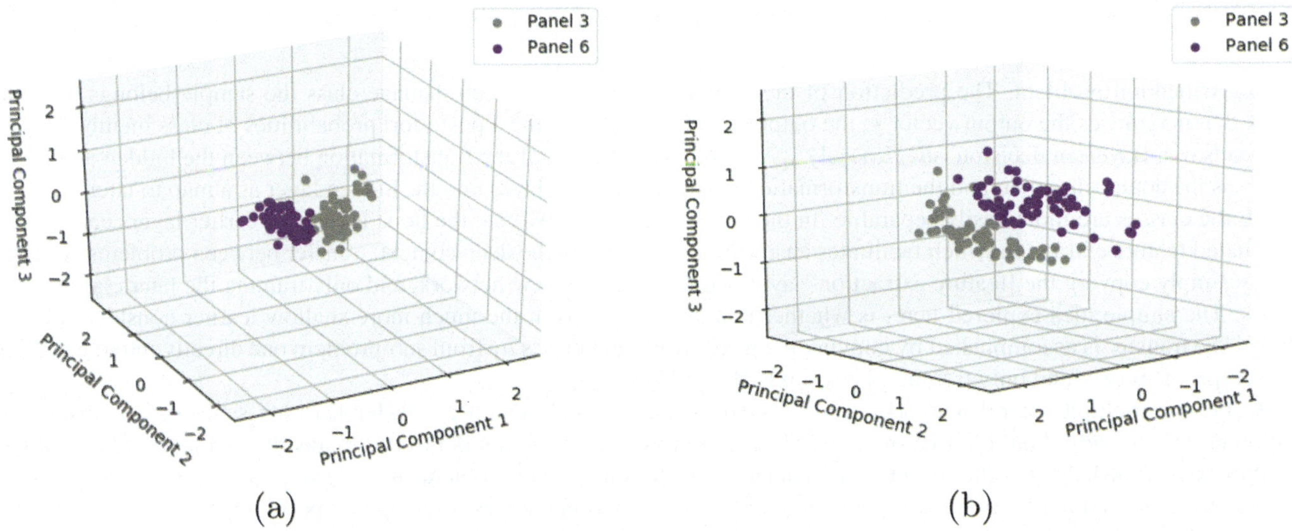

Fig. 5.7 Principal components of original features of the second dataset (**a**) and transformed features (**b**)

The training histories of the two models can be seen in Fig. 5.8. It is clear that the loss history of the model with transformed data (blue and cyan lines) converges faster, especially in the initial part of the training, and it also reaches a lower minimum value for the loss function in the same number of training epochs. This can be explained by looking at the effect of the learnt transformation on the data. In Figs. 5.6 and 5.7 this effect is illustrated. (Note that the points are different from those in Fig. 5.4, because principal component analysis was performed this time on the normalised data in the interval $[-1, 1]$ for the neural network training). The transformation spreads out the points of the original problem (first dataset) in order to make their separation by the decision layer easier; however, it is clear that it also accomplishes the same result on the second dataset. The points in Fig. 5.7b are spread out compared to the initial points and thus, their separation by the single layer neural network is easier. Furthermore, the points lay further away from the required decision boundary and this explains

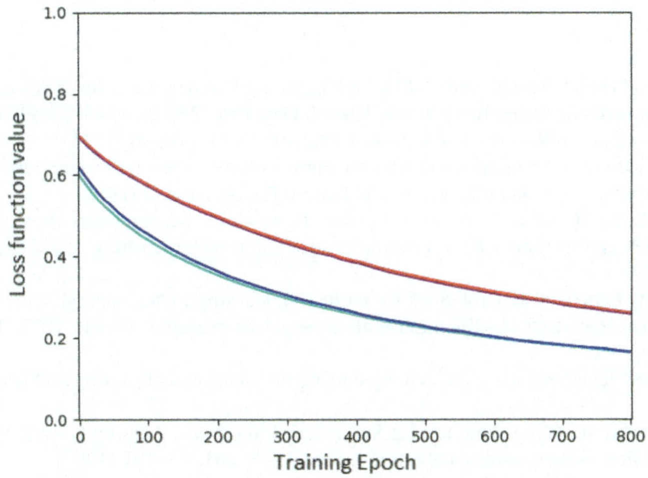

Fig. 5.8 Loss histories of transferred model: train(blue), validation(cyan) and model trained on initial data: train(red), validation(magenta)

both the faster training convergence and the lower minimum achieved. In contrast to the transformation of the first dataset, in the second dataset, the transformation does not concentrate points of the same class in specific areas of the feature space (Fig. 5.7). In Fig. 5.6b the points are both spread and concentrated closer according to the class they belong. This probably means that only a part of the physics of the problem is transferred in the second problem through this specific transformation.

5.5 Discussion and Conclusions

For the SHM classification (location) problem considered here, splitting the dataset into two subsets contributed to increasing the classification accuracy by a small percentage. This result was explained by the lesser effect that the small panel removals had on the novelty index features. This issue arose because the points representing these classes were close to each other and also points from other classes – those corresponding to large panel removal/damage. By considering the two damage cases as different problem, perfect accuracy was achieved in the task of classifying damage to the small panels, and there was also a small increase in the performance of the classifier tasked to identify the more severe damage states.

An attempt at a crude form of transfer learning was also investigated. Having trained the neural network classifier on the first dataset of the seven damage cases, transfer of knowledge to the second sub-problem was considered. This was accomplished by copying the first two layers of the first classifier – the 'feature extraction' layers – directly into the second classifier and only training the connections from the hidden layer to the output. The result is not particularly profound; the transfer does allow a good classifier, even with the smaller set of trainable parameters, but is not as good as training the network from scratch. The result is interesting, because it is clear that the source network is carrying out a feature clustering and cluster separation on the source data, that is still useful when the target data are presented. This suggests that the main issue with the small-panel damage classification is that the data are masked by the close presence of the large-panel data. Separating out the small-panel is the obvious answer. The results are interesting because they illustrate in a 'toy' example, how the early layers in deeper networks are manipulating features automatically in order to improve the ultimate classification step. The other benefit of the separation into sub-problems, was the faster convergence of the network training.

Acknowledgments The authors would like to acknowledge Graeme Manson for providing the data used. Moreover, the authors would like to acknowledge the support of the Engineering and Physical Science Research Council (EPSRC) and the European Union (EU). G.T is supported by funding from the EU's Horizon 2020 research and innovation programme under the Marie Sklodowska-Curie grant agreement DyVirt (764547). The other authors are supported by EPSRC grants EP/R006768/1, EP/R003645/1, EP/R004900/1 and EP/N010884/1.

References

1. Farrar, C.R., Worden, K.: Structural Health Monitoring: A Machine Learning Perspective. John Wiley & Sons ltd. West Sussex (2012)
2. Pan, S.J., Yang, Q.: A survey on transfer learning. IEEE Trans. Knowl. Data Eng. **22**(10), 1345–1359 (2009)
3. Bishop, C.M.: Neural Networks for Pattern Recognition. Oxford University Press, Oxford (1995)
4. Oquab, M., Bottou, L., Laptev, I., Sivic, J.: Learning and transferring mid-level image representations using convolutional neural networks. In: Proceedings of the IEEE Conference on Computer Vision and Pattern Recognition (2014)
5. Shin, H.-C., Roth, H.R., Gao, M., Lu, L., Xu, Z., Nogues, I., Yao, J., Mollura, D., Summers, R.M.: Deep convolutional neural networks for computer-aided detection: CNN architectures, dataset characteristics and transfer learning. IEEE Trans. Med. Imaging. **35**(5), 1285–1298 (2016)
6. Bingel, J., Søgaard, A.: Identifying beneficial task relations for multi-task learning in deep neural networks. CoRR, abs/1702.08303 (2017)
7. Fung, G.P.C., Yu, J.X., Lu, H., Yu, P.S.: Text classification without negative examples revisit. IEEE Trans. Knowl. Data Eng. **18**(1), 6–20 (2005)
8. Al-Mubaid, H., Umair, S.A.: A new text categorization technique using distributional clustering and learning logic. IEEE Trans. Knowl. Data Eng. **18**(9), 1156–1165 (2006)
9. Rytter, A.: Vibrational Based Inspection of Civil Engineering Structures. PhD thesis, Aalborg University (1993)
10. Worden, K.: Structural fault detection using a novelty measure. J. Sound Vib. **201**, 85–101 (1997)
11. Worden, K., Manson, G., Feilelr, N.R.J.: Damage detection using outlier analysis. J. Sound Vib. **229**, 647–667 (2000)
12. Worden, K., Manson, G.: The application of machine learning to structural health monitoring. Philos. Trans. R. Soc. A: Math. Phys. Eng. Sci. **365**, 515–537 (2007)
13. Manson, G., Worden, K., Allman, D.J.: Experimental validation of a structural health monitoring methodology: part III. Damage location on an aircraft wing. J. Sound Vib. **259**, 365–385 (2003)
14. Worden, K., Manson, G., Hilson, G., Pierce, S.G.: Genetic optimisation of a neural damage locator. J. Sound Vib. **309**(3–5), 529–544 (2008)
15. Tarassenko, L.: Guide to Neural Computing Applications. John Wiley and Sons Inc, New York (1998)

George P. Tsialiamanis studied at the school of Civil Engineering at the National Technical University of Athens during the period 2012–2017. His specialisation is on Structural Engineering and his thesis was about high performance surrogate models for stochastic finite element problems. After his studies he had an internship at Beta-CAE Systems SA (2017–2018), on applications of machine learning techniques in various engineering problems. More specifically, machine learning was used to predict the results of car crash test simulations. At the moment he is a PhD student at the University of Sheffield in the DyVirt program funded by the Marie Sklodowska-Curie EU Horizon 2020 research and innovation programme. The PhD is focused on decision support systems for structural health monitoring (SHM) and also in building a unifying framework for the various components of Dynamic Virtualization.

Chapter 6
An Ontological Approach to Structural Health Monitoring

George P. Tsialiamanis, David J. Wagg, I. Antoniadou, and K. Worden

Abstract In the current work, an ontological framework for structural health monitoring (SHM) is discussed. Ontologies are used in disciplines like knowledge engineering and natural language processing, but their structure and goals also fit the purposes of SHM. In SHM projects – as in all projects – many problems arise during knowledge sharing and application. Ontologies can deal with these problems and at the same time have more benefits for SHM processes, as their modularity may assist in extending and transferring knowledge. An SHM-specific ontology is constructed here and described; It contains many objects that can be used in the procedure of monitoring structures. The ontology can also be used as a database to store data acquired, but also serves as a knowledge-base for the current discipline's algorithms and methods. Further, having close connections to object-oriented programming, ontologies straightforwardly facilitate software development and reusability of their components. Certainly, the ontology can be used to save time during the application of SHM, but also can be applied to improve performance of existing methods, by finding within the ontology the best algorithm to fit the purpose of each method.

Keywords Structural health monitoring (SHM) · Ontologies · Database · Knowledge-base · Knowledge engineering

6.1 Introduction

In modern societies, everyday activities are becoming increasingly dependent on structural and mechanical systems. These systems have a design life which is heavily dependent on the external conditions that they will be subjected to throughout their operation. Since it is critical that they survive their design while remaining operational and safe, a framework to ensure both operability and safety is required. With this purpose in mind, *structural health monitoring* (SHM) can be employed. SHM refers to the process of implementing a damage detection strategy for aerospace, civil or mechanical engineering infrastructure [1]. Application of SHM to systems can be partitioned into the following steps [2]: (1) observation of the system during its operation, (2) data acquisition from the system and, (3) extraction of features that are sensitive to damage and determine the current state of system's health. The steps of SHM can be implemented in many ways. More specifically, the final two steps of feature extraction and current state evaluation have been implemented with various methods that come from different scientific disciplines like signal processing, machine learning, physical modelling, etc. All these methods interact with each other and have their advantages and disadvantages. Being motivated by this and by the fact that in many projects problems arise from poor communication between various project members and difficulty in knowledge sharing [3], this paper is proposing a connecting framework for these components and for sharing knowledge within the SHM process.

The proposed framework here is an *ontological* one. Ontologies are used in many fields including computer science, semantics and natural language processing. The Ontology's purpose of sharing knowledge, developing software modules and interoperability between different projects fits the needs of SHM and can be exploited to improve performance of such applications. In the current work, an SHM ontology is constructed and discussed in the context of using it to better understand its component functions: using it as a database, to apply SHM techniques more efficiently and to implement software according to the ontology.

Although applicable to SHM in a broad sense, an ontology could be particularly useful in a Population-based SHM (PBSHM) setting [4–8], where the goal is to develop general inference tools across a population. Here an ontology would

G. P. Tsialiamanis (✉) · D. J. Wagg · I. Antoniadou · K. Worden
Dynamics Research Group, Department of Mechanical Engineering, University of Sheffield, Sheffield, UK
e-mail: g.tsialiamanis@sheffield.ac.uk; david.wagg@sheffield.ac.uk; i.antoniadou@sheffield.ac.uk; k.worden@sheffield.ac.uk

B. Dilworth (ed.), *Topics in Modal Analysis & Testing, Volume 8*, Conference Proceedings of the Society for Experimental Mechanics Series, https://doi.org/10.1007/978-3-030-47717-2_6

identify useful, and perhaps overlooked, connections between objects such as Irreducible Element (IE) models and Attributed Graph (AG) representations of structures [5, 6] and appropriate knowledge transfer methods like transfer learning [7] and 'forms' [4, 7]. This may provide benefits in highlighting appropriate methods for each data source such that destructive phenomenon like negative transfer in transfer learning are avoided.

A similar approach has been outlined in previous work [9] to exploit the advantages of using ontologies in the scope of verification and validation (V&V) and system identification, fields that also have many interacting components.

6.2 Components of the Ontology

Ontologies have many definitions. The one that is preferred here is given in [10]: "An ontology is a specification of a conceptualization". This means that the ontology is a description of knowledge on a specific domain that can be helpful in sharing and explaining, storing and reusing/transferring it to similar projects and domains.

Ontology construction is usually carried out using ontology languages like OWL [11]. To facilitate ontology construction, software with user interfaces exist e.g. Protege [12] and GATE. The specific ontology construction software used herein was Protege, developed at Stanford University in collaboration with the University of Manchester; it was chosen for its convenience in editing the ontology and defining the components. In Fig. 6.1 the interface of Protege is shown, on the left various pieces of the ontology are listed and on the right a specific class is being edited. Protege has many capabilities and flexibility in editing and creating ontologies like defining variable types according to one's needs, automatic clustering of objects following axioms, etc.

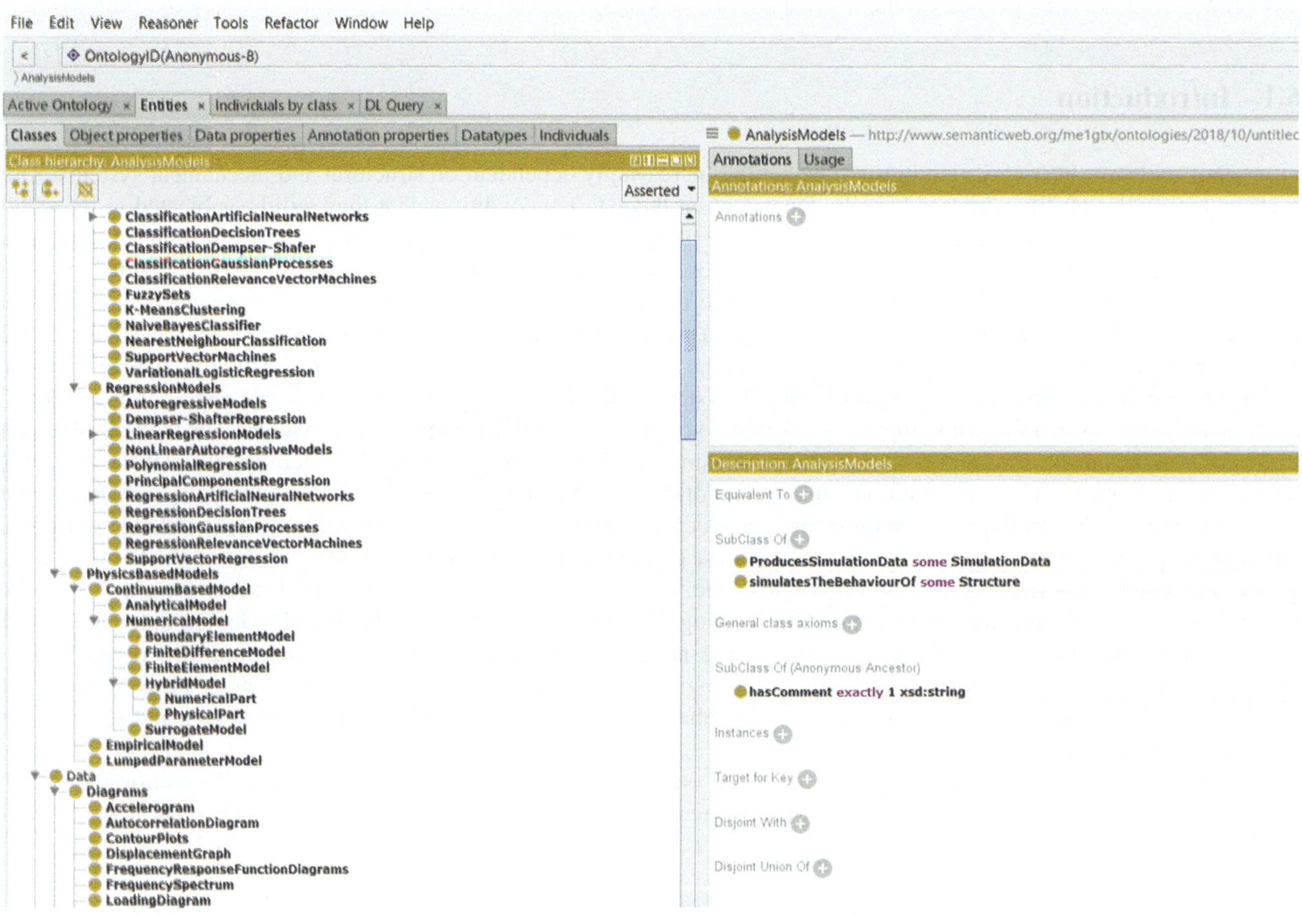

Fig. 6.1 Protege ontology editor

The procedure of constructing the ontology is executed through construction of its components. The components constituting an ontology are:

1. **Individuals**: are members of the ontology. Individuals are instances of the classes that they belong to and have corresponding characteristics.
2. **Classes**: are sets that include instances with similar characteristics. Classes are partitioned in subclasses and all the instances belonging to them share the properties of the superclass.
3. **Connections**: are the relationships connecting classes. The defined connections show the way that two or more classes may be related. The name of the connection is the semantic definition of the interaction between connected classes.
4. **Attributes**: are characteristics of individuals; they refer to a quantitative characteristic of an individual and their values are not always the same between individuals of the same class.
5. **Properties**: are the connections between individuals and attributes. A property of an individual reveals an attribute that may differentiate it from similar individuals. As with the connections, the property name defines the semantics of the individual-attribute relationship.
6. **Annotations**: are strings including definitions of objects. They can be thought as an attribute being a string that describes the functionality of an individual or a class.

Having defined an ontology through the components above, a hierarchy of classes and subclasses is created. This is also called a taxonomy and is the result of the ontology having connected different classes only through connections defining the relationship of subclass and superclass. Following this path, an initial tree structure was accomplished. This is the fundamental structure of the in-hand ontology. Connections linking classes with different relationships were defined thereafter. In the following sections, both the fundamental structure and the complete structure will be described.

6.3 Main Ontology Structure

In order to facilitate the shaping of the ontology, a taxonomy was initially defined comprising of the most important superclasses of instances used in SHM. These superclasses were chosen to be:

1. analysis models;
2. data;
3. data processing;
4. physical parts;
5. SHM methods.

The first four classes contain algorithms and data types that are used in many disciplines and can be easily transferred into other ontologies. The final class contains the SHM methods that may be used in the procedure of monitoring structures and must be connected with instances in other classes, since these methods use instances of all other superclasses. One can think of the final class as the main goal and connecting component of the ontology (Fig. 6.2).

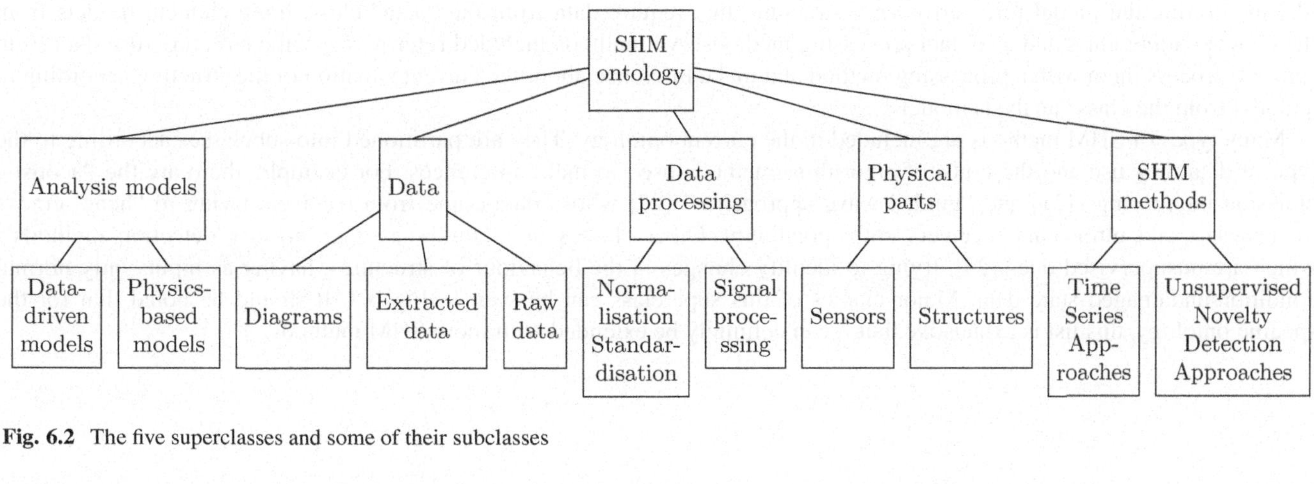

Fig. 6.2 The five superclasses and some of their subclasses

To further explain the content of the aforementioned classes, their description and some of their subclasses will be presented subsequently. The "analysis models" class contains anything that is used to analyse a structure. Since many types of models are used in various structural health monitoring techniques, like monitoring the modal assurance criterion (MAC) [13] or neural networks [14] to classify/localise damage. The first partition is into the classes of "physical models" and "data-driven models". "Data-driven models" is a class that mainly contains machine learning techniques. These are models that simulate physics, but infer their output using existing data from structures. These models are partitioned mainly into *regression* models and *classification* models. The first of these classes is composed of models predicting values of variables (e.g. natural frequencies, displacements, etc.) and are more commonly used in simulating the behaviour of structures, whilst the second one is often used to discriminate data corresponding to damaged or undamaged states of the structure. Some of the included models are neural networks [15], K-means clustering [16], support vector machines [17], autoassociative neural networks [18] etc. Physics-based models are constructed by studying and explaining the physics of problems. Two major subclasses of this class are the analytical and numerical models. Numerical models contain finite element models, surrogate models, lumped mass models etc; they are the most often-used ones, since they can be simulated using computers. On the other hand, the analytical models are solved using calculations and are not perhaps as common as the numerical ones, since engineering problems often preclude analytical solutions; however, they are still included in the ontology for completeness.

The "Data" superclass contains all the data that are acquired from structures. The data are partitioned in classes according to their type. For example, acceleration, displacement and velocity data, which belong to the time domain, but also data in the frequency domain. Furthermore, the data are also separated into data from sensors ("raw data"), data from processing the sensor data ("processed data") and data from any models ("simulation data"). "Data" is probably the class that can be more easily transferred into ontologies describing other domains, since most processes nowadays produce data and their analyses are based on data. Additionally, data can be transferred from one SHM application to another. Analysing data from a structure and making inference about them, can help in understanding the behaviour of similar structures or materials used in another SHM application. Some subclasses included in this class are "accelerograms", "contour plots", "mode shapes", "displacement simulation data", "displacement sensor data", etc.

The next superclass is related to methods that are used to process the data. These methods mainly belong to the discipline of signal processing, from the simple and ubiquitous Fourier transform [19] and signal statistics extraction, to more complicated methods like wavelet decomposition of signals [20]. These methods are vital for SHM, because they may reveal unseen features of the signals that are sensitive to damage. It is quite common that a damaged state can be spotted by observing a spectrum (which is produced by performing a Fourier transform on the acceleration signal of a sensor) rather than the acceleration time-history itself. Methods included in this class are: "signal smoothing algorithms", "Hilbert transform", "principal component analysis", etc.

The fourth superclass is the "physical parts" that the all methods refer to. This class simply contains instances of the structures that are monitored and the sensors placed on them. The instances are just objects referring to an existing structure. The structures belonging to this class can be partitioned in substructures for a more detailed and modular representation of a greater object. For example a wind turbine can be partitioned into its blades and the tower and even deeper, the blade can be partitioned into the cell and the stiffeners existing inside. This class serves the purpose of registering all monitored physical objects but also separating them into classes to facilitate search for data from a specific structure or type of structures.

Last but not least, is the class containing all methods that are used in SHM. This class is the most important one, as it uses components from every other class and combines them into methods that monitor a structure. Some of the methods here are trivial, such as monitoring the maximum value of a sensor signal; but there are also much more sophisticated methods, like monitoring the modal force error on a structure, that require data from the "data" class, finite element models from the corresponding class and also data processing methods. All methods included refer to a specific structure, use data from sensors, process them with a processing method and make inference about the current situation of the structure according to a model from the class "analysis models".

Many types of SHM methods are included in the current ontology. They are partitioned into subclasses according to the type of data they use and the types of algorithms/methods used to make inferences. For example, there are the "acoustic emission monitoring" [21] and "guided wave approaches" [22] whose data come from receivers trying to "hear" cracks in materials and ultrasound receivers correspondingly. Other classes of methods are the "novelty detection methods", which are unsupervised ones [23], trying to identify changes in the behaviour of structures, having as inputs only normal condition/undamaged state data. Major classes of this superclass can be seen in Fig. 6.3. It should be noted that for the specific ontology, this list is exhaustive, but it can definitely be extended with more SHM methods.

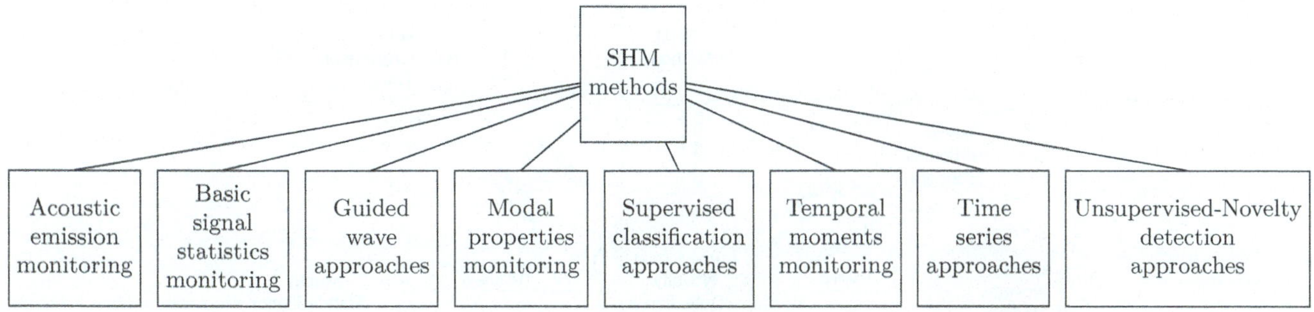

Fig. 6.3 SHM "methods" class

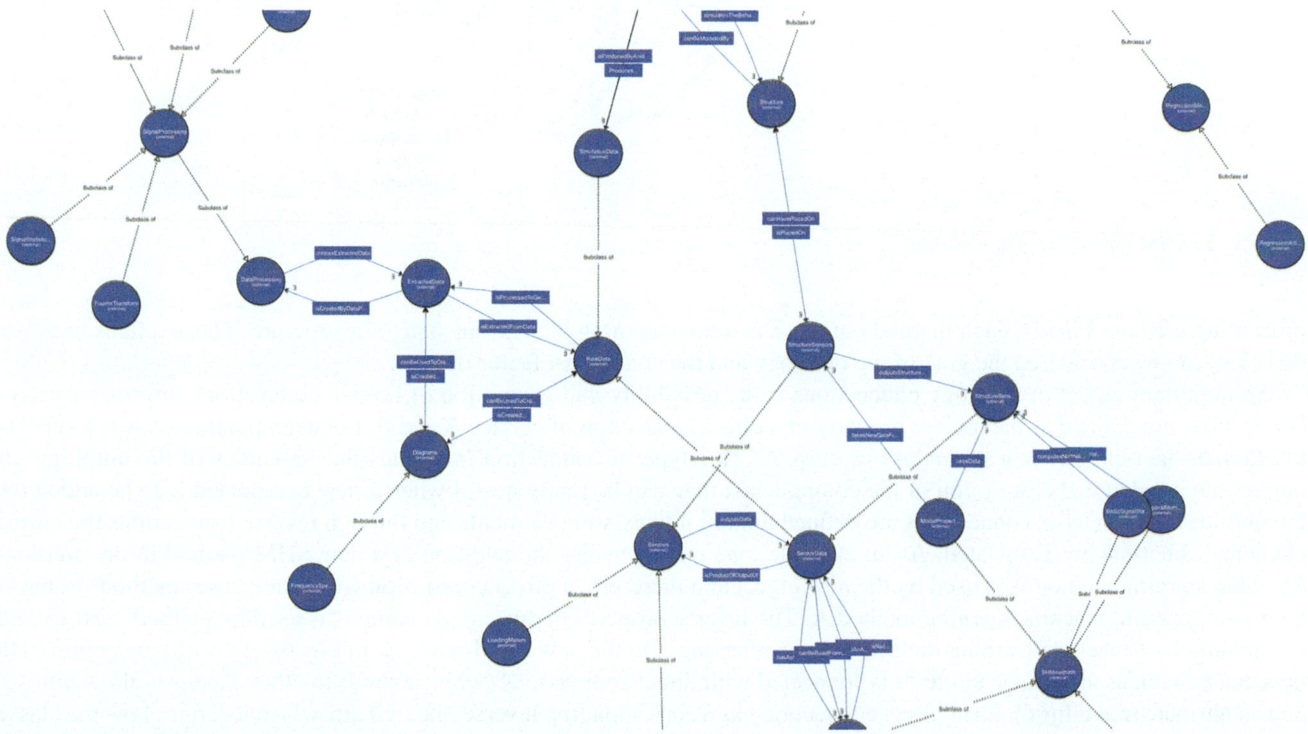

Fig. 6.4 Schematic overview of the ontology

6.4 Connections

Having defined a tree structure for the ontology and included all the components that are used in SHM procedures, further connections have to be defined in order to specify more analytically the interaction of these components. This is achieved by adding connections in the ontology. The name of the connection, as mentioned above, is the semantic definition of the connection; it explains how two objects are connected within the ontology for the purpose of SHM. So far the defined connections are only of type "is a subclass of". For example, "data-driven models" is a subclass of "analysis models".

Connections are added to explain the functionality of components. Data-driven models use data so they are connected to the type of data that they use with a connection named "uses data of type". Furthermore, all types of models produce data and are connected to them with a connection called "produces data". Following this logic, all the components gradually get connected to other ones and form a more complicated diagram that represents the existing knowledge about SHM. The result can be seen in Fig. 6.4. It is clear that adding these connections to the architecture makes it much more complicated than the tree that was the initial structure (part of it is shown in Fig. 6.2).

As expected, the "SHM methods" class is connected to most of the other classes. All methods have to use one or more analysis models, they all refer to a structure, use sensors, exploit data taken from sensors and process the data using a data

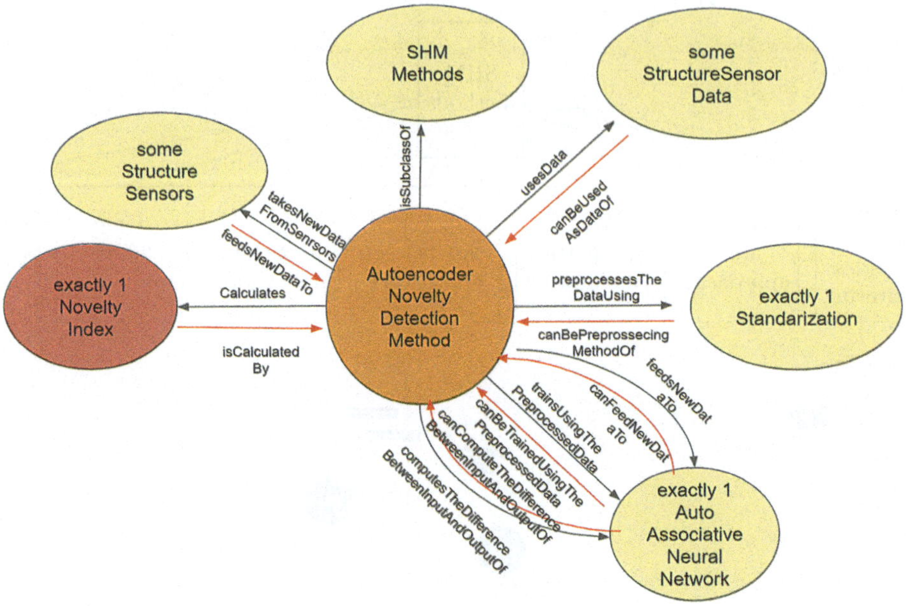

Fig. 6.5 An SHM novelty detection method

processing method. Finally, each method outputs a result concerning the current state of a structure. This explains both why this class can be considered the goal of the ontology and the connective factor of it.

An important aspect of ontology connections is the possibility and generation of inverse connections. Inverse or reverse connections are defined as the inverse property of a direct connection of classes A and B. For example if class A is a subclass of class B, then class B is a superclass of class A. This type of connection integrates the semantics of the ontology and defines a more detailed description of the components; they can be really useful when a new component is to be added into the ontology. In that case, connections are defined linking it to existing elements and through reverse connections the current elements' functionality is updated. As an example, one may consider the addition of a new SHM method in the ontology. Machine learning method A is used by the new object and therefore, a direct connection with name "uses method" connects the newcomer and machine learning method A. The inverse property in this case is named "is used by method" and extends the definition of machine learning method A by connecting it to the new SHM method. In Fig. 6.5 a novelty detection SHM data can be seen as the orange circle. It is connected with direct connections (white arrows) to other classes (yellow ellipses) and an attribute (red ellipse). Each direct connection has a corresponding inverse one (red arrows) that defines how the classes in yellow circles affect the class under examination.

6.5 Aspects of the Ontology

Having constructed the ontology, some of its uses or aspects will be examined. The first and most straightforward one is using the ontology to share knowledge. It is clear that the fabricated network is really effective in explaining the functionality of its components and their interaction. Apart from the annotations explaining each object's function, one is lead by the connections to other components directly connected to the one that is of interest, and learns their functionality too, gaining further intuition in the methods. Furthermore, within a project, collecting knowledge in the ontology and sharing it facilitates everyone's understanding in each other's part. For example, if a method is developed by a member of a group and it gets included in the ontology by defining connections with existing pieces of knowledge, methods and structures, it gets automatically explained to other members by observing it as a part of the ontology. This can solve difficulties in communication, misunderstandings and boost a group's productivity.

Further to helping groups in understanding and communicating, ontologies can be used in the same scope to increase the efficiency of SHM applications, to automate them and expand existing knowledge. It is clear that the ontology is quite modular. Using a regression algorithm for an SHM method means that another one can be used and maybe different algorithms can yield better results than the initial one. A method's connection to a specific algorithm almost certainly means that it can be connected to algorithms in the same superclass as the existing one. Following this scheme algorithms that optimise SHM methods can be found, and new pieces of knowledge are produced, since the newly assembled SHM method may be something that has never been applied before. An example of the above is having an algorithm of "supervised classification approaches" class connected to "classification neural networks" to perform the classification. The neural network class is a subclass of "classification models" just like "classification support vector machines". The neural networks then can be replaced by support vector machines and an alternative way of applying the existing "supervised classification approaches" method has been created. This procedure is shown in Fig. 6.6.

Another aspect of the ontology is that it can assist in developing software about the topic it describes. Ontologies are tightly connected to object-oriented programming [24]. Classes exist in both cases and objects of a class inherit properties form their parents. Implementing knowledge from an ontology in combination with an object-oriented scheme is much easier than trying to directly implement all methods as different software pieces. Clustering in classes encourages reusability and makes software more understandable. Different modules created in a project can be transferred and used in other projects, both in

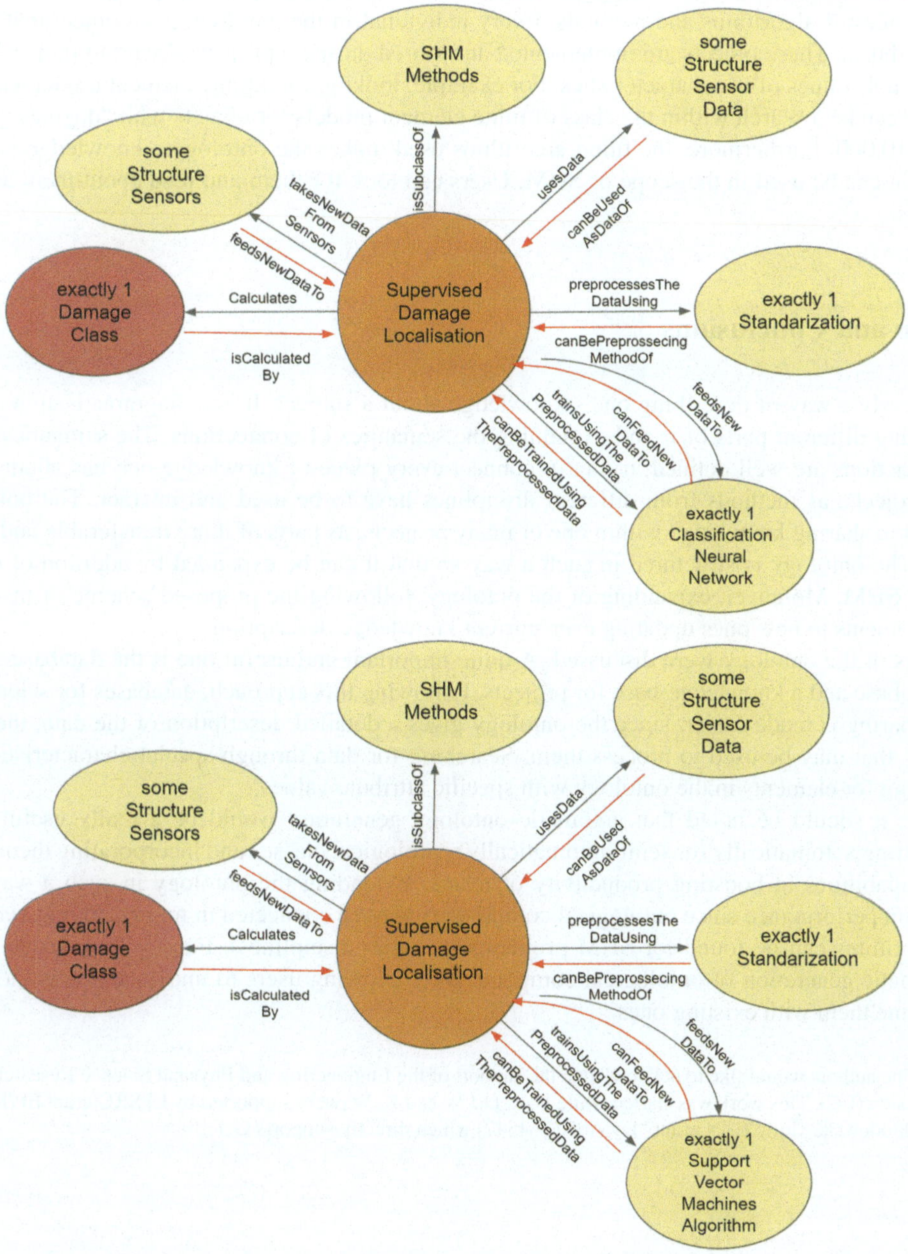

Fig. 6.6 Initial method of supervised damage classification using neural networks (top) and alternative method created by swapping neural networks for support vector machines (bottom)

the case of software modules and as ontological classes. Connections define clearly the interaction between software objects and help developers understand the way all components should be combined to achieve a specific result. This approach can be considered as another way of increasing productivity within a group, by facilitation of all software implementations needed.

The database aspect of the ontology is also a straightforward one. The "Data" class contains all data coming from both real-life structures and simulation models. Connections between these data and the structures and sensors they come from form a filing system or a database. Users of the ontology can at anytime find data coming from objects of interest, since connections exist between them and are produced by their data. At the same time, data are partitioned into categories according to what they represent; for example displacement, acceleration, frequency response functions, probability density functions of events, etc. Moreover, data are connected to processing methods that potentially transform them and users can have access directly to these methods and process the data in hand. This detailed description of data also assists exchanging data in different projects and disciplines.

Apart from direct consideration of the ontology as a database described in the previous paragraph, it can also be considered a database for instances of algorithms and methods. Every individual in the ontology is an object that interacts with other objects and has attributes. These objects are implemented and stored during a project. Trying to find an object can be easier if it is searched through values of it's characteristics. For example, looking for a finite element model with more than 10,000 degrees of freedom can be a search within the class of finite element models for objects with "degrees of freedom" attribute value greater than 10,000. Furthermore, the filing algorithms used makes the ontology a knowledge-base, including every possible method that can be used in the scope of SHM. Users can look for them and read about them and their interactions with other instances.

6.6 Discussion and Conclusions

The ontology is clearly a way of describing one's knowledge about a subject. It is a diagramatisation of intuition within a discipline, connecting different parts of it and explaining the semantics of connections. The semantics of components and those of their interactions are well defined, trying to connect every piece of knowledge one has about a subject. It is well suited for SHM projects, as methods from different disciplines have to be used and interact. The ontological framework proposed can assist in sharing knowledge within one or many projects, as parts of it are transferable and may be fit for other purposes as well. The ontology is structured in such a way so that it can be expanded by addition of more elements used under the scope of SHM. Moreover, expanding of the ontology, following the proposed scheme of inverse properties, will connect existing elements to new ones updating even current knowledge description.

Different aspects of the ontology were discussed. A quite important and useful one is the database aspect. An ontology potentially is a database and a knowledge-base for projects. Following this approach, databases for scientific projects can be created and data sharing is made easier, since the ontology gives a detailed description of the data, the source of data and also the algorithms that may be used to process them. Searching for data through special characteristics that one needs is facilitated by looking for elements in the ontology with specific attribute values.

As future work, it should be noted that automatic ontology generation would be a really useful tool in the current framework. Generating automatically (or semi-automatically) ontological classes and incorporating them within the ontology will increase its capabilities in boosting productivity of teams. Extending the ontology in such a way will also assist in optimising algorithm performance since ontological components may be connected in more different ways as it extends, and more efficient algorithms can be found for SHM problems (or other disciplines). Finally, knowledge extension would be benefited by automatic generation of ontological components by allowing users to understand new methods and how they are used and combine them with existing ones.

Acknowledgments The authors would like to acknowledge the support of the Engineering and Physical Science Research Council (EPSRC) and the European Commission (EC). This work was started while I. A., D.J.W and K. W. were supported by EPSRC grant EP/K003836/2, and finished under the EC Marie Skłodowska-Curie ETN grant "DyVirt" (764547) which directly supports G.T.

References

1. Farrar, C.R., Worden, K.: Structural Health Monitoring: A Machine Learning Perspective. John Wiley & Sons Ltd, West Sussex, United Kingdom (2012)
2. Farrar, C.R., Worden, K.: New Trends in Vibration Based Structural Health Monitoring. In: An introduction to structural health monitoring, pp. 1–17. Springer, Vienna (2010)
3. Uschold, M., Gruninger, M.: Ontologies: principles, methods and applications. Knowl. Eng. Rev. **11**, 93–136 (1996)
4. Bull, L.A., Gardner, P.A., Gosliga, J., Rogers, T.J., Haywood-Alexander, M., Dervilis, N., Cross, E.J., Worden, K.: Towards a population-based structural health monitoring. Part I: homogeneous populations and forms. In: Proceedings of IMAC XXXVIII – the 38th International Modal Analysis Conference, Houston (2020)
5. Gosliga, J., Gardner, P.A., Bull, L.A., Dervilis, N., Worden, K.: Towards a population-based structural health monitoring. Part II: heterogeneous populations and structures as graphs. In: Proceedings of IMAC XXXVIII – the 38th International Modal Analysis Conference, Houston (2020)
6. Gosliga, J., Gardner, P.A., Bull, L.A., Dervilis, N., Worden, K.: Towards a population-based structural health monitoring. Part III: graphs, networks and communities. In: Proceedings of IMAC XXXVIII – the 38th International Modal Analysis Conference, Houston (2020)
7. Gardner, P.A., Worden, K.: Towards a population-based structural health monitoring. Part IV: Heterogeneous populations, matching and transfer. In: Proceedings of IMAC XXXVIII – the 38th International Modal Analysis Conference, Houston (2020)
8. Lin, W., Worden, K., Cross, E.J.: Towards a population-based structural health monitoring. Part VII: EoV fields: environmental mapping. In: Proceedings of IMAC XXXVIII – the 38th International Modal Analysis Conference, Houston (2020)
9. Antoniadou, I., Barthorpe, R.J., Worden, K.: Towards an ontology for verification and validation in structural dynamics. In: Proceesdings 26th International Conference on Noise and Vibration Engineering (ISMA2014), Leuven, pp. 1171–1178 (2014)
10. Gruber, T.R.: A translation approach to portable ontology specifications. Knowl. Acquis. **5**, 199–220 (1993)
11. Knublauch, H., Oberle, D., Tetlow, P., Wallace, E.: A semantic web primer for object-oriented software developers. W3C Working Group Note 9th March 2006, W3C (2006)
12. Musen, M.A.: The protégé project: a look back and a look forward. AI Matters **1**, 4–12 (2015)
13. Allemang, R.J.: The modal assurance criterion–twenty years of use and abuse. Sound and vibration. **37**(8), 14–23 (2003)
14. Bishop, C.M.: Neural Networks for Pattern Recognition. Oxford University Press, Great Clarendon Street, Oxford (1995)
15. Bishop, C.M.: Pattern Recognition and Machine Learning. Spring Street, New York, NY, USA (2006)
16. Macqueen, J.: Some methods for classification and analysis of multivariate observations. In: Proceedings of 5th Berkeley Symposium on Mathematical Statistics and Probability, pp. 281–297 (1967)
17. Cortes, C., Vapnik, V.: Support-vector networks. Mach. Learn. **20**, 273–297 (1995)
18. Kramer, M.A.: Nonlinear principal component analysis using autoassociative neural networks. AIChE J. **37**, 233–243 (1991)
19. MacKay, D.J.C.: Inference and Learning Algorithms. In: Information Theory. Cambridge University Press, Cambridge England (2002)
20. Chui, C.K.: An Introduction to Wavelets. Academic, San Diego, California (1992)
21. Hensman, J.H., Worden, K., Eaton, M., Pullin, R., Holford, K.M., Evans, S.L.: Spatial scanning for anomaly detection in acoustic emission testing of an aerospace structure. Mech. Syst. Signal Process. **25**, 2462–2474 (2011)
22. Fuentes, R., Cross, E.J., Ray, N., Dervilis, N., Guo, T., Worden, K.: In-process monitoring of automated carbon fibre tape layup using ultrasonic guided waves. In: Special Topics in Structural Dynamics, vol. 6, pp. 179–188. Springer (2017)
23. Dervilis, N., Choi, M., Taylor, S.G., Barthorpe, R.J., Park, G., Farrar, C.R., Worden, K.: On damage diagnosis for a wind turbine blade using pattern recognition. J. Sound Vib. **333**, 1833–1850 (2014)
24. Kindler, E., Krivy, I.: Object-oriented simulation of systems with sophisticated control. Int. J. Gen. Syst. **40**, 313–343 (2011)

George P. Tsialiamanis studied at the school of Civil Engineering at the National Technical University of Athens during the period 201–2017. His specialisation is on Structural Engineering and his thesis was about high performance surrogate models for stochastic finite element problems. After his studies he had an internship at Beta-CAE Systems SA (2017–2018), on applications of machine learning techniques in various engineering problems. More specifically, machine learning was used to predict the results of car crash test simulations. At the moment he is a PhD student at the University of Sheffield in the DyVirt program funded by the Marie Sklodowska-Curie EU Horizon 2020 research and innovation programme. The PhD is focused on decision support systems for structural health monitoring (SHM) and also in building a unifying framework for the various components of Dynamic Virtualization.

Chapter 7
Passive Aeroelastic Tailored Wing Modal Test Using the Fixed Base Correction Method

Natalie Spivey, Rachel Saltzman, Carol Wieseman, Kevin Napolitano, and Benjamin Smith

Abstract In modal testing and finite element model correlation, analysts desire modal results using free-free or rigid boundary conditions to ease comparisons of test versus analytical data. It is often expensive both in cost and schedule to build and test with boundary conditions that replicate the free-free or rigid boundaries. Static test fixtures for load testing are often large, heavy, and unyielding, but do not provide adequate boundaries for modal tests because they are dynamically too flexible and often contain natural frequencies within the frequency range of interest of the test article. Dynamic coupling between the test article and test fixture complicates the model updating process because significant effort is required to model the test fixture and boundary conditions in addition to the test article. If there were a way to correct the modal results for fixture coupling, then setups used for other structural testing could be adequate for modal testing. In the case described in this paper, a partial static loads testing setup was used, which allowed significant schedule and cost savings by eliminating a unique setup for a modal test. A fixed base correction technique was investigated during modal testing of a flexible wing cantilevered from part of a static test fixture.

The technique was successfully used to measure the wing modes de-coupled from the dynamically active test fixture. The technique is promising for future aircraft applications, but more research is needed.

Keywords Modal Test · Ground Vibration Test · Fixed Base · Passive Aeroelastic Tailored Wing

Nomenclature

a	acceleration
accel	accelerometer
AFRC	Armstrong Flight Research Center
CFAST	a NASTRAN element that connects two shell elements to provide the joint stiffness of the connecting elements
CReW	Calibration Research Wing
DOF	degrees of freedom
f	external force
fwd	forward
FBC	Fixed Base Correction
FEM	finite element model
FLL	Flight Loads Laboratory
FRF	frequency response function

N. Spivey (✉) · R. Saltzman
Armstrong Flight Research Center, National Aeronautics and Space Administration, Lilly, CA, USA
e-mail: natalie.d.spivey@nasa.gov; rachel.saltzman@nasa.gov

C. Wieseman
Langley Research Center, National Aeronautics and Space Administration, Hampton, VA, USA
e-mail: carol.d.wieseman@nasa.gov

K. Napolitano
ATA Engineering, Inc, San Diego, CA, USA
e-mail: kevin.napolitano@ata-e.com

B. Smith
Aurora Flight Sciences, Fairborn, OH, USA
e-mail: Smith.Benjamin@aurora.aero

B. Dilworth (ed.), *Topics in Modal Analysis & Testing, Volume 8*, Conference Proceedings of the Society for Experimental Mechanics Series, https://doi.org/10.1007/978-3-030-47717-2_7

GVT	ground vibration test
Hz	Hertz
iso	isometric
k	structural stiffness
lb	pound
LE	leading edge
m	mass
MAC	modal assurance criterion
MIF	Mode Indicator Function
NASA	National Aeronautics and Space Administration
NMIF	Normal Mode Indicator Function
OML	outer mold line
PAT	Passive Aeroelastic Tailored
PSMIF	Power Spectrum Mode Indicator Function
TE	trailing edge
W1B	wing 1st bending
W2B	wing 2nd bending
W3B	wing 3rd bending
W4B	wing 4th bending
W5B	wing 5th bending
W6B	wing 6th bending
W7B	wing 7th bending
W1T	wing 1st torsion
W2T	wing 2nd torsion
W1F/A	wing 1st fore/aft
W2F/A	wing 2nd fore/aft
W3F/A	wing 3rd fore/aft
W4F/A	wing 4th fore/aft
WLTF	Wing Loads Test Fixture
x	displacement
ω	frequency

7.1 Introduction

New aircraft structures often require static and dynamic structural ground testing to validate the analytical structural finite element models (FEMs) used in determining airworthiness. Static and dynamic ground tests require different boundary conditions, which result in two different costly and specialized test setups. Therefore, it would be beneficial if a modal survey could be conducted while a test article is mounted in a static test fixture for a structural loads test, allowing for two traditionally separate structural tests to be performed using one test fixture. This paper discusses an effort to apply a fixed base correction technique to measure fixed base modes from a test article mounted to part of a dynamically active static test fixture.

The Flight Loads Laboratory (FLL) at the National Aeronautics and Space Administration (NASA) Armstrong Flight Research Center (AFRC) (Edwards, California) specializes in both structural modal testing and loads calibration testing of aerospace research structures [1]. To facilitate the loads calibration test on the Passive Aeroelastic Tailored (PAT) Wing, a Wing Loads Test Fixture (WLTF), shown in Fig. 7.1, was designed.

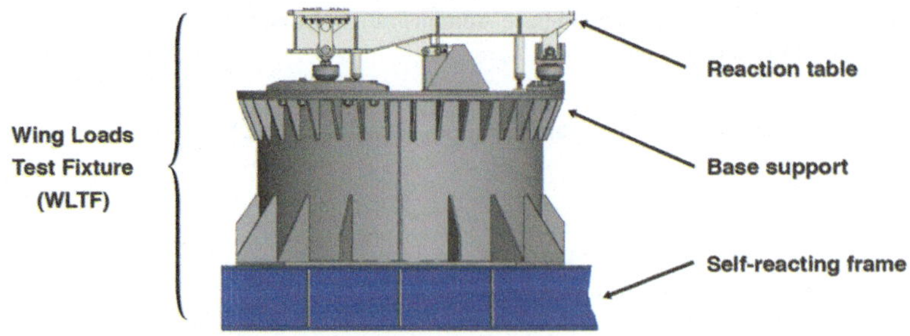

Fig. 7.1 Side view of the dynamically active static Wing Loads Test Fixture

The PAT Wing - a carbon-epoxy high-aspect-ratio wing of an approximately 39-ft semi-span - was built to investigate a new composite technology known as tow-steering to increase aeroelastic efficiencies [2–4] and underwent a modal test in the FLL. The modal test of the high-aspect-ratio, tow-steered wingbox was conducted to validate the FEM. The objective of the modal test was to measure the primary frequencies, mode shapes, and damping up to the Wing 1st Torsion (W1T) mode (expected to be approximately 55 Hz). To streamline the modal test and save significant project resources of time, cost, and schedule, the modal test used the same hardware configuration as was used for the follow-on loads testing, with the wing cantilevered out from the WLTF table.

This setup differs from a standard modal test setup. The boundary conditions of the wing mounted onto the WLTF table were not ideal for modal testing because truly rigid boundary conditions were not required for static loads testing. Finding the analytical connection stiffness of how the wing was physically mounted would be a very difficult analytical task; however, by using the Fixed Base Correction (FBC) test method, the table-mounted boundary conditions were analytically fixed for the modal test.

The uniqueness of FBC methodology compared to traditional modal tests is that it requires an equal number of independent drive point inputs (that is, shakers) as base mode shapes to remove. The result is that many shaker inputs are required; the number depends on the complexity of the base which is desired to be "fixed." The FBC method allows for test articles to be tested with non-ideal modal testing boundary conditions that can normally complicate testing and drive up cost and schedule. While a traditional ground vibration test (GVT) only requires shakers be attached to the wing or test article, the FBC method also requires multiple shakers be attached to the mounting fixture. The fixture excitation accelerations are used as references when calculating frequency response functions (FRFs) instead of using the traditional shaker forces as references. This FBC strategy analytically removes and de-couples enough of the fixture response from the wing in order to "fix" the fixture and aid in comparing modal ground test results to FEM modal results. More detail regarding the background of the FBC method is presented below. This paper details the second time that the FBC technique has been applied to aeronautics applications. The FLL previously conducted a test on the Calibration Research Wing (CReW); this was a pathfinder GVT for the PAT Wing test [5]. Using the CReW static loads testing setup for the GVT and implementing FBC allowed significant schedule and cost savings by eliminating a unique setup for a modal test.

7.2 Theory/Correction Methodology

There exists considerable literature discussing how to extract fixed base modes from structures, mainly satellite-related structures, mounted on shake tables [6–14]. These methods require two different approaches to extract fixed base modes from structures mounted on flexible shake tables. One method applies a constraint equation to measured mass-normalized mode shapes to generate fixed base modes [15]. The advantage of using mass-normalized modes is that a large number of shakers do not necessarily need to be mounted on the base, which simplifies the test setup. The accuracy of this method, however, depends on how well a linear combination of the measured modes can represent the fixed base modes. If the measured test modes don't span the space of the true fixed base modes, then this method will not be able to accurately estimate them. The method also requires well-excited modes so that modal mass can be accurately calculated. A second method, hereafter called the Fixed Base Correction method, is the focus of this paper and uses base accelerations as references to calculate the FRFs associated with a fixed base [16, 17]. The FRFs are then post-processed to extract fixed based modes of the test article.

The FBC method can be illustrated with a simple spring-mass 2-degrees-of freedom (2-DOF) system, as shown in Fig. 7.2.

Applying Newton's second law, the equation of motion for an undamped system in the frequency domain is as shown in Eq. (7.1):

$$\begin{bmatrix} -\omega^2 m_1 + k & -k \\ -k & -\omega^2 m_2 + 2k \end{bmatrix} \begin{Bmatrix} x_1 \\ x_2 \end{Bmatrix} = \begin{Bmatrix} f_1 \\ f_2 \end{Bmatrix} \tag{7.1}$$

where m is the mass, ω is the frequency, k is the structural stiffness, x is the displacement, and f is the external force. The subscripts 1 and 2 refer to blocks 1 and 2, respectively. It should be noted that the 2-DOF system above is the same k value, but the FBC method can be further generalized for different structural stiffness values.

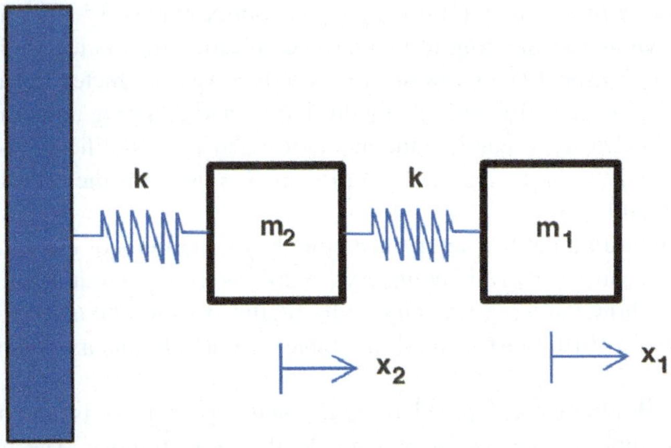

Fig. 7.2 Spring-mass two-degrees-of-freedom system

The FRF for traditional modal testing is calculated using the shaker forces applied to DOF 1 and DOF 2 as references to obtain the full system response. The results of using these traditional FRFs are referred to in this paper as the "uncorrected" results, as shown in Eq. (7.2):

$$a_1 = \left[\frac{-\omega^2\left(-\omega^2 m_2 + 2k\right)}{\left(-\omega^2 m_2 + 2k\right)\left(-\omega^2 m_1 + k\right) - k^2} \; \frac{-\omega^2 k}{\left(-\omega^2 m_2 + 2k\right)\left(-\omega^2 m_1 + k\right) - k^2} \right] \begin{Bmatrix} f_1 \\ f_2 \end{Bmatrix} \tag{7.2}$$

where a is the acceleration.

When implementing the FBC, however, if the force at DOF 1 and the acceleration at DOF 2 are used as references, then the resulting FRFs are associated with a structural system with dynamics associated with DOF 2 fixed, as shown in Eq. (7.3):

$$a_1 = \left[\frac{-\omega^2}{-\omega^2 m_1 + k} \; \frac{k}{-\omega^2 m_1 + k} \right] \begin{Bmatrix} f_1 \\ a_2 \end{Bmatrix} \tag{7.3}$$

Furthermore, the FRF associated with the force applied at DOF 1 is equivalent to an FRF associated with DOF 2 being fixed. This property is exploited in the FBC method by using drive point accelerations, instead of the traditionally used shaker forces, on the test fixture as references when calculating the FRF.

The key necessity of the FBC method is at least one independent excitation source, usually modal shakers, for each degree of freedom that is desired to be fixed. Therefore, FBC modal testing requires multiple shakers used on the test fixture in addition to the test article. Although not described in this paper, the FBC technique could also use constraint shapes as references when the number of independent sources is larger than the number of independent DOF of the test fixture [16]. The fundamental FBC strategy is to use shaker accelerations as references, rather than the traditional shaker forces, when calculating FRFs. Personnel at ATA Engineering, Inc. (San Diego, California) have implemented the FBC modal methodology into their IMAT™ (Interface between MATLAB®, Analysis and Test) software (MATLAB is a registered trademark of The MathWorks, Natick, Massachusetts).

The fixed base corrected FRF can be calculated directly using shaker accelerations on the fixture and shaker forces on the test article as references [18], or by performing a partial inversion of the baseline FRFs that have been calculated using all shaker forces as references [6, 17]. In fact, the results are equivalent if measured forces are used as basis vectors when calculating the FRF directly [19].

One advantage of calculating the FRF directly is that doing so removes the requirement to mount load cells to the shakers on the test fixture. The advantage of performing a partial inversion of the FRF matrix is that boundary conditions can be changed quickly by changing which DOF are to be inverted.

One potential disadvantage of the FBC method is that the measured damping values of the FBC modes have been observed to be slightly different from expected values; sometimes very lightly damped modes may even be calculated to have slightly negative damping. In these cases, it may be better to report the damping values of the mode from the uncorrected test data that best align with the FBC mode. Analytically, the damping values calculated from the FBC method should be accurate. Further study is needed to understand how the FBC method affects damping measurements.

7.3 Test Description

The following sections describe the PAT Wing test article, the finite element model, and the modal testing details. The PAT Wing modal testing was conducted in the summer of 2018 with the intent to use the FBC method.

7.3.1 Test Article

The PAT Wing test article, depicted in Fig. 7.3, is a carbon-epoxy, semi-span, flexible right wingbox designed and manufactured by Aurora Flight Sciences (Manassas, Virginia) ("Aurora") using a composite technology called tow steering. The test article is a 27% scale model of the NASA undeflected Common Research Model (uCRM) with a high aspect ratio of 13.5, 36.8-degree wing sweep, and approximately 39-ft semi-span.

The wingbox consists of two skins; both the upper and the lower surfaces are of the tow-steered wingskins. The forward and aft spars are primarily carbon-fiber composite. An outboard section of the forward spar was replaced with approximately 12 ft of aluminum due to manufacturing difficulties. The wing contains 58 composite ribs connected to the wingskins. Along the wingspan there are 14 load lugs: seven on the leading edge (LE) spar and seven on the trailing edge (TE) spar). The load lugs are permanently installed to the wingbox for the load testing; see Fig. 7.4. Along the TE spar there is a Yehudi break between lugs 1 and 2. At the wing root there are two large steel reaction plates (forward/LE and aft/TE) each containing two three-inch reaction pins. The root reaction plates are mounted to attachment hardware connecting to the WLTF table. The wingbox was assembled using numerous fasteners and, in some locations, bonding agents.

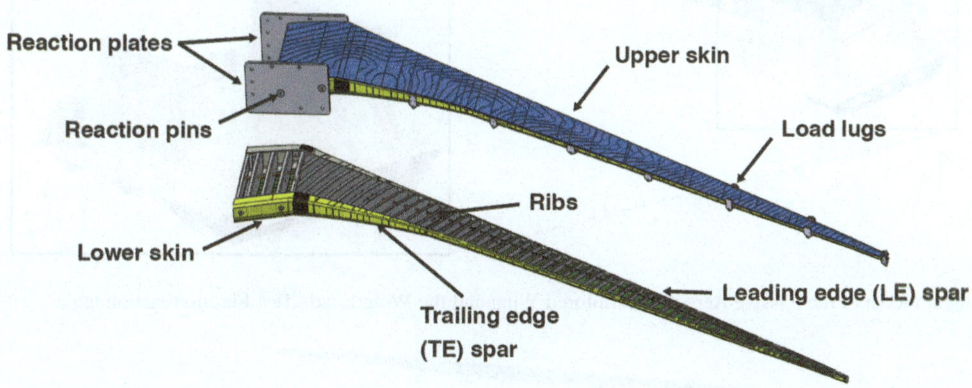

Fig. 7.3 The Passive Aeroelastic Tailored Wing Test Article

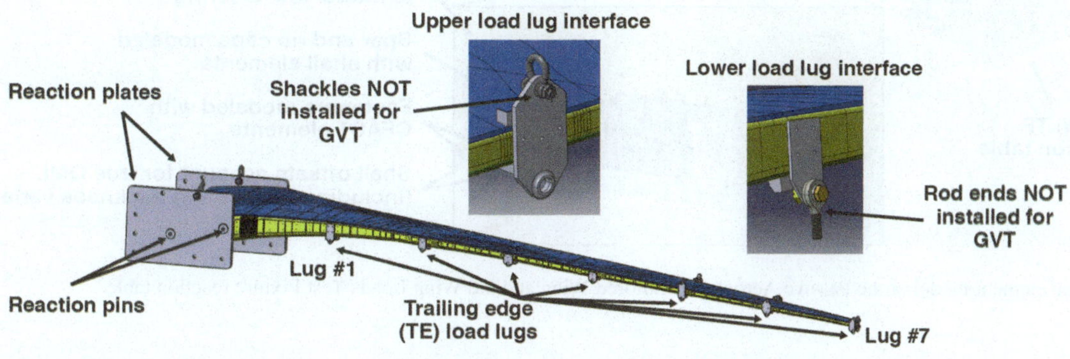

Fig. 7.4 Load lugs on the leading edge and the trailing edge of the Passive Aeroelastic Tailored Wing

A weight and balance test was performed prior to the wing being shipped from Aurora to NASA AFRC. The wing weighed 2621 lb with the installed reaction plates, the internal strain gage instrumentation, and the wire bundle.

While the root reaction plates are adequate for static testing because they carry static loads properly, they are not adequate for measuring fixed base modes because plates are dynamically flexible in the out-of-plane direction.

7.3.2 *Finite Element Model*

Data from both the modal and loads tests were used to validate the FEM analytical models, modeling techniques, and assumptions used for the towed-steering technology. Personnel of Aurora and of the NASA Langley Research Center (Hampton, Virginia) participated in a combined effort to create an MSC Nastran™ (MSC Software, Newport Beach, California) FEM of the PAT Wing along with the wing reaction plates, attachment hardware, and WLTF reaction table, shown in Fig. 7.5. Unique material orientations were assigned to each wingskin element to account for the spatially varying tow-steering paths. Skins consisted of a laminate with 62.5% of the tows following the local tow-steering path and the balance of the laminate consisting of plies offset −45 deg., +45 deg., and 90 deg. from the local tow-steering path. Homogenized laminate properties corresponding to this ply fraction were assigned to the skin elements based on unnotched tensile testing performed on representative tow-steering coupons.

The FEM modeled both rib and spar caps with shell elements, and fasteners were modeled utilizing discrete CFAST elements. Element offsets were applied to all skin elements to position them at the as-measured OML, which includes deviations from the nominal design due to local variations in liquid shim thickness between the skins and spar and rib caps in Fig. 7.6.

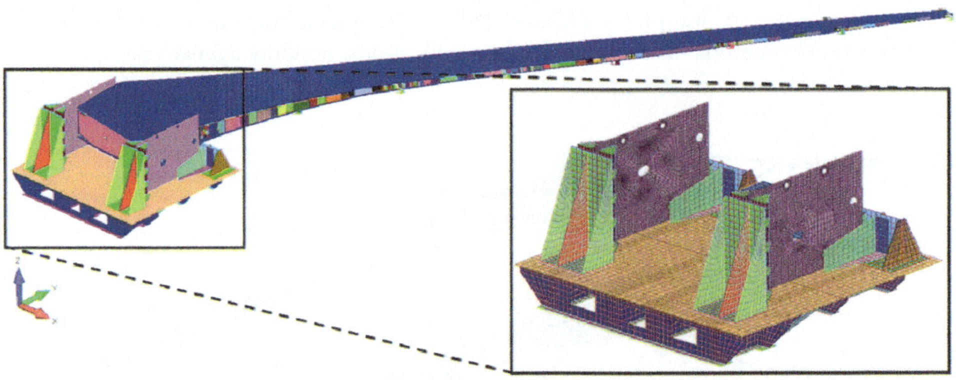

Fig. 7.5 Finite element model of the Passive Aeroelastic Tailored Wing and the Wing Loads Test Fixture reaction table

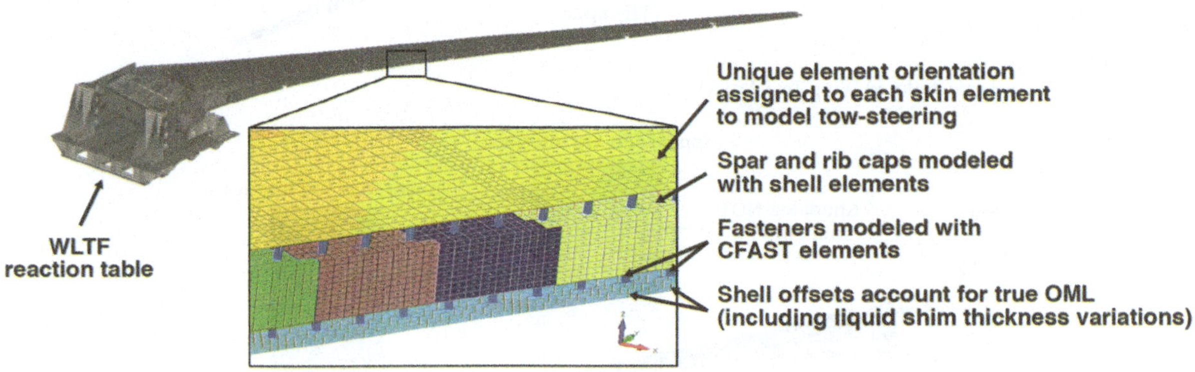

Fig. 7.6 Finite element model of the Passive Aeroelastic Tailored Wing and the Wing Loads Test Fixture reaction table

During the wing fabrication, some unexpected defects were found and multiple fixes were made to ribs and spars. Aurora subsequently updated the FEM to incorporate differences of the as-built wing. Non-structural masses were updated to account for as-measured part and assembly masses. This FEM was used for test predictions and post-test comparisons. Currently there are no plans to correlate the as-built FEM with test results.

7.3.3 Modal Test Setup

The PAT Wing modal test using the FBC method took place July 10–12, 2018 in the NASA AFRC FLL high bay. The original modal test setup plan for the FBC PAT Wing, shown in Fig. 7.7, was to perform the test with the wing installed on the dynamically active WLTF, as the wing would be for the loads testing. In this test configuration, the PAT Wingtip would be approximately 124 inches above the Flight Loads Laboratory floor.

Upon further review of the test plan and understanding of the FBC technique, it was determined that the modal test could be simplified by setting the WLTF table directly on the FLL floor (see Fig. 7.8), rather than installing the reaction table on top of the WLTF base support. Using FBC made this approach possible because it analytically removed the effects of the reaction table and the hardware below it. The reaction table being set directly on the FLL floor also significantly simplified the shaker setup, because the wingtip now was only approximately 48 inches, rather than approximately 124 inches, above the FLL floor. This further simplification allowed cost savings by increasing access to the wing; attaching the shakers at lower heights also prevented adding additional flexibility and potential errors into the test. The reaction table on the FLL floor was supported by four retractable feet and secured with a strap to floor tracks as shown in Fig. 7.9. The wing was cantilevered from the reaction table by securing the two wing reaction plates to attachment hardware connected to the table.

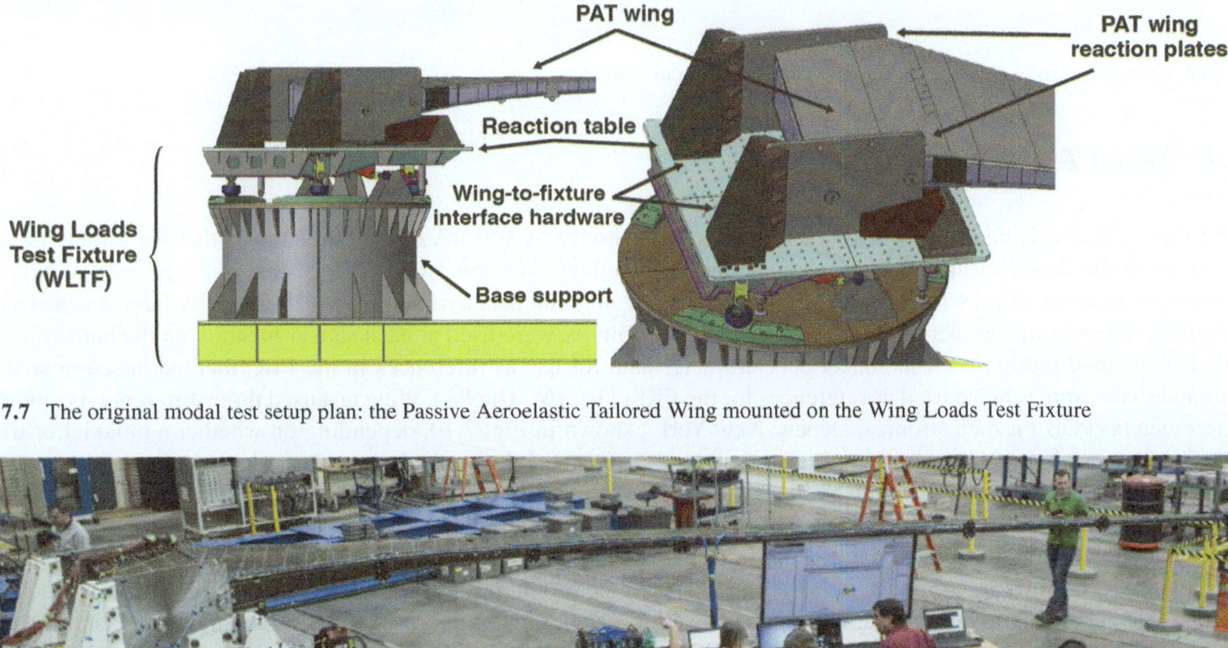

Fig. 7.7 The original modal test setup plan: the Passive Aeroelastic Tailored Wing mounted on the Wing Loads Test Fixture

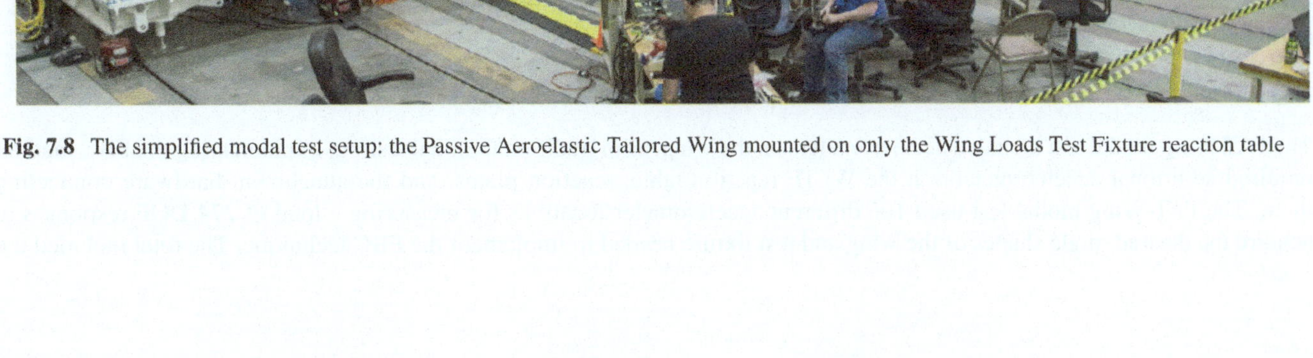

Fig. 7.8 The simplified modal test setup: the Passive Aeroelastic Tailored Wing mounted on only the Wing Loads Test Fixture reaction table

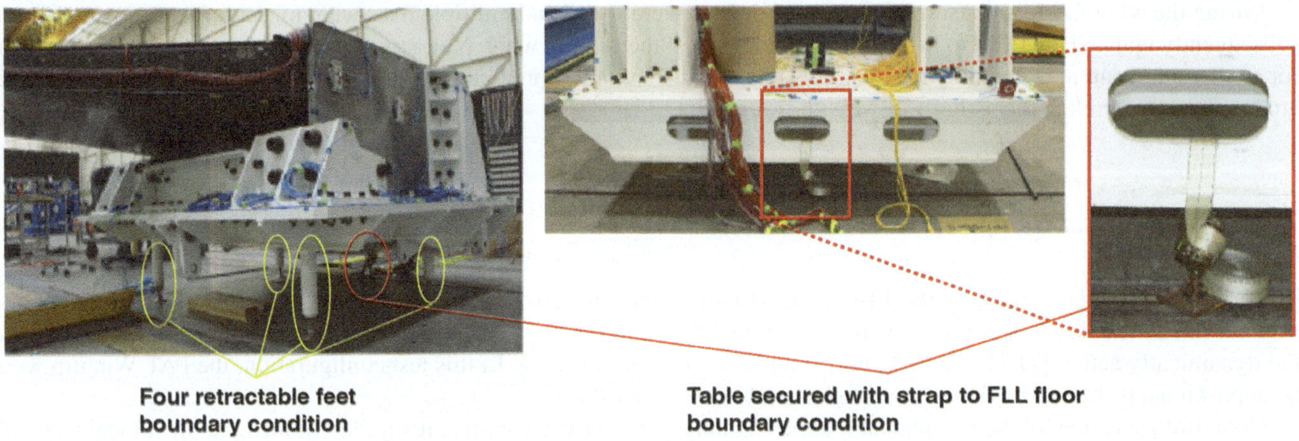

Fig. 7.9 Boundary conditions of the Wing Loads Test Fixture reaction table on the Flight Loads Laboratory floor

Fig. 7.10 Ground test accelerometers used for modal testing (not to scale)

7.3.4 Modal Test Instrumentation

Modal testing normally requires accelerometers with a sensitivity of 100 mV/g distributed over the test article and force transducers at the shaker locations. To implement the FBC method, additional 100 mV/g accelerometers were added on the hardware being fixed along with a small handful of seismic uniaxial accelerometers, which typically have a sensitivity of 1000 mV/g. The seismic accelerometers with the higher sensitivity were used at each shaker location on the hardware being fixed. This method produced clean shaker accelerometer data for use as references in the FBC method, as compared with traditional shaker forces being used as references for the FRFs [15, 16]. The PAT Wing test used three different types of modal accelerometers (PCB Piezoelectronics, Depew, New York), shown in Fig. 7.10, depending on whether a uniaxial or triaxial accelerometer was desired to measure a certain number of DOF at each location along with the seismic accelerometers at the fixed shaker locations.

For every shaker attached to the reaction table or wing reaction plates, a reference seismic accelerometer in the direction of the shaker excitation along with a force transducer attached to the shaker stinger were used to measure the excitation input. Figure 7.11 shows an example of the seismic accelerometer and force transducer shaker setup that was used on the reaction table. The wingtip shaker did not require a seismic accelerometer and used a traditional modal accelerometer and force transducer because the force was used as a reference when calculating the FRF; see Fig. 7.12.

7.3.5 Modal Test Accelerometer Layout

The PAT Wing modal test included accelerometers on the wing, as in traditional modal testing; implementing the FBC method required additional accelerometers on the WLTF reaction table, reaction plates, and the attachment hardware connecting them. The PAT Wing modal test used 106 different accelerometer locations for measuring a total of 274 DOF responses to acquire the desired mode shapes of the wing and test fixture needed to implement the FBC technique. The total included the

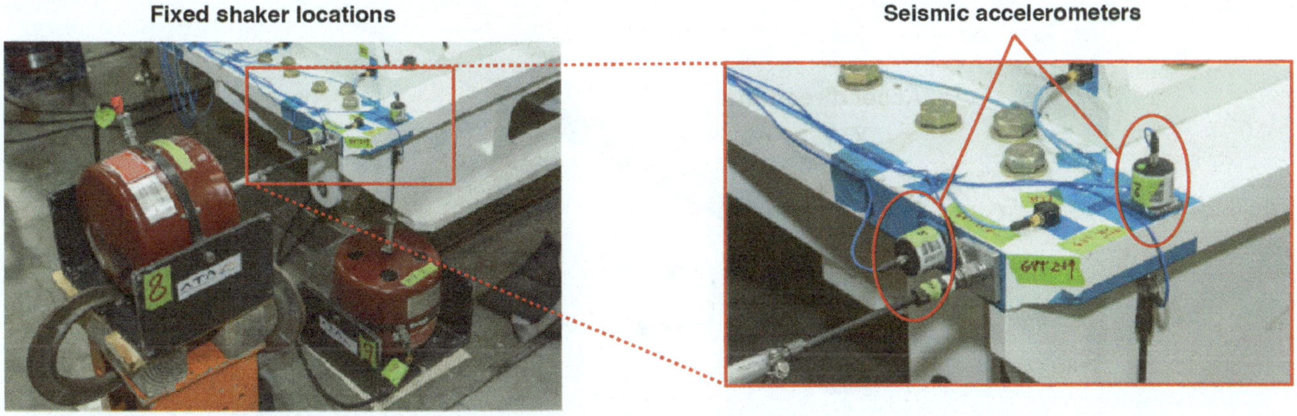

Fig. 7.11 Typical shaker setup on the Wing Loads Test Fixture reaction table using seismic accelerometers

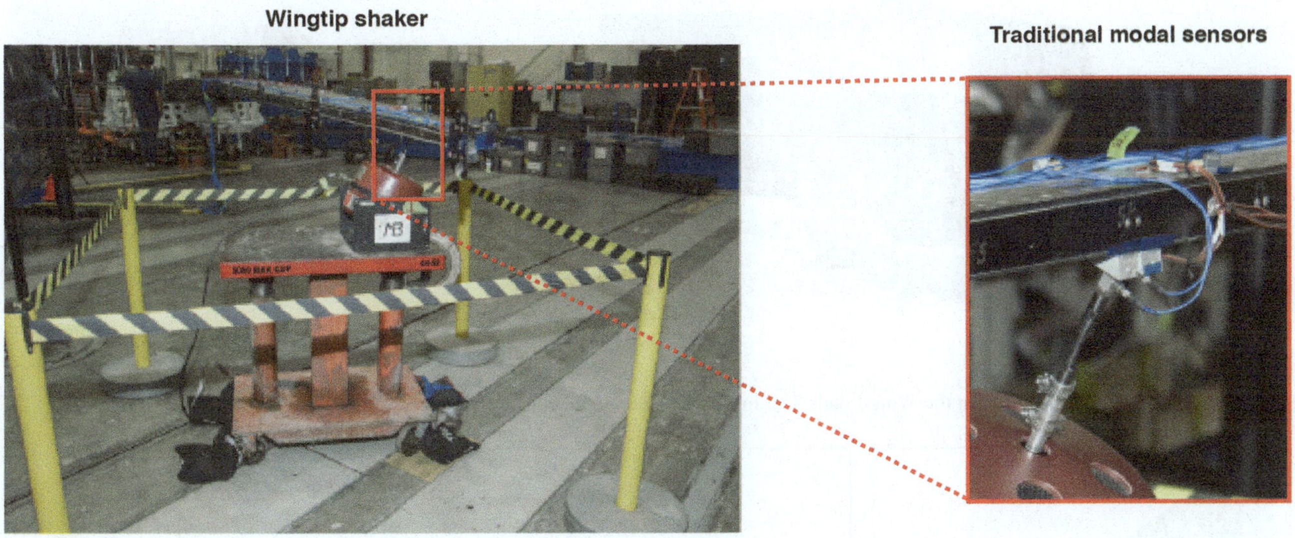

Fig. 7.12 Wingtip shaker setup using traditional modal accelerometer and force transducer

accelerometer responses for the one wingtip shaker along with each fixed shaker location; these were later used as reference for the FBC. The data acquisition system also included the 14 shaker force transducers measured as references. Accordingly, a total of 288 channels were recorded with the data acquisition system for each test run.

Of the 106 total locations there were 31 accelerometer locations on the wing (see Fig. 7.13), which had triaxial accelerometers to measure a total of 87 DOF for the wing. The placement of the wing accelerometers was the same as for any traditional modal test; sensors should be placed to adequately observe and differentiate modes of the structure.

The remaining 75 locations were on the WLTF reaction table, attachment hardware, and the wing reaction plates to perform the FBC calculations. The majority of these locations used triaxial accelerometers for a total of 187 DOF measured on the hardware being fixed; see Fig. 7.14 (some accelerometer locations are not visible in the figure).

The coordinates of the 106 accelerometer locations were used to create the test display model shown in Fig. 7.15. The test display model was used to visualize the test mode shapes.

7.3.6 *Modal Test Shaker Layout*

The FBC technique requires multiple independent drive points (that is, shakers) be mounted to both the WLTF reaction table and the PAT Wing test article. The shaker layout depends on where the FBC technique is trying to fix the boundary conditions.

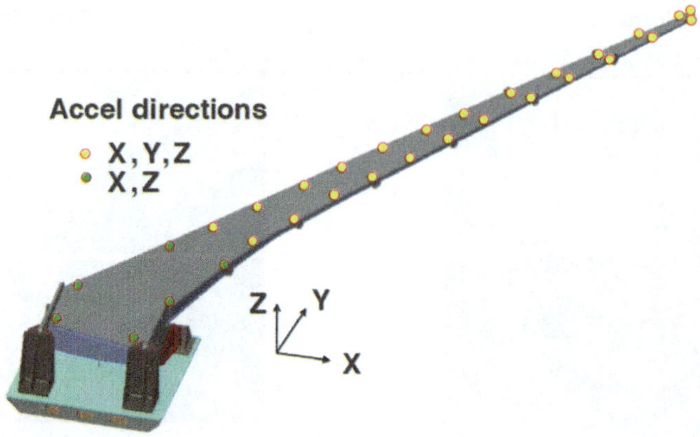

Fig. 7.13 Accelerometer locations on the Passive Aeroelastic Tailored Wing

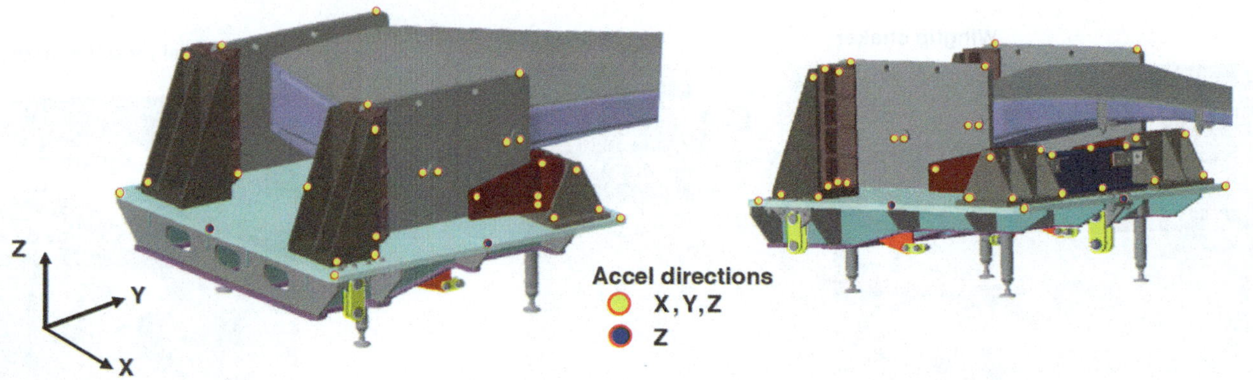

Fig. 7.14 Accelerometer locations on the Wing Loads Test Fixture reaction table, attachment hardware, and wing reaction plates

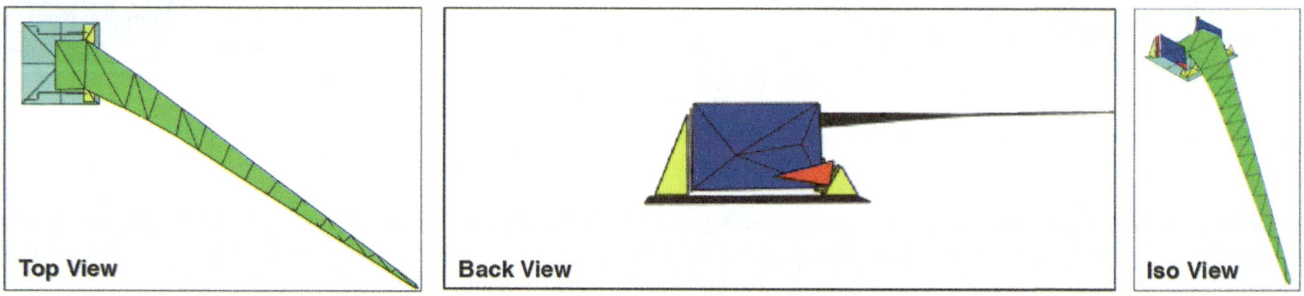

Fig. 7.15 The Passive Aeroelastic Tailored Wing modal test display model

There must be at least as many independent sources as there are independent boundary deformations of the hardware desired to be fixed in the test article frequency range of interest. The PAT Wing modal test included an effort to fix the reaction table by adding more shakers to improve the fixed base modes. For each shaker configuration, one shaker was always positioned on the wingtip as for traditional modal testing, and multiple other shakers were positioned around the WLTF reaction table and connecting attachment hardware.

During the PAT Wing modal test three different shaker configurations (see Fig. 7.16) were attempted with the FBC method to fix different hardware to improve the fixed base modes:

- 10 Shakers: nine shakers on the reaction table, one shaker on the wingtip
  - Vertical wingtip excitation

Fig. 7.16 Shaker configurations on the Passive Aeroelastic Tailored Wing

- 12 Shakers: two shakers added on aft triangular brackets (fore/aft)
  - Vertical wingtip excitation
- 14 Shakers: two shakers added on wing root reaction plates (fore/aft)
  - Vertical and fore/aft wingtip excitation

The direction of the shakers on the reaction table and connecting hardware are important and essentially eliminate the effect of the hardware moving in each shaker direction; see Fig. 7.17. A few different shaker configurations were attempted to improve the fixed base modes to fix the reaction table. The final shaker layout consisted of 14 total shakers with one wingtip shaker plus 13 shakers around the reaction table and connecting hardware, as shown in Fig. 7.18. This method fixed the reaction table and connecting hardware enough to decouple the wing modes.

The placement of the shakers around the WLTF was adjusted to excite primary base modes and maximize the capability of the FBC to decouple the base modes from the wing modes. The shakers used were MB Dynamics (Cleveland, Ohio) Modal 110-lb and Modal 50-lb electromagnetic shakers. Higher shaker forces were required on the base because it was stiffer than the wing which required less force at the wingtip. The wingtip shaker force was approximately 0.7 lb RMS for the various tests; the base shaker forces varied between 3 and 5 lb RMS.

To compare the modal test FBC results with the above described shaker configurations to the FEM pre-test prediction results, analytical boundary conditions were placed on the FEM. Figure 7.19 shows how each FEM component was fully fixed. The FEM boundary conditions on all of the nodes that were rigid in all 6 DOF are shown in cyan. The fully-fixed FEM used for the pre-test modal analysis was the sum of the fixed boundary conditions on the reaction table, the aft two and forward two triangular brackets of the attachment hardware, and both wing reaction plates.

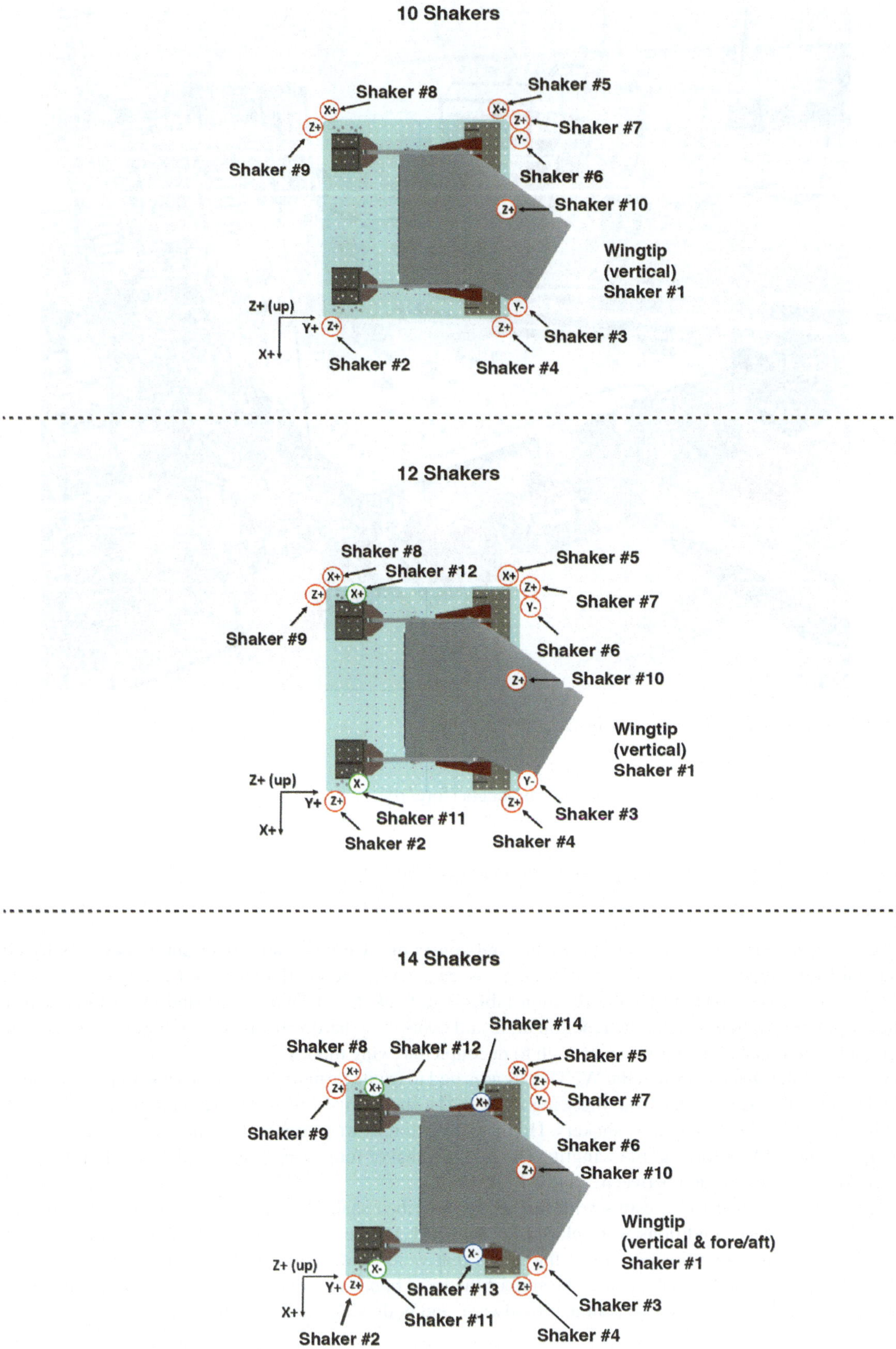

Fig. 7.17 Shaker locations on the Passive Aeroelastic Tailored Wing

Fig. 7.18 The Passive Aeroelastic Tailored Wing modal test shaker layout for the Fixed Base Correction Method

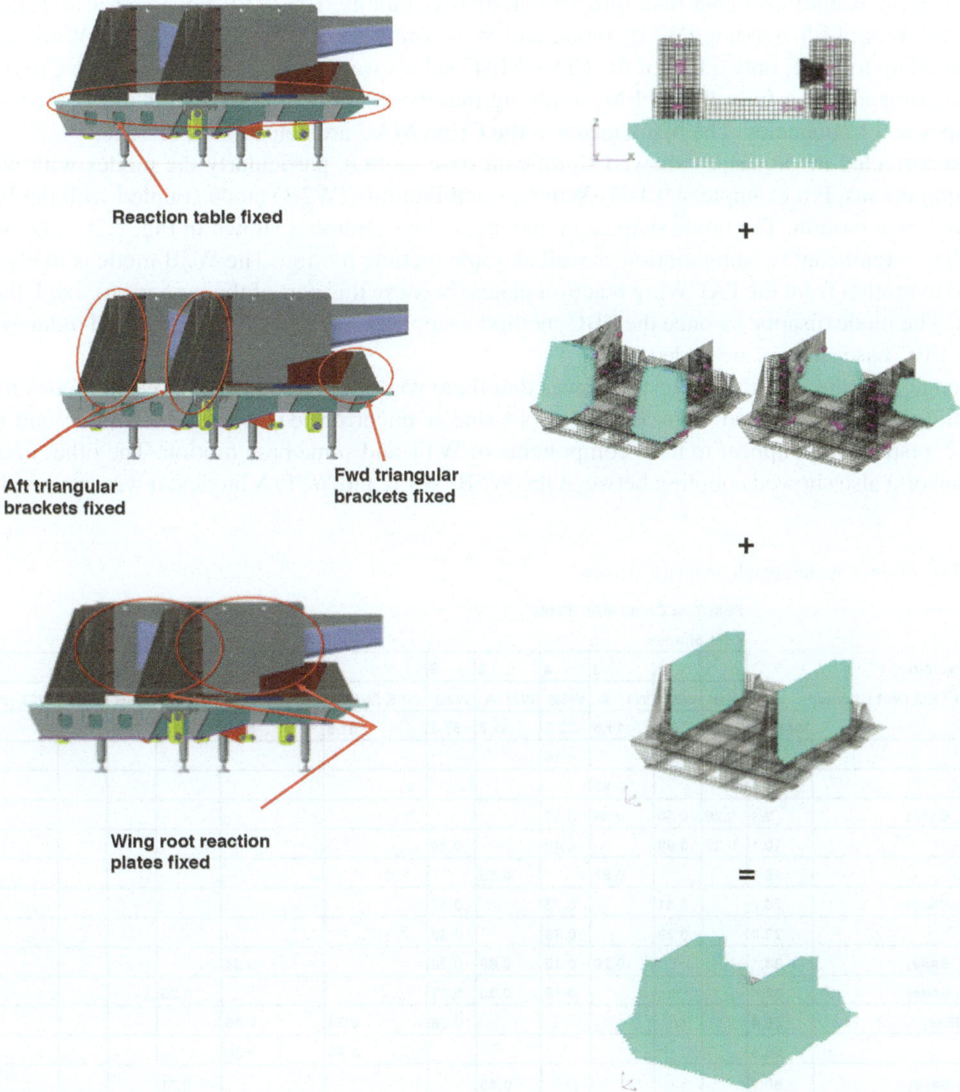

Fig. 7.19 Finite element model fully-fixed boundary conditions

7.4 Results

A total of 25 test runs were performed during the PAT Wing modal test using up to 14 shakers simultaneously to ensure the primary wing modes up to the W1T were cleanly extracted. The following PAT Wing modal results show that the FBC modes were successfully extracted using a total of 14 shakers.

7.4.1 Uncorrected Results

The uncorrected results using traditional modal testing techniques for the 10-shaker, 12-shaker, and 14-shaker configurations were all very similar. The 10-shaker and 12-shaker tests had vertical wingtip excitations, as did one of the 14-shaker tests. The other 14-shaker test had fore/aft wingtip excitation; the results of this test are shown in FEM/Test Cross MAC (modal assurance criterion) Table 7.1. The main difference between the tests was that the WIT mode was difficult to pick up in tests that had vertical wingtip excitations rather than fore/aft wingtip excitations. The W1T mode was also difficult to pick up due to coupling with the Wing fifth Bending (W5B) mode and Wing third Fore/Aft (W3F/A) mode. While all of the expected modes were captured up to W1T, only a few of the Cross MAC values were above 0.9, or 90%. The Cross MAC values also were not diagonal compared to a fully-fixed FEM, implying that the uncorrected test results did not accurately capture the correct mode shapes and frequencies. The blank terms in the Cross MAC are below 0.15, or 15%.

Many of the uncorrected mode shapes showed significant base motion, particularly the modes with wing fore/aft (F/A) and torsion (T) components. For example, a 9.1-Hz Wing second Bending (W2B) mode coupled with the Wing first Fore/Aft (W1F/A) mode and base motion. The mode shape and undeflected wireframe is shown in Fig. 7.20. The bottom plate of the base is shown to have significant twisting motion as well as some rocking motion. The W2B mode is likely coupling with the W1F/A mode due to motion from the PAT Wing reaction plates; because this part of the base is not fixed, the base is coupling with wing modes. The mode disappears once the FBC method is applied, because the FBC method reduces the motion of the reaction plates as FBC base shakers are added.

Another challenge with the uncorrected test data was that there was a lot of coupling between modes in the 50- to 60-Hz range. This coupling can be discerned from Table 7.1 (14-shaker uncorrected), as the W5B mode and the W3F/A mode (Modes 10 and 12, respectively) appear to have components of W1T and some base motion. The other uncorrected tests (10 shakers and 12 shakers) also showed coupling between the W5B, W1T, and W3F/A modes; it was not possible to find a W1T

Table 7.1 Cross MAC of uncorrected results with 14 shakers

Uncorrected — FEM/Test Cross MAC Table
14 Shakers — FEM Shapes
Fore/Aft Wingtip Excitation
Fully Fixed Pretest FEM (Not Updated)

Test Shapes			1	2	3	4	5	6	7	8	9	10	11	12	13	14
			W1B	W2B	W1F/A	W3B	W2F/A	W4B	W5B (W1T)	W1T (W5B)	W3F/A	W6B	W2T	W2T (W4F/A)	W7B	W4F/A
		MAC	3.4	10.4	11.3	22.5	31.7	37.2	51.8	55.2	64.3	76.8	92.9	95.3	103.1	115.9
1	W1B	3.5	0.99	0.30		0.16										
2	W1F/A (Base)	5.1			0.83											0.17
3	W2B (W1F/A, Base)	9.1	0.26	0.50	0.34	0.17										
4	W2B	10.1	0.32	0.98		0.40		0.19								
5	W2F/A (Base)	16.5			0.87		0.53									0.34
6	W3B (W2F/A, Base)	20.2		0.31		0.73		0.37								
7	W3B (Base)	22.0		0.28		0.88		0.35								
8	W2F/A (W4B, Base)	34.1			0.20	0.15	0.66	0.26		0.21						0.21
9	W4B (W2F/A, Base)	35.4				0.18	0.30	0.71			0.29					0.27
10	W5B (W1T, Base)	50.4						0.26	0.23	0.35						
11	W1T (Base)	56.5							0.70	0.30						
12	W3F/A (W1T, Base)	60.1					0.45				0.71					0.50
13	W6B (W3F/A, Base)	67.8					0.21				0.77					0.43
14	W6B (W4F/A, Base)	81.2										0.85			0.25	0.21
15	W4F/A (Base)	87.8									0.48					0.55
16	W4F/A (Base)	95.6									0.42					0.53
17	W2T (Base)	98.9											0.91	0.18		
18	W7B	N/A														

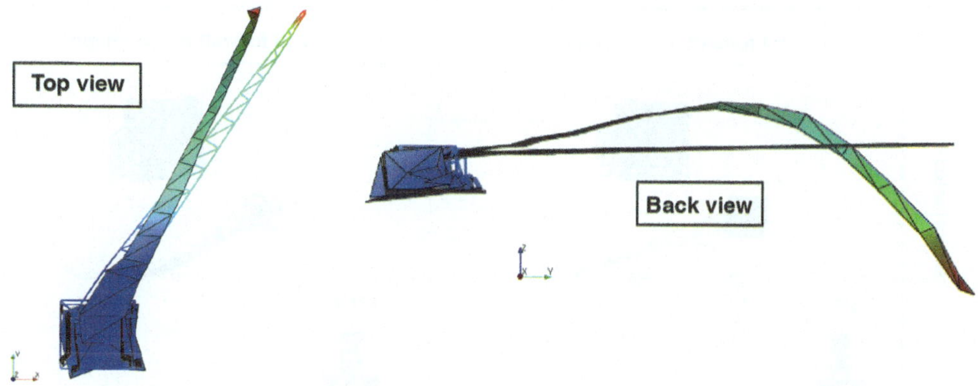

Fig. 7.20 Uncorrected 9.1-Hz Wing 2nd Bending (Wing 1st Fore/Aft) mode with significant base motion (undeflected shape depicted by wireframe)

Table 7.2 Cross MAC of Fixed Base Corrected configuration with 10 shakers

Fixed Base Corrected
10 Shakers
Vertical Wingtip Excitation
Fully Fixed Pretest FEM (Not Updated)

FEM/Test Cross MAC Table — FEM Shapes

Test Shapes		MAC	1 W1B	2 W2B	3 W1F/A	4 W3B	5 W2F/A	6 W4B	7 W5B (W1T)	8 W1T (W5B)	9 W3F/A	10 W6B	11 W2T	12 W2T (W4F/A)	13 W7B	14 W4F/A
		MAC	3.4	10.4	11.3	22.5	31.7	37.2	51.8	55.2	64.3	76.8	92.9	95.3	103.1	115.9
1	W1B	3.5	0.99	0.31		0.16										
2	W2B	10.2	0.33	0.99		0.40		0.18								
3	W1F/A	10.7			0.94		0.24									0.22
4	W3B	21.2		0.35		0.99		0.40								
5	W2F/A	29.5			0.43		0.95				0.20					0.44
6	W4B	35.1				0.33		0.95		0.20						
7	W5B (W1T, Base)	52.1								0.26				0.15	0.32	
8	W1T (W3F/A, Base)	55.6					0.27		0.20	0.41	0.32					0.21
9	W3F/A (W1T, Base)	57.7			0.17		0.23		0.41		0.19					
10	W6B (W3F/A, Base)	75.6									0.56	0.19				0.35
11	W6B (W4F/A, Base)	78.9								0.16		0.81				0.29
12	W4F/A (W2T, Base)	88.8									0.35					0.46
13	W4F/A (W2T, Base)	95.9									0.33					0.53
14	W2T (Base)	98.8											0.94	0.15		
15	W7B (Base)	105.5										0.25			0.84	

mode from the 10-shaker and 12-shaker uncorrected tests. Utilizing the FBC method enabled decoupling these sorts of mode shapes.

7.4.2 Fixed Base Corrected Results with 10 Shakers

A buildup testing approach was used to add shakers around the WLTF base to analytically fix it. The tests were initially started with 10 shakers, nine of which were on the WLTF reaction table; the 10th shaker was on the wingtip in the vertical direction as described above. The Cross MAC table for the 10-shaker FBC is Table 7.2. As compared with the uncorrected Cross MAC in Table 7.1, the first six mode shapes are now diagonal, indicating that the test results better match the FEM results. The Cross MAC results are actually better than expected; the first six flexible modes have very high (above 0.9, or 90%) diagonal Cross MAC values.

Table 7.2 also shows that there are now only 15 modes up to the Wing 7th Bending (W7B) mode (as opposed to the 18 modes up to W7B in Table 7.1), thus some of the redundant base modes were removed by applying FBC. The W7B mode was also able to be found by applying FBC; this mode was not found in the uncorrected Cross MAC Table 7.1.

The Wing 1st Bending (W1B) and W2B modes did not show a lot of change after implementing FBC, which implies that the non-ideal modal test setup boundary condition was already stiff enough in the vertical direction to capture these modes. The FBC did, however, appear to significantly stiffen the W1F/A and Wing 2nd Fore/Aft (W2F/A) modes, increasing the

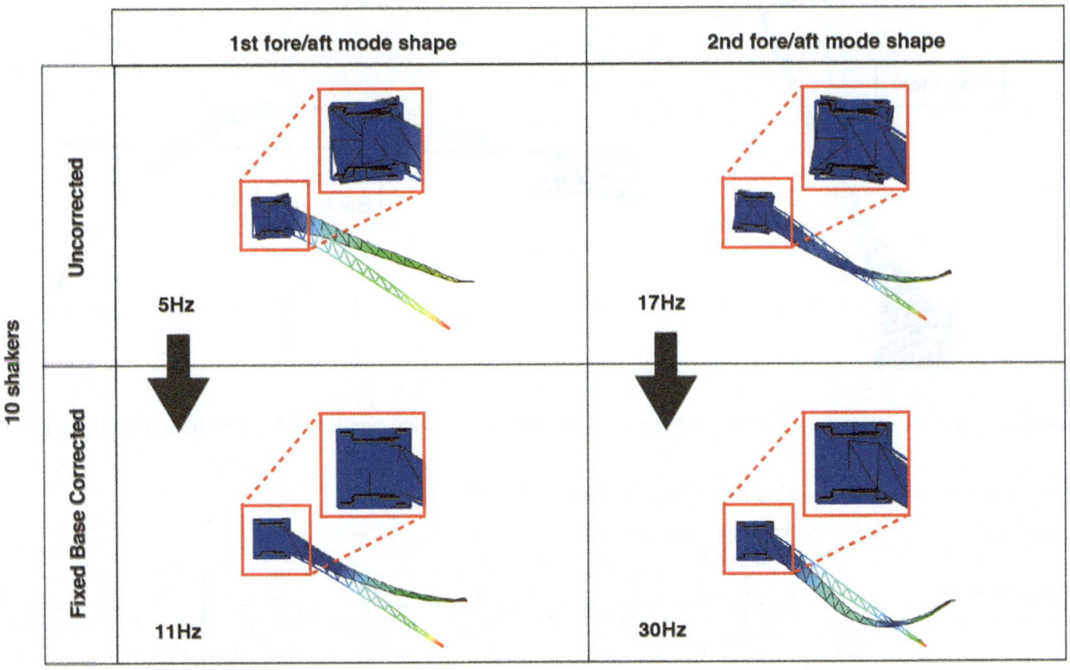

Fig. 7.21 10-shaker comparison of uncorrected versus Fixed Base Corrected for Fore/Aft modes (undeflected shape depicted by wireframe)

Cross MAC values of these modes from approximately 0.85 (85%) to approximately 0.95 (95%), providing confidence that the FBC method is matching the test results to the FEM results.

The W1F/A and W2F/A mode shapes are shown in Fig. 7.21 comparing uncorrected and 10-shaker FBC test results. The FBC method significantly reduced, and almost eliminated, any base rotation, which can be seen as the blue lines in the insets in the figure. The wing shapes also show some improvement as well, since the FBC mode shapes have higher relative wingtip displacements at the wingtip than do the uncorrected mode shapes. Figure 7.21 also shows how applying the FBC method significantly stiffened the F/A modes, increasing the W1F/A frequency from 5 Hz to 11 Hz and increasing the W2F/A frequency from 17 Hz to 30 Hz. These results suggest that the FBC method has already significantly improved the quality of the GVT data gathered with only 10 shakers. The FBC method had the greatest effect on the F/A modes in the 10-shaker configuration.

7.4.3 *Fixed Base Corrected Results with 12 Shakers*

After the 10-shaker FBC tests, an additional two shakers (shaker number 11 and number 12) were added on the attachment hardware known as the aft triangular brackets in the F/A direction, as seen in Fig. 7.17. These two shaker locations were chosen because the PAT Wing reaction plates showed significant F/A deflection. This method separated some of the coupled modes and added multiple more diagonal modes on the Cross MAC table, as can be seen in Table 7.3. One highlight is that the W5B mode, W1T mode, and W3F/A mode were decoupled to some degree. This decoupling was very difficult to accomplish with only 10 shakers using the FBC method (Table 7.2); it was also difficult to accomplish without using the FBC method (Table 7.1). The W5B mode, the W1T mode, and the W3F/A mode finally could show up on the Cross MAC diagonal with values of 0.6 (60%) or better; it is likely that these values were not higher due to some remaining motion in the base. It is also notable that adding the two shakers reduced the number of modes from 15 (Table 7.2) to 13 (Table 7.3), again reducing the number of redundant base modes. The 12-shaker FBC configuration was a significant improvement over the previous 10-shaker FBC configuration, which illustrates how crucial it is to the FBC method that an adequate number of shakers are added in the correct directions to fix the base modes up to the frequencies that are desired to be measured in the test article.

Adding the two F/A shakers also allowed the Wing 6th Bending (W6B) mode to show up on the Cross MAC diagonal. As can be seen in the 10-shaker FBC Cross MAC table (Table 7.2), one of the test W6B modes appeared to couple with the FEM W3F/A mode due to the F/A motion of the PAT Wing reaction plates. Figure 7.22 compares the W6B mode shapes of the 10-shaker and 12-shaker FBC datasets to show how adding the two F/A shakers removed the W3F/A coupling by removing some of the base motion in the F/A direction.

Table 7.3 Cross MAC of Fixed Base Corrected configuration with 12 shakers

Fixed Base Corrected
12 Shakers FBC
Vertical Wingtip Excitation
Fully Fixed Pretest FEM (Not Updated)

FEM/Test Cross MAC Table
FEM Shapes

Test Shapes			1	2	3	4	5	6	7	8	9	10	11	12	13	14
			W1B	W2B	W1F/A	W3B	W2F/A	W4B	W5B (W1T)	W1T (W5B)	W3F/A	W6B	W2T	W2T (W4F/A)	W7B	W4F/A
		MAC	3.4	10.4	11.3	22.5	31.7	37.2	51.8	55.2	64.3	76.8	92.9	95.3	103.1	115.9
1	W1B	3.5	0.98	0.34		0.17										
2	W2B	10.1	0.34	0.98		0.39		0.18								
3	W1F/A	10.9			0.96		0.25									0.23
4	W3B	21.3		0.35		0.99		0.41								
5	W2F/A	29.7			0.42		0.95				0.19					0.44
6	W4B	35.1				0.32		0.95		0.19						
7	W5B (W1T)	52.2						0.19	0.60	0.30		0.23				
8	W1T (W3F/A, Base)	57.1							0.42	0.59						
9	W3F/A (Base)	58.2					0.47				0.81					0.47
10	W6B (Base)	77.4									0.15	0.71			0.21	
11	W4F/A (Base)	81.4					0.18				0.54	0.23				0.69
12	W2T (Base)	98.5											0.92	0.18		
13	W7B (Base)	106.8												0.18	0.71	

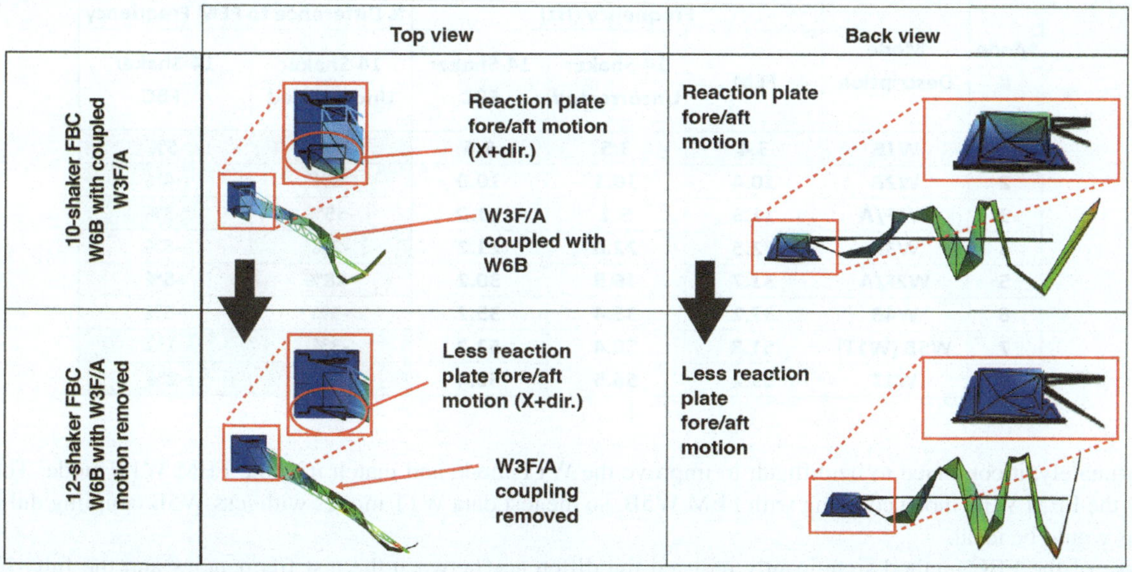

Fig. 7.22 Comparison of Fixed Base Corrected Wing 6th Bending mode shapes with and without wing Fore/Aft coupling (10 versus 12 Shakers)

7.5 Fixed Base Corrected Results with 14 Shakers

Due to the remaining F/A motion in the base, two more shakers were added to the wing root reaction plates (Fig. 7.16 and Fig. 7.17) to remove most of the remaining base motion in the test article frequency range of interest. While the objective of the GVT was to accurately capture modes up to W1T, it was desired to examine several higher frequency modes to evaluate how well the FBC method could work with non-ideal modal boundary conditions.

There was not a significant change in the Cross MAC table between the 12-shaker and 14-shaker tests that used FBC. The 14-shaker FBC table is presented as Table 7.4. The main difference is that the W6B mode showed improvement with the Cross MAC value increasing from 0.71 (71%) to 0.88 (88%). The W7B mode also improved from 0.71 (71%) to 0.82 (82%). It is promising that these modes improved, which shows how the FBC method continued to remove more base motion as more shakers are added in the correct directions and locations on the base.

Table 7.4 Cross MAC of Fixed Base Corrected configuration with 14 shakers

Fixed Base Corrected
14 Shakers
Fore/Aft and Vertical Wingtip Excitation
Fully Fixed Pretest FEM (Not Updated)

FEM/Test Cross MAC Table
FEM Shapes

Test Shapes	Wingtip Excitation			1	2	3	4	5	6	7	8	9	10	11	12	13	14
				W1B	W2B	W1F/A	W3B	W2F/A	W4B	W5B (W1T)	W1T (W5B)	W3F/A	W6B	W2T	W2T (W4F/A)	W7B	W4F/A
			MAC	3.4	10.4	11.3	22.5	31.7	37.2	51.8	55.2	64.3	76.8	92.9	95.3	103.1	115.9
1	Fore/Aft	W1B	3.6	0.99	0.33		0.17										
2	Vertical	W2B	10.0	0.29	0.98		0.40		0.19								
3	Fore/Aft	W1F/A	11.0			0.94		0.24									0.21
4	Fore/Aft	W3B	21.2		0.34		0.99		0.41								
5	Fore/Aft	W2F/A	30.2			0.41		0.96				0.18					0.43
6	Fore/AFt	W4B	35.2				0.32		0.95		0.20						
7	Vertical	W5B (W1T)	52.2						0.20	0.69	0.21		0.21				
8	Vertical	W1T	56.4							0.40	0.57						
9	Vertical	W3F/A (W1T)	59.1					0.46			0.15	0.73					0.46
10	Vertical	W6B (Base)	77.4										0.88			0.23	
11	Vertical	W4F/A (W2T, Base)	88.6			0.16		0.21				0.50		0.17	0.17		0.70
12	Vertical	W2T (Base)	98.6											0.90	0.20		
13	Vertical	W7B (Base)	106.4										0.17			0.82	

Table 7.5 Frequency percent difference with respect to finite element model between 14-shaker uncorrected and 14-shaker Fixed Base Corrected

Mode #	Mode Description	Frequency (Hz)			% Difference to FEM Frequency	
		FEM	14-Shaker Uncorrected	14-Shaker FBC	14-Shaker Uncorrected	14-Shaker FBC
1	W1B	3.4	3.5	3.6	3%	5%
2	W2B	10.4	10.1	10.0	-3%	-4%
3	W1F/A	11.3	5.1	11.0	-55%	-3%
4	W3B	22.5	22.0	21.2	-2%	-6%
5	W2F/A	31.7	16.5	30.2	-48%	-5%
6	W4B	37.2	35.4	35.2	-5%	-5%
7	W5B (W1T)	51.8	50.4	52.2	-3%	1%
8	W1T	55.2	56.5	56.4	2%	2%

Unfortunately, it continued to be difficult to improve the W1T mode and match it to the FEM W1T mode. The cause is probably the FEM W1T mode coupling with FEM W5B, so the test data W1T modes with less W5B coupling did not match as well as would be ideal.

The use of the FBC method significantly reduced the difference between the test frequencies and the fully-fixed FEM frequencies, as can be seen in Table 7.5. The frequencies with a percent difference under 5% are shaded green, the frequencies with a percent difference under than 10% are shaded orange, and the frequencies with a percent difference above 10% are shaded red. Only the modes up to the W1T test objective are shown. Five of the uncorrected redundant base modes were excluded from this table in order to simplify the comparison between uncorrected and FBC. The F/A modes benefitted the most from correction. The W1F/A mode percent difference dropped by about 50%, which also corresponded to a large frequency shift from 5 Hz to 11 Hz. Modes other than F/A did not exhibit frequencies that looked significantly different. The main thing that the FBC method improved for these modes was cleaning up the test wing mode shapes to better match the FEM mode shapes, which is reflected by the Cross MAC table values increasing as more shakers were added while using the FBC method. Some values, however, did not increase significantly due to remaining small amounts of base motion.

Applying the FBC method can either increase or decrease the frequency of an uncorrected mode. If there is an inertial boundary condition effect (which is common with shake tables), the frequency tends to decrease as FBC is applied. If there is a stiffness boundary condition effect (which is common with static structures), the frequency tends to increase as FBC is applied.

A comparison of the W7B modes between the 12-shaker and 14-shaker FBC tests shows a little improvement in the base stiffness, as can be seen in Fig. 7.23. While both modes were relatively clean after adding so many base shakers, the 14-shaker configuration reduced the F/A motion of the reaction plates which reduced the Wing 4th Fore/Aft (W4F/A) motion that the W7B mode shapes were experiencing.

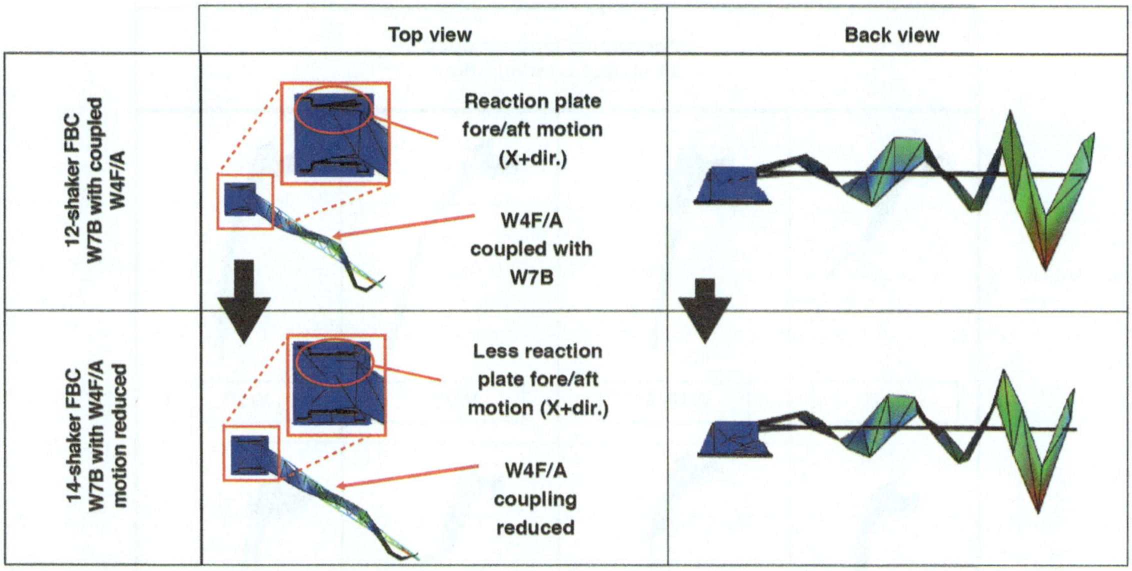

Fig. 7.23 Comparison of Fixed Base Corrected Wing 7th Bending mode shapes from 12-shaker versus 14-shaker: Wing 4th Fore/Aft coupling reduced

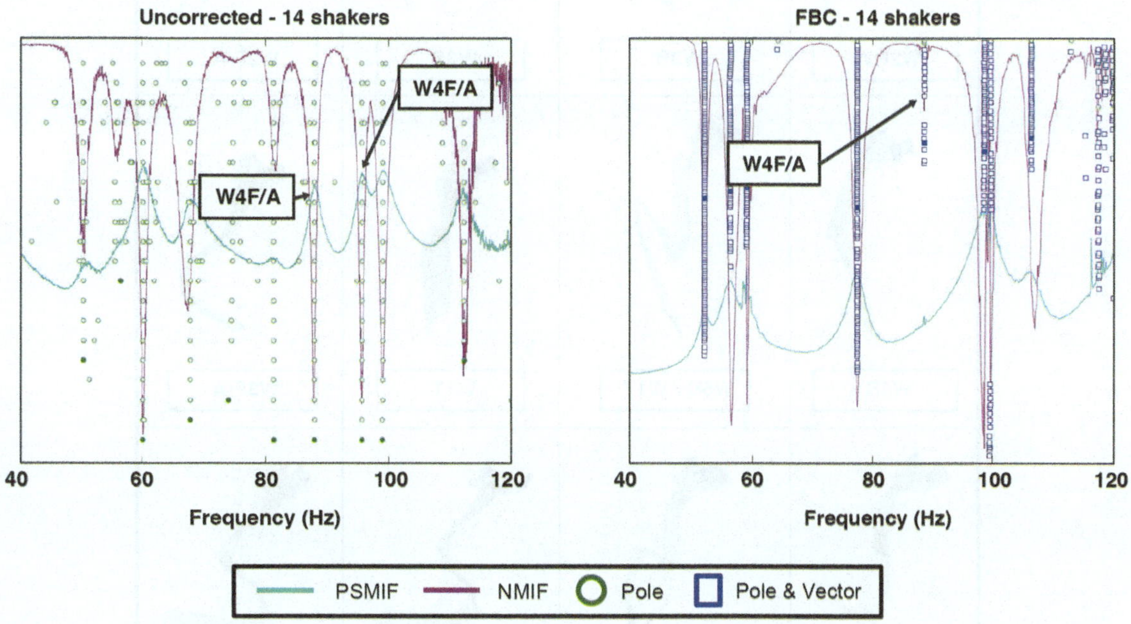

Fig. 7.24 Comparison of mode indicator functions for 14-shaker uncorrected versus 14-shaker Fixed Base Corrected

One of the primary benefits of applying the FBC method was reducing the base motion that coupled with various wing modes. One example was the W4F/A mode. In the uncorrected post-processing Mode Indicator Functions (MIFs), there were two peaks associated with W4F/A: one at 87 Hz, and the other at 95 Hz. Both mode shapes looked similar and the MIF peaks were about the same size, making it difficult to determine which was the true W4F/A mode. In contrast, there was only one MIF peak associated with W4F/A in the FBC post-processing at 88 Hz, as can be seen in Fig. 7.24.

In future tests, FBC results could potentially be improved by adding more accelerometers on the wing and base. There could also be more optimization performed when choosing locations for the base shakers. The PAT Wing GVT shows that the FBC method is promising, although more research is needed for aeronautics applications.

All of the uncorrected 14-shaker mode shapes can be seen in Fig. 7.25. Only an abbreviated version of the mode shape names is shown in Fig. 7.25; the complete uncorrected test mode shape names can be seen in Table 7.1.

All of the FBC 14-shaker mode shapes can be seen in Fig. 7.26.

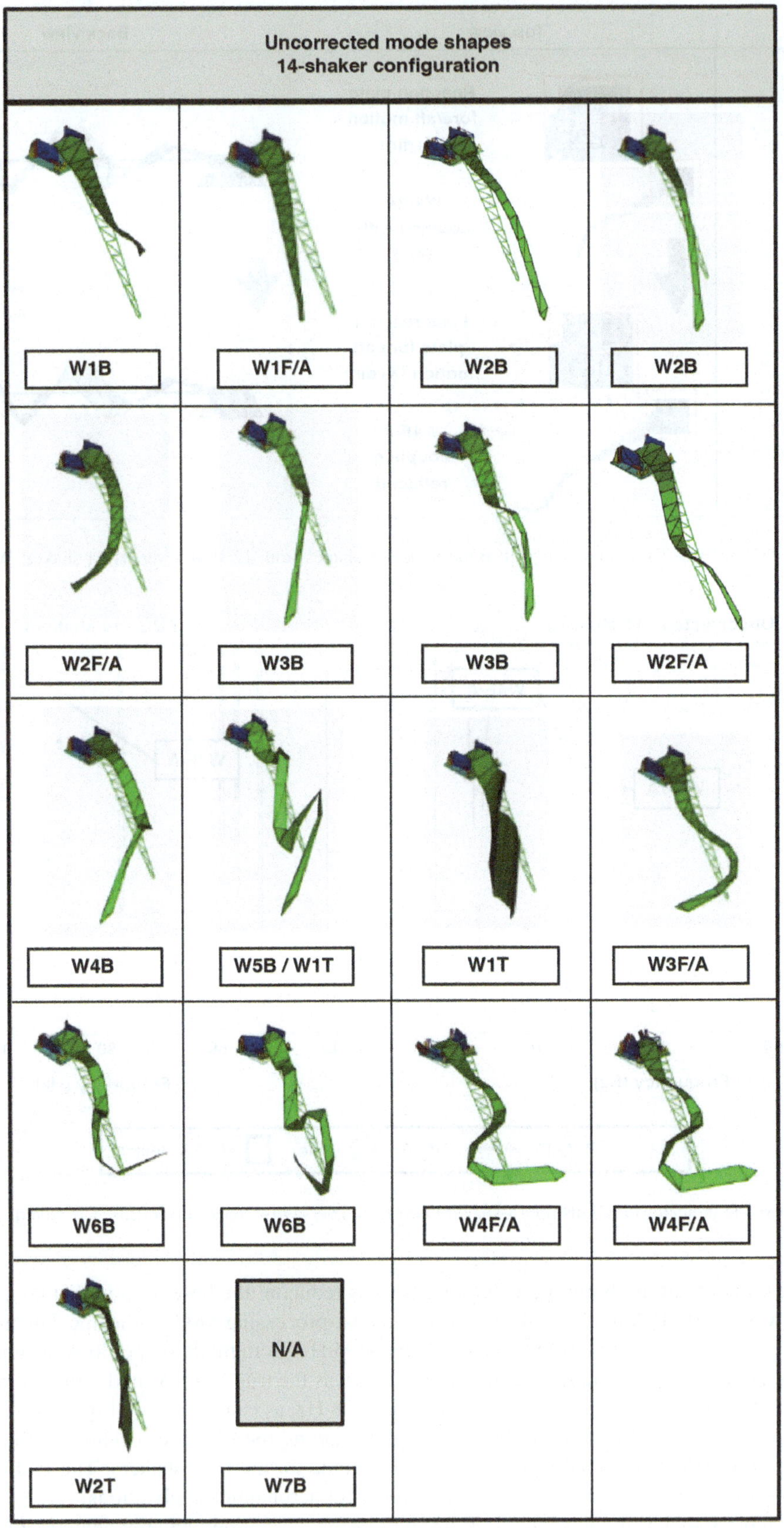

Fig. 7.25 Isometric mode shape pictures: uncorrected 14-shaker configuration

Fig. 7.26 Isometric mode shape pictures: fixed base corrected 14-shaker configuration

7.6 Summary

Passive Aeroelastic Tailored Wing ground vibration test results show the feasibility of using the Fixed Base Correction (FBC) method to decouple the wing and test fixture modes for a long flexible wing mounted to a dynamically active static test fixture. The test frequencies and mode shapes of the wing better matched fixed boundary condition finite element model predictions by using this method. The FBC technique is implemented by applying an excitation to the desired "fixed" boundary hardware with multiple independent sources (that is, shakers) where there are at least as many independent sources as there are independent boundary deformations in the test article frequency range of interest. The FBC method then uses the shaker boundary accelerations (measured by seismic accelerometers) as independent references when calculating frequency response functions. This FBC method has the potential to change how modal testing is traditionally performed and can save money and schedule time by eliminating an independent setup for modal testing. The FBC results also produce test results with reliable and comparable boundary conditions to replicate in and compare with analytical models.

Acknowledgements The authors gratefully acknowledge the NASA Advanced Air Transport Technology (AATT) project for the funding support to accomplish the Passive Aeroelastic Tailored (PAT) Wing modal testing. The PAT Wing is under the AATT project of the NASA Advanced Air Vehicles Program (AAVP). We also thank the NASA Armstrong Flight Loads Laboratory technicians and mechanics for their assistance in setting up and performing the ground vibration test. The NASA authors acknowledge Aurora Flight Sciences for designing and fabricating the towed-steered wingbox, and ATA Engineering, Inc. for assisting with the testing and guidance through the post-test data analysis using the fixed base correction method.

References

1. Armstrong Flight Loads Laboratory. https://www.nasa.gov/centers/armstrong/research/Facilities/FLL. Accessed 17 Oct 2019
2. Brooks, T.R., Martins, J.R.R.A., Kennedy, G.J.: High-Fidelity Aerostructural Optimization of a High Aspect Ratio Tow-steered Wing. AIAA-2016-1179. (2016)
3. Brooks, T.R., Martins, J.R.R.A., Kennedy, G.A.: High-Fidelity Multipoint Aerostructural Optimization of a High Aspect Ratio Tow-steered Composite Wing. AIAA-2017-1350. (2017)
4. Brooks, T.R., Kenway, G.K.W., Martins, J.R.R.A.: Benchmark aerostructural models for the study of transonic aircraft wings. AIAA Journal. **56**(7), (2018)
5. Spivey, N., Miller, K., Saltzman, R., Napolitano, K.: Modal testing of a flexible wing on a dynamically active test fixture using the fixed base correction method. IFASD 2019-108. (2019)
6. Crowley, J.R., Klosterman, A.L., Rocklin, G.T., Vold, H.: Direct structural modification using frequency response functions. In: Proceedings of the 2nd International Modal Analysis Conference, Orlando, Florida, pp. 58–65 (1984)
7. Beliveau, J.G., Vigneron, F.R., Soucy, Y., Draisey, S.: Modal parameter estimation from base excitation. J. Sound Vib. **107**(3), 435–449 (1986)
8. Imregun, M., Robb, D.A., Ewins, D.J.: Structural modification and coupling dynamic analysis using measured FRF data. In: Proceedings of the 5th International Modal Analysis Conference, London, England, pp. 1136–1141 (1987)
9. Carne, T.G., Martinez, D.R., Nord, A.R.: A comparison of fixed-base and driven-base modal testing of an electronics package. In: Proceedings of the 7th International Modal Analysis Conference, Las Vegas, Nevada, pp. 672–679 (1989)
10. Fullekrug, U.: Determination of effective masses and modal masses from base-driven tests. In: Proceedings of the 14th International Modal Analysis Conference, Dearborn, Michigan, pp. 671–681 (1996)
11. Sinapius, J.M.: Identification of Free and Fixed Interface Normal Modes by Base Excitation. In: Proceedings of the 14th International Modal Analysis Conference, Dearborn, Michigan, pp. 23–31 (1996)
12. Mayes, R.L., Bridgers, L.D.: Extracting fixed base modal models from vibration tests on flexible tables. In: 27th Conference and Exposition on Structural Dynamic 2009 (IMAC XXVII), Vol. 2, Orlando, Florida, pp. 957–970 (2009)
13. Allen, M.S., Mayes, R.L.: Recent advances to estimation of fixed-interface modal models using dynamic substructuring. In: Dynamics of Coupled Structures, Volume 4, Proceedings of the 36th IMAC, A Conference and Exposition on Structural Dynamics, Orlando, Florida, pp. 157–170 (2018)
14. Napolitano, K.L., Yoder, N.C.: Fixed base FRF using boundary measurements as references – analytical derivation. In: Topics in Modal Analysis I, Volume 5, Proceedings of the 30th IMAC, A Conference on Structural Dynamics, Jacksonville, Florida, pp. 299–308 (2012)
15. Mayes, R.L., Rohe, D.P., Blecke, J.: Extending the frequency band for fixed base modal analysis on a vibration slip table. In: Topics in Experimental Dynamic Substructuring, Volume 2, Proceedings of the 31st IMAC, A Conference on Structural Dynamics, Garden Grove, California, pp. 287–297 (2013)
16. Napolitano, K.L., Yoder, N.C., Fladung, W.A.: Extraction of fixed-base modes of a structure mounted on a shake table. In: Topics in Experimental Dynamic Substructuring, Volume 2, Proceedings of the 31st IMAC, A Conference on Structural Dynamics, Garden Grove, California, pp. 299–309 (2013)
17. Napolitano, K.L.: Fixing degrees of freedom of an aluminum beam by using accelerometers as references. In: Proceedings of the 37th IMAC, Orlando, Florida, pp. 53–60 (2019)

18. Yoder, N.C., Napolitano, K.L.: Fixed base FRF using boundary measurements as references - experimental results. In: Topics in Modal Analysis I, Volume 5, Proceedings of the 30th IMAC, A Conference on Structural Dynamics, Jacksonville, Florida, pp. 309–317 (2012)
19. Napolitano, K.L.: Using singular value decomposition to estimate frequency response functions. In: Topics in Modal Analysis & Testing, Volume 10, Proceedings of the 34th IMAC, San Diego, California, pp. 27–43 (2016)

Chapter 8
Nonlinear Normal Mode Estimation with Near-Resonant Steady State Inputs

Michael Kwarta and Matthew S. Allen

Abstract Nonlinear normal modes (NNMs) have been widely used for understanding and characterizing the motion of nonlinear structures, yet current methods to measure them experimentally are time-consuming and not always reliable. Since the structural nonlinearities usually occur when the sample oscillates at high amplitudes, specimens can be damaged or at least develop fatigue cracks when the testing is lengthy. Moreover, the interaction between the shaker and the structure can lead to distortions of the excitation force and can impact the quality of the measured test data. In our previous work, we proposed an NNM estimation algorithm that can help to overcome the issues mentioned above. The approach uses near-resonant data together with an algorithm based on the Single Nonlinear Resonant Mode (SNRM) method to then estimate the NNM backbone. The SNRM algorithm, in its original form, requires vibration modes to be well-separated and assumes no internal resonances between them. This work proposes a possible modification to the algorithm that will allow the modal coupling to be detected as well. The final version of the algorithm will be first tested with data generated numerically using a reduced model of a curved beam experiencing modal interactions. Then the method will be used to estimate the NNMs of a curved steel beam that exhibits significant modal interactions. The results will be validated against those obtained using well-established testing approaches.

Keywords Nonlinear system identification · Single nonlinear resonant mode Method · Modal coupling · Nonlinear normal modes · Nonlinear modal analysis

8.1 Overview of the Basic SNRM Algorithm

The authors' prior work, presented in [1, 2], used the Single Nonlinear Resonant Mode method to predict the Nonlinear Normal Mode backbone of a mechanical system experiencing very limited modal coupling. That algorithm is based on the SNRM equation (8.1), which was first proposed in [3].

$$V^{\text{meas}} \approx V^{model} = \left.\frac{\mathbf{\Phi}_j \mathbf{\Phi}_j^T \mathbf{F}\Omega}{\tilde{\omega}_{0,j}^2 - \Omega^2 + 2i\tilde{\zeta}_j\tilde{\omega}_{0,j}\Omega}\right|_{\substack{\text{pt. of max.}\\ \text{deflection}}} + \sum_{\substack{k=1\\k\neq j}}^{N_{lin}} \left.\frac{\mathbf{\Phi}_k \mathbf{\Phi}_k^T \mathbf{F}\Omega}{\omega_{0,k}^2 - \Omega^2 + 2i\zeta_k\omega_{0,k}\Omega}\right|_{\substack{\text{pt. of max.}\\ \text{deflection}}}, \tag{8.1}$$

where:

- V is the complex amplitude of the velocity signal,
- Ω is the forcing frequency,
- $\mathbf{\Phi}_i$, $\omega_{0,i}$, ζ_i are the mode shape, natural frequency and modal damping ratio of the i-th mode, respectively,
- $\mathbf{F}$ is a vector giving the spatial distribution of the sinusoidal excitation force,
- j is the index of the dominant mode,
- N_{lin} denotes the number of relevant linear modes, and
- the quantities marked $(\sim)$ vary with the vibration level.

M. Kwarta (✉) · M. S. Allen
Department of Engineering Mechanics, UW–Madison, Madison, WI, USA
e-mail: kwarta@wisc.edu; matt.allen@wisc.edu

B. Dilworth (ed.), *Topics in Modal Analysis & Testing, Volume 8*, Conference Proceedings of the Society for Experimental Mechanics Series, https://doi.org/10.1007/978-3-030-47717-2_8

To identify a mechanical system experiencing modal coupling, Eq. (8.1) has to be modified. A discussion on how this might be done is presented in the next sections. The concepts proposed herein are motivated by the measurements collected in several numerical tests.

8.2 Nonlinear Resonant Steady State Response Analysis

The modifications proposed here are motivated by the results collected in a numerical simulation of a single input Force Appropriation test. This test was performed on a simulation model of a *curved beam* with clamped-clamped boundary conditions. The beam was created using 400 shell elements resulting in a total of 3030 DOFs and was reduced to a 2-mode ICE-ROM including modes 1 and 2.

The backbone curve of the first Nonlinear Normal Mode of the beam is shown in Fig. 8.1. It consists of three segments that were computed separately because the response of the beam is unstable in the vicinity of pairs of points (1G, 2A) and (2E, 3A), which are marked in the figure. The authors suspect that the structure experiences internal resonance near pair (1G, 2A). The reason for instability near the backbone's point of minimum frequency is at this moment unknown and will be investigated.

The response of the nonlinear part of the mechanical system at point 3D (also marked in Fig. 8.1), decomposed into modal velocities $\dot{\mathbf{q}}(t)$, is presented in Fig. 8.2 and Table 8.1. Figure 8.2 shows time responses of modal velocities, while in Table 8.1

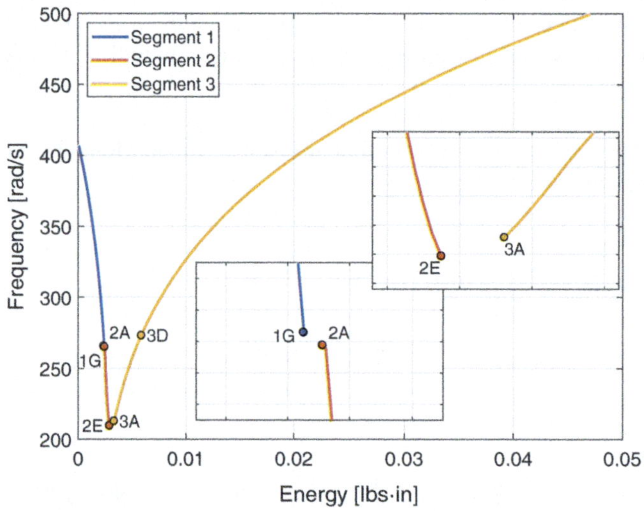

Fig. 8.1 Segments of the NNM backbone curve presented on frequency-energy plot

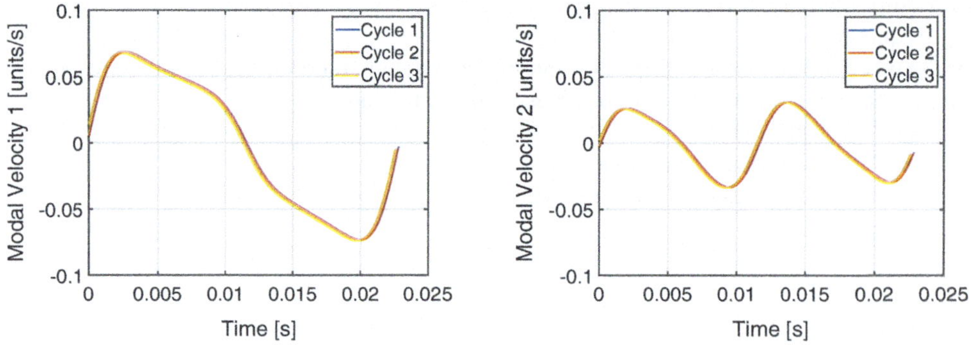

Fig. 8.2 Modal velocity signals, $\dot{q}_k(t)$, $k \in \{1, 2\}$, of the mechanical system oscillating at point 3D (marked in Fig. 8.1) during three consecutive cycles ($T = \frac{2\pi}{\Omega}$) (The cycles almost overlay)

Table 8.1 Fourier coefficients magnitudes of the first five harmonics of modal velocities presented in Fig. 8.2 and their ratio to the maximal coefficient value (expressed in %). Values marked with blue correspond to the modes/harmonics which are considered to exhibit modal coupling

	Ω		2Ω		3Ω		4Ω		5Ω	
$\dot{q}_1(t)$	**6.77e-02**	**(100.00)**	**1.59e-02**	**(23.45)**	**1.37e-02**	**(20.29)**	3.52e-03	(5.19)	2.58e-03	(3.81)
$\dot{q}_2(t)$	1.04e-03	(1.54)	**2.84e-02**	**(41.89)**	1.87e-03	(2.76)	**6.48e-03**	**(9.57)**	9.61e-04	(1.42)

the magnitudes of their Fourier coefficients are presented and compared with one another indicating which modes/harmonics participate the most in the system's response. The quantities $\dot{\mathbf{q}}(t)$ and $\mathbf{v}_j^{nl}(t)$ are defined in Eqs. (8.2) and (8.3), respectively.

$$\dot{\mathbf{q}}(t) = \begin{bmatrix} \dot{q}_1(t) \\ \dot{q}_2(t) \\ \vdots \end{bmatrix} = \mathbf{\Phi}^{-1}\mathbf{v}_j^{nl}(t) \tag{8.2}$$

Table 8.1 shows that the modal coupling at point 3D takes place between five modes/harmonics. Namely, between mode one occurring as first, second and third harmonics and mode two occurring as second and fourth harmonics. These modes/harmonics are also dominant in the steady-state response of the structure oscillating in the vicinity of point 3D. Thus it might be possible to estimate the NNM parameters based on the near-resonant measurements, even if the structure experiences modal coupling. The next section presents an overview of a concept that could be used to modify the original SNRM formulation so that it can successfully identify a nonlinear mechanical system using near-resonant response data such as that shown here.

8.3 Discussion on the SNRM Algorithm Extension

A generalized form of the SNRM model function is presented in Eq. (8.3). Rather than expressing the response using a single complex amplitude, as in Eq. (8.1), it considers the motion over a certain time. Hence, one could include the sub- or higher-harmonics in the system's response (as e.g. indicated in Table 8.1). This form also brings the model closer to the original Nonlinear Normal Modes definition, which introduces them as (non-necessarily synchronous) periodic motions of the conservative system [4].

$$\mathbf{v}^{meas}(t) = \mathbf{v}_j^{nl}(t) + \underbrace{\mathrm{Re}\left\{\sum_{\substack{k=1\\k\neq j}}^{N_{lin}} \frac{\mathbf{\Phi}_k\mathbf{\Phi}_k^T\mathbf{F}\Omega\, e^{i\Omega t}}{\omega_{0,k}^2 - \Omega^2 + 2i\zeta_k\omega_{0,k}\Omega}\right\}}_{\mathbf{v}_j^{lin}(t)}, \tag{8.3}$$

where:

- $\mathbf{v}^{meas}(t)$ is the full-field velocity response of the structure (measured experimentally or numerically),
- $\mathbf{v}_j^{nl}(t)$ is the full-field nonlinear velocity response of the structure oscillating near the j-th NNM and
- $\mathbf{v}_j^{lin}(t)$ is a term responsible for modeling the response of the system far from the j-th NNM.

One of the possible concepts of how to express the quantity $\mathbf{v}_j^{nl}$ is shown in Eq. (8.4). This formula allows for modeling the response with several modes and/or harmonics. Additionally, it has certain similarities to the nonlinear term from Eq. (8.1), which facilitates a physical interpretation of the new quantities introduced in (8.4).

$$\mathbf{v}_j^{nl}(t) = \mathrm{Re}\left\{\sum_{n=1}^{N_{nl}} \frac{\mathbf{\Phi}_j(n)\mathbf{\Phi}_j^T(n)\,\mathbf{F}\,\Omega\, e^{i[h_j(n)]\Omega t}}{\tilde{\omega}_{0,j}^2 - \Omega^2 + 2i\,\tilde{\zeta}_j(n)\,\tilde{\omega}_{0,j}\Omega}\right\} \tag{8.4}$$

The new quantities introduced in Eq. (8.4) are defined as follows:

- N_{nl} indicates how many coupling terms should be considered in the system identification process,
- $\tilde{\zeta}_j(n)$ is the damping ratio corresponding to mode $\boldsymbol{\Phi}_j(n)$ which is expected to occur in the system response. If $\tilde{\zeta}_j(n)$ is a large number than the mode $\boldsymbol{\Phi}_j(n)$ is negligible in the system's response.
- $h_j(n)$ indicates if the mode $\boldsymbol{\Phi}_j(n)$ vibrates with the forcing frequency Ω ($h_j(n) = 1$), or if it appears as a sub- or higher harmonic ($h_j(n) \neq 1$).

In the case study presented in the previous section, which focuses on the motion near the first NNM ($j = 1$), the quantities introduced above should be given the following values: $N_{nl} = 5$, $\boldsymbol{\Phi}_1(n) = [\boldsymbol{\Phi}_1\ \boldsymbol{\Phi}_1\ \boldsymbol{\Phi}_1\ \boldsymbol{\Phi}_2\ \boldsymbol{\Phi}_2]$, $h_1(n) = [1\,2\,3\,2\,4]$ and $\tilde{\zeta}_1(n) = [\tilde{\zeta}_{1,1}\ \tilde{\zeta}_{1,2}\ \tilde{\zeta}_{1,3}\ \tilde{\zeta}_{1,4}\ \tilde{\zeta}_{1,5}]$. The damping ratios of the coupled modes ($\tilde{\zeta}_{1,k},\ k \in \{1, \ldots, 5\}$) could be modeled as unknown functions of vibration level, with values known when the system vibrates at low amplitudes. At low vibration levels, the response of the structure is dominated by the underlying linear system. Thus, $\tilde{\zeta}_{1,1} = \tilde{\zeta}_1^{lin}$ and $\tilde{\zeta}_{1,k},\ k \in \{2, \ldots, 5\}$ should be given large enough values, so that the contribution of the their pseudo-modes to the system's response can be treated as negligible.

The extension to the SNRM algorithm discussed briefly in this section is one of several possible concepts the authors are currently investigating. The final version of the model function (8.4) and the discussion on its correctness from the physical standpoint as well as the ability to capture modal interactions experienced by the oscillating structure will be presented at the conference.

8.4 Conclusion and Future Work

This work briefly discussed one possible modification to the SNRM algorithm that would enable it to capture modal coupling. The main goal of this extension is to estimate the Nonlinear Normal Mode backbone curve and additionally detect the modal interaction. The authors are currently investigating variations on the model function in (8.4) in order to determine which to implement in the final version of the system identification algorithm.

In future work, the method will be tested numerically using a ROM of a curved beam, which experiences significant modal coupling. Then the algorithm will be used to identify the NNMs of a curved steel beam that exhibits modal interactions. The results will be validated against those obtained using well-established testing approaches.

Acknowledgments This work was supported by the Air Force Office of Scientific Research, Award # FA9550-17-1-0009, under the Multi-Scale Structural Mechanics and Prognosis program managed by Dr. Jaimie Tiley. The authors would also like to thank Joseph Hollkamp from the Air Force Research Laboratory's Structural Sciences Center, for his insightful comments and suggesting several improvements to this work.

References

1. Kwarta, M., Allen, M.S., Hollkamp, J.J.: An interpolation algorithm to speed up nonlinear modal testing using force appropriation. In: Mains, M.L., Dilworth, B.J. (eds.) Topics in Modal Analysis & Testing, vol. 8. Springer International Publishing, pp. 193–196 (2021)
2. Kwarta, M., Allen, M.S.: NNM Backbone Curve Estimation with Near-Resonant Steady State Inputs. Available at "https://sd.engr.wisc.edu/wp-content/uploads/sites/709/2019/10/SNRM_report_01.pdf" (unpublished). Technical report (2019)
3. Szemplińska-Stupnicka, W.: The modified single mode method in the investigations of the resonant vibrations of non-linear systems. J. Sound Vib. **63**(4), 475–489 (1979)
4. Rosenberg, R.M.: Normal modes of nonlinear dual-mode systems. J. Appl. Mech. **27**, 263–268 (1960)

Michael Kwarta is a graduate student in Mechanical Engineering at the University of Wisconsin – Madison. His recent research focuses on proposing new or improving existing methods used in nonlinear system identification.

Chapter 9
Automatic Modal Parameter Identification with Methods of Artificial Intelligence

Maik Gollnick, Daniel Herfert, and Jan Heimann

Abstract The increasing use of operational and experimental modal analysis provides experienced and inexperienced users always face new challenges. Automated evaluation, as a tool for distinguishing between physical and mathematical poles and for removing uncertainties in the accuracy of results, is becoming increasingly important (Jenny Lau, et al.: "Automatic modal analysis: reality or myth?"). This work deals specifically with the description of extreme values and their automated selection in indicator functions. The developed method is based on methods of pattern recognition and artificial intelligence and does not require any further parameters and or expert knowledge for later execution. The functional values of the indicator functions are represented by a simple feature vector with small dimensions and classified by a "Support Vector Machine" (SVM). Since the characteristics describe the function course, they are independent of the measurement parameters, such as sampling rate or frequency spacing. To increase robustness, existing methods from the field of "computer vision" were adapted and new ideas developed. Due to the versatile applicability of the method, an enormous database is required for the learning process. Therefore, a method for the creation of a training basis with synthetic data was developed. This covers a very wide range of applications and special cases, such as modes that are close together or strongly damped. In addition, the influence of inaccuracies, such as sensor noise, was modelled.

Keywords Modal analysis · Identification · Artificial intelligence · Peak detection

Nomenclature

N_R	Number of reference frequency response functions
N_F	Number of spectral lines (frequencies)
ω	Frequency (rad/s)
$[H(\omega)]$	Frequency response function matrix
$[T]$	Eigenvector matrix
$[\Lambda]$	Eigenvalue matrix (diagonal)
$[U]$	Left singular vector matrix (unitary)
$[\Sigma]$	Signular value matrix (diagonal)
$[V]$	Right singular vector matrix (unitary)
Λ_k	kth eigenvalue
Σ_k	kth singular value
$l(\omega)$	Left prominence feature at frequency ω
$r(\omega)$	Right prominence feature at frequency ω
$h(\omega)$	Relative height feature at frequency ω
$CMIF_k$	kth complex mode indicator function
$[]^H$	Hermitian (conjugate transpose) of matrix

M. Gollnick · D. Herfert (✉) · J. Heimann
Department of Structural Dynamics/Pattern Recognition, Society for the Advancement of Applied Computer Science, Berlin, Germany
e-mail: gollnick@gfai.de; herfert@gfai.de

B. Dilworth (ed.), *Topics in Modal Analysis & Testing, Volume 8*, Conference Proceedings of the Society for Experimental Mechanics Series, https://doi.org/10.1007/978-3-030-47717-2_9

9.1 Introduction

The automatic determination of modal parameters is an essential point to perform a modal analysis even without much previous knowledge of the user. Furthermore, automatic reproducible determination is very important for monitoring applications that also use modal parameters. This work deals specifically with the description of extreme values and their automated selection in indicator functions. This approach tries to use the same parameters to determine the modal parameters that a human would use for evaluation. Thereby the indicator functions are represented by a simple feature vector with small dimensions and classified by a SVM [2]. The features were selected so that they describe the function independently of measurement parameters. This means, for example, that they are independent of sampling rate or frequency spacing.

This approach for automatic determination of modal parameters is integrated in the vibration analysis software WaveImage Modal [6]. It thereby offers the use of experimental modal analysis not only for experts, but also for beginners and completes the package to perform a dynamic structural analysis. In addition to the intuitive acquisition of data using the recorder module, the entire modal analysis process is simplified and combined in a single software package. Also, the modal parameters obtained in this way can be used directly to improve the FE simulation via model updating. In addition to experimental modal analysis, the software also provides operational modal analysis, operating deflection shapes, order analysis and a large number of options for signal processing. WaveImage is specialized in processing large amounts of data of several Gigabytes (e.g. Laser Doppler Measurements).

The automatic modal parameter identification is validated in this publication exclusively by examples of experimental modal analysis. However, it can also be used in the software for applications of operational modal analysis.

9.2 Background

The method presented in this work is based on procedures and their approaches from the field of machine vision and image comprehension. The sense and content of the data is achieved by the description and classification via representing object features. The characteristics are determined as generally as possible and independent of external parameters, so that the recognition of the same object under changed conditions is guaranteed. This approach has been transferred to modal analysis for finding peaks in indicator functions. The peak locations are described and classified by general properties.

The procedure is applicable to many types of indicator functions and is not specialized for a particular type. In addition, the developed method serves only as pre-processing for modal analysis, since the semantics of a peak (physical or complex mode) are not considered.

Due to the widespread use of the "Complex Mode Indicator Function" (CMIF) [3] in the field of modal analysis, this function was chosen as a reference example. Furthermore, the CMIF method offers an easy and efficient way to determine modal parameters of a complex object.

The first formulation of the CMIF process was based on the eigenvalue decomposition of the following matrix:

$$H(\omega)^H H(\omega) = T(\omega)\, \Lambda(\omega)\, T(\omega)^H \tag{9.1}$$

It quickly became clear that singular value decomposition (SVD) of the FRF matrix was a more practical approach and that it could dispense the matrix multiplication $H(\omega)^H H(\omega)$. Thus, the CMIF method for the FRF matrix $H(\omega)$ is defined by each frequency ω:

$$H(\omega) = U(\omega)\, \Sigma(\omega)\, V(\omega)^H \tag{9.2}$$

The matrix $U(\omega)$ forms the left singular vectors (the approximated mode shapes) and the matrix $V(\omega)$ the right singular vectors (the approximated modal participation factors). The k CMIF curves are given by the singular values $\Sigma_k(\omega)$ for the number of reference FRFs N_R at each frequency ω. The reference to the eigenvalue decomposition is given by:

$$\sqrt{\Lambda_k(\omega)} = \Sigma_k(\omega), k = 1, 2, ..N_R \tag{9.3}$$

The natural frequencies and modes close to each other can be identified by peaks in the different CMIF curves. But not every peak corresponds to a mode, since they can also be caused by noise, leakage effects, non-linearity, or the cross eigenvalue effect [7].

9.3 Peak Descriptor

For the automatic recognition of peaks in indicator functions, three phases are passed through: the pre-processing of the available data, the compilation of feature vectors and the classification. In the first phase, all data-improving preprocessing steps are possible. Since the procedures are often related to a use case, prior knowledge of the possible characteristics of the peaks is always required. For example the window size of a median filter is always data specific and not generally usable. In order to guarantee the general applicability of this method, this paper has only decided to remove the linear trend.

In the second phase, a feature vector is constructed to distinguish peaks from any other function value. The three features are: left $l(\omega_i)$ and right $r(\omega_i)$ prominence and the relative height $h(\omega_i)$ of a peak at frequency ω_i.

$$l\left(\omega_i\right) = \left|\left\{\omega_j < \omega_i \mid CMIF_k\left(\omega_j\right) \leq CMIF_k\left(\omega_i\right)\ and\ \nexists l\ with\ j < l < i\ and\ CMIF_k\left(\omega_l\right) > CMIF_k\left(\omega_i\right)\right\}\right| / N_F \tag{9.4}$$

$$r\left(\omega_i\right) = \left|\left\{\omega_j > \omega_i \mid CMIF_k\left(\omega_j\right) \leq CMIF_k\left(\omega_i\right)\ and\ \nexists l\ with\ j < l < i\ and\ CMIF_k\left(\omega_l\right) > CMIF_k\left(\omega_i\right)\right\}\right| / N_F \tag{9.5}$$

$$h\left(\omega_i\right) = \max\left(\min_{\omega_j < \omega_i} CMIF_k\left(\omega_j\right), \min_{\omega_j > \omega_i} CMIF_k\left(\omega_j\right)\right) \tag{9.6}$$

The left prominence describes the number of consecutive function values that are smaller to the left of the currently viewed frequency. The right prominence accordingly the smaller values on the right hand side. So this counter is independent of frequency spacing because it is normalized by the number of frequency lines for the respective indicator function (see Fig. 9.1).

As shown in Fig. 9.1, peaks are represented by a high left and right prominence. In contrast, noise peaks produce small values for both features or are distributed along one of the feature axes. This example shows two weak points. On one hand, both features are always related to the position of the peak within the considered frequency band and on the other hand, modes close to each other are difficult to separate, since the stronger ones dominates the weaker ones. If the peak is moved from frequency zero to the nyquist frequency, the behavior in the feature space can be seen in Fig. 9.2.

The Fig. 9.2 clearly shows that the position of a peak is described by a straight line in the feature space. For the separability between peaks and noise, it means that a peak cannot occur at the left or right edge of the frequency band, if it is to be reliably distinguished from a noise peak. Therefore, it must be ensured during the measurement that a sufficiently large frequency interval is covered and that the significant peaks are not at the edge.

For better separability, the relative height of a maximum has been defined as the difference between the current peak and the maximum of the left and the right minimum value is calculated (see Fig. 9.3). Finally, the altitude value is scaled to the

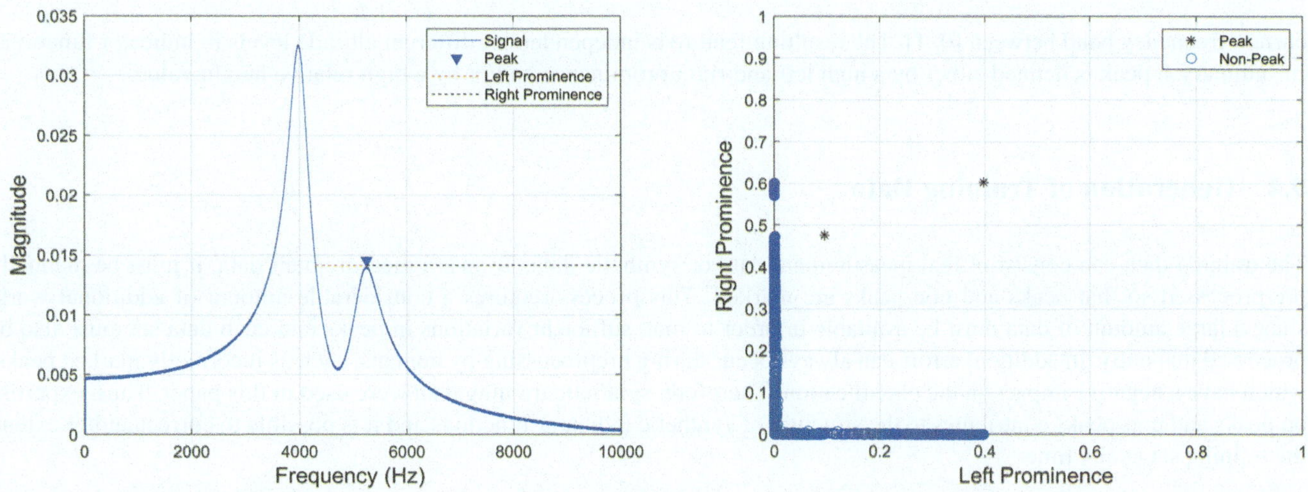

Fig. 9.1 Left: Example of an indicator function with two peaks to show the left and right peak prominence. Right: Indicator function transformed to feature space

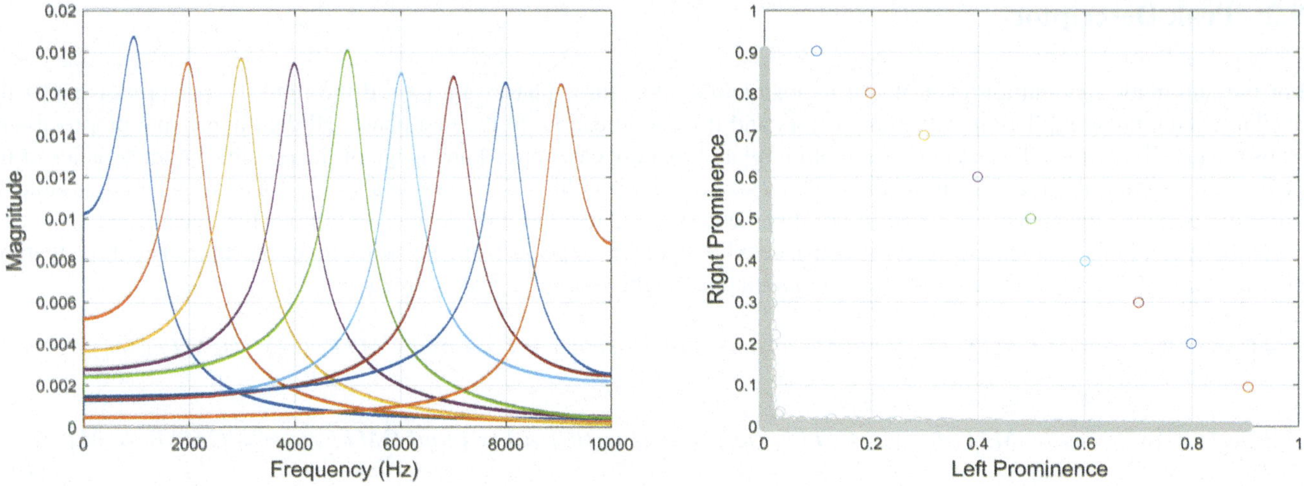

Fig. 9.2 Left: Moving peak from 1 kHz to 9 kHz. Right: Behavior of moving peaks in the feature space

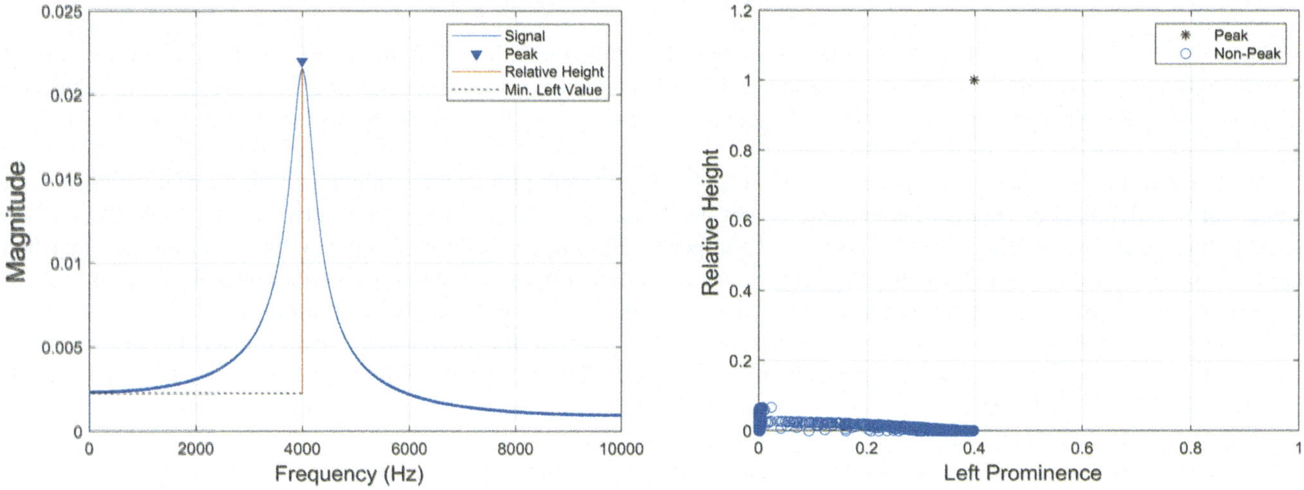

Fig. 9.3 Left: Example of an indicator function with one peak to show the relative height. Right: Indicator function transformed to feature space

current frequency band between [0..1]. The resulting feature is independent of different altitude levels in indicator functions. In summary, a peak is defined either by a high left and right prominence and/or by a high relative height value.

9.4 Generation of Training Data

The training data can consist of real measurement data or synthetic data. To insert measurement data, it must be manually pre-processed so that peaks and non-peaks are marked. This process requires a considerable amount of additional work, since a large amount of data must be available in order to map sufficient variations in peaks and each data set must also be processed manually. In addition, errors can always occur during preprocessing by humans, such as incorrectly marked peaks, which have a negative impact on the classification. Therefore, synthetic training data were used in this paper. Thus, expertise on peaks and non-peaks contributes to the structure of synthetic indicator functions and it is possible to correct and/or extend the training set at any time.

For the synthetic indication functions, the decaying sinusoidal oscillations were generated by the following equation:

$$y = \sum_{i=1}^{n} A_i \cdot e^{-\frac{t}{\tau}} \cdot \sin\left(2 \cdot \pi \cdot f_i \cdot t\right) \tag{9.7}$$

The parameters amplitude $\boldsymbol{A_i}$, decay $\boldsymbol{\tau}$ and frequency $\boldsymbol{f_i}$ were randomly generated. To generate the time vector and to restrict the frequencies f_i, the sampling rate, the number of frequency lines and the number of frequencies n contained in the signal were set to a fixed value. Since the features are independent of these parameters, a random assignment is not necessary. Finally, white noise was added.

9.5 Classification

The basic design of a "Support Vector Machine" (SVM) [2] requires a training set that defines which class belongs to which data set. The training objects are presented as feature vectors in a vector space. In our case, this is the $\mathbb{R}^3$. The assignment of the training data to classes is realized by a hyperplane, which acts as a separation plane. The minimal distance of the sets of vectors to the plane is maximized. This should guarantee a better separability of non-trivial data.

For objects that are not linearly separable, the hyperplane is transferred to a higher dimensional space. In this space, with a sufficiently high dimension, the training data can again be separated linearly. When transforming back into a space with a smaller dimension, the linear hyperplane becomes a non-linear, possibly non-contiguous hyperplane. The so-called "kernel trick" is used to reduce the complexity and the calculation load for the outward and backward transformation. Kernel functions are used to describe the hyperplane in the low and the high dimensional spaces. With this step, the support vectors at the edge of the hyperplane are sufficient to fully describe the outer boundaries of the classes.

By generating synthetic indicator functions, the size of the training data quickly increases, since we try to cover every special case. In addition, a clear linear separation is not always guaranteed. Noise or other simulated disturbances for example can shift positions of non-peaks from one class to another. These effects can be minimized by using an SVM with suitable kernel functions.

To get these suitable kernel functions, the used SVM was optimized with regard to the training data using an "Iterative Single Data Algorithm" (ISDA) [4] and the assumption of 5% outliers.

9.6 Measurement Setup

Two structures were measured and simulated as part of the publication [5]. These are a flange and a UAV rotor blade.

The vibration response of a stainless steel flange (120 mm × 14 mm) to a force excitation was investigated. The flange was attached to a frame with a rubber band. Thus free boundary conditions can be assumed on the entire surface. The flange was excited by a shaker (PCB SmartShaker with integrated power amplifier, model K2007E01) via a thin stinger (diameter approx. 2 mm). In order to get a force transmission only normal to the surface the stinger with force sensor (PCB type 208C02) was mounted on the back side at the outer edge using glue, see Fig. 9.4 (left). The system response of the flange to the force excitation was measured with a 3D-Laser Doppler Vibrometer (3D-LSV) measurement system (PSV-500, Polytec). As test signal a periodic chirp was used. Based on simulation results, the frequency range of the excitation was limited to 4–12 kHz. The surface velocities in all three directions (*x, y,z*) were scanned using an unevenly distributed mesh grid with more than 200 measuring points over the whole surface of the flange. The correlation between excitation and response leads to the individual frequency response function for each measuring point. This is used to identify the modal parameters.

The system response of a carbon fiber reinforced polymer (CFRP) rotor blade was investigated. The rotor blade (274 mm length) was fixed at its root on one side in order to realize the actual installation situation. For this purpose the rotor blade was mounted to a frame. The light structure was excited broadly by an automatic modal hammer (WaveHit, gfai tech GmbH) to avoid additional mass coupling by a sensor. In order to achieve a sufficient force transmission the location of the impact have been chosen to be on the free end of the blade tip, see Fig. 9.4 (right). The modal hammer, as a full automatic device, was synchronized with 3D-LSV data acquisition system, which measured the individual system response in all three directions (*x, y, z*) at 170 unevenly distributed measuring points on the whole surface of the structure.

Fig. 9.4 Left: Setup for the Experimental Modal Analysis on a flange. The structure with free boundary conditions is excited by a shaker between 4–12 kHz. Acquisition of the system response via 3D-LSV. Right: Setup for the Experimental Modal Analysis on a rotor blade. The structure fixed on one side was excited broadly by an automatic modal hammer (WaveHit, gfai tech GmbH). Acquisition of the system response via 3D-LSV

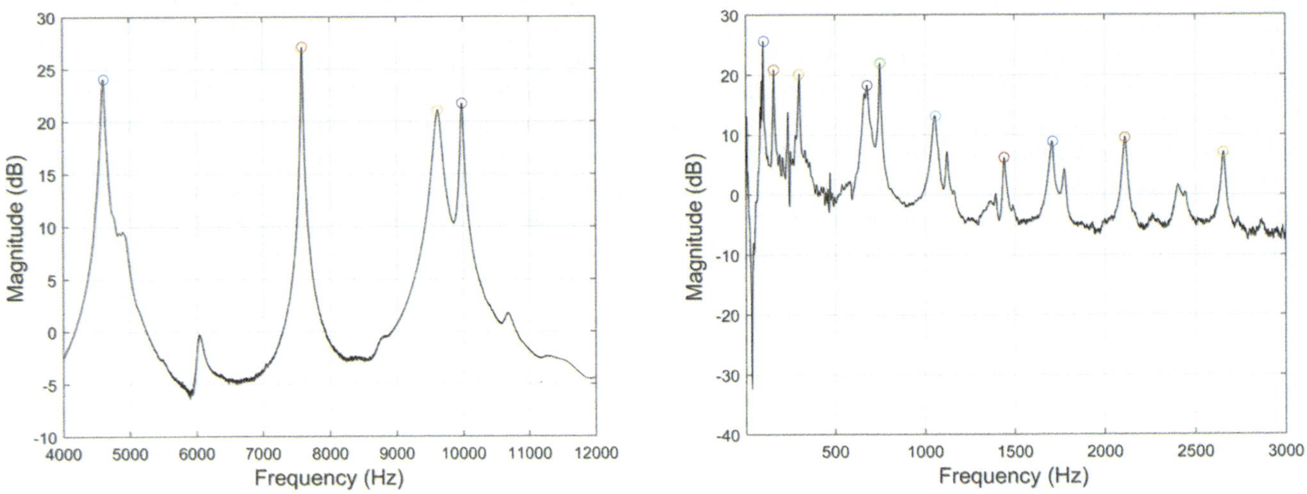

Fig. 9.5 Left: First CMIF curve of a stainless steel flange. Right: First CMIF curve of an UAV rotor blade. Classified peaks are marked

To create the training data, indicator functions with a sampling rate of 20 kHz and a frequency spacing of 0.0005 Hz were generated. The amplitude, decay and frequencies of the sinusoidal oscillation were randomly generated and white noise was added. The training set consisted of a total of 6044 peaks and 99,940 non-peaks. The classification was applied to the CMIF curves of the stainless steel flange and rotor blade. Both objects were excited at one reference point, shaker and modal hammer, respectively. This meant that only one CMIF curve was available for the evaluation of both measurement objects (see Fig. 9.5).

Figure 9.5 shows the results of the classification for the stainless steel flange and rotor blade using the first CMIF curve. It becomes clear that nearly all peaks are found correctly by the applied method. Peaks with too low height and to low left or right prominence are classified as non-peak. The frequency 6043 Hz for the stainless steel flange and the frequencies 1023 Hz, 1775 Hz and 2405 Hz for the rotor blade were not detected as peaks. This result is due to the fact that these peak characteristics were not taken into account during the learning process. The amount of training must therefore be adjusted with regard to these variants.

In summary, it becomes clear that peaks and non-peaks in the feature space are distinct from each other (see Figs. 9.6, 9.7 and 9.8) and that the respective features complement each other.

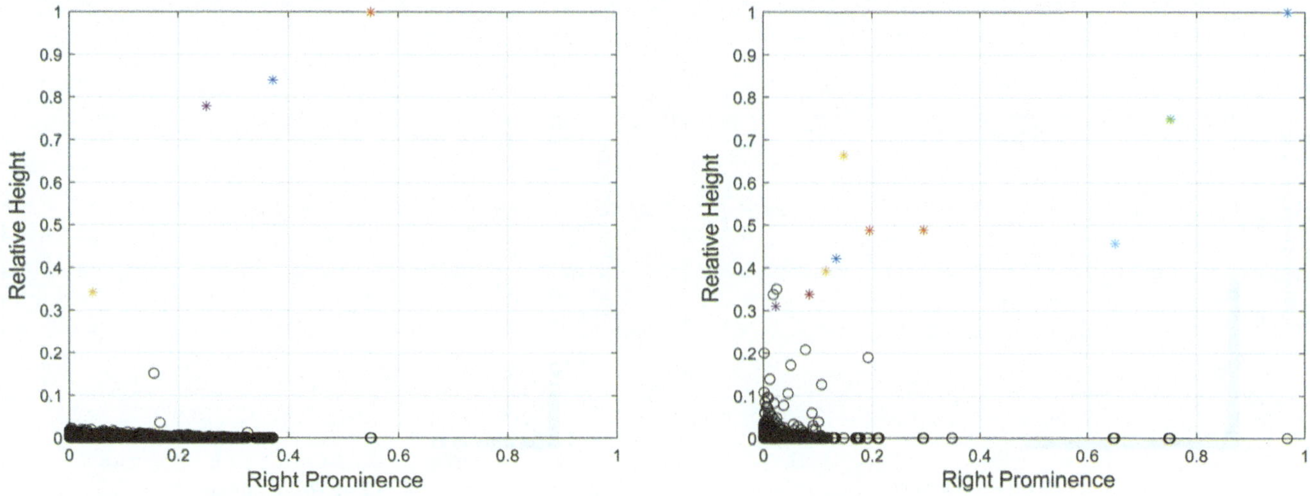

Fig. 9.6 Left: Indicator function (stainless steel flange) transformed to feature space. Right: Indicator function (UAV rotor blade) transformed to feature space. Relative height and right prominence plane is shown

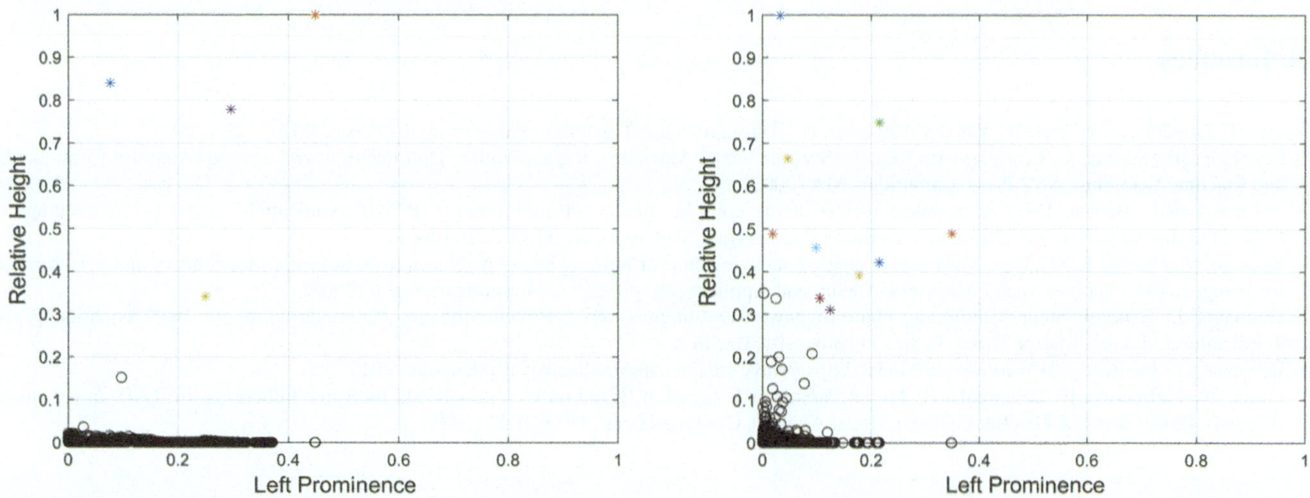

Fig. 9.7 Left: Indicator function (stainless steel flange) transformed to feature space. Right: Indicator function (UAV rotor blade) transformed to feature space. Relative height and left prominence plane is shown

9.7 Conclusion

In this paper a new machine learning based approach for automatic peak finding is presented. For this purpose a feature vector consisting of the left and right peak prominence and relative height was presented. In addition, an approach for the synthetic creation of learning data was described as an alternative to use real data. With this method, a lot of training data was generated quickly and flexibly. All possible peak characteristics are mapped and disturbances such as sensor noise were modelled. Furthermore, the use of synthetic data reduces the additional effort involved in compiling training examples and avoids possible errors by the user. For classification, an SVM was optimized with the "Iterative Single Data Algorithm" and under the assumption of 5% outliers. The results of the classification procedure were presented by 3D -Laser Doppler Vibrometer measurements on a stainless steel flange and a reinforced polymer (CFRP) rotor blade. The CMIF curves were used as indicator functions. For further investigations, modal indicators, such as the average or modal coherence, can be added as characteristics, to better adapt the solution to the application of modal analysis.

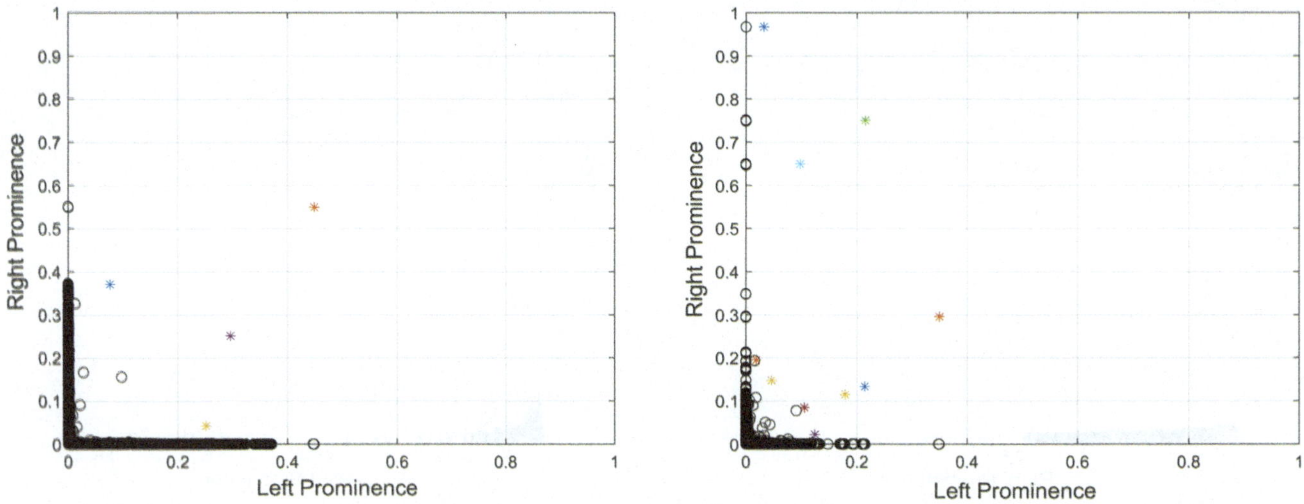

Fig. 9.8 Left: Indicator function (stainless steel flange) transformed to feature space. Right: Indicator function (UAV rotor blade) transformed to feature space. Relative right prominence and left prominence plane is shown

References

1. Lau, J., Lanslots, J., Peeters, B., Van der Auweraer, H.: Automatic modal analysis: reality or myth? Proce. IMAC. **25**, (2007)
2. Schölkopf, B., Smola, A.: Learning with Kernels: Support Vector Machines, Regularization, Optimization, and beyond (Adaptive Computation and Machine Learning). MIT Press, Cambridge, MA (2002)
3. Allemang, R.J., Brown, D.L.: A complete review of the complex mode indicator function (CMIF) with applications. In: Proceedings of ISMA2006 Int. Conference on Noise and Vibration Engineering, Sept 2006, pp. 3209–3246 (2006)
4. Kecman, V., Huang, T.-M., Vogt, M.: Iterative single data algorithm for training kernel machines from huge data sets: theory and performance. In: Wang, L. (ed.) Support Vector Machines: Theory and Applications, pp. 255–274. Springer, Berlin (2005)
5. Heimann, J.: Experimentelle Validierung eines Eigenwert-Optimierers sowie Erweiterung des Parameterraums zur Berücksichtigung von Temperatureinflüssen. Master Thesis Technical University, Berlin
6. Homepage of the Software WaveImage. Modal. https://wave-image.com/modalanalysis/?lang=en, 2019
7. Leurs, W., Deblauwe, F., Lembregts, F.: Modal Parameter Estimation Based on Complex Mode Indicator Functions. In: LMS Intl. Leuven, Belgium, Proceedings of 11th International Modal Analysis Conference, pp. 1035–1041 (1993)

Chapter 10
A Single Step Modal Parameter Estimation Algorithm: Computing Residues from Numerator Matrix Coefficients of Rational Fractions

Nimish Pandiya, Christian Dindorf, and Wim Desmet

Abstract The current state-of-the-art modal parameter estimation algorithms follow a two-step procedure to estimate from measurements, the modal parameters in the form of complex natural frequencies (poles), participation and modal vectors, and modal scaling factors. The current work investigates the use of previously neglected matrix-coefficients of the numerator polynomial in the rational fraction matrix description model for computing the residues for each of the poles identified. While the denominator polynomial describes the global characteristics of the system, the local characteristics are included in the numerator polynomial and residues and residuals may be extracted by appropriate mathematical manipulations. The procedure is labeled as a "single-step" algorithm mainly because the least-squares fitting using the measured frequency response functions is carried out only once. The proposed method is applied to a lumped mass multi-degree-of-freedom system where the frequency response function matrix is truncated in its output degrees of freedom to mimic a realistically measured multiple-input multiple-output frequency response function matrix. The parameters are validated against the traditional two-step approach using the accuracy of the reconstructed frequency response functions and several existing model validation techniques. The results indicate that the proposed algorithm yields an accurate model of the dynamic system under test.

Keywords Modal parameter estimation · Rational fraction model · Partial fraction model · Residue estimation · Stabilization chart

10.1 Introduction

Modal parameter estimation (MPE) refers to system identification of vibrating structures by fitting a known parametric model to measured frequency (FRF) or impulse (IRF) response functions. The parameters computed, i.e. complex natural frequencies, mode-shapes and participation vectors, and modal scaling factors [1], describe the dynamics of the structure under test in a linear regime. As such, MPE finds applications in analyses of vibrating structures like finite-element model verification and validation, reduced-order modeling, sub-structuring etc. With a long history [2] of successfully proposed approaches associated with it, the maturity of this branch of research may be gauged by the current research focus being on statistical approaches [3], operational modal parameter estimation [4] and minimizing user interaction during the MPE process [5, 6]. Applicability to industrial (and noisy) data is the driving force behind investigations and much effort has been put into handling and tracking of measured functions and validation of results from realistically damped systems.

N. Pandiya (✉)
Center of Competence for Vibration, Robert Bosch GmbH, Stuttgart, Germany

Department of Mechanical Engineering, KU Leuven, Leuven, Belgium
e-mail: nimish.pandiya@student.kuleuven.be; nimish.pandiya@de.bosch.com

C. Dindorf
Center of Competence for Vibration, Robert Bosch GmbH, Stuttgart, Germany
e-mail: christian.dindorf@de.bosch.com

W. Desmet
Department of Mechanical Engineering, KU Leuven, Leuven, Belgium

DMMS Lab, Flanders Make, Leuven, Belgium
e-mail: wim.desmet@kuleuven.be

B. Dilworth (ed.), *Topics in Modal Analysis & Testing, Volume 8*, Conference Proceedings of the Society for Experimental Mechanics Series, https://doi.org/10.1007/978-3-030-47717-2_10

The current standard commercial algorithms (e.g. PolyMax [7]) employ a two-step process for the computation of the said parameters. The Unified Matrix Polynomial Approach (UMPA) [8] provides a general framework for various algorithms and is also used in this work. UMPA utilizes measured FRFs (or IRFs) to compute a rational fraction model (Equation 10.3) by minimizing residuals in a least-squares sense. The complex natural frequencies and participation vectors are obtained in the first step by carrying out an eigen-value decomposition on a companion matrix derived from the denominator matrix-coefficients. The numerator coefficients on the other hand are eliminated [7] or left unused [8]. After the selection of valid poles (typically done by the analyst using a stabilization/consistency chart [9]), the calculation of the residue and the residuals from the pole-residue representation of FRFs (Equation 10.11) is accomplished in a second step. The modal model is often validated by using metrics like modal assurance criterion (MAC) [10], mean phase correlation (MPC), mean phase deviation (MPD) and synthesis correlation coefficient [9, 11].

The present research aims at answering a more basic question – can the numerator coefficients be manipulated to make the complete residue information available during the pole selection phase, to be included in the stabilization chart?

10.2 Theoretical Background

A left matrix-factor description (MFD) is used to ascribe a rational fraction model to measured FRF data (Equation 10.1). The complete process is transferable to a right MFD form, but for the sake of simplicity, only one development is discussed.

$$\Big[H(s)\Big]_{N_o \times N_i} = \left[\sum_{i=0}^{m}\Big[[\alpha_i]s^i\Big]\right]^{-1}_{N_o \times N_o}\left[\sum_{i=0}^{m-1}\Big[[\beta_i]s^i\Big]\right]_{N_o \times N_i} + \left[\sum_{i=-n_l}^{n_u}\Big[[R_i]s^i\Big]\right]_{N_o \times N_i} \tag{10.1}$$

Here, $[H(s)]$ represents the transfer function ($s = j(2\pi f)$, for measured frequency response functions, f being the sampled frequency in Hz). Its size is defined by the number of outputs (N_o) and the number of inputs (N_i) for each frequency line observed (N_f). The denominator polynomial model order is indicated by m, while n_l and n_u indicate the order of the polynomials used to represent the lower and upper residuals respectively.

The description in Equation 10.1 is in terms of complex frequency. The equation holds for N_f number of frequencies on the positive imaginary frequency axis and N_f on the negative imaginary frequency axis. The reader is referred to the concept of characteristic space in [2]. Naturally, since the negative frequencies are not measured, they are simply assigned as complex conjugates of the positive imaginary axis in the complex plane. The FRFs are also assigned as complex conjugates of the corresponding positive frequency FRFs.

The terms in Equation 10.1 can be rearranged to result in the system shown in Equation 10.2, which is effectively the rational fraction model with additional numerator coefficients for inclusion of residual terms. This is discussed in detail by Fladung [12] and hereafter, the values $n_l = 2, n_u = 0$ are used for physical inertia and stiffness residuals.

$$\left[\sum_{i=0}^{m}\Big[[\alpha_i]s^i\Big]\Big[H(s)\Big]\right] = \left[\sum_{i=-n_l}^{m+n_u}\Big[[\hat{\beta}_i]s^i\Big]\right] \tag{10.2}$$

To avoid calculation of the trivial solution, without any loss in generality, the highest alpha coefficient is assumed to be the identity matrix. Equation 10.3 for each frequency ($N_\omega = 2N_f$ frequency lines) is then be used to set up an over-determined system of linear equations, which is solved in a least squares sense to recover the matrix coefficients. The complete setup of the over-determined least-squares problem is shown in Appendix B of [8] for a right MFD formulation, which is simply the transposed version of the left MFD equation.

$$\left[\sum_{i=0}^{m-1}\Big[[\alpha_i]s^i\Big]\Big[H(s)\Big]\right] - \left[\sum_{i=-n_l}^{m+n_u}\Big[[\hat{\beta}_i]s^i\Big]\Big[I\Big]\right] = \left[-\Big[I\Big]\Big[H(s)\Big]s^m\right] \tag{10.3}$$

The inversion of the Vandermonde-type matrix obtained by arranging the measurements in Equation 10.3 is an ill-conditioned problem and often a Z-transform approach is used to improve the conditioning. By substituting $s = exp(2j\pi f\Delta t)$, the complex two-sided FRF is wrapped around a unit circle such that any increase in the frequency values is essentially a rotation in the complex plane. Here, f is the frequency under transformation and Δt is the discrete time interval obtained from the Nyquist criterion. Employing orthogonal polynomials [8] for frequency tranformation results is another approach to improve conditioning, but is computationally more expensive.

Once the least squares solution for the rational fraction matrix-coefficients (both $[\alpha_i]$ and $[\hat{\beta}_i]$) is computed, the residue and the numerator matrix-coefficients may be separated using a simple, fully determined de-convolution solution as shown in [12]. Hence for each model order iteration, the complete set of $[\alpha]$, $[\beta]$, $[\hat{\beta}]$ and $[R]$ coefficients is available i.e. Equation 10.4 is fully defined.

$$[H(s)] = [\alpha(s)]^{-1}[\beta(s)] + [R(s)] = [\alpha(s)]^{-1}[\hat{\beta}(s)] \tag{10.4}$$

Where,

$$[\alpha(s)] = \left[\sum_{i=0}^{m}\left[[\alpha_i]s^i\right]\right]_{N_o \times N_o}$$

$$[\beta(s)] = \left[\sum_{i=0}^{m-1}\left[[\beta_i]s^i\right]\right]_{N_o \times N_i}$$

$$[R(s)] = \left[\sum_{i=-n_l}^{n_u}\left[[R_i]s^i\right]\right]_{N_o \times N_i}$$

$$[\hat{\beta}(s)] = \left[\sum_{i=0}^{m-1}\left[[I][\beta_i]s^i\right]\right]_{N_o \times N_i} + \left[\sum_{i=0}^{m}\left[[\alpha_i]s^i\right]\right]_{N_o \times N_o}\left[\sum_{i=-n_l}^{n_u}\left[[R_i]s^i\right]\right]_{N_o \times N_i}$$

Normally, the numerator matrix-coefficients remain unused and are discarded [8] or they are expressed in the form of denominator matrix-coefficients [7] to improve the speed of the least-squares solution. The denominator coefficients are used to construct a companion matrix, and its eigen-values and eigen-vectors (state-vectors of the order of the polynomial) are calculated. These poles and vectors are then used to build the stabilization chart.

However, instead of utilizing anything but the denominator matrix-coefficients, all the available information is used to compute the residues. The algorithm from Vu [13, 14] allows the inverse of a square matrix-coefficient polynomial to be described as a ratio of its adjoint (matrix-coefficient) polynomial and the characteristic (monic) equation (Equation 10.5).

$$[\alpha]^{-1}_{N_o \times N_o} = \frac{\left[\sum_{i=0}^{(N_o-1)m}\left[[\alpha_i^+]_{N_o \times N_o} s^i\right]\right]}{\sum_{i=0}^{N_o m} d_i s^i} = \frac{[\alpha^+]}{d} \tag{10.5}$$

The algorithm requires recursive computation of a new set of matrices as defined in Equation 10.6 and coefficients as defined in Equation 10.7 for the nth model order.

$$[B_{d,c}] = \sum_{v=1}^{c}\sum_{w=1}^{d}\left([B_{d-w,c-v}][\alpha_w][\alpha_0]^{v-1}\right) \tag{10.6}$$

Where, $c = 1, 2, \ldots, N_o$; $d = 0, 1, \ldots, N_o * n$; *and*, $[B_{0,c}] = [\alpha_0]^c$

$$b_{j,i} = \frac{1}{j}\sum_{c=1}^{j}\sum_{d=0}^{i}\left((-1)^{c-1}b_{j-c,i-d} * trace([B_{d,c}])\right) \tag{10.7}$$

Where, $i = 0, 1, 2, \ldots, N_o * n$

The coefficients required in Equation 10.5 are then computed using the Equations 10.8 and 10.9.

$$d_i = a_{N_o,i} \tag{10.8}$$

$$[\alpha_i^+] = \sum_{c=0}^{N_o-1} \sum_{d=0}^{i} \left((-1)^c b_{N_o-1-c,i-d} * trace([B_{d,c}]) \right) \tag{10.9}$$

The eigen-solution using the companion matrix leads to the same poles as the roots of the characteristic equation [15]. In other words, the characteristic equations of a square polynomial matrix and the companion matrix constructed using its matrix-coefficients is identical. The monic denominator-polynomial can hence be used to to obtain the eigen-frequencies present in the frequency range of interest.

The FRF matrix (or generally the transfer function matrix) is expressed using Equations 10.4 and 10.5 as:

$$[H(s)]_{N_o \times N_i} = \frac{[\alpha^+(s)]_{N_o \times N_o}[\beta(s)]_{N_o \times N_i}}{d(s)_{1\times 1}} + [R(s)]_{N_o \times N_i} \tag{10.10}$$

The residues for each of the poles are computed by substituting the partial-fraction model (Equation 10.11) of the FRF in Equation 10.10.

$$\Big[H(s)\Big]_{N_o \times N_i} = \sum_{r=0}^{N_r} \left(\frac{[A_r]_{N_o \times N_i}}{s - \lambda_r} + \frac{[A_r]^*_{N_o \times N_i}}{s - \lambda_r^*} \right) + \left[\sum_{i=-n_l}^{n_u} [R_i] s^i \right]_{N_o \times N_i} \tag{10.11}$$

In conjunction with Equation 10.17, UMPA uses this equation system for the second least-squares step for estimating the scaling factors and modal vectors [8]. Equation 10.12 is over-determined using the frequencies available according to the frequency range selected.

$$\Big[H(s)\Big]_{N_o \times N_i} = \left[\Big[L_r\Big]_{N_o \times 2N_r} \left[\frac{1}{s - \lambda_r} \right]_{2N_r \times 2N_r} \right] \Big[Q_r \psi_r\Big]^H_{2N_r \times N_i} + \left[\sum_{i=-n_l}^{n_u} [R_i] s^i \right]_{N_o \times N_i} \tag{10.12}$$

By multiplying the resulting equation throughout by the factor $(s - \lambda_r)$, the equation for the residue $[A_r]$ of the rth pole emerges (Equation 10.13).

$$\Big[A_r\Big]_{N_o \times N_i} = \lim_{s \to \lambda_r} \left(\Big[H(s)\Big] * (s - \lambda_r) \right) = \lim_{s \to \lambda_r} \left(\left[\frac{[\alpha^+(s)][\beta(s)]}{d(s)} + [R(s)] \right] * (s - \lambda_r) \right) \tag{10.13}$$

Since λ_r is a root of the polynomial represented by $d(s)$, the limit takes an indeterminate form (the numerator and denominator both vanish at λ_r). However, since the denominator is now a monic polynomial, L'Hôpitals rule may easily applied to compute the limit for the residue.

$$\Big[A_r\Big]_{N_o \times N_i} = \left(\frac{[\alpha^+(s)][\beta(s)]}{d'(s)} \right) \Bigg|_{s=\lambda_r} \tag{10.14}$$

Hence, the residue is calculated for each pole at each model order using Equation 10.15. This added information is beneficial in applying an additional filter to the stabilization chart during the pole selection stage to test for consistency of residues between successive model order iterations.

$$\Big[A_r\Big]_{N_o \times N_i} = \frac{\left[\sum_{i=0}^{(N_o-1)m} \left[[\alpha_i^+] s^i \right] \right] \left[\sum_{i=0}^{m-1} \left[[\beta_i] s^i \right] \right]}{\sum_{i=1}^{N_o m} i d_i s^{i-1}} \Bigg|_{s=\lambda_r} \tag{10.15}$$

The residue in Equation 10.15 is evaluated in the Z-domain and it must be scaled back to the complex-domain of the original FRFs so that it may be used to reconstruct FRFs. The scaling factor is derived in Equation 10.16 and is dependent on the Z-domain pole.

$$\left[A_r\right]_{(j\omega)} = \frac{\left[A_r\right]_{(s)}}{s_r * \Delta t} \tag{10.16}$$

It is also noteworthy that the residue is of unity rank. The rows of the residue matrix are the modal vectors scaled by appropriate entries from the participation vectors. Therefore, the row-wise (or column-wise) MAC is expected to be unity, indicating complete linear dependency of the residue matrix on each row/column. The structure of the residue matrix is shown in Equation 10.17.

$$\left[A_r\right]_{N_o \times N_i} = Q_r [L_r]_{N_o \times 2N_r} [\psi_r]^H_{2N_r \times N_i} \tag{10.17}$$

Here, Q_r is the modal scaling factor, $[L_r]$ is the matrix of modal participation vectors and $[\psi_r]$ is the matrix of mode shape vectors. H denotes the Hermitian (complex-conjugate transpose) operation on a matrix.

The comparison of the residue matrices obtained from the proposed methodology and the "traditional" process can hence be accomplished using MAC at each pole. Additionally, MPC and MPD metrics [16] that are applied to modal vectors can also be applied to the residue matrix after reshaping it into a vector.

A summary of the current modal parameter estimation process, alongside the proposed method, is shown in Fig. 10.1.

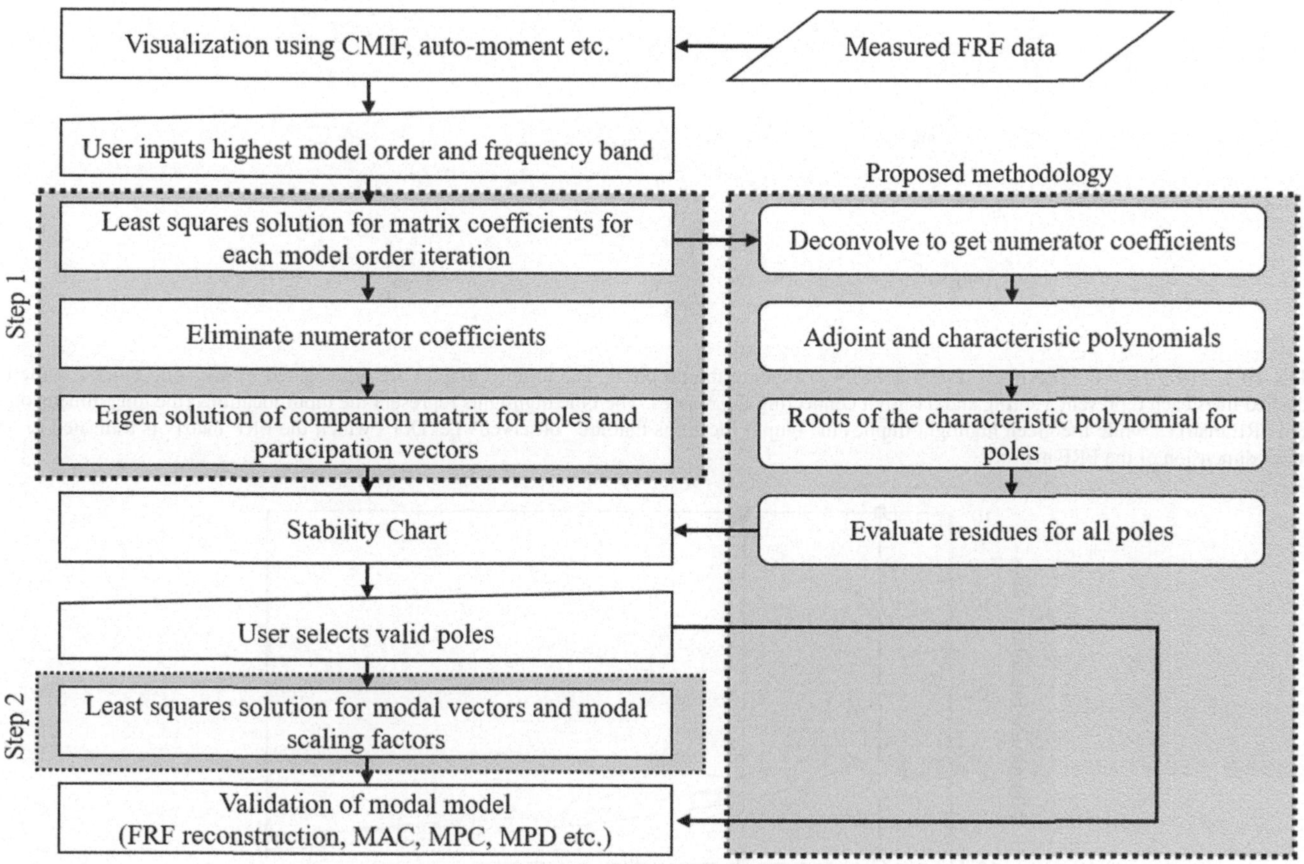

Fig. 10.1 The basic modal parameter estimation algorithm along with proposed modifications to bypass the second stage

10.3 Analysis

The proposed method was evaluated on a theoretical system which is shown in Fig. 10.2. The lumped parameter model's FRF matrix was constructed using complete information about the masses, stiffness coefficients and proportional-damping coefficients. The square FRF (9 × 9) matrix was then truncated to a rectangular size (($N_o =$) 4 outputs and ($N_i =$) 9 inputs) to mimic experimental matrix sizes.

Figure 10.3 shows the Complex Mode Indicator Function (CMIF) used to select a suitable frequency range of interest for the MPE process. The model order (m) is a user-input and for this case ranged from 2 to 10. The proposed method was applied post the least squares estimate of the matrix-coefficients. Using the traditional process, the stabilization chart (Fig. 10.4), was constructed to select valid poles and was limited to a participation vector consistency check (blue diamonds). Due to the proposed approach, the calculation of residues for all computed poles allowed for an additional consistency check (for residues). This was applied to the pole results to obtain the stabilization chart shown in Fig. 10.5. Table 10.1 shows the tolerances used to construct these stabilization charts.

From a comparison of the Figs. 10.4 and 10.5, it is observed that the poles showing consistency in participation vector estimates also show consistency in their residue estimates. Although this is expected for the current case due to absence of noise, it must be noted that this extra consistency parameter may be useful towards plotting *very* clear stabilization charts in practical situations and for autonomous parameter estimation. It has been shown that inclusion of the residue information

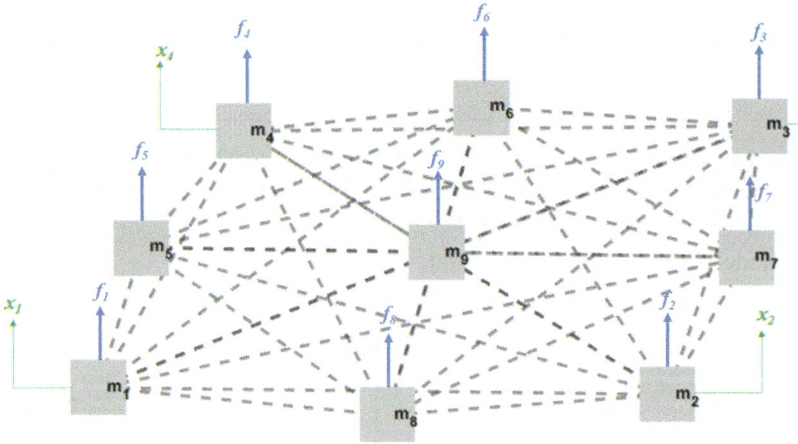

Fig. 10.2 The 9-DoF lumped mass system used for evaluating the proposal. The lumped masses are represented by the solid blocks, while the dashed lines each represent a spring and dash-pot connecting the masses. The blue highlights represent the input locations (the long dimension of the FRF matrix), while the green highlight implies the output locations that are "observed" i.e. DoFs which the FRF matrix is truncated to (the short dimension of the FRF matrix)

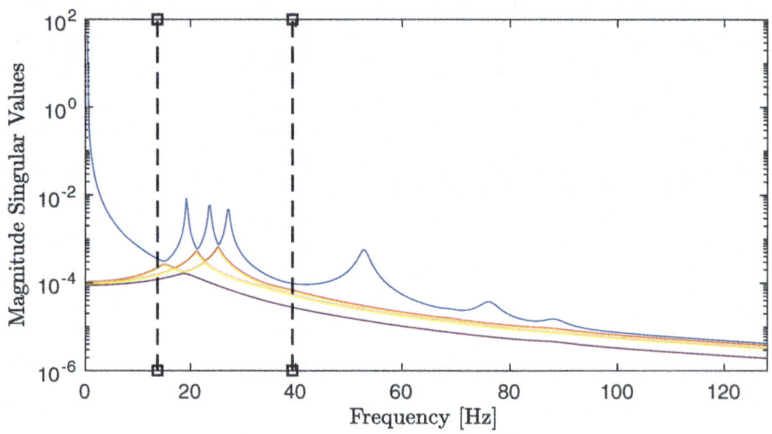

Fig. 10.3 Complex Mode Indicator Function for the system under test. The bands limit the number of frequencies used for the least squares solution from 13.75 to 40 Hz

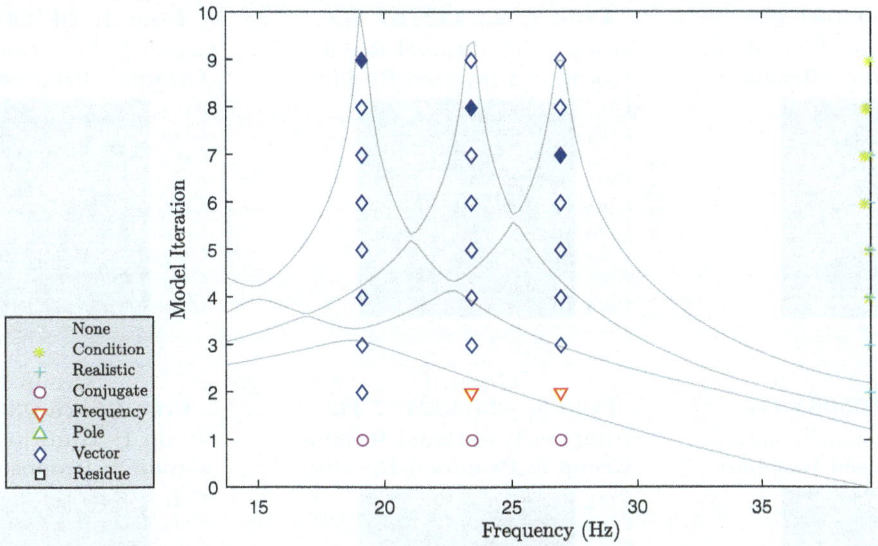

Fig. 10.4 Stabilization with participation vector comparison. Solid symbols indicate the selected poles

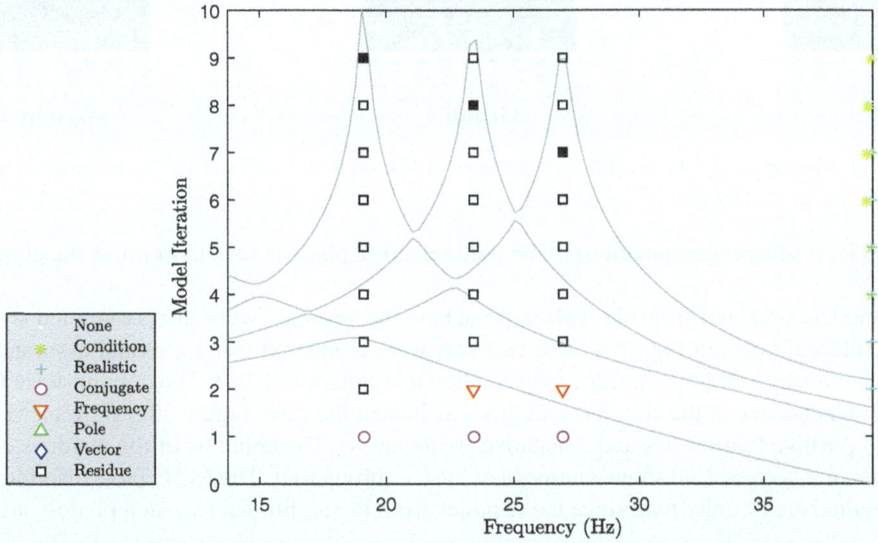

Fig. 10.5 Stabilization chart with residue comparison between iterations. The solid symbols indicate the selected poles and are numerically the same as the ones shown selected in Fig. 10.4

Table 10.1 Tolerances utilized for the construction of the stabilization chart. The legends in Figs. 10.4 and 10.5 may be referred for the symbols used [8]

Consistency metric	Tolerance	Description
None	–	(Initial assignment)
Condition	$<10^{10}$	Condition number check
Realistic	Negative real part	Filter out positively damped poles
Conjugate	0.1%	Conjugate check
Frequency	0.5%	Imaginary part of pole
Pole	1%	Complex pole
Vector	5%	Participation vector
Residue	5%	Scaled residue

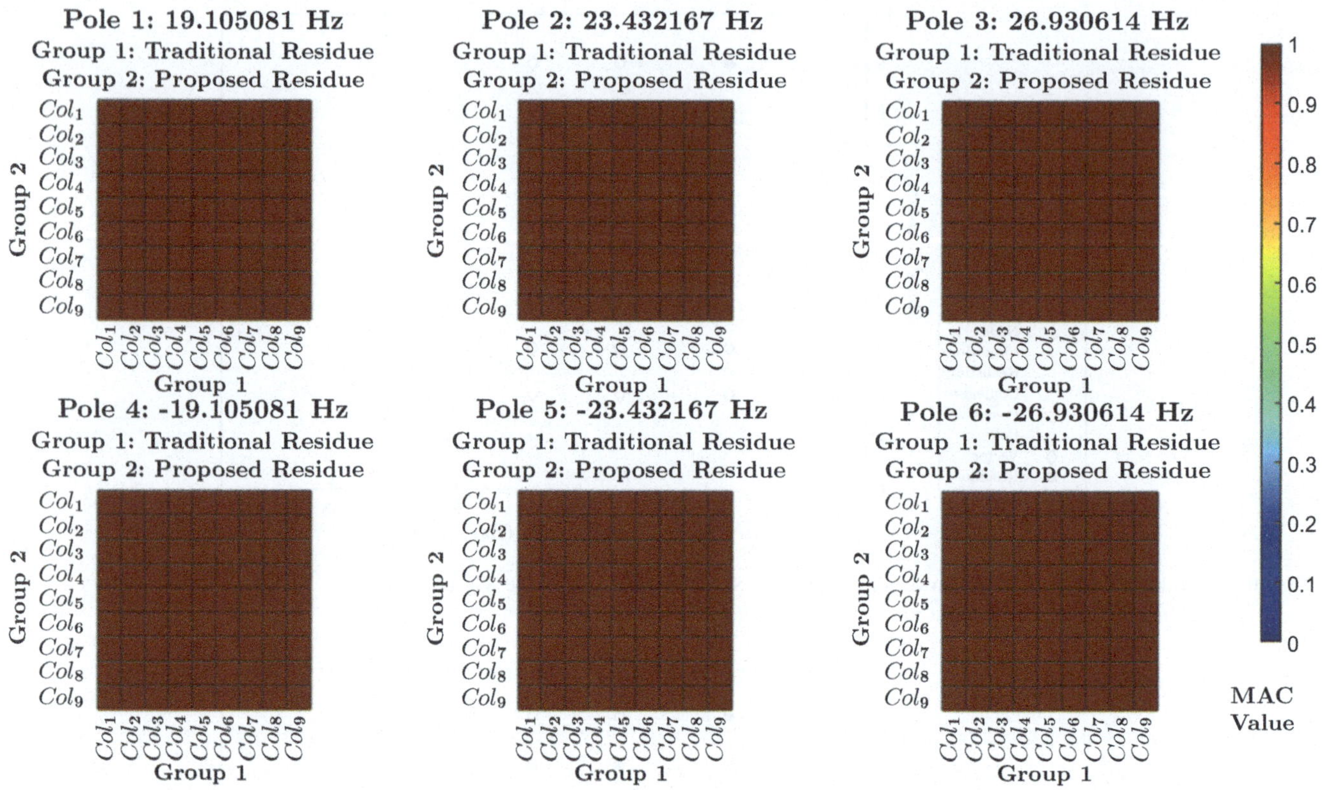

Fig. 10.6 MAC matrix for dependence of columns of the residue matrix obtained from traditional and the proposed process for the three poles

(from the second-step least squares computation) in the pole selection phase is seen to improve the clarity of the stabilization chart [16].

To compare the models obtained from the two approaches, the residues were also computed using the second least-squares step for the selected poles in Fig. 10.4. The two sets were compared using a modal assurance criterion approach. This is possible due to the structure of the residue matrix, shown in Equation 10.17. The column-wise MAC was expected to show complete linear dependence of the columns and this was indeed the case. Figure 10.6 shows a MAC plot of residue for each of the 6 poles (3 positive frequencies and 3 negative frequencies). The columns of the residues correspond to the long dimension (4 elements in 9 vectors) and show a unity MAC value throughout. The MAC plots also indicate that the residues from the proposed method are of unity rank since the residues from the traditional two-step process are forced to be of unity rank.

The FRFs were then reconstructed using the residue from the traditional as well as the proposed approach, and plotted against the original FRFs (Fig. 10.7, for example). It may be seen from the reconstructed FRF curves that the proposed process identifies the same modal model as traditional process. Of course, a deviation between the two modal models is to be expected in presence of noise in the measurements.

The residues, after being reshaped as vectors were also plotted in the complex plane (Fig. 10.8) and the mean phase correlations indicated that both the methods resulted in real modes. The low deviations from the mean phase for all the poles and the imaginary nature of the residues both indicate a correct estimation of the modal model.

It is worthwhile to mention that there is no restriction on the complexity or scaling of the modal residue under the proposed scheme. There are various scaling strategies for the participation vectors that have been applied to clear up the stabilization

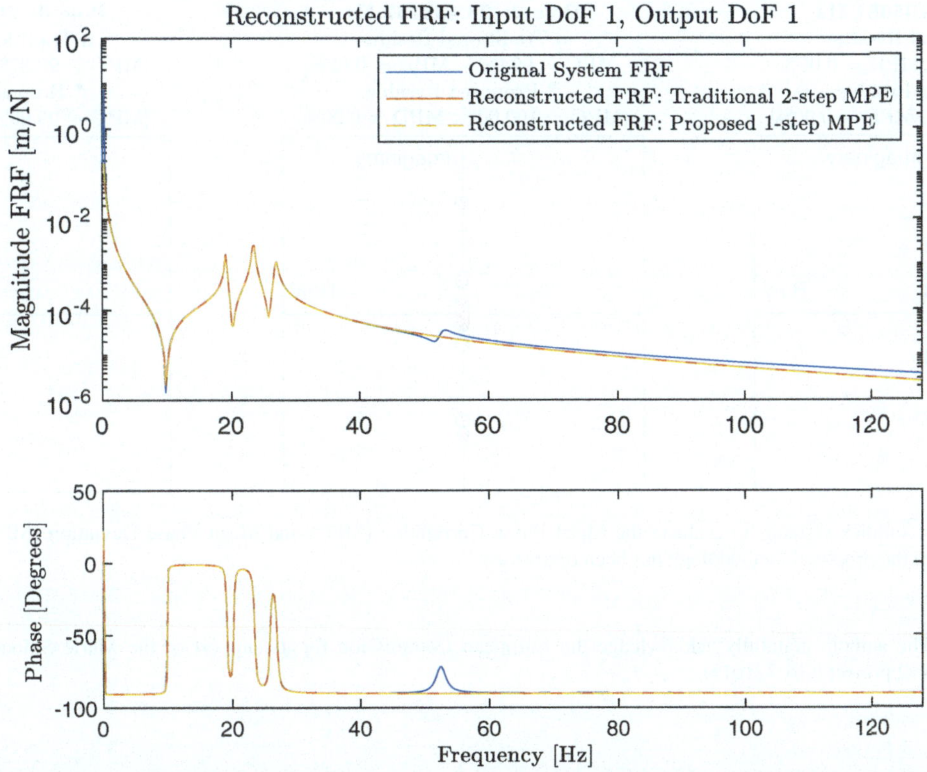

Fig. 10.7 Comparison of reconstructed FRFs using the traditional two-step approach and the proposed one-step approach against the original FRF for input 1 and output 1

diagram and ease the process of pole selection for the user through a clean stabilization chart [16]. From Fig. 10.8 it can also be seen that the residue matrix obtained in the two cases is practically identical.

10.4 Conclusions

In the presented work, a "single-step" modal parameter estimation algorithm (Direct Estimation of Residues from Rational-fraction Polynomials, or DERRP) is proposed. The procedure is based on the computation of the adjoint matrix-coefficient polynomial and the characteristic equation from a rational fraction matrix-description of the measured FRF matrix. It is shown that the proposed algorithm makes the residues available for comparison during the plotting of the stabilization chart. Modal parameters are estimated using the proposed and the traditional two-step approach for a multi-degree-of freedom lumped mass system and compared using existing model-validation techniques. The results indicate that the estimated parameters are accurate and completely define the modal model of the system under investigation.

The merit of the proposal lies in the fact that complete residue information is available to be used at the pole selection phase. Hence, the poles that are extracted from the stabilization chart convey a greater statistical confidence. Additionally, the pole selection phase is essentially transformed to allow model validation. With the (scaled) residue information for selected poles already available, the need for the second least-squares solution step is eliminated. The algorithm is expected to be helpful in effectively automating the pole selection process and hence the complete MPE process.

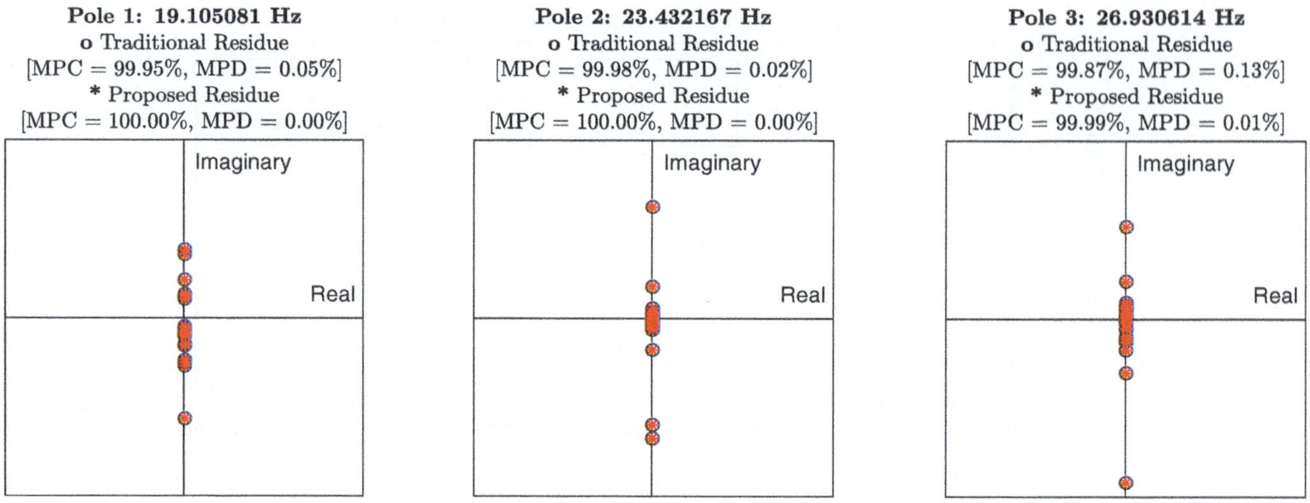

Fig. 10.8 Plot of the complex residues to evaluate the Mean Phase Correlation (MPC) and Mean Phase Deviation (MPD). No scaling of the residues resulting from the proposed methodology has been performed

Acknowledgments The authors gratefully acknowledge the European Commission for its support of the Marie Sklodowska-Curie program through the ETN PBNv2 project (GA 721615).

References

1. Ewins, D.J.: Modal Testing: Theory, Practice, and Application. Mechanical Engineering Research Studies. Engineering Dynamics Series, 2nd edn. Research Studies Press, Baldock (2000). ISBN: 0863802184
2. Brown, D.L., Allemang, R.J.: Review of spatial domain modal parameter estimation procedures and testing methods. In: Proceedings of the 27th International Modal Analysis Conference (IMAC), Orlando, p. 23 (2009)
3. Guillaume, P., Verboven, P., Vanlanduit, S.: Frequency-domain maximum likelihood identification of modal parameters with confidence intervals. In: Proceedings of the International Seminar on Modal Analysis, Katholieke Universiteit Leuven, vol. 1 (1998)
4. Reynders, E.: System identification methods for (operational) modal analysis: review and comparison. Arch. Comput. Methods Eng. **19**(1), 51–124 (2012)
5. Phillips, A.W., Allemang, R.J., Brown, D.L.: Autonomous modal parameter estimation: methodology. In: Modal Analysis Topics, vol. 3, pp. 363–384. Springer, New York, NY (2011)
6. Lanslots, J., Rodiers, B., Peeters, B.: Automated pole-selection: proof-of-concept and validation. In: Proceedings of the ISMA International Conference on Noise and Vibration Engineering, Leuven. Citeseer (2004)
7. Peeters, B., Auweraer, H.V.D., Guillaume, P., Leuridan, J.: The PolyMAX frequency-domain method: a new standard for modal parameter estimation? Shock Vib. **11**(3–4), 395–409 (2004)
8. Allemang, R.J., Brown, D.L.: A unified matrix polynomial approach to modal identification. J. Sound Vib. **211**(3), 301–322 (1998)
9. Heylen, W., Lammens, S., Sas, P.: Modal analysis theory and testing. KUL. Faculty of engineering. Department of Mechanical Engineering. Division of Production Engineering, Machine Design and Automation (1997). ISBN: 907380261X
10. Allemang, R.J., Brown, D.L.: A correlation coefficient for modal vector analysis. In: Proceedings of the 1st international modal analysis conference, SEM Orlando, vol. 1, pp. 110–116 (1982)
11. Allemang, R.: Experimental modal analysis: lecture notes. Structural Dynamics Research Laboratory, University of Cincinnati (2013)
12. Fladung, W.A. Jr: A generalized residuals model for the Unified Matrix Polynomial Approach to frequency domain modal parameter estimation. PhD thesis, University of Cincinnati (2001)
13. Vu, K.M.: An extension of the Faddeev's algorithms. In: 2008 IEEE International Conference on Control Applications (2008)
14. Vu, K.M.: Pencil characteristic coefficients and their applications in control. IEEE Proc. Control Theory Appl. **146**(5), 450–456 (1999)
15. Kailath, T.: Linear Systems, vol. 156. Prentice-Hall, Englewood Cliffs (1980)
16. Phillips, A.W., Allemang, R.J.: Application of modal scaling to the pole selection phase of parameter estimation. In: Structural Dynamics, vol. 3, pp. 499–518. Springer, New York, NY (2011)

Nimish Pandiya received his Masters from Cincinnati in 2017 and is currently pursuing his PhD at KU Leuven, in collaboration with Robert Bosch GmbH, under the European Marie Curie Program Pass-by-Noise for electric powertrains. His research is focused on modal methods for load predictions for power-electronics sub-components.

Chapter 11
Improved Expansion Results Using Regularized Solutions

Chris Beale, Ryan Schultz, and Deborah Fowler

Abstract Traditional expansion techniques utilize a modal projection wherein modal response is estimated based on a generalized inverse of measurements at a sparse set of degrees of freedom. Those modal response estimates are then used to project out to a larger set of degrees of freedom, resulting in predicted responses at more points or even full- field. As with any generalized inverse problem, the results are sensitive to noise and conditioning of the inverted matrix. While much has been done to improve numerics of matrix inversion problems in the context of input estimation or source identification problems, little has been done to improve the numerics of inverse solutions in expansion problems. This work presents numerical correction or regularization techniques applied to expansion problems using both simple and complex example structures. The effects of degree of freedom selection and noise are explored. Improved expansion results are obtained using straightforward regularization techniques, meaning higher accuracy responses can be obtained at expansion degrees of freedom with no change in the sparse set of measurements.

Keywords Expansion · Modal filter · Projection · Regularization · Inverse problems

11.1 Introduction

Expansion techniques enable responses to be estimated at unmeasured degrees of freedom (DOFs) based on responses at measured DOFs. Therefore, expansion can be used to better quantify the dynamics of a complex system despite the lack of instrumentation or inability to instrument deeply embedded components. The larger set of unmeasured DOF (*n-set*) are estimated by applying a projection operation on the smaller set of measured DOFs (*a-set*). The projection operation requires an inverse or pseudo-inverse on a matrix of projection vectors, which can introduce errors in the expanded results depending on the form of the matrix. When a proper selection of measured *a-set* DOFs are used and minimal noise exists in the responses, the projection vector matrix will be well-formed and accurate expansion results can be expected. However, the measured *a-set* DOFs cannot always be properly selected (due to channel number limitations or spatial limitations) and the measured responses will always exhibit some level of noise contamination. As a result, the projection vector matrix can be poorly conditioned (formed), which propagates and magnifies error (noise) from the measured *a-set* DOFs out to the larger *n-set* of expanded DOFs. In order to mitigate the effects of poorly conditioned matrices in the expansion process, this paper investigates the use of regularization in the expansion process to enhance the accuracy of expansion results.

System equivalent reduction expansion process (SEREP) is one expansion technique which works well for a variety of systems in various environments. SEREP utilizes a transformation matrix comprised of the modal vectors of the system. This transformation matrix effectively transforms the measured physical responses into modal space and then projects the

Sandia National Laboratories is a multimission laboratory managed and operated by National Technology & Engineering Solutions of Sandia, LLC, a wholly owned subsidiary of Honeywell International Inc., for the U.S. Department of Energy's National Nuclear Security Administration under contract DE-NA0003525.

C. Beale · R. Schultz (✉)
Engineering Sciences, Sandia National Laboratories, Albuquerque, NM, USA
e-mail: rschult@sandia.gov

D. Fowler
Department of Mechanical Engineering, University of Massachusetts Lowell, Lowell, MA, USA

B. Dilworth (ed.), *Topics in Modal Analysis & Testing, Volume 8*, Conference Proceedings of the Society for Experimental Mechanics Series, https://doi.org/10.1007/978-3-030-47717-2_11

modal responses back to physical space at a larger set of response DOFs. A pseudo-inverse is used to develop transformation matrix, which implies that the technique is susceptible to the selection of the measured *a-set* DOF and the presence of noise in the measured responses.

Two systems, one simple and one complex, are used to demonstrate expansion and investigate the effects of gauge placement (DOF selection) and noise on the accuracy of the expansion results with and without regularization. While these are contrived systems implemented as finite element models, they exhibit characteristics similar to the systems encountered in practice making them proper candidates for investigating the factors that affect the accuracy of the expansion results. The simple system, representative of a tuning fork, is used to demonstrate how gauge placement affects the conditioning of the expansion transformation matrix and accuracy of the expansion results. The complex system consists of three coupled components and is analyzed with a fixed-set of gauge locations (*a-set* DOF) where one component is entirely unmeasured. This represents typical field instrumentation configurations where certain components cannot be fully instrumented due to space limitations, cabling limitations, or other practical considerations. This complex system is used to examine the effects of noise on expansion results, and to demonstrate how regularization techniques can be used to improve expansion results when there is noise on the *a-set* DOF responses, which is typical of real measurements.

These models demonstrate several important characteristics of expansion problems. First, the test engineer must choose the locations of the *a-set* DOFs (measured gauge locations) carefully to maximize the information which is used in the expansion process. Next, the quality of the measurements should be high to avoid noise propagation and amplification issues during the expansion process. If noise is present and non-trivial, regularization techniques such as Tikhonov regularization or singular value regularization can be used to significantly improve expansion results by minimizing the noise amplification effects in the expansion process.

11.2 Theory

Generally, expansion involves a transformation from the set of known responses (*a-set* DOFs) to a larger set of unknown responses (*n-set* DOFs), via a transformation matrix such as

$$\{x_n(t)\} = [T]\{x_a(t)\}, \tag{11.1}$$

where the known and unknown responses are $\{x_a\}$ and $\{x_n\}$, respectively, and the transformation matrix is $[T]$. This transformation matrix can take many forms and in a fundamental sense is simply a mechanism for relating response at a few points to the response at many points, often via some shape-based spatial relationship. The *a-set* and *n-set* responses can be time histories, as shown in Eq. 11.1, frequency domain quantities such as linear spectra and cross- power spectral density (CPSD) matrices, or mode shape vectors. Expansion of linear spectra is identical to Eq. 11.1, except the response quantities are linear spectra vectors:

$$\{X_n(\omega)\} = [T]\{X_a(\omega)\}, \tag{11.2}$$

where the known and unknown linear spectra are $X_a(\omega)$ and $X_n(\omega)$, respectively. Expansion of mode shape matrices from the *a-set* DOF, $[U_a]$, to the *n-set* DOF, $[U_n]$, is also similar:

$$[U_n] = [T][U_a]. \tag{11.3}$$

Expansion of CPSDs at the *a-set* DOF, $[S_{aa}]$, to the *n-set* DOF, $[S_{nn}]$, is simply an outer product of Equation 11.2 because a CPSD is an outer product of linear spectra vectors:

$$[S_{nn}(\omega)] = [T][S_{aa}(\omega)][T]^T. \tag{11.4}$$

11.2.1 SEREP Expansion

The SEREP method [1] uses a transformation matrix developed from the mode shape vectors of the system. During the transformation, a modal projection is performed on the *a-set* responses, $\{x_a\}$, to estimate the modal responses, $\{p\}$, using a pseudo-inverse of the mode shape matrix at the *a-set* DOFs, $[U_a]$. The modal responses are then projected back to physical space, but at the *n-set* DOFs resulting in the expanded responses $\{x_n\}$. The SEREP transformation matrix is given by

$$[T] = [U_n]\,[U_a]^+, \tag{11.5}$$

where the $[U_a]^+$ represents the generalized inverse of the *a-set* DOF mode shape matrix. Substituting Eq. 11.5 into Eq. 11.1 demonstrates the modal projection-based expansion method in SEREP more clearly:

$$\{x_n\} = [T]\,\{x_a\} = [U_n]\,[U_a]^+\,\{x_a\} = [U_n]\,\{p\}, \tag{11.6}$$

where $\{p\}$ is a vector of modal responses. The modal responses are related to the physical responses via the mode shape matrix:

$$\{x\} = [U]\,\{p\}. \tag{11.7}$$

As with any pseudo-inverse operation, the pseudo inverse of the *a-set* DOF mode shape matrix, $[U_a]$, is subject to error propagation which can become significant if the form of the $[U_a]$ matrix is poor. That is, the estimate of the modal response can be affected by the form of the matrix. A well-conditioned $[U_a]$ matrix will result in an accurate projection into modal space, estimate of the modal responses, and projection out to physical space at the *n-set* DOFs. However, a poorly-conditioned $[U_a]$ matrix can result in a poor projection into modal space, inaccurate estimate of the modal responses, and thus a poor estimate of the responses in physical space at the *n-set* DOFs. A typical measure of the quality of the form of a matrix is the condition number which is the ratio of the largest and smallest singular values [2, 3]. A large condition number is indicative of a poorly-formed matrix and a small condition number is indicative of a well-formed matrix. It is essentially a measure of the independence of the rows or columns of a matrix. In the case of SEREP expansion, it would be a measure of the independence of the rows or columns of the mode shape matrix at the *a-set* DOF. These issues are universal in matrix inverse problems and various techniques, called regularization, exist for improving the pseudo-inverse operation to improve the accuracy of these types of problems.

11.2.2 Theory of Regularization in Direct Pseudo-Inverse Solutions

Regularization is used to modify the inverted matrix to reduce the effects of poor form and/or conditioning. Common techniques are Tikhonov and singular value regularization. In both these methods, the matrix is perturbed slightly to improve the condition number and reduce the errors which result from a direct inverse solution using that matrix. Changing the matrix too much can result in additional error, so there is a balance of the amount of perturbation, or regularization, applied and the improvement in the results.

The typical Moore-Penrose pseudo-inverse of a matrix $[A]$ is given by

$$[A]^+ = \left[[A]^T\,[A]\right]^{-1}[A]^T. \tag{11.8}$$

Tikhonov regularization can be applied to the pseudo-inverse adding a diagonal matrix $\lambda^2[I]$ to Eq. 11.8:

$$[A]^+ = \left[[A]^T\,[A] + \lambda^2\,[I]\right]^{-1}[A]^T, \tag{11.9}$$

where the value λ is the regularization parameter. The value of λ dictates the amount of regularization. A very small value of λ would result in little regularization as the argument $[[A]^T[A] + \lambda^2[I]]$ is nearly equal to $[[A]^T[A]]$. Conversely, a large value of λ results in more regularization as the argument is very different from $[[A]^T[A]]$. Very large values of λ can perturb the argument too much, resulting in erroneous estimates of $[A]^+$. Careful choice of the regularization value is necessary to clean up numerical issues with the matrix while not over-regularizing and introducing new errors into the solution.

Singular value regularization uses a slightly different approach. First singular value decomposition (SVD) is used to represent the matrix $[A]$ in terms of two unitary matrices, $[U_\Sigma]$ and $[V_\Sigma]$, and a diagonal matrix, $[S_\Sigma]$. $[U_\Sigma]$ and $[V_\Sigma]$ are the left and right singular vector matrices, respectively, and $[S_\Sigma]$ contains the singular values along the diagonal. A pseudo-inverse of $[A]$ can be obtained using this SVD form by:

$$[A]^{+} = [U_{\Sigma}][S_{\Sigma}]^{-1}[V_{\Sigma}]^{H}. \tag{11.10}$$

Regularization can be applied by perturbing the smallest singular values, which has the effect of improving the condition number and reducing the error propagation effects of inverting a poorly-conditioned matrix. The smallest singular values can either be set to some larger values or set to zero, which has the effect of removing or truncating the smallest singular values and vectors. These regularization techniques will be applied to expansion problems in the following sections.

11.3 Models Used for Demonstrating Expansion Effects

Two models were used to demonstrate the effects of gauge location and noise on expansion results and investigate regularization as a technique to achieve higher accuracy results. The models are presented in order of complexity starting with a simple two beam system, labeled the tuning fork model, and leading into an intricately coupled three beam system, labeled the coupled components model.

11.3.1 Simple Model: Tuning Fork

The tuning fork model is schematically depicted in Fig. 11.1 and all the associated properties are listed in Table 11.1. The model is comprised of two beams, a 20 inch base beam fixed at one end that is coupled to a 10 inch top beam at the mid-length of the base beam. The tuning fork model was defined to focus on the effects of gauge placement on the expansion results and how the results can be improved through regularization. The model serves to show that even in the case of a simple model in a controlled analytical environment with minimal noise, the accuracy of the expansion results can be severely depreciated by a poorly conditioned transformation matrix.

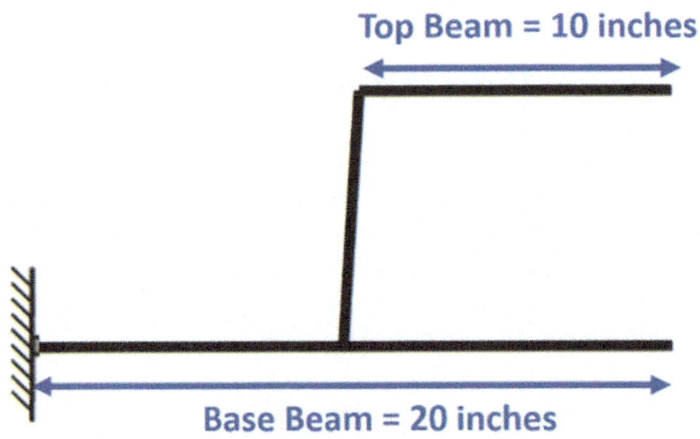

Fig. 11.1 Schematic depiction of the tuning fork model

Table 11.1 Tuning fork model properties

Overall properties	Parameter	Value	Units
	Young's Modulus	1.00E+07	psi
	Density	2.59E-04	slug • in^{-3}
	Width	0.50	in
	Height	0.25	in
Base beam	Parameter	Value	Units
	Length	20	in
	Nodes	41	
Top beam	Parameter	Value	Units
	Length	10	in
	Nodes	21	

11.3.2 Complicated Model: Coupled Components

The coupled components model is schematically depicted in Fig. 11.2 and all the associated properties are listed in Table 11.2. The model is comprised of three beams intricately coupled together to serve as a practical example of a system-subsystem-component assembly. The system beam, beam A, is 140 inches long and connected to ground by soft springs at each end. The subsystem, beam B, is 50 inches long and coupled at both ends to the system beam A 45 inches from either end. The component, beam C, is 10 inches long and coupled at one end to the subsystem beam B 90 inches from the origin, or 45 inches down the length of beam B.

The coupled components model resembles a structure commonly encountered in environments engineering analysis, in which several subsystems and components are coupled to, or within, a larger system. Obtaining the response of the subsystems and components is a challenge because of limited number of acquisition channels, inability to instrument deeply embedded subsystems and small components, and noise contamination. Expansion is a viable approach to appropriately characterize the response of non-instrumented components but is susceptible to noise. This model serves to demonstrate how well regularization can enhance the expansion results of a representative engineering system under the limitations of practical instrumentation.

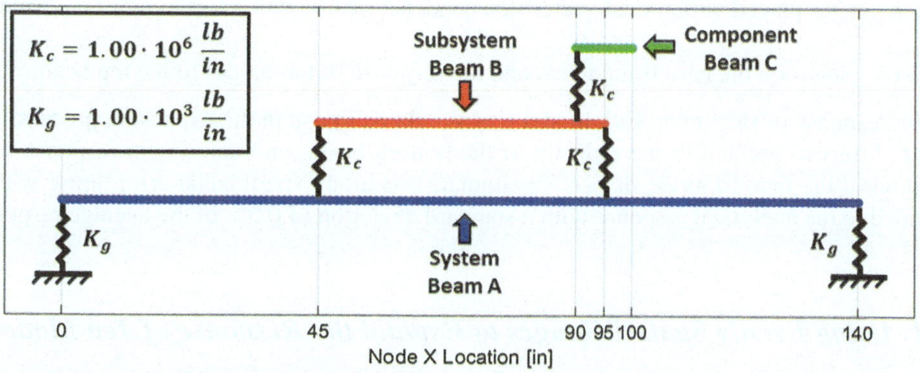

Fig. 11.2 Schematic depiction of the coupled components model

Table 11.2 Coupled components model properties

Overall properties	Parameter	Value	Units
System Beam A	Young's Modulus	1.00E+07	psi
	Density	2.59E-04	slug • in^{-3}
	Width	3.00	in
	Height	1.50	in
	Wall thickness	0.188	in
	Coupling Springs	1.00E+06	lb • in^{-1}
	Ground Springs	1.00E+03	lb • in^{1}
	Parameter	Value	Units
Subsystem Beam B	Length	140	in
	Nodes	141	
	Parameter	Value	Units
Component Beam C	Length	50	in
	Nodes	51	
	Parameter	Value	Units
	Length	10	in
	Nodes	11	

11.4 Demonstration of the Effects of Gauge Location on Expansion Results

Modal expansion accuracy relies on a well-conditioned mode shape matrix to reduce sensitivity to noise. Prior to data collection, conditioning is improved by properly placing gauges to maximize modal vector independence. Gauges are often placed in symmetric grid patterns for ease in visualization and bookkeeping, but these configurations often have poor conditioning. This can be remedied by using many more sensors than modes. However, most tests have limited channels available, making efficient gauge placement the preferred solution. Sensor selection techniques to optimize modal vector independence are often employed to maximize the number of modes characterized with a limited number of gauges. The condition number of the mode shape matrix can be used to evaluate sensor placement and determine the optimal test geometry for the desired application.

To demonstrate the sensitivity of SEREP expansion to the number and placement of sensors, three cases are presented. Case 1 demonstrates the detrimental effect of using minimal, poorly placed gauges. Case 2 uses the same number of gauges as Case 1 but places them using techniques that optimize modal vector independence. Case 3 shows that accurate expansion of test data is infeasible without regularization techniques unless gauges are placed on each active component. Each case is summarized as follows:

- Case 1: Using ten evenly spaced gauges to expand the response of ten modes
- Case 2: Using ten gauges placed with modal vector independence optimization techniques to expand the response of ten modes
- Case 3: Using forty gauges on the base beam to expand the response of ten modes to the top beam

The time domain response of the tuning fork model was simulated to use in the expansion process. The analytical input pulse shown in Fig. 11.3 was applied to the right tip of the bottom beam, and the system response was calculated using a modal superposition of the first 10 mode shapes. To simulate common experimental error, noise with a standard normal distribution was added to the analytical response with a standard deviation of 0.5% of the average response magnitude.

11.4.1 Case 1: Using Evenly Spaced Gauges to Expand the Response of Ten Modes

If the number of gauges available approaches the number of modes needed for expansion, placing the gauges in an even grid results in poor conditioning. To demonstrate this effect, equal numbers of gauges and modes are used in the expansion process, with the gauges evenly spaced on the tuning fork model. Figure 11.4 shows the *a-set* DOF, which was defined as the translational DOF of every sixth node for both beams. This resulted in a condition number of 90, which is higher than desired for optimal expansion. The first ten modes of the system were used to calculate the SEREP transformation matrix,

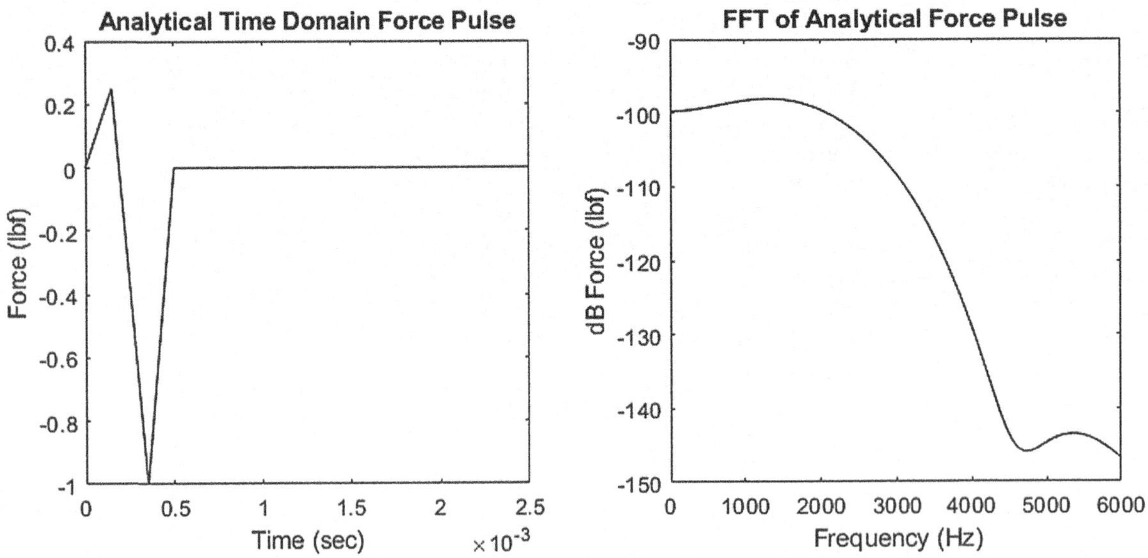

Fig. 11.3 Excitation pulse applied to the two-beam model, in the time (left) and frequency (right) domains

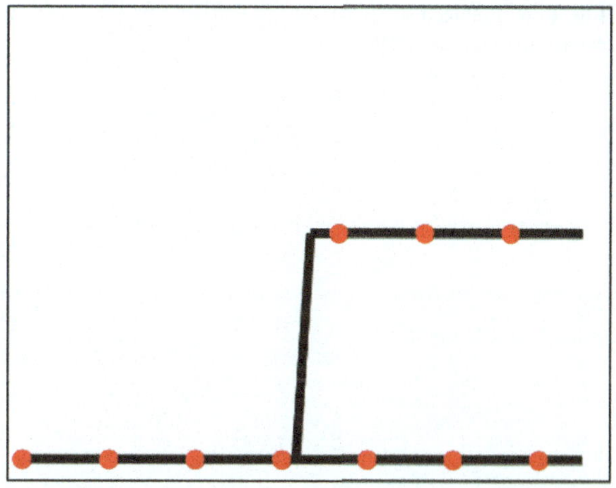

Fig. 11.4 Reduced set of 10 evenly spaced *a-set* DOF

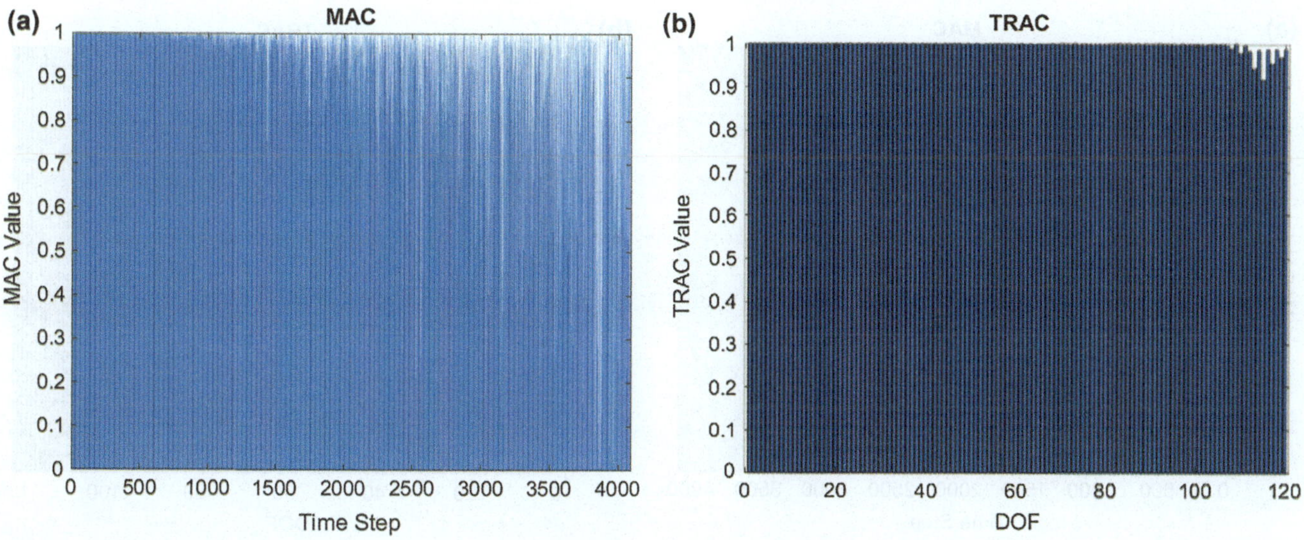

Fig. 11.5 MAC (left) and TRAC (right) comparing expanded responses to the truth responses

then the system response was expanded to full field. Figure 11.5 shows the MAC comparing the expanded response to the truth response over all points at each time step, and the TRAC comparing the responses at each DOF. The MAC and TRAC indicate that some error was propogated through the expansion process, particularly effecting the end of the top beam in the latter third of the time response. This is a result of the noise added to the data propogating through the poorly conditioned inverted mode shape matrix.

While an even grid of points is problematic when the number of *a-set* DOF is close to the number of modes, an even grid can still be used if the number of sensors is significantly increased. A follow-on example is presented to shown this, where a reduced set of twenty *a-set* DOF was defined as the translational degree of freedom of every third node on both beams. This resulted in an evenly spaced set of *a-set* DOF, shown in Fig. 11.6, for which the mode shape matrix had a condition number of 1.9. This condition number is significantly less than the previous example using ten *a-set* DOF. Consequently, the MAC and TRAC values shown in Fig. 11.7 are all approximately one, indicating high correlation and an accurate expansion.

11.4.2 Case 2: Using Ten Gauges Placed with Effective Independence or Condition Number Optimization to Expand the Response of Ten Modes

For situations when available channels are limited, it is possible to significantly reduce the number of gauges needed to preserve independence of the modal vectors by using an optimization algorithm to select sensor locations. This paper

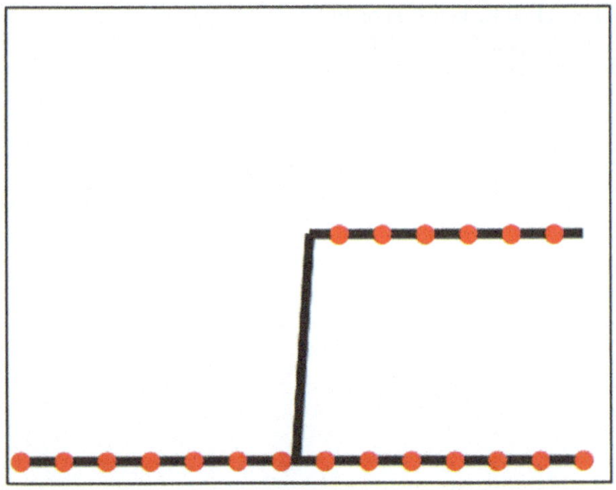

Fig. 11.6 Reduced set of 20 evenly spaced *a-set* DOF

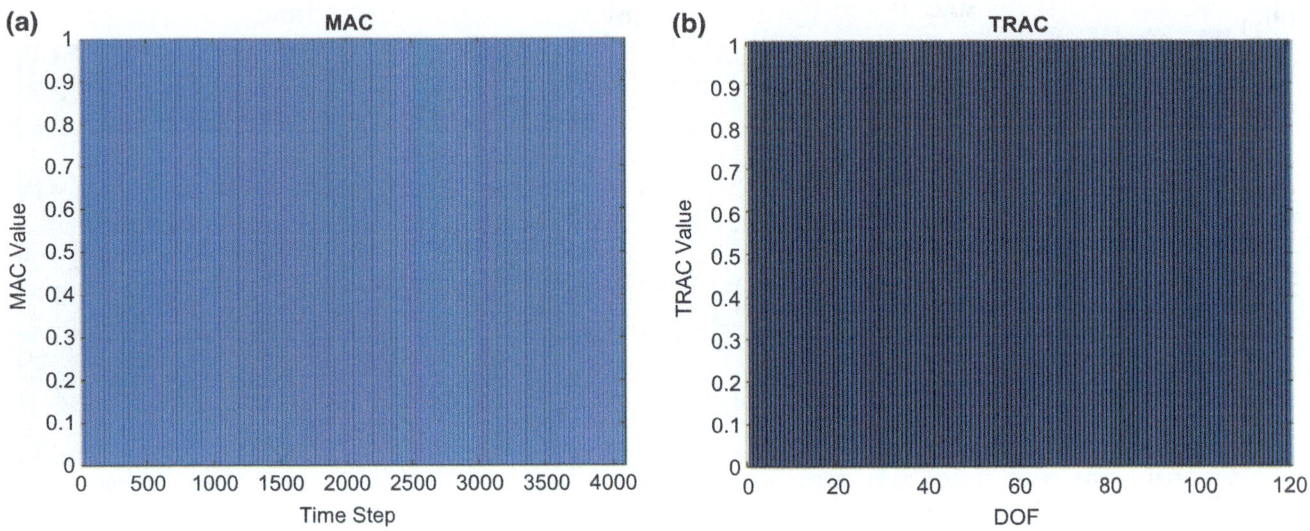

Fig. 11.7 MAC (left) and TRAC (right) comparing expanded responses to the truth responses

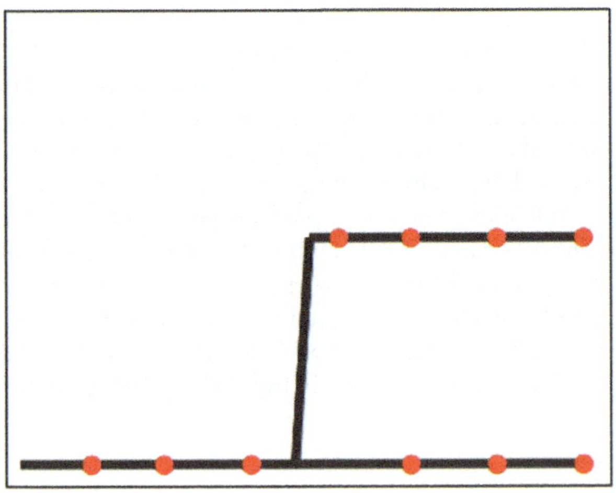

Fig. 11.8 Reduced set of 10 *a-set* DOF chosen using effective independence

demonstrates two methods for sensor selection: effective independence and condition number optimization. For this simplified case, both techniques produce good results.

Ten gauges were placed using the effective independence method, shown in Fig. 11.8. Although the number of gauges used is identical to the first example in Case 1, a condition number of 3.2 is obtained as opposed to 90 from case 1 because

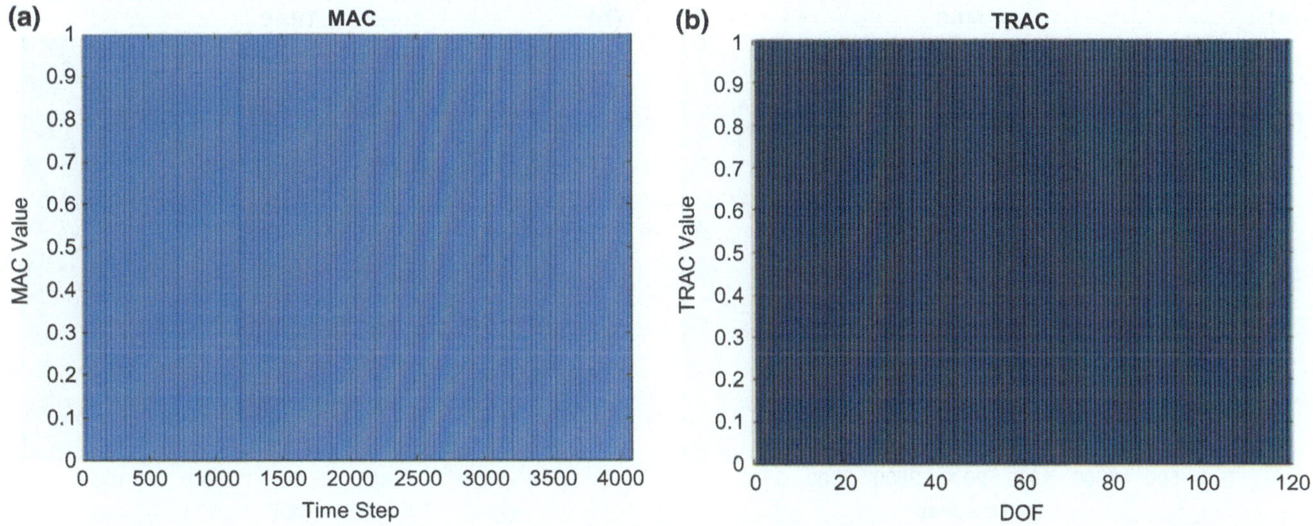

Fig. 11.9 MAC (left) and TRAC (right) comparing expanded responses to the truth responses

Fig. 11.10 Reduced set of 10 *a-set* DOF chosen using effective independence

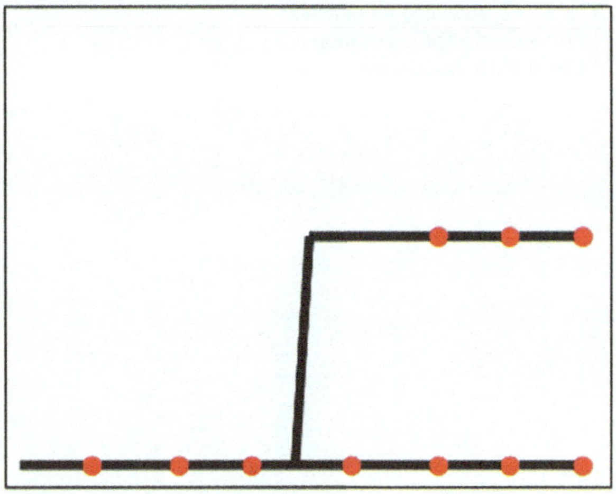

the gauges were located more appropriately. The MAC and TRAC shown in Fig. 11.9 indicate a highly accurate expansion, as would be expected based on the condition number.

This case was repeated using the condition number optimization method to place the ten sensors, shown in Fig. 11.10. This resulted in a condition number of 4.4, and a corresponding highly accurate expansion demonstrated in Fig. 11.11. These results show that both techniques are effective at placing gauges to optimize for independence of the modal vectors, which improves the accuracy of the expansion results when compared to a symmetric or grid-like test geometry.

11.4.3 Case 3: Using Forty Gauges on the Base Beam to Expand the Response of Ten Modes to the Top Beam

When designing a test setup, there are often many components that are of interest and limited gauges for instrumentation. For accurate expansion without the use of regularization techniques, each component of interest must be instrumented with at least one sensor. To illustrate this, all the translational DOF of the bottom beam are used as the *a-set* DOF and no top beam DOF are included, shown in Fig. 11.12. This results in a condition number of 13,040, which is four orders of magnitude higher than the previous cases with accurate expansion. Correspondingly, the MAC and TRAC shown in Fig. 11.13 show extremely poor expansion results for the behavior of the top beam. Notably, the expansion results are very poor even while

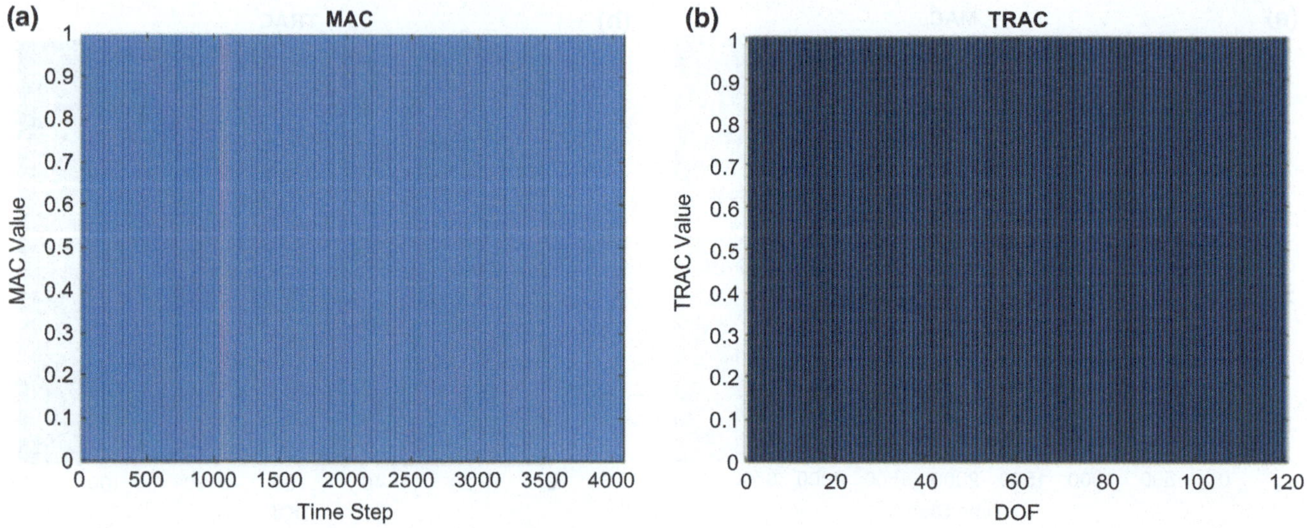

Fig. 11.11 MAC (left) and TRAC (right) comparing expanded responses to the truth responses

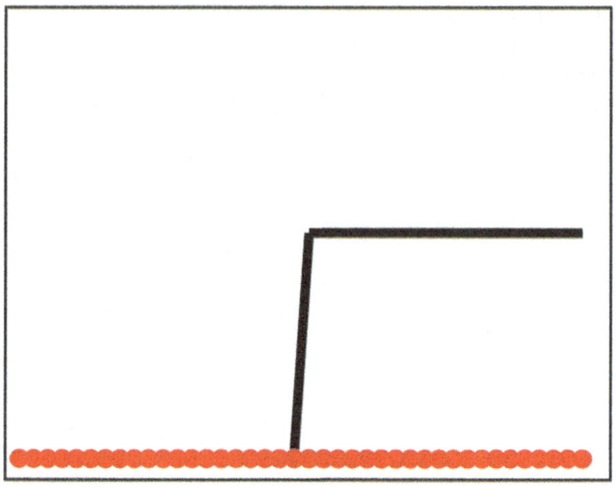

Fig. 11.12 Reduced set of *a-set DOF* defined as all translation DOF of the bottom beam

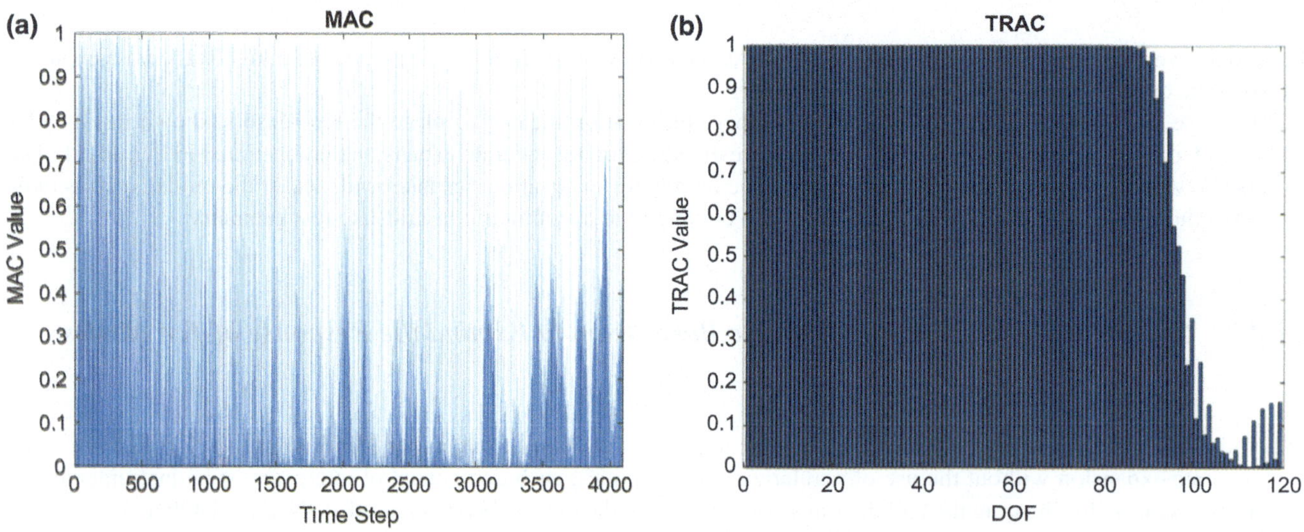

Fig. 11.13 MAC (left) and TRAC (right) comparing expanded responses to the truth responses

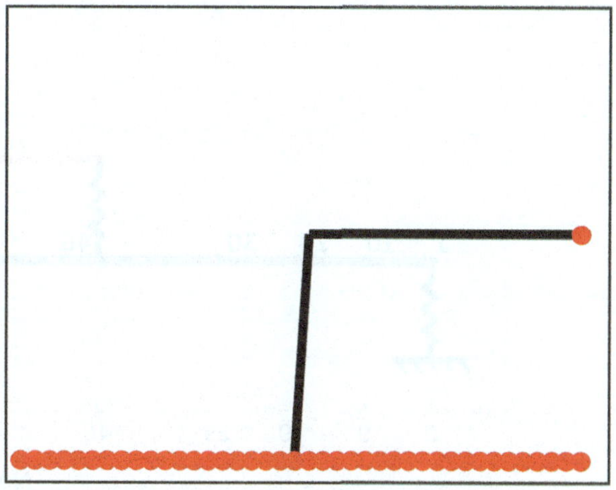

Fig. 11.14 Reduced *a-set* DOF defined as all translation DOF of the bottom beam and one translational DEF on the top beam

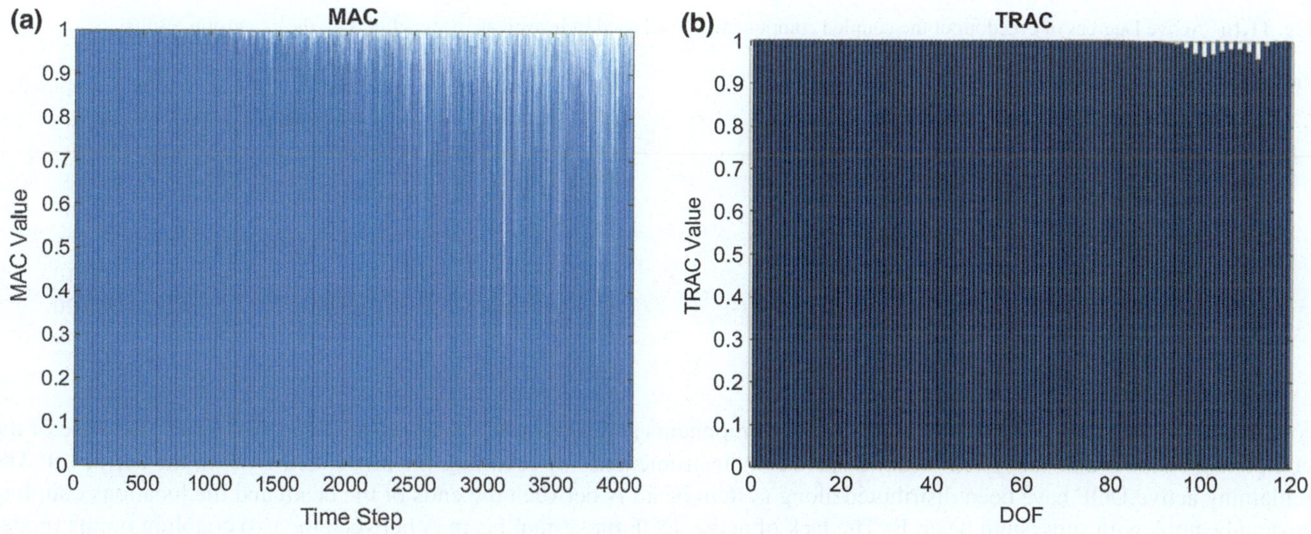

Fig. 11.15 MAC (left) and TRAC (right) comparing expanded responses to the truth responses

using four times as many gauges as needed in Case 2, demonstrating that intelligent gauge placement is more important than the total number of gauges used.

The expansion results can be dramatically improved with a single additional gauge placed on the top beam. The *a-set* DOF is shown in Fig. 11.14, with a condition number of 230. The MAC and TRAC shown in Fig. 11.15 indicate that there is still some substantial error in the expansion results, but they are significantly less compared to the previous case where the condition number was 13,040. Instrumenting every active component on a structure is pivotal if full- field expansion results are desired.

11.5 Demonstration of the Effects of Noise on Expansion Results

11.5.1 Model Configuration (Active Degrees of Freedom)

The effects of noise on expansion results were investigated using the coupled components model. A single set of active (*a-set*) DOF were used, shown in Fig. 11.16, that resemble a practical mapping of instrumentation used on a real engineering structure. The number of active DOF were limited to 12 replicating a lack of available channels commonly experienced in the field. None of the active DOF have been placed on component beam C and only a single DOF has been placed on subsystem beam B. In most cases, it is impractical to instrument deeply embedded components and.

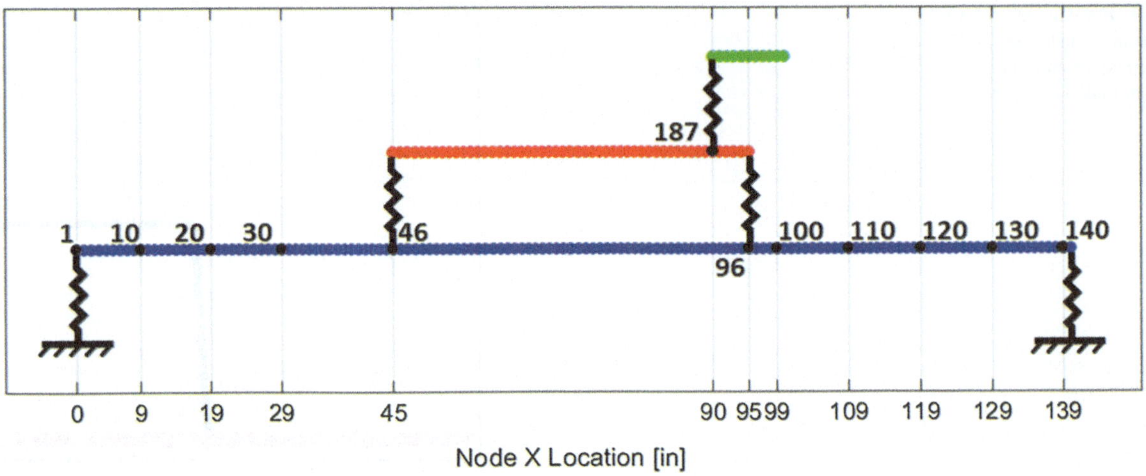

Fig. 11.16 Active Degrees of Freedom of the coupled components model used to investigate noise effects on the expansion results

Table 11.3 Noise severity test cases

	Case	Amplitude
Mode shape	1	0.10
	2	0.20
	3	0.40
	4	0.80
Time response	1	0.005
	2	0.010
	3	0.020
	4	0.040

subsystems, and the response at the base of the components and subsystems is more desirable. Therefore, the base of the component beam C and subsystem beam B has been "instrumented" by retaining the associated DOF in the active set. The remaining active DOF have been distributed along system beam A between the ends of the beam and the locations coupling system beam A with subsystem beam B. The lack of active DOF on system beam A between the two coupling points (nodes 46 and 96) could be due to lack of clearance between the system and subsystem, or an inability to properly instrument the structure between those two locations.

11.5.2 Noise Demonstration Test Matrix

Noise is a natural and unavoidable phenomenon experienced in all experimental measurements. Even the slightest bit of noise can have detrimental effects on the expansion results. To investigate how the severity of noise, and noise in general, effects the expansion results a series of test cases were developed and executed. A total of 8 test cases, defined in Table 11.3, were performed split evenly into 4 noise cases of varying severity for each type of expansion, mode shape and time domain.

All noise was defined as a zero mean Gaussian distributed random variable. The severity of the noise was adjusted by modifying the maximum amplitude of the noise. Due to the sensitivity of the expansion results to noise on the *a-set* response, generating a new realization of noise each test case could yield drastically different expansion results. Therefore, a single realization of noise was used and scaled to consider different levels of severity for each type of expansion. Each test case was performed by simply adding the noise of the specified severity to the mode shape values or time responses of the active DOF set.

11.5.3 Noise Contaminated Expansion Results

Mode Shape Expansion

Mode shape expansion was performed with the active DOF set defined in Sect. 11.5.1 (Fig. 11.16) for all noise severity cases defined in Sect. 11.5.2 (Table 11.3). Only the first ten modes were considered in the expansion process. The quality of the results was quantified by computing the MAC of the expanded shapes with a set of reference shapes. The reference shapes correspond to the analytical mode shapes superimposed with the noise of the considered test case. The MAC matrices computed for each noise severity case is presented in Fig. 11.17. Inspecting the MAC matrices along the diagonals of each test case, the values exhibit a clear trend of decreasing MAC value with increasing noise severity. Starting from the lowest severity of noise, the MAC values of modes 4 (0.09) and 6 (0.76) are poor. By the third level of noise severity the MAC values along the diagonal are all below 0.9 indicating a poor correlation.

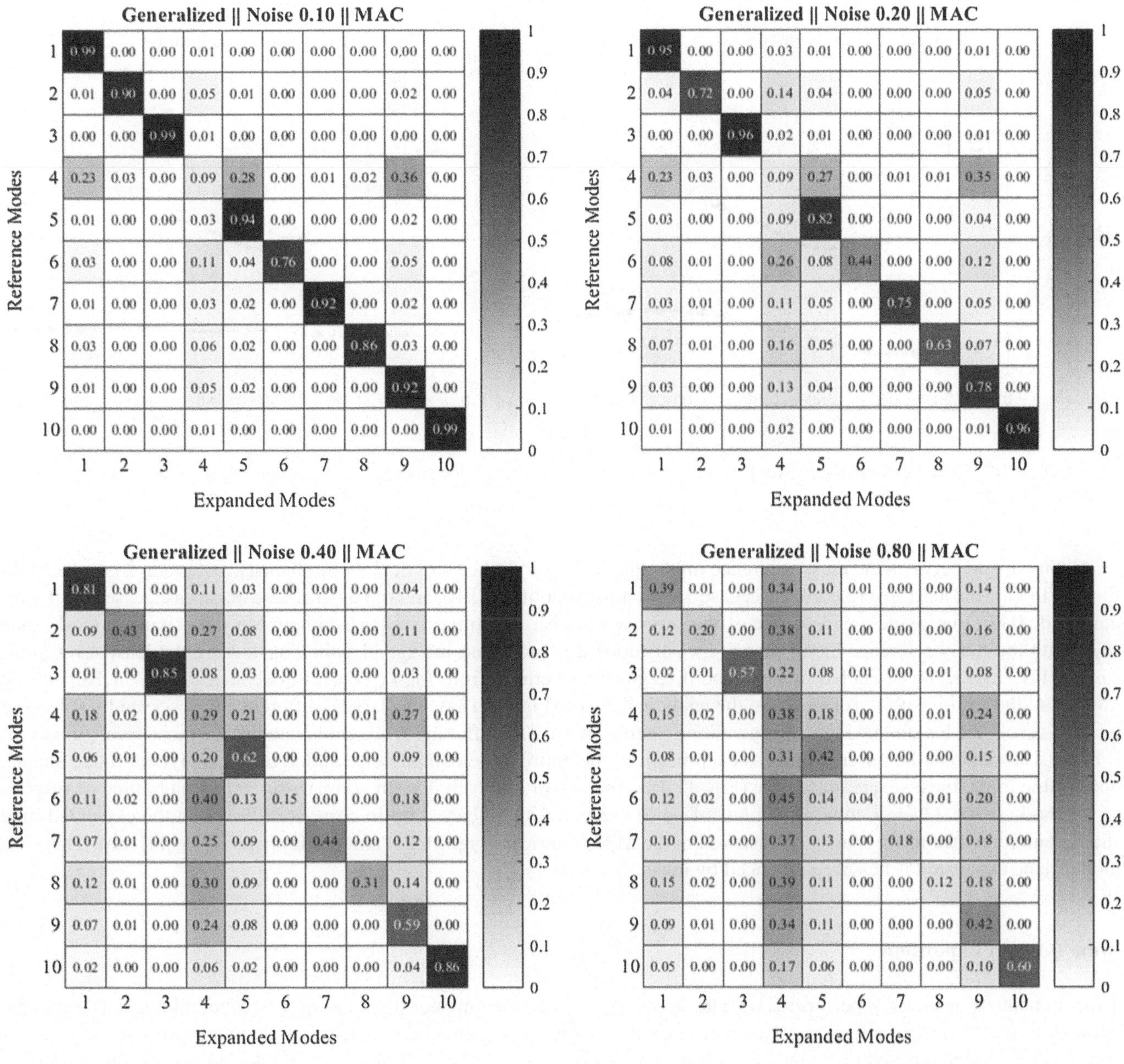

Fig. 11.17 MAC matrices computed from the expanded noisy mode shapes for each test case using the standard generalized inverse approach with no regularization

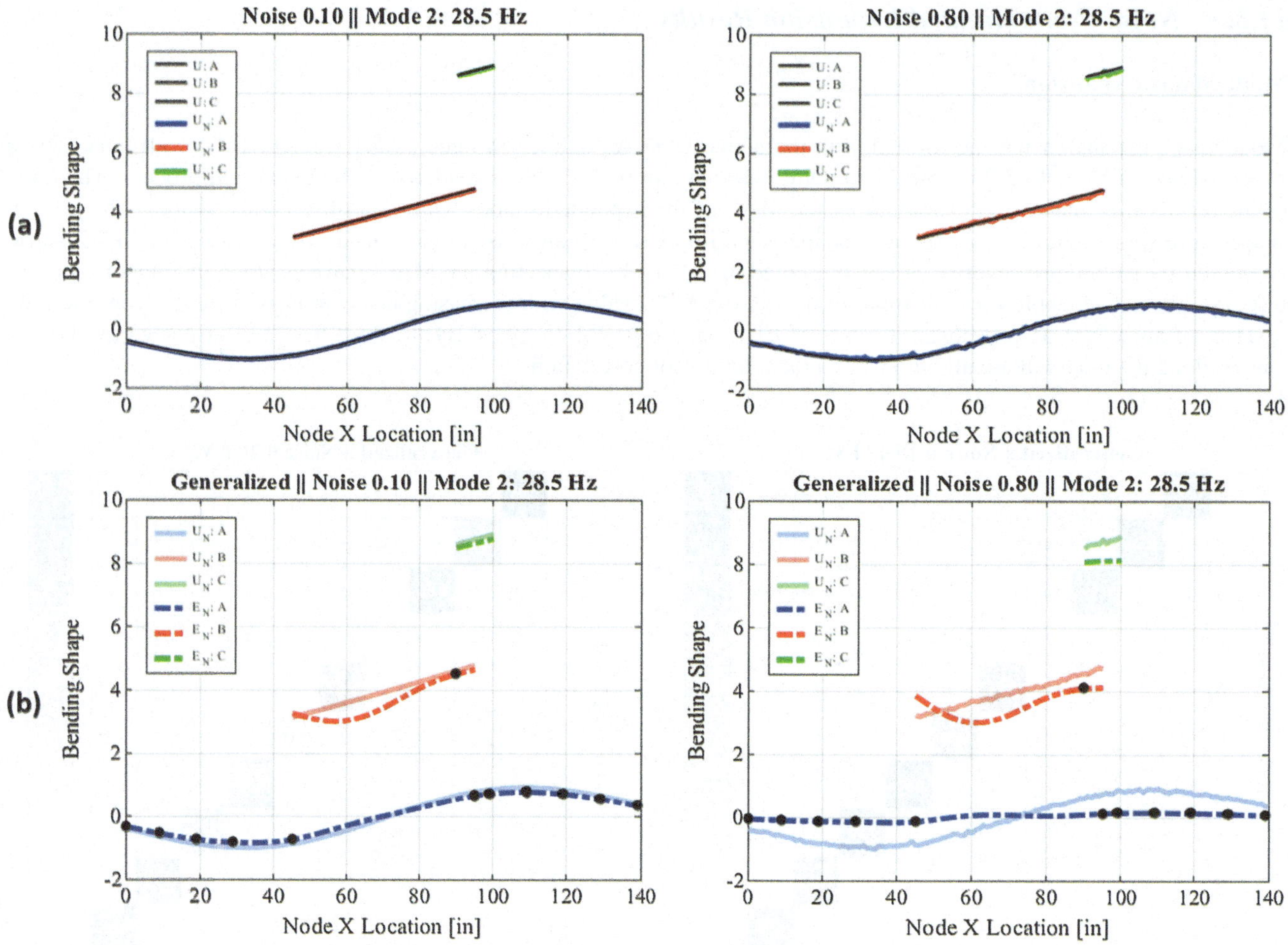

Fig. 11.18 Mode shape 2 comparison between the (**a**) noisy reference shape with the analytical mode shape and (**b**) expanded shape with the noisy reference shape

A comparison between the noisy reference mode shape (U_N) with the analytical mode shape (U) of mode 2 is presented in Fig. 11.18a for the least severe noise case, case 1 with noise amplitude 0.10, and the most severe noise case, case 4 with noise amplitude 0.80, to provide a perspective of the severity of noise. A similar comparison between the expanded mode shape (E_N) with the noisy reference mode shape (U_N) of mode 2 is presented in Fig. 11.18b. Considering the least severe noise case in Fig. 11.18, the expanded mode shape diverges considerably from the reference mode shape (Fig. 11.18b) despite how small the additive noise is relative to the analytical shape (Fig. 11.18a). The largest difference is observed by subsystem beam B across the length lacking instrumentation (active DOF). The differences are enunciated in the most severe noise case (Fig. 11.18b) and can be observed across each beam in the entire system. The noise added to the analytical mode shape is discernible in the most severe noise case (Fig. 11.18a) but is still relatively low in comparison to the magnitude of response of the mode shape. Overall, minor additions of noise induced a significant drop in correlation between the expanded mode shapes and reference mode shapes as observed in the MAC matrices (Fig. 11.17) and mode shapes (Fig. 11.18b). Therefore, mode shape expansion is heavily influenced by noise.

Time Domain Expansion

Time domain expansion was performed with the active DOF set defined in Sect. 11.5.1 (Fig. 11.16) for all noise severity cases defined in Sect. 11.5.2 (Table 11.3). The time domain response was obtained through modal superposition of the coupled components model exposed to a triangular pulse function at node 40. The sampling and excitation parameters used in the modal superposition solution are listed in Table 11.4. A schematic of the modal superposition model and the response of the input force in the time and frequency domain is shown in Fig. 11.19. The input forcing function was defined to provide

Table 11.4 Modal superposition parameters used with the coupled components model

Overall properties	Parameter	Value	Units
	Sample rate	4000	Hz
	Sample size	3000	
	Sample time	0.75	s
	Pulse length	0.005	s
	Pulse amplitude	10	lbf

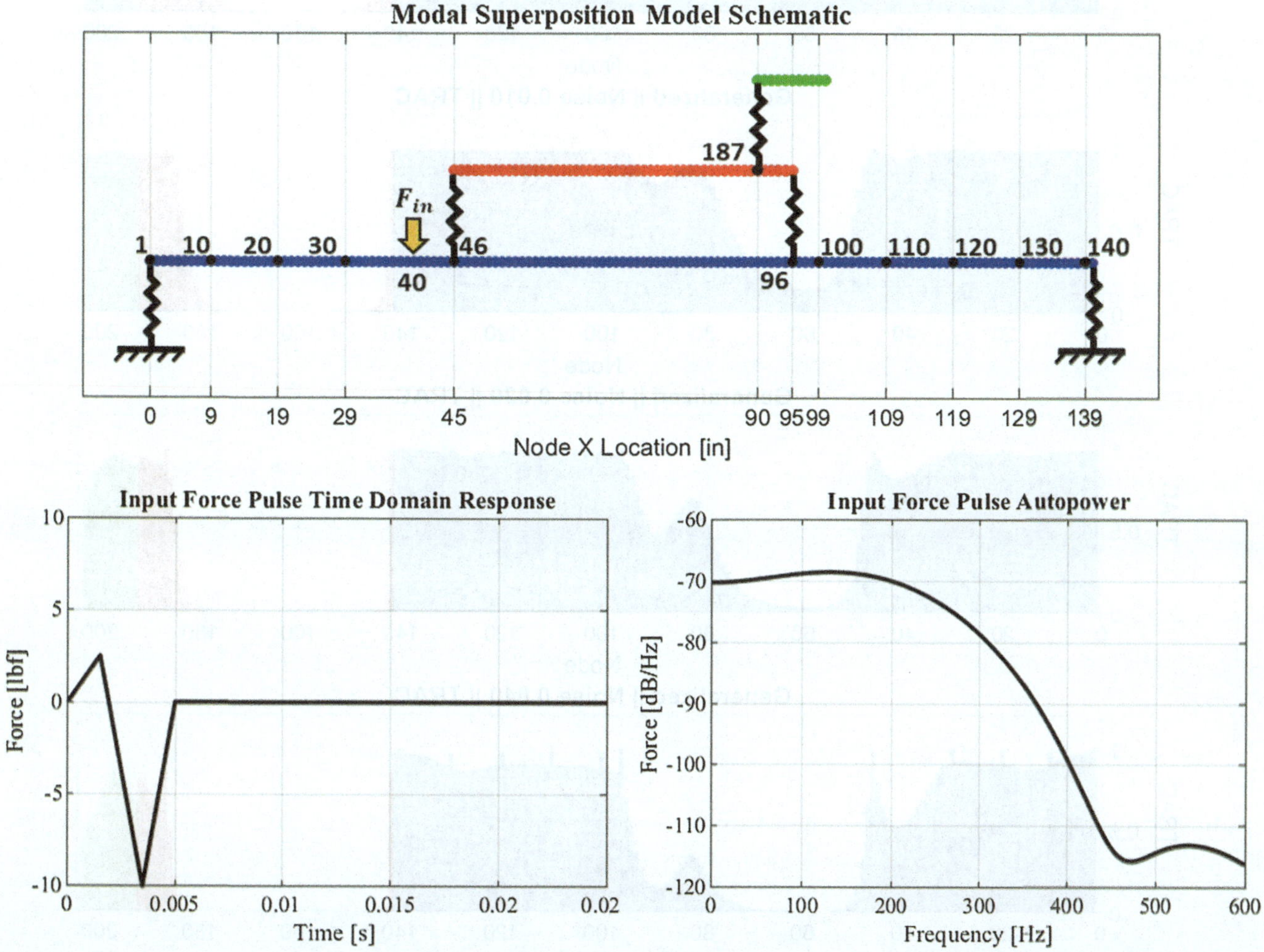

Fig. 11.19 Modal superposition schematic of the coupled components model and the response of the input force in the time and frequency domain

adequate excitation up to 400 Hz to cover the first ten modes of the coupled components model (7–389 Hz), which were solely considered in the expansion process.

The quality of the results was quantified by computing the TRAC of the expanded time responses with the reference time responses. The reference time responses correspond to the analytical time domain responses superimposed with the noise of the considered test case. The TRAC values of all DOF computed for each noise severity case is presented in Fig. 11.20. Inspecting the TRAC values of each test case for all DOF, the values exhibit a clear trend of decreasing TRAC value with increasing noise severity. Starting from the lowest severity of noise, the TRAC values of nodes 50 through 76 (mid-span of system beam A between the coupling points to subsystem beam B) and 143 through 184 (first 40 inches of subsystem beam B) are low. Lower TRAC values at the aforementioned nodes, as opposed to all other nodes, are likely due to the lack of instrumentation. The TRAC values have dropped below 0.9 for most nodes in component beam C by the third level of severity, and for most nodes in system beam A by the fourth level of severity indicating a poor correlation.

The effects of noise on the expanded time domain responses are shown using node 202 as an example. Node 202 corresponds to the second to last node of the component beam C, shown in Fig. 11.21a. The results of the least severe noise case, case 1 with noise amplitude 0.005, are provided in Fig. 11.21b and the results of the most severe noise case,

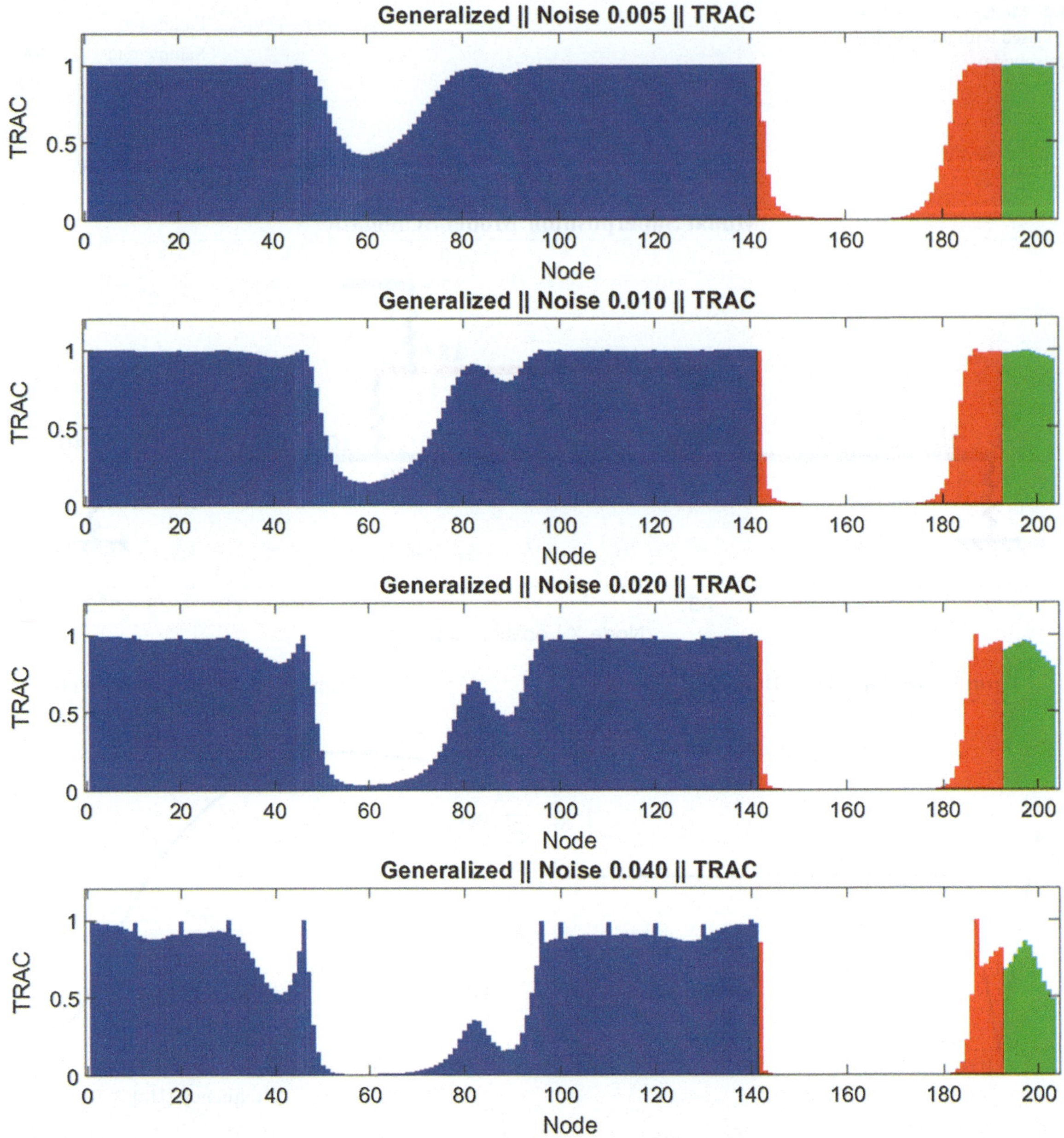

Fig. 11.20 TRAC values computed from the expanded noisy time responses for each test case using the standard generalized inverse approach with no regularization

case 4 with noise amplitude 0.040, are provided in Fig. 11.21c. A comparison of the noisy reference response (U_N) with the analytical response (U) is shown for each noise case to give a perspective of the noise severity. Similarly, a comparison of the expanded response (E_N) with the noisy reference response (U_N) is shown for each noise case to show the influences of noise on the expanded results. Considering the least severe noise case (Fig. 11.21b), the noise exhibited on the reference response is negligible relative to the analytical response. The resultant expanded response exhibits a slight increase in noise compared to the noisy reference response, but otherwise correlates well with a TRAC of 0.99. In the most severe noise case (Fig. 11.21c), the noise exhibited on the reference response is discernable, but still small relative to the analytical response. Despite the low relative amplitude of the noise, the expanded results exhibit a large amount of noise resulting in a poor TRAC value of 0.54. Similar to the observations extracted from mode shape expansion, time domain expansion is heavily influenced by noise.

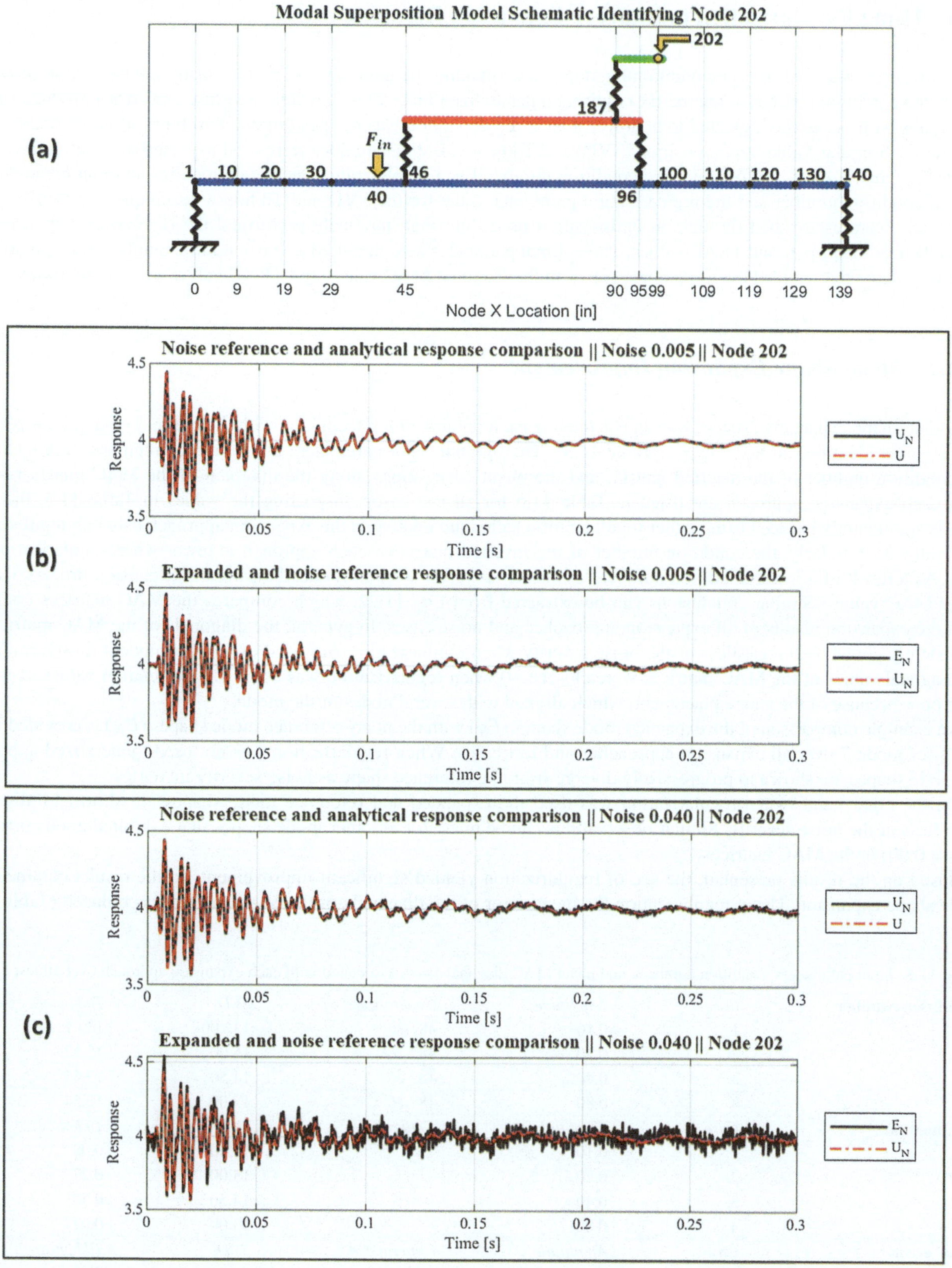

Fig. 11.21 (**a**) Model schematic identifying node 202 and time response comparisons between the (**b**) noisy reference time response with the analytical time response and (**c**) expanded time response with the noisy reference time response

11.6 Using Regularization to Improve Expansion Results

Regularization was used as a preconditioning step prior to taking the generalized inverse during the expansion process to reduce the condition number of the matrix requiring a generalized inverse. As a result, the noise and error introduced during the expansion process was expected to be lower than if regularization was not performed. Two types of regularization were considered, Singular Value Decomposition (SVD) and Tikhonov. Both techniques were used to perturb the matrix, increasing low-valued matrix elements to reduce the condition number. Each regularization technique requires an input parameter, the desired condition number and the regularization parameter value for the SVD and Tikhonov techniques, respectively. The parameters were established through an optimization procedure that iteratively performed the regularized expansion and stored the resultant MAC and TRAC values. The optimal parameter was identified as the condition number and regularization parameter value that yielded the lowest mean error in the diagonal MAC values and TRAC values over all iterations.

11.6.1 Mode Shape Expansion Improvement

The mode shape expansion process was repeated using each regularization technique with the optimal input parameters over all test cases identified in Sect. 11.5.2 (Table 11.3). The optimal input parameters used for each regularization technique, the condition number of the inverted matrix, and the mean error values along the diagonal of the MAC matrix resultant from each expansion approach are listed in Table 11.5 for all test cases. Inspecting the values in Table 11.5, the MAC error is significantly reduced using either regularization technique relative to the expansion approach using no regularization (generalized). Similarly, the condition number of the inverted matrix of each approach is lower when regularization was used. As a result of a lower condition number, less noise and error were introduced during the expansion process yielding better MAC values. Similar conclusions can be extracted from Fig. 11.22, which compares the MAC matrices computed from the expansion results of all expansion approaches and noise cases. In general, the diagonals of the MAC matrices are consistently above 0.80 regardless of the noise severity when regularization was used except for mode 4. Furthermore, the off-diagonal values of the MAC matrices were closer to 0 when regularization was used. The correlation values of mode 4 were poor because of the gauge placement, which aligned with several nodes of the mode.

An example comparison of the expanded mode shapes (E_N) with the noisy reference mode shapes (U_N) is provided in Fig. 11.23 for mode 7 over all expansion approaches and test cases. When regularization was not used (generalized approach), the mode shapes are shown to progressively diverge from the reference shape as noise severity increases.

On the other hand, when regularization was used the expanded and reference mode shapes are shown to align well regardless of the noise severity. Similar observations can be made for all other mode shapes that exhibited good correlation (above 0.90) in the MAC matrices.

Based on the results presented, the use of regularization yielded significant improvements to the results obtained from mode shape expansion. Using regularization the mean error of the diagonals of the MAC matrix was reduced (Table 11.5),

Table 11.5 Input parameters, condition numbers, and mean MAC diagonal errors associated with each expansion approach over all test cases

Condition Number	*Case*	*Amplitude*	*Generalized*	*SVD*	*Tikhonov*
	1	0.10	487.26	132.00	133.11
	2	0.20	487.26	15.00	15.52
	3	0.40	487.26	14.50	13.44
	4	0.80	487.26	13.00	12.54
Parameters	*Case*	*Amplitude*	*Generalized*	*SVD*	*Tikhonov*
	1	0.10	–	132.00	0.08
	2	0.20	–	15.00	0.27
	3	0.40	–	14.50	0.50
	4	0.80	–	13.00	0.67
MAC error	*Case*	*Amplitude*	*Generalized*	*SVD*	*Tikhonov*
	1	0.10	0.16	0.08	0.08
	2	0.20	0.29	0.10	0.10
	3	0.40	0.46	0.11	0.11
	4	0.80	0.67	0.14	0.13

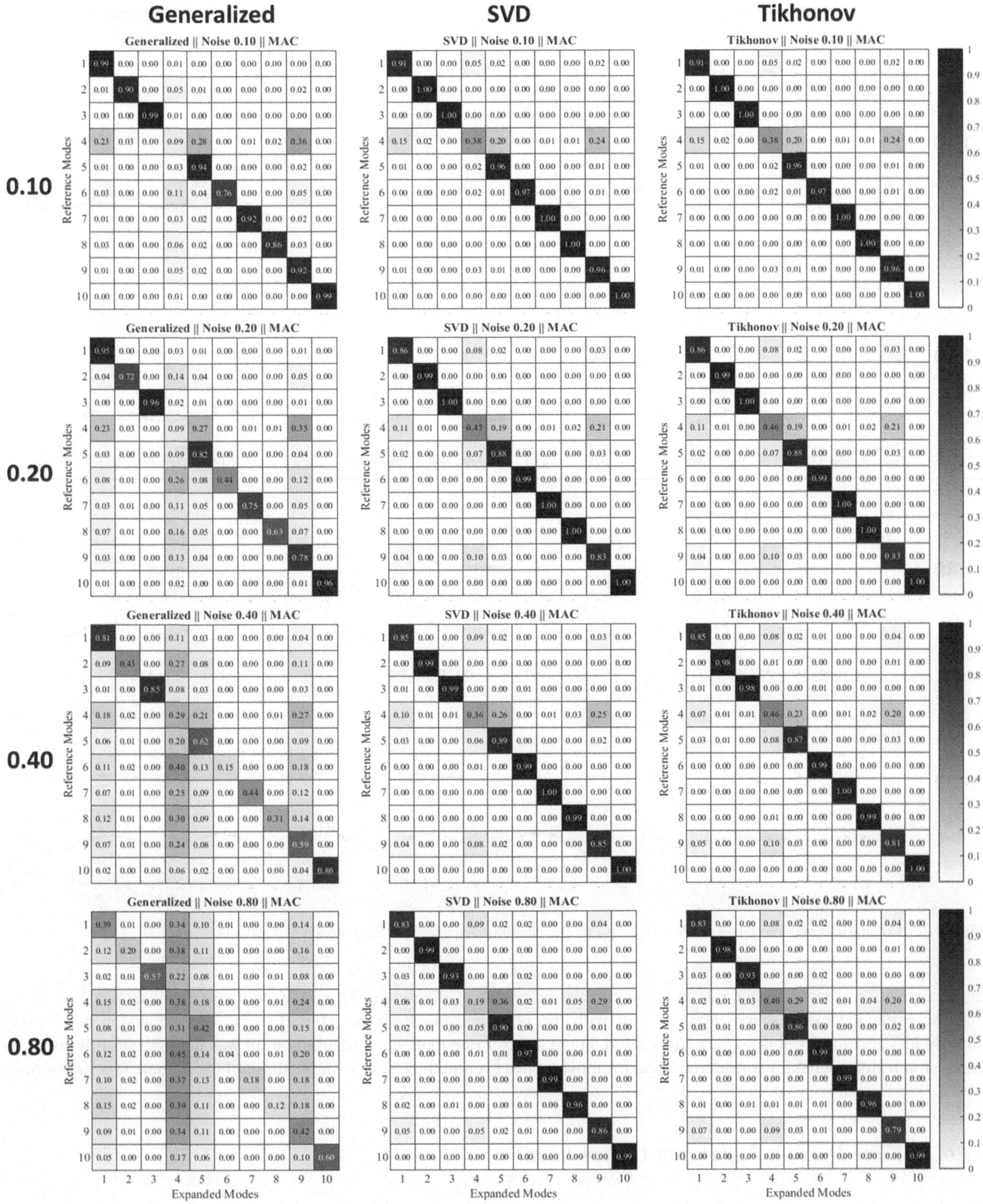

Fig. 11.22 MAC matrices computed from the expanded mode shapes for all expansion approaches and test cases

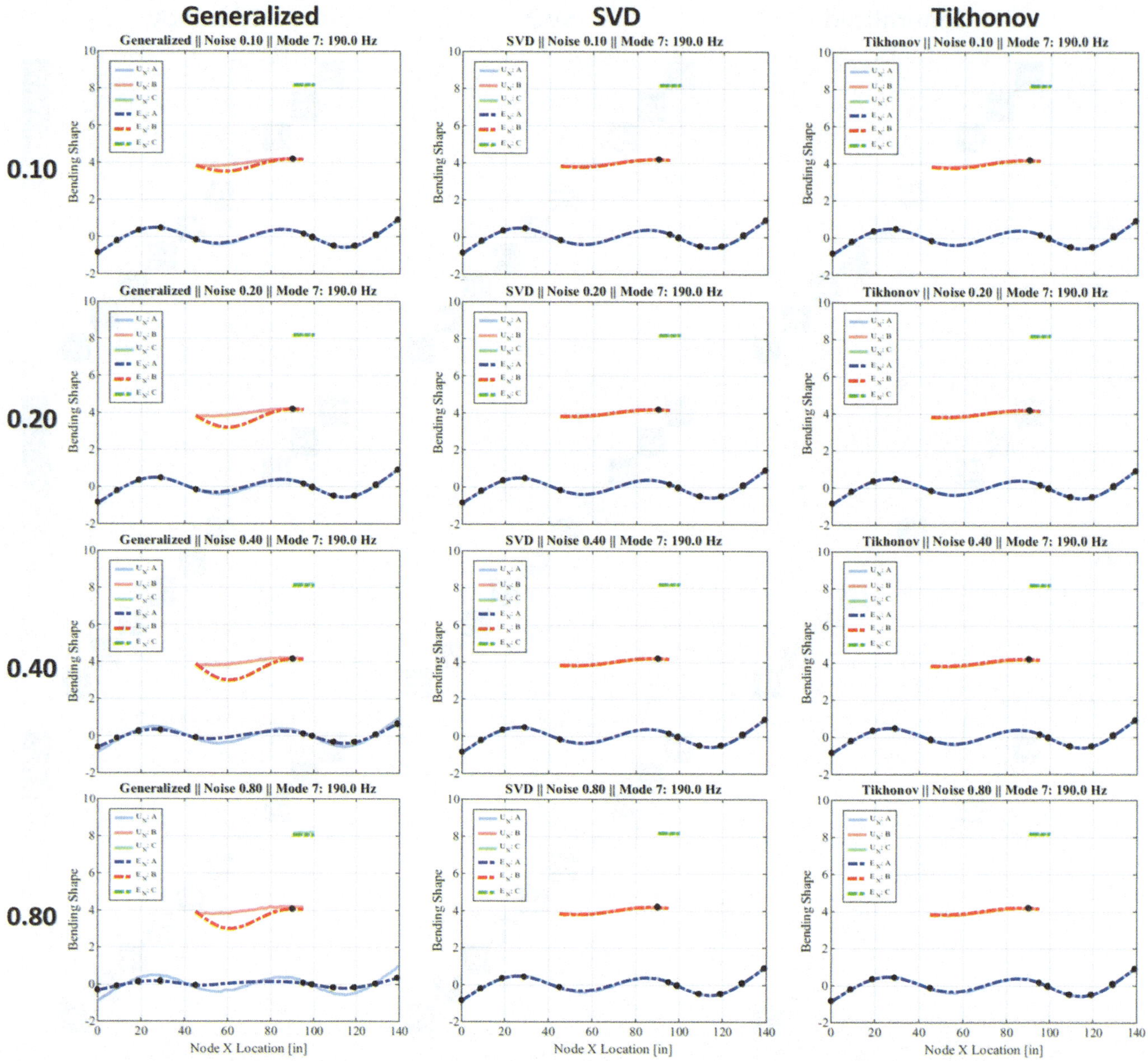

Fig. 11.23 Comparison between the expanded and noisy reference mode shape 6 for each approach and test case

good correlation along the diagonals and poor correlation on the off-diagonals of the MAC matrix were observed (Fig. 11.22), and the expanded mode shapes aligned well with the reference mode shapes (Fig. 11.23). The expansion results were slightly less accurate as noise severity increased, but at a lesser rate than when regularization was not used. Furthermore, the expansion results were still considered good for most modes even in the test case that exhibited the highest severity of noise.

11.6.2 Time Domain Expansion Improvement

The time domain expansion process was repeated using each regularization technique with the optimal input parameters over all test cases identified in Sect. 11.5.2 (Table 11.3). The optimal input parameters used for each regularization technique, the condition number of the inverted matrix, and the mean error values of the TRAC for all nodes resultant from each expansion approach are listed in Table 11.6 for all test cases. Inspecting the values in Table 11.6, the TRAC error is significantly reduced using either regularization technique relative to the expansion approach using no regularization (generalized). Similarly,

Table 11.6 Input parameters, condition numbers, and mean TRAC errors associated with each expansion approach over all test cases

Condition number	*Case*	*Amplitude*	*Generalized*	*SVD*	*Tikhonov*
	1	0.005	487.26	12.80	13.09
	2	0.010	487.26	9.80	11.81
	3	0.020	487.26	4.40	13.87
	4	0.040	487.26	3.40	25.64
Parameters	*Case*	*Amplitude*	*Generalized*	*SVD*	*Tikhonov*
	1	0.005	–	12.80	0.57
	2	0.010	–	9.80	0.80
	3	0.020	–	4.40	1.36
	4	0.040	–	3.40	2.84
TRAC error	*Case*	*Amplitude*	*Generalized*	*SVD*	*Tikhonov*
	1	0.005	0.24	0.07	0.07
	2	0.010	0.31	0.10	0.09
	3	0.020	0.39	0.16	0.14
	4	0.040	0.49	0.28	0.26

the condition number of the inverted matrix of each approach is lower when regularization was used. Because of a lower condition number, less noise and error were introduced during the expansion process yielding better TRAC values. Similar conclusions can be extracted from Fig. 11.24, which compares the TRAC values computed for each node from the expanded results of all expansion approaches and noise cases. The TRAC values computed from the expanded results obtained with regularization are generally always larger than the TRAC values obtained when regularization was not used. The TRAC values obtained from the regularized results are considered good (>0.9) for all nodes of system beam A in the first two test cases and for all nodes of component beam C in the first three test cases. Regardless of the expansion approach or noise severity, the TRAC values of most nodes in subsystem beam B were poor. The poor correlation is likely due to the lack of active DOF (instrumentation) local to the nodes exhibiting poor correlation. Similar conclusions can be drawn regarding the poor correlation observed between nodes 50 through 76 of system beam A for all expansion results obtained without regularization and all expansion results obtained with regularization in test cases 3 and 4.

An example comparison of the expanded time response (E_N) with the noisy reference response (U_N) is provided in Fig. 11.25 for nodes 70 (mid-length of system beam A) and 202 (second last node of component beam C) in test cases 2 (noise amplitude of 0.010) and 4 (noise amplitude of 0.040) for all expansion approaches. All comparisons of the time responses obtained when regularization was not used (generalized approach) exhibit larger errors and noise-like behavior that increases drastically as the noise amplitude increases. On the other hand, the time responses obtained when regularization was used are generally in good agreement with the reference responses regardless of the node or noise severity. The expanded time responses obtained with regularization by node 70 with a noise amplitude of 0.040 has the lowest correlation but is a significant improvement relative to the results obtained without regularization. Furthermore, as the noise severity increases the error is introduced at a lesser rate in the expansion process when regularization was used. Similar observations can be made for all other nodes that exhibited good correlation (TRAC value above 0.90).

Based on the results presented, the use of regularization yielded significant improvements to the results obtained from time domain expansion. Using regularization the mean error of the TRAC values were reduced (Table 11.6), good correlation between the expanded time responses and references time responses of most nodes were observed (Fig. 11.24), and the expanded time responses aligned well with the reference time responses (Fig. 11.25) in most cases. The expansion results were less accurate as noise severity increased, but at a lesser rate than when regularization was not used. Furthermore, the expansion results were still considered good for most nodes up until the most severe level of noise was considered.

11.6.3 Additional Considerations

Significant improvements were obtained by incorporating regularization in the expansion process when the mode shapes (Sect. 11.6.1) or time responses (Sect. 11.6.2) were exposed to noise. The amount of regularization used generally increased as the amount of noise contamination increased. This was observed in Table 11.5 and Table 11.6 by the decrease in the condition number wanted of the SVD approach and increase in the regularization parameter value of the Tikhonov approach as the noise severity increased. Although the amount of regularization increased, the optimal regularization parameters

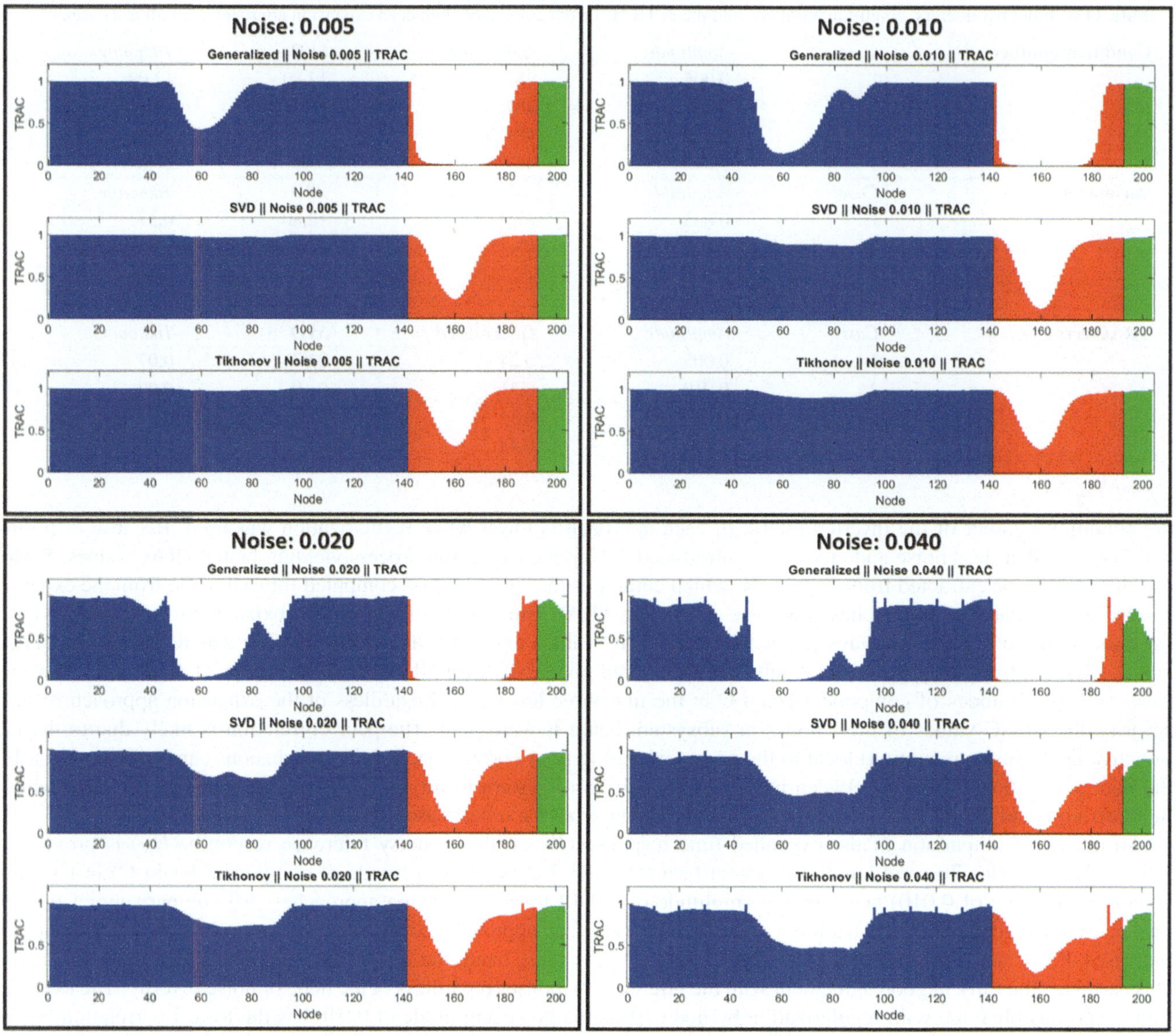

Fig. 11.24 TRAC values of all nodes computed from the expanded time responses for all expansion approaches and test cases

generally yielded an inverted matrix having a condition number within the range of 3–25. The only case that yielded a higher condition number was during the mode shape expansion analysis for the lowest noise severity. This is likely because the noise was too insignificant to have a notable effect, as explained by the good correlations that were achieved without regularization using an inverted matrix with a condition number of ~487.

The expanded results obtained using regularization were significant but were not enough to yield good correlations for the shapes of all modes and time responses of all nodes. In the most severe noise cases for both expansion types, several diagonal MAC values and numerous TRAC values were low (below 0.9). Therefore, it is important to note that regularization can provide considerable improvements to the results but cannot completely resolve the issue of noise contamination in the expansion process. The data should be thoroughly assessed and appropriate engineering judgement should be used before deciding to apply regularization, and expansion in general, on noise contaminated data.

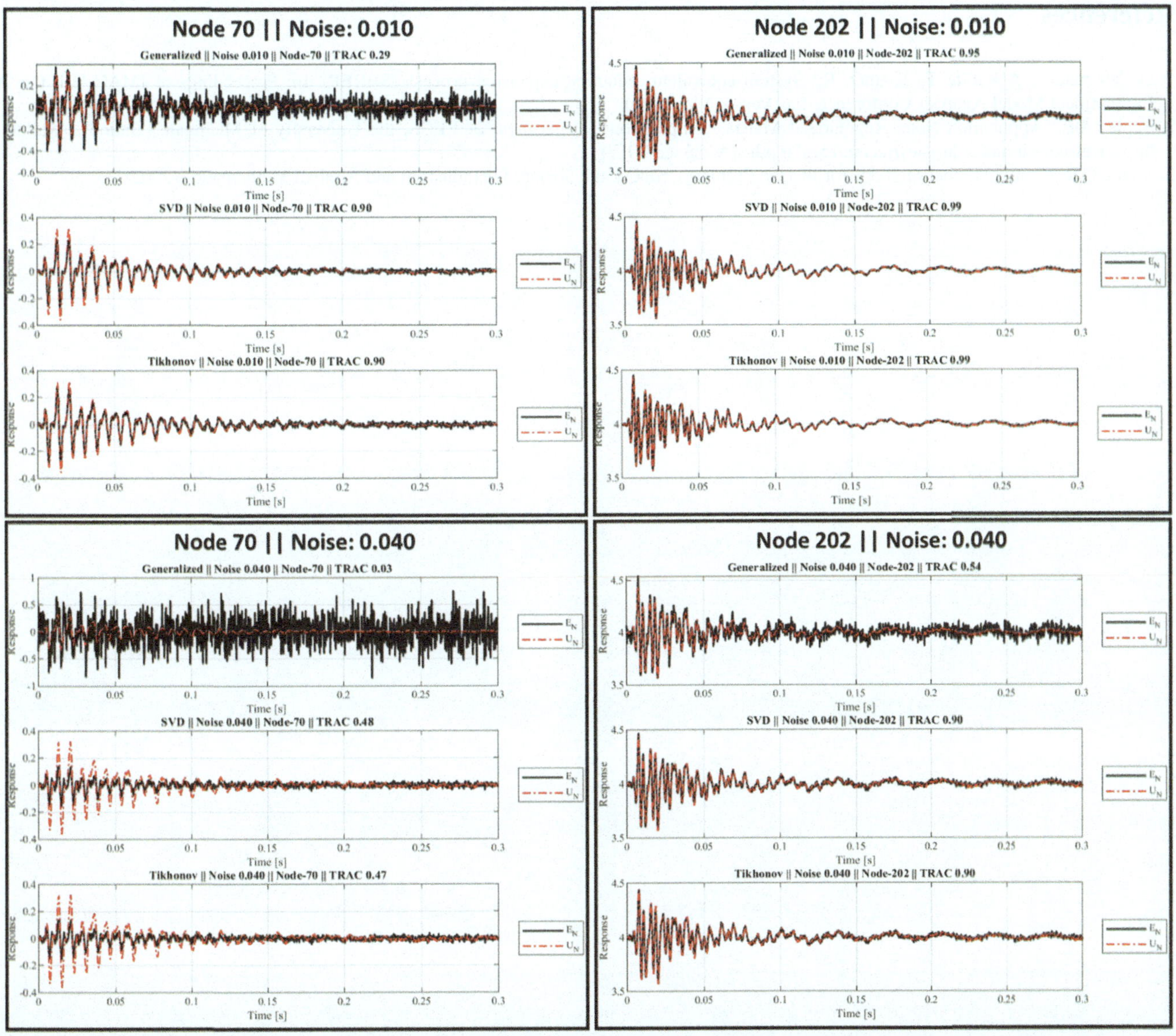

Fig. 11.25 Comparison between the expanded and noisy reference time responses of nodes 70 and 202 for each approach in test cases 2 (noise amplitude of 0.010) and4 (noise amplitude of 0.040)

11.7 Conclusions

Expansion is a very useful tool for making the most of measured data. However, as was shown here, the expansion results are sensitive to various factors including the location of gauges and noise on the measured data. Much of the error which results from an expansion process is due to error propagation and amplification in the pseudo-inverse used in forming the expansion transformation matrix. As with any matrix inverse operation, inverting a poorly-conditioned matrix will result in errors. Conditioning is greatly improved by carefully selecting measurement locations which are independent and cover as much of the structure as possible. However, there are many cases where optimal gauge locations cannot be used because of practical, logistical considerations. In those cases, regularization can be an effective method for improving expansion results. Here, significant improvements in both time and mode shape expansion was achieved by using Tikhonov and singular value regularization techniques, though care had to be taken in selecting the appropriate amount of regularization.

References

1. O'Callahan, J., Avitabile, P., Riemer, R.: System equivalent reduction expansion process (SEREP). In: Proceedings of IMAC VII, the 7th International Modal Analysis Conference, Las Vegas, NV (1989)
2. Yagle, A.E.: Application Note: Regularized Matrix Computations. Department of EECS, the University of Michigan. [Online]. Available: http://web.eecs.umich.edu/~aey/recent/regular.pdf. [Accessed 2017]
3. Vogel, C.R.: Computational Methods for Inverse Problems. SIAM, the Society for Industrial and Applied Mathematics (2002)

Chapter 12
Expansion of Coupled Structural-Acoustic Systems

Ryan Schultz, Dagny Beale, and Ryan Romeo

Abstract Expansion is useful for predicting response of un-instrumented locations and has traditionally been applied to structures alone. However, there are a range of hollow structures where the influence of the acoustic cavity affects the structural response, and the structural response affects the acoustic response. This structural-acoustic coupling results in a gyroscopically coupled system with complex modes. Though more complicated than modes of a structure alone, the modes of the coupled structural-acoustic system can be used as the basis vectors in an expansion process. In this work, complex modes of a model of a coupled structural-acoustic system are used to expand from a sparse set of structural and acoustic response degrees of freedom to a larger set of both structural and acoustic degrees of freedom. The expansion technique is demonstrated with a finite element model of a hollow cylinder with simulated displacement and pressure measurements, and expansion is studied for both modal and transient responses. Though more nuanced than traditional structure-only expansion problems, the displacement and pressure response of a coupled structural-acoustic system can be expanded using the coupled-system modes.

Keywords Expansion · Coupled system · Acoustoelastic · Structural-acoustic · Modes

12.1 Motivation and Theory

One major challenge in experimental testing is that only a few discrete points on an object of interest are measured but the response of the full object is usually desired. Expansion is a method that can take the limited number of measurements and some knowledge of the mode shapes of the object to predict the response at a larger number of points, and is especially useful for structures with components that cannot easily be measured. Traditionally, expansion has only been applied to solid structures with mode shapes comprised of displacement, velocity, or acceleration responses. However, there are many objects of interest that are hollow structures that contain an acoustic cavity, such as rocket fairings and automobiles. The responses of these coupled structural-acoustic systems depend on both structural modes and acoustic pressure modes. This work applies traditional expansion to a coupled system to show that expansion is possible in coupled structures, and to determine the challenges unique to coupled expansion.

Expansion from a set of measurements at the a-DOF to a set of unmeasured responses at the n-DOF is achieved by multiplying the measured responses, $\{x_a\}$, by a transformation matrix, $[T]$, as in Eq. 12.1.

$$\{x_n\} = [T]\,\{x_a\}\,. \tag{12.1}$$

Sandia National Laboratories is a multimission laboratory managed and operated by National Technology and Engineering Solutions of Sandia, LLC., a wholly owned subsidiary of Honeywell International, Inc., for the U.S. Department of Energy's National Nuclear Security Administration under contract DE-NA0003525.

R. Schultz (✉) · D. Beale · R. Romeo
Sandia National Laboratories, Albuquerque, NM, USA
e-mail: rschult@sandia.gov

B. Dilworth (ed.), *Topics in Modal Analysis & Testing, Volume 8*, Conference Proceedings of the Society for Experimental Mechanics Series, https://doi.org/10.1007/978-3-030-47717-2_12

A matrix of mode shapes from a test at the a-DOF, $[E_a]$, can similarly be expanded to shapes at the n-DOF by Eq. 12.2.

$$[E_n] = [T]\,[E_a]\,. \tag{12.2}$$

With System Equivalent Reduction Expansion Process (SEREP) expansion method, the method that will be used in this paper, the transformation matrix is composed of mode shape matrices at the a- and n-DOF, $[U_a]$ and $[U_n]$ respectively [1], as in Eq. 12.3.

$$[T] = [U_n]\,[U_a]^+. \tag{12.3}$$

These shape matrices typically come from a model, which allows the n-DOF to be at any of the model DOF. A pseudo-inverse of the $[U_a]$ matrix, $[U_a]^+$, projects the a-DOF measurements to modal space, giving the modal responses. These modal responses are then projected back to physical space via a multiplication by the $[U_n]$ shape matrix, giving the expanded response.

Extending this expansion method to coupled systems is straightforward. Consider a system comprised of two components, 1 and 2. Let system 1 be a structural system with displacements in the *x*, *y*, and *z* directions, contained in the structural mode shape matrix $[U_1]$. Let system 2 be an acoustic domain with a pressure response, *p*, contained in the acoustic mode shape matrix $[U_2]$. The measured mode shapes at the a-DOF of the components are $[U_{a,\,1}]$ and $[U_{a,\,2}]$ and the expanded mode shapes at the n-DOF are$[U_{n,\,1}]$ and $[U_{n,\,2}]$. Then, the coupled system transformation matrix can be formed with mode shape matrices comprised of the shape matrices from each component:

$$[T] = \begin{bmatrix} U_{n,1} \\ U_{n,2} \end{bmatrix} \begin{bmatrix} U_{a,1} \\ U_{a,2} \end{bmatrix}^+. \tag{12.4}$$

This is nothing more than a partitioning of the system DOF in terms of the DOF of each component.

The unique aspect of the coupled structural-acoustic system comes from the fact that this is a gyroscopically-coupled system, which results in complex modes. In fact, some of the modes of these systems will be structurally-dominant, meaning the mode is mostly structural response, and some modes will be acoustically-dominant, meaning the mode is mostly acoustic response. Additionally, some modes may be coupled, with similar contributions from structural and acoustic DOFs. This coupling is called acoustoelastic coupling and has some interesting effects on the system dynamics, including peak splitting [2, 3]. This paper aims to use complex modes of a model of a coupled structural-acoustic system in an expansion problem by simply using the complex shape matrices for the component matrices shown in Eq. 12.4 and see how the expanded response compares to the actual response at a chosen set of un-measured DOFs.

Computing these coupled system modes with a finite element model is non-trivial because it is a quadratic eigenvalue problem (QEVP), which presents an additional challenge in this work. Here, the coupled system modes were computed using the QEVP solution in Sandia National Laboratories Sierra/Structural Dynamics solver [4]. While there are several QEVP methods available, here the so-called SA_eigen method is used. The SA_eigen method uses the uncoupled, free-free modes of the two subdomains (structural component and acoustic component) to form a reduced size coupled system which is then solved in full. Then, the solution of the reduced system is projected back to full space to give an estimate of the coupled system modes. This method is effective, though is sensitive to the uncoupled component modes and modal truncation.

12.2 Modal Expansion of a Coupled Structure

A Finite Element (FE) model was created for the coupled expansion analysis that consisted of a cylindrical structure enclosing an acoustic fluid. The cylindrical structure was chosen to be 8-inch outer diameter, 24-inch long aluminum cylinder with 0.5-inch thick walls and end caps on both ends. The bottom end cap was solid, while the top end cap has a 2-inch hole in the center. The inside cavity of the cylindrical structure was modeled as an acoustic cavity filled with air, as shown in Fig. 12.1. The dark circle shown on the top of the acoustic body is just a product of the mesh appearance; the acoustic domain is a solid cylinder, and the mesh is consistent with the structural domain mesh. The dimensions of the coupled structure were designed to create acoustic and structural modes that have similar frequencies and inherently couple. The model was meshed using 62,191 hexahedral elements. The overlapping nodes between the acoustic and the structural domains were merged to create the coupled system. The input to the QEVP solver asked for a target of 40 structural modes and 80 acoustic modes for

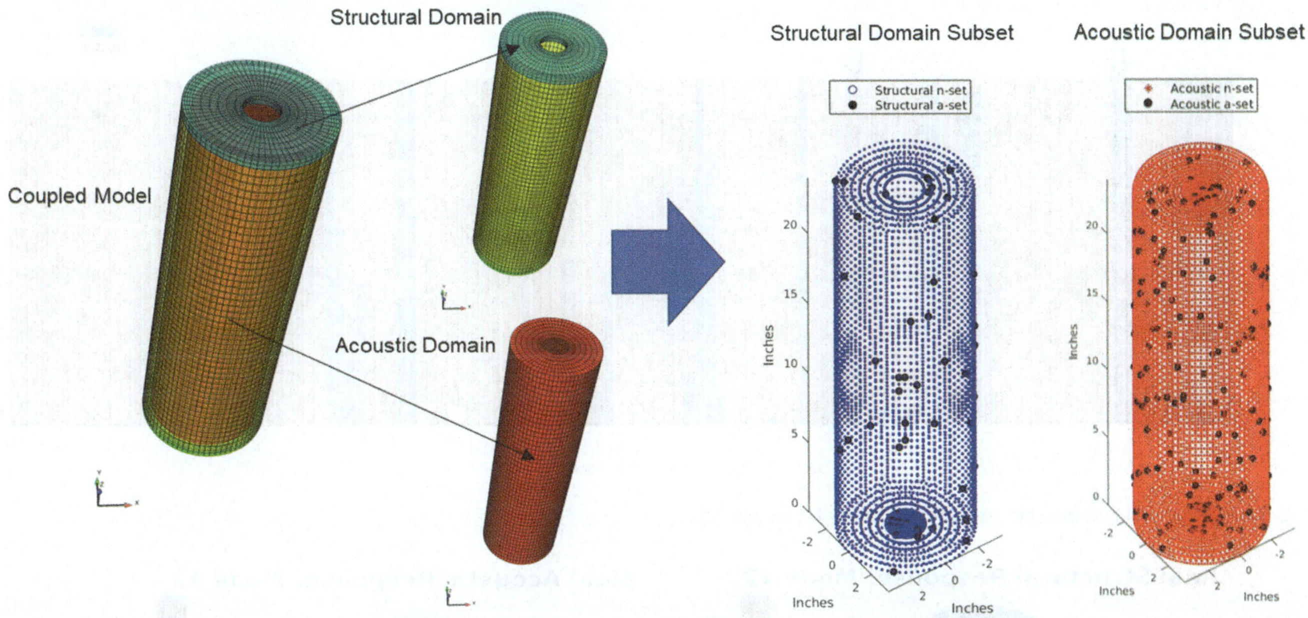

Fig. 12.1 Coupled model with structural and acoustic domains (left) and chosen subsets of the structural and acoustic domains for analysis (right)

a total of 120 modes. The results of this solver are given as complex-valued modes that contain both a real and an imaginary portion.

After the solution was found, subsets of the structural and acoustic domains were chosen to perform the expansion analysis. The internal surface of the aluminum cylinder and the external surface of the acoustic domain were the chosen subsets. The chosen subsets of the acoustic and structural domains will be referred to as the n-set of the model. A sample of 50 points on the solid cylinder and 160 points on the acoustic domain within the n-set points were then chosen to create the a-set points. The n-set points are shown in Fig. 12.1 as three-dimensional scatter plots of each domain, and the a-set points are marked as black dots. The a-set points will be considered as the "measurement" points while the n-set points will be considered as the "full-field" response. To make the a-set "measurements" more representative of actual measurements, low-level random noise at 1% of the response mean was added to the a-set responses to simulate imperfect data.

Three modes will be used to study expansion in this paper: a mode dominated by structural response, a mode dominated by acoustic response, and a mode with somewhat equal contributions of structural and acoustic responses. The modes were chosen by plotting the contributions of structural and acoustic responses calculated in the FE result file, as shown in Fig. 12.2. The contribution of each type of response to each mode was given as a value from zero to one, with higher values indicating a higher contribution to the mode.

From the contribution factors shown in Fig. 12.2, mode 12 was chosen to be the representative structurally-dominated mode, mode 38 was chosen to be the representative acoustically-dominated mode, and mode 66 was chosen to be the representative coupled mode. The real and imaginary portions of each mode are shown, respectively, in Figs. 12.3, 12.4, and 12.5. Take note of the color scale in each plot, as some responses look significant but are actually very small with respect to the dominant motion. In the structural mode in Fig. 12.3, both the real and imaginary structural modes have a strong response while the acoustic response is small and has a similar shape to the structural mode. In the acoustic mode in Fig. 12.4, imaginary part of the acoustic mode dominates the response. In the coupled mode in Fig. 12.5, the real part of the structural mode and the imaginary part of the acoustic mode both contribute strongly to the response.

To perform the expansion, a transformation matrix was required that could map between both structural and acoustic modes. All solution modes, including rigid body modes, were included in the expansion. The transformation matrix was computed by assembling each mode shape variable (x, y, and z for the structural response and p for the acoustic pressure response) into a universal mode shape matrix, and then calculating the transformation matrix using Eq. 12.4. The results of expansion are shown in the next three figures.

The structure-dominated mode 12 expansion is shown in Fig. 12.6, with the major contributors to the mode highlighted. The structural expansion matches closely with the actual structural mode in shape and magnitude. The expanded acoustic mode, however, does not resemble the actual acoustic mode and the expanded acoustic response has a higher amplitude than the actual acoustic amplitude. This is likely due to the noise that was added to make the response more realistic. The acoustic

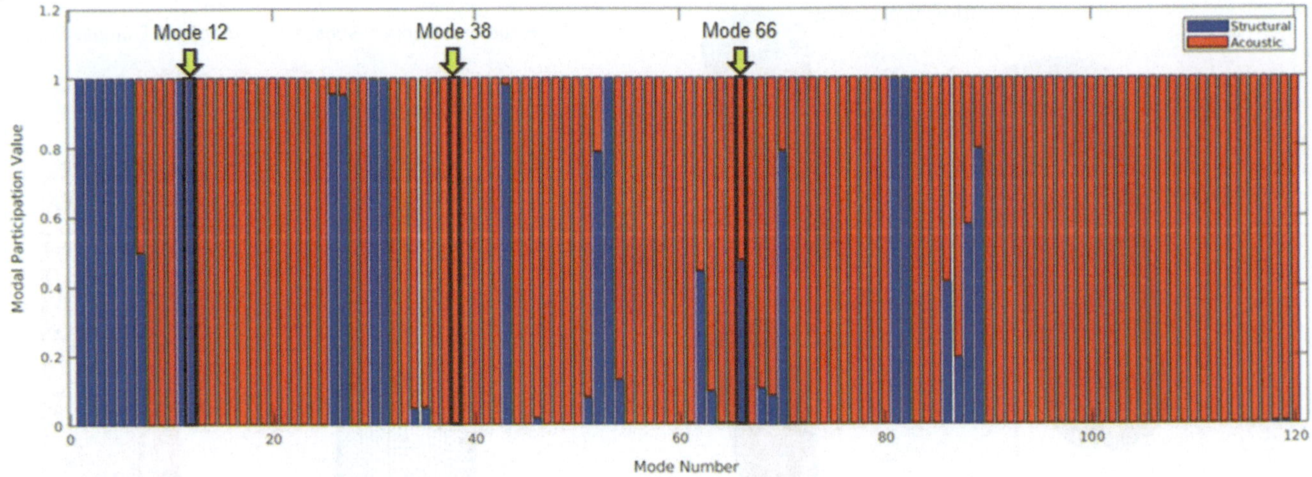

Fig. 12.2 Structural and acoustic contribution values for each mode

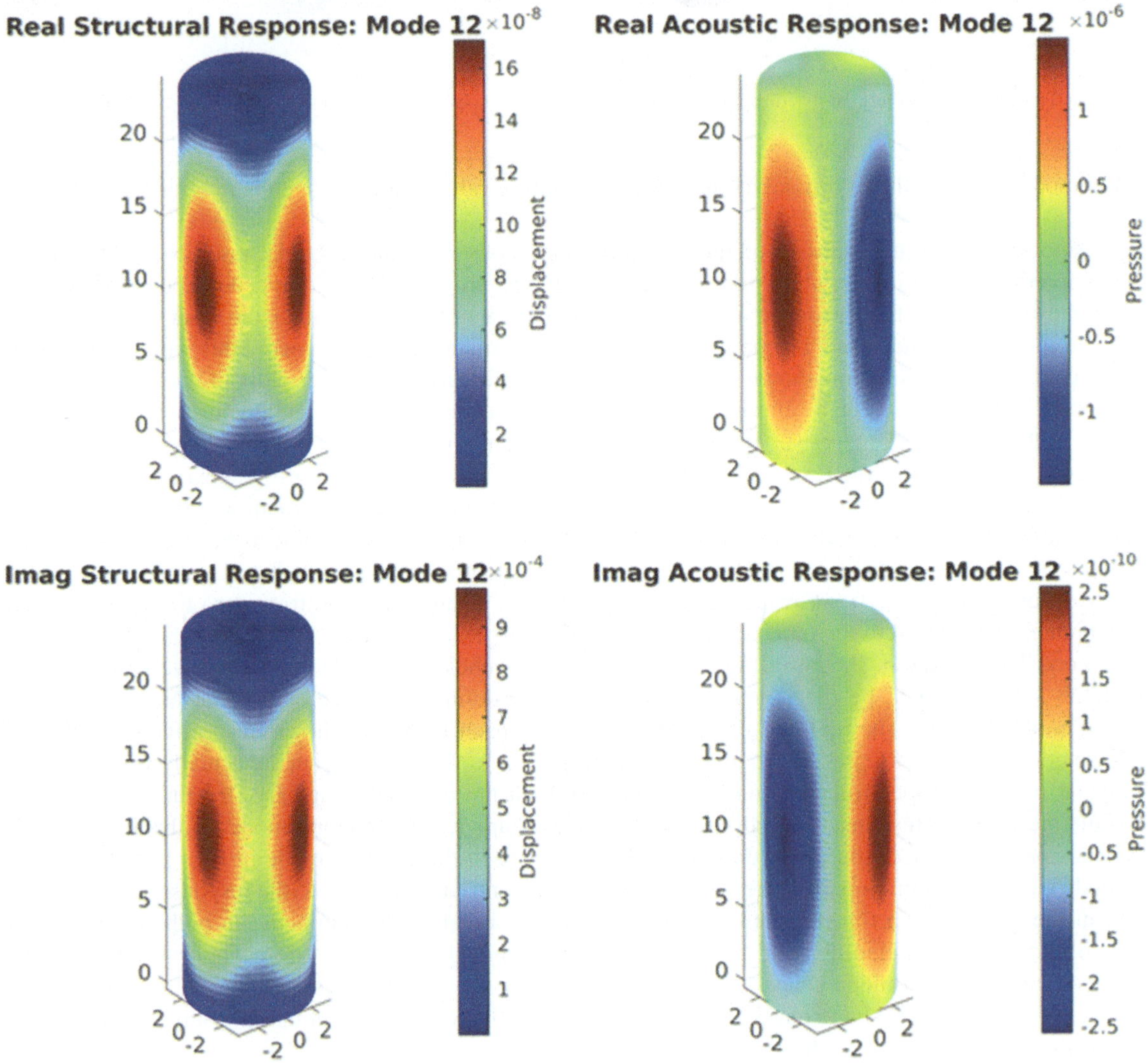

Fig. 12.3 Representative structurally-dominated mode (mode 12, 1047 Hz)

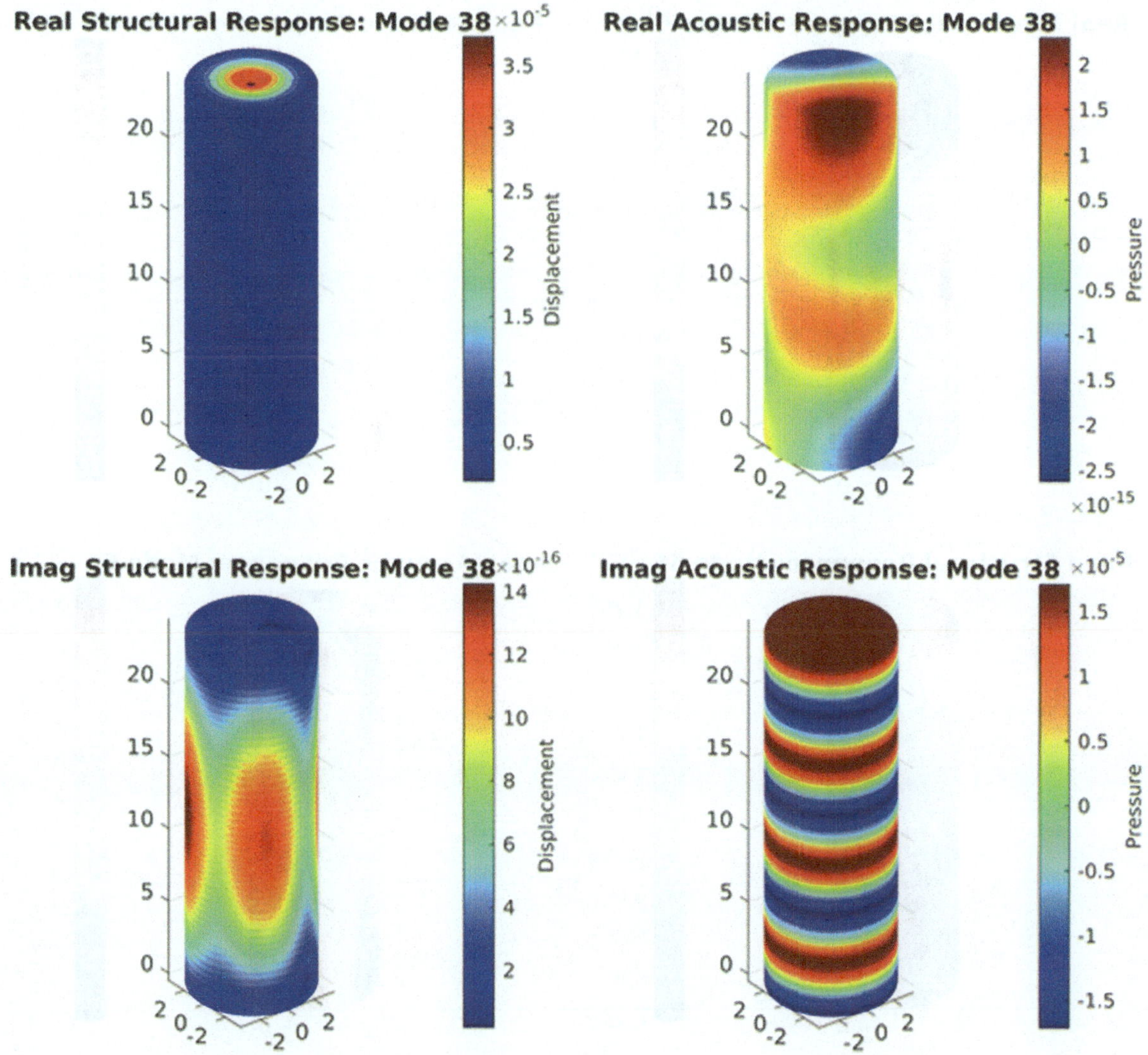

Fig. 12.4 Representative acoustically-dominated mode (mode 38, 1969 Hz)

response is essentially nonexistent, so the addition of noise masks the insignificant acoustic response. When expansion is performed, noise is being expanded, so the compared acoustic responses are not expected to look the same.

The acoustic-dominated mode 38 expansion is shown in Fig. 12.7, with the major contributors to the mode highlighted. The imaginary acoustic expansion (the dominant response in this mode) matches closely with the actual mode in shape and magnitude. The expanded real portion of the structural mode is also a fairly close match to the real structural response, with some slight differences in amplitude. The expanded imaginary structural response and the expanded real acoustic response do not look similar to their counterparts. However, again this is likely due to the small, insignificant response values that were poorly expanded in the presence of noise.

The coupled mode 66 expansion is shown in Fig. 12.8, with the major contributors to the mode highlighted. The expanded real portion of the structural mode is very similar to the real structural response. The expanded imaginary acoustic response is similar to the actual imaginary acoustic response, with some differenced around the center of the cylinder. These differences could be due to mode truncation. Although more acoustic than structural modes were included in the expansion, the coupled mode is fairly high in frequency, and the mode set used for expansion might not span the space of the acoustic response. Again, the very small valued portions of the expanded mode (imaginary structural response and real acoustic response) did not closely resemble the actual mode.

Overall, the coupled expansion was shown to be a success for the dominant portions of the mode shapes. Expansion using complex modes generated from the QEVP solver was able to be used in traditional SEREP equations. However, some of the

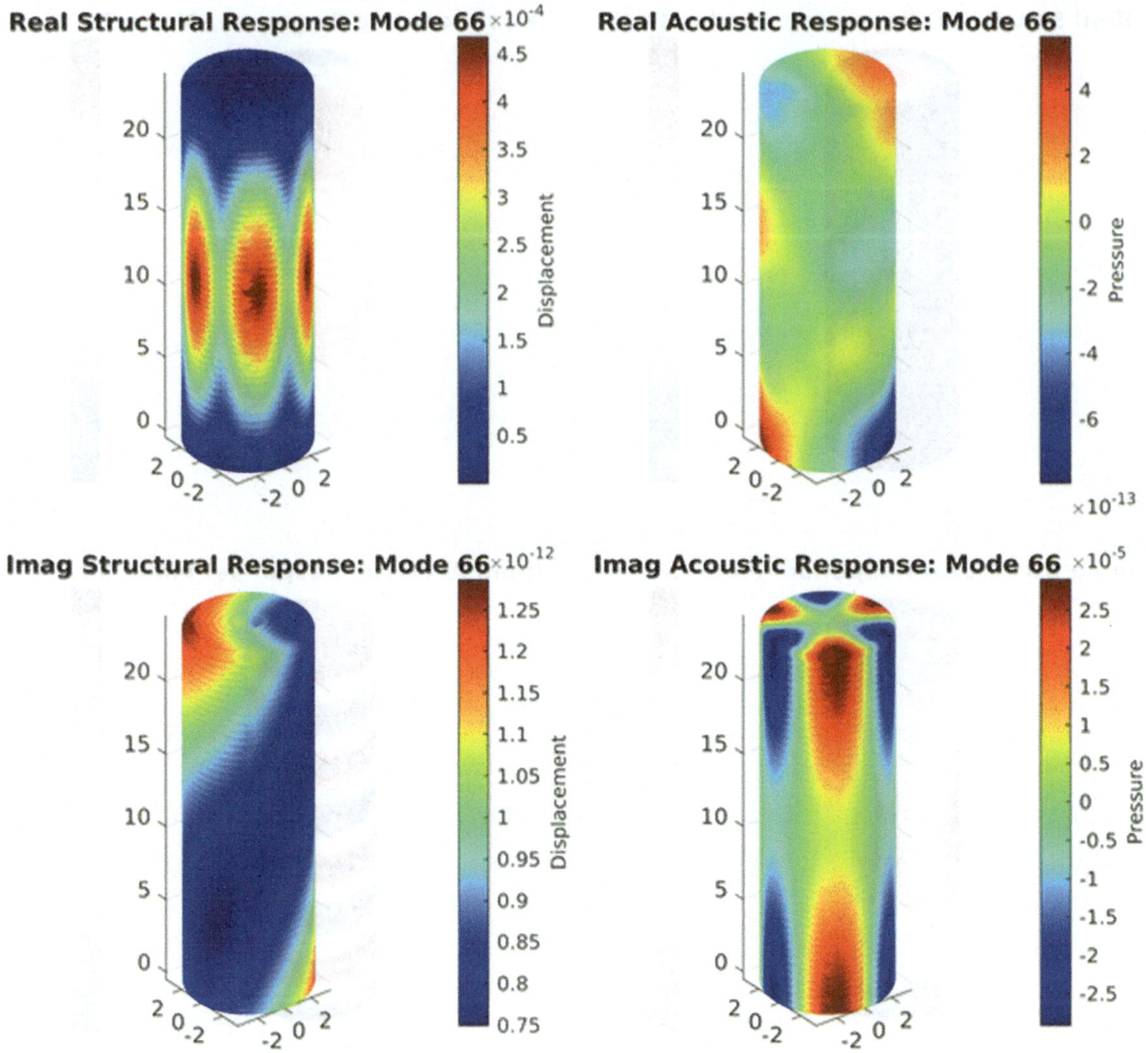

Fig. 12.5 Representative coupled mode (mode 66, 2608 Hz)

smaller modal response portions were not captured during the modal expansion. This analysis has shown that SEREP modal expansion is applicable to coupled systems.

12.3 Coupled Expansion of a Transient Response

In the previous section, modal expansion of a coupled structural-acoustic system was shown. Here, the FE model of the shell and its air cavity is used to demonstrate the expansion of transient response data. A 0.4 ms haversine pulse force was applied in the radial direction to the shell node shown in Fig. 12.9. Then, the response of the shell and the air cavity was captured at many nodes distributed throughout, as shown in Fig. 12.9. A subset of these DOF were assigned to the a-DOF and the remaining to the n-DOF. Ideally, some optimal selection method, such as effective independence [5], would have been used to choose the a-DOF from the structure and acoustic component, but here locations were randomly selected. The 84 structure nodes have acceleration response in the radial, tangential, and axial directions (three DOF per node), and the 168 acoustic nodes only have acoustic pressure response. The a-DOF consisted of the radial response at 30 structure nodes and the pressure response at 50 acoustic nodes. While the number of a-set DOF was higher than would be desired for practical implementation, fewer a-set nodes did not achieve good expansion results in this implementation. It is believed that improved results, using fewer a-DOF could be achieved if the a-DOF were chosen with some optimum gauge selection algorithm.

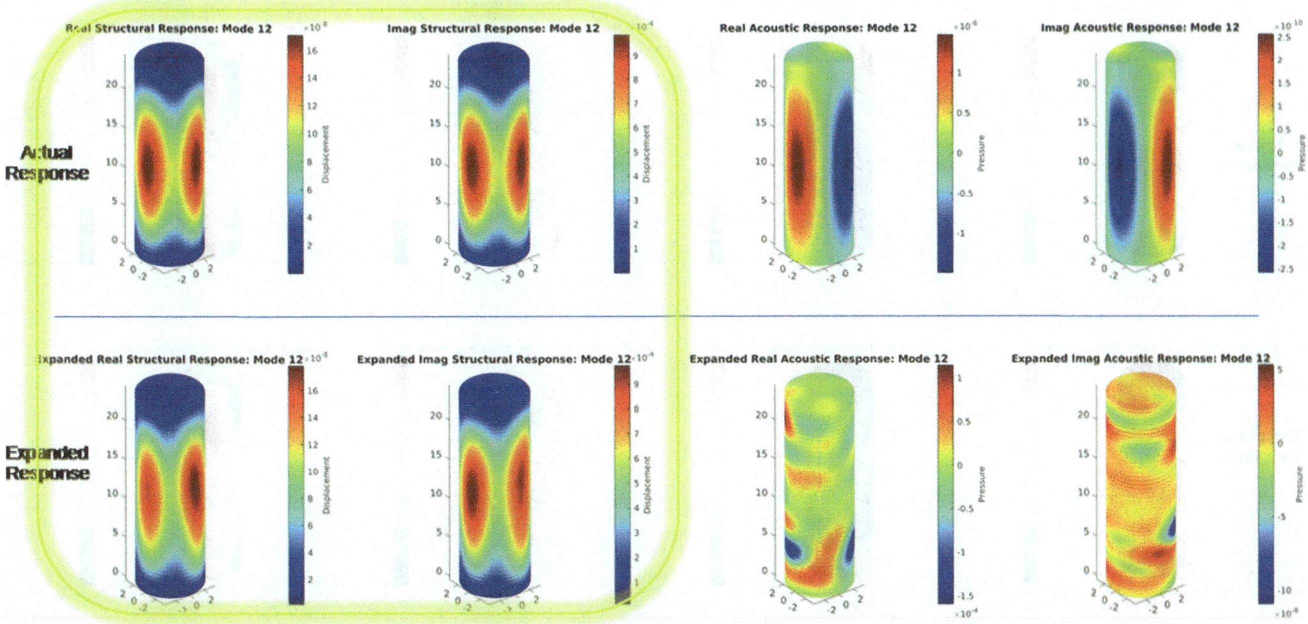

Fig. 12.6 Structural mode expansion (mode 12, 1047 Hz) from left to right: real structural mode, imaginary structural mode, real acoustic mode, and imaginary acoustic mode

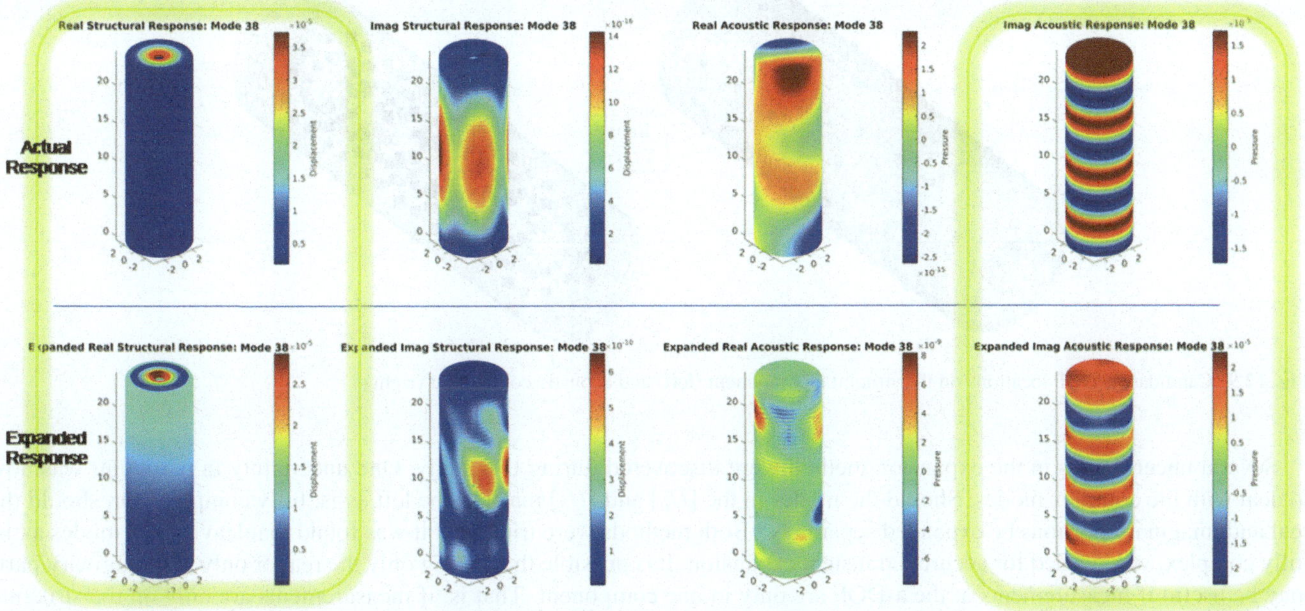

Fig. 12.7 Acoustic mode expansion (mode 38, 1969 Hz) from left to right: real structural mode, imaginary structural mode, real acoustic mode, and imaginary acoustic mode

This expansion problem, just like all expansion problems, was sensitive to mode selection and truncation. Through preliminary expansion simulations, it was observed that omitting the rigid body modes produced an improved solution. In addition, preliminary analysis showed that modes up to 2500 Hz (the first 80 modes) were required to get a good match to the actual response at the n-DOF. While there are only a few structural modes in this bandwidth, there are many acoustic modes and coupled structural-acoustic modes. As such, a high number of modes were required. For other coupled systems with fewer acoustic modes, fewer total system modes and thus fewer a-DOF may be required.

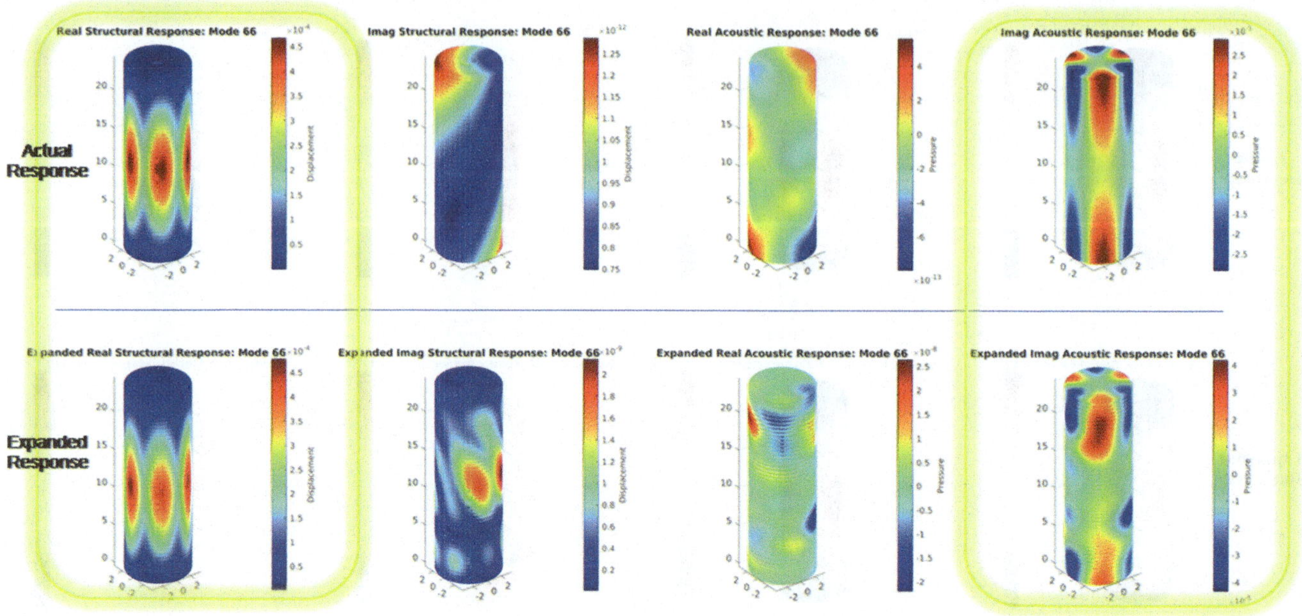

Fig. 12.8 Coupled mode expansion (mode 66, 2608 Hz) from left to right: real structural mode, imaginary structural mode, real acoustic mode, and imaginary acoustic mode

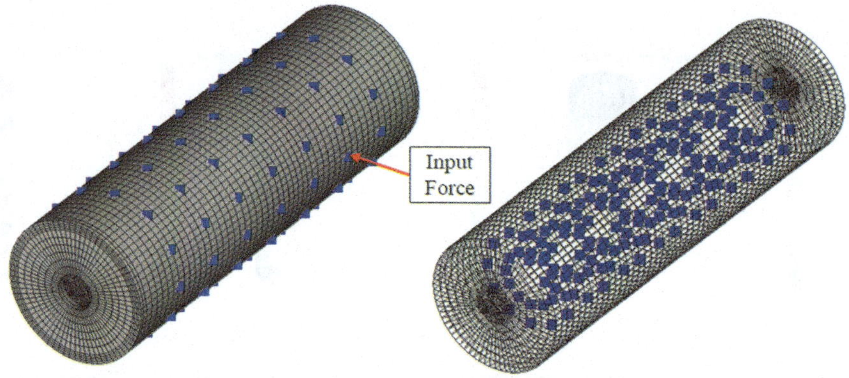

Fig. 12.9 Candidate a-DOF locations on the structural component (left) and acoustic component (right)

Several uncertainties in the expansion method were discovered during this study. One uncertainty in particular was how to deal with the complex modes. Should the modes in the $[U_a]$ and $[U_n]$ matrices be left as-is, fully complex? Or, should the real and imaginary portions be expanded separately? Both methods were tried, and it was found that leaving the modes as-is, fully complex, was needed for accurate transient expansion. It is possible that taking only the real or only the imaginary parts may be useful if measurements at the a-DOF are only in one component. That is, if measurements are only on the structure and the expansion is to some acoustic DOF, then the $[U_a]$ matrix may benefit by being forced to be purely real or imaginary depending on the coupled nature of each mode. This is a possible avenue for future research.

Expansion of the transient response to a forced impulse was calculated by using all modes but the rigid body modes, and calculating the transformation matrix with the complex mode shapes. The expansion results at two n-DOF are compared to the actual simulated responses in terms of the time response in Figs. 12.10 and 12.11. Two of the structure and two of the acoustic DOFs are shown for brevity; the results from these DOFs is representative of overall quality of the expansion. Although differences are seen in both comparison plots, the expanded time responses follow the same general trends as the actual time responses.

The magnitude of the linear spectra of the time responses were also compared and are shown in Figs. 12.12 and 12.13. The spectra comparison plot shows that the expanded responses deviate more at higher frequencies, which could be an indicator of modal truncation. Perhaps the addition of more modes, or a more logical selection of modes could improve the expansion results.

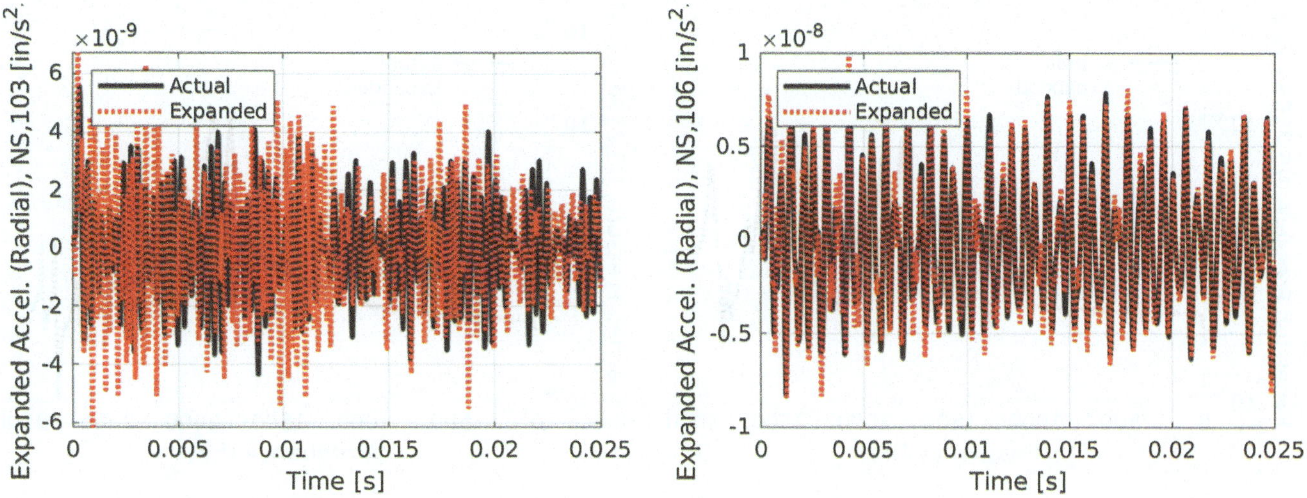

Fig. 12.10 Expanded time responses at two example structural DOF

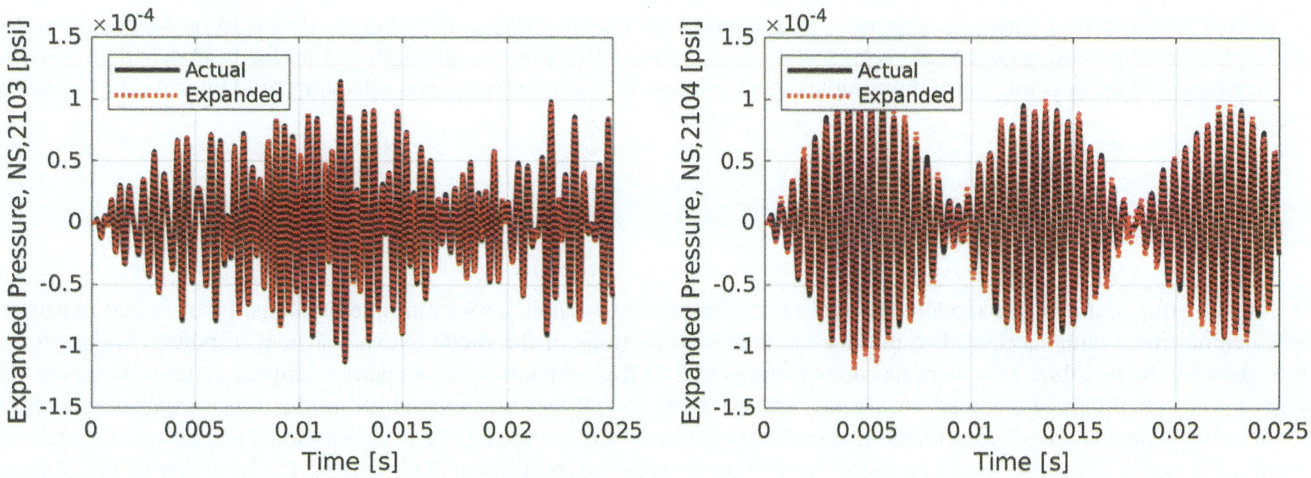

Fig. 12.11 Expanded time responses at two example acoustic DOF

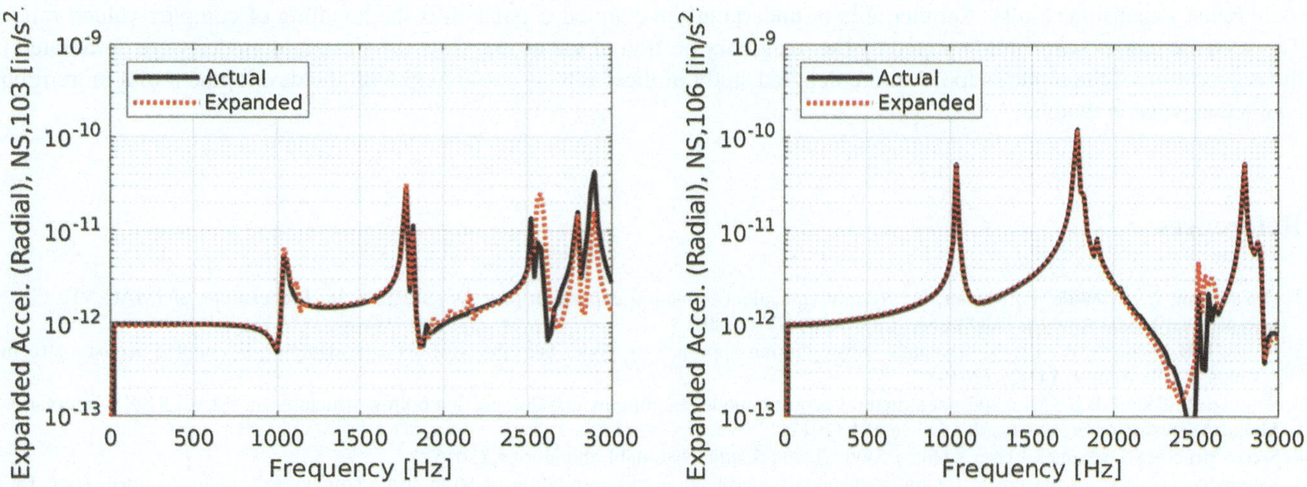

Fig. 12.12 Expanded time responses at two example structural DOF converted to linear spectrum magnitude

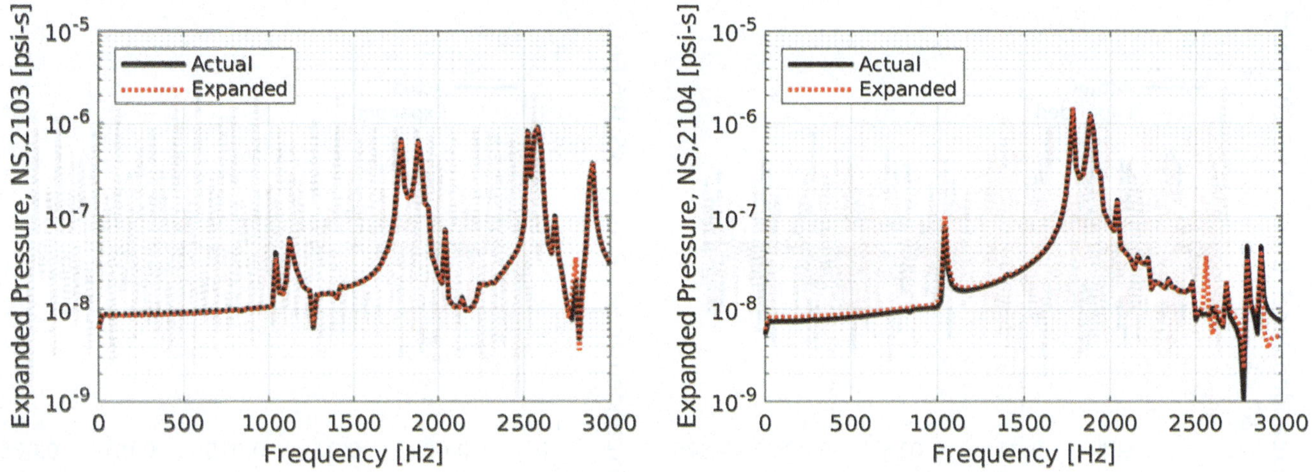

Fig. 12.13 Expanded time responses at two example acoustic DOF converted to linear spectrum magnitude

SEREP expansion of transient responses on a coupled structural-acoustic system was shown to be fairly successful. The expanded time and spectra were representative of the actual responses, although not perfectly matched. Additional understanding of the best way to handle complex mode expansion and improved mode selection could improve the expansion capability.

12.4 Discussion and Conclusions

A coupled structural-acoustic model was used to demonstrate expansion across multiple domains. First, modal expansion was demonstrated with successful results for the dominant portions of the mode. Next, transient response expansion was also shown to be possible. This work has demonstrated that SEREP expansion works across coupled systems, however, this work has also highlighted the need for a better understanding of the expansion variables. In the modal expansion analysis, modal truncation could have contributed to the differences between the expanded and actual modal responses. A systematic method for modal selection would be useful in for future expansion processes. Additionally, the inclusion of insignificant modes in the expansion process could be reducing the accuracy of the expansion. Perhaps only including acoustic dominated modes for the acoustic domain expansion and only including structurally dominated modes for structural expansion could yield better expansion results. Another area of uncertainty in coupled expansion is the handling of complex-valued modes. The acoustic modes seemed dominated by the imaginary portion of the mode while the structural modes were dominated by the real portion of the mode shape. A more detailed study of these effects could help with the development of a more robust coupled expansion method.

References

1. O'Callahan, J., Avitabile, P., Riemer, R.: System equivalent reduction expansion process (SEREP). In: Proceedings of IMAC VII, the 7th International Modal Analysis Conference, Las Vegas, NV (1989)
2. Davis, R.B., Schultz, R.: Using a dynamic substructuring approach to model the effects of acoustic damping in coupled acoustic-structure systems. J. Vib. Acoust. **141**(2), (2019)
3. Schultz, R., Pacini, B.R.: Mitigation of structural-acoustic mode coupling in a modal test of a hollow structure. In: IMAC-XXXV International Modal Analysis Conference, Garden Grove, CA (2017)
4. Sierra Structural Dynamics – User's Notes. Sierra Team, Sandia National Laboratories, (2015)
5. Kammer, D.C.: Sensor placement for on-orbit modal identification and correlation of large space structures. J. Guid. Control. Dyn. **14**(2), 251–259 (1991)

Chapter 13
Expansion Methods Applied to Internal Acoustic Problems

Ryan Schultz and Dagny Joffre

Abstract Expansion techniques have been used for many years to predict the response of un-instrumented locations on structures. These methods use a projection or transformation matrix to estimate the response at un-instrumented locations based on a sparse set of measurements. The transformation to un-instrumented locations can be done using modal projections or transmissibilities. Here, both expansion methods are implemented to demonstrate that expansion can be used for acoustic problems, where a sparse set of pressure measurements, say from a set of microphones in a cavity or room, are used to expand and predict the response at any location in the domain. The modal projection method is applied to a small acoustic cavity, where the number of active modes is small, and the transmissibility method is applied to a large acoustic domain, where the number of active modes is very large. In each case, expansion is shown to work well, though each case has its benefits and drawbacks. The numerical studies shown here indicate that expansion could be accurate and therefore useful for a wide range of interior acoustic problems where only sparse measurements are available, but full-field information is desired, such as field reconstruction problems, or model validation problems.

Keywords Expansion · Modal projection · Interior acoustics · Frequency response function · Transmissibility

13.1 Introduction

Expansion is a process where sparse measurements are used to infer the response at other locations in the domain. Most typically, this is done with structures using accelerometer measurements at various locations on the structure and the System Equivalent Reduction Expansion Process (SEREP) to expands from those sparse accelerometer measurements to other locations of interest on the structure [1]. Often, a model of the structure is used to create the two mode shape matrices needed for SEREP. The first shape matrix is at the measured response locations, here called the a-set degrees of freedom (DOFs).

The second shape matrix is at the expanded response locations. The expansion could be to all locations, the n-DOFs, or to a subset of other locations, the b- or c-DOFs. The shape matrixes need to have enough modes to span the space of the content in the measured responses and the number of a-set DOFs must be greater than the number of modes in the matrix. As such, if the response is broadband or involves many active modes, many a-set DOFs are required.

Just as expansion is useful for structural problems, expansion could be useful for many types of acoustic systems. For example, expansion could be used to take a small number of microphone measurements in the inside of an aircraft fuselage and provide the acoustic field at all locations within the fuselage. Expansion could also be used to determine the as-tested pressures on a test article in an acoustic environmental test. In that case, the acoustic domain would be the test chamber, the measured a-set DOFs would be a small number of microphones in the chamber, and the expanded, b-set DOFs would be the points on the surface of a test article. In that way, the tested pressure loads on the test article could be known in a

Sandia National Laboratories is a multimission laboratory managed and operated by National Technology and Engineering Solutions of Sandia, LLC., a wholly owned subsidiary of Honeywell International, Inc., for the U.S. Department of Energy's National Nuclear Security Administration under contract DE-NA0003525.

R. Schultz (✉) · D. Joffre
Sandia National Laboratories, Albuquerque, NM, USA
e-mail: rschult@sandia.gov

B. Dilworth (ed.), *Topics in Modal Analysis & Testing, Volume 8*, Conference Proceedings of the Society for Experimental Mechanics Series, https://doi.org/10.1007/978-3-030-47717-2_13

full-field sense. These use cases motivated this work, which uses simulations of big and small systems to exercise expansion of acoustic domains and assess how typical structural expansion techniques may be used for a range of acoustic problems.

A span of acoustic domain sizes represents a span of modal density regimes. A small interior acoustic domain, such as an automobile cabin or the payload bay of a rocket, perhaps may only have tens of modes in a typical bandwidth of interest (e.g. below 2 kHz). Conversely, a large interior acoustic domain, such as a large auditorium or reverberation test chamber, can have tens of thousands of modes in the same bandwidth. Thus, these represent very different types of dynamic systems which require different expansion approaches. For the small domain with a small number of active modes, SEREP could be very practical; the number of a-set DOFs (measurements) is tractable. However, the large domain, with a very large number of active modes cannot use SEREP expansion as the number of a-set DOFs would be impractical. Here, expansion of a large domain is accomplished instead with a transmissibility approach wherein the frequency response function (FRF) matrices are formed between the a-set DOF outputs and inputs as well as between the b-set DOF outputs and inputs. Then, the transmissibility matrix relating the b-set DOF outputs to the a-set DOF outputs is formed. In that way, the measured a-set DOF outputs can be used to expand to the b-set DOF outputs. With a transmissibility approach, there only needs to be at least as many a-set DOF as independent inputs, so it can work in very high modal density regimes.

In this paper, the theory of these two expansion approaches will be presented briefly. Then, SEREP will be used to expand acoustic responses inside a model of a cavity of a shell test article, which is an example of a small domain with low modal density. Next, transmissibilities will be used to expand acoustic responses in a model of a reverberation chamber, which is an example of a large domain with high modal density. Unfortunately, no test data was available at the time of this writing to demonstrate these techniques with actual measurements but results of these numerical example problems are promising and should motivate an experimental demonstration in the future.

13.2 Theory

SEREP expansion uses a modal projection to convert a set of measured DOFs, the a-set DOFs, to a set of expanded DOFs, the n-set DOFs. This is accomplished using a transformation matrix, $[T]$:

$$\{x_n(t)\} = [T]\{x_a(t)\}. \tag{13.1}$$

The transformation matrix is formed using a set of mode shape matrices, $[U_a]$, $[U_n]$, at the a- and n-DOFs, respectively. A pseudo-inverse of the $[U_a]$ mode shape matrix projects the measured a-set DOFs into modal space, giving the modal responses. Those modal responses are then projected back to physical space at the n-set DOFs through the $[U_n]$ mode shape matrix. The SEREP transformation matrix is:

$$[T] = [U_n][U_a]^+, \tag{13.2}$$

where the superscript $[\cdot]^+$ indicates a pseudo-inverse of the shape matrix. While Eq. 13.1 shows expansion of time responses, frequency domain quantities can also be expanded in the same way. For example, linear spectra at a-DOFs, $\{X_a\}$, can be expanded to n-DOFs, $\{X_n\}$, with an analogous equation:

$$\{X_n(\omega)\} = [T]\{X_a(\omega)\}. \tag{13.3}$$

Expansion can also be achieved using transmissibility functions, which are a ratio of two FRFs. The first FRF relates the measured responses at the a-DOFs to the input. This FRF is computed (if using the H1 FRF estimator) based on the CPSD of the outputs and inputs, $[S_{ai}]$, and the CPSD of the inputs, $[S_{ii}]$:

$$[H_{ai}] = [S_{ai}]\left[S_{ii}^+\right]. \tag{13.4}$$

The FRFs at the n-DOFs can be similarly computed, giving $[H_{ni}]$ With the two FRFs relating outputs at a- and n-DOFs to some inputs i, the transmissibility matrix can be formed which relates the response at the n-DOFs to the response at the a-DOFs [2]:

$$[T_{na}] = [H_{ni}][H_{ai}]^+. \tag{13.5}$$

With the transmissibility matrix, $[T_{na}]$, the measured response at the a-DOFs can be used to estimate the response at the n- DOFs with:

$$\{X_n\} = [T_{na}]\{X_a\}. \tag{13.6}$$

Note that the number of a-DOFs required for SEREP must be equal to or greater than the number of modes in $[U_a]$ and $[U_n]$. Similarly, there is a minimum number of a-DOFs for the transmissibility expansion, where the number of a-DOFs must be equal to or greater than the number of inputs to the system. Also, with the transmissibility approach, the input locations and directions must be known and consistent. That is, the input locations used to determine the transmissibility matrix must be the same as the actual input locations in any measurement used to expand. If the input locations change, the results will not be valid. These techniques are general to any linear system and here are simply applied to acoustic systems to demonstrate their usefulness for these types of problems.

13.3 SEREP Expansion for Acoustic DOF in a Small Domain Model

Acoustic finite element models were created for the small shell cavity and the large reverberation chamber. The shell cavity model is based on a piece of test hardware shown in Fig. 13.1. The interior air cavity is cylindrical with a 7 inch diameter and 24 inch length. The model was meshed using hexahedral elements with the element size of 0.5 inch chosen to give accurate results to at least 2 kHz. A 1 ms haversine input is provided at the edge of one end of the cavity as an acoustic velocity on a set of element faces, shown in Fig. 13.1. The pressure response due to this input is measured at the a-set DOF and then used to expand to the other DOF in the cavity. The first task was to determine the a-set DOF which can best measure the response due to a set of modes in the chosen bandwidth of 1 kHz. This frequency range includes the first 20 modes, so at least 20 a-set.

DOF are needed. Here 25 DOF were chosen from a set of 168 candidate locations using the effective independence (EFI) algorithm [3]. The candidate locations and the chosen a-set locations are shown in Fig. 13.2.

SEREP expansion of this acoustic domain is demonstrated two ways. First, the pressure modes of the cavity are expanded. In this case, the cavity modes from the finite element model at the a-set DOF were used to expand to all the DOF in the model. To make this expansion of "perfect" model data more realistic, random noise was added to the a-set "measured" shapes. As shown in Fig. 13.3, the expansion of the acoustic pressure mode shapes of the cavity worked properly with this set of modes and a-set DOF. The expanded mode shapes are nearly identical to the actual mode shapes at all the DOF in the cavity.

Next, expansion of transient pressure response is demonstrated by simulating the transient response of the cavity due to the 1 ms haversine input at the corner of the cylinder. Again, some random noise was added to the measured a-DOF response to simulate the effects of having imperfect response data. While the response of all the DOF could be used in the expanded b-set DOF as in the mode shape expansion case, just two nodes specified in Fig. 13.4 were used in the b-set to simplify visualization. The expansion results are shown in terms of the time response, in Fig. 13.5, and the power spectral density

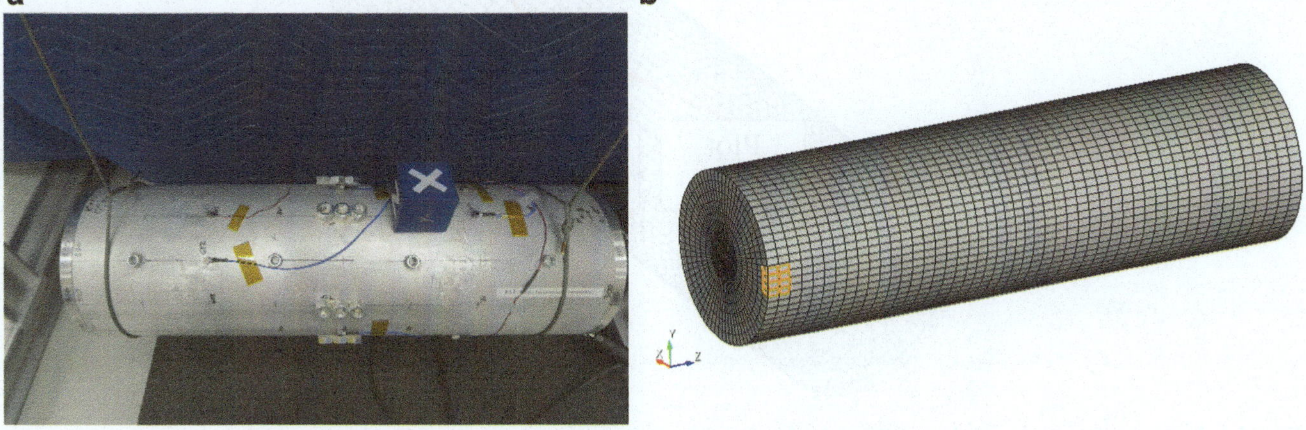

Fig. 13.1 Shell test article hardware (left) and finite element mesh of the interior acoustic cavity (right)

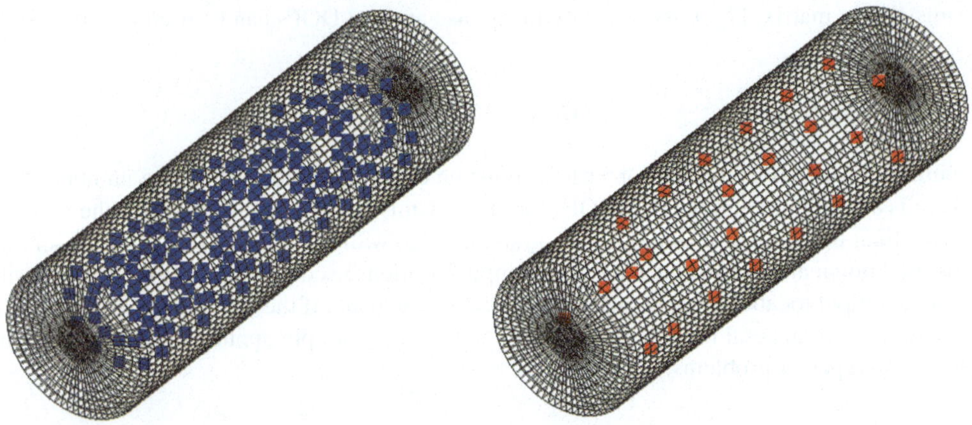

Fig. 13.2 168 Candidate measurement DOF (left) and 25 a-set DOF selected with EFI (right)

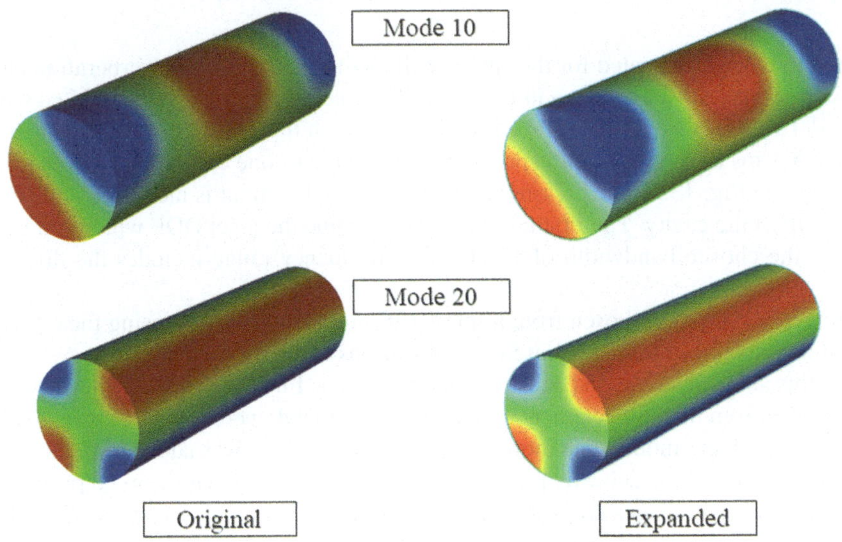

Fig. 13.3 Example mode shapes of the shell cavity. Left: Actual mode shapes. Right: Expanded mode shapes

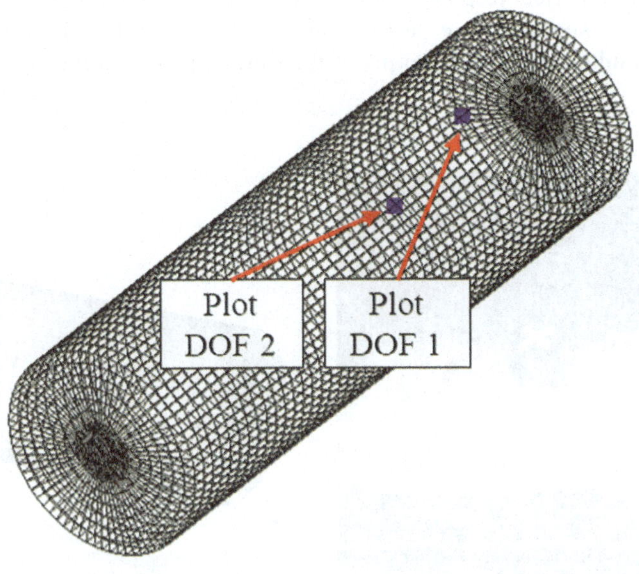

Fig. 13.4 Locations of two expanded b-set DOF in the interior of the cavity

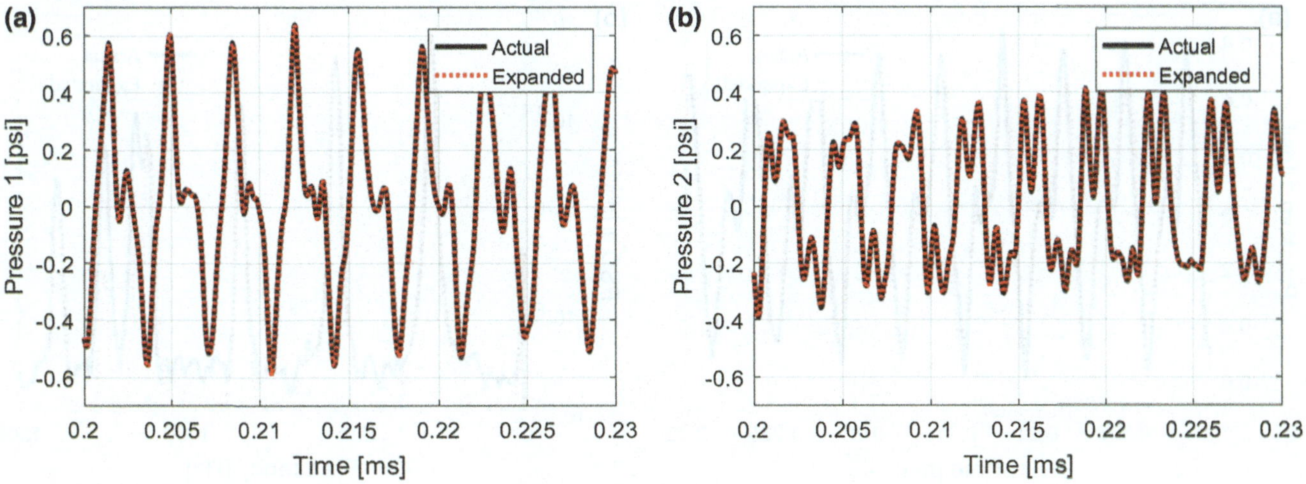

Fig. 13.5 Time expansion shown as time histories at the two expanded b-set DOF

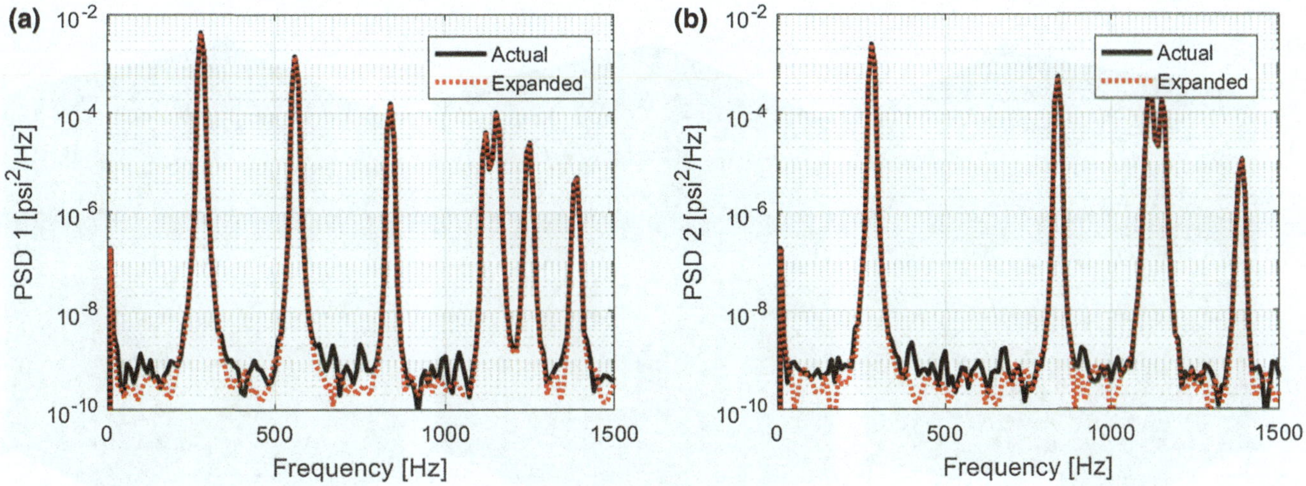

Fig. 13.6 Time expansion shown as PSDs at the two expanded b-set DOF

(PSD) of that time response, in Fig. 13.6. Using either metric, the expanded response of the two b-set DOF in this acoustic domain match very well with the actual response, indicating that SEREP is a useful method for acoustic expansion of not only mode shapes, but transient responses as well. A typical artifact of mode-based expansion methods is mode truncation errors that are caused by not including enough modes in the mode shape matrices used to make the SEREP transformation matrix. The effects of mode truncation are demonstrated in Fig. 13.7 where just three modes were used to populate the transformation matrix. This is exactly the expected behavior, with the expanded response matching well at low frequencies and not as well at higher frequencies.

The results from SEREP expansion of the acoustic response in terms of modes and time responses show that SEREP is effective for internal acoustic problems as well as structures. While this isn't surprising, it is useful to see the technique demonstrated with this atypical application in mind. Just like with structural system expansion, acoustic system expansion is sensitive to problem setup considerations such as mode truncation, as demonstrated here, and also factors such as a-set DOF and mode selection. For systems with a small number of active modes, such as some automobile cabins or small rocket payload sections, this expansion technique could be effective in generating full-field responses from a small number of measurements.

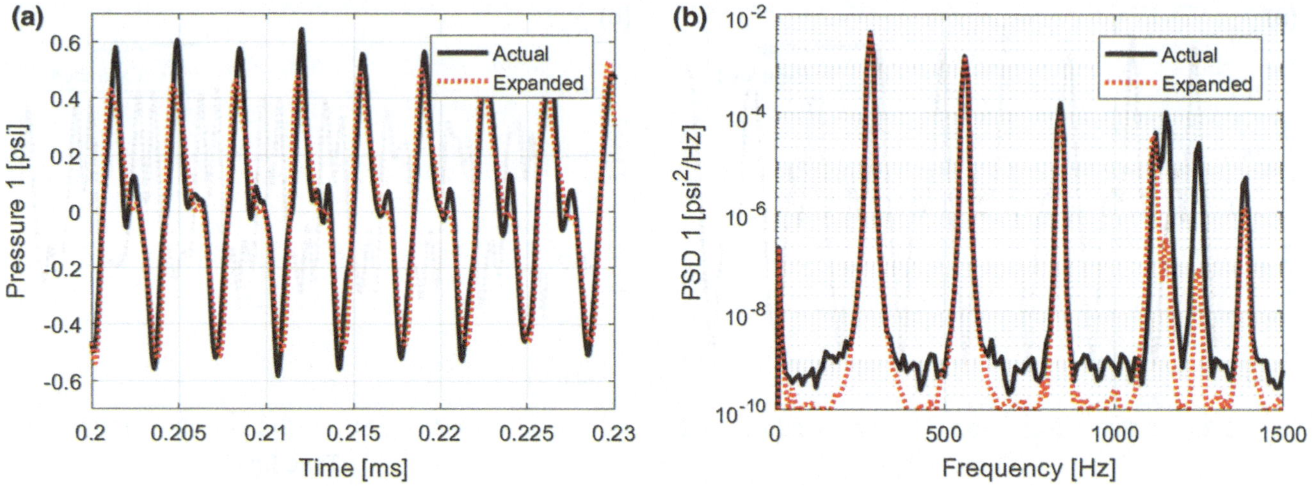

Fig. 13.7 Example expanded response results with mode truncation, using only modes 1,2, and 3 in the expansion. Left: Time response at one b-set DOF. Right: PSD at one b-set DOF

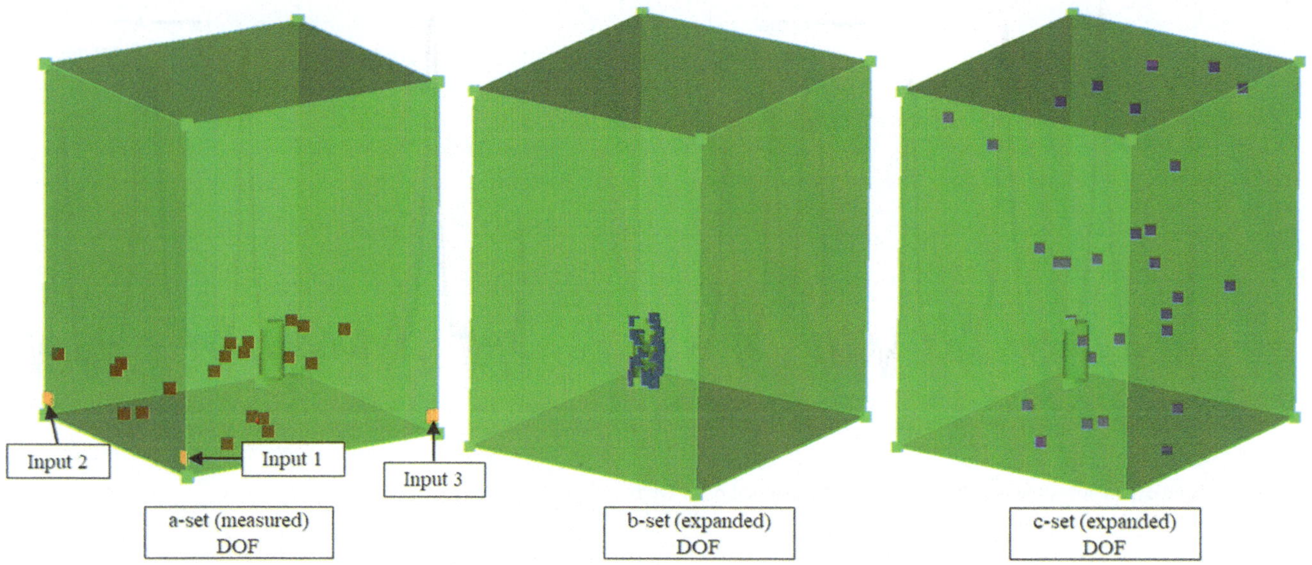

Fig. 13.8 Acoustic chamber model showing the three input locations and a-set DOF (left), b-set DOF located on the surface of a cylindrical test article (center) and c-set DOF distributed throughout the chamber volume (right)

13.4 Transmissibility Expansion for Acoustic DOF in a Large Domain Model

Expansion of a large domain using transmissibilities is demonstrated using a model of a 21x25x30 foot reverberation chamber. Due to the size of this domain, the element size was limited to 4 inches, which limits the maximum frequency of the analysis to around 500 Hz. Three independent inputs were provided by surface velocities at small patches in three of the bottom corners of the chamber model, which could represent three loudspeaker inputs, as shown in Fig. 13.8. Locations of 20 a-set DOF are shown as red dots in the left image of Fig. 13.8. These locations were chosen at random in this case, rather than with the EFI algorithm. Two sets of expansion DOF were chosen, representing two different use cases. The first, the b-set DOF, represents points on the wetted surface of a cylindrical test article located in one corner of the chamber, useful for cases where the as-tested pressure on a test article is desired. The second, the c-set DOF, are just random points throughout the entire chamber domain, which provides an indication of the quality of the expansion over the entire space. These two sets of expansion DOF are shown in the center and right images in Fig. 13.8.

The chamber has modes, just like the shell cavity, which can be viewed in terms of a pressure mode shape. Figure 13.9 shows two such modes of the chamber. The difference between the shell cavity and this chamber is the modal density. Where

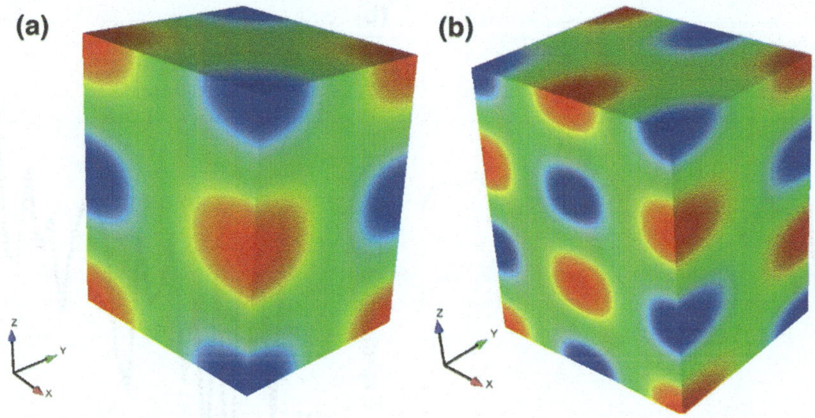

Fig. 13.9 Example pressure mode shapes of the acoustic chamber

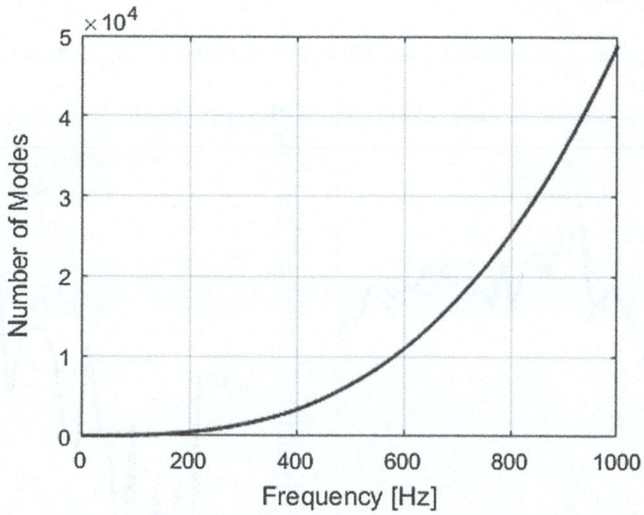

Fig. 13.10 Modal density of the chamber shown as the number of modes below each frequency from 0 to 1000 Hz

the shell cavity has just 4 modes below 1000 Hz, the chamber has more than 40,000 modes, as shown in Fig. 13.10. This extreme number of modes is what motivated the use of transmissibilities instead of a modal projection for expansion of this system.

Three independent, white noise inputs were applied at the three surface patches in terms of an acoustic velocity and used in a transient simulation of the chamber model to obtain the pressure response at the a-, b-, and c-DOFs. Next, the pressure response/input surface velocity FRFs were created using time responses of the pressure at the a-, b-, and c-DOF and the applied surface velocity. To get accurate FRF estimates, 30 s of simulated response was needed. With approximately 0.1% Rayleigh damping assigned to the air in the model, it took around 4-5 s for the levels to become stationary in the chamber. Also, it was observed that the FRF estimates were sensitive to the block size used. With a short block size, such as 1 or 2 s, the FRFs were not converged and resulted in inaccurate expanded responses. Five to seven seconds was found to give acceptable results in this case, though further study of this phenomenon is needed. The FRFs were used to create transmissibility matrices relating the response at the b- and c- DOFs to the a-DOFs due to these three input locations. Next, the PSD of the a-DOF response of this transient simulation was used along with the transmissibility matrices to expand to the b- and c-DOF. Results of expansion of two example b- and c-DOF are shown in Figs. 13.11 and 13.12. The response at all these locations is matched very well over this frequency range. So, with 20 a-set DOF this transmissibility-based expansion works well. Next, the a-set was reduced to determine how well the expansion works with a very reduced set of DOF. Figure 13.13 shows expansion results at one b-set DOF and one c-set DOF using just four a-set DOF. Even with the number of a-set DOF being thousands of times less than the number of modes, this expansion works properly. It should be noted that three a-set DOF was also used, with decent results, though increasing to four provided more accurate results.

The similarities in the expanded and actual response in Figs. 13.11, 13.12, and 13.13 highlight the benefit of using transmissibility for expansion of this high modal density system. Often the acoustic pressure on a test article during an

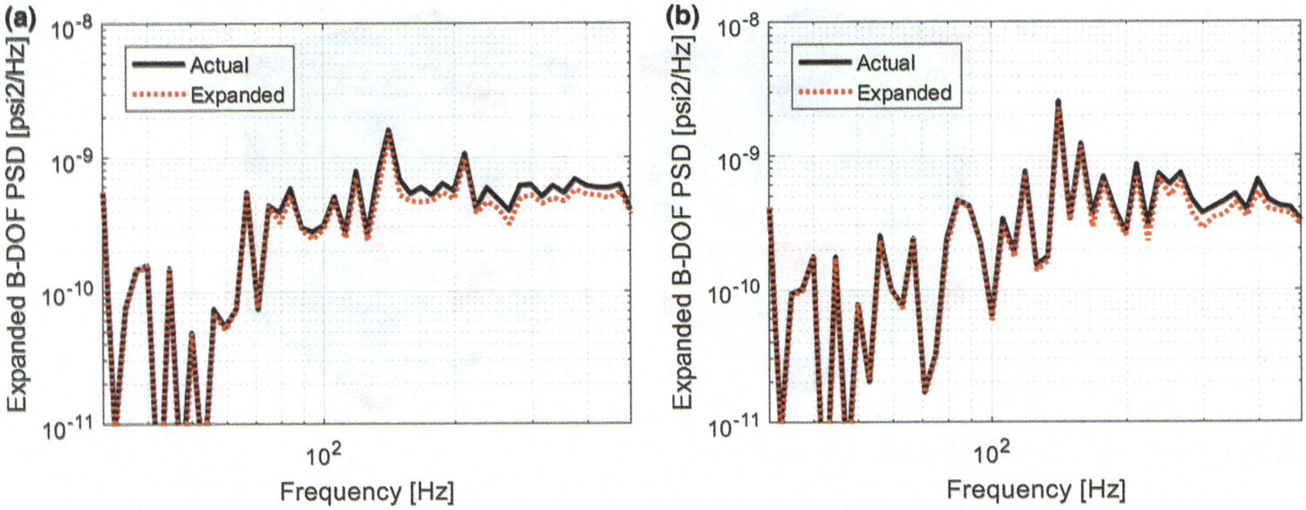

Fig. 13.11 Actual and expanded response at the expanded b-set DOF located on the surface of a cylindrical test article, shown in 12th octaves

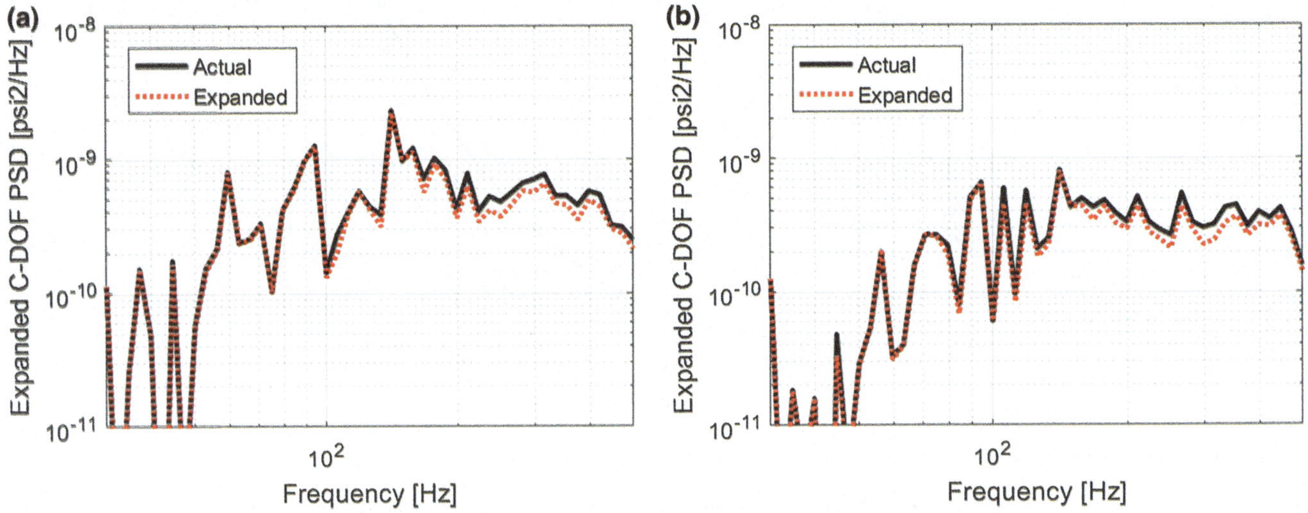

Fig. 13.12 Actual and expanded response at the expanded c-set DOF distributed throughout the chamber volume, shown in 12th octaves

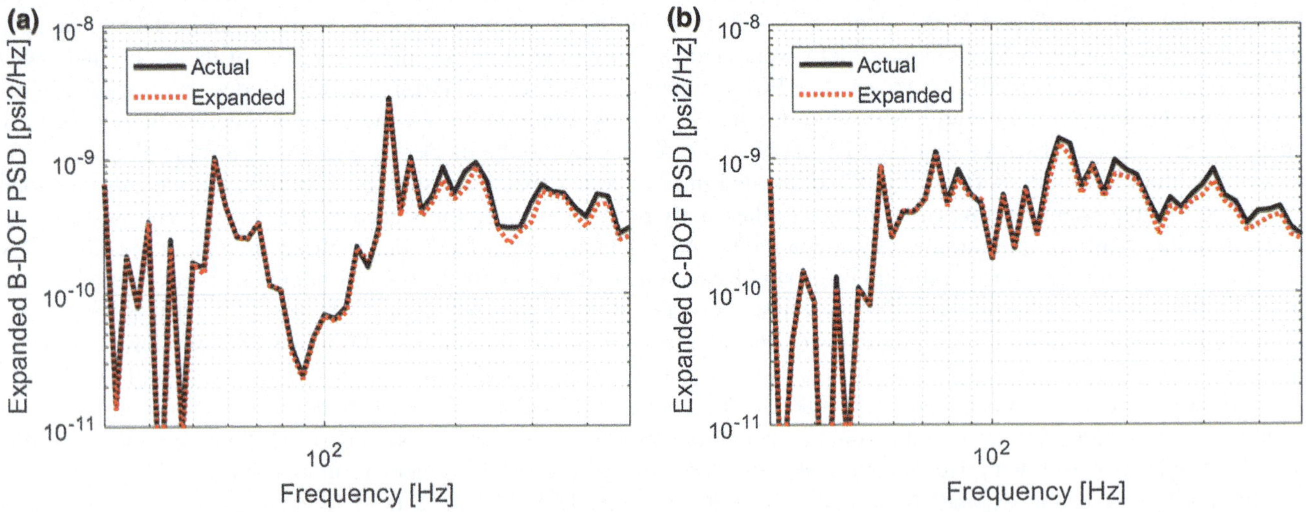

Fig. 13.13 Response of one b-set DOF (left) and one c-set DOF (right) using expansion with just 4 a-set DOF

acoustic environment test in a reverberation chamber is desired but cannot be measured. This transmissibility expansion technique was able to take a very small number of measurements and accurately expand to many locations on the surface of a test article or anywhere else in the chamber.

13.5 Discussion and Conclusions

Expansion techniques can make limited measurements much more useful, providing response estimates at any location on the structure, or as shown here, in the acoustic domain. Modal projection approaches, such as SEREP, can work very well provided there are enough a-set DOF for the active modes in the response. SEREP allows for shapes, time responses, or frequency quantities such as linear spectra to be expanded with no knowledge of the input forces (or their locations) required. As shown in the shell cavity example here, SEREP is also effective for interior acoustic problems with low modal density.

For problems with very large numbers of active modes, such as the reverberation chamber model used in this work, SEREP is not appropriate as the number of measurements (a-set DOF) is impractically large. Instead, transmissibilities could be used for expansion of these types of systems. The benefit of a transmissibility approach is that a much smaller number of a-set DOF are required, because the number of a-set DOF only needs to be as large as the number of independent inputs to the system. In the example shown here, four a-set DOF were used to expand response of a three-input system with thousands of modes in the bandwidth. The downside of the transmissibility approach is that the input DOF must be known.

Overall, this work provided two simple examples which represent two different, but typical, acoustic systems for which expansion would be a useful tool. The two expansion methods demonstrated here each provided accurate response estimates, indicating expansion of acoustic domains is possible, and behaves just like the expansion of structures. Future efforts will try to experimentally validate these findings using tests of small and large acoustic domains.

References

1. O'Callahan, J., Avitabile, P., Riemer, R.: System equivalent reduction expansion process (SEREP). In: Proceedings of IMAC VII, the 7th International Modal Analysis Conference, Las Vegas, NV (1989)
2. Ewins, D.J.: Modal Testing: Theory, Practice and Application, vol. 2. Research Studies Press Ltd, Hertfordshire (2000)
3. Kammer, D.C.: Sensor placement for on-orbit modal identification and correlation of large space structures. J. Guid. Control. Dyn. **14**(2), 251–259 (1991)

Chapter 14
Scaling an OMA Modal Model of a Wood Building Using OMAH and a Small Shaker

Osama Abdeljaber, Michael Dorn, and Anders Brandt

Abstract Operational modal analysis, OMA, results in unscaled mode shapes, since no forces are measured. Yet, obtaining a scaled modal model, i.e. knowing the modal mass of each mode (assuming proportional damping), is essential in many cases for structural health monitoring and load estimation. Several methods have therefore recently been developed for this purpose. The so-called OMAH method is a recently developed method for scaling OMA models, based on harmonic excitation of the structure. A number of frequencies are excited, one by one, and for each frequency, one or more frequency response values are calculated, that are then used for estimation of the modal masses of each mode, and residual effects of modes outside the frequency of interest. In the present paper, measurements were made on a four-story office building which was excited with a small, 200 N sine peak electrodynamic shaker. It is demonstrated that this small shaker was sufficient to excite the building with a force level of approx.. 1.8 N RMS close to the first eigenfrequency of the building, which was sufficient to produce harmonic response across the building. Reliable modal masses were possible to obtain within an accuracy of 6%. This demonstrates the feasibility of the OMAH method.

Keywords Operational modal analysis · OMAH method · Mode shape scaling

14.1 Introduction

Operational modal analysis (OMA) is a powerful technique for extracting modal properties (i.e. natural frequencies, mode shapes, and damping ratios) from the ambient vibration response of a dynamic system. Unlike experimental modal analysis (EMA), OMA is suitable for testing large civil structures since it does not require measuring the forces acting on the structure. This, however, leads to unscaled mode shapes that are inadequate for many structural health monitoring, model calibration, and vibration mitigation applications [1].

Several methods have been proposed for scaling of OMA mode shapes, some of which require carrying out OMA repeatedly with different mass and stiffness layouts [2–4]. Another approach, referred to as OMA with exogenous input (OMAX), requires applying a known broadband excitation on the structure [5, 6]. A third approach is to create an accurate finite element (FE) model of the structure and rely on the mass matrix of the model to scale the mode shapes [7]. All these methods are interesting and applicable for certain cases. However, the following practical issues can be identified here:

1. Conducting OMA with several mass/stiffness configurations can be laborious and time-consuming. Furthermore, the additional masses required for large structures can be infeasible [4].
2. Carrying out OMAX on large structures requires large and expensive actuators to apply the required level of broadband excitation.
3. The third approach is highly dependent on the accuracy of the FE model.

In an attempt to overcome these drawbacks, the so-called OMAH method was recently introduced [8]. OMAH is similar to the aforementioned OMAX method except that it employs mono-harmonic excitations. The main advantage of OMAH is

O. Abdeljaber · M. Dorn
Department of Building Technology Faculty of Technology, Linnaeus University, Växjö, Sweden
e-mail: osama.abdeljaber@lnu.se; michael.dorn@lnu.se

A. Brandt (✉)
Department of Technology and Innovation, University of Southern Denmark, Odense M, Denmark
e-mail: abra@iti.sdu.dk

B. Dilworth (ed.), *Topics in Modal Analysis & Testing, Volume 8*, Conference Proceedings of the Society for Experimental Mechanics Series, https://doi.org/10.1007/978-3-030-47717-2_14

that narrowband harmonic excitations can be readily generated using inexpensive shakers [9]. Also, as demonstrated in [8], signal processing techniques can be utilized to extract the vibration response under harmonic excitations from the measured signals even when the signal-to-noise (SNR) ratio is very low. This means that a small shaker is potentially sufficient for carrying out OMAH on large structures.

So far, OMAH has been tested only on relatively small structures (a staircase and a helipad) [8, 10]. Therefore, the aim of the study presented in this paper is to investigate the feasibility of employing a small and inexpensive shaker for scaling the mode shapes of a large civil structure using OMAH.

14.2 OMAH Theory

The frequency response function FRF between the displacement at degree-of-freedom (DOF) p and a force at DOF q (i.e. the receptence between p and q) can be written as:

$$H_{p,q}\left(j\omega\right)=\sum_{r=1}^{N_m}\frac{\psi_r^p\psi_r^q}{m_r\left(j\omega-s_r\right)\left(j\omega-s_r^*\right)} \tag{14.1}$$

where r is the mode number, N_m is the total number of modes considered, ω is the angular frequency, m_r is the modal mass of mode r, ψ_r^p and ψ_r^q are the mode shape coefficients of mode r at p and q, respectively, and the superscript $*$ indicates complex conjugation. The pole associated with mode r can be expressed in terms of the corresponding eigenfrequency ω_r and the damping ratio ζ_r as follows:

$$s_r=-\zeta_r\omega_r+j\omega_r\sqrt{1-\zeta_r^2} \tag{14.2}$$

Assuming well-separated modes, the FRF around the eigenfrequency ω_r can be estimated as:

$$H_{p,q}\left(j\omega\right)\big|_{\omega\approx\omega_r}\approx\frac{\psi_r^p\psi_r^q}{m_r\left(j\omega_r-s_r\right)\left(j\omega_r-s_r^*\right)} \tag{14.3}$$

The unscaled mode shapes ψ_r along with the poles s_r can be identified using OMA. Scaling OMA modes requires estimating the modal masses m_r, which is the objective of OMAH. To do so, a harmonic excitation $F_q(jw_{\text{ex}})$ is applied on a certain DOF q at some frequency w_{ex} around the eigenfrequency ω_r of the desired mode. The response $X_p(jw_{\text{ex}})$ under this excitation is measured at a certain location p. Note that both q and p should be picked from the DOFs considered in OMA. Next, the receptence at w_{ex} is calculated as $H_{p,\,q}(j\omega_{ex})=X_p(jw_{\text{ex}})/F_q(jw_{\text{ex}})$. The modal mass corresponding to this mode can then be estimated as:

$$m_r\approx\frac{\psi_r^p\psi_r^q}{H_{p,q}\left(j\omega_{\text{ex}}\right)\left(j\omega_{\text{ex}}-s_r\right)\left(j\omega_{\text{ex}}-s_r^*\right)} \tag{14.4}$$

Once the modal masses are identified, the scaled FRF between any two DOFs can be easily computed as given in Eq. (14.1). For more details about OMAH, the reader is referred to [8] and [10], where also more general techniques, that take all modes into account simultaneously, are presented.

14.3 Structure and Instrumentation

The test structure in this study was a four-story wood building located in Växjö, Sweden (Fig. 14.1). The total area of the building is about 5700 m^2 in total, separated into about 3700 m^2 of office space and approximately 2000 m^2 of restaurants and conference rooms. Since July 2018, temperature, humidity, wind speed, displacement, and vibration at multiple points have been continuously monitored using a network of sensors distributed across the building.

Fig. 14.1 The test structure

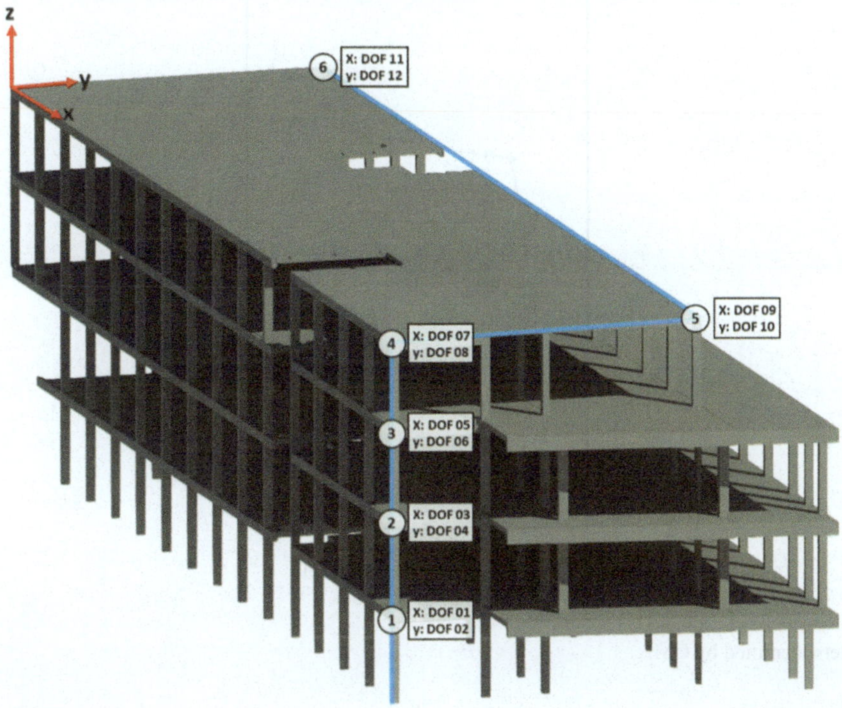

Fig. 14.2 Measurement locations and DOFs

A total of 12 geophones (Sunfull, PS-4.5B) were installed to monitor vibrations at the six locations given in Fig. 14.2. As shown in Fig. 14.3, each pair of geophones was placed inside a small box which was attached to the load bearing structure at the desired location to measure the dynamic response in x and y. Six ADC cards were provided to operate the 12 geophones. The ADC card has four channels, each with a digital filter (Maxim Integrated, MAX7401), a sample/hold amplifier (Texas Instruments, LF298) and a 24-bit A/D converter (Texas Instruments, ADS1255). An amplification circuit was also employed to amplify the geophone outputs. Among the six ADC cards, a single card was configured as "Master" and used to send out synchronization pulses to the other ones.

Additionally, a weather station (Davis Vantage Pro 2) was installed on the roof to measure outdoor temperature, humidity, rain, wind speed and wind direction, and air pressure. A specially designed sensor card was used to acquire data from the weather station.

Fig. 14.3 Installation of geophones. (**a**) Two geophones inside a box. (**b**) The box attached to the structure

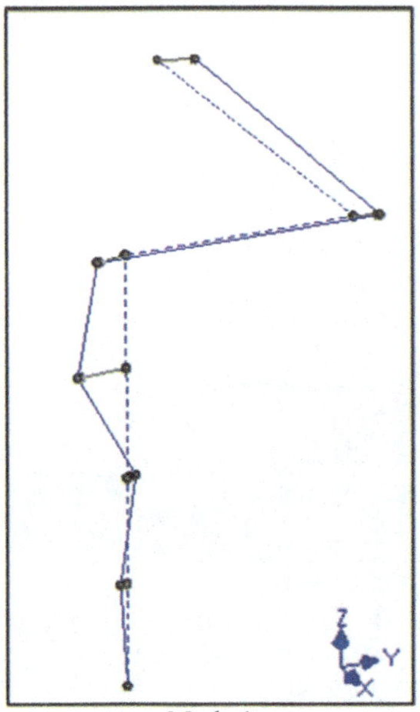

Fig. 14.4 Modal parameters identified by OMA

14.4 Parameter Extraction by OMA

The ambient vibration signals measured by the geophones were utilized to identify the modal properties of the test structure. A total of 189 vibration datasets were examined in terms of both wind speed and operational deflection shape (ODS) quality. As a result, a single dataset was chosen for carrying out OMA. The chosen dataset consists of 12 geophone signals recorded for approximately 61 min at a sampling frequency of 120 Hz.

The geophone signals were calibrated and converted to velocity signals according to the method described in [11]. A high-pass filter with a cutoff frequency of 1 Hz was applied to eliminate unwanted low frequency components. The filtered signals were then downsampled from 120 Hz to 10 Hz, since the frequency range of interest in this work is only up to 5 Hz. Next, a complete correlation matrix was computed by considering the 12 DOFs as references. The resulting correlation matrix was used to estimate the Hankel matrix required for OMA.

OMA was then carried out in Matlab [12] using the function 'ir2pmitd' available in ABRAVIBE Toolbox for Noise and Vibration Analysis [13]. This function uses the Multi-reference Ibrahim Time Domain (MITD) method [14] to extract the modal parameters corresponding to the first mode of the building. The resulting eigenfrequency, damping ratio, and mode shape are presented in Fig. 14.4.

14.5 Mode Shape Scaling by OMAH

To obtain the measurements required for OMAH, a small 200 N sine peak electrodynamic shaker (LDS V406) was used to apply harmonic excitations at DOF 6 (i.e. third floor, +y direction). As shown in Fig. 14.5, the shaker was placed at the corner near location 3 in Fig. 14.2 without mounting it to the floor. An additional mass of 1 kg was attached to the moving part of the shaker, resulting in a total moving mass of 1.2 kg. Therefore, the inertial force exerted by the shaker can be estimated by multiplying the acceleration of the moving part by 1.2 kg. Three accelerometers (Kistler 8772A5) were installed to measure the vibration response of the moving part of the shaker as well as the shaker's trunnion and the wall in the direction of excitation.

Using this setup, the building was harmonically excited at its first natural frequency $\omega_{ex} = 3.3$ Hz with a force level of approximately 1.8 N RMS. The acceleration response was recorded for a duration of 800 s, equivalent to approximately 2640 periods of the eigenfrequency. The sampling frequency was set to 40.96 Hz.

The acquired time-domain signals were used to compute the linear (root mean square, RMS) spectra of the response of the three accelerometers. Fig. 14.6 shows that the acceleration spectra measured at the wall and trunnion during the first OMAH test are in good agreement around the excitation frequency ω_{ex}. This indicates that the dynamics between that trunnion and the wall are insignificant (i.e. the shaker actually excited the building and was not only sliding on the carpet).

The FRF (i.e. the accelerance) $H_{6,\,6}(j\omega)$ between the wall and the moving part of the shaker was measured and then scaled to displacement/force. The value of the FRF at the excitation frequency $H_{6,\,6}(j\omega_{ex})$ was used to scale the mode shape (i.e. estimate the modal mass of the first mode) according to Eq. (14.4).

Fig. 14.5 OMAH setup

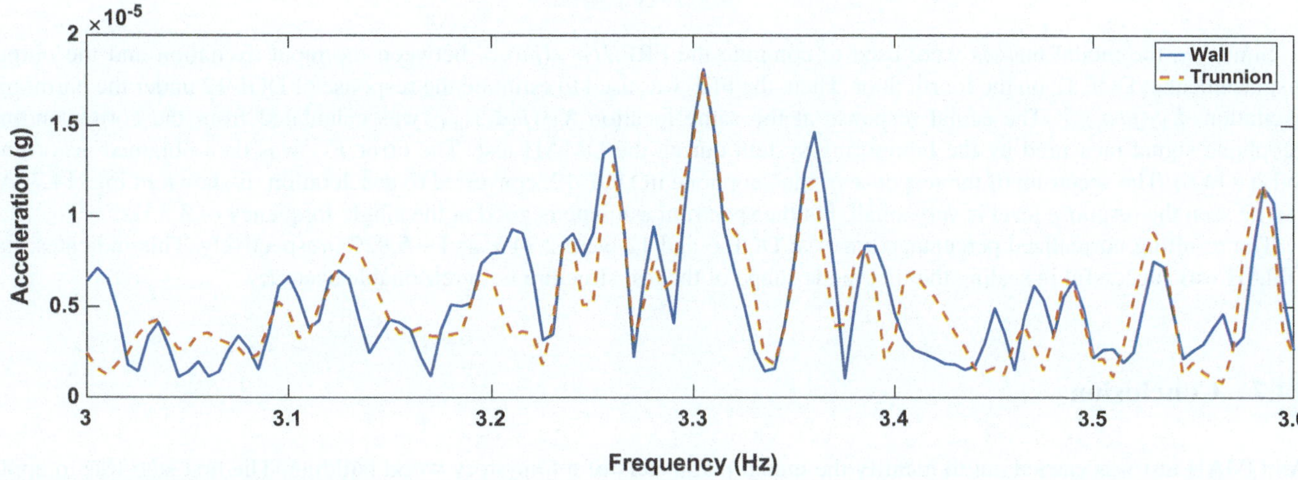

Fig. 14.6 Comparison between the response measured by the accelerometer on the wall and that measured at the shaker's trunnion. Note that it is only the peak at 3.3 Hz that is of interest

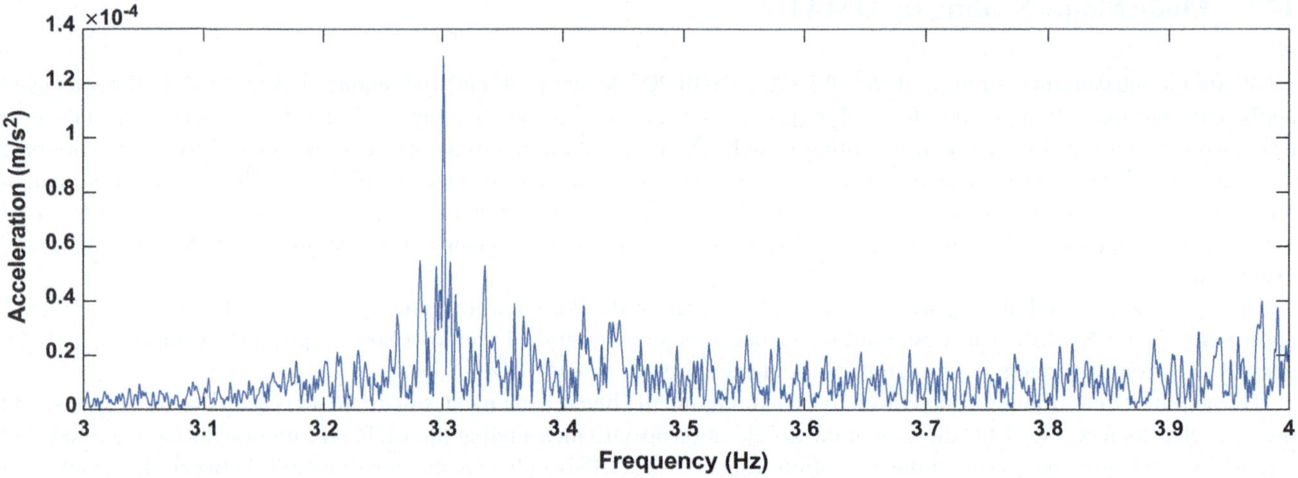

Fig. 14.7 Linear RMS spectrum of geophone signal in DOF 12, converted to acceleration

14.6 Verification of OMAH Result

The estimated modal mass was plugged into Eq. (14.1) to synthesize the FRF $H_{6,6}(j\omega)_{syn}$ between an input force at DOF 6 and an output displacement at the same DOF. The FRF was then used to estimate the RMS of the response under a harmonic force at the excitation frequency, $X_6(j\omega_{ex})_{est}$ as follows:

$$X_6(j\omega_{ex})_{est} = F_6(j\omega_{ex})_{meas} \times H_{6,6}(j\omega_{ex})_{syn} \tag{14.5}$$

where $F_6(j\omega_{ex})_{meas}$ is the measured force obtained from the acceleration of the moving mass of the shaker.

The actual (i.e. measured) RMS level of the displacement response at $X_6(j\omega_{ex})_{meas}$ was then computed by processing the signal measured by the accelerometer on the wall. To verify the modal mass estimated by OMAH, the normalized percentage error between the estimated $X_6(j\omega_{ex})_{est}$ and the actual $X_6(j\omega_{ex})_{meas}$ response was calculated as:

$$E_6 = \frac{X_6(j\omega_{ex})_{est} - X_6(j\omega_{ex})_{meas}}{X_6(j\omega_{ex})_{meas}} \times 100 \tag{14.6}$$

Similarly, the modal masses were used to compute the FRF $H_{12,6}(j\omega)_{syn}$ between the input excitation and the output displacement at DOF 12 on the fourth floor. Then, the FRF was used to estimate the response of DOF 12 under the harmonic excitation, $X_{12}(j\omega_{ex})_{est}$. The actual response at the same location $X_{12}(j\omega_{ex})_{meas}$ was calculated from the corresponding geophone signal measured by the monitoring system during the OMAH test. The error E_{12} was then obtained according to Eq. (14.6). The spectrum of the response of the geophone in DOF 12, converted to acceleration, is shown in Fig. 14.7. As can be seen the response level is very small, but the spectrum estimate is good at the single frequency of 3.3 Hz.

The resulting normalized percentage errors at DOF 6 and 12 were 2.74% and −5.82%, respectively. This indicates that OMAH was successful in scaling the first mode shape of the test structure with reasonable accuracy.

14.7 Conclusion

An OMAH test was carried out to identify the modal parameters of a four-story wood building. The first step was to apply OMA to extract the first natural frequency, damping ratio, and the unscaled mode shape from the ambient vibration response of the building. A small shaker was then used to excite the structure at its first eigenfrequency with a force of approx. 1.8 N RMS. The force applied by the shaker along with the acceleration measured at the excitation location were used to estimate the corresponding modal mass and scale the mode shape. The estimated modal mass was used to synthesize the harmonic response at two locations. To evaluate the accuracy of the scaling results, the synthesized response was compared to the measured one and found to be within 6% of the measured responses. This comparison demonstrated that a small and inexpensive shaker is sufficient for scaling the mode shapes of a large civil structure using OMAH.

Acknowledgements Funding provided by Växjö kommun, VKAB and Videum allowed to install the measurement devices and to perform monitoring. The authors would like to thank colleagues and collaborators that helped to put up install the sensors and the measurement network, namely Bertil Enquist and Jonas Klaeson, Linnaeus University, Department of Building Technology, as well as Per Finander, SAAB AB. Also thanks to Åsa Bolmsvik, now at Skanska AB, for helping with the shaker measurements.

References

1. Brandt, A., Berardengo, M., Manzoni, S., Cigada, A.: Using three-parameter sine fitting for scaling mode shapes with OMAH. In: IOMAC 2017 – 7th International Operational Modal Analysis Conference, pp. 297–302 (2017)
2. Parloo, E., Verboven, P., Guillaume, P., Van Overmeire, M.: Sensitivity-based operational mode shape normalisation. Mech. Syst. Signal Process. **16**, 757–767 (2002). https://doi.org/10.1006/mssp.2002.1498
3. López-Aenlle, M., Fernández, P., Brincker, R., Fernández-Canteli, A.: Scaling-factor estimation using an optimized mass-change strategy. Mech. Syst. Signal Process. **24**, 1260–1273 (2010). https://doi.org/10.1016/j.ymssp.2009.06.011
4. Bernal, D.: Modal scaling from known mass perturbations. J. Eng. Mech. **130**, 1083–1088 (2004). https://doi.org/10.1061/(ASCE)0733-9399(2004)130:10(1255)
5. Cara, J.: Computing the modal mass from the state space model in combined experimental-operational modal analysis. J. Sound Vib. **370**, 94–110 (2016). https://doi.org/10.1016/j.jsv.2016.01.043
6. Reynders, E., Degrauwe, D., De Roeck, G., Magalhães, F., Caetano, E.: Combined experimental-operational modal testing of footbridges. J. Eng. Mech. **136**, 687–696 (2010). https://doi.org/10.1061/(ASCE)EM.1943-7889.0000119
7. Aenlle, M.L., Brincker, R.: Modal scaling in operational modal analysis using a finite element model. Int. J. Mech. Sci. **76**, 86–101 (2013). https://doi.org/10.1016/j.ijmecsci.2013.09.003
8. Brandt, A., Berardengo, M., Manzoni, S., Cigada, A.: Scaling of mode shapes from operational modal analysis using harmonic forces. J. Sound Vib. **407**, 128–143 (2017). https://doi.org/10.1016/j.jsv.2017.06.033
9. Brandt, A., Berardengo, M., Manzoni, S., Vanali, M., Cigada, A.: Overview of the new Omah technique for scaling OMA mode shapes. Sound Vib. **52**, 18–22 (2018). https://doi.org/10.32604/sv.2018.03872
10. Brandt, A., Berardengo, M., Manzoni, S., Vanali, M., Cigada, A.: Global scaling of operational modal analysis modes with the OMAH method. Mech. Syst. Signal Process. **117**, 52–64 (2019). https://doi.org/10.1016/j.ymssp.2018.07.017
11. Brincker, R., Bolton, B., Brandt, A.: Calibration and processing of geophone signals for structural vibration measurements. Conf. Proc. Soc. Exp. Mech. Ser. **3**, 1375–1379 (2011). https://doi.org/10.1007/978-1-4419-9834-7_121
12. The Mathworks Inc.: MATLAB – MathWorks. www.Mathworks.com/products/matlab (2019)
13. Brandt, A.: ABRAVIBE–A MATLAB toolbox for noise and vibration analysis and teaching. http://www.abravibe.com (2018)
14. Allemang, D.R.J., Brown, D.D.L.: Experimental modal analysis and dynamic component synthesis. Struct. Dyn. Res. Lab. Univ. Cincinnati, Ohio. **I**, 1–123 (1987)

Chapter 15
Quantitative Study on the Modal Parameters Estimated Using the PLSCF and the MITD Methods and an Automated Modal Analysis Algorithm

Silas Sverre Christensen, Stefano Manzoni, Marcello Vanali, Alfredo Cigada, and Anders Brandt

Abstract There are many advanced algorithms used to estimate modal parameters. In this paper, the modal parameters extracted from the Poly-reference Least Squares Complex Frequency (PLSCF) algorithm and the Multi-reference Ibrahim Time Domain (MITD) algorithm, are compared. The former, is widely used in the industry and is known to produce almost crystal clear stabilization diagrams with barely any spurious pole estimates. The latter, is less common and the stabilization diagrams typically contain some spurious pole estimates. An Automated Modal Analysis (AMA) algorithm, that utilizes the statistical representation of the pole estimates combined with a number of decision rules based on the Modal Assurance Criteria (MAC), is employed, to detect probable physical poles. Simulated data from a Plexiglas plate is used in the study. Results indicate that the absolute bias error associated with the modal parameter estimates output by the PLSCF algorithm is higher than the bias error related to the modal parameter estimates output by the MITD algorithm. It was not conclusive which of the two methods that had the lowest random error. It should also be mentioned that, while the MITD algorithm could process all references and responses, the PLSCF algorithm relied strongly on a delicate selection of representative references and that not too many references were used.

Keywords Automated operational modal analysis · Automated modal analysis · Poly-reference least squares complex frequency · Multi-reference Ibrahim time domain · Damping

15.1 Introduction

Modal parameters describe the dynamic properties of structures and can aid in the development of sophisticated methods for validating and updating models as well as monitoring structures during operation. There are many advanced algorithms that can be used to estimate modal parameters. Among the most popular are: Multi-reference Ibrahim Time Domain (MITD) [1, 2], Poly-reference Time Domain (PTD) [3], Eigensystem Realization Algorithm (ERA) [4], Stochastic Subspace Identification (SSI) [5], Poly-reference Least Squares Complex Frequency (PLSCF) [6], Frequency Domain Decomposition (FDD) [7] and Polyreference Frequency Domain (PFD) [8, 9].

In this paper we shall look at two of the methods mentioned, namely the PLSCF and MITD algorithms. The former is widely used in the industry and especially popular for its ability to produce very clear stabilization diagrams that produce almost no spurious information. The latter is less common and produces less clear stabilization diagrams. It may be convenient for a modal analyst when interpreting stabilization diagrams that they are clear, but there is no guarantee that a clear stabilization diagram yields the best modal parameter estimates. The time spend to interpret a stabilization diagram and select the most probably physical poles may also be inconsistent depending on the appointed operator. There are many methods available in the literature that automates this process, among some are: The K-Means (or Fuzzy-Means)

S. S. Christensen (✉) · A. Brandt
Department of Technology and Innovation, University of Southern Denmark, Odense M, Denmark
e-mail: ssv@iti.sdu.dk; abra@iti.sdu.dk

S. Manzoni · A. Cigada
Department of Mechanical Engineering, Politecnico di Milano, Milan, Italy
e-mail: stefano.manzoni@polimi.it; alfredo.cigada@polimi.it

M. Vanali
Department of Engineering and Architecture, Università degli Studi di Parma, Parma, Italy
e-mail: marcello.vanali@unipr.it

B. Dilworth (ed.), *Topics in Modal Analysis & Testing, Volume 8*, Conference Proceedings of the Society for Experimental Mechanics Series, https://doi.org/10.1007/978-3-030-47717-2_15

Clustering algorithm [10, 11] and the Agglomerative Hierarchical Clustering algorithm [12, 13]. Some methods also utilizes the statistical representation of the pole estimates [14–16]. In this paper an automated modal analysis (AMA) algorithm that extends from the statistical representation of the pole estimates is used. The method is extended by adding a number of decision rules based on the Modal Assurance Criterion (MAC).

The structure of this paper is as follows. In Sect. 15.2 the general outline of the PLSCF and the MITD algorithms are described. Simulated data from a Plexiglas plate are presented in Sect. 15.3 along with the Automated Modal Analysis (AMA) algorithm. The modal parameter estimates output by the two methods are presented in Sect. 15.4, and their dependencies are discussed in Sect. 15.5. A summary of the concluding remarks are found in Sect. 15.6.

15.2 Theory

Any frequency response matrix can be decomposed into a sum of system poles, s_r, containing damping ratios, ζ_r, and natural frequencies, f_r, as well as a residue matrix, $\mathbf{A_r}$, which carry information about the mode shape vectors, $\boldsymbol{\psi}$, given as

$$\mathbf{H}(j\omega) = \sum_{r=1}^{N} \frac{\mathbf{A}_r}{j\omega - s_r} + \frac{\mathbf{A}_r^*}{j\omega - s_r^*} = \sum_{r=1}^{2N} \frac{\mathbf{A}_r}{j\omega - s_r} \tag{15.1}$$

Note that the rightmost term is but a renumbering where the sum goes to 2N instead of N, in this way, every second residue matrix is the complex conjugate of the previous residue. Information about the mode shape is retained in the numerator, while the denominator contains information about the system poles. The basic equation of an individual frequency response, H_{pq}, can also be written as a fraction of polynomials

$$H_{pq}(\omega_i) = \frac{U_p(\omega_i)}{F_q(\omega_i)} = \frac{\beta_n(s_i)^n + \beta_{n-1}(s_i)^{(n-1)} + \cdots + \beta_0(s_i)^0}{\alpha_m(s_i)^m + \alpha_{m-1}(s_i)^{(m-1)} + \cdots + \alpha_0(s_i)^0} \tag{15.2}$$

where ω_i, is some measured frequency and $s_i = j\omega_i$, is the generalized frequency of the system while α and β denotes the polynomial coefficients. The order of the numerator is typically two less than the order of the denominator, i.e. $n = m - 2$. Rearranging Equation (15.2) and introducing the full frequency response matrix along with the polynomial coefficient matrices $\boldsymbol{\alpha}_k$ and $\boldsymbol{\beta}_k$ yields

$$\sum_{k=0}^{m} \boldsymbol{\alpha}_k (s_i)^k \mathbf{H}(\omega_i) = \sum_{l=0}^{n} \boldsymbol{\beta}_l (s_i)^k \tag{15.3}$$

Equation (15.3) is the frequency domain formulation using frequency response matrices. A similar expression can be derived for the time domain where impulse responses, correlation functions or random decrement signatures may be used. The force coefficients are zero, which gives

$$\sum_{k=0}^{m} \boldsymbol{\alpha}_k \mathbf{h}(t_{i+k}) = 0 \tag{15.4}$$

The polynomial coefficients for the frequency domain and the time domain are given in Equations (15.5) and (15.6)

$$\boldsymbol{\alpha}_m s^m + \boldsymbol{\alpha}_{m-1} s^{m-1} + \boldsymbol{\alpha}_{m-2} s^{m-2} + \cdots + \boldsymbol{\alpha}_0 = 0 \tag{15.5}$$

$$\boldsymbol{\alpha}_m z^m + \boldsymbol{\alpha}_{m-1} z^{m-1} + \boldsymbol{\alpha}_{m-2} z^{m-2} + \cdots + \boldsymbol{\alpha}_0 = 0 \tag{15.6}$$

15.2.1 The Poly-reference Least Squares Complex Frequency (LSCF) Algorithm

The Poly-reference Least Squares Complex Frequency (PLSCF) algorithm was first formulated in [6]. It is a poly-referenced version of the Least Squares Complex Frequency (LSCF) algorithm [17], which is the frequency domain alternative to the Least Squares Complex Exponential (LSCE) algorithm [18]. The poly-referenced version of LSCE is commonly known as the Poly-reference Time Domain (PTD) algorithm [3]. The PolyMAX algorithm is similar to the PLSCF algorithm [19], yet the exact similarities are not know since the internals of the commercial implementation are not known. This paper will adopt the formulation given in [6] and [19].

The initial step of the PLSCF algorithm is to use a Right Matrix Fraction Description (RMFD) model, which is obtained by post mutiplying Equation (15.3) by $\sum_{k=0}^{m} \boldsymbol{\alpha}_k^{-1}(j\omega_i)^k$ yielding

$$\mathbf{H}(\omega_i) = \sum_{l=0}^{n} \boldsymbol{\beta}_l (j\omega_i)^k \sum_{k=0}^{m} \boldsymbol{\alpha}_k^{-1}(j\omega_i)^k \tag{15.7}$$

where the matrix polynomial $\boldsymbol{\alpha}_k(j\omega_i)^k$ and $\boldsymbol{\beta}_l(j\omega_i)^k$ have $\mathrm{N_o} \times \mathrm{N_i}$ and $\mathrm{N_o} \times \mathrm{N_o}$ coefficients, respectively. Here $\mathrm{N_o}$ denote the number of responses while $\mathrm{N_i}$ denote the number of references. The number of references are typically lower than the number of responses. For the RMFD model the number of eigenvalues equal $\mathrm{mN_i}$.

The basic concept of the PLSCF algorithm is to solve a least squares problem by minimising a cost function from an error function that is derived from Equation (15.7). These steps are not included in this paper, but are found in [6] and [19], where it is shown that the error function is only linear for $\mathrm{N_i} = 1$ and otherwise non-linear. This is undesirable as $\mathrm{N_i}$ typically is higher than one. By some algebraic manipulation, also left out in this paper, the non-linear error function can be approximated by a linear error function. A cost function can then be formulated based on the linearised error function. The cost function is minimized by setting its derivatives with respect to the unknown polynomial coefficients, θ, equal to zero. This gives us the following reduced normal equations

$$2Re \begin{bmatrix} \mathbf{X}_1^H\mathbf{X}_1 & \mathbf{0} & \cdots & \mathbf{X}_1^H\mathbf{Y}_1 \\ \mathbf{0} & \mathbf{X}_2^H\mathbf{X}_2 & & \mathbf{X}_2^H\mathbf{Y}_2 \\ \vdots & & \ddots & \vdots \\ \mathbf{Y}_1^H\mathbf{X}_1 & \mathbf{Y}_2^H\mathbf{X}_2 & \cdots & \sum_{k=1}^{N_iN_o} \mathbf{Y}_k^H\mathbf{Y}_k \end{bmatrix} \begin{Bmatrix} \boldsymbol{\beta}_1 \\ \boldsymbol{\beta}_2 \\ \vdots \\ \boldsymbol{\beta}_{N_iN_o} \\ \boldsymbol{\alpha} \end{Bmatrix} = 2Re(\mathbf{J}^H\mathbf{J})\boldsymbol{\theta} = \mathbf{0} \tag{15.8}$$

with J being the Jacobian matrix where $\mathrm{X_k}$ and $\mathrm{Y_k}$ depend on the polynomial basis function as well as the estimated frequency response matrices. A weighting function that accounts for the quality of the measured frequency responses matrices can also be included in Equation (15.8).

By setting the $\boldsymbol{\beta}_1 \ldots \boldsymbol{\beta}_{N_iN_o}$ polynomial coefficients equal to zero, the $\boldsymbol{\alpha}$ polynomial coefficients can be solved. Hereafter $\boldsymbol{\beta}_1 \cdots \boldsymbol{\beta}_{N_iN_o}$ may be found. In this way, the denominator of Equation (15.2), which contains the pole estimates are determined first. Then the $\boldsymbol{\beta}_1 \ldots \boldsymbol{\beta}_{N_iN_o}$ polynomial coefficients, i.e. the modal participation factors (or mode shape vectors) can be estimated.

A common way of solving the polynomial coefficient $\boldsymbol{\alpha}$ is by using the companion matrix formulation, given as

$$\mathbf{C} = \begin{bmatrix} -\boldsymbol{\alpha}_{m-1} & -\boldsymbol{\alpha}_{m-2} & \cdots & -\boldsymbol{\alpha}_0 \\ \mathbf{I} & \mathbf{0} & \cdots & \mathbf{0} \\ \mathbf{0} & \mathbf{I} & \cdots & \mathbf{0} \\ \vdots & & & \vdots \\ \mathbf{0} & \cdots & \mathbf{I} & \mathbf{0} \end{bmatrix} \tag{15.9}$$

To solve Equations 15.5 and 15.6, it is normal to set $\boldsymbol{\alpha}_m = 1$. It was, however, found in [20], that by utilizing a low order normalization, i.e. setting $\boldsymbol{\alpha}_0 = 1$ in Equation 15.5, very clear stabilization diagrams were obtained when using the PLSCF algorithm.

The PLSCF algorithm presented does not use Singular Value Decomposition (SVD), and it should not be needed, if an appropriate number of references is selected and the number of references are limited to only a few references [20]. It was pointed out in [21] that the PTD algorithm (time domain alternative to the PLSCF algorithm) produces bad stabilization diagrams when the number of references exceeds more than 3.

A Left Matrix Fraction Description (LMFD) model can also be used, but that would yield mN_o eigenvalues, and since $o >> i$, the use of the SVD (or any other equation condensation) is then advised.

15.2.2 The Multi-reference Ibrahim Time Domain (MITD) Algorithm

The Multi-reference Ibrahim Time Domain (MITD) algorithm [2] is a multi-reference version of the Ibrahim Time Domain algorithm [1]. This method operates in time domain thus requiring impulse response functions, correlation functions or random decrement signatures as inputs. Correlation functions are used in the present analysis. The correlation function matrix, $\mathbf{R}(\tau)$, is defined as

$$\mathbf{R}(\tau) = \mathbb{E}\left[\mathbf{y}(t)\mathbf{y}^T(t+\tau)\right] \tag{15.10}$$

where τ denotes the time lag given in seconds and $\mathbf{y}(\mathrm{t}) = \left[\mathrm{y}_1(\mathrm{t}), \mathrm{y}_2(\mathrm{t}), \ldots, \mathrm{y}_{\mathrm{M}(\mathrm{t})}\right]^{\mathrm{T}}$ is the response vector consisting of M measurements. This means that the diagonal elements in $\mathbf{R}(\tau)$ represent the autocorrelation functions while the off-diagonal elements represent the cross-correlation functions. For the MITD method it is common to gather all the information at different time lags in a block Hankel matrix, given by

$$\mathbf{H_{nm}}(\tau) = \begin{bmatrix} [\mathbf{R}(\tau))] & [\mathbf{R}(\tau) + \Delta t)] & \cdots & [\mathbf{R}(\tau) + (m-1)\Delta t)] \\ [\mathbf{R}(\tau) + \Delta t)] & [\mathbf{R}(\tau) + 2\Delta t)] & \cdots & [\mathbf{R}(\tau) + m\Delta t)] \\ \vdots & \vdots & \ddots & \vdots \\ [\mathbf{R}(\tau) + (n-1)\Delta t)] & [\mathbf{R}(\tau) + n\Delta t)] & \cdots & [\mathbf{R}(\tau) + (n+m-2)\Delta t)] \end{bmatrix} \tag{15.11}$$

Equation (15.11) can be decomposed into a mode shape matrix $\boldsymbol{\Psi}$, a diagonal exponential pole matrix $e^{\mathbf{S}_r t}$ and a modal participation matrix $\mathbf{L}^T$

$$H_{mn}(\tau) = \boldsymbol{\Psi} e^{\mathbf{S}_r t}\mathbf{L}^T \tag{15.12}$$

The block Hankel matrix given in Equation (15.11) is also formulated for $\tau + \Delta t$, i.e. $\mathbf{H_{nm}}(\tau + \Delta t)$. Using these two expressions for the block Hankel matrix and by applying some algebraic manipulation, an eigenvalue problem can be constructed. Since this eigenvalue problem is very large, it is condensed using the SVD, a tool that has shown to be powerful in removing redundant information. The modal parameters are then extracted by solving the reduced eigenvalue problem.

15.3 Methodology

Modal parameter estimates were extracted from an experimental dataset of a rectangular Plexiglass plate using the MITD algorithm. The modal parameter estimates were subsequently used to produce a simulated dataset, from which modal parameter estimates have been extracted using the PLSCF and the MITD algorithms.

The Plexiglass plate measures 533 × 321 × 20 mm and is described by 35 degrees of freedom (DOFs) in a 7 by 5 grid. See [22] for details on the experimental study whose data were also used in this paper. The modal parameters from the experimental dataset were extracted using the MITD algorithm and an Automated Modal Analysis (AMA) algorithm. The AMA algorithm, described in [23], starts with constructing a statistical representation of the pole estimates for varying model orders. It is then complemented by a number of decision rules based on the Modal Assurance Criteria (MAC). The general outline of the AMA algorithm is as follows

1. Choose a modal parameter estimation method capable of producing poles and mode shape vectors at varying model orders
2. Estimate poles at mode shape vectors at various model orders
3. Initiate the statistical representation of the pole estimates
4. Introduce an occurrence threshold and temporarily exclude modal parameter estimates below said threshold
5. Compute the MAC for each remaining bin and remove modal parameter estimates whose MAC value is poor in comparison to the MAC values of the majority of poles in a given bin
6. Combine adjacent bins whose MAC values are similar
7. Add any adjacent modal parameters that did not pass step (4) to a bin if the MAC values are similar
8. Compute mean values and standard deviation for each remaining bin

The AMA algorithm has shown successful in detecting structural modes from datasets on several real structures, e.g. the Little Belt Suspension Bridge, the Heritage Court Tower and a Ro-Lo ship [23]. The modal parameter estimates from the experimental dataset using the AMA algorithm, were then used to simulated 300 s forced response sequences at a sampling frequency of 5000 Hz using superposition and digital filter theory [24]. Gaussian white noise processes were used as input in all 35 DOFs with a noise level corresponding to 0.01% of the standard deviation of the forced response.

The simulated data were input into the PLSCF algorithm and the MITD algorithm to estimate modal parameters. A model order of 100 was used for both algorithm, to ensure that many modal parameters were output.

For the PLSCF algorithm three of the four corner points were used as references of the 35 responses. The correlation function estimates used to compute the spectral densities were postmultiplied by an exponential window so that the correlation function value at the end of the measurement time was reduced to 0.01%. This is a common procedure used to reduce the bias error added from truncations in the time domain [25]. A total of 100 averages were used to compute the spectral density estimates. The damping ratio added from applying an exponential window was subtracted from the estimated damping ratios output by the PLSCF algorithm.

For the MITD algorithm an unbiased Welch estimator with a blocksize of 512 samples, was used to compute the correlation functions for all 35 measurement channels. The first 15 time lag values were removed from all correlation functions to suppress measurement noise [26]. Exactly 90 time lag values from each of the 1225 correlation functions were used as input into the MITD algorithm. At this point the correlation functions had almost fully decayed.

For the modal parameter estimates output by both algorithm, only poles originating from an under-damped system were considered, i.e. having low damping and positive frequency. Furthermore, damping ratio estimates above 10% were discarded since the damping ratio of the plexiglass plate does not exceed 3.5% for any of the modes of interest.

15.4 Results

The modal parameters estimated using the two algorithms are presented in the following. The stabilization diagrams are seen in Figs. 15.1 and 15.2.

In the present work, when using the PLSCF algorithm, it was crucial that a representative number of references were chosen and that the number of references were limited to only a few. The most clear stabilization diagram were obtained by choosing three of the four corner points as references, which was also mentioned in Sect. 15.3. By choosing more references, the number of spurious poles present in the stabilization diagrams would rise. It was attempted to use 35 references, which rendered the stabilization diagram completely indecipherable for the AMA algorithm to interpret. This observation is related to the fact that the number of eigenvalues estimated equals mN_i, where m is the number of frequency lines and N_i denotes the number of references. By increasing the number of references more eigenvalues are estimated as a multiple of the number of frequency lines. Therefore, if too many references (e.g. SVD) are used, equation condensation should be employed, while, when only a few references are used equation condensation should not be required.

For the PLSCF algorithm nine modes were identified, which correspond to the number of modes used to simulate the data. It is quite clear that the frequency component of the pole estimates stabilize for increasing model order and that barely any spurious pole estimates are present. This is in line with previous observations, that very clear stabilization diagrams are output, when using the PLSCF algorithm, that was presented in Sect. 15.2. Upon zooming onto the first two and closely spaced modes in the same plot, it is seen, up to a model order of 20, that the model parameters are not stabilizing well on the frequency axis. Above a model order of 50 it appears that the estimates are strongly frequency stable. In the intermediate rage, model order ranging from 20 to 50, the frequency component of the model parameter estimates are slightly skewed to the right, yielding higher frequency estimates. It should be clarified that in Fig. 15.1, the horizontal dashed black line corresponds to the occurrence threshold that was described in step (4), see Sect. 15.2. Modal parameter estimates marked

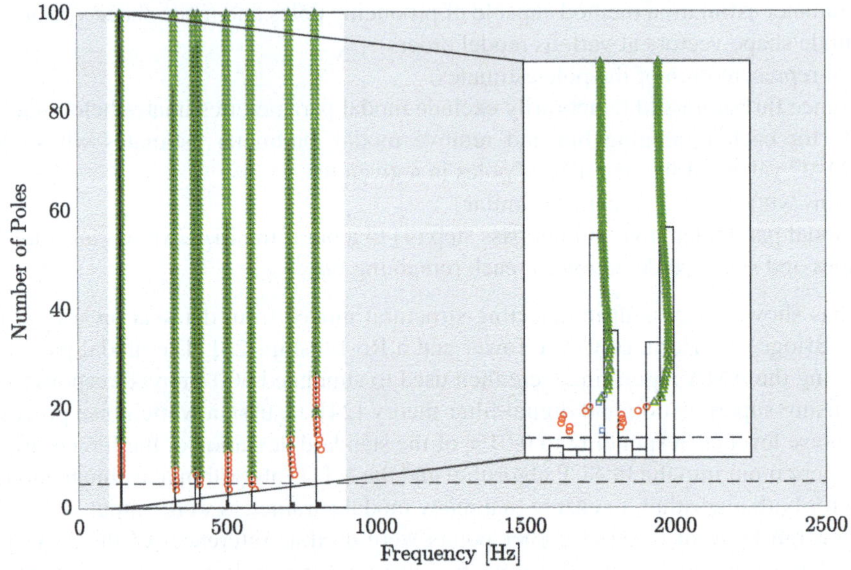

Fig. 15.1 Stabilization diagram overlaid by its probability mass function based on modal parameter estimates derived using the PLSCF algorithm on simulated data. ○ circle – bins with pole estimates that are below the horizontal dashed black line and outside the white patched area; □ square – poor MAC valued pole estimates; △ triangle – stable pole estimates with similar MAC values

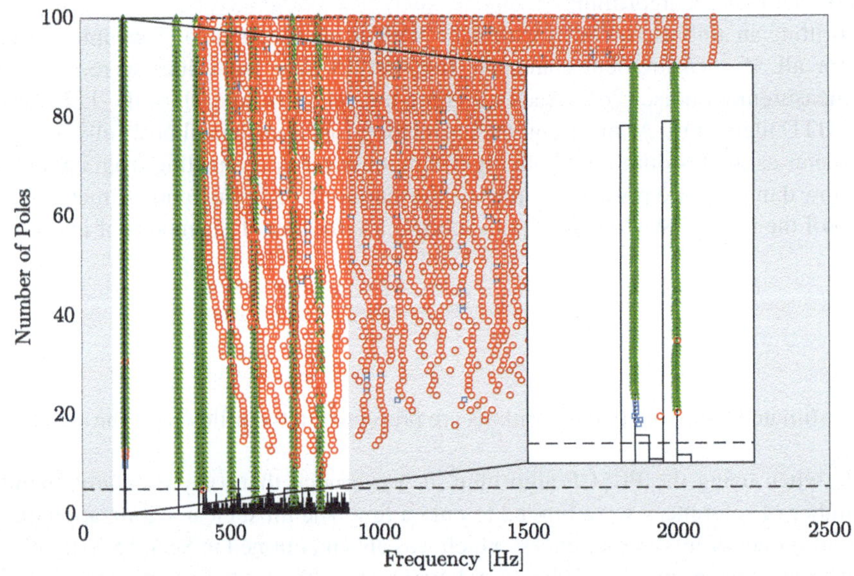

Fig. 15.2 Stabilization diagram overlaid by its probability mass function based on modal parameter estimates derived using the MITD method on simulated data. ○ circle – bins with pole estimates that are below the horizontal dashed black line and outside the white patched area; □ square – poor MAC valued pole estimates; △ triangle – stable pole estimates with similar MAC values

as red circles does not satisfy step (4) and step (7), while modal parameter estimates denoted as blue squares does not satisfy step (5), see also Sect. 15.2. The remaining modal parameter estimates (and the bin they belong to), marked as green triangles, are strongly frequency stable and have high MAC similarity. Mean values and standard deviations are computed for the estimates in each of the combined bins.

When looking at Fig. 15.2 the amount of spurious information is abundant, and in direct contrast to what was seen Figure in 15.1, where barely any spurious information was present. It is however seen that almost no spurious information is present at the first four modes. It should be mentioned that the same nine modes that was found when using the PLSCF algorithm were also found using the MITD algorithm. When zooming onto the first two modes, it is observed that below a model order of 50 the frequency component of the pole estimates are more stable than those found when using the PLSCF algorithm. At model orders above 50 it appears that the two algorithms produced similar results.

Table 15.1 Mean values and standard deviations of the model parameter estimates output by the PLSCF algorithm. The relative difference on the mean value in comparison to the true value are also reported

	Mean		Std		True		Rel. Diff. Mean		
Mode [-]	f [Hz]	ς [%]	f [Hz]	ς [%]	f [Hz]	ς [%]	f [%]	ς [%]	Count [-]
1	141.41	3.14	0.049	0.101	141.90	3.17	0.34	1.07	84
2	142.45	2.88	0.065	0.051	142.58	3.06	0.09	5.88	87
3	323.59	2.72	0.061	0.007	323.25	2.68	−0.11	−1.38	92
4	393.20	2.54	0.023	0.004	393.04	2.60	−0.04	2.38	93
5	408.95	2.50	0.025	0.016	408.73	2.53	−0.05	1.22	94
6	504.17	2.50	0.022	0.004	503.79	2.45	−0.08	−2.00	89
7	583.84	2.42	0.018	0.008	583.63	2.46	−0.04	1.75	94
8	712.05	2.39	0.056	0.029	712.01	2.42	0.00	1.40	87
9	799.69	2.18	0.038	0.044	800.09	2.22	0.05	1.58	76

Table 15.2 Mean values and standard deviations of the model parameter estimates output by the MITD algorithm. The relative difference on the mean value in comparison to the true value are also reported

	Mean		Std		True		Rel. Diff. Mean		
Mode [-]	f [Hz]	ς [%]	f [Hz]	ς [%]	f [Hz]	ς [%]	f [%]	ς [%]	Count [-]
1	141.89	3.14	0.018	0.010	141.90	3.17	0.00	1.07	89
2	142.67	3.09	0.018	0.016	142.58	3.06	−0.07	−0.98	87
3	323.26	2.68	0.005	0.002	323.25	2.68	0.00	0.11	94
4	392.82	2.57	0.102	0.031	393.04	2.60	0.05	1.23	93
5	408.88	2.53	0.067	0.024	408.73	2.53	−0.04	0.04	91
6	503.89	2.45	0.134	0.027	503.79	2.45	−0.02	0.04	93
7	583.57	2.50	0.291	0.139	583.63	2.46	0.01	−1.50	90
8	712.14	2.40	0.168	0.025	712.01	2.42	−0.02	0.99	96
9	800.03	2.26	0.700	0.336	800.09	2.22	0.01	−2.03	88

Both methods were able to output a large number of frequency and MAC stable modal parameter estimates for the nine modes identified for different model orders. These estimates are now included in a quantitative study. In Tables 15.1 and 15.2, the number of modal parameter estimates, their mean values and standard deviations are presented. Also the mean values are compared to the true modal parameters that was used for the simulated dataset.

Upon comparing the results from the two Tables 15.1 and 15.2, it is seen that for the first three modes, the standard deviations of the modal parameter estimates output by the PLSCF algorithm are highest. Quite the opposite is seen for the last six modes, where the standard deviations of the modal parameter estimates output by the MITD algorithm are highest. When comparing the mean values of the estimates to the true values, it is evident that the relative differences, in absolute terms, are largest for the modal parameter estimates output by the PLSCF algorithm.

15.5 Discussion

We have established, that the mean values of the damping ratio estimates output by the PLSCF algorithm are further from the true values, in comparison to the mean values of the damping ratio estimates output by the MITD algorithm. Although the damping ratio estimates, in absolute terms, are largest for the modal parameter estimates output by the PLSCF algorithm, it can not be concluded that the PLSCF algorithm consistently over- or underestimates the damping ratio estimates. Nor can this be said about the damping ratio estimates output by the MITD algorithm. It appears that this bias error is dependent on the mode in question. It is known that a bias error is introduced as a result of time discretization [25]. In other words, spectral leakage is the result of applying a window in the time domain prior to using the Fourier transform. Therefore a bias error is always present for any frequency domain modal parameter estimation method, including the PLSCF algorithm. However, for methods operating in the time domain, including the MITD algorithm, there is also a bias error present. This bias error is associated with the number of time lag values used in the correlation function estimates, whose optimum, is known to be different for each mode [27]. Therefore, whether the PLSCF algorithm or the MITD algorithm are used, there will always be a bias error present. The exponential window that was used in conjunction with the PLSCF algorithm, see Sect. 15.3, was

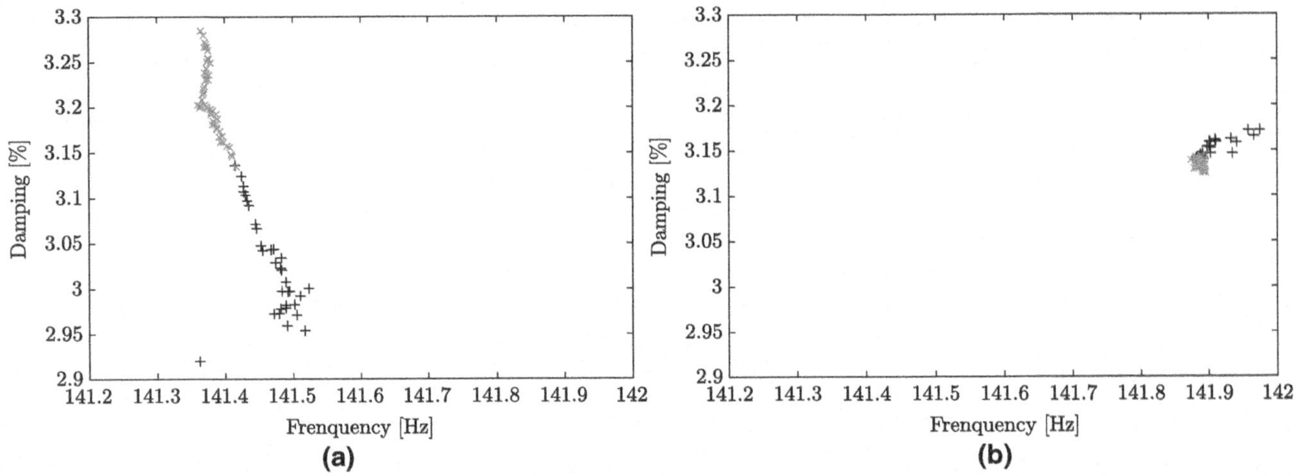

Fig. 15.3 Frequency-damping ratio plot for the first mode using the modal parameters estimated from (**a**) PLSCF algorithm and (**b**) MITD algorithm. The black + (plus) denotes the lower half of the modal parameter estimates while the gray × (cross) represents the upper half of the modal parameter estimates as seen in the stabilisation diagrams, Figs. 15.1 and 15.2

defined so that the correlation function value at the end of the measurement time was reduced to 0.01%. By increasing or decreasing this value by a multiple of 10, no noticeable differences were observed in the modal parameter estimates. Also for the MITD algorithm it was attempted to vary the number of time lag values used in the correlation function estimates from 70 to 130, which had minimal impact on the modal parameter estimates.

The random error for the damping ratio estimates is not definitively lower whether the PLSCF or the MITD algorithm are used. For instance high standard deviations are attributed to the damping ratio estimates for the fourth, seventh and the ninth mode when using the MITD algorithm. The standard deviation associated with the damping ratio estimates, for the first mode, is high when using the PLSCF algorithm. By taking a closer look at the frequency-damping ratio plot for this mode, seen in Fig. 15.3, it is observed that the frequency and damping ratio estimates follow a trend. Upon further inspection it is seen that the pole estimates for the upper half (grey cross) of the stabilization diagram have lower standard deviation than those for the lower half (black plus). For the MITD algorithm the frequency and damping ratio estimates are neatly clustered, but it is also observed that the upper half of the estimates are clustered better than the lower half. Since the random error is dependent on the measurement duration, it was attempted to double the measurement time from 300 to 600 s, which would allow twice the number of averages. This barely had any impact on the standard deviations of modal parameter estimates output by either of the two algorithms.

It was mentioned in Sect. 15.2 and in Sect. 15.4, that the PLSCF algorithm is sensitive to the number of references chosen. A study [28] using the MITD algorithm showed that the modal parameter estimates were almost unaffected whether all or only a few references were used when constructing the block Hankel matrix. This is expected since the MITD algorithm employs SVD, and thus removes redundant information, making the selection of poorly located references irrelevant as long as optimal reference locations are available. It was also reported in the study that by choosing fewer references a higher variance was obtained on the modal parameter estimates. Although the PLSCF algorithm uses fewer references than the MITD algorithm, the standard deviations of the modal parameters output were not much different.

The frequency-damping ratio plot for the fourth mode is seen in Fig. 15.4. As for the first mode it is seen that the upper half of the modal parameter estimates have the lowest variation when using the PLSCF algorithm. For the MITD algorithm the complete opposite is seen, compared to the behaviour observed for the first mode, that the modal parameter estimates for the lower half of the stabilization diagram have the lowest variation. For both methods, still considering the fourth mode, it is seen that a number of potential outliers are present. The outliers seem to be more profound for the modal parameter estimates output by the MITD algorithm. Potential outliers were also present for the modal parameter estimates output for the seventh and the ninth mode when using the MITD algorithm. This naturally contributes to higher standard deviations, and may also affect the mean values. By omitting the four pole estimates furthest to the right in Fig. 15.4a and those seven pole estimates furthest to the left in Fig. 15.4b, the variation of pole estimates appear to be similar for the two methods. However, the PLSCF algorithm still underestimates the damping ratio more than the MITD algorithm does.

Outlier detection is outside the scope of this paper, but it seems that when using a AMA algorithm, outlier detection should be employed. When an operator interprets a stabilisation diagram, that person may also assess whether one or more

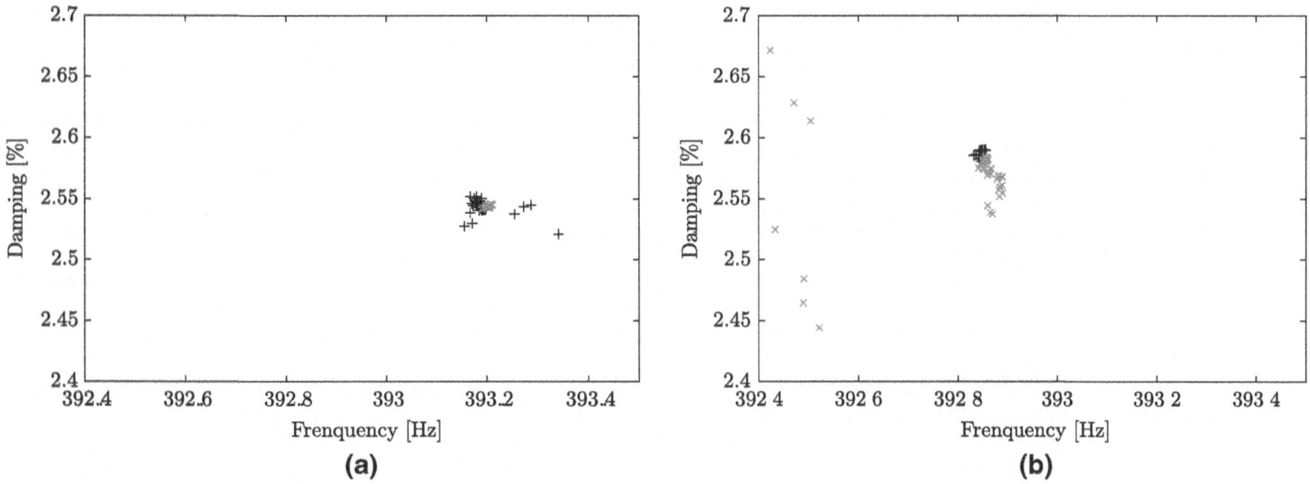

Fig. 15.4 Frequency-damping ratio plot for the fourth mode using the modal parameters estimated from (**a**) PLSCF algorithm and (**b**) MITD algorithm. The black + (plus) denotes the lower half of the modal parameter estimates while the gray × (cross) represents the upper half of the modal parameter estimates as seen in the stabilization diagrams, Figs. 15.1 and 15.2

modal parameter estimates are deviating strongly from the majority of pole estimates. The operator would then discard those estimates in a similar manner as most outlier detecting methods would.

15.6 Conclusions

Simulated data from a Plexiglas plate were used to estimate modal parameters using the Poly-reference Least Squares Complex Frequency (PLSCF) and the Multi-reference Ibrahim Time Domain (MITD) algorithms. An Automated Modal Analysis (AMA) algorithm based on the statistical representation of the pole estimates and complemented by a number of decision rules based on the Modal Assurance Criterion (MAC) were employed to ensure that probably physical modes were output. Upon comparing the damping ratio estimates from the two methods it was found that modal parameter estimates output by the PLSCF algorithm had higher bias, in absolute terms, in comparison to the modal parameter estimates output by the MITD algorithm. The nature of the random error associated with the damping ratio estimates was ambiguous, however, for some modes governed by outliers. It should also be mentioned that, while the MITD algorithm, can handle excess amounts of data due to the ability of the Singular Value Decomposition (SVD) to remove redundant information, the PLSCF relied on a careful selection of references and that only a few references were chosen.

Acknowledgments The work presented is supported by the INTERREG 5A Germany-Denmark program, with funding from the European Fund for Regional Development.

References

1. Ibrahim, S.R., Mikulcik, E.C.: A method for the direct identification of vibration parameters from the free response. Shock Vib. Bull. **47**, 183–198 (1977)
2. Fukuzono, K.: Investigation of multiple-reference Ibrahim Time Domain modal parameter estimation technique. Master's thesis, University of Cincinnati (1986)
3. Vold, H., Kundrat, J., Rocklin, G.T., Russell, R.: A multi-input modal estimation algorithm for mini-computers. In: Society for Automotive Engineers International Congress and Exposition, pp. 815–821 (1982)
4. Juang, J., Richard, P.S.: An eigensystem realization algorithm for modal parameter identification and model reduction. J. Guid. Control Dyn. **8**, 620–627 (1984)
5. Van Overschee, P., De Moor, B.: Subspace Identification for Linear Systems – Theory – Implementation – Applications. Kluwer Academic Publisher (1996) https://www.springer.com/gp/book/9781461380610
6. Guillaume, P., Verboven, P., Vanlanduit, S., Auweraer, H., Peeters, B.: A poly-reference implementation of the least-squares complex frequency-domain estimator. In: International Modal Analysis Conference (2003)

7. Brincker, R., Zhang, L., Andersen, P.: Modal identification from ambient responses using frequency domain decomposition. In: International Modal Analysis Conference, pp. 625–630 (2000)
8. Leuridan, J.M.: Some direct parameter model identification methods applicable for multiple input modal analysis. Ph.D. thesis, University of Cincinnati (1984)
9. Zhang, L., Kanda, H., Brown, D.L., Allemang, R.J.: A polyreference frequency method for modal parameter identification. In: American Society of Mechanical Engineers Design Engineering Technical Conference, pp. 1237–1245 (1985)
10. Ubertini, F., Gentile, C., Materazzi, A.L.: Automated modal identification in operational conditions and its application to bridges. J. Eng. Struct. **46**, 264–278 (2012)
11. Rainieri, C., Fabbrocino, G.: Development and validation of an automated operational modal analysis algorithm for vibration-based monitoring and tensile load estimation. Mech. Syst. Signal Process. **60–61**, 512–534 (2015)
12. Devriendt, C., Magalhães, F., Weijtjens, W., De Sitter, G., Cunha, A., Guillaume, P.: Structural health monitoring of offshore wind turbines using automated operational modal analysis. J. Struct. Health Monitor. **13**(6), 644–659 (2014)
13. Cardoso, R., Cury, A., Barbosa, F.: A robust methodology for modal parameters estimation applied to SHM. Mech. Syst. Signal Process. **95**, 24–41 (2017)
14. Scionti, M., Lanslots, J., Goethals, I., Vecchio, A., Van der Auweraer, H., Peeters, B., De Moor, B.: Tools to improve detection of structural changes from in-flight flutter data. In: International Conference on Recent Advances in Structural Dynamics (2004)
15. Allessandro, F.: Modal parameters estimation in the Z-domain. Mech. Syst. Signal Process. **23**, 217–225 (2009)
16. Bakir, P.G.: Automation of the stabilization diagrams for subspace based system identification. J. Expert Syst. Appl. **38**(12), 14390–14397 (2011)
17. Auweraer, H., Guillaume, P., Verboven, P., Vanlanduit, S.: Application of a fast-stabilizing frequency domain parameter estimation method. ASME J. Dyn. Syst. Meas. Control **123**(4), 651–658 (2001)
18. Brown, D.L., Allemang, R.J., Zimmerman, R., Mergeay, M.: Parameter estimation techniques for modal analysis. SAE Trans. **88**(1), 828–846 (1979)
19. Peeters, P., Auweraer, H., Guillaume, P., Leuridan, J.: The PolyMAX frequency-domain method: a new standard for modal parameter estimation? Shock Vib. **11**(3–4), 395–409 (2004)
20. Cauberghe, B.: Applied frequency-domain system identification techniques in the field of operational modal analysis. Ph.D. thesis, Vrije University Brussel (2003)
21. Guillaume, P., Verboven, P., Vanlanduit, S.: Frequency-domain maximum likelihood identification of modal parameters with confidence intervals. In: Proceedings of the International Conference on Noise and Vibration Engineering (1998)
22. Orlowitz, E., Brandt, A.: Comparison of experimental and operational modal analysis on a laboratory test plate. Measurement **102**, 121–130 (2017)
23. Christensen, S.S., Brandt, A.: Automatic operational modal analysis using statistical modelling of pole locations. In: Proceedings of the International Conference on Noise and Vibration Engineering (2018)
24. Ahlin, K., Magnevall, M., Josefsson, A.: Simulation of forced response in linear and nonlinear mechanical systems using digital filters. In: Proceedings of the International Conference on Noise and Vibration Engineering (2006)
25. Brandt, A.: Noise and Vibration Analysis – Signal Analysis and Experimental Procedures. Wiley (2011) https://onlinelibrary.wiley.com/doi/book/10.1002/9780470978160
26. Orlowitz, E., Brandt, A.: Influence of noise in correlation function estimates for operational modal analysis. In: Proceedings of International Modal Analysis Conference (2018)
27. Christensen, S.S., Brandt, A.: Parameter study of statistics of modal parameter estimates using automated operational modal analysis. In: Proceedings of the 37th International Modal Analysis Conference (2019)
28. Orlowitz, E.: Effects of sensor count on damping estimates from operational modal analysis. In: Proceedings of International Modal Analysis Conference, vol. 8, pp. 357–364 (2019)

Silas Sverre Christensen finished his masters in Structural Engineering in 2015 after which he got employed at a structural engineering consultancy company (name: COWI). In the summer 2017 he pursued a PhD. The tittle of the PhD project is: "Vibration Monitoring of Offshore Structures using Virtual Sensing techniques".

Chapter 16
Empirical Models for the Health Monitoring of High-Rise Buildings: The Case of Palazzo Lombardia

Marta Berardengo, Francescantonio Lucà, Stefano Manzoni, Marcello Vanali, and Daniele Acerbis

Abstract This paper deals with the monitoring of Palazzo Lombardia, one of the tallest skyscrapers in the city of Milan. The monitoring system collects both dynamic and static data in order to assess the health condition of the whole structure. Accelerometers are used for collecting dynamic data, while static data are obtained from clinometers. Moreover, environmental conditions are acquired by employing different further transducers.

The aim of this paper is to develop an empirical model to predict the values of the first eigenfrequencies of the structure as a function of the environmental conditions. Such a model can be then used to assess the health condition of the building. Indeed, deviations between the values expected from the model and those identified by means of an operational modal analysis of the acceleration data may indicate the occurrence of a structural change. However, it is not straightforward to build the mentioned empirical model because of some difficulties in managing the environmental data. As an example, it is hard to understand which temperature must be considered since it changes significantly along the structure. The paper shows how to overcome these problems, avoiding the use of the environmental data and employing, in place of them, the data coming from the clinometers and accelerometers. This approach allows to build the mentioned model successfully. Different empirical models are compared in the paper, evidencing which one is able to describe the trends of the eigenfrequencies at best. Furthermore, the uncertainty associated to the predictions of the model is also discussed. The whole work has been carried out on data acquired by the monitoring system for a time duration of almost 2 years.

Keywords Structural health monitoring · Empirical model · Operational modal analysis · Environmental effects · Eigenfrequency model

16.1 Introduction

This work deals with a project related to Structural Health Monitoring (SHM) of Palazzo Lombardia, one of the tallest skyscrapers recently built in Milan [1–4]. SHM is a discipline which aims at evaluating the safeness of civil and mechanical systems in an automated and reliable way, a problem which has attracted the interest of many researchers in recent times. Thanks to SHM it is possible to switch from a time-based to a condition-based maintenance approach. Indeed, an on-line evaluation of the data coming from the monitoring system allows to detect anomalous behaviours of the structure and consequently to start the maintenance procedures. This gives a clear economic advantage, but also a more reliable method for evaluating the health condition of the structure. Being the raw data automatically elaborated by the SHM algorithms, the human operator has just to check if the output of the algorithms themselves seem to indicate an anomaly in the system.

M. Berardengo
Department of Mechanical, Energy, Management and Transportation Engineering, Università degli Studi di Genova, Genoa, GE, Italy
e-mail: marta.berardengo@unige.it

F. Lucà · S. Manzoni (✉) · D. Acerbis
Department of Mechanical Engineering, Politecnico di Milano, Milan, MI, Italy
e-mail: francescantonio.luca@polimi.it; stefano.manzoni@polimi.it; daniele.acerbis@mail.polimi.it

M. Vanali
Department of Engineering and Architecture, Università degli Studi di Parma, Parma, PR, Italy
e-mail: marcello.vanali@unipr.it

B. Dilworth (ed.), *Topics in Modal Analysis & Testing, Volume 8*, Conference Proceedings of the Society for Experimental Mechanics Series, https://doi.org/10.1007/978-3-030-47717-2_16

16.2 Background

Normally, in the context of SHM, dynamic parameters such as eigenfrequencies and mode shapes are identified from the dynamic response of the structure in operational conditions [5–7]. Operational Modal Analysis (OMA) is an output-only approach which enables the estimation of modal parameters exploiting natural excitation (e.g. wind, traffic) [8–10]. The quality of the identification is dependent on the signal-to-noise ratio and the frequency resolution of the data used for the analysis [3]. Generally, an incipient damage produces a slight change of modal parameters. Therefore, it is fundamental to limit the uncertainty affecting the results of OMA in order to be able to detect a small anomaly in modal parameters. This can be obtained by properly carrying out the signal processing and setting the OMA input parameters (e.g. number of averages for calculating the power- and cross-spectra, frequency resolution) [3, 8, 11].

Beyond the uncertainty of the modal identification process, another issue which can affect the damage detection strategy is the effect of environmental conditions. Modal parameters, especially eigenfrequencies, can strongly change because of a variation of the environmental conditions such as temperature, humidity, wind speed etc. If these variables cannot be filtered out from the damage identification process, two scenarios could occur: the variation of the modal parameters generated by the environment could be recognized as a damage without any reason, or the variation due to the environmental conditions could cover a damage that is actually present in the structure.

In this context, this paper proposes a study based on the data collected from the monitoring system installed at Palazzo Lombardia and aimed at developing an empirical model for predicting the eigenfrequencies of the structure as functions of the environmental parameters. Palazzo Lombardia is made by 42 floors (plus three floors underground) and it is equipped with a continuous monitoring system collecting data from several accelerometers and clinometers. Since the monitored floors are just 5 (see Fig. 16.1a), it is not convenient to use mode shapes for SHM purposes. This is why the eigenfrequencies have been selected as damage features focusing on the effects of the environmental conditions. However, the next step is to take into account also the mode shapes because the number of sensors could be increased in the next future with the consequent possible use of mode shapes for SHM purposes as well.

In order to use the eigenfrequencies as structural change detectors, their trends must be monitored in time, and deviations from the expected behavior could indicate structural modifications. In order to perform this type of analysis, a model relating the eigenfrequency trends and the main influencing variables (exception made for possible damages, of course) must be derived. The main variables that must be taken into account are related to the environmental conditions, mainly the temperature, sun exposure and the wind speed and direction. Indeed, the temperature and sun exposure are shown to be able to influence the eigenfrequency values [12–15], as well as the wind (when the wind speed is high enough) [3] due to possible non-linear behaviours related to the building oscillations and aeroelastic effects.

However, measuring temperature and wind features is not enough for developing such a model. Indeed, several issues arise. As an example, measuring temperature requires to understand how many sensors must be used and in which locations they must be placed. The problem gets more and more complicated when the height of the structure increases. Indeed, it is difficult to understand which temperature values (one or more than one) must be considered, e.g. at the top of the building, at the ground floor, on a side exposed/non-exposed to the sun light.

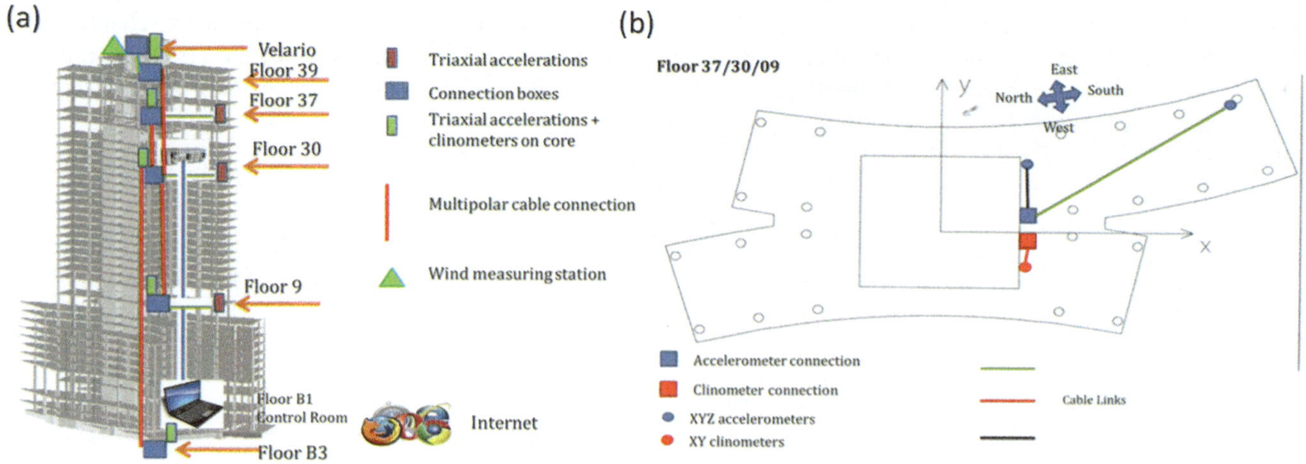

Fig. 16.1 Instrumented floors (floor B3 is the third floor underground) (**a**) and sensor layout on some instrumented floors (**b**)

Table 16.1 Nominal values of the first two eigenfrequencies of the building (estimated by means of a preliminary experimental modal analysis) and corresponding mode shapes

	Mode 1	Mode 2
Nominal eigenfrequency value [Hz]	0.32	0.40
Mode type	Bending (East-West direction)	Bending (North-South direction)

In order to overcome this problem, the authors proposed in [3] to use data related to the static and dynamic behavior of the building and somehow connected to the environmental conditions. In order to account for the temperature and sun exposure effects, the signals coming from clinometers placed in the building were used. Indeed, these environmental variables are able to affect the static configuration of the structure, that can be deducted by the signals collected by the clinometers. Furthermore, the signals coming from clinometers placed at the lower floor of the building (third floor underground) can be used to detect changes in the static configuration of the structure due to adjustments of the foundations (e.g. due to possible changes of the groundwater level). Furthermore, the effect of the wind can be taken into consideration by calculating the root-mean-square (RMS) value of the acceleration signal sensed by the accelerometer with the highest mode shape component for the considered eigenfrequency. This parameter allows to account for the wind speed and direction at the same time. Obviously, the signal must be preliminarily subjected to a filtering operation in order to consider just its components in the frequency range of the mode considered.

With this approach, the authors proposed a preliminary empirical model [3] correlating the eigenfrequency values of the first two modes (see Table 16.1) to the accelerometer and clinometer signal features mentioned above. This paper proposes a refined analysis, comparing different empirical models and showing the capability of these models to predict the trend of the eigenfrequency values. This is addressed in the next two sections.

16.3 Empirical Model

Many different models have been investigated. The most general law used to correlate the inclination values and the acceleration RMS to the i-th eigenfrequency value f_i was the following:

$$f_i = b_0 + b_1 I_{-3}^{x} + b_2 I_{-3}^{y} + b_3 \left(I_{-3}^{x}\right)^2 + b_4 \left(I_{-3}^{y}\right)^2 + b_5 \frac{1}{I_{-3}^{x}} + b_6 \frac{1}{I_{-3}^{y}} + b_7 I_{30}^{x} + b_8 I_{30}^{y}$$
$$+ b_9 \left(I_{30}^{x}\right)^2 + b_{10} \left(I_{30}^{y}\right)^2 + b_{11} \frac{1}{I_{30}^{x}} + b_{12} \frac{1}{I_{30}^{y}} + b_{13} \sqrt{A_{RMS}^{i}} + b_{14} \exp\left(\pm A_{RMS}^{i}\right) + b_{15} \log\left(A_{RMS}^{i}\right) \quad (16.1)$$

where the b_k (with $k = 1, \ldots, 15$) coefficients of the model are the unknown coefficients to be determined, I_{-3}^{y} is the mean value of the measured inclination around the y axis (see Fig. 16.1b) at the third floor underground, I_{30}^{y} is the mean value of the measured inclination around the y axis at the 30-th floor, I_{-3}^{x} is the mean value of the measured inclination around the x axis (see Fig. 16.1b) at the third floor underground, I_{30}^{x} is the mean value of the measured inclination around the x axis at the 30-th floor, and A_{RMS}^{i} is the RMS of the measured acceleration signal coming from the accelerometer with the highest mode shape component for the i-th mode (it changes mode by mode, the accelerometer used in this case is always at the 37-th floor), with the acceleration signal band-pass filtered around the i-th eigenfrequency in order to filter out the effects of the other modes. The I and A indexes were calculated on a time base equal to 2 h, with an overlap of 1 h [3]. Therefore, a new value of the I and A indexes is available every hour. Before calculating the I and A indexes, all the signals undergo to a reliability check in order to discard corrupted data.

All the terms in Eq. (16.1) are related to specific behaviours noticed in the experimental trends of the f_i values estimated by means of OMA (f_i^{exp} in the following). The model of Eq. (16.1), and sub-models where some of the terms in Eq. (16.1) are neglected, have been taken into account to relate the f_i^{exp} values to the I and A indexes [16]. It must be noticed that also models with past values of the I indexes were considered. This allowed to take into consideration also possible effects of the thermal inertia. Since these additions did not provide any benefit, these further models with the past values are not discussed here.

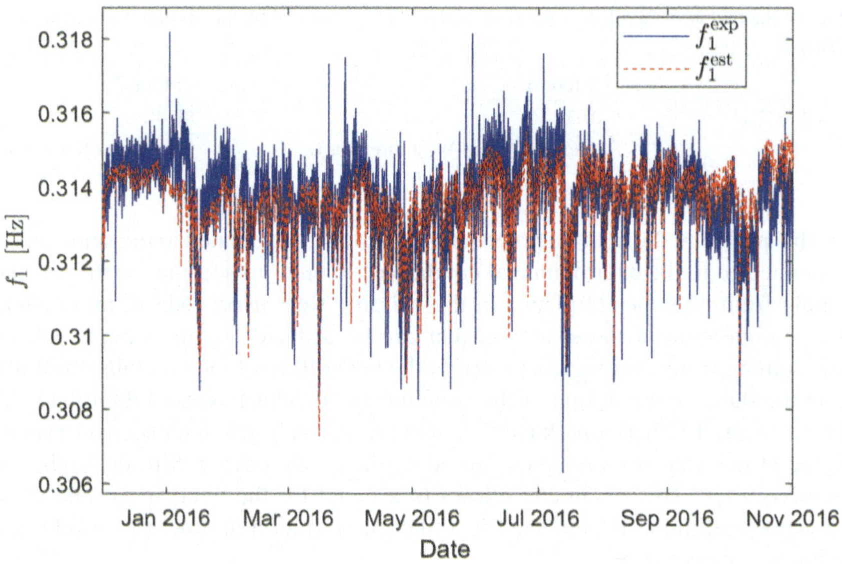

Fig. 16.2 Trend of f_1^{est} and f_1^{exp}

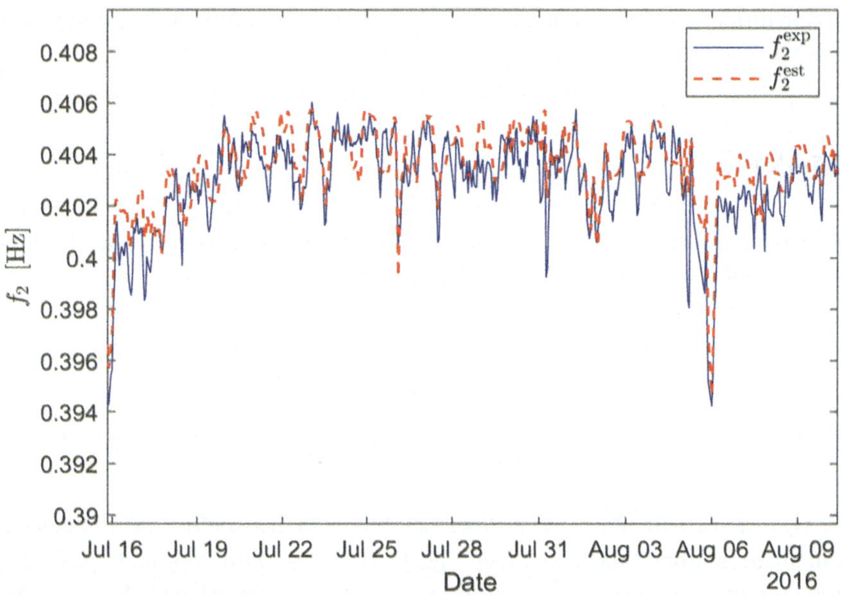

Fig. 16.3 Trend of f_2^{est} and f_2^{exp}

After a series of check related to the significance of the models (e.g. analysis of the trend of the residues), many of the them were discarded and that providing the highest fitting capability was found to be:

$$f_i = b_0 + b_1 I_{-3}^{x} + b_2 I_{-3}^{y} + b_7 I_{30}^{x} + b_8 I_{30}^{y} + b_{13}\sqrt{A_{RMS}^{i}} \tag{16.2}$$

With the model of Eq. (16.2) (using data from October 2015 to October 2016), it was then possible to estimate the building eigenfrequencies (f_i^{est} in the following). Figure 16.2 shows a comparison between f_i^{est} and f_i^{exp} for the first mode. It is evident that there is a good match between the trends. The same applies for the second mode (see Fig. 16.3 for a comparison on a shorter time-window that allows for a more detailed comparison).

Moreover, Fig. 16.4 shows (on a limited time base in order to make the plot clear) the trends of f_i^{est} and f_i^{exp} with the corresponding confidence intervals (95% level). The confidence interval for f_i^{exp} was derived in [3], while that for f_i^{est} can be calculated as explained in [16].

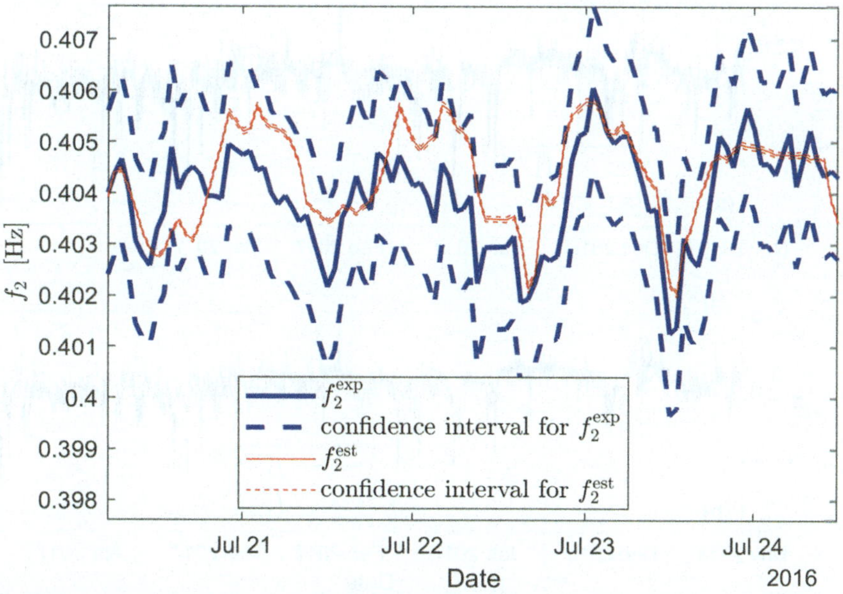

Fig. 16.4 Trend of f_2^{est} and f_2^{exp} with the corresponding confidence intervals (95% confidence level)

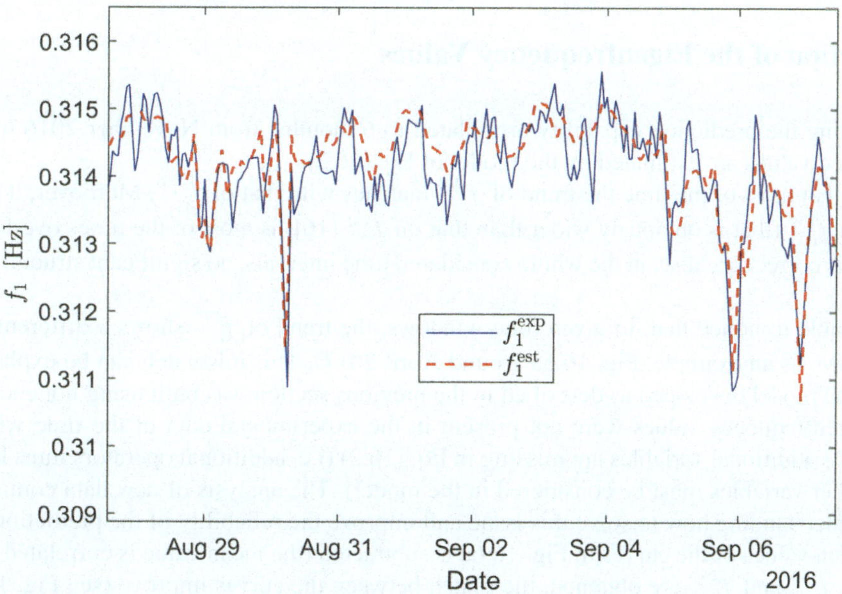

Fig. 16.5 Trend of f_1^{est} and f_1^{exp}

Finally, it is noticed that the use of the root square for the terms related to A (see Eq. (16.2)) is able to properly describe the sudden drops of the eigenfrequency values due to an increase of vibration related to an increase of the wind action (see, as an example, the peaks downward in Fig. 16.5).

The model of Eq. (16.2) for the two eigenfrequencies of Table 16.1 has then been used to predict the values of f_i (f_i^{pre} in the following) according to the values of the I and A indexes measured from November 2016 to April 2017 which were not previously used to derive the model (see the next section). A significant difference between f_i^{pre} and f_i^{exp} could, indeed, indicate a structural change.

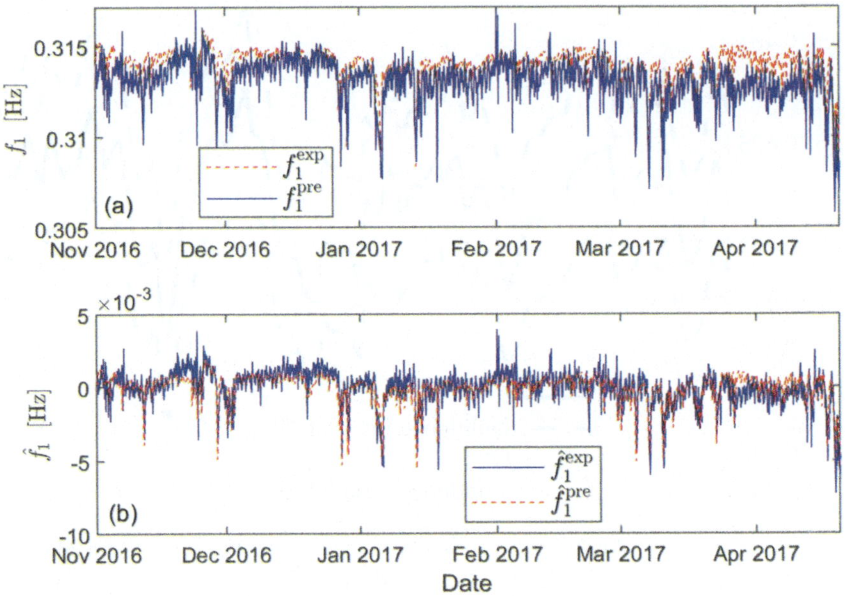

Fig. 16.6 Trends of f_1^{est} and f_1^{exp} (**a**) and trends of $\hat{f}_1^{est}$ and $\hat{f}_1^{exp}$ (**b**)

16.4 The Prediction of the Eigenfrequency Values

The data used for testing the prediction capability are related to 6 months, from November 2016 to April 2017, while the corresponding predicted values are estimated by the model of Eq. (16.2).

Figure 16.6 shows that most of the time the trend of f_i^{pre} matches with that of f_i^{exp}. Moreover, it has been found that the confidence interval on f_i^{pre} (that is obviously wider than that on f_i^{est} [16]) is most of the times overlapped to the confidence interval on f_i^{exp}. This is correct because, in the whole considered time intervals, no significant structural changes are expected to have occurred.

However, it is possible to notice that, in given time windows, the trend of f_i^{pre} shows a different mean value compared to the trend of f_i^{exp} (see, as an example, Fig. 16.6a around April 2017). This mismatch can be explained with two different possible reasons: (i) the model developed as described in the previous section was built using not enough data and long-term oscillations of the eigenfrequency values were not present in the experimental data of the time window October 2015 to October 2016, (ii) some additional variables are missing in Eq. (16.2) (i.e. additional operators must be used for the the I and A indexes or new further variables must be considered in the model). The analysis of new data coming from the monitoring system will help in understanding how to solve this issue and improve the reliability of the prediction model.

However, if the mean values of the curves in Fig. 16.6a is subtracted (the mean value is correlated with long-term trends), and the new variables $\hat{f}_i^{pre}$ and $\hat{f}_i^{exp}$ are obtained, the match between the curves improve (see Fig. 16.6b). Moreover, it has been found that the confidence intervals of $\hat{f}_i^{pre}$ and $\hat{f}_i^{exp}$ are always superimposed. Therefore, the model related to $\hat{f}_i$ can be already employed for a comparison with experimentally identified eigenfrequency values for SHM purposes.

16.5 Conclusion

The paper has dealt with the use of empirical models for SHM purposes for Palazzo Lombardia. Particularly, the model aims at describing the trend of the first eigenfrequencies of the building as functions of given indexes. These indexes are calculated relying on dynamic and static signals measured by the monitoring system. Indeed, these indexes are able to summarize the effects of different environmental variables. The prediction of the eigenfrequency trends is almost in accordance with the experimental trends, even if the description of long-term drifts must be improved.

References

1. Cigada, A., Mola, E., Mola, F., Stella, G., Vanali, M.: Dynamic behavior of the palazzo Lombardia tower : comparison of numerical models and experimental results. J. Perform. Constr. Facil. **28**(3), 491–501 (2014). https://doi.org/10.1061/(ASCE)CF.1943-5509.0000431
2. Busca, G., Cigada, A., Mola, E., Mola, F., Vanali, M.: Dynamic testing of a helicopter landing pad: comparison between operational and experimental approach. J. Civ. Struct. Heal. Monit. **4**, 133–147 (2014)
3. Berardengo, M., Busca, G., Grossi, S., Manzoni, S., Vanali, M.: The monitoring of palazzo Lombardia in Milan. Shock Vib. **2017**, 8932149 (2017). https://doi.org/10.1155/2017/8932149
4. Berardengo, M., Cigada, A., Manzoni, S., Vanali, M.: Design and Installation of a Permanent Monitoring System for Palazzo Lombardia in Milano, Italy. In: VII European Congress on Computational Methods in Applied Sciences and Engineering – ECCOMAS Congress 2016 – 5–10 June 2016, ISBN: 9786188284401, Crete Island (Greece), pp. 3640–3651 (2016). https://doi.org/10.7712/100016.2062.9244
5. Devriendt, C., Presezniak, F., De Sitter, G., Vanbrabant, K., De Troyer, T., Vanlanduit, S., Guillaume, P.: Structural health monitoring in changing operational conditions using tranmissibility measurements. Shock Vib. **17**, 651–675 (2010). https://doi.org/10.3233/SAV-2010-0556
6. Zhu, X., Hao, H.: Development of an integrated structural health monitoring system for bridge structures in operational conditions. Front. Struct. Civ. Eng. **6**, 321–333 (2012). https://doi.org/10.1007/s11709-012-0161-y
7. Ubertini, F., Gentile, C., Materazzi, A.L.: Automated modal identification in operational conditions and its application to bridges. Eng. Struct. **46**, 264–278 (2013). https://doi.org/10.1016/j.engstruct.2012.07.031
8. Rainieri, C., Fabbrocino, G.: Operational Modal Analysis of Civil Engineering Structures. Springer, New York (2014)
9. Peeters, B., Van Der Auweraer, H., Guillaume, P., Leuridan, J.: The PolyMAX frequency-domain method : a new standard for modal parameter estimation? Shock Vib. **11**, 395–409 (2004). https://doi.org/10.1155/2004/523692
10. Brincker, R., Zhang, L., Andersen, P.: Modal identification of output-only systems using frequency domain decomposition. Smart Mater. Struct. **10**, 441–445 (2001)
11. Brandt, A.: Noise and Vibration Analysis – Signal Analysis and Experimental Procedures. Wiley, Chichester (2011)
12. Shi, H., Worden, K., Cross, E.J.: A regime-switching cointegration approach for removing environmental and operational variations in structural health monitoring. Mech. Syst. Signal Process. **103**, 381–397 (2018). https://doi.org/10.1016/j.ymssp.2017.10.013
13. Bull, L.A., Rogers, T.J., Wickramarachchi, C., Cross, E.J., Worden, K., Dervilis, N.: Probabilistic active learning: an online framework for structural health monitoring. Mech. Syst. Signal Process. **134**, 106294 (2019). https://doi.org/10.1016/j.ymssp.2019.106294
14. Sen, D., Erazo, K., Zhang, W., Nagarajaiah, S., Sun, L.: On the effectiveness of principal component analysis for decoupling structural damage and environmental effects in bridge structures. J. Sound Vib. **457**, 280–298 (2019). https://doi.org/10.1016/j.jsv.2019.06.003
15. Wah, W.S.L., Owen, J.S., Chen, Y.T., Elamin, A., Roberts, G.W.: Removal of masking effect for damage detection of structures. Eng. Struct. **183**, 646–661 (2019). https://doi.org/10.1016/j.engstruct.2019.01.005
16. Montgomery, D.: Design and Analysis of Experiments, 7th edn. Wiley, New York (2009)

Chapter 17
Towards Population-Based Structural Health Monitoring, Part II: Heterogeneous Populations and Structures as Graphs

Julian Gosliga, Paul Gardner, Lawrence A. Bull, Nikolaos Dervilis, and Keith Worden

Abstract Information about the expected variation in the normal condition and various damage states of a structure is crucial in structural health monitoring. In an ideal case, the behaviour associated with each possible type of damage would be known and classification would be possible. However, it is not realistic to obtain data for every possible damage state in an individual structure. Examining a population of structures gives a much larger pool of data to work with. Machine learning can then potentially allow inferences across the population using algorithms from transfer learning.

The degree of similarity between structures determines the level of possible knowledge transfer between different structures. It is also useful to quantify in which ways two structures are similar, and where these similarities lie. This information determines whether or not certain the transfer learning approaches are applicable in a given situation. It is therefore necessary to develop a method for analysing the similarities between structures. First, it must be decided which properties of the structure to use when measuring the similarity. For example, comparing 3D CAD models or Finite Element models is not a suitable approach, since these contain a lot of irrelevant information. It is better to abstract this information into a form that contains only the relevant information.

This paper proposes Irreducible Element (IE) models, which are designed to capture the features that are crucial in determining whether or not transfer learning is possible. This information is then converted into an Attributed Graph (AG). The Attributed Graph for a structure contains the same information as the Irreducible Element model; however, the graph carries this information as a list of attributes attached to nodes. Organising the information in this manner makes it easier for graph-matching algorithms to perform a comparison between two structures. This comparison can then be used to generate a measure of similarity between the two structures and determine the most appropriate transfer learning method.

Keywords Population-based structural health monitoring · Irreducible element model · Attributed graph

17.1 Introduction

Moving beyond detecting damage in a structure, to locating and diagnosing the type of damage, raises a *supervised leaning* problem, which requires data relating to each particular type of damage (also known as the damage states of the structure). There is a large cost associated with obtaining information on the damage states of a particular engineering structure. *Population-based SHM* (PBSHM) seeks to reduce this cost by developing methods to allow the sharing of data between structures. The concept of PBSHM has been introduced in [1, 2]. If the population of structures is *homogeneous* (i.e. composed of nominally-identical structures), then it may be possible to establish a normal condition which is common across the population of structures. Even if the population is *heterogeneous* (composed of disparate structures), it may be possible to use information regarding the damage state from one structure to diagnose the same type of damage across the population. The technology that appears most likely to allow the transfer of information across populations is found in the machine learning discipline of *Transfer Learning*.

Within a strongly homogeneous population of structures, all structures have nominally-identical materials, geometry, and topology. (*Topology* in this case refers to how the individual parts which make up the structure are attached to one another.) This usually implies that these are structures of the same make and model, and that the only variation within the population is due to manufacturing uncertainty. In addition, structures in a *strongly homogeneous* population must all have the same

J. Gosliga (✉) · P. Gardner · L. A. Bull · N. Dervilis · K. Worden
Dynamics Research Group, Department of Mechanical Engineering, University of Sheffield, Sheffield, UK
e-mail: j.gosliga@sheffield.ac.uk; p.gardner@sheffield.ac.uk; l.a.bull@sheffield.ac.uk; n.dervilis@sheffield.ac.uk; k.worden@sheffield.ac.uk

B. Dilworth (ed.), *Topics in Modal Analysis & Testing, Volume 8*, Conference Proceedings of the Society for Experimental Mechanics Series, https://doi.org/10.1007/978-3-030-47717-2_17

boundary conditions. It is often the case that populations do not meet all these criteria, but are still homogeneous to some degree. This situation is typically the case for wind farms, and so far there has been work taking the first steps towards PBSHM in the Lillgrund wind farm [3]. There are many ways of performing PBSHM for a homogeneous population with one approach being to represent a homogeneous population with a single model called a form [4], and another approach being the use of transfer learning.

For a population of heterogeneous structures, each structure can have completely different materials, geometry, or topology. Within such a population, it is necessary to assess, on a case-by-case basis, whether or not damage data from one structure can be used to diagnose damage in another. For example, it may not be possible to use damage data from an aeroplane to locate damage in a bridge; however, damage data from one aeroplane may prove useful when diagnosing damage in a different aeroplane. In addition, the correct tools for making these diagnostic inferences will depend on the level of similarity between the structures [5]. Therefore, it is necessary to develop a method for consistently and quantitatively assessing the degree of similarity between structures.

The first step in assessing the similarity between structures, in the framework proposed here, involves creating an *Irreducible Element* (IE) representation of the structure, which is closely tied to the physics of the structure. This IE representation is then converted to an *Attributed Graph*. The procedure for generating an IE representation of a structure, along with a brief description of how to generate the Attributed Graph is included in [6]. This paper will expand on the choice of properties included in the IE representation, and how these help to determine the level of inference possible between structures.

The AG representation makes it possible to quantitatively assess the degree of similarity between structures. Attributed Graphs have previously been used in manufacturing in [7, 8], to measure the degree of similarity in the geometry of manufactured parts. The attributes of the graph provide an efficient method for assessing any differences in the materials or geometry of the structures. The graph directly reflects the topology of the structure it represents and so can be used to assess any differences in the topology between structures. This paper will also describe the general procedure for converting the IE representation of a structure into an Attributed Graph, and highlight how a computer implementation of the ideas can be designed, with reference to the language *Python*.

Section 17.2 will define the IE models for both an aeroplane and a wind turbine, with a discussion of how the geometry, topology (where joints form part of the description of topology), and material properties of the IE model affect knowledge transfer between two structures. Geometry is discussed in Sect. 17.2.1. Topology and joints are discussing in Sects. 17.2.2 and 17.2.2, respectively. Material properties are discussed in Sect. 17.2.3. The procedure for creating a AG from the IE model is described in Sect. 17.3. The approach for including boundary conditions in the IE and AG is described in Sect. 17.4. Formal definitions for homogeneous and heterogeneous populations are provided in Sect. 17.5.

17.2 Irreducible Element Representations of Structures

To determine whether or not two structures share similarities which allow inferences to be made, it is not necessary to consider every property or dimension of the structure. For example, in order to determine geometric similarity (to a certain resolution), it would be computationally inefficient to compare 3D models—such as Finite Element (FE) models—of structures directly. Such 3D models contain a large amount of redundant information (i.e. the detailed construction of the mesh in an FE model) that have no strong effect on the overall geometry of the structure. It is far more useful to consider which properties and dimensions are significant and abstract this information.

The proposed solution is an IE representation of the structure. For a given structure, an IE model attempts to simplify the geometry of various structural components into shapes that have well-defined dynamic behaviour, such as beams or plates, as illustrated in Fig. 17.1. The motivation for this is the belief that, for example, all plates exhibit similar behaviour (given a similar set of boundary conditions) and therefore would respond to damage in a similar way, or would have similar damage features. In the case where two IEs have the same shape and dimensions, it can be assumed that the behaviour exhibited is identical for a given set of boundary conditions. However, in reality it is likely that the geometry will be somewhat more complex and matches will not be exact. The complexity of the geometry and the degree of uncertainty within the match will influence the certainty with which inferences can be made. The IE representation can be used to determine which inference tools are appropriate, as well as the level of information that can be inferred. For example, if the method used to transfer information is transfer learning, the structures would need to be similar enough that the issue of *negative transfer* did not arise.

The IE representation captures important information about the geometry, topology, and materials in a structure that make it possible to determine the level of homogeneity between two structures. These aspects of a structure tie into the level

Fig. 17.1 Irreducible Element representations seek to remove dimensions from a model which do not significantly affect the bulk dynamic properties of a structure, or have no relevance in determining the level of information transfer possible

of information transfer between two structures, based on Rytter's hierarchy [9]. (The relation between differences in the IE representations of a structure and the level of inference possible are discussed further in [5].) The most extreme case would be if the IE representations of two structures were *identical*, which realistically would only occur when comparing a structure with itself. Within a population of structures that are all the same make and model, such as a wind farm, the IE representations would be *nominally-identical*. This condition implies that there is some probability distribution over the dimensions and material properties due to variation in manufacturing. If the population is described as strongly homogeneous, this also implies that the boundary conditions for the two structures are the same, for example all turbines in a wind farm with foundations on the same part of the seabed. A population where the IE representations are at least nominally-identical is a homogeneous population, and the idea of a *form* can be used in this case [4], in addition to other transfer learning methods.

Geometry, topology and materials have been chosen here as the main characteristics to use when judging the level of homogeneity of structures. Geometry is important when determining the overall dynamic behaviour of structures. If it is possible to assume that the dynamic behaviour is the same between two structures, then (given the same environmental conditions) any difference in the behaviour of the two structures must be due to damage. This can be seen in the Lilligrund windfarm example [3]. If the topology of two structures is the same, then it is possible to assume that any damage location labels can be applied in a consistent manner on both of them. This observation is true even if the entire structure is not similar. If two structures share a common substructure, then it may be possible to infer damage location between the two structures, provided damage occurs within this common substructure. This idea works because within this common substructure, feature and location labels should be consistent. Including materials in the description means one can determine whether inferences on the type of damage is possible. If two structures are made from the same materials, then the extent and classification labels for damage will be consistent, enabling the use of transfer learning. The following sections will detail the relevance of these properties further.

17.2.1 Geometry

Since it may not be possible to obtain complete geometrical information for all structures, it is necessary to define a hierachy of properties which define the IE to increasing levels of precision. This can be seen in Tables 17.1 and 17.2, where the coarsest level of description is the *geometry* class, followed by *shape*. In the full IE model there will be a further level where the shape is fully defined with the major dimensions, for example length, breadth and width. This hierachy of properties encodes the fact that certain transfer learning approaches are valid at a coarser level than others, with some able to deal with different parameters in the source and target domains when transfer learning is applied [5]. Similarities become better defined at a lower level of the hierarchy and improve the chance of success knowledge transfer. Using the example of a beam, certain inferences will be possible across all beams, some will require that the cross-section is the same, and other inferences rely on the geometry matching exactly.

The first four proposed geometry classes are: *beam*, *plate*, *shell*, and *complex* (and these will be sufficient for the purposes of the discussion within this paper). Beam and shell are fairly self-explanatory and correspond to the definitions used in multi-body modelling. The class of plate is introduced to differentiate between shells which enclose a volume and those that do not. There are three main cases where the complex class becomes necessary. Firstly, the complex class is used to

Table 17.1 List of elements and their properties for Turbine 1

Element designations for Turbine 1				
Name	Element ID	Material	Geometry	Shape
Rotor blade	A	FRP	Beam	Aerofoil
Rotor blade	B	FRP	Beam	Aerofoil
Rotor blade	C	FRP	Beam	Aerofoil
Rotor hub	D	FRP	Complex	Rotor hub
Nacelle	E	FRP	Shell	Cuboid
Tower section 1	F	Metal	Beam	Cylindrical
Tower section 2	G	Metal	Beam	Cylindrical
Tower section 3	H	Metal	Beam	Cylindrical
Foundation	I	Concrete	Plate	Cylindrical
Name	Element ID	Boundary	–	–
Footing	1	Ground	–	–

Table 17.2 List of elements and their properties for Aeroplane 1

Element designations for Aeroplane 1				
Name	Element ID	Material	Geometry	Shape
Fuselage	A1	FRP	Shell	Truncated cone
Fuselage	A2	FRP	Beam	Cylindrical
Fuselage	A3	FRP	Shell	Cone
Wing 1	B	FRP	Beam	Aerofoil
Pylon 1	C	FRP	Complex	Pylon
Engine 1	D	Assembly	Shell	Cylinder
Pylon 2	E	FRP	Complex	Pylon
Engine 2	F	Assembly	Shell	Cylinder
Wing 2	G	FRP	Beam	Aerofoil
Pylon 3	H	FRP	Complex	Pylon
Engine 3	I	Assembly	Shell	Cylinder
Pylon 4	J	FRP	Complex	Pylon
Engine 4	K	Assembly	Shell	Cylinder
Vert stabiliser 1	L	FRP	Beam	Aerofoil
Vert stabiliser 2	M	FRP	Beam	Aerofoil
Horz stabiliser	N	FRP	Beam	Aerofoil
Front landing gear	O	Assembly	Complex	Assembly
Rear landing gear	P	Assembly	Complex	Assembly
Name	Element ID	Boundary	–	–
Tarmac	1	Ground	–	–

represent the components within the structure that could in theory be broken down further into simpler IEs,[1] but doing so would mean creating multiple elements out of a single structural component. An example of this can be seen in the elements for the aeroplane in Table 17.2, where the fuselage has been broken into separate elements in order to better define the geometry. However, when it comes to similarity matching, there are benefits to defining it as a single complex element. Using aeroplanes as an example, it may be beneficial to define the fuselage as a single element so that each aeroplane now shares a common feature: a fuselage element attached to two wing elements. Secondly, the complex class can be used in the case where it is beneficial to represent a collection of components as a single element, for example the landing gear in Table 17.2 is represented as a single complex element. The third case where the *complex* class is useful is when there exists an element which is not easily described by a simple shape, for example the rotor hub on a wind turbine in Table 17.1.

[1]There might appear to be a contradiction in the terminology here, in the idea that an *irreducible* element might be decomposed. However, the term is used here in the sense that an IE model can provide a description of *sufficient* complexity for the purposes of matching structures, but the user may choose a higher resolution.

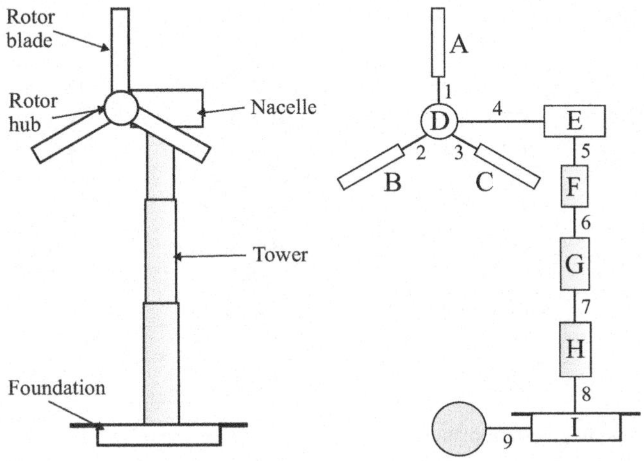

Fig. 17.2 An expanded IE representation of a wind turbine with the elements labelled A to I and the connections labelled 1 to 9. The shaded node is a special element representing the ground, where the boundary condition is defined in the attributes for Joint 9

17.2.2 *Topology*

The topology of the structure plays a large part in determining the dynamic behaviour of a structure. The topology here is determined by physical connections between the various elements, as shown in Fig. 17.2. With reference to Rytter's hierachy [9], the topology has a strong link to damage location. Structures without corresponding topology cannot have fully consistent location labels.

The complex class provides flexibility to simplify the topology of structures where it is believed that the topology is not significant for the problem. For example, if the application was comparison of the overall dynamics of an aeroplane, it may be useful to simplify the landing gear or the fuselage into a single element so that differences in overall configuration, such as engine placement are emphasised. Alternatively, if the application was locating damage within the fuselage, it would be beneficial to break this complex element into constituent parts.

Joints

Joints are classed as capturing topological information since they define where connections between elements occur; they also have the same effect on the label consistency as topology [5]. Example lists of joints and their properties can be found in Tables 17.3 and 17.4 for a wind turbine and an aeroplane respectively. The element set describes which elements the joint forms a connection for. This is crucial in recovering the topology of the structure. The coordinates define the geometrical location of the joint; the joint coordinate aids in constructing physical models from the IE representation. These coordinates define the mid-point of the joint and are determined by examining the geometry of the structure. The joint type describes some of the key physical aspects of the joint, such as the type of joint and the degrees of freedom that are constrained. The joint type and restricted degrees of freedom are again hierarchical, as it may not be possible to obtain this information in every case. It should also be noted that for static joints it is assumed that all degrees of freedom are constrained and so this property is not specified for static joints. The properties contained within the joints are important both for constructing the Attributed Graph, as well as for creating physical models of the structure.

Anticipating the next section a little, the *nodes* in the Attributed Graphs will represent the IEs, the *edges* will represent the joints between them.

17.2.3 *Materials*

The topology and geometry of the structure are most strongly linked with the transfer of damage detection and damage location labels. The material properties, on the other hand, determine the assessment and classification labels. A hierarchical set of materials properties are used to describe the materials within an IE, as again it cannot be guaranteed that detailed

Table 17.3 List of joints and their properties for Turbine 1

Joint designations for Turbine 1					
Joint ID	Element set	Coordinate	Type	Disp. DoF	Rot. DoF
1	A, D	8, 15, 235.75	Bearing	[x, y, z]	[y, z]
2	B, D	8, 14, 254	Bearing	[x, y, z]	[y, z]
3	D, E	10, 15, 253	Bearing	[x, y, z]	[y, z]
4	D, C	8, 16, 254	Bearing	[x, y, z]	[x, y]
5	E, F	15, 15, 250	Bearing	[x, y, z]	[x, y]
6	F, G	15, 15, 183	Bolted	–	–
7	G, H	15, 15, 105	Bolted	–	–
8	H, I	15, 15, 5	Bolted	–	–
9	I, 1	15, 15, 0	Soil	–	–

Table 17.4 List of joints and their properties for Aeroplane 1

Joint designations for Aeroplane 1					
Joint ID	Element set	Coordinate	Type	Disp. DoF	Rot. DoF
1	A1, A2	34.2, 14.68, 5.165	Perfect	–	–
2	A2, A3	34.2, 60.96, 5.165	Perfect	–	–
3	A2, B	32.2, 29.79, 2.89	Lug	–	–
4	B, C	13.2, 42.67, 4.74	Complex	–	–
5	C, D	13.2, 40.17, 4.74	Complex	–	–
6	B, E	23.2, 30.79, 3.57	Complex	–	–
7	E, F	23.2, 28.29, 3.57	Complex	–	–
8	A2, G	36.2, 29.79, 2.89	Lug	–	–
9	G, H	45.2, 30.79, 3.57	Complex	–	–
10	H, I	45.2, 28.29, 3.57	Complex	–	–
11	G, J	55.2, 42.67, 4.74	Complex	–	–
12	J, K	55.2, 40.17, 4.74	Complex	–	–
13	A3, L	33.2, 68.58, 7.55	Lug	–	–
14	A3, M	35.2, 68.58, 7.55	Lug	–	–
15	A3, N	34.2, 64.58, 9.16	Lug	–	–
16	A1, O	34.2, 7.75, 1.75	Complex	–	–
17	A2, P	34.2, 29.67, 1.75	Complex	–	–
18	O, 1	34.2, 7.75, 0	Plane	[z]	[x, y]
19	P, 1	34.2, 29.67, 0	Plane	[z]	[x, y]

information on materials used will be available for each structure. The coarsest level of detail will be the *material* class (ceramics, metals). The next level of detail will be a specific description of the material within the class, for example brass and steel belong to the material class of metals. The finest level of detail will be the specific properties of the material, for example Young's Modulus and density. Material grade may also form part of the attributes. Once again, to be confident of an exact match between two structures, all of the material properties must be known down to the bottom of the hierarchy, otherwise there will be a degree of uncertainty.

The material properties determine whether it is possilbe to make inferences using damage assessment and classification labels between two structures. Two materials of the same material class will experience similar failure modes, and could be expected to exhibit a similar response to a particular type of damage giving more confidence in the classification of damage. However, to build confidence that the assessment (extent) of the damage will be the same between two structures, the material would ideally be the same. For example, aluminium and steel will both suffer corrosion (damage classification), but there is little guarantee that the change in material properties will be the same for the same extent of corrosion. If the materials are identical, then transferring damage assessment and classification labels is trivial.

17.3 Producing an Attributed Graph

For the Attributed Graph (AG) the same information is embedded as for the IE representation, but whereas the properties for the IE representation are organised in a tabular format to improve readability by humans, the information in the Attributed Graph is organised so that it can be more efficiently processed by a graph-matching algorithm [10].

The structure of the elements and joints is extracted from the properties in the tables to create the graph itself. A graph G consists of a set of nodes V and edges E. The set of nodes contains all of the elements from the IE representation. The element set for each joint is used to construct the set of edges. One can define the graph of the turbine G_{turbine} from the set of nodes $V_{\text{turbine}} = \{A, B, C, D, E, F, G, H, I, 1\}$ and set of edges $E_{\text{turbine}} = \{(A, D), (B, D), (D, E), (D, C), (E, F), (F, G), (G, H), (H, I), (I, 1)\}$. The graph is represented as a Python dictionary, which is a datatype where values are stored with a lookup key, e.g. dictionary = {'key': 'values'}. The dictionary representation of a graph contains the list of nodes (labelled with their IE designation as shown in Fig. 17.3) and the values for each node in the dictionary are a list of neighbouring nodes.

One can define a dictionary for a subgraph G_1 of the turbine graph G_{turbine} shown in Fig. 17.3:

```
graph = {'A' : ['D']
         'B' : ['D']
         'C' : ['D']
         'D' : ['A', 'B', 'C', 'E']
         'E' : ['D']}
```

where the curly brackets are used to specify that the datatype will be 'dictionary'. The quotation marks are used to denote strings, which are used as node labels in the code. The square brackets are used to specify a 'list'. The edges of the graph can be obtained by examining the dictionary 'key' (to the left of the colon) and creating a pair with any single entry in the list of neighbours (to the left of the colon). For example, the edge (A, D) can be created by pairing the dictionary key A with its neighbour D. Alternatively, the equivalent edge (D, A) could be created by pairing the dictionary key D with its neighbour A.

Examining the subgraph G_1 shown in Fig. 17.4 with the full turbine graph G_{turbine} shown in Fig. 17.3, it is clear that this is indeed a valid subgraph. By defining the node set $V_1 = \{A, B, C, D, E\}$ and edge set $E_1 = \{(A, D), (B, D), (D, E), (D, C)\}$ for the subgraph, it can be stated that $V_1 \subseteq V_{\text{turbine}}$ and $E_1 \subseteq E_{\text{turbine}}$. Since E_1

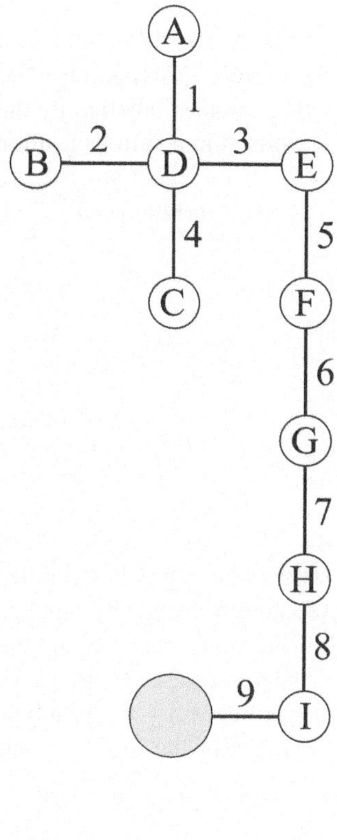

Fig. 17.3 The resulting Attributed Graph from the IE representation shown in Fig. 17.2. It is possible to see how the topology is defined by the element set in the joints

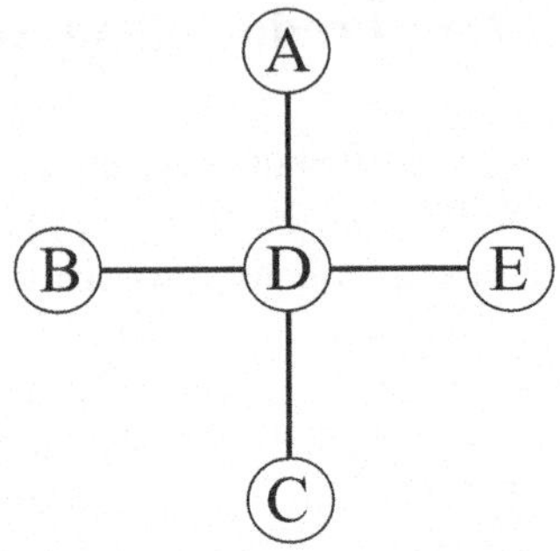

Fig. 17.4 A sample subgraph of Fig. 17.3

contains all edges from E_{turbine} that have both endpoints in V_1, G_1 is an *induced* subgraph of G_{turbine}. For determining similarity between engineering structures, induced subgraphs are the only subgraphs of interest, since to preserve topology it is necessary to include all connections that exist within the two parent graphs when examining any sub-structures. In addition, only connected subgraphs will be considered, since in a structure, every part must be physically connected to at least one other part of the structure.

The dictionary data type in Python is also used to store the properties from the IE representation as node and edge attributes. This datatype allows the graph-matching algorithm to easily query the attributes of a given node, since the node label can be used as a dictionary key. The organisation of the information for elements consists of using the element ID as the dictionary key, and the element attributes are stored as a list, with nested lists containing the geometrical and material attributes. For example the node attributes for the elements in Table 17.1 appear as:

```
turbine.elements = {‘A’ : [‘FRP’,      [‘Beam’, ‘Aerofoil’]],
                    ‘B’ : [‘FRP’,      [‘Beam’, ‘Aerofoil’]],
                    ‘C’ : [‘FRP’,      [‘Beam’, ‘Aerofoil’]],
                    ‘D’ : [‘FRP’,      [‘Complex’, ‘Rotor hub’]],
                    ‘E’ : [‘FRP’,      [‘Shell’, ‘Cuboid’]],
                    ‘F’ : [‘Metal’,    [‘Beam’, ‘Cylindrical’]],
                    ‘G’ : [‘Metal’,    [‘Beam’, ‘Cylindrical’]],
                    ‘H’ : [‘Metal’,    [‘Beam’, ‘Cylindrical’]],
                    ‘I’ : [‘Concrete’, [‘Plate’, ‘Cylindrical’]],
                    ‘1’ : [‘Ground’]}
```

For edges, the label is exchanged with the element set. This is because when looking up edges in resulting subgraphs, the easiest way to query if an edge exists between node v_1 and node v_2 is to check whether $(v_1, v_2) \in E$ and so naturally edges are best labelled by their node pair. The coordinates and joint information (for the turbine joints in Table 17.3) is then organised in a similar fashion to the node attributes, where the hierachical information is contained within a nested list:

```
turbine.joints = { (‘A’, ‘D’) : [‘1’, [8, 15, 235.75], [‘Bearing’, [‘x’, ‘y’, ‘z’], [‘y’, ‘z’]]],
                   (‘B’, ‘D’) : [‘2’, [8, 14, 254],    [‘Bearing’, [‘x’, ‘y’, ‘z’], [‘y’, ‘z’]]],
                   (‘C’, ‘D’) : [‘3’, [8, 16, 254],    [‘Bearing’, [‘x’, ‘y’, ‘z’], [‘y’, ‘z’]]],
                   (‘D’, ‘E’) : [‘4’, [10, 15, 253],   [‘Bearing’, [‘x’, ‘y’, ‘z’], [‘x’, ‘y’]]],
                   (‘E’, ‘F’) : [‘5’, [15, 15, 250],   [‘Bearing’, [‘x’, ‘y’, ‘z’], [‘x’, ‘y’]]],
                   (‘F’, ‘G’) : [‘6’, [15, 15, 183],   [‘Bolted’]],
                   (‘G’, ‘H’) : [‘7’, [15, 15, 105],   [‘Bolted’]],
                   (‘H’, ‘I’) : [‘8’, [15, 15, 5],     [‘Bolted’]],
                   (‘I’, ‘1’) : [‘9’, [15, 15, 0],     [‘Soil’]]}
```

where the regular parentheses represent a *tuple* in Python. While organising the information in this way reduces the readability from a human perspective, it makes it easier for a graph-matching algorithm to search for node or edge attributes. Furthermore, there are no empty rows or columns when presented with an incomplete set of attributes.

Given the set of nodes V and edges E for a structure, it is possible to plot the graph using Python. Plotting the graphs is not necessary for graph comparison, but can aid in assessing whether or not the code is producing valid subgraphs. An example of the graphs produced using the Python code is shown in Fig. 17.5. From these two graphs of the turbine and

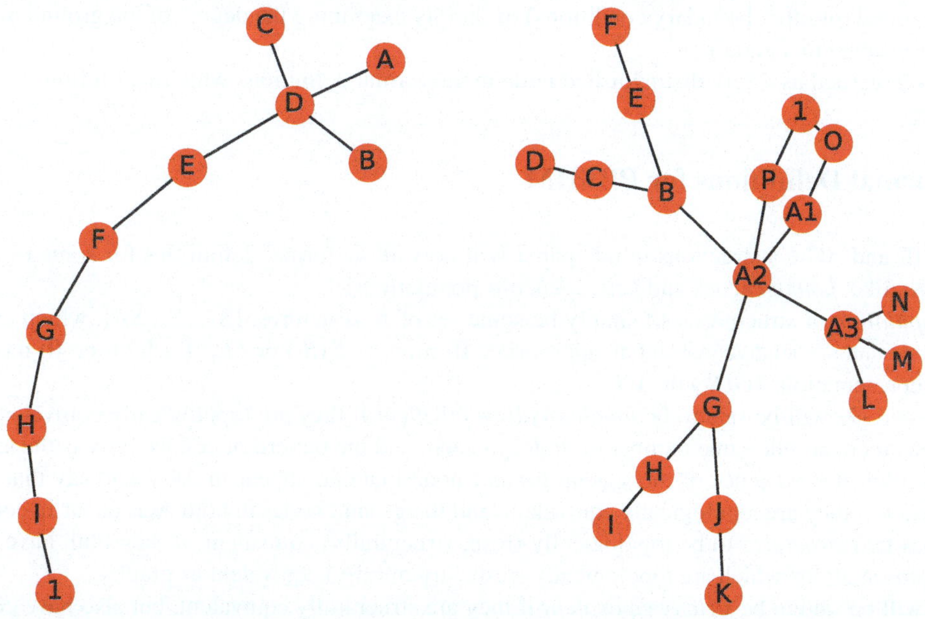

Fig. 17.5 The graphs produced from list of elements in Tables 17.2 and 17.1 and list of joints in Tables 17.4 and 17.3. The graph shown on the left is from the wind turbine (compare with Fig. 17.3) and the graph on the right is from the aeroplane

the aeroplane, it is possible to visually determine that G_1 (Fig. 17.4) is a valid subgraph (compare the subset of nodes $\{A, B, C, D, E\}$ in the turbine with subset of nodes $\{A2, A3, N, M, L\}$ in the aeroplane) of both G_{turbine} and $G_{\text{aeroplane}}$. Using the node attributes, it is possible to align the aerofoil elements in each node set, giving the corresponding node set for the common subgraph $V_1 = \{(A, N), (B, M), (C, L), (D, A3), (E, A2)\}$. This node re-labelling requires an update of the edge set $E_1 = \{((A, N), (D, A3)), ((B, M), (D, A3)), ((C, L), (D, A3)), ((E, A2), (D, A3))\}$.

Once an AG is defined, it immediately induces a basic modal model – the *Canonical Modal Model* (CMM). This is constructed as a lumped-mass model as follows: first, one replaces each node (IE) in the model by a lumped mass computed using the appropriate node attributes (i.e. density and dimensions). Secondly, one replaces each edge by a spring with stiffness computed using the edge attributes. Taking care to preserve ground nodes, this procedure defines a basic model from which modal invariants (natural frequencies) can be computed.

17.4 A Note on Boundary Conditions

Because boundary conditions are essentially a type of joint, they are defined using special edges, which also determine which coordinate degrees of freedom are actually constrained. In fact, a complete boundary condition will be specified by an edge-node pair; defining the boundary condition and the object to which the boundary is connected. There are two main cases to deal with here; the first is when the connection is to 'ground', the second is when the structure is connected to another structure which is deemed to be distinct. Boundary conditions imposed by the same root cause can be represented by the same node. For example, in the case of the aeroplane described in Tables 17.2 and 17.4, both sets of landing gear as restricted by the tarmac (ground), as the aeroplane sits on the runway.[2] As such, these two boundary conditions can be represented by the same sort of edge, connected to the same node, which represents the runway (ground). The ground also features as part of a boundary condition for the wind turbine in Table 17.1, although as can be seen from Table 17.3, the wind turbine has a permanent static connection to ground. The same type of boundary condition can manifest in different ways depending on how the structure is connected, and so a complete match in boundary conditions requires that both the type of node, and the edge connected to the boundary are the same.

The justification for this is partially physical – all connections to 'ground' are modelled as rigid connections to the earth – but also practical. Having single edge-ground combinations for boundary conditions makes it possible to calculate the number

[2]If the aircraft is stationary, this joint represents a static friction; if the aircraft is in motion, one would need to consider the joint as a rolling friction; if the aircraft were in flight, the joint would disappear

of connections to ground (or other boundary conditions) by simply examining the degree of the ground nodes. (Degree is the number of edges connected to a node.)

'Ground' will be regarded as a very distinguished node in the formal definitions which will follow.

17.5 Some Formal Definitions for PBSHM

The details of the IE and AG constructions in this paper will now allow formal definitions for some of the terms that have been used up to now, like *homogeneous* and *heterogeneous* populations.

First of all a *population* of structures will simply be some set of N structures, $\{S_1, \dots, S_N\}$, which exist in reality. With each structure S_i, one can associate/construct an appropriate IE model $IE(S_i)$ or IE_i. Each of the IE models will induce an Attributed Graph representation $AG(S_i)$ or AG_i.

Two graphs G_1 and G_2 will be said to be *topologically equivalent* if they are topologically equivalent, when considered as simple graphs, i.e. there are the same number of nodes in each, and the pattern of connectivity is the same.

Now, one can exploit the existence of the special *ground* node that can appear in AGs, and say that the two graphs are *structurally equivalent* if they are topologically equivalent, and the ground nodes in both AGs occur in corresponding places.

Now, one defines two *structures* to be topologically (resp. structurally) equivalent, if they both have IE representations, that induce AG representations which are topologically (resp. structurally) equivalent as graphs.

Two structures will be said to be *strictly equivalent* if they are structurally equivalent, but also have corresponding IEs at corresponding nodes in the AG; this means that the attribute vectors for the nodes must have the same entries, with the same physical meanings.

Now, a population of structures $P = \{S_i\}$ is *topologically (resp. structurally, strictly) homogeneous*, if all structures in P are pairwise topologically (resp. structurally, strictly) equivalent. If *any* pair of structures fail in equivalence, the population P is *heterogeneous*.

Consider a strictly homogeneous population $P = \{S_i : i = 1, \dots, N\}$, where each structure S_i has parameters representing the node attributes, denoted by $\underline{\theta}^i_j$, where j is the node index, and parameters $\underline{\theta}^i_{jk}$, where j and k indicate that the edge is between nodes j and k. By the strict homogeneity assumption, the parameters in the vectors for each structure are in one-to-one correspondence in terms of physical meaning. A population now, will be termed *fully homogeneous* if the attributes can be considered to be random draws from distributions $p_j\underline{\theta}_j)$ and $p_{jk}(\underline{\theta}_{jk})$, which are *independent* of the structure. Clearly, this definition encompasses the idea that a homogeneous population is comprised of structures that are exactly the same make and model, with variations arising only from manufacturing/embodiment variations.

In order to fully motivate the idea of a *form*, it may be necessary to add a stronger condition, even than this. For example, one might say that a population is *strongly homogeneous* if it is homogeneous, and the attribute densities p_j and p_{jk} are unimodal, or convex etc.

A wind farm with all the turbines the same model, but subject to manufacturing variations would be expected to be strongly homogeneous, assuming that some parameterised foundation model was appropriate to all the turbines.

These definitions are part of the essential machinery that will be needed to assess if transfer of inferences is likely to succeed between two different structures. Part VI of the papers in this sequence, will return to this issue [5].

17.6 Conclusions

The properties of a pair of IE models of two structures determines the level of inference possible between the structures. This is due to the fact that consistent damage labels are required for transfer learning. Matching topology and element geometry gives consistent location labels. Matching the material properties for elements gives consistent damage classification and assessment labels. Therefore, when looking for exact matches, it is important to consider all three aspects of a particular structure: topology, materials and geometry. Accordingly, these three categories are captured in the information contained in the AG. The nodes contain attributes which correspond to these properties in the IE, however the information is arranged in a way that makes it more accessible to any graph-matching algorithms. Joints carry important topological information, particularly in terms of connections to ground, which is necessary for not only creating the AG, but also when searching for common subgraphs. By carefully choosing the information and data structure, it is possible to compare the AG for two structures in order to determine the most appropriate transfer learning tool. Comparing the attributes of the graphs (which correspond to the properties of the IE model) allows one to group structures based on the level of knowledge transfer possible between any two given structures.

Acknowledgments The authors would like to thank the UK EPSRC for funding through the Established Career Fellowship EP/R003645/1 and the Programme Grant EP/R006768/1.

References

1. Antoniadou, I., Dervilis, N., Papatheou, E., Maguire, A.E., Worden, K.: Aspects of structural health and condition monitoring of offshore wind turbines. Philos. Trans. R. Soc. A Math. Phys. Eng. Sci. **373**, 1–14 (2015)
2. Worden, K., Cross, E.J., Dervilis, N., Papatheou, E., Antoniadou, I.: Structural health monitoring: from structures to systems-of-systems. IFAC-PapersOnLine, pp. 1–17 (2015)
3. Papatheou, E., Dervilis, N., Maguire, A.E., Antoniadou, I., Worden, K.: A performance monitoring approach for the novel Lillgrund offshore wind farm. IEEE Trans. Indus. Electron. **62**, 6636–6644 (2015)
4. Bull, L.A., Gardner, P.A., Gosliga, J., Maguire, A.E., Campos, C., Rogers, T.J., Haywood-Alexander, M., Dervilis, N., Cross, E.J., Worden, K.: Towards population-based structural health monitoring, part I: Homogeneous populations and forms. In: Proceedings of IMAC XXXVIII – the 38th International Modal Analysis Conference, Houston (2020)
5. Gardner, P.A., Worden, K.: Towards population-based structural health monitoring, part IV: Heterogeneous populations, matching and transfer. In: Proceedings of IMAC XXXVIII – the 38th International Modal Analysis Conference, Houston (2020)
6. Gosliga, J., Worden, K.: A general representation for assessing the similarity of structures. In: Proceedings of the 12th International Workshop on Structural Health Monitoring, Palo Alto (2019)
7. El-Mehalawi, M., Allen Miller, R.: A database system of mechanical components based on geometric and topological similarity, part I: representation. Comput. Aided Des. **35**, 83–94 (2001)
8. El-Mehalawi, M., Allen Miller, R.: A database system of mechanical compoennts based on geometric and topological similarity, part II: indexing, retrieval, matching, and similarity assessment. Comput. Aided Des. **35**, 94–105 (2001)
9. Rytter, A.: Vibrational based inspection of civil engineering structures. Ph.D. thesis, Deptartment of Building Technology and Structural Engineering, Aalborg University (1993)
10. Gosliga, J., Gardner, P.A., Bull, L.A., Dervilis, N., Worden, K.: Towards population-based structural health monitoring, part III: Graphs, networks and communities. In: Proceedings of IMAC XXXVIII – the 38th International Modal Analysis Conference, Houston (2020)

Julian Gosliga completed his Masters degree in Mechanical Engineering at the University of Sheffield. He then went on to complete a PhD on Energy Harvesting within the same department. Julian now works as a Research Associate for the Dynamics Research Group at the University of Sheffield, focusing on population-based structural health monitoring.

Chapter 18
Developing a Correlation Criterion (SpaceMAC) for Repeated and Pseudo-repeated Modes

Pranjal M. Vinze, Randall J. Allemang, and Allyn W. Phillips

Abstract One of the most important factors in the validation of a finite element model is whether the modal vectors obtained from the finite element solution match sufficiently with the modal analysis results of the part or a test prototype. Modal Assurance Criteria (MAC) (Allemang, Brown, A correlation coefficient for modal vector analysis. In: Proceedings of the international modal analysis conference, pp 110–116, 1982) is usually a very effective way to check this condition. However, in case of repeated modal frequencies and vectors, MAC can give misleading results. Hence, there is a need for a method that could indicate how well the finite element method-based estimates for the repeated modes correlate with the modal analysis modes. This work is an attempt to develop a correlation criterion between a set of repeated roots from a finite element method solution and an experimental modal analysis solution.

Building on a low dimensional modal vector example, a vector subspace-based approach was identified to help properly define the solution to a characteristic equation with repeated roots. This analogy was extended to higher dimension modal vector cases and vector subspace or hyper planes were identified as a way to model a repeated mode case.

Similar to MAC consistency of the solution was considered as ideal way to establish correlation. But in this case the consistency of the solution subspace was found to be more important than that of normalized modal vectors. The smallest principal angle between the two solution subspaces was identified as a way of measuring the consistency. The criterion, referred to as spaceMAC, was developed as a function of this angle such that the range of the criterion is 0–1, similar to MAC was defined as 1- sin(θ) where θ is the principal angle between the two solution subspaces. This criterion was tested with two datasets 2001 Circular Plate Dataset and 2014 Circular Plate Dataset.

Keywords Modal Assurance Criterion (MAC) · Repeated modes · Correlation · Modal vector · Vector subspaces

18.1 Introduction

Modal Assurance Criterion (MAC) is the most common way of measuring the consistency of the estimates of modal vectors when the lengths of the modal vectors are same. It can be used in a 'black box' manner where the user can input the modal vectors and get a direct indicator of correlation between various modal vectors. However, there is a drawback in this approach. MAC can give misleading results in case of repeated and pseudo-repeated vectors. A usual workaround that is used in these cases is to observe a wireframe animation or to observe the mode shape for the modes in question. Then it can be checked if any shape can be rotated in space to get the shape for some other mode.

Apart from using the wireframe animation, some methods exist to detect and correlate repeated roots. For example, it has been shown that by taking the singular value decomposition (SVD) of unity scale factor (USF), which is defined similarly to modal scale factor (MSF), the number of repeated roots can be found [1]. A method developed for correlating repeated modes is shown in [2], where the test mode is expressed as a linear combination of all possible combinations of analytic modes. The cross-MAC values for all these combinations and the test mode shape are calculated. From these cross-MAC values it is found which combination is a repeated mode combination. A new method has been developed based on vector subspace theory to address some limitations of existing methods and make the criteria more 'black-box' approach friendly.

P. M. Vinze (✉) · R. J. Allemang · A. W. Phillips
Department of Mechanical Engineering, College of Engineering and Sciences, University of Cincinnati, Cincinnati, OH, USA

Structural Dynamics Research Laboratory, College of Engineering and Sciences, University of Cincinnati, Cincinnati, OH, USA
e-mail: vinzepm@mail.uc.edu; allemarj@ucmail.uc.edu; philliaw@ucmail.uc.edu

B. Dilworth (ed.), *Topics in Modal Analysis & Testing, Volume 8*, Conference Proceedings of the Society for Experimental Mechanics Series, https://doi.org/10.1007/978-3-030-47717-2_18

A FEM analytical solution from ANSYS was extracted and downsampled to the experimental modal vectors for existing datasets (2001 Circular Plate Dataset and 2014 Circular Plate Dataset). These two datasets were used to test the correlation criteria.

18.2 Background

18.2.1 Repeated Roots

A repeated modal frequency condition is said to exist when the modal frequencies of two or more modes occur at the same frequency. This condition in experimental analysis exhibits equal complex modal frequencies, that is both the modal frequency and the modal damping are equal. It is important to correctly estimate repeated modes to get an accurate model of the system. Therefore, it is important to have a method to estimate the certainty in the estimates of repeated modes. Analytically, modal vectors are calculated from Eq. 18.1a whereas experimental results correspond to Eq. 18.1b

$$\left(s^2\,[\mathrm{M}] + [\mathrm{K}]\,\{\mathrm{X}\}\right) = \{0\} \tag{18.1a}$$

$$\left(s^2\,[\mathrm{M}] + \mathrm{s}\,[\mathrm{C}] + [\mathrm{K}]\,\{\mathrm{X}\}\right) = \{0\} \tag{18.1b}$$

This is why the experimental modal vectors are complex valued while analytical modal vectors are real valued. In the case of repeated roots when the analytically evaluated eigenvalue is substituted in Eq. 18.1a, the equation is rank deficient by the number of repeated roots at that eigenvalue. If a set of Nr repeated roots are present at the eigenvalue λr, Nr physical coordinates have to be assumed in the calculation of the eigenvector. This has to be repeated Nr times, once for each repeated eigenvalue. A number of Nr physical coordinates can be selected and they can be assigned infinite number of arbitrary combinations of constants. Based on our selection of sets of these coordinates and their values, different eigenvectors are obtained. This is why normalized eigenvectors for repeated modes are not unique like regular eigenvectors. Experimentally, the repeated modes will probably occur as closely spaced frequencies. The multi-degree of freedom (MDOF) FRF is expressed as a linear superposition of many single degree of freedom systems, each belonging to a mode. To be able to detect repeated modes experimentally, it is important to have at a minimum, the number of independent columns in the FRF matrix equal to the number of repeated mdoes in the largest repeated mode set that is to be identified. It is usually better to have more columns (references) in the FRF matrix than that.

A modal vector that belongs to a repeated mode set is orthogonal to other modal vectors but not necessarily to the modal vectors that are from the same repeated mode set. In case of a repeated mode of multiplicity 2, two independent vectors will be required to describe the motion of plate at that frequency, but these two vectors are not unique. Since the complex valued frequencies may not be perfectly equal in the case of repeated modes in experimental analysis, it can be problematic when regular modes occur at closely spaced frequencies. This condition is called a pseudo-repeated mode.

18.2.2 Modal Vectors as Vector Spaces and Subspaces

A vector subspace is a nonempty set of vectors, on which operations of vector addition and scalar multiplication are defined, such that they follow ten basic axioms stated in [3]. Vector spaces can contain within them other vector spaces. Such a vector space is said to be a vector subspace. A vector space H, which is a subset of a vector space V, is said to be the subspace of V if,

1. The zero vector of V belongs to H
2. For all vectors $\mathbf{u}$ and $\mathbf{v}$ in H, $\mathbf{u} + \mathbf{v}$ belongs to H
3. For all vectors $\mathbf{u}$ in H and all scalars a, $a\mathbf{u}$ belongs to H

It can be checked with the axioms in [3], that the eigenspace corresponding to modal vectors is a vector space. In case of repeated modes, the modes are not unique and any linear combination of the modes of a repeated set can be said to be a mode in the repeated set. Also, any mode multiplied by a scalar still describes the same mode. These two considerations mean that a repeated mode set is a vector subspace of the eigenspace corresponding to the characteristic equation, that is Eq. 18.1a for analytical solution and Eq. 18.1b for experimental solution.

18.2.3 Modal Vectors as Vector Spaces and Subspaces

The angle between subspaces is an important concept when trying to correlate subspaces. This angle is easy to visualize in a subspace made in a R^3 space. This means the vector length is 3. The subspace in this case would be a plane. If the estimates of this plane are consistent the angle would be small. The minimum principal angle between two subspaces of equal vector length X and Y is given by,

$$\cos(\theta) = \max\left(\max\left(\left|x^H y\right|\right)\right) \tag{18.2}$$

where x and y are unit vectors in the X and Y subspaces respectively. For computation of this minimal angle, MATLAB function subspace used. This function first finds the orthonormal bases for both sets of vectors. The orthonormal bases are swapped to maintain order of rank. It then makes use of the sine projection to calculate the angle.

18.2.4 Existing Correlation Methods

Correlation methods like MAC and shape difference indicator (SDI) [8] do not directly work on repeated modes. MAC is defined by the Eq. 18.3,

$$MAC_{qr} = \frac{\{\phi_r\}^H \{\psi_q\} \{\psi_q\}^H \{\phi_r\}}{\{\phi_r\}^H \{\phi_r\} \{\psi_q\}^H \{\psi_q\}} \tag{18.3}$$

Figure 18.1 shows the cross-MAC values for 2001 Circular Plate Dataset between experimental and finite element method solutions. In case of 12th and 13th modes it can be seen that the cross-MAC values are not close to unity on the diagonal or

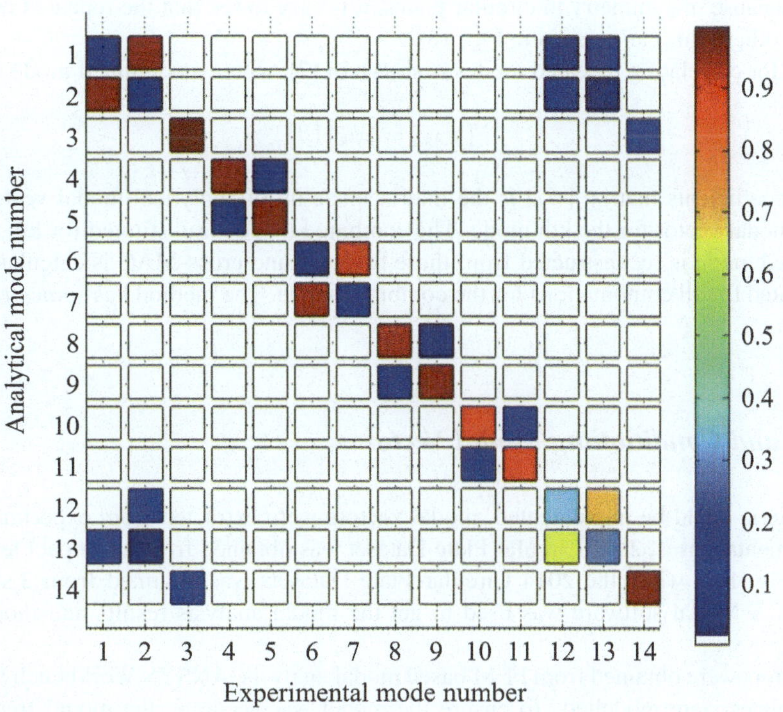

Fig. 18.1 Cross-MAC between experimental and analytical modes, 2001 Circular Plate Dataset

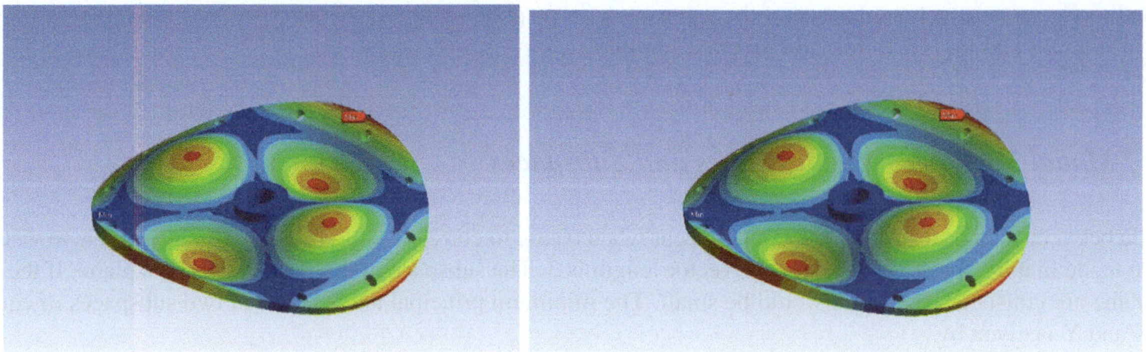

Fig. 18.2 Repeated roots – 12th and 13th mode, 2001 Circular Plate Dataset

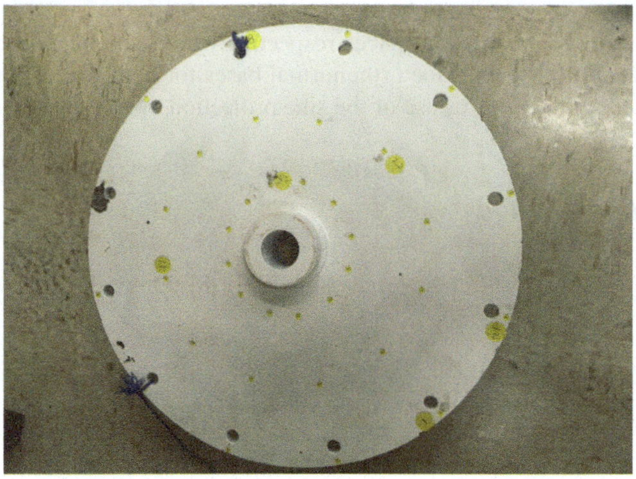

Fig. 18.3 2001 plate

on the cross diagonal. Because of symmetry in circular plates, it is easy to see that the repeated modes are actually spatially rotated versions of each other. This can be seen in Fig. 18.2.

A method developed for correlating repeated modes is shown in [2], where a combined mode is first set up as

$$[\psi_{Ak}] = [\psi_k]\{\beta\} \tag{18.4}$$

where {β} is the linear coefficients matrix, $[\psi_k]$ is the matrix made up of analytical modal vectors for a set of modes and $[\psi_{Ak}]$ is the combined modal vector for the kth mode. The combined mode is substituted for kth test mode and β values are calculated. The combined mode is reconstructed from these β values and cross-MAC is calculated between combined and test modes. This is repeated for all combinations for the combined mode. The method has some drawbacks that are discussed in Sect. 18.3.1.

18.2.5 Extracting and Conditioning Modal Vector

Before a correlation criteria could be implemented, modal vectors were extracted from experimental and analytical modal analysis. For the experimental cases, 2001 Circular Plate Dataset was obtained from an impact testing setup with 36 impact locations as shown in Fig. 18.3 while the 2014 Circular Plate Datasets was obtained from a shaker test dataset with 30 accelerometer locations. X-Modal software was used to get the modal analysis results and modal vectors were extracted (Fig. 18.4).

Analytical modal vectors were obtained from FEM based modal analysis. ANSYS Workbench was selected as the analysis method. Both circular plates were modelled. To ensure the model was accurate, the modal frequencies and vectors were

Fig. 18.4 2014 plate

compared. The modal vector comparison was done using the cross-MAC. Modal vectors obtained directly were of bigger size (vector length) compared to experimental vectors. These vectors were then down sampled to the number of nodes present in experimental vectors. Nodes from analytical solutions that were closest to the experimental nodes were selected. Note that the modal frequency between analytical and experimental models cannot be used without first comparing the modal vectors to be certain that the modal vectors are correctly matched pairs of sets.

18.3 Analysis

18.3.1 Implementing Existing Methods

Repeated mode correlation as shown by Walther et al. [2] can be easily implemented in MATLAB. A set of modes was selected from the complete set of analytical modal vectors. One vector was then selected from the complete set of experimental modal vectors. Preferably, the selected experimental mode should be such that the corresponding analytical mode exists in the selected set of analytical modes. The selected analytical set can then be used to get a combined analytical mode as in the following equation

$$[\psi_{Ak}] = [\psi_k]\{\beta\} \tag{18.5}$$

This combined mode was then substituted with kth test mode $[\psi_{Ak}]$. The coefficient vector {β} was then calculated for the selected $[\psi_k]$ according to the equation

$$\{\beta\} = [\psi_k]^T\{\psi_{Tk}\} \tag{18.6}$$

The combined mode was then reconstructed from Eq. 18.5 by utilizing the {β} matrix. The MAC between the *k*th test mode $[\psi_{Tk}]$ and the corresponding combined mode was calculated. This calculation was then repeated for all sets of $[\psi_k]$. MAC values close to unity suggests the selected set of analytic modes used for that combined mode may be repeated mode set. Some of results are shown in Table 18.1 to show the two problems with this approach. The two problems are

1. The high correlation between the combination [12, 13, 14] and the 12*th* experimental mode suggests that the repeated mode set is [12, 13, 14]. This is clearly not true as can be seen from the Fig. 18.5, where 12th and 13th can be seen to be similar whereas the 14th mode is a very different modal shape as can be seen in Fig. 18.6.

Table 18.1 MAC between combined and experimental modes

Experimental mode number	Analytical modes	MAC
12	12,13	0.972
12	12,13,14	0.977
12	12,10	0.867
12	12,11	0.867
12	12,10,14	0.871
12	13,12	0.972
12	13,11	0.848

Fig. 18.5 Repeated roots – 12th and 13th mode 2001 Circular Plate Dataset

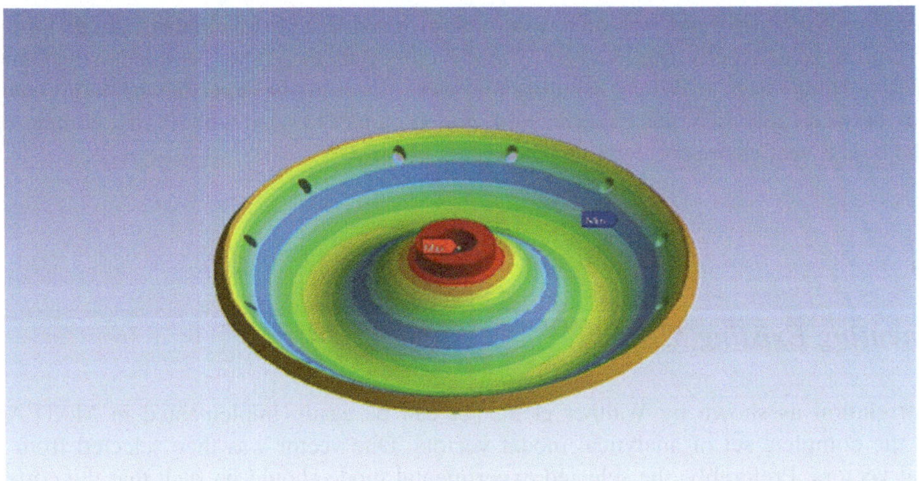

Fig. 18.6 14th mode 2001 Circular Plate Dataset

2. The high correlation between the combination [12, 10, 14] and the 12th experimental mode is more problematic as it suggests that high correlation may be obtained whenever the analytical set contains the one experimental mode it is being correlated with. High correlation with the [12, 10] also presents the same problem.

18.3.2 SPACEMAC Implementation

For comparing vector spaces, a criterion was developed. The concept of principal angle between subspaces, θ, was used to estimate the similarity of the vector subspaces. The criterion as calculated as $1 - |sin(\theta)|$ also known as the coverse sine function. A $cos(\theta)$ function can not be used because it shows very low sensitivity to the angle at low angular values. Figure 18.7 shows the flowchart for the method of implementing SpaceMAC.

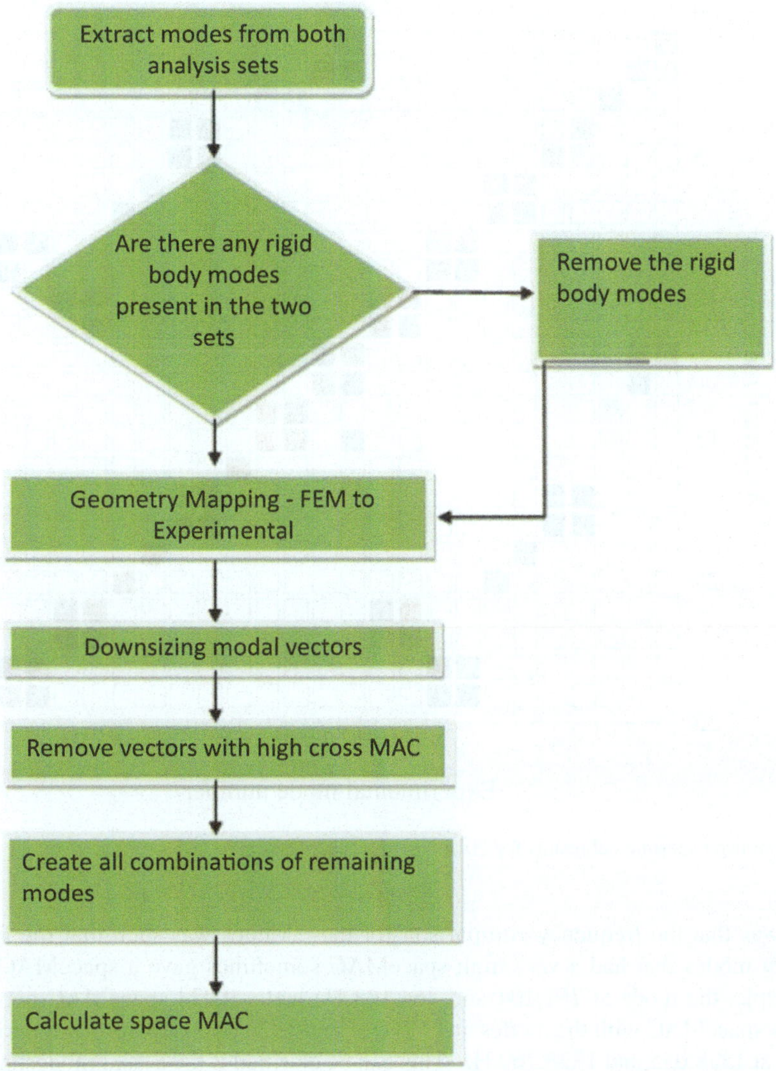

Fig. 18.7 Flowchart for implementing spaceMAC

18.3.3 Results

When two different sets of experimental results were compared for both plates, one important observation was that the repeated modal vectors were numerically consistent to begin with. The cross-MAC is shown in Fig. 18.8. In such cases spaceMAC will not be required. One factor that might be affecting this is the real dominant normalization procedure that is applied to modal vector during residue estimation. SpaceMAC would still be useful for FEM to experimental comparison.

The spaceMAC method correlated one modal vector space of a selected multiplicity and another modal vector space of the same multiplicity. This is important as the drawback of the correlation method described in Sect. 18.3.1 is that it correlates a space with a vector. This gives space-MAC an advantage over this method. For example, if the 12th, 13th and 14th modes for 2001 Circular Plate Dataset are taken as in Sect. 18.3.1 it is easier to detect and correlate repeated modes with spaceMAC. But it has to be made sure that the search for repeated modes must be made in an increasing order of multiplicity of repeated modes. Table 18.3 shows the space-MAC values of multiplicity equal to 2. The cross-MAC threshold was selected as 0.9. The cross-MAC values can be seen in Fig. 18.9 for 2001 Circular Plate Dataset. Space MAC values above 0.8 can be considered a good enough correlation. It can be seen in the Table 18.2 that there are some space-MAC values around 0.6.

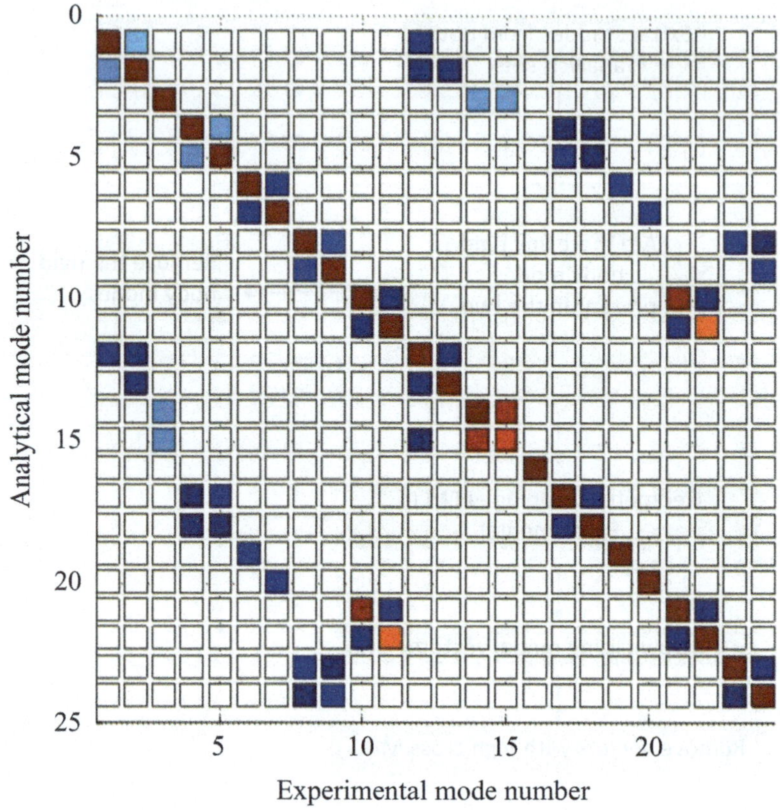

Fig. 18.8 Cross-MAC between two experimental results for 2014 Circular Plate Dataset

Another observation was that the frequency comparison for these values was such that the frequencies were far apart. Also, two sets (of two) of modes that had a very high spaceMAC sometimes gave a spaceMAC of about 0.6 when group as a quadruple. For example, the mode at 761.104 and 764.163 Hz had a 0.924 spaceMAC between them and both these modes showed a very low spaceMAC with the modes at 1223.011 and 1224.080 Hz while they showed a spaceMAC value of around 0.6 for the modes at 1328.055 and 1328.769 Hz. This can be clarified if the actual modeshape is observed (animated). Figures 18.9, 18.10, 18.11, 18.12, 18.13 and 18.14 show the modeshapes at these six frequencies. As can be seen from these figures, the spaceMAC values around 0.6 seems to be because of spatial aliasing phenomenon. But clearly the spatial aliasing was not strong enough to affect the cross-MAC values. SpaceMAC was observed to have higher sensitivity to spatial aliasing effects.

Following an increasing order of selected multiplicity is an important factor in spaceMAC implementation. SpaceMAC calculations for multiplicity 3 were done. Some sample values are show in Table 18.3 below. As is already known this set of modal vectors does not contain any repeated modes of multiplicity 3. But some high values can be seen. Sets of 3 modes with spaceMAC$_3$ values of around 0.6 usually result from spatial aliasing of a pair of repeated mode with another mode. Similar effects can be observed when calculating spaceMAC$_4$.

The MAC threshold was varied from 0.7 to 0.95 but the spaceMAC values remained unchanged. If MAC value threshold is set too low it could eliminate some repeated mode pairs so it is important to set the threshold depending on which mode have to be checked.

Table 18.2 SpaceMAC$_2$ values for 2001 Circular Plate Dataset

Number	Frequency 1	Frequency 2	Principal angle	spaceMAC
1	362.337	363.648	0.044	0.956
2	362.337	761.104	1.255	0.049
3	362.337	764.163	1.255	0.049
4	362.337	1223.011	1.257	0.049
5	362.337	1224.080	1.255	0.049
6	362.337	1328.055	1.256	0.049
7	362.337	1328.769	1.255	0.049
8	362.337	2322.186	1.266	0.049
9	362.337	2322.555	1.195	0.070
10	363.648	761.104	1.239	0.054
11	363.648	764.163	1.239	0.054
12	363.648	1223.011	1.254	0.050
13	363.648	1224.080	1.239	0.054
14	363.648	1328.055	1.239	0.054
15	363.648	1328.769	1.239	0.054
16	363.648	2322.186	1.183	0.074
17	363.648	2322.555	1.247	0.052
18	761.104	764.163	0.077	0.924
19	761.104	1223.011	1.253	0.050
20	761.104	1224.080	1.173	0.078
21	761.104	1328.055	0.385	0.624
22	761.104	1328.769	0.388	0.622
23	761.104	2322.186	0.387	0.623
24	761.104	2322.555	0.405	0.606
25	764.163	1223.011	1.253	0.050
26	764.163	1224.080	1.173	0.078
27	764.163	1328.055	0.395	0.615
28	764.163	1328.769	0.395	0.615
29	764.163	2322.186	0.397	0.614

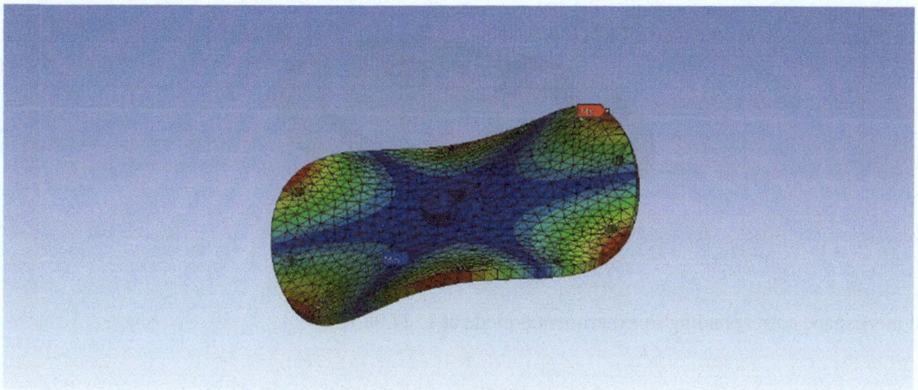

Fig. 18.9 Analytical modeshape corresponding to experimental mode at 761.104 Hz

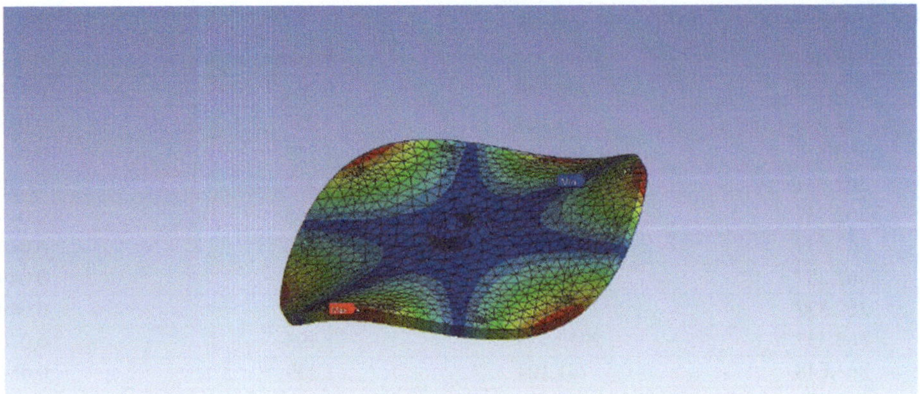

Fig. 18.10 Analytical modeshape corresponding to experimental mode at 764.163 Hz

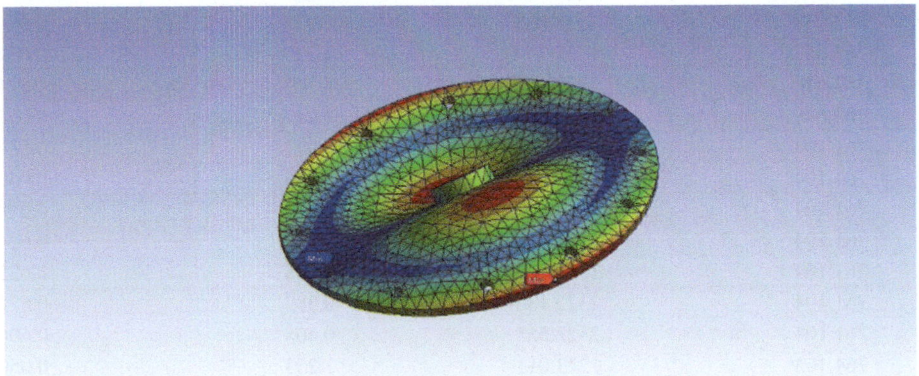

Fig. 18.11 Analytical modeshape corresponding to experimental mode at 1223.01 Hz

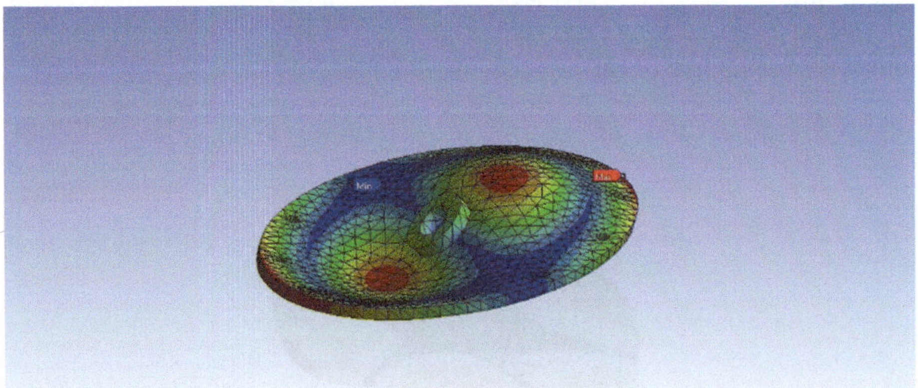

Fig. 18.12 Analytical modeshape corresponding to experimental mode at 1224.08 Hz

18.4 Conclusion

In comparison to the method described in the paper by Walther et al. [2], the main advantage spaceMAC had is that it compares two modal vector spaces instead of a modal vector space with a modal vector. This means if an increasing order of multiplicity is followed when searching and correlating repeated mode sets, a clearer indication of which sets of modes are repeated sets can be obtained. Once the sets of modal vectors that belong to a repeated set have been identified and the vector spaces have been shown to have a high correlation, the {β} coefficient matrix method can be used to get experimental modal vectors that have high MAC value with the analytical modal vectors. A high sensitivity to spatial aliasing was observed in spaceMAC. However, the spaceMAC values obtained due to spatial aliasing were around 0.6 whereas the spaceMAC values obtained for repeated modes were higher than 0.8. Also, since the modal frequencies are normally quite different, it is easy

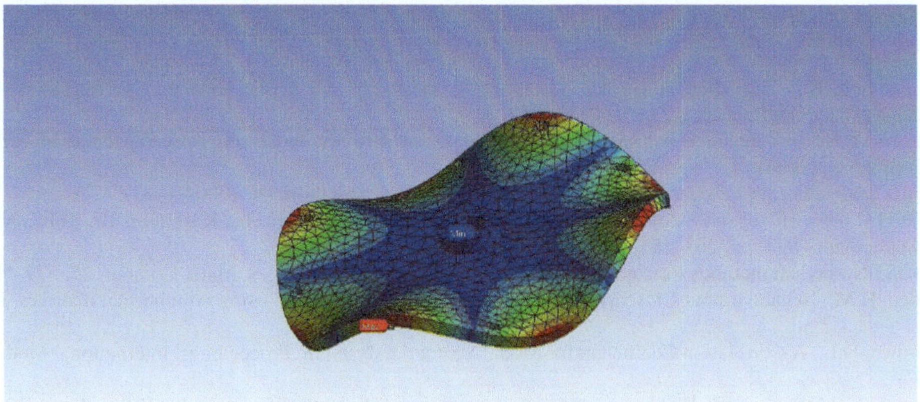

Fig. 18.13 Analytical modeshape corresponding to experimental mode at 1328.05 Hz

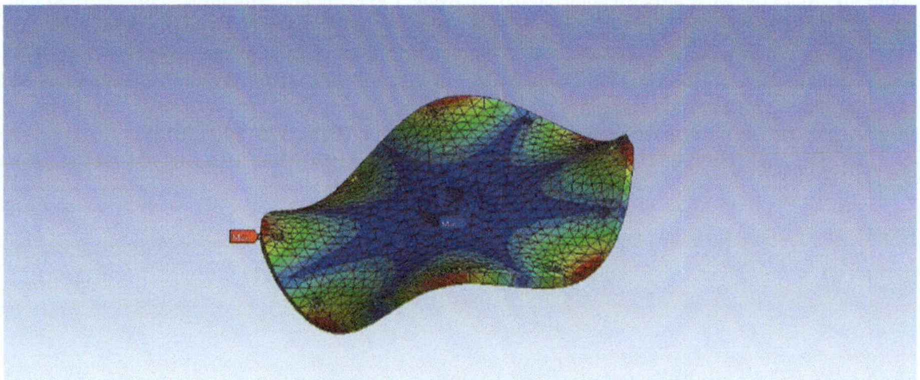

Fig. 18.14 Analytical modeshape corresponding to experimental mode at 1328.77 Hz

Table 18.3 spaceMAC$_3$ values

Number	Frequency1 (Hz)	Frequency2 (Hz)	Frequency3 (Hz)	spaceMAC
1	362.337	363.648	761.104	0.624
2	362.337	363.648	764.163	0.615
3	362.337	363.648	1223.011	0.050
4	362.337	761.104	764.163	0.049
5	362.337	761.104	2322.186	0.046
6	362.337	363.648	2322.186	0.647
7	362.337	363.648	2322.555	0.631

to eliminate those cases as being possible repeated roots cases based upon frequency alone. It has been observed that MAC values between two different experimental results belonging to the same experimental setup were close to unity even for repeated modes. This observation was contrary to expectation. It was the same for different force functions used, signal processing methods and number of reference points.

When compared to another method for correlation of closely spaced modes, the results obtained with spaceMAC were clearer as long as an increasing order of multiplicity was followed. This was largely because spaceMAC works with correlation of two modal vector spaces whereas the other method compares a space with a vector. The effect of variation of MAC threshold was also explored. It has been observed that it did not directly affect the spaceMAC calculation for a particular set of modes.

References

1. Phillips, A.W., Allemang, R.J.: Utilizing the Modal Assurance Criterion (MAC) in the Detection of Repeated Roots. In: Proceedings, International Modal Analysis Conference, pp. 665–671 (1996)
2. Walther, H.P., Kmetyk, L.N., Holzmann, W.A., Segalman, D.: Modal Correlation with Closely Spaced Modes. In: Proceedings, International Modal Analysis Conference (2004)
3. Beezer, R.A.: A First Course in Linear Algebra. University of Pugetsound, Washington, DC (2004)
4. Wedin, P.A.: On angles between subspaces of a finite dimensional inner product space. In: Kagstrom, B., Ruhe, A. (eds.) Matrix Pencils, Lecture Notes in Mathematics 973, pp. 263–285. Springer, Berlin (1983)
5. Bjorck, A., Golub, G.: Numerical methods for computing angles between linear sub-spaces. Math. Comput. **27**, 579–594 (1973)
6. Horn, B.K.P., Hilden, H.M.: Negahdaripour Closed-form solution of absolute orientation using orthonormal matrices. J. Opt. Soc. Am., 1127–1135 (1998)
7. Allemang, R.J., Brown, D.L.: A Correlation Coefficient for Modal Vector Analysis. In: Proceedings, International Modal Analysis Conference, pp. 110–116 (1982)
8. Richardson, S., Tyler, J., McHargue, P., Richardson, M.: A new measure of shape difference. In: Proceedings, 32nd International Modal Analysis Conference, pp. 71–79 (2014)
9. Allemang, R.J.: The modal assurance criterion – twenty years of use and abuse. Sound Vib., 14–21 (2003)
10. Allemang R. J., Vibrations: Analytical and Experimental Modal Analysis, Experimental Vibrations, University of Cincinnati (2017)
11. Allemang, R.J., Phillips, A.W.: Un-weighted and Weighted Versions of the Modal Assurance Criterion (MAC) for Evaluation of Modal Vector Contamination In: Proceedings, International Modal Analysis Conference, Topics in Modal Analysis I, vol. 7, pp 173–180 (2014)
12. Hiremanglur, R.: Real-Normalization of Experimental Complex Modal Vectors with Modal Vector Contamination. Master's thesis, University of Cincinnati (2014)
13. Knyazev, A.V., Zhu, P. Angles between subspaces and their tangents. J. Numer Math., 325–340 (2013)
14. Ipsen, I.C.F., Meyer, C.D.: The angle between complementary subspaces. Am. Math. Mon. **102**(10), 904–911 (1995)

Chapter 19
Subsecond Model Updating for High-Rate Structural Health Monitoring

Michael Carroll, Austin Downey, Jacob Dodson, Jonathan Hong, and James Scheppegrell

Abstract High-rate monitoring, condition assessment, and control of structural systems experiencing high-rate dynamics is challenging due to these structures experiencing events of timescales below 10 ms. Examples of structures that require high-rate structural health monitoring include spacecraft, hypersonic vehicles, ballistic packages, and active barriers for blast mitigation. Subsecond model updating techniques for this unique class of structures must be capable of tracking the system through rapidly changing input forces and time-varying structural parameters. Moreover, any methodology designed for high-rate structural health monitoring must account for the challenges associated with high-speed data measurement and model updating in its formulation. This work presents and experimentally validates a subsecond model updating methodology for enabling high-rate structural health monitoring of a structure that undergoes system-level changes (i.e., damage) while accounting for uncertainties in the measurements, model, and system. To achieve this, a parallelized residual minimization model updating technique is implemented on an FPGA where model parameters are drawn from a continuously updated parameter pool. The parameter pool is updated based on previous system states and known upcoming events (e.g., impacts). In this work, the DROPBEAR experimental test bed at the Air Force Research Laboratory is used to validate the proposed methodology for a one-degree-of-freedom system with a continuously changing boundary condition. Results demonstrate that a continuously changing boundary condition can be successfully tracked at time intervals of 10 ms or less. Computational speed, prediction accuracy as a function of model size, and the role of measurement noise are examined in this work.

Keywords Real-time model updating · Finite elements analysis · Structural health monitoring · High-rate dynamics

19.1 Introduction

The development of real-time modeling methods for structures undergoing high-rate dynamics is driven by the demand to develop active structures. Examples of active structures that experience high-rate dynamics include active blast mitigation barriers, ballistics packages, and hypersonic vehicles [1–4]. The end goal of such development is the monitoring and mitigation of high-speed impacts and shock events on structures in real-time. For example, a 1 ms delay in a decision made for a ballistic package traveling at Mach 5 would result in the package being 1.7 meters down range. Due to the timescales considered, the monitoring and condition assessment of structures must occur at the single-digit millisecond timescale.

Due to the unmodeled nature of the structures operating in these harsh environments that are currently unprecedented, a model updating technique cannot rely on pre-calculated data sets. This paper presents an algorithm capable of real-

M. Carroll
Department of Mechanical Engineering, University of South Carolina, Columbia, SC, USA

A. Downey · M. Carroll (✉)
Department of Civil and Environmental Engineering, University of South Carolina, Columbia, SC, USA

Department of Mechanical Engineering, University of South Carolina, Columbia, SC, USA
e-mail: austindowney@sc.edu

J. Dodson
Air Force Research Laboratory, Wright-Patterson AFB, OH, USA
e-mail: jacob.dodson.2@us.af.mil

J. Hong · J. Scheppegrell
Applied Research Associates, Albuquerque, NM, USA
e-mail: jhong@ara.com

B. Dilworth (ed.), *Topics in Modal Analysis & Testing, Volume 8*, Conference Proceedings of the Society for Experimental Mechanics Series, https://doi.org/10.1007/978-3-030-47717-2_19

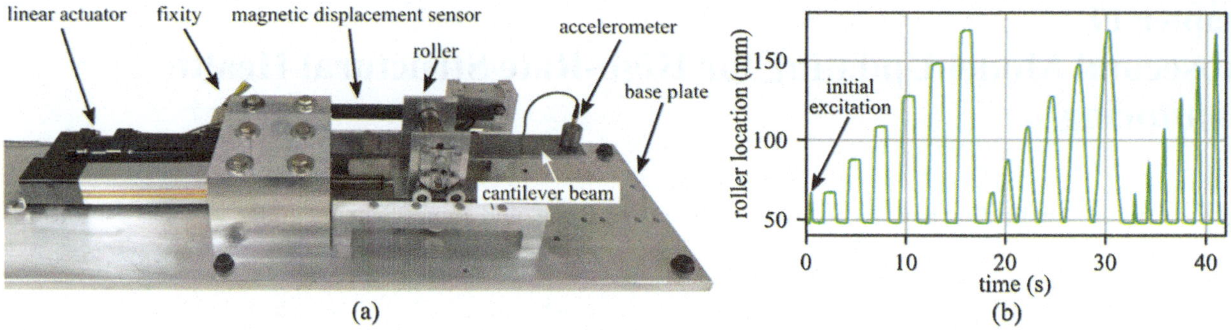

Fig. 19.1 The Dynamic Reproduction of Projectiles in Ballistic Environments for Advanced Research (DROPBEAR) experimental test bed showing: (**a**) key components; and (**b**) the test profile of the roller movement used in this work

time model updating and experimentally validates the algorithm using a test structure. The algorithm leverages an error minimization technique in the frequency domain to develop a methodology that can operate with the desired single-digit millisecond timescales while producing accurate estimation of parameter changes in the test structure.

19.2 Background

The DROPBEAR (Dynamic Reproduction Of Projectiles in Ballistic Environments for Advanced Research) experimental setup was first introduced by Joyce et al. [5] and is presented in Fig. 19.1a. It was designed with two user-controlled parameters: a detachable mass and a continuously variable roller cart, where the detachable mass represents a discrete system change (i.e. damage) and the movable roller a system change. For the experiments conducted in this work, the roller cart was used to simulate a progressive change in the structure and the mass drop was not utilized.

The experimental test bed features a large, rectangular aluminum base securely affixed to a base plate. This plate serves as the mounting base for the clamp housing in which the cantilever beam is secured. The beam is made from steel and dimensioned with a width of 51 mm (2 in), a free length of 350 mm (13.81 in), and a thickness of 6.3 mm (0.25 in). One single-axis PCB Piezotronics 393B04 accelerometer was attached to the beam at 337 mm (13.27 in) from the primary clamp (Fig. 19.1a). The accelerometer was connected to a 24-bit IEPE ADC NI-9234 housed in a National Instruments (NI) cDAQ-9172 eight-slot chassis. A NI PXIe-8880 controller placed in the chassis was used for real-time data acquisition and processing using custom LabVIEW codes.

A Parker Drive linear actuator is used to control the location of the cart, and can be programmed with predetermined courses of motion [6]. Once programmed, the actuation of the cart was triggered using NI-LabVIEW through a digital output signal from a NI PXI-6133. The excitation of the beam was generated by the cart motion along the beam length as the sole input force into the system (Fig. 19.1). A LabVIEW program acquired and saved response data from an accelerometer. The estimated cart position was verified through a Honeywell Sensing and Productivity Solutions magneto-resistive linear transducer, model number SPS-L225-HALS. The analog voltage input from the linear transducer was acquired through LabVIEW via a 14-bit ADC NI PXI-6133 coupled with an NI TB-2709. MATLAB and Python were employed for post-processing of data to perform modal analysis.

The development of a numerical study for the changing parameters based on a finite element analysis (FEA) code revealed how the natural frequencies and mode shapes change linearly with changes in the cart position, as presented in Fig. 19.2 and discussed in detail in [7]. Due to the redundancy in the natural frequency of the beam during the cart's full travel along the beam, the roller movement was limited to the range of 48 mm to 168 mm from the fixed end.

The proposed algorithm is presented in Fig. 19.3. The algorithm functions by acquiring data from the accelerometer, performing a fast Fourier transform (FFT), and generating a value for the first frequency. Additionally, an Euler-Bernoulli FEA (Finite Element Analysis) model, implemented in LabVIEW, is used as the model to be updated in real-time. The FEA model has a specified number of unique cart locations at which the natural frequency can be calculated. To enable real-time model-updating without pre-calculated data sets, these FEA models are solved in real-time. Modal analysis is used to calculate the structure's natural frequency, where the damping term (C) can be disregarded due to an insignificant contribution of damping within the system. Therefore, we employ the following free vibration equation of motion:

$$\boldsymbol{M}\ddot{\boldsymbol{x}} + \boldsymbol{K}\boldsymbol{x} = \boldsymbol{0} \tag{19.1}$$

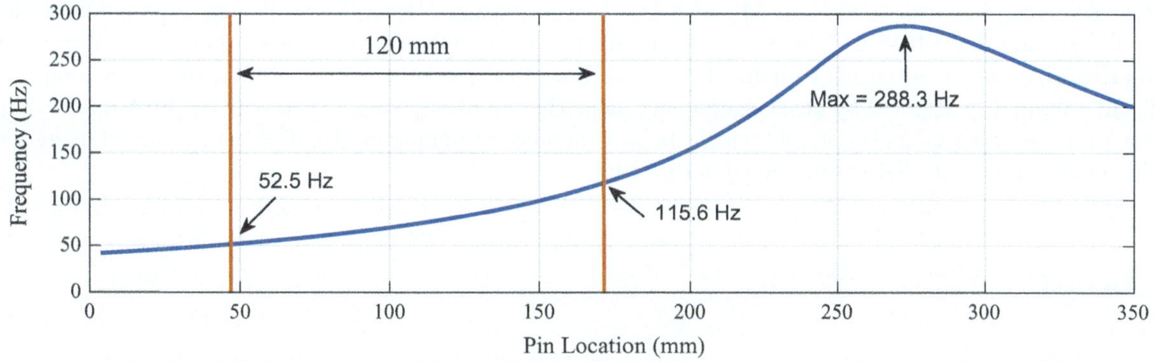

Fig. 19.2 Comparison of first frequency mode with changing pin location. The mode is from the finite element model using Python. For the experimental data, initial and final frequencies for the pin location are shown

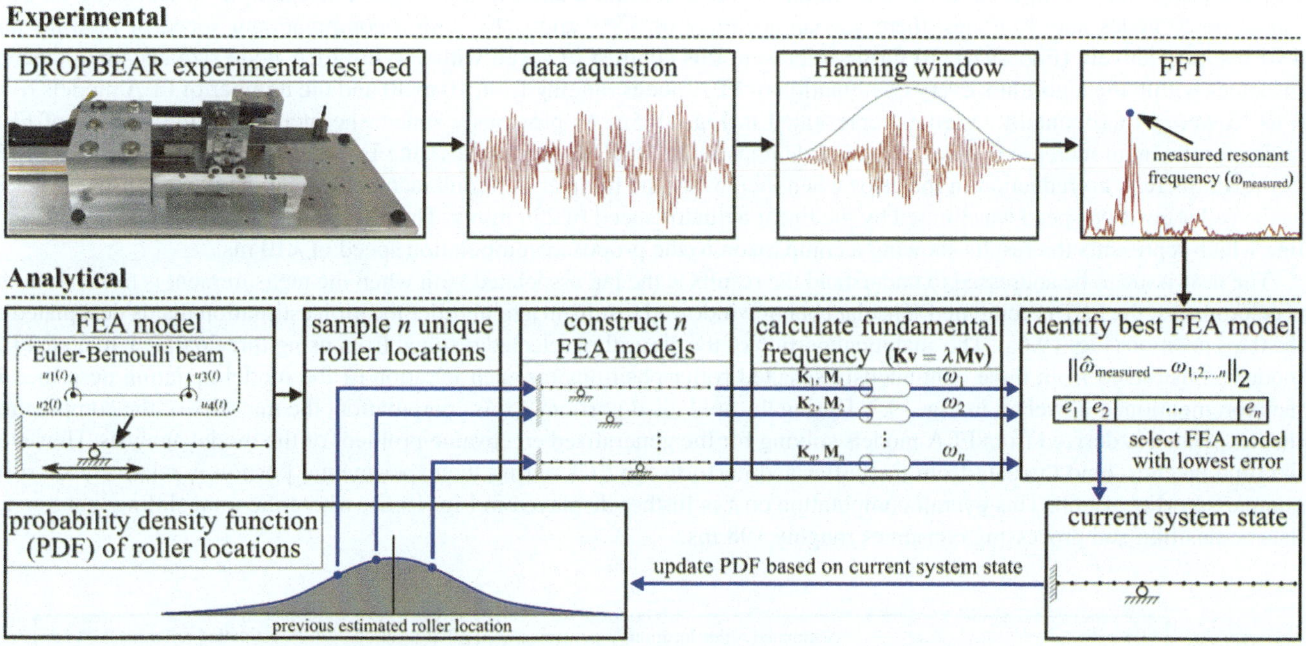

Fig. 19.3 Diagram showing the flow of the millisecond model updating method

This solution can be further simplified by making the assumption that for the harmonic solution $x_n = \Phi_n \sin(\omega_n t)$ where Φ represents the mode shapes of the solution and ω_n is the undamped natural frequency. Upon substitution, where $\mathbf{\Phi} = \mathbf{0}$ is a trivial solution, the equation becomes:

$$\left(-\omega^2 M + K\right) \mathbf{\Phi} = \mathbf{0} \tag{19.2}$$

The solution of interest is where the natural frequencies $\boldsymbol{\omega}$ and modes $\mathbf{\Phi}$ that satisfy the generalized eigenvalue problem, where $\lambda_n = \omega_n^2$, results in the equation:

$$\boldsymbol{K}\mathbf{\Phi} = \boldsymbol{\lambda}\boldsymbol{M}\mathbf{\Phi} \tag{19.3}$$

Within this equation, $\boldsymbol{\lambda}$ is the diagonal matrix of eigenvalues and $\mathbf{\Phi}$ are the associated eigenvectors where $\omega_n = \sqrt{\lambda_n}$. Therefore, the undamped natural frequency can be solved for using the generalized eigenvalue problem and selecting the lowest valued eigenvalue.

For each sampled pin location, a unique **M** and **K** matrix was generated. In order to calculate each pin location's unique first natural frequency, LabVIEW's Cholesky factorization and Jacobi Method of decomposition were employed to calculate the appropriate eigenvalues associated with the FEA model. Each FEA model's unique first frequency was compared against the FFT's first frequency acquired by the accelerometer data. The pin location associated with the FEA model frequency of least error would become the mean around which the next number of unique pin locations are centered using a Gaussian probability density function (PDF) as presented in Fig. 19.3.

19.3 Analysis

The results for the experimental test when implementing the DROPBEAR test bed will be discussed with consideration to real-time model updating techniques described in the preceding background section of this paper.

When increasing the number of nodes in the FEA model, there is an increase in the accuracy of the position estimation. However, this increase in accuracy results in an increase in computation time as well. For example, when using an FEA model of 20 nodes and 40 nodes, there was an accuracy of 3.9% and 2.9% with a computational speed of 0.86 ms and 4.04 ms, respectively (Fig. 19.4). To further represent this contrast of speed with varying FEA node size and simultaneous instances within the algorithm, every combination of FEA nodes ranging from 10 to 40 and the number of FEA models from 3 to 15 were experimentally tested as represented in Fig. 19.5a. As previously stated, the increase in the number of FEA nodes results in an increase in error reduction demonstrated as MAE (Mean Absolute Error) and represented in Fig. 19.5b. However, there is no reduction in the error when increasing the number of simultaneous FEA models (number of particles) as the roller position speed was limited by the linear actuator speed of 250 mm/s. The red lines in Fig. 19.5 indicate the 5 ms line which represents the results showing a comparison to the proposed computation speed of <10 ms.

The next issue to be addressed to understand the results is the lag associated with when the measurement is taken from the accelerometer, the FFT is computed. The data being windowed results in a final roller location estimation that is designated as the 0 ms position (Fig. 19.6a). The analytical portion of the algorithm selected resulted from using three parallel 40 node FEA models constructed from three continuously selected roller positions for each iteration of the model updating described in the aforementioned flowchart in Fig. 19.3. During the analytical portion of this computation, the majority of the computation time expended is derived from FEA models solving for the generalized eigenvalue problem of the modal analysis. However, the time taken to load the data from the buffer and perform the FFT during the experimental portion is relatively short by comparison (Fig. 19.6b. This overall computation cost is further displayed in Fig. 19.6c, where the total sliding window for data acquisition and processing consumes roughly 198 ms.

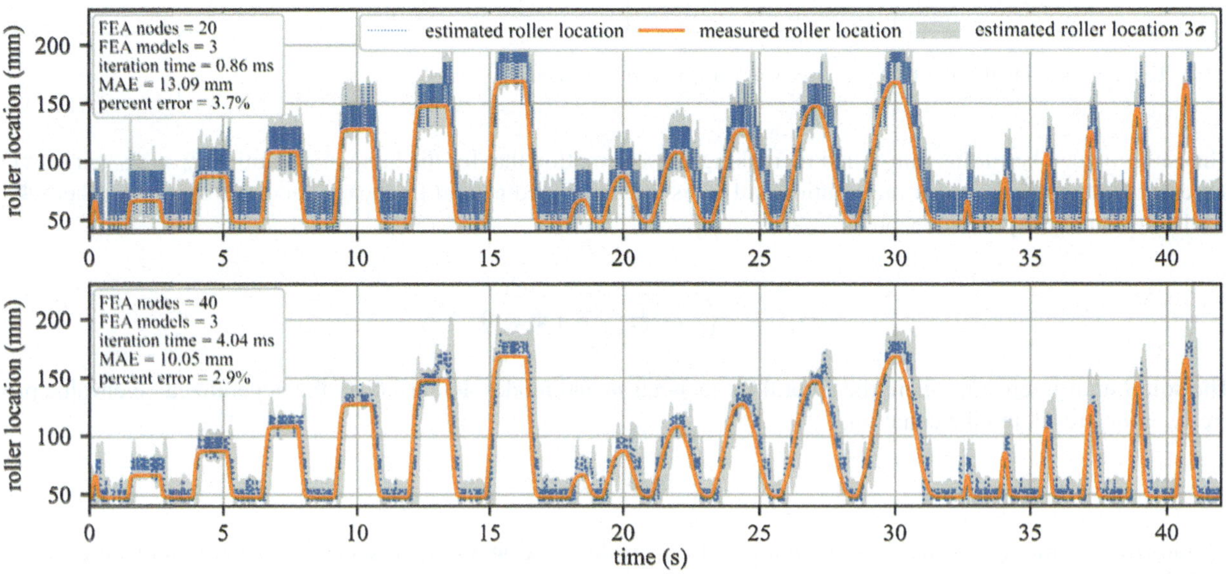

Fig. 19.4 20 and 40 node FEA models, each with three parallel instances of FEA models being solved, are shown with temporal data responses

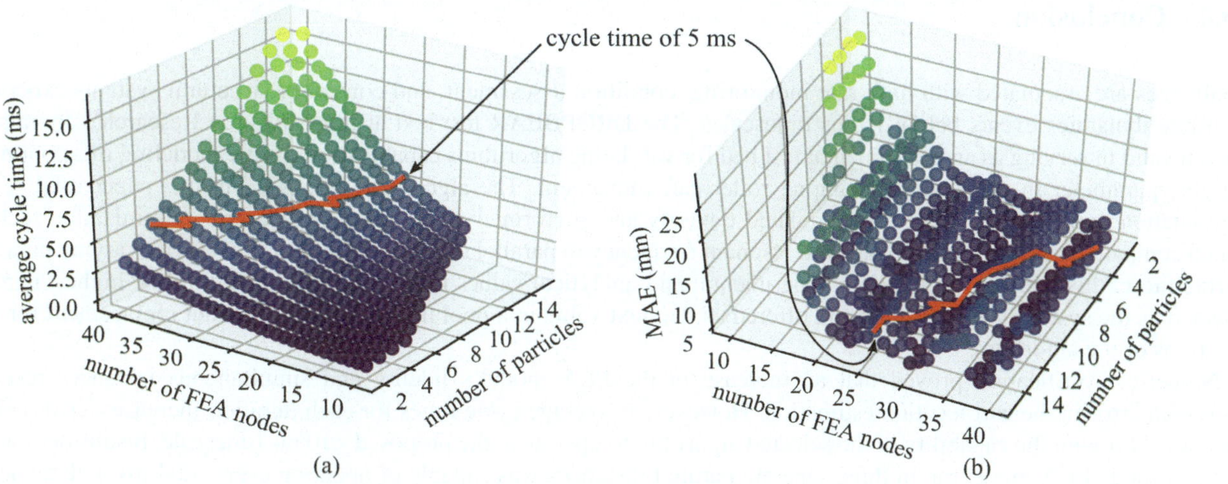

Fig. 19.5 233 Results of automated tests displaying the: (**a**) average iteration time for each possible combination, (**b**) mean absolute error for each combination. Note: There is a reversal of the axes numbering to aid plotting

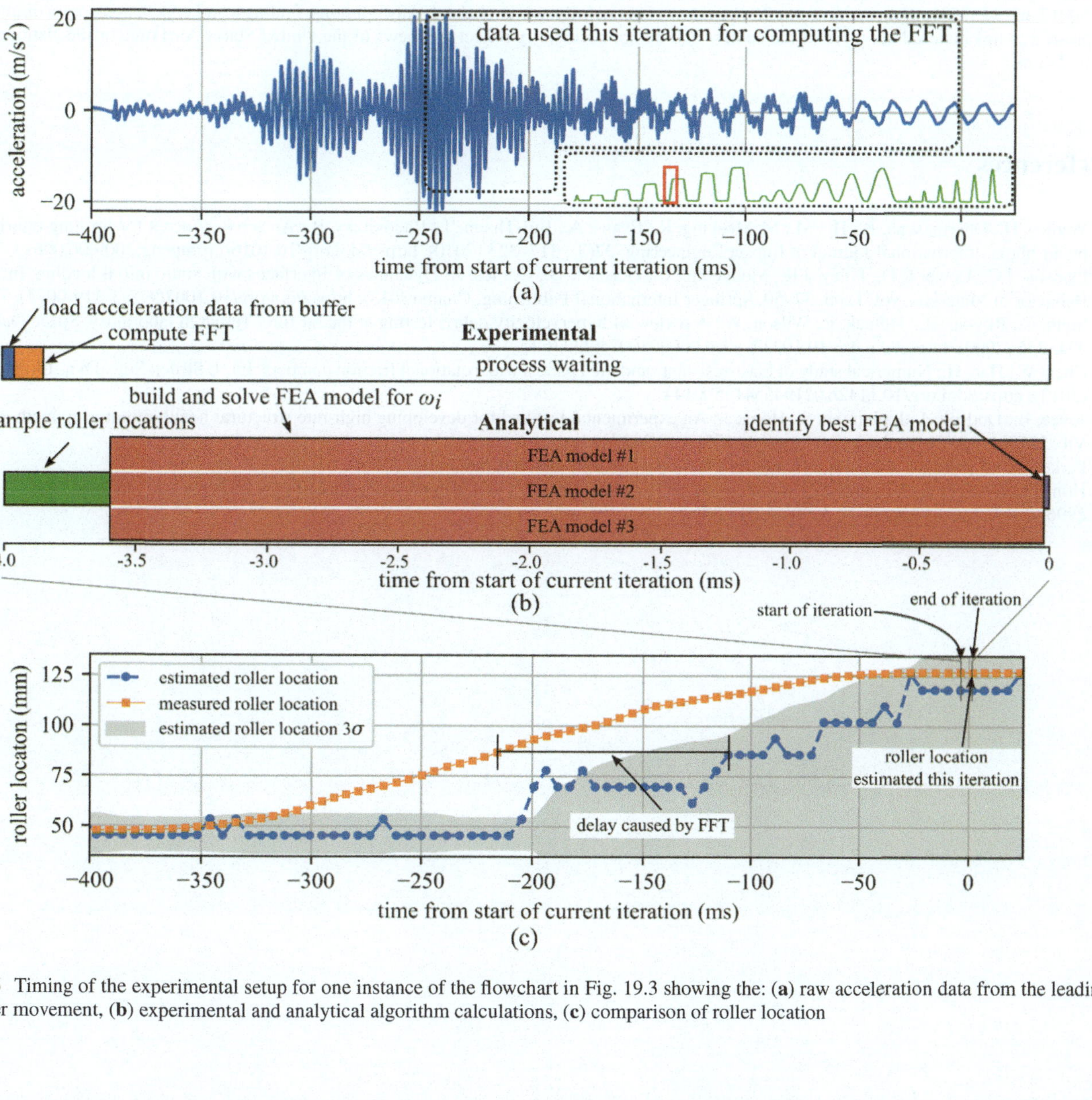

Fig. 19.6 Timing of the experimental setup for one instance of the flowchart in Fig. 19.3 showing the: (**a**) raw acceleration data from the leading step roller movement, (**b**) experimental and analytical algorithm calculations, (**c**) comparison of roller location

19.4 Conclusion

Challenges are associated with high-rate monitoring, condition assessment, and control of structural systems experiencing high-rate dynamics events below 10 ms timescales. The DROPBEAR test bed at the Air Force Research Laboratory has proven valid in serving as an experimental test bed for validating algorithms capable of real-time modeling of a continuously variable parameter in the form of changing roller cart movement. The algorithms discussed in this paper used real-time computation without the use of pre-calculated datasets and performed a frequency-based error minimization technique, which compared an experimentally derived resonant frequency to parallel instances of a FEA model with varying parameters. Furthermore, this comparison between experimental and analytical values resulted in the value with the highest agreement selected as the system's current state and offered as the next value in a pool that is sampled without replacement for further iterations of the algorithm.

Numerical verification proved that an increase in the FEA model's fidelity and simultaneous instances resulted in decreased error in the cart location estimation. However, as average cycle times for each iteration increases, error reduction decreases. Proving the capability of the selected algorithm to operate at the proposed <10 ms timescale, results demonstrated that a 40-node FEA model ran in three separate parallel instances was capable of updating every 4.04 ms with an accuracy of 2.9%.

Acknowledgements This material is based upon work supported by the United States Air Force through the Air Force Research Laboratory Summer Internship Program and AFRL/RWK contract number FA8651-17-D-0002 and partly by the National Science Foundation Grant No. 1850012 and 1937535. The support from these agencies is gratefully acknowledged. Any opinions, findings, and conclusions or recommendations expressed in this material are those of the authors and do not necessarily reflect the views of the United States Air Force or the National Science Foundation.

References

1. Wadley, H., Dharmasena, K., He, M., McMeeking, R., Evans, A., Bui-Thanh, T., Radovitzky, R.: An active concept for limiting injuries caused by air blasts. International Journal of Impact Engineering. **37**(3), 317–323 (2010). https://doi.org/10.1016/j.ijimpeng.2009.06.006
2. Dodson, J.C., Lowe, R.D., Foley, J.R., Mougeotte, C., Geissler, D., Cordes, J.: Dynamics of interfaces with static initial loading. In: Dynamic Behavior of Materials, vol. 1, pp. 37–50. Springer International Publishing, Cham (2013). https://doi.org/10.1007/978-3-319-00771-7_5
3. Stein, C., Roybal, R., Tlomak, P., Wilson, W.: A review of hypervelocity debris testing at the air force research laboratory. Space Debris. **2**(4), 331–356 (2000). https://doi.org/10.1023/b:sdeb.0000030024.23336.f5
4. Chen, W., Hao, H.: Numerical study of blast-resistant sandwich panels with rotational friction dampers. Int. J. Struct. Stab. Dyn. **13**(06), 1350014 (2013). https://doi.org/10.1142/s0219455413500144
5. Joyce, B., Dodson, J., Laflamme, S., Hong, J.: An experimental test bed for developing high-rate structural health monitoring methods. Shock. Vib. **2018**, 1–10 (2018). https://doi.org/10.1155/2018/3827463
6. Parker Servo Systems. onexia.com/parker/pdf/ONExia-Parker-P-Series-Catalog.pdf (2016)
7. Hong, J., Dodson, J., Laflamme, S., Downey, A.: Transverse vibration of clamped-pinned-free beam with mass at free end. Appl. Sci. **9**(15), 2996 (2019). https://doi.org/10.3390/app9152996

Chapter 20
Phase Quadrature Backbone Curve for Nonlinear Modal Analysis of Nonconservative Systems

Martin Volvert and Gaëtan Kerschen

Abstract Nonlinear normal modes (NNMs) are often used to predict the backbone of resonance peaks in nonlinear frequency response functions. Regardless of the definition considered, one important limitation remains, i.e., the NNMs require a multi-point, multi-harmonic external forcing to be excited. To address this limitation, the present study proposes a new definition of NNMs, termed phase quadrature backbone curve (PQBC). The advantage of PQBC is that an actual solution of the nonlinear frequency response obtained under mono-point, mono-harmonic external forcing is followed for which phase quadrature of a selected degree of freedom is achieved. Additionally, super and subharmonic resonance peaks can be captured by adapting the phase quadrature condition. Finally, no post-processing is required to get amplitude-forcing relations of the NNMs. Isolated responses can, in turn, be predicted from these relations.

Keywords Nonlinear vibrations · Harmonic forcing · Nonlinear modal analysis · Resonances

20.1 Introduction

NNMs are used for modal analysis of nonlinear systems. Considering the unforced and undamped system, they can be defined as *a (nonnecessarily synchronous) periodic motion* [1]. To be excited, NNMs require a multi-point, multi-harmonic external forcing equal to the dissipative terms and in phase quadrature with each harmonic component of the displacement, as shown in [2].

In the presence of damping, the aforementioned definition can no longer be used. To address this issue, a new definition of NNMs accounting for linear and nonlinear damping was proposed by Krack [3], in which periodicity is enforced by a mass-proportional damping term, i.e., a fictitious force. Phase quadrature is obtained between each of the displacement harmonics and the fictitious force term. However, a multi-point, multi-harmonic forcing is still required for exciting the considered NNM, which is in contrast with the fact that nonlinear frequency responses are usually obtained under mono-point, mono-harmonic forcing.

In this study, a new NNM definition is proposed, which relies on the phase quadrature between a mono-point, mono-harmonic forcing and the first harmonic component of the displacement of the excited degree of freedom. This definition can also be extended to super and subharmonic resonance peaks by imposing phase quadrature between a higher or lower harmonic component of the displacement and of the external forcing.

20.2 Phase Quadrature Backbone Curve

Similar to what was achieved in [3], the basic idea of the PQBC definition is to introduce a fictitious external forcing to enforce periodicity. However, unlike [3], the fictitious forcing, which plays the role of mono-point and mono-harmonic forcing, only involves the first harmonic component of the velocity at the selected degree of freedom:

$$M\ddot{x} + C\dot{x} + Kx + f_{nl}(x, \dot{x}) - \xi I_l \dot{x}_{1h} = 0 \tag{20.1}$$

M. Volvert (✉) · G. Kerschen
Department of Aerospace and Mechanical Engineering, University of Liège, Liège, Belgium
e-mail: m.volvert@uliege.be; g.kerschen@uliege.be

B. Dilworth (ed.), *Topics in Modal Analysis & Testing, Volume 8*, Conference Proceedings of the Society for Experimental Mechanics Series, https://doi.org/10.1007/978-3-030-47717-2_20

where $1h$ stands for first harmonic, I_l is an $n \times n$ matrix filled with 0 s and one 1 on the l-th diagonal term, selecting the excited degree of freedom, and ξ is the pseudo-amplitude of the fictitious forcing. In addition, super and subharmonics backbone curves can also be predicted by carefully selecting the harmonic component of the velocity in the fictitious forcing. Specifically, for super and subharmonic resonances, higher and lower than 1 harmonic components of the displacement are in phase quadrature with the forcing. Eq. (20.1) thus becomes:

$$M\ddot{x} + C\dot{x} + Kx + f_{nl}(x, \dot{x}) - \xi I_l \dot{x}_{nh} = 0 \tag{20.2}$$

for the n-th superharmonic resonance, and:

$$M\ddot{x} + C\dot{x} + Kx + f_{nl}(x, \dot{x}) - \xi I_l \dot{x}_{\frac{1}{n}h} = 0 \tag{20.3}$$

for the n-th subharmonic resonances.

20.3 Numerical Results

The PQBC definition is first demonstrated using a simple Duffing oscillator:

$$m\ddot{x} + c\dot{x} + kx + k_{nl}x^3 = 0 \tag{20.4}$$

The results are displayed in Fig. 20.1. In addition to the fundamental resonance, the system presents super and subharmonic resonances, the loci of which are all accurately predicted using the PQBC NNMs. It should be noted that the subharmonic resonance only exists from a certain amplitude of the external forcing.

Second, the method is applied to a two-degree-of-freedom system:

$$\begin{cases} m_1\ddot{x}_1 + (c_1 + c_{12})\dot{x}_1 - c_{12}\dot{x}_2 + (k_1 + k_{12})x_1 - k_{12}x_2 + k_{nl}x_1^3 = F\sin\omega t \\ m_2\ddot{x}_2 + (c_2 + c_{12})\dot{x}_2 - c_{12}\dot{x}_1 + (k_2 + k_{12})x_2 - k_{12}x_1 = 0 \end{cases} \tag{20.5}$$

The undamped NNMs of this system feature several modal interactions. However, in practice, they can be annihilated by damping. Using PQBC, it is possible to accurately determine whether modal interactions are still present or not in the presence of damping. Specifically, the PQBC NNMs in Fig. 20.2 clearly indicate that the modal interaction is eliminated by damping, which is confirmed by the calculation of the nonlinear frequency responses.

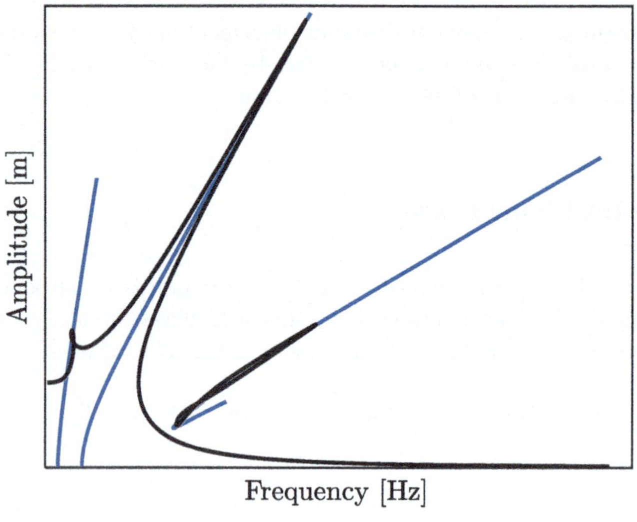

Fig. 20.1 Nonlinear frequency response (in black) of the Duffing oscillator and PQBC NNMs (in blue)

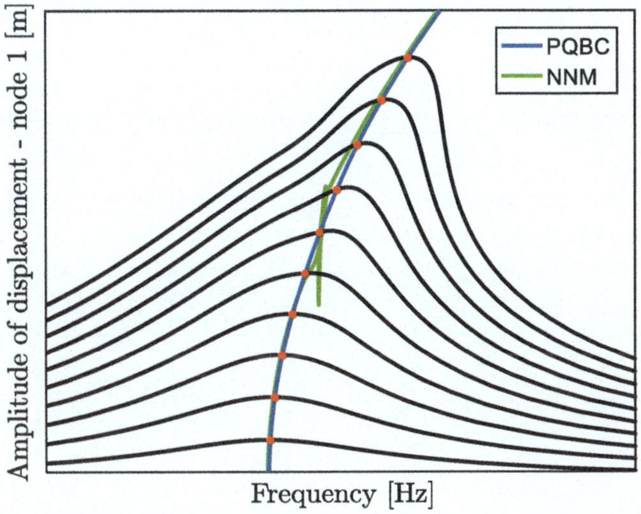

Fig. 20.2 Comparison between undamped NNMs (green) and PQBC NNMs (blue). The red dots represent the phase quadrature points of the nonlinear frequency responses (black)

20.4 Conclusion

This study has introduced a new definition of NNMs, termed phase quadrature backbone curve (PQBC). PQBC is such that phase quadrature is enforced between the excited degree of freedom and the external forcing. It has been shown that PQBC can calculate accurately the loci of fundamental, super and subharmonic resonances.

References

1. Kerschen, G., Peeters, M., Golinval, J.C., Vakakis, A.: Nonlinear normal modes, part I: a useful framework for the structural dynamicist. Mech. Syst. Signal Process. **23**, 170–194 (2009)
2. Peeters, M., Kerschen, G., Golinval, J.C.: Dynamic testing of nonlinear vibrating structures using nonlinear normal modes. J. Sound Vib. **330**, 486–509 (2011)
3. Krack, M.: Nonlinear modal analysis of nonconservative systems: extension of the periodic motion concept. Comput. Struct. J. **154**, 59–71 (2015)

Chapter 21
Preliminary Results of Vibration Measurements on a Wind Turbine Test Bench

Jesper Berntsen and Anders Brandt

Abstract In this paper some preliminary measurements aimed at determining the dynamics of a test bench are described. The test bench for which measurements were applied, is a Highly Accelerated Life-Time (HALT) tester used for testing wind turbine nacelles. The HALT tester is a large and complex machine with nine actuators that can provide tilt and yaw moments up to 25 MNm to emulate forces from the wind turbine blades. The dynamics of the test bench are unknown, and the measurements that were carried out should provide an understanding of the dynamics, which later should lead to fatigue assessment of vital components of the test benches. Accelerometers and geophones were placed at strategic positions around the HALT tester, both on the steel structure and concrete foundation. Experimental vibration data were obtained both with impact excitation and recordings during operation. From analysis of data obtained during operation, resonances at 1.5 and 3.9 Hz were found. These resonances could potentially be low enough to influence the life time of the test bench. Several resonances from operation measurements could be directly compared to the ones from impact excitation. A comparison between accelerometers and geophones were made, which showed the geophones in general performed best for the tests. For the impact test two sizes of impact hammers were used and the results showed that the larger sledgehammer produced the best results.

Keywords Experimental vibration analysis · Impact test · Operational measurement · Highly accelerated life time tester · Wind industry

21.1 Introduction

This paper describes the process and preliminary results from measurements of a highly complex test bench. The aim of the measurements is to provide a better understanding of its dynamics. The test bench on which measurements were applied, is a Highly Accelerated Life-Time (HALT) tester used for testing wind turbine nacelles, see Fig. 21.1. The HALT tester is a large and complex machine which has a Hexapod with nine actuators that can provide tilt and yaw moments up to 25 MNm used to simulate forces from the wind turbine blades. Two large 8 MW electrical motors drive the shaft of the machine and can produce up to 14 MNm of torque. All these forces are used to simulate the lifetime of a nacelle in about half a year. The attached unit undergoing tests is referred to as Device Under Test (DUT). Lindø Offshore Renewables Center (LORC), who owns the HALT tester, has a keen interest in knowing the expected lifetime of the machine. To determine the lifetime of the HALT test bench, dynamic properties of the test bench is vital to obtain the correct estimations. The test bench is running at a low rotational speed, which means that to get a dynamic amplification of the machine's displacement at its resonances, the frequency of these must be low. The dynamics of the test bench is currently unknown and to get an understanding of these, measurements were carried out. Two types of experimental vibration tests were conducted, impact test and measurements with operational excitation. Two test types were used to compare results between the two in order to find similarities and differences that could help obtain more information on the machine. As mentioned, the HALT tester is a large and complex

J. Berntsen (✉)
Department of Technology and Innovation, University of Southern Denmark, Odense M, Denmark
Lindø Offshore Renewables Center, Munkebo, Denmark
e-mail: jbe@iti.sdu.dk

A. Brandt
Department of Technology and Innovation, University of Southern Denmark, Odense M, Denmark
e-mail: abra@iti.sdu.dk

B. Dilworth (ed.), *Topics in Modal Analysis & Testing, Volume 8*, Conference Proceedings of the Society for Experimental Mechanics Series, https://doi.org/10.1007/978-3-030-47717-2_21

Fig. 21.1 HALT test bench

structure which makes the task of producing satisfactory impact measurements difficult. The test bench consists of many components. The main parts are welded together and connected by various joints. Operational measurements are usually preferred over impact and shaker tests on large structures such as bridges, buildings and wind turbine towers. It can be difficult to get enough excitation from impact tests and shakers for large structures are very expensive. Operational excitation often provides a more thorough excitation of the structure. In addition to the two used measurement techniques, two different sensor types were used for the measurements and are shown in this article. For the impact tests two sizes of impact hammers were applied in order to explore differences between the two and deciding which was better for the tests.

21.2 Pretest

As a first step to obtain a better understanding of the dynamics of the HALT tester, some impact pretests were conducted. These tests included four accelerometers, two placed on the bracket of an actuator and two on the concrete wall. A medium sized impact hammer was used for excitation. The placement of the accelerometers can be seen in Fig. 21.3 as degree of freedom (DOF) 7, 8, 9 and 10. The purpose of the sensor placements was to investigate the behavior of the concrete and for the actuators to see if the measurement quality at these points were sufficient to extract stiffness, which at a later stage would be used in a finite element analysis. The point of impact was performed both on the actuator and concrete. The sampling rate of the experiments was 1652 Hz (the lowest sampling rate of NI-9234), and measurement duration of 150 and 30 s between each impact. These tests were performed with a small measurement setup to gain fast results that could give indications of the behavior of the structure.

The hardware and software used for the pretests were:

- 4 accelerometers – Dytran 3097 A3T, 500 mV/g, 4.3 g.
- Medium size impact hammer – Kistler 9728a20000.
- DAQ AC/DC coupling – NI-9234.
- DataLogger – Matlab GUI for logging data.
- ImpactGui – Matlab GUI for data processing of impact test [1].

Time data from the tests where imported into the ImpactGui and processed according to [2]. Blocksize of 2K and exponential window of 0.01%, were used. The Frequency response functions (FRF) were obtained from auto spectras and cross spectras of input and output, which where used for calculation of the H_1 estimator. The H_1 estimator is used for all impact tests as the FRFs. The results from one of the accelerometers on the concrete are shown in Fig. 21.2. The sensors on the actuators did not produce useful information, with bad FRF and coherence. The accelerance FRF in Fig. 21.2 shows the lowest visual peak at 12 Hz, but the peaks are in general not well defined. The coherence is approximately one from

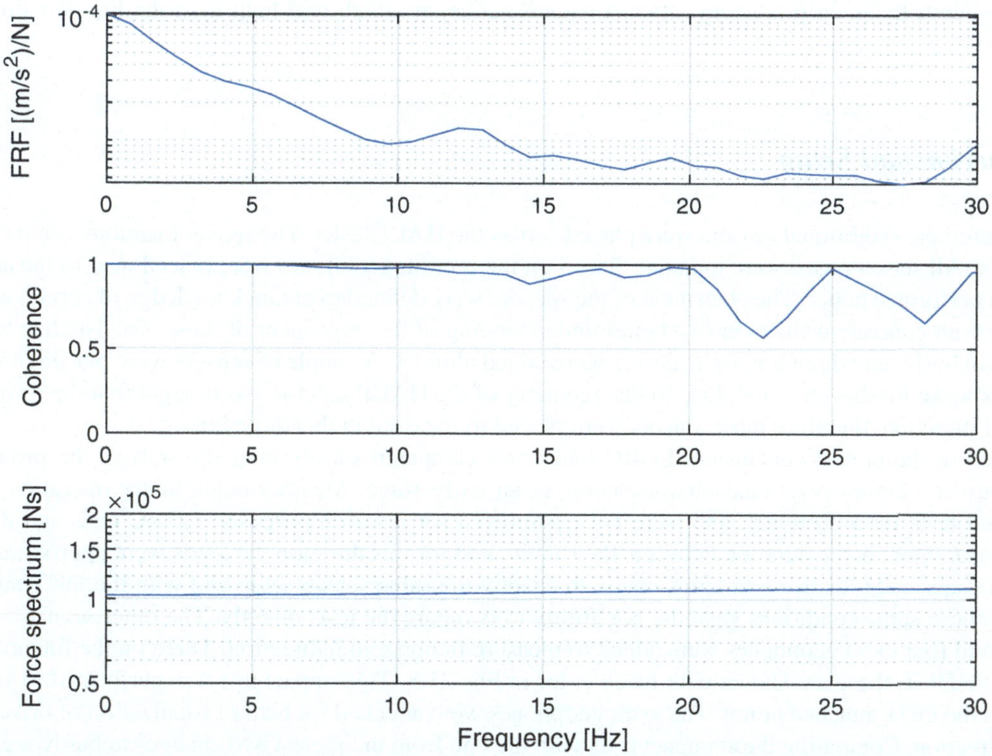

Fig. 21.2 Accelerance, coherence and force spectrum from impact test with accelerometer on concrete in DOF 10. Impact location was close to DOF 9 approximate 3 m away

0–5 Hz, but the FRF has no peaks in this frequency range, which could be caused by poor measurement quality. For the low frequencies before the first peak, the magnitude of the accelerance is decreasing instead of increasing from zero as it should. This is an error due to the measurements and is not of physical value, the first frequencies approximately below 3 Hz should therefore be disregarded when examining the plots. This phenomena are noticed in all FRFs from the impact tests. For the pretest the poor coherence and FRF could be caused by low excitation from the medium sized hammer.

21.2.1 Conclusion from Pretest

The pretest results could indicate that the accelerometers did not receive enough excitation to produce reasonable FRF and coherence, and this was with impact location close to the sensor. That meant some adjustments had to be made in order to produce better measurements. Two adjustments that could be made were trying different sensors and a larger hammer to produce better excitation. Shortly after the pretest measurements, LORC had planned tests which was a convenient opportunity to get operational measurements at an early stage of the project. Based on the pretest, another sensor should be applied for these tests to have a better chance of producing satisfactory measurements. The low frequencies were of interest which meant the sensor should perform well at that range. The laboratory at the university had geophones in storage which are favorable at low frequencies due to their low noise floor. These were chosen for the next measurements alongside the accelerometers. To improve the next impact tests a larger impact hammer should be tried, because the hypothesis was that the excitation level was too low, and a larger hammer would produce a larger impulse.

21.3 Experimental Measurements

Based on the pretest results, additional sensors and another impact hammer were used for the next tests. Many measurements were performed and can in general be split into the three following cases: Operational measurements with the DUT attached, impact test with the DUT attached and impact test without the DUT attached. These experiments will be described and examined closer with focus on producing satisfactory measurement which will help describe the dynamics of the HALT tester.

21.3.1 Experimental Setup

For these measurements additional sensors were placed across the HALT tester. The sensor locations can be seen in Fig. 21.3 with white color. All sensors used were uniaxial. Based on the pretest, geophones were placed next to the accelerometers to compare which performed better. The placement of the sensors were defined to obtain knowledge of certain areas. Geophones were placed on both concrete walls to have a better understanding of the movement of these. To visualize how the Hexapod would displace in horizontal direction, four sensors were added along it. A couple of sensors were placed in vertical direction to obtain the response for this direction. Due to the geometry of the HALT tester it was thought that the lowest modes would be in horizontal direction, therefore most sensors were placed to measure in this direction.

As mentioned in the pretest conclusion, LORC had a test campaign shortly after the start of the project which was a good opportunity to obtain operational measurements at an early stage. Measurements under operation were conducted when LORC had tests for a costumer, and were not customized for research purposes. LORC tests and therefore also the operational measurement had a duration between 10–15 min, and for this duration the loads were approximately constant.

Impact tests were made for two scenarios, one with exactly the same sensor setup and with the DUT attached as for the operation test. These same conditions gave the possibility to compare the tests directly. The other scenario did not have the DUT attached and four extra geophones were added to measure in horizontal direction. Three on the linkarm and one on the attachment to the DUT, these are shown with black color in Fig. 21.3. This impact test was performed to see the difference between having the DUT attached or not. The extra geophones were attached for better visualization of drive train movement in horizontal direction. Comparing these impact tests with the one from the pretest two changes in hardware were made. The two changes were the addition of another sensor type (geophones), and a larger impact hammer was used (sledgehammer). These two changes would be examined with a comparison between the pretest and these impact tests. The measurement duration were 120 and 15 s between impact, which is less than in the pretest because it was seen that the impulse died out

Fig. 21.3 Sensor placement. A is accelerometer, G is geophone and the number is the DOF. G11 and G12 are geophones on the other concrete wall placed in the same way as G9 and G10 for symmetry. The ones with white color show the original sensor setup for test with the DUT attached. The black color sensor location shows the additional four geophones added on the linkarm and at the attachment to the DUT. On this picture they have not been placed yet, but the marks are where they eventually were placed

fast. Even shorter time between impacts could have been used, but 15 s were chosen to have more options on variation of the blocksize.

All measurements were sampled with 1000 Hz and acquired with the data acquisition hardware:

- Accelerometer – Dytran 3097 A3T, 500 mV/g, 4.3 g.
- Geophone – SS-4.5N, 28.8 V/m/s, 86 g.
- NI PXIe-1073 chassis with 3 NI PXIe-4497.
- Sledgehammer – Dytran 5803AT (Impact test).
- DataLogger – Matlab GUI for logging data.
- ImpactGui – Matlab GUI for data processing of impact test [1].

When performing measurements on a big machine like the HALT there are many surrounding effects that could influence the measurements. There are electrical components, hydraulic components and other surroundings that could have an influence. These surroundings have to be kept in mind when preparing the measurements setup and examining the data.

21.3.2 Data Processing and Analysis

For data analysis of the geophones the signal had to be transferred from mV to acceleration, to be compared with the accelerometers. This was done in frequency domain by calculating the FRF for the raw signal and then first transferring mV to velocity by dividing the FRF for the time data with the FRF for the geophones. These calculations were made because the geophone FRF has a slope below its natural frequency of 4.5 Hz. The next step was multiplying by $j\omega$ to produce accelerance FRF. The same procedure with some changes was done for the PSDs.

Operational Measurements

LORC's test campaign included different load scenarios with variations in loads and in the beginning of each test the loads were always ramped. This meant that the data could be non-stationary, the time data had to be analyzed and the most stationary part was used to calculate power spectral densities (PSDs). The tests were analyzed and most tests showed the same tendencies. The time data, probability density and root mean square (RMS) shown in Fig. 21.4 are for DOF 10. The other DOFs have been analyzed in the same fashion, which had some variations. To check whether the time data are normally distributed, one of the measure which was taking was creating a histogram with a normal probability density function superimposed, see Fig. 21.4b. From the histogram and other tools like kurtosis it was seen that the time data for DOF 10 were close to being normal distributed. Inspecting statistical moments among others for the time data which were divided into 50 segments for these calculations, show some peaks in the data when looking at for example mean, RMS, skewness and

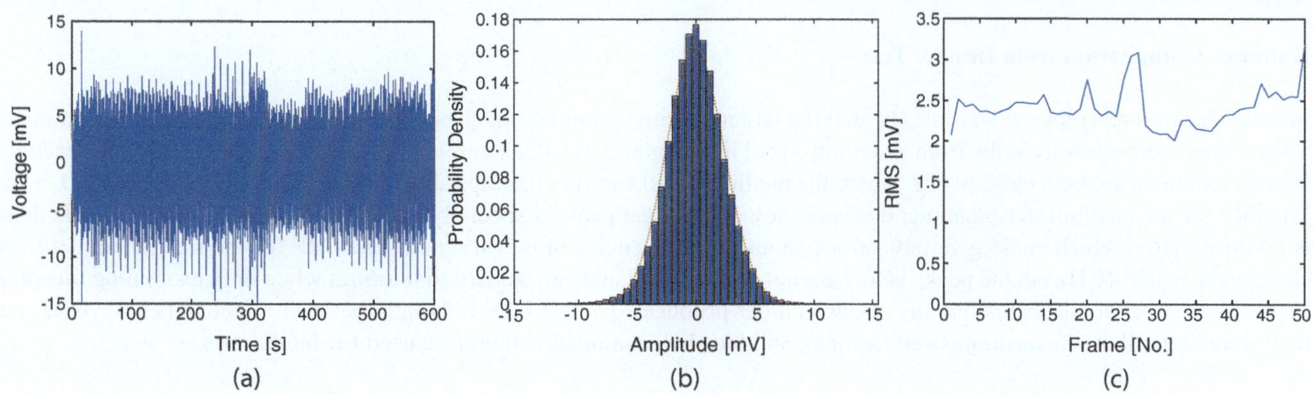

Fig. 21.4 Examination of time data. (**a**) shows the selected time data for which statistical methods were applied. In (**b**) a histogram is plotted with a normal probability density function superimposed. The time data was segmented into 50 frames and RMS of each frame was calculated to form the plot in (**c**)

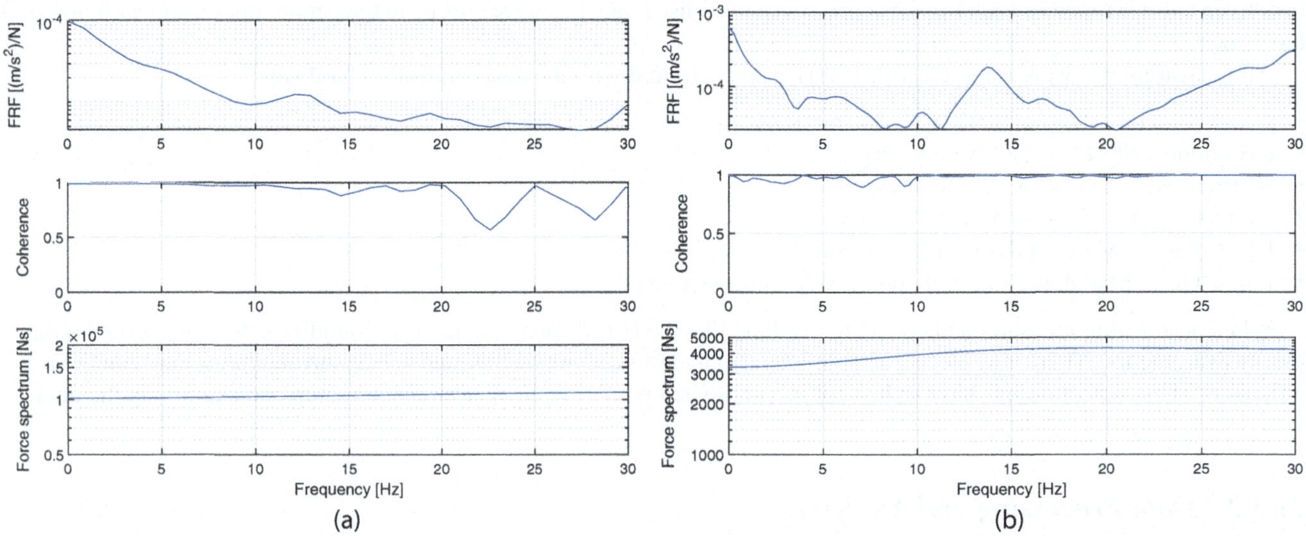

Fig. 21.5 Frequency response function, coherence and force spectrum for an accelerometer in DOF 10 which is on the concrete wall. Two different impact hammers were used, in (**a**) a medium sized impact hammer and in (**b**) a sledgehammer

kurtosis; the plot for RMS values is shown in Fig. 21.4c. The time data showed a small offset in mean which was removed before further processing.

After removing the mean, the PSDs were calculated with Welch's method, with a blocksize of 4K and 291 averages with overlap of 50%. To determine the blocksize, the blocksize was increased until the peak did not increase with increased blocksize, to remove bias error. To obtain a low random error, multiple averages should be calculated which was done with 291 averages.

Impact Tests

The data processing and analysis for these impact tests were the same as for the pretest. Computation of FRF, coherence and force spectra were done in the ImpactGui, and the H_1 estimator were used for the FRFs. Blocksize was 4k and the exponential window was 0.01%.

21.3.3 Results and Discussion

Hammer Comparison from Impact Test

In a number of impact tests it was investigated if a larger hammer would produce better results than a medium sized hammer. The comparison between results from a medium sized hammer and a sledgehammer is shown in Fig. 21.5. The coherence for the two hammers are both close to one, where the medium sized hammer has dips above 20 Hz but is a little better for 0–5 Hz. The FRF for the medium sized hammer does not show many clear peaks. Especially in the low frequencies below 10 Hz there is no information which could give indications about the frequencies of the structure. The FRF for the sledgehammer have clear peaks above 10 Hz but the peaks below are not well defined and can be hard to interpret when only examining this plot. From this comparison it seems that the sledgehammer produces the best FRF, although they were unclear below 10 Hz, but they were better than the medium sized hammer, and the sledgehammer is therefore used for future tests.

Sensor Comparison from Impact Test

Based on the pretest, geophones were added on the structure beside the accelerometers. A comparison between these two can be seen in Figs. 21.6 and 21.7. Two different impact tests were conducted where the impact position changed. In Fig. 21.6

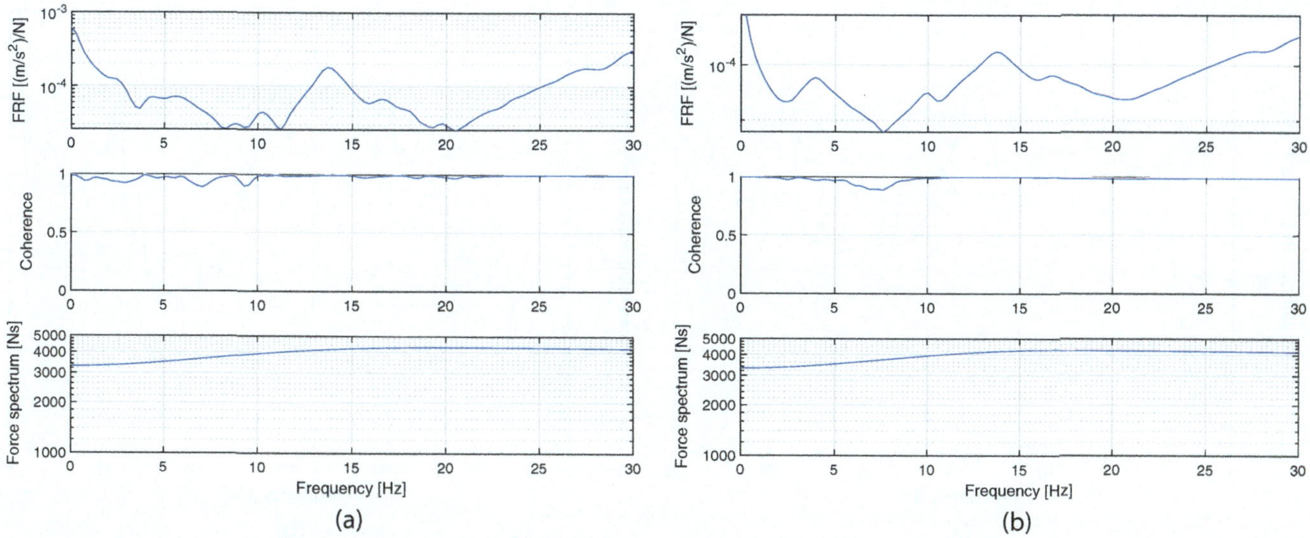

Fig. 21.6 Frequency response function, coherence and force spectrum for an accelerometer (**a**) and geophone (**b**) both in DOF 10 with impact close to DOF 9 approximate 3 m away. For this impact test the DUT was attached

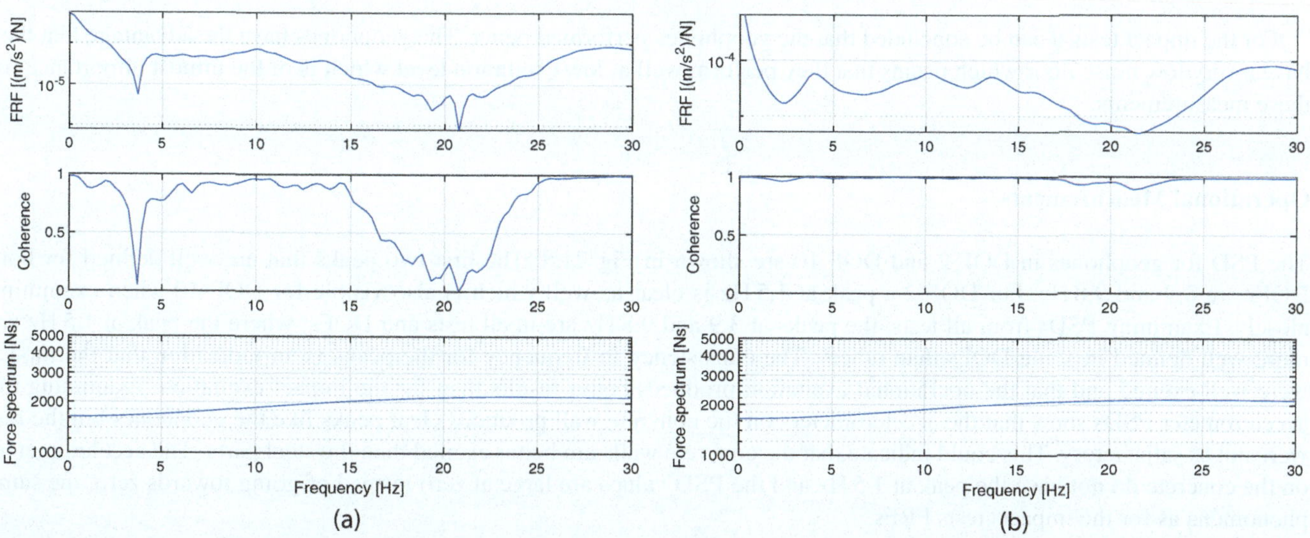

Fig. 21.7 Frequency response function, coherence and force spectrum for an accelerometer (**a**) and geophone (**b**) both in DOF 10 with impact at DOF 12 11 m away. For this impact test the DUT was not attached

impact was conducted 3 m away from DOF 10 close to DOF 9, and in Fig. 21.7 impact location was at DOF 12, which is 11 m away. For the excitation close to the sensor, see Fig. 21.6, the coherence is close to one both for the accelerometer and the geophone, where the geophone is a little better in the low frequencies. The force spectrum is identical for both sensors and are constant through the frequencies of interest. From the FRF it is seen that the peaks are much clearer for the geophones than the accelerometers where it is hard to determine the low frequency peaks. The geophone shows the first peak at 3.9 Hz and the second at 10.1 Hz.

When examining the impact test with excitation far away, see Fig. 21.7, it can be seen that the coherence for the geophone is almost one for frequencies up to 15 Hz which are the frequencies of interest. The coherence for the accelerometer is bad with many large dips over a large frequency interval. The FRF for accelerometers does not show clear peaks, compared to the geophones which show peaks at 3.9 and 10.1 Hz as the first two eigenfrequencies. When comparing the two tests with difference in excitation placement, it is seen that the accelerometer FRF and coherence becomes much worse, where the geophones are not affected significantly. Another difference between the two tests was that the DUT was attached in the second test with excitation far away. This difference is not visual in the FRF, where the first three peaks have the same frequencies.

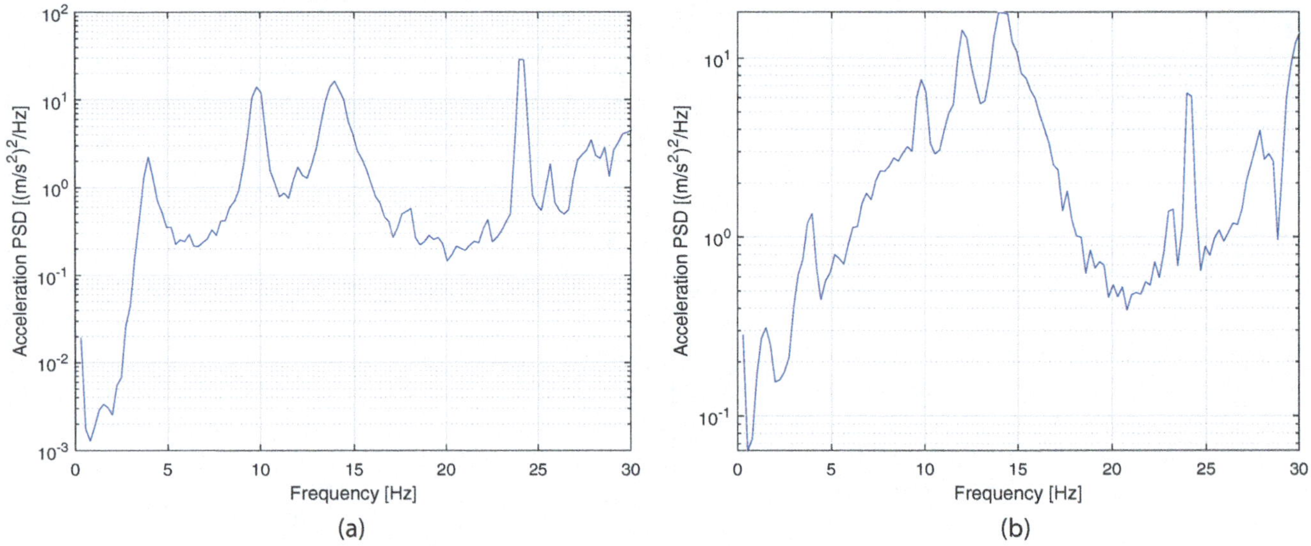

Fig. 21.8 PSD from operational measurements for geophone in DOF 10 (**a**) and DOF 2 (**b**)

For the impact tests it can be concluded that the geophones performed better. The geophones have the advantage that they have a very low noise floor which means that they perform well at low excitation level which is of the utmost importance for these measurements.

Operational Measurements

The PSD for geophones in DOF 2 and DOF 10 are shown in Fig. 21.8. The first two peaks that are well defined for both DOFs are 3.9 and 9.8 Hz. For DOF 2 a peak at 1.5 Hz is clear as well which is also visible for DOF 10 when examining closely. Examining PSDs from all tests, the peaks at 3.9 and 9.8 Hz are in all tests and DOFs, where the peak at 1.5 Hz are more well defined in some DOFs than others. The consistency in frequency for the peaks is an indication that the modes were well excited and that the operational excitation produces better results than for the impact excitation. Examining the accelerometer PSDs show that the accelerometers on the concrete wall produced clear peaks like the geophones but the rest were not as satisfactory. This could indicate that the concrete walls are better excited than the steel parts. The accelerometers on the concrete do not have the peak at 1.5 Hz and the PSD values are large at zero instead of going towards zero, the same phenomena as for the impacts tests FRFs.

Comparing Impact Test to Operational Measurements

From analysis of FRF's for all the sensors on HALT it is seen that the frequency for the second peak show variation depending on sensor placement. The second peak varies between two frequencies 8 and 10 Hz, where the 8 Hz is mostly shown on the Hexapod and 10 Hz is more visual on the concrete, with some exceptions. The results from the linkarm, which connects the Hexapod to the motors, predominantly show the same as for the concrete. Both these frequencies are shown on most plots where one is dominant and the other is a little variation in the FRF. The shape of the peaks in the FRFs are round and not a clear apex, this introduces small deviations when determining the peaks frequency. This means that there in some cases almost are two points defining the peak, which can cause a shift of one frequency increment in frequency. Part of the reason for the round peaks is added damping from the applied exponential window during data processing, but even with this off the peaks are not as clear as for the PSDs. When comparing the peaks between the two types of experiments in Fig. 21.9 it is clear that the peaks from the PSD are well defined at 3.9 and 9.8 Hz compared to impact tests. When comparing the impact tests peak frequencies with the operational measurements, they are not identical but the peak at 3.9 Hz is the same in most cases. The second peak for impact excitation 8 or 10 Hz is close to the third peak for the operational measurements which is at 9.8 Hz. The operational measurements show another peak lower than 3.9 Hz, at 1.5 Hz. This is not visual on any impact tests. For operational excitation the frequencies do not vary between the sensors or tests. Operational excitation produce a

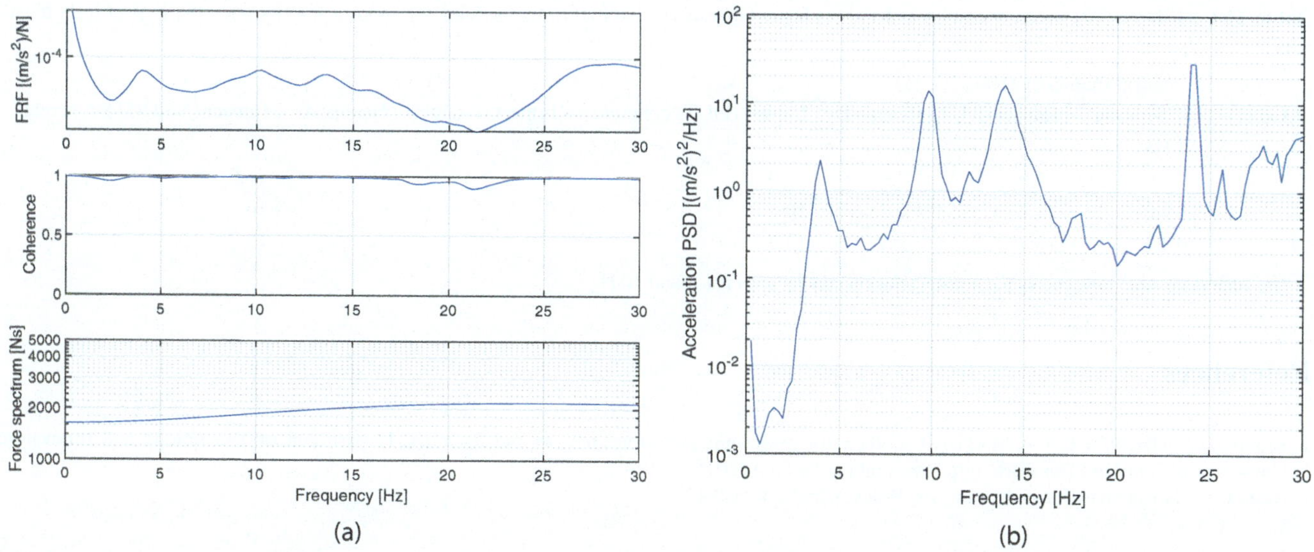

Fig. 21.9 Frequency response function, coherence and force spectrum from impact test and PSD from operational measurements for a geophone in DOF 10 which is on the concrete wall

more thorough excitation of the modes which is also shown in the plots where the peaks are more defined. The frequency values for operational measurements are for that reason thought to be more precise.

For impact tests when the DUT is not attached, the peak at 13.7 Hz is shown for the concrete walls but not for the rest of the sensors. When the DUT is attached both for operational measurements and impact tests, the peak at 13.7 Hz are visual for all sensors.

The vertical sensor at DOF 5 showed the same peaks at 3.9, 9.8 and 13.7 Hz for the operational measurements as were seen the horizontal sensors. This could indicate a rigid body mode for the concrete which had both horizontal and vertical movement. Another aspect that supports assumption that the low frequencies are rigid body modes, were that the eigenfrequencies both with and without a DUT remain the same for the experimental vibration tests. This supports the assumption that the frequencies could be rigid body modes of the entire concrete structure. If it is not a rigid body mode it means that the DUT does not change these low frequency modes for the HALT tester. This has to be examined further, and more measurements have to be performed on the concrete in order to determine if it is a rigid body mode.

21.4 Conclusion

The HALT tester is a large and complex structure for which satisfactory measurements are difficult to obtain. Especially impact tests were difficult to conduct due to the size and complexity of the machine. Different adjustments were made to improve the tests. Two types of sensors were used and for impact tests two sizes of impact hammers were tried. A comparison between the hammers was made and it was concluded that the sledgehammer performed best with better coherence and a FRF with better details.

Accelerometers and geophones were compared and showed that the geophones produced the best results. The geophones showed a better coherence overall. It especially outperformed the accelerometer when the excitation level was low, due to the low noise floor of the geophone. The FRF for the geophones had better defined peaks which were easier to interpret. For operational measurement the accelerometers on the concrete wall showed clear peaks where the rest performed poorly. This indicates that the concrete foundation was better excited than the steel parts during operation.

Frequencies for the first 3–4 peaks are shown in Table 21.1. Operational measurements showed an additional peak compared to the impact tests at a lower frequency at 1.5 Hz. The peak at 3.9 Hz can be compared across the two test types and is consistent at 3.9 Hz, both with the DUT attached and without it. This could indicate that this frequency could be a rigid body mode for the concrete foundation, which has to be examined further. The peak after the 3.9 Hz varies depending on test and which sensors for the impact tests. The peak at 13.7 Hz is consistent in frequency but is not visual on the Hexapod when the DUT is not attached.

Table 21.1 Difference between operation and impact tests. Operational excitation frequencies are more stable than for the impact tests for all sensors

	Natural frequency [Hz]				
Mode [-]	Operation	Impact (DUT) Hexapod	Impact (DUT) concrete	Impact (No DUT) Hexapod	Impact (No DUT) concrete
1	1.5	–	–	–	–
2	3.9	3.9	3.9	3.9	3.9
3	9.8	–*	10	8	10
4	13.7	13.7	13.7	–	13.7

* The FRF does not show peaks but some variation can be seen for 8 and 10 Hz

References

1. Brandt, A.: ABRAVIBE – A MATLAB toolbox for noise and vibration analysis and teaching, Department of Technology and Innovation, University of Southern Denmark. http://www.abravibe.com (2018)
2. Brandt, A.: Noise and Vibration Analysis. Wiley, Chichester (2011)

Jesper Berntsen Background as a mechanical engineer. Worked at year at an engineering consultancy before taking a master's degree in construction at the University of Southern Denmark. Started as an industrial PhD student in March, shortly after finishing the master's degree.

Chapter 22
Feasibility for Damage Identification in Offshore Wind Jacket Structures Through Numerical Modelling of Global Dynamics

Mark Richmond, Ursula Smolka, and Athanasios Kolios

Abstract Operational modal analysis is currently being researched to detect damages in offshore wind turbine structures. To make practical considerations, such as where to place sensors, it is valuable to know how different damages present themselves in the structural dynamics and what kind of damages can be detected given certain detectability limits.

Through conducting natural frequency analysis, the modal response of a structure to damages are explored and resulting considerations for damage detection are discussed. A state of the art, detailed, numerical design model of the Wikinger jacket structure is used to investigate damage detection. Aspects of the model are changed to simulate damage, including member damage, scour, corrosion and more. The resulting modal parameters are calculated, these parameters are compared to those from an unaltered structure and metrics are calculated including frequency change, modal assurance criterion (MAC).

It is found that when the frequencies of two modes intersect, there is a large change in the mode shape of the two modes from the undamaged case and the choice of which modes to compare between healthy and damaged can change the results significantly. The findings from each simulated damage are presented.

Keywords Offshore wind jacket structure · Natural frequencies · Modal shapes · Damage identification · Finite element model

22.1 Introduction

Data on failed structures is typically not available, the being able to identify failure when it occurs is of significant economic benefit. To overcome this lack of data, researchers have modeled structures to simulate damage. Some researchers aimed to detect damage in their model in the presence of measurement noise; for example, Malekzehtab et al. used mode shapes and natural frequency in an objective function, combined with a penalty term to avoid false positives [1]. Liu et al. found, even with 3% noise, modal flexibility could still effectively detect damage [2]. Wang et al. investigated their approach in the presence of temperature variation and used Modal Strain Energy Decomposition to detect damage [3] while Xu et al. used a residual strain energy-based approach [4].

This work differs from others in that a detailed design model is used which includes more complex aspects of the structure and allows for a more detailed investigation. This work includes the mass and moment of inertia for the nacelle and blades, an aspect not present in many other studies, but which is important in the modal properties of a wind turbine structure. Both element damage and scour are simulated allowing for direct comparison of their impacts on the same structure.

There are four levels of damage identification: damage detection, damage localization, damage severity assessment and damage consequence/progression [5]. Detecting a damage is beneficial because while the turbine continues to operate, the remaining fatigue life is used much more rapidly. By detecting the damage and parking the turbine, the excess consumption of fatigue life can be reduced. By localizing the damage, time and resources can be saved in its repair. With the quantification and assessment of damage, decisions regarding the growth and operational impact can be made. A great deal of knowledge is required to achieve all four levels. The use of virtual sensing, along with pre-existing detailed design models, can a cost-effective way of improving the detection and assessment of damage. Operational modal analysis (OMA) can provide

M. Richmond (✉) · A. Kolios
Naval Architecture Ocean and Marine Engineering, University of Strathclyde, Glasgow, UK

M. Richmond · U. Smolka
Offshore Wind, New Services Department, Ramboll, Germany
e-mail: mark.richmond@strath.ac.uk; ursula.smolka@ramboll.com

B. Dilworth (ed.), *Topics in Modal Analysis & Testing, Volume 8*, Conference Proceedings of the Society for Experimental Mechanics Series, https://doi.org/10.1007/978-3-030-47717-2_22

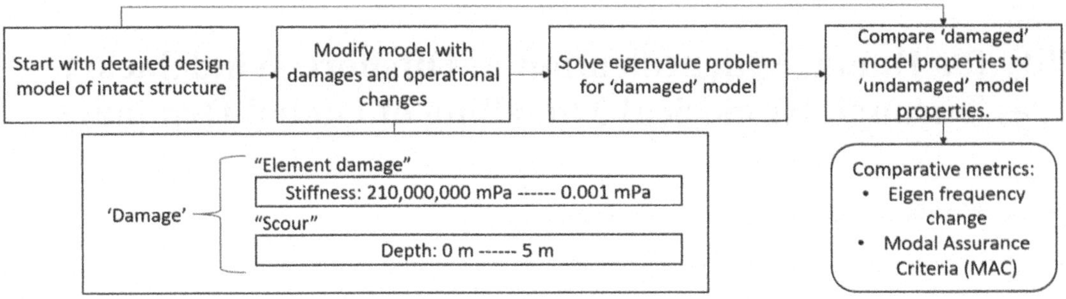

Fig. 22.1 Diagram of damage process used to map damages implemented in a structural model with structural response caused by a change in the structure

information about the structure, and numerically solving the eigenvalue problem can provide information on how the structure responds to damage cases. The aim of this research is to investigate how the use of a numerical model can improve understanding of the impact of damage on the structure and aid those conducting OMA to detect, identify and quantify damage as it occurs. The approach used is shown in Fig. 22.1.

22.2 Background

The structure being is a the four-legged jacket from the Wikinger wind farm. These jackets are at a water depth of between 39–42 meters. At their base they are 23 m square, they are 62 meters tall and support 5 MW turbines [6]. Each leg is fixed with 40 m piles [7]. The rotor diameter is 135 m.

Dynamic analysis is conducted using Ramboll's proprietary, in-house software, Ramboll Offshore Structural Analysis (ROSA). In this case, version 53, the latest at the time, is used [8]. ROSA, the main part of the ROSAP package, is a beam element-based design software. The ROSAP package, and the module ROSA, has been used for the design of the majority of offshore wind turbine foundations worldwide [9]. The first five mode shapes are investigated. The first two are first order modes, one where the tower sways fore-aft and the other side-side, these have natural frequencies around 0.3 Hz. The third is a torsional mode where the tower oscillates about the verticle axis, its natural frequency is calculated to be around 1.3 Hz. The 4th and 5th modes are 2nd order, tower swaying modes, fore-aft and side-side, both with frequencies around 1.4 Hz.

A damage metric used in this study is the modal assurance criterion (MAC). MAC calculates indicates the level of consistency between two vectors $\{\varphi_A\}$ and $\{\varphi_X\}$. The equation is as follows [10]:

$$MAC\,(r,q) = \frac{\left|\{\varphi_A\}_r{}^T\{\varphi_X\}_q\right|^2}{\left(\{\varphi_A\}_r{}^T\{\varphi_A\}_r\right)\left(\{\varphi_X\}_q{}^T\{\varphi_X\}_q\right)} \tag{22.1}$$

Some environmental factors which are expected to occur are included at an average value expected over the life of the structure, such as marine growth, scour, corrosion and some others [11, 12]. Scour is investigated as a process which impacts the system's modes and is different from element damage. Scour is varied from 0 to 5.5 m, which in this case is removing the top layers, exposing some length of the piles.

Researchers who have developed damage detectability models by implementing some damage into an FEA structure have typically limited their investigation to joint damage [1]. The way that joint damage has typically been implemented is through the reduction of either joint or element stiffness. Within many equations used to derive the equations of motion for the beam members, the effect from a change in stiffness and from area are proportional, as is seen in the Lagrange equation. The assumption that a change in stiffness is proportional to a change in area due to the growth of a crack is not always strictly realistic, but for investigating the first five natural frequencies this assumption has been made by other authors [3, 13].

Table 22.1 MAC values for first five modes with stiffness in leg joint reduced to 100,000 Pa

E = 100,000 Pa	1	2	3	4	5
1	0.999	0.000	0.164	0.124	0.000
2	0.000	0.999	0.001	0.001	0.132
3	0.164	0.000	1.000	0.444	0.010
4	0.123	0.001	0.443	1.000	0.001
5	0.000	0.132	0.010	0.001	1.000

Healthy mode along the horizontal direction and damaged along the vertical. Top five values shaded in grey

Table 22.2 MAC values for first five modes with stiffness in leg joint reduced to 100 Pa

E = 100 Pa	1	2	3	4	5
1	0.014	0.985	0.013	0.089	0.126
2	0.955	0.015	0.136	0.055	0.003
3	0.009	0.158	0.109	0.523	0.454
4	0.000	0.124	0.020	0.222	0.995
5	0.205	0.001	0.910	0.359	0.010

Healthy mode along the horizontal direction and damaged along the vertical. Top five values shaded in grey

22.3 Analysis

An observation made in the course of this work is that quantifying the damage based on changes to the modal properties requires consideration for how the modes are compared. If the first five modes of the healthy structure are compared to the first five modes of the damaged structure then this results in 25 possible comparisons, only five of which provide any value. The diagonal elements can be used, that is to compare mode 1 healthy to mode 1 damaged and so on, but with significant enough damage, the correct order to compare might change.

To illustrate this point, a damage is modelled half way up one of the legs of the structure. This damage is not rotationally symmetric and so when looking at fore-aft and side-side mode pairs, this will affect one more than the other. As a result of this unsymmetrical damage, the natural frequencies of similar modes, with frequencies which are very close, change at different rates. Consequentially a mode with a frequency higher than another, changes to having a frequency lower than that mode. This damage is modelled, as described previously, by reducing the elastic modulus of the element. The MAC values resulting from this are shown, with a modulus of 100,000 Pa in Table 22.1 and a modulus of 100 Pa shown in Table 22.2.

The MAC values indicate the similarity of the damaged modes to the unaltered modes. With a low level of damage, close to undamaged, all the diagonal elements of the table are close to 1, indicating consistency. Conversely, with a high level of damage, none of the diagonal elements are close to 1, but off-diagonal elements are.

There are several practical implications of this observation. This is expected to be the typical case, rather than a fringe case, as wind turbine jackets typically have these paired fore-aft and side-side modes which have very close frequencies. For simply detecting the damage this is not of much consequence, but it is critical for damage identification or anything higher as it can change the conclusions about what damage has occurred. There is no time-history to connect the modes, they are separate observations in time, and so a method is needed to correlate the observations for the sake of comparison. In this study, maximizing the MAC values with the Hungarian algorithm [14] is used.

22.4 Results

Figures 22.2 and 22.3 show the normalized change in natural frequency and the MAC values respectively, given the loss of each joint for the first five modes, this does not include any leg damage, only cross-members. The normalized frequency is the change from the intact frequency, which is typically a negative change, divided by the intact frequency. Modes 1 and 2, the tower fore-after and side-side modes, are not affected by damage to the cross members, but the damage does impact higher order modes in this model. Based on field experience, frequency changes larger than 4% and a MAC value less than 0.8 are detectable, in that it can be discerned clearly from the standard variation caused by environmental and other factors, and similar limits have been found by other authors [3, 4]. However, this is only a preliminary detectability limit. Lower joints have more effect than higher ones, as has been found in other studies [3]. From these graphs, it can be concluded that

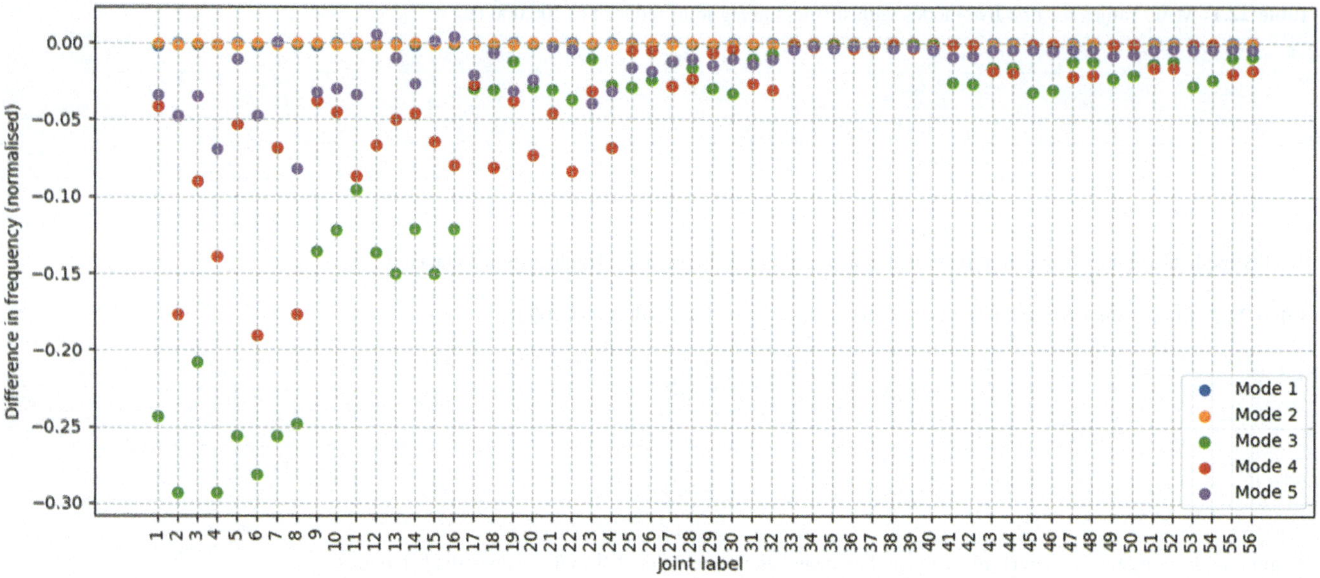

Fig. 22.2 Normalized change in natural frequency for each joint on the structure and the first five modes given damage in only that joint. Joints lower down in the structure are further left

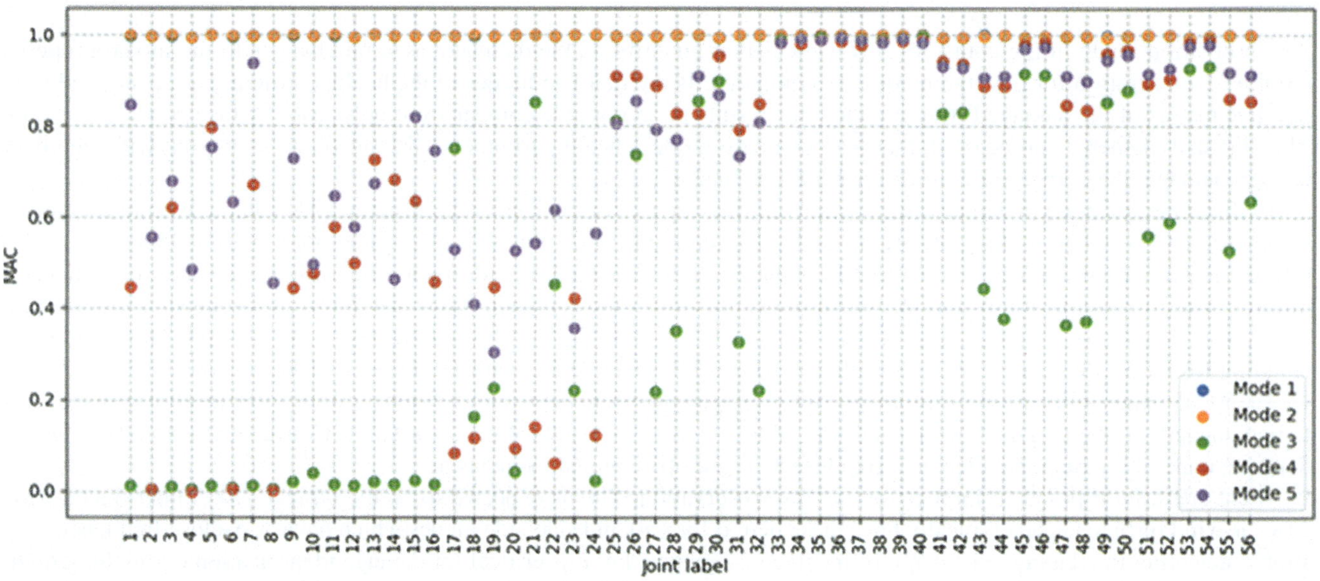

Fig. 22.3 MAC values for each joint on the structure and the first five modes given damage in only that joint. Joints lower down in the structure are further left

more damages can be detected with MAC than with change in frequency and that damage to horizontal members (33–40) cannot be detected with either approach.

Jacket structures and monopiles respond differently to scour; others have found that the first mode frequency changes significantly with scour, around 5% for a scour/diameter ratio of 1 [15] for a monopile, however, the results presented here in Fig. 22.4 show that for this jacket structure the first mode frequency is almost unaffected and the MAC doesn't go below 0.996. Modes 3, 4 and 5 are more affected by this change. At just under 4 m scour depth, the frequencies of the 4th and 5th modes cross each other, resulting in a dip in the MAC values for these modes. The damage index calculated for this case is almost completely linear, fitting a linear trendline gives an increase in damage index by 0.22 per meter. A certain level of scour is expected, so over time from installation, the natural frequencies will reduce [16].

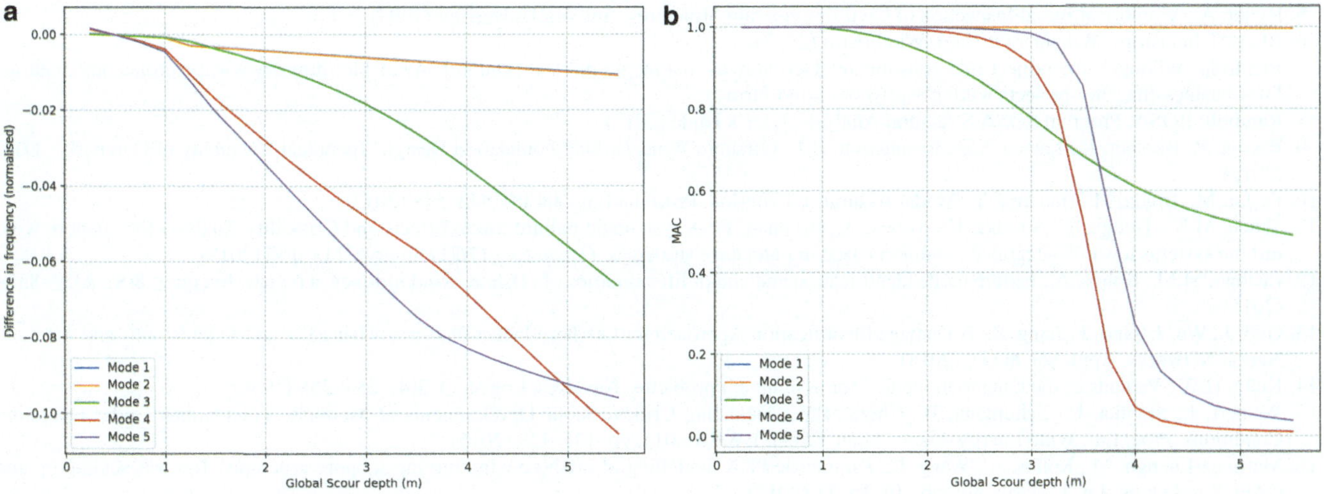

Fig. 22.4 Effect of gradually increased depth of global scour, compared to 0 scour depth. Modes compared based on maximized MAC (**a**) Difference in natural frequency, (**b**) MAC value. Scour depth in meters for the 23 m x 23 m jacket

22.5 Conclusion

The feasibility of detecting a range of damages, based on simulated damages to a numerical model, was investigate. An eigenvalue solution using a detailed design model was conducted for the sake of calculating damage metrics at these conditions. The model is more detailed than other models typically used in similar studies, such as including the mass and moment of inertia of the nacelle and blades and environmental factors expected in the life of the structure. The effects simulated included not only damage to structural elements but also scour. The metrics used are ones which can theoretically also be determined from accelerometer values, which are natural frequency changes, MAC values.

Non-symmetric damage and the nature of mode shape typical to wind turbine jacket structures often results in a change of frequency order which needs to be accounted for. In such cases, to characterize the damage, comparison between intact and damaged can be done based on highest MAC values rather than frequency order. For damage to elements, it is found that lower elements change mode shape and frequency more than higher ones. For non-joint effects which are symmetric on the structure such as global scour, the change in frequencies and the MAC values follow a consistent order, given moderate levels of damage. This can make distinguishing these types of damage difficult.

This research shows how different types of damages can be assessed using structural models which can then be used in the future with OMA approaches to distinguish different damage types. In the present study, cross-member damage as well as scour were investigated. Future research will look at many more damage types as well as further considerations in the assessment of the changes in modal parameters resulting from the implementation of damage into the model.

Acknowledgements This project has received funding from the European Union's Horizon 2020 research and innovation program under grant agreement No. 745625 (ROMEO) ("Romeo Project" 2018) (https://www.romeoproject.eu). The dissemination of results herein reflects only the author's view and the European Commission is not responsible for any use that may be made of the information it contains.

Furthermore, this work was also supported by the sponsorship of University of Strathclyde for pursuing Doctoral Studies within the Centre for Doctoral Training in Renewable Energy Marine Structures - REMS (http://www.rems-cdt.ac.uk/)..

References

1. Malekzehtab, H., Golafshani, A.A.: Damage detection in an offshore jacket platform using genetic algorithm based finite element model updating with noisy modal data. Procedia Eng. **54**, 480–490 (2013)
2. Liu, K., Yan, R.-J., Guedes Soares, C.: Damage identification in offshore jacket structures based on modal flexibility. Ocean Eng. **170**, 171–185 (Dec. 2018)
3. Wang, S., Zhang, M., Li, H.: Damage localization of an offshore platform considering temperature variations. Math. Probl. Eng. **2015**, 1–10 (2015)
4. Xu, M., Wang, S., Li, H.: A residual strain energy based damage localisation method for offshore platforms under environmental variations. Ships Offshore Struct., no. December 2018, 2018

5. Rytter, A.: Vibrational Based Inspection of Civil Engineering Structures. Aalborg Universitet (1993)
6. BLADT Industries: Wikinger Offshore Wind Farm (2017)
7. Iberdrola: Wikinger, the project that consolidates Germany as a strategic market. [Online]. Available: https://www.iberdrola.com/about-us/lines-business/flagship-projects/wikinger-offshore-wind-farm
8. Ramboll: ROSA, Program ROSA Structural Analysis, User's guide (2019)
9. Passon, P., Branner, K., Larsen, S.E., Rasmussen, H.J.: Offshore Wind Turbine Foundation Design. Technical University of Denmark – DTU (2015)
10. Pastor, M., Binda, M., Harčarik, T.: Modal Assurance Criterion. Procedia Eng. **48**(48), 543–548 (2012)
11. Scheu, M.N., Tremps, L., Smolka, U., Kolios, A., Brennan, F.: A systematic Failure Mode Effects and Criticality Analysis for offshore wind turbine systems towards integrated condition based maintenance strategies. *Ocean Eng.***176**(February), 118–133 (2019)
12. Luengo, M.M., Kolios, A.: Failure mode identification and end of life scenarios of offshore wind turbines: a review. Energies. **8**(8), 8339–8354 (2015)
13. Guo, J., Wu, J., Guo, J., Jiang, Z.: A Damage Identification Approach for Offshore Jacket Platforms Using Partial Modal Results and Artificial Neural Networks. Appl. Sci. **8**(11), (2018)
14. Kuhn, H.W.: Variants of the hungarian method for assignment problems. Nav. Res. Logist. Q. **3**(4), 253–258 (1956)
15. Weinert, J., Smolka, U., Schümann, B., Cheng, P.W.: Detecting Critical Scour Developments at Monopile Foundations Under Operating Conditions. Proc. Eur. Wind Energy Assoc. Annu. Event, EWEA 2015, pp. 135–139 (2015)
16. Martinez-Luengo, M., Kolios, A., Wang, L.: Parametric FEA modelling of offshore wind turbine support structures: Towards scaling-up and CAPEX reduction. Int. J. Marine Energy. **19**, 16–31 (2017)

Chapter 23
Use of Operational Modal Analysis to Identify Systems with Oscillatory Masses

Lasse Førde Thunbo, Niklas Carl Ørum-Nielsen, Tobias Friis, Sandro D. R. Amador, Evangelos Katsanos, and Rune Brincker

Abstract Structures experience frequently moderate or even excessive structural vibrations due to their exposure to dynamic actions such as wind- and/or wave-induced loads. To mitigate the amplitude of those oscillations, passive damping systems can be used and their efficient performance is dependent on their tuning to the vibration properties of the structure, in which they have been installed. The Tuned Liquid Damper (TLD), being commonly a container including some kind of liquid, has been already used to damp excessive vibrations of different structural systems. However, there are structures that are expected to carry containers of varying size filled with liquids during their lifetime and hence, this inevitable, vibration-wise interaction between the structure and the container may eventually affect, either favorably or adversely, the structural response. The offshore platforms, being usually equipped with a tank at its top for the temporary storage of the extracted oil, constitute an example of structures with oscillatory mass at the top. Hence, it is rather appealing to investigate the dynamic behavior of this coupled system (offshore platform and the oil tank at its deck) and especially, to identify its modal properties. To do so, a thorough study, including both experimental and numerical investigation respectively, is undertaken and the robust Operational Modal Analysis is applied herein to estimate the effect of the oscillatory mass of the TLD on the dynamic properties of the offshore platform.

Keywords Operational modal analysis · Random vibrations · Tuned liquid damper · Experimental investigation

23.1 Introduction

The trend of using high strength and lightweight materials in structural systems often leads to an increasing flexibility and lack of inherent damping of structures that, in turn, may amplify their susceptibility to various sources of loads like, for example, the wind- and-wave induced excitations. In order to reduce the risk of potential structural failures, external damping devices have been widely used to mitigate excessive vibrations. Along these lines, a simple and economical passive damping device is a Tuned Liquid Damper (TLD), which is known to work effectively for structural systems that are subjected to lateral excitations while responding mainly in the low frequency range.

The performance of a TLD, being commonly a container including some kind of liquid, is dependent on how well its own dynamic properties are tuned to the dynamic properties of the structural system per se. In other words, the appropriate interaction between the vibration of the structure and the self-vibration of the TLD defines the effectiveness of this passive damping system to reduce the response of the structure. However, there are structures that are expected to carry containers of varying size filled with liquids during their lifetime. Hence, such an inevitable, vibration-wise interaction between the structures and the containers may eventually affect, either favorably or adversely, the structural response. Along these lines, the offshore platforms are typical examples of structures that carry at their top tanks for the temporary storage of the extracted oil. Hence, it is rather appealing to investigate reliably this vibration-driven interaction between the oil tank and the offshore platform, the latter being continuously exposed to a multi-hazard environment consisting, among others, excessive wind and wave lateral loads. Such an investigation requires to scrutinize the dynamic properties of the coupled system by identifying its modal properties (frequencies, mode shapes and damping ratios). To do so, the robust Operational Modal Analysis (OMA) can be used assuming that vibration responses for both the structure and the tank are available.

L. F. Thunbo · N. C. Ørum-Nielsen · T. Friis · S. D. R. Amador · E. Katsanos (✉) · R. Brincker
Technical University of Denmark, Kongens Lyngby, Denmark
e-mail: s143874@student.dtu.dk; s143857@student.dtu.dk; tofri@byg.dtu.dk; sdio@byg.dtu.dk; vakat@byg.dtu.dk; runeb@byg.dtu.dk

B. Dilworth (ed.), *Topics in Modal Analysis & Testing, Volume 8*, Conference Proceedings of the Society for Experimental Mechanics Series, https://doi.org/10.1007/978-3-030-47717-2_23

Several studies have already proven that the OMA-based identification techniques are solid and can be robustly used to identify dynamic properties of structural systems responding mainly in the linear regime. However, the use of OMA to identify the dynamic properties of a structure with TLD can be questioned since the latter is widely considered as a nonlinear system with an infinite number of degrees of freedom that the dynamic properties and mainly the associated damping are highly dependent on the external loading and, in case of offshore structures, are associated with the occurrence of wave breaking and other surface contamination that may lead to a turbulent flow [1]. As it concerns the numerical modelling of the TLD per se, it has been treated as a single degree of freedom (SDOF) system [2, 3] while the sloshing frequency of the liquid of the TLD has been determined by using the Lamb's linear theory of liquid surface motion [4]. Despite the convenience that this simplified approach is associated with, its experimental validation has been only seen to be marginal. Additional studies have focused on the structures being equipped with TLD by investigating the mass ratio (between the mass of the structure and the one of the TLD) that needs to be tuned appropriately in order to have the maximum reduction of the structural vibrations. For example, Jin-Ting Wang et. al [5] determined an optimal range for the mass ratio to be between 1% and 4%. Several studies have also been conducted in order to improve the damping effect of the TLD, by implementing a filter screen across the middle of the container or by using a sloped bottom TLD, which additionally reduces the possibility of the "stiffness hardening" effect of the liquid [1, 6].

Based on the discussion made above, plenty of research effort has been dedicated to elaborate the dynamic properties of the TLD per se. However, the effect of the TLD on the dynamic properties of the structural system, which carries this passive damping device, has attracted until now marginal research work. Along these lines, the current study, including both experimental investigation and numerical analysis, scrutinizes the effect of a TLD on the dynamic properties of a structural system with closely spaced modes. Especially, a linear, SDOF-based approach is adopted herein to tune a rectangular TLD to the first structural mode of a scaled steel model that represents an offshore platform with an oil-tank mounted at its top. The entire structural system is excited by random loading at low excitation amplitude with white noise characteristics. The response of the platform is measured by three accelerometers while the water elevation of the TLD is measured by four laser sensors. The platform is analyzed with and without the TLD installed through an experimental investigation, where the modal parameters are identified by the use of a time-domain, OMA-based identification technique. By conducting this study, two main research questions need to be addressed:

- Is possible to identify experimentally the effect of the TLD on the modal properties of the structural system by applying the OMA techniques?
- Is the experimental identification of the sloshing frequency of the TLD in agreement with the one estimated by using the theoretical linear formula and the associated SDOF-based approach?

23.2 Theoretical Background

23.2.1 Operational Modal Analysis

The basic idea behind the identification technique known as OMA is to characterize the modal parameters, such as the natural frequencies, f_n, damping ratios, ζ_n, and the mode shapes, $\mathbf{b}_n$, of an linear structural system through the identification of its natural vibration modes from its operating response. The modal parameters can be identified by the Auto Regressive – Poly Reference technique (AR/PR) that works in the time domain. This specific method utilizes estimated correlation functions (CF), being considered as free decays, with the use of AR models that contain the physical properties of the system. Using the free decays expressed in discrete time, an equation describing the free response properties of the system can be formed.

$$\mathbf{y}(n) - \mathbf{A}_1\mathbf{y}(n-1) - \mathbf{A}_2\mathbf{y}(n-2) - \cdots - \mathbf{A}_{na}\mathbf{y}(n-na) = 0 \tag{23.1}$$

where the free decay, $\mathbf{y}(n)$, contains $n = 1, 2, \ldots np$ number of samples and $\mathbf{A_1}, \ldots, \mathbf{A_{na}}$ are the AR matrices. An estimate of the AR polynomial coefficients is then obtained by solving an overdetermined problem with the use of linear least squares (LS) regression. The estimation of the AR matrices enables the modal parameters to be identified by forming the corresponding companion matrix, $\mathbf{A}_C$, and performing an eigenvalue decomposition (EVD) of the companion matrix.

$$\mathbf{A}_C\, \mathbf{\Phi} = \mathbf{\Lambda}\, \mathbf{\Phi} \tag{23.2}$$

from where the discrete time poles, $\mu = e^{\lambda \Delta t}$, can be found in the diagonal matrix $[\boldsymbol{\Lambda}]$, while the discrete eigenvectors, $\boldsymbol{\varphi}_d$, are contained in the matrix $[\boldsymbol{\Phi}]$. In addition to the time-domain identification, a qualitative inspection of the system information contained in the measured signals is performed in the frequency domain. The method is based on the Frequency Domain Decomposition (FDD) technique, in which a singular value decomposition of the spectral density (SD) matrix is performed and the modal peaks of the singular values are detected. More elaborate details about the identification techniques and estimation of CF matrices can be found in [7].

23.2.2 *Tuned Liquid Damper*

A TLD is a economical passive damping system that is expected to reduce the lateral vibrations of a structural system responding in the low frequency range. The TLD, typically, consists of a tank being filled with a liquid and its vibration will counteract the vibration of the structural system per se. In general it is known that the TLD has quite similar effect on a structural system as a tuned mass damper and therefore, in order for the TLD to work as a damping device it has to be tuned to a specific mode which highly depends on the geometry of the tank and water depth.

A rectangular TLD is used in this study in order to suppress the vibrations of the scaled platform and the tuning process is simplified by a linear approach that is used herein to estimate the dynamic properties of the TLD. The simplification is based on the fact that the structural system is subjected to low amplitude excitation and hence, a linear behaviour is expected to be seen. As described by Ramos [2], the fundamental sloshing frequency, f_w, of the TLD can be determined by a linear approach using Lamb's linear theory of liquid surface motion [4] yielding Eq. (23.3).

$$f_w = \frac{1}{2\pi}\sqrt{\frac{\pi\, g}{L}\tanh\left(\frac{\pi\, h_0}{L}\right)} \tag{23.3}$$

where g is the gravitational acceleration, L is the length of the TLD in the tuning direction of the rectangular tank and h_0 is the still water depth. Since low amplitude excitation is considered, only a part of the mass in the TLD will participate in the oscillation. Housner [3] has developed a linear model of a rectangular tank that consists of a rigid attached mass to the structure, m_b, and a active mass, m_a, that will participate in the damping process which are determined by Eqs. (23.4) and (23.5) respectively.

$$m_b = m\,\frac{\tanh(\sqrt{3}\,(a/h_0))}{\sqrt{3}\,(a/h)} \tag{23.4}$$

$$m_a = m\left(\frac{1}{3}\sqrt{\frac{5}{2}}\,\frac{a}{h_0}\tanh\left(\sqrt{\frac{5}{2}}\,\frac{h_0}{a}\right)\right) \tag{23.5}$$

where m is the total liquid mass in the TLD and $a = L/2$. The two main tuning parameters are therefore the tuning ratio, Ω, and mass ratio, μ.

$$\Omega = \frac{f_w}{f}, \quad \mu = \frac{m_a}{m_j} \tag{23.6}$$

where m_j is the modal mass of the structure. The optimal mass ratio is considered to be between 1% and 4%, while the optimal tuning rate is $\Omega = 1$ [5].

23.3 Experimental Model

The experimental setup is a simplified and scaled model of an offshore platform consisting of three stacked steel plates supported by four steel columns. The dimensions of the experimental model with the TLD installed is illustrated in Fig. 23.1. One single plate has the specific dimensions of 450 × 450 × 5 mm, leading to a total thickness of the platform's deck being equal to 15 mm. The plates are connected to four steel columns with four A4-70 bolts connecting each column to the deck. The columns are made of 850 mm tall SHS steel profiles with the cross-sectional dimensions 15 × 15 × 10 mm. At the

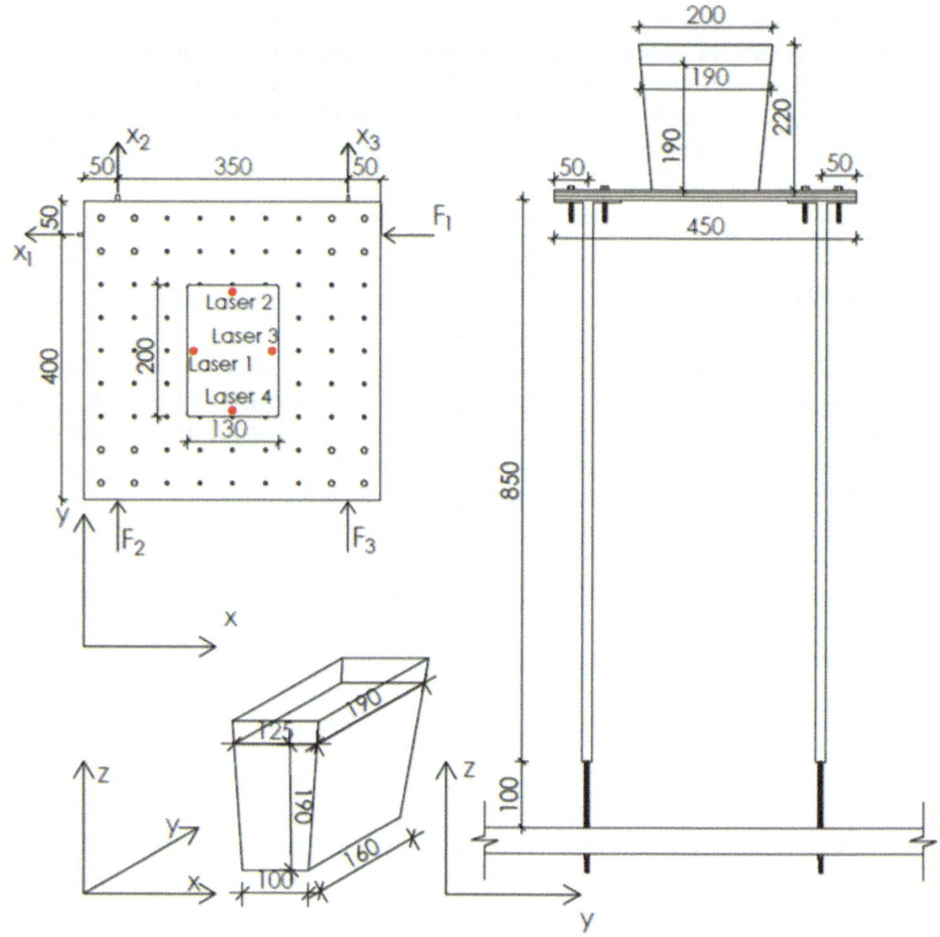

Fig. 23.1 Experimental model seen from above with the TLD attached and measurements points (top left corner), dimensions of the TLD (bottom left corner) and dimensions of the platform seen from the front with the TLD installed on top (middle)

supports of each leg the inner part of the columns consist of a thread, from where a threaded steel rod with a diameter of 5 mm connects the columns to the supporting plate. The threaded steel rod has a free length of 100 mm, leading to a total height of 965 mm for the experimental model. The steel rod is connected to a 40 mm thick wooden plate by four nuts where two of them are counter nuts. The supports can therefore be described as fixed in the translational directions with rotational springs, where the wooden plate acts as the foundation for the experimental model.

The platform is excited by a pneumatic actuator system which induces random loading with white noise characteristics. The load is applied through three nozzles located 20 cm from the platform as illustrated in Fig. 23.2, while the load is efficiently transmitted to the platform by three square aluminum plates with the dimensions 50 × 50 × 2 mm. The locations of the load are denoted as F_1, F_2 and F_3 in Fig. 23.1. In order to distribute the random loading with approximately the same magnitude on both sides of the platform, the two nozzles aiming in the y-direction have a diameter of 1 mm while the single nozzle in the x-direction has a diameter of 1.5 mm. The response of the platform is measured by three accelerometers and their location is denoted as x_1, x_2 and x_3 in Fig. 23.1. The three accelerometers are glued on the second of the three steel plates that together constitute the platform as illustrated in Fig. 23.2, to ensure that the accelerometers capture the overall response. Furthermore, the accelerometers are located at a distance of 50 mm from the edges of the top plates.

Experimental investigation with and without the TLD is conducted with four laser sensors placed on top of the platform and above the water container. The laser sensors, used only for measuring the motion of the water in the TLD, are attached to aluminum profiles with double sided tape and two strings, where the aluminum profiles are connected to the platform with bolts. The laser sensors, raised 10 cm above the water surface, can measure the water deflections with an accuracy of ±3.5 cm. The laser sensors record the water elevation in the middle of each side of the rectangular TLD as depicted in Fig. 23.1.

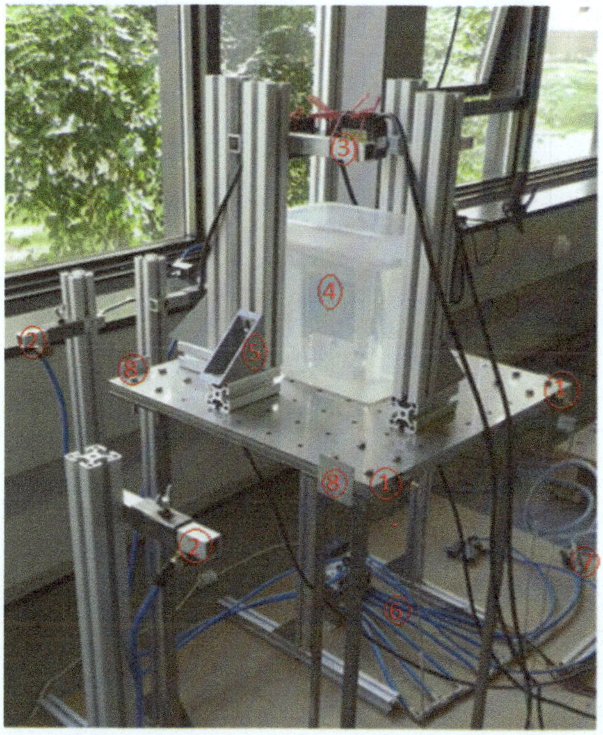

Number	Function
1	Accelerometers
2	Nozzles
3	Laser Sensors
4	TLD
5	Equipment holding the lasers
6	Air hose
7	Valve to open and close the air
8	Aluminum plates

Fig. 23.2 Experimental setup (left), surface in the TLD (top right) and definitions of the numbering

Table 23.1 Final dimensions of the TLD with corresponding properties in the longitudinal and transversely direction

	L	b^*	h_0	m	f_w	m_a	m_b
	[m]	[m]	[m]	[kg]	[Hz]	[kg]	[kg]
Longitudinal direction	0.19	0.125	0.19	4.51	2.02	1.18	3.64
Transverse direction	0.125	0.19	0.19	4.51	2.49	0.78	4.08

* b is the width of the container

The rectangular TLD container is made of plastic and it is located on top of the steel plates and in the middle of the deck, where it is secured with double-sided tape. The container is filled with 19 cm of water, leading to the width and depth at still water level being 125×190 mm as depicted in Fig. 23.1. Polystyrene with humid paper is placed on top of the water surface from where the laser light is reflected back to the sensors. This type of surface acts as a floating roof, where wave breaking, turbulence and other contamination of the water surface is prevented due to the higher stiffness of the material. The properties of the TLD in both the longitudinal and transversely direction is listed in Table 23.1, yielding a mass ratio between the (modal) mass of the first mode of the platform and the active mass of the TLD equal to $\mu = 3.93\%$ while the tuning ratio is set to be equal to $\Omega = 0.99$.

It is seen that the active mass is only 26% of the total mass and this is due to the fact that the TLD is considered to belong to the deep water regime since $h_0/L > 0.15$. This is highly dependent upon the dimensions of the container, which therefore could be further optimized by changing its geometry leading, probably, to the TLD being considered in the shallow water regime, $h_0/L < 0.15$. The latter would yield a larger damping effect and a more optimal use of the TLD per se since higher amount of the total mass would become active [8].

23.4 Results

The experimental investigation is conducted with and without the TLD installed on top of the platform. The response of the platform is measured by three accelerometers which are converted to displacement by double integration in the frequency domain, while the four laser sensors capture the relative water elevation of the TLD. The signals are over-sampled during

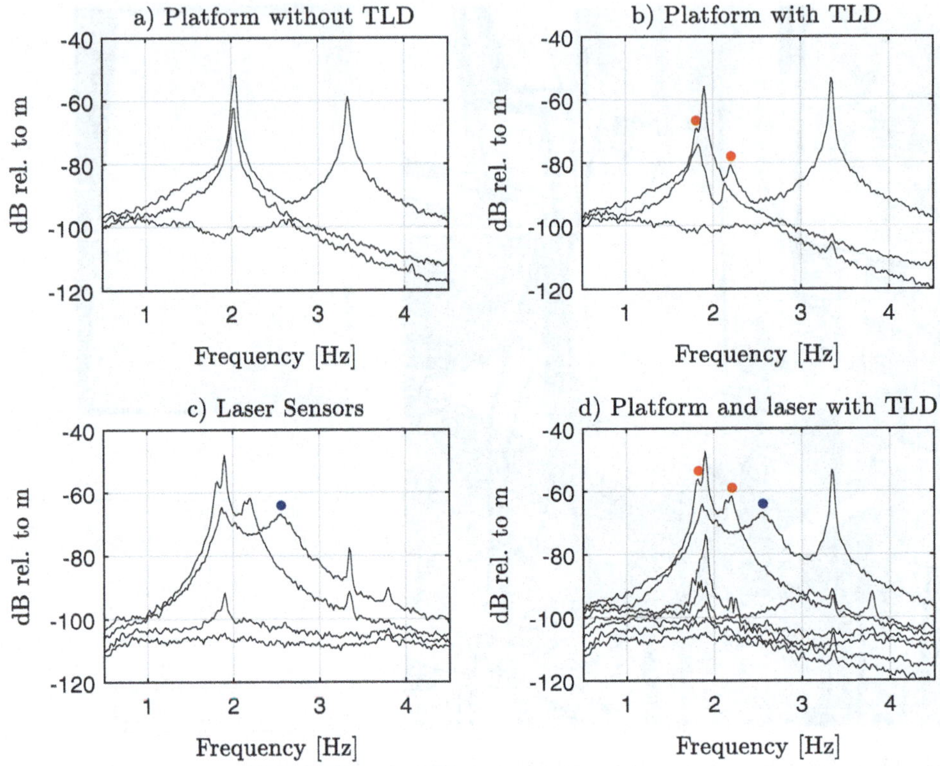

Fig. 23.3 Singular values of the SD matrix for the response of: (**a**) the platform without the TLD, (**b**) platform with the TLD, (**c**) water elevation measured by the laser sensors and (**d**) total structural system including the platform and TLD

data acquisition and decimated down to a cut-off frequency of 8.6 Hz. Furthermore, to isolate the modes of interest, a band-pass filter is applied to the response obtained both from the accelerometers and the laser sensors. The total response of the platform and TLD are used in the aforementioned OMA identification techniques, providing the graphical representation of the singular values of the SD matrix, the identification of the modal characteristics and the schematic illustration of the corresponding mode shapes. It is notable that the SD matrix is obtained by the Welch averaging method with a Hanning window and a 50% overlap.

23.4.1 Identification of Modal Properties

The singular values of the SD matrices obtained from the experiment with and without the TLD are depicted in Fig. 23.3. From the first plot in Fig. 23.3a, which represents the response from the three accelerometers without the TLD, two singular value lines appear to peak in the frequency band of approximately 2–3.5 Hz. The first and second peak clearly indicate the presence of two closely-spaced translational modes around 2 Hz, while the third peak indicates the presence of a third torsional mode at 3.5 Hz.

Figure 23.3b contains the response from the three accelerometers with the TLD, where the singular value lines now indicate the presence of four modes. The first three peaks in the frequency band of 1.8–2.2 Hz indicate the presence of three translational modes, where the first and third peak represents the first mode of the system without the TLD being split into two modes with similar mode shapes (highlighted with red dots). By considering the two new modal peaks, it is observed that the tuning is not optimal since the difference between the amplitude of the peaks is approximately 11 db, where the optimal case would result in the same amplitude yielding a plateau [9]. As previously observed the third singular value line is more flat than the first two, and a torsional mode is observed at 3.5 Hz.

In Fig. 23.3c, the measurements of the water elevation from the four laser sensors are depicted, meaning that the four signals yields four singular value lines. An interaction between the TLD and the platform is observed as three peaks are depicted in the frequency band of 1.8–2.2 Hz as previously observed. However, the occurrence of a fourth peak at approximately 2.5 Hz is only observed when considering the singular value lines of the response from the laser sensors,

Table 23.2 Natural frequencies, f_n, and damping ratios, ζ_n, of the experiment without and with the TLD. The subscript "w/o" refer to the case without the TLD installed

		Mode1	Mode2	Mode3	Mode4	Mode5
$f_{n,w/o}$	[Hz]	2.014	2.045	3.347	–	–
$\zeta_{n,w/o}$	[%]	0.41	0.21	0.31	–	–
f_n	[Hz]	1.811	1.899	2.192	2.497	3.352
ζ_n	[%]	1.19	0.33	2.16	5.45	0.14

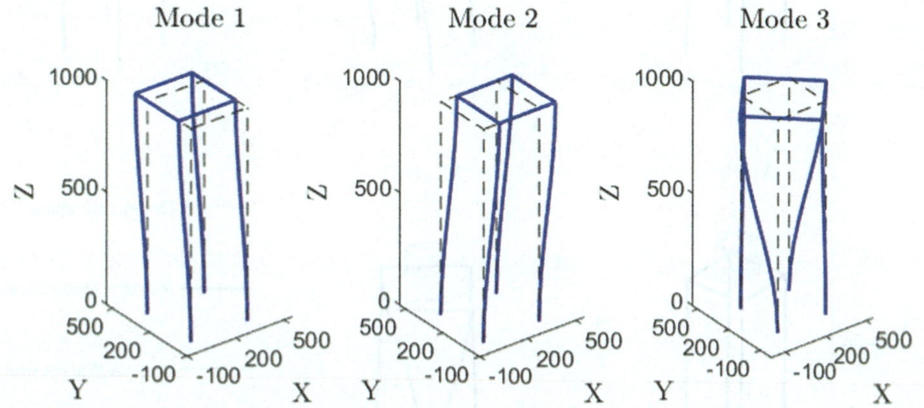

Fig. 23.4 First three mode shapes of the experimental model without TLD

meaning that this mode is considered to be the second asymmetric mode of the TLD (highlighted with the blue dot). The first and third peak depicts the presence of the first and third translational mode with possibly similar mode shapes, while the second and fourth peak indicates the second and fourth translational mode also with similar mode shapes.

The fourth plot of the singular values of the SD matrix in Fig. 23.3d is obtained by combining the response from the accelerometers with the response from the laser sensors. It therefore consists of 7 signals yielding 7 singular values at every frequency line and the identification of 5 modes with the fourth mode being the second asymmetric mode of the TLD.

The results from the AR/PR OMA-related identification technique is presented in Table 23.2, where the upper and lower part present the identified modal parameters of the platform without and with the TLD respectively. Regarding the identification of the scaled structural system with the TLD at its top, mode 1 and 3 correspond to the first mode of the structural system without the TLD, being split into two modes. Furthermore, it is observed that a lower frequency is identified for mode 2, due to the additional mass that is primarily related to the TLD while the mass of the measurement equipment (laser sensors) may also contribute to lowering the frequencies. The frequency of the fourth mode, considered as the second asymmetric mode of the TLD, resembles the theoretical sloshing frequency in the transversely direction as presented in Table 23.1. The fifth mode yields approximately the same frequency as the torsional mode before mounting the TLD. In terms of the damping ratios it is seen that they all increase when the TLD is attached to the platform, except for the torsional mode where the damping seems to decrease.

The identified mode shapes of the structural system without and with the TLD are presented in Figs. 23.4 and 23.5. The mode shapes are normalized in relation to the largest real component and projected on planes parallel to the platform sides. The mode shapes with the TLD are scaled 1:4, meaning that the mode of the platform is multiplied by a factor of 4 for illustrative purposes, since the movement of the TLD is much larger than the platform itself.

From Fig. 23.4 it observed that the first two modes of the scaled model without the TLD are translational ones in the y- and x-direction respectively, while the third mode is a torsional mode. When considering the scaled offshore platform model with the TLD at its top, the mode shapes indicate that the TLD is in-phase with the platform's vibration for mode 1 and out-of-phase for mode 3. Moreover, it is observed that the TLD moves with the platform at the second mode, while the fourth mode reflects the second asymmetric mode of the TLD in its transverse direction. The fifth torsional mode shape illustrates only the vibration of the platform, even though this is not the case, which is a result of the laser sensors not measuring the water elevation in the corners of the TLD. The lack of optimal tuning is again illustrated by considering the shapes of modes 3 and 1, for which the platform movement is smaller for mode 3 compared to mode 1.

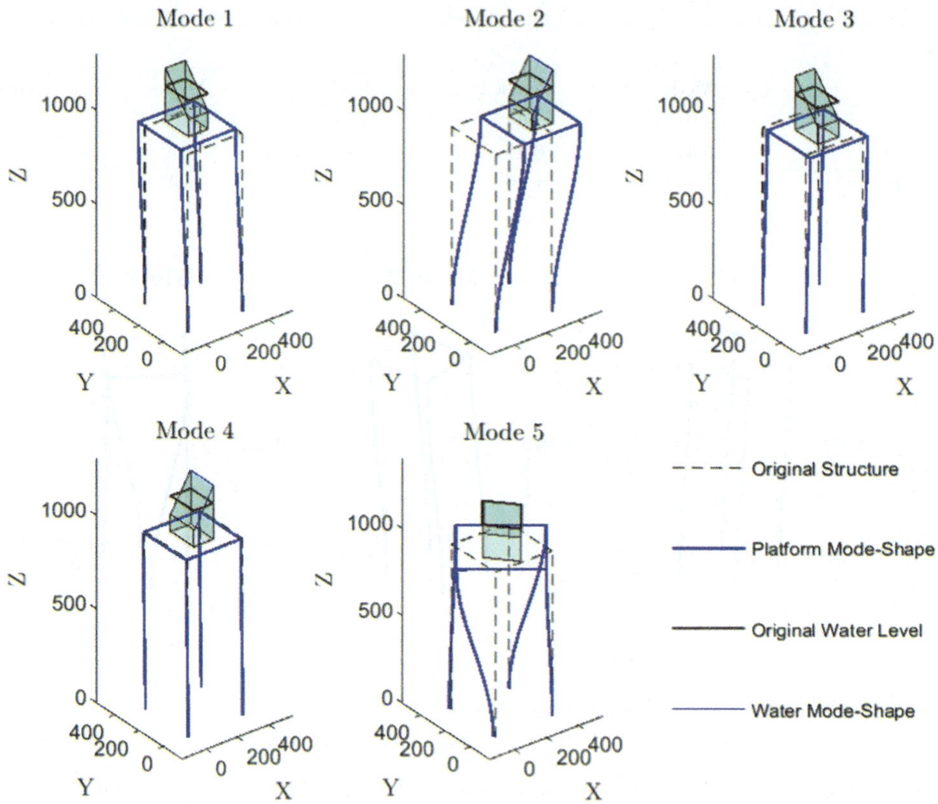

Fig. 23.5 First five mode shapes of the experimental model with the attached TLD

23.4.2 Estimation of Sloshing Frequency and Damping Ratio

A simple experiment is conducted in order to assess the theoretical sloshing frequency as stated by Eq. (23.3), while the damping ratio of the sloshing motion is also estimated. The experiment is performed on a rigid table with the TLD alone and one laser sensor measuring the water elevation in the middle of the longitudinal and transversely direction of the container. Since it is only the sloshing frequency and damping ratio of the TLD that are of interest in this investigation, one side of the tank is lifted and dropped to obtain a free decay of the water surface in the direction of interest. The obtained free decay vibrations in both directions are illustrated in Fig. 23.6.

The sloshing frequency of the TLD is determined by detecting every second zero-crossing of the above response from which the period can be obtained. Knowing the period, T_w, of the liquid sloshing motion, the sloshing frequency can be readily determined by $f_w = 1/T_w$. Furthermore, the damping ratio, ζ_w, is determined by following the widely known logarithmic decrement, δ, that can be calculated by considering two consecutive peak displacement responses, a_j and a_{j+1}, from the free decay of the TLD per se.

$$\delta = \ln\left(\frac{a_j}{a_{j+1}}\right) = \frac{2\pi\,\zeta_w}{\sqrt{1-\zeta_w^2}} \tag{23.7}$$

The resulting sloshing frequencies and estimated damping ratios from the simple experiment are presented together with the theoretical values from Eq. (23.3) in Table 23.3.

It is clear from Table 23.3 that the theoretical sloshing frequencies defined by Eq. (23.3) show high compliance with the experimental results in both the longitudinal and transversely direction.

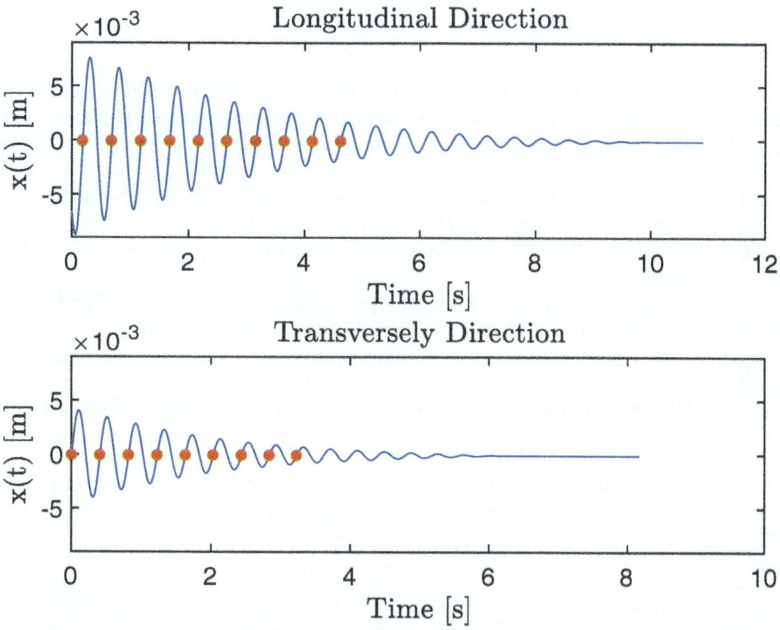

Fig. 23.6 Free decay of the measured relative displacement of the water surface in its longitudinal and transversely direction, including every second zero-crossing to indicate the period, T_w

Table 23.3 Theoretical and experimental sloshing frequencies and damping ratios of the TLD in the longitudinal and transversely direction where subscript "t" indicate the theoretical values and "exp" indicate the experimental values

Excitation dir.		Longitudinal	Transversely
$f_{w,t}$	[Hz]	2.023	2.499
$f_{w,exp}$	[Hz]	2.014	2.460
Δf_w	[%]	0.45	1.56
$\zeta_{w,exp}$	[%]	2.62	3.64

23.5 Conclusion

This study investigates the effect of a TLD, implemented herein by a water container, on the dynamic behaviour of a scaled offshore platform model. Such a TLD, mounted at the top of the scaled steel model, represents the oil tanks that are commonly located at the deck of offshore platforms in order to store temporarily the extracted oil. Especially, the main objective of the study is to investigate whether the dynamic properties of the coupled system (i.e., the offshore platform model and the TLD) can be accurately identified by the use of the OMA techniques. To this end, the offshore platform model, with and without the TLD at its top, is excited by a well designed pneumatic actuator system that induces time-varying excitations with white noise characteristics. The response of the platform is measured by three accelerometers and the response of the TLD is measured by four laser sensors. Moreover, a simple experiment that induces free decays of the TLD per se is performed to verify the analytically-defined sloshing frequency of the water included in the TLD. The study reveals the following conclusions:

- The application of the OMA-based identification technique is found to provide reliable identification results for the steel scaled model and the liquid container at its top, the latter acting as a TLD.
- The analytically-defined sloshing frequency of the TLD is found to fit very well with the experimental findings.
- It is finally concluded that the TLD can be reliably tuned to a specific mode by using a linear and SDOF-based approach.

Acknowledgments The authors acknowledge the funding received from the Centre for Oil and Gas – DTU/Danish Hydrocarbon Research and Technology Centre (DHRTC).

References

1. Tait, M.J., El Damatty, A.A., Isyumov, N.: The dynamic properties of a tuned liquid damper using and equivalent amplitude dependent tuned mass damper. In: Structural Specialty Conference of the Canadian Society for Civil Engineering, vol. 1, pp. 10 (2002)
2. Ramos, L.J.S.T.: Study and simulation of the synchronized pedestrian footstep load and passive damping solutions for footbridges. Ph.D.-thesis, University of Porto (2011)
3. Housner, G.W.: Dynamic pressures on accelerated fluid containers. Bull. Seismol. Soc. Am. **47**, 15–24 (1957)
4. Lamb, H.: Hydrodynamics. The University Press, Cambridge (1932)
5. Wang, J.T., Gui, Y., Zhu, F., Jin, F., Zhou, M.X.: Real-time hybrid simulation of multi-story structures installed with tuned liquid damper. Struct. Control Health Monitor. **23**(7), 17 (2016)
6. Gardarsson, S., Yeh, H., Reed, D.: Behavior of sloped-bottom tuned liquid dampers. Eng. Mech. **3**, 7 (2001)
7. Brincker, R., Ventura, C.: Introduction to Operational Modal Analysis. Wiley, Chichester (2015)
8. Li, H.N., Yi, T.H., Jing, Q.Y., Huo, L.S., Wang, G.X.: Wind-induced vibration control of dalian international trade mansion by tuned liquid dampers. Math. Probl. Eng. **2012**, 21 (2012)
9. Krenk, S.: Dynamics of Structures. Department of Mechanical Engineering, Civil Engineering, Technical University of Denmark (2013)

Evangelos Katsanos is Assistant Professor at the Dept. of Civil Engineering of the Technical University of Denmark. His research interests are focused on structural dynamics, operational modal analysis and earthquake engineering.

Chapter 24
OMA-Based Modal Identification and Response Estimation of a Monopile Model Subjected to Wave Load

Jóhan Bech Húsgard, Frederik Alexander Hvelplund Uhre, Bruna Silva Nabuco, Renata Grabowsky, Sandro Amador, Evangelos Katsanos, Erik Damgaard Christensen, and Rune Brincker

Abstract The Operational Modal Analysis (OMA) has shown potential to provide reliable identification of offshore structures, being exposed to harsh environmental conditions that may degrade their integrity over the designed lifetime. Therefore, this study aims to employ OMA techniques to estimate the response of a scaled structural model that represents an offshore monopile-supported platform exposed to irregular wave loads. An experimental investigation was undertaken including a polycarbonate monopile of 0.90 m height that supports a steel deck. The monopile structure, considered to be fixed at its bottom, was placed inside a wave flume and the wave-induced vibration responses were measured along with the wave elevations. The recorded responses were used to conduct OMA-based identification of the modal properties of the scaled experimental model. To support the experimental investigation, a parallel track of numerical study was followed by developing a finite element (FE) model of the scaled structure. The FE model was appropriately updated to achieve a better matching between the experimental and numerical modal properties. The experimental measurements from wave gauges were also used along with the linear wave theory to estimate the wave kinematics, the latter being calculated by the Wheeler modification. Such a wave kinematics calculation enabled estimating the hydrodynamic forces or the so-called Morison forces while conventional assumptions were made for the definition of the necessary hydrodynamic coefficients, namely the drag, C_D, and inertia, C_M, coefficients of the Morison equation. The calculated forces were eventually applied to the FE model and the numerically obtained displacements were compared with the ones derived through the experimental investigation. This comparative assessment was further enriched by accounting for several data sets, being associated with varying combinations between significant wave height and peak period. The discrepancy that was found between the simulated and measured responses (i.e., displacements) constitutes an indication that the use of the code-recommended and widely adopted values for Morison's drag and inertia coefficients, namely $C_D = 1.0$ and $C_M = 2.0$, can yield conservatism to the calculation of the hydrodynamic forces that, in turn, may over-estimate the response of the structural systems being exposed to wave loads. Further research is, though, required to strengthen the validity of the current findings.

Keywords Operational modal analysis · Structural dynamics · Morison forces · Wave simulation

24.1 Introduction

Over the last fifty years there has been an increasing demand for offshore structures worldwide in order, mainly, to facilitate the needs of the Oil and Gas Industry. Since offshore structures are being constructed in waters and exposed to harsh environmental conditions that can threaten their structural integrity and survivability, there is an increasing demand for knowledge in the field. In 1950, Morison et al. [1] developed a formulation to evaluate the hydrodynamic forces that are exerted to the submerged structural components. Those forces were determined by introducing two non-dimensional coefficients, the drag and the inertia coefficients, C_D and C_M, being dependent on the velocity and acceleration of the water particles respectively [1]. In 1958, these coefficients were experimentally estimated by Keulegan and Carpenter [2] for cylindrical elements with different diameters exposed to a range of flow kinematics. These experiments led to the definition of the widely known Keuleguan-Carpenter (KC) number, which is used nowadays to verify if the hydrodynamic forces,

J. B. Húsgard · F. A. H. Uhre · B. Nabuco · S. Amador · E. Katsanos (✉) · R. Brincker · E. Damgaard Christensen
DTU, Kgs.Lyngby, Denmark
e-mail: s143860@student.dtu.dk; s133930@student.dtu.dk; brunan@byg.dtu.dk; sdio@byg.dtu.dk; vakat@byg.dtu.dk; runeb@byg.dtu.dk

R. Grabowsky
DNV GL, London, UK

B. Dilworth (ed.), *Topics in Modal Analysis & Testing, Volume 8*, Conference Proceedings of the Society for Experimental Mechanics Series, https://doi.org/10.1007/978-3-030-47717-2_24

imposed to a submerged structural element, is either drag- or inertia-dominated. Later, in 1999, Wolfram and Naghipour [3] investigated various approaches to determine the Morison's drag and inertia coefficients for multiple cylindrical elements with rough surfaces by undertaking experiments in a wave basin while the related responses were processed both in the time and frequency domain respectively. The current state-of-practice recognizes also the key role of both the drag and inertia coefficient for estimating the hydrodynamic forces on the submerged elements of an offshore structure. Along these lines, the recommended practice DNVGL-RP-C205 [4] accounts for both these two coefficients and the recommended values, being widely in several industrial applications, are in order of $C_D = 1.0$ and $C_M = 2.0$ for cylindrical structural members.

Despite the research advances already made in estimating the hydrodynamic forces of the submerged structural elements, the definition of the pertinent drag and inertia coefficients is still associated with uncertainty that can propagate through the subsequent analysis phase and eventually, affect adversely the design of new offshore structures and/or the assessment of the existing ones. To this end, the current study aims to re-visit the estimation of the drag and inertia coefficients by combining the robust Operational Modal Analysis (OMA) technique [5], being lately applied for modal identification of offshore structures [6, 7], and numerical (simulation-based) investigation. More specifically, this paper describes the experimental investigation of a bottom-fixed monopile structure being exposed to irregular waves through a wave flume testing campaign. The experimental study allowed for measuring both the wave elevations, produced by the generated waves, and the corresponding structural response of the monopile model. The wave-induced response was also calculated numerically by developing a finite element (FE) model of the monopile that was updated on the basis of the experimentally identified, via the use of OMA, modal properties of the test model. The updated FE model was subsequently subjected to wave loads estimated by Morison formulation. Linear wave theory is applied to determine the horizontal wave kinematics from the measured sea elevation and the Wheeler modification is used to describe the kinematics in the splash zone.

By doing so, this paper aims, mainly, to:

- Identify the modal properties (i.e., natural frequencies, mode shapes and damping ratios) of the monopile scaled model by OMA;
- Simulate the structural response (i.e., displacement) of the monopile model with the use of OMA and the linear wave theory while considering also the recommended values for Morison's drag and inertia coefficients;
- Compare the simulated and measured responses to conclude whether the use of recommended values for Morison's drag and inertia coefficients yields, eventually, over-estimated hydrodynamic forces that, in turn, provide conservative response results.

24.2 Theory

The purpose of this section is to provide a brief overview of the theoretical background used herein to perform the current study.

24.2.1 Operational Modal Analysis

The modal properties of an experimental model were identified herein by using OMA, which is a robust and reliable method to estimate the modal properties of a system under operating conditions [5]. OMA is carried out under the hypothesis that the structure is subjected to excitation with approximately white noise characteristics, i.e., broad band properties where all relevant frequencies are being excited. Within the framework of OMA, the modal parameters of a vibrating structure can be identified by using either time-domain or frequency-domain identification techniques. Along these lines, the Time Domain Poly Reference (TDPR) method, being a widely adopted technique to estimate the modal properties (i.e., the natural frequencies, damping ratios and mode shapes) of a system, considers the Correlation Functions as free decays. The specific modal identification technique follows the principles shortly outlined in the following (Brincker and Ventura [5]): the Correlation Function matrix is first established and the Hankel matrices are calculated. A companion matrix is then computed and by solving the eigenvalue problem, the modal parameters can be estimated. When using the TDPR technique, noise modes might be falsely identified. In order to include only the modes deemed to be the physical ones, the time-domain signals, including the structural response, need to be appropriately processed and filtered.

24.2.2 Wave Simulation and System Response

Ocean waves are random in height, length, shape and speed of propagation and usually fit to a wave spectrum, $S_\eta(\omega)$. By monitoring the wave elevation $\eta(t)$, the linear wave theory can be applied in order to provide the wave amplitude A and the random phase φ. In this way, the horizontal water particle velocity $u(t, z')$ and acceleration $du/dt(t, z')$ can be obtained by [4]:

$$u(t, z') = \sum_{i=1}^{N} A_i \cos(\omega_i t - k_i x + \varphi_i)\omega_i \frac{\cosh(k_i(z' + d))}{\sin(k_i d)} \tag{24.1}$$

$$\frac{du}{dt}(t, z') = -\sum_{i=1}^{N} A_i \cos(\omega_i t - k_i x + \varphi_i)\omega_i^2 \frac{\cosh(k_i(z' + d))}{\sin(k_i d)} \tag{24.2}$$

where N is the number of increments of the one-sided wave spectra, ω is a discrete frequency increment, t is the time history, k is the wave number, x is the distance between the wave gauge and the structure, d is the water depth and z' is the vertical coordinate positive upward. The wave amplitude is related to the wave spectrum $A_i = \sqrt{2S_\eta(\omega_i)\Delta\omega}$, where $\Delta\omega = \omega_i - \omega_{i-1}$ is the discrete frequency increment. The wave number k_i can be found from the dispersion relation [8]:

$$\omega_i^2 = k_i\, g\, \tanh(k_i\, d) \tag{24.3}$$

where g is the gravitational acceleration. The wave kinematics are calculated from the seabed up to the mean sea level. Since the instantaneous surface elevation is changing continuously, Wheeler [9] described how the kinematics can be determined by stretching or compressing the wave profile. Wheeler stretching is the most commonly accepted stretching approach [8] and it has been applied herein by computing the wave kinematics from $-d \leq z' \leq 0$ and stretching or compressing the profile to the true sea profile $-d \leq z(t) \leq \eta(t)$ [4, 8]:

$$z(t) = z' \frac{d + \eta(t)}{d} + \eta(t) \tag{24.4}$$

Wave loads are calculated using the Morison equation, where the force is found as the sum of the inertia force proportional to the acceleration and the drag force proportional to the square velocity [1, 4, 8, 10]:

$$F(t, z(t)) = \rho C_M A \frac{du}{dt}(t, z(t)) + \frac{1}{2}\rho C_D D u(t, z(t))|u(t, z(t))| \tag{24.5}$$

where ρ is the water density, A is the cross sectional area and D is the cross sectional diameter. It is notable that in this formula the loads are applied with respect to the actual sea elevation described by the vertical coordinate $z(t)$ from Eq. (24.4). In order to verify if the Morison forces are drag or inertia dominated, the dimensionless KC-number is calculated as [2, 10]:

$$KC = \frac{U_M T_p}{D} \tag{24.6}$$

where U_M is taken as the 95%-fractile particle velocity and T_p is the peak period of the wave spectra. Small KC-numbers indicate inertia dominated forces, while large KC-numbers indicate drag dominated forces [8].

The deformations of the structure can be described by the second order differential equation of motion. To simulate the system response, the solution to the equation of motion in the frequency domain is adopted [5, 11].

$$\mathbf{Y}(\omega) = \left(\sum_{n=1}^{N_m} \mathbf{b}_n\, H_n(\omega)\, \mathbf{b}_n^T \right) \mathbf{X}(\omega) \tag{24.7}$$

where $\mathbf{Y}(\omega)$ is the response in the frequency domain, N_m is the number of considered modes, $\mathbf{b}$ is the mode shape, $H(\omega)$ is the frequency response function (FRF) and $\mathbf{X}(\omega)$ is the external force vector. An important property of the used simulation method is that it does not introduce any bias on the physical properties of the system [5].

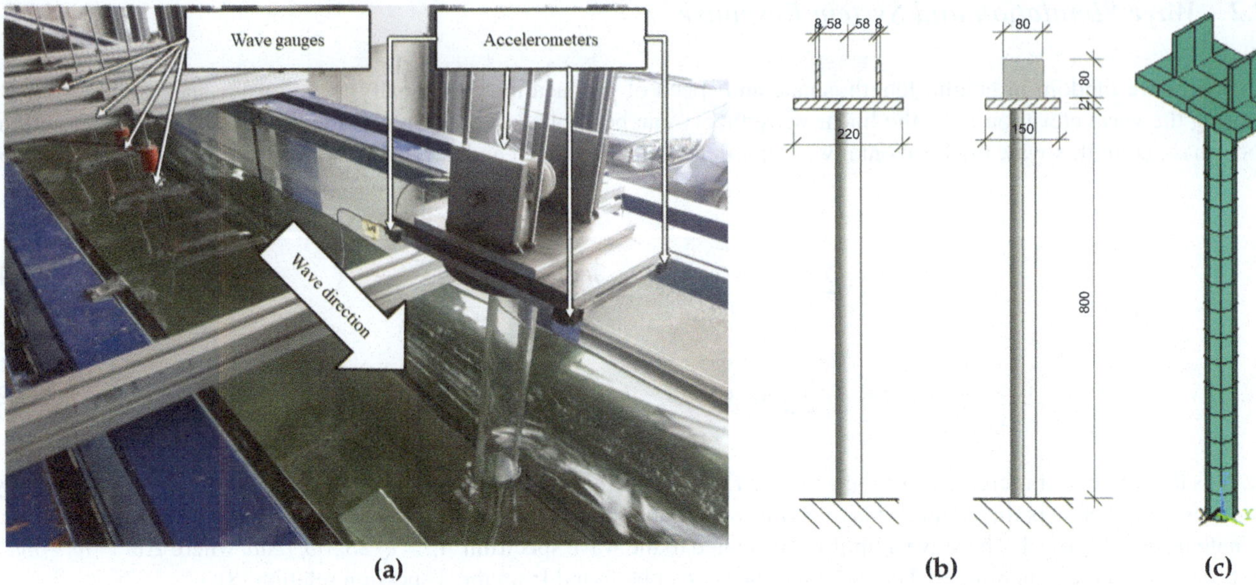

Fig. 24.1 (a) Experimental setup in wave flume, (b) schematic illustration and (c) FE-model of monopile model

Table 24.1 Main characteristics of the monopile model

Water depth [m]	Diameter monopile [m]	Height [m]	Scale [−]	Mass on top [kg]
0.60	0.05	0.90	1:70	6.17

Table 24.2 Significant wave height and peak period for the datasets for the monopile model

Dataset	[−]	1	2	3	4	5	6	7	8	9
H_s	[m]	0.087	0.122	0.075	0.065	0.023	0.093	0.078	0.038	0.079
T_p	[s]	1.31	1.63	1.30	1.24	0.98	1.31	1.16	0.98	1.12

24.3 Experimental Setup

An experimental monopile structure, made of polycarbonate material, was created as a 1:70 scaled model of a real offshore structure. The experimental setup and schematic illustration of the model are shown in Fig. 24.1a, b. The model was subjected to irregular wave loads in a wave flume 25 m long, 0.6 m wide and 0.8 m high. Four wave gauges were placed along the wave flume before the waves reached the structure. Additionally, four triaxial accelerometers were installed on the steel plate (i.e.,deck) of the monopile model, providing a total of 12 measured degrees of freedom (DOFs). Due to rigid body motion of the steel plate, only six DOFs were eventually used for the purpose of identifying the modal properties of the experimental model. The main characteristics of the monopile model is shown in Table 24.1.

The wave flume input was defined by the JONSWAP spectrum [4]. The significant wave height, H_s, and peak period, T_p, have been defined by scaling wave measurements taken on the real structure. In total, 9 datasets with varying H_s and T_p have been genarated during 20 min each—see Table 24.2.

24.4 Numerical Analysis

In this section the numerical model, the loads and other relevant considerations are described. Especially, the FE model is briefly outlined while the description of the loads follows. Lastly, a short overview of the response simulation is provided.

Table 24.3 Type of material, Young's modulus, material density and Poison ratio of FE model

	Type of material [−]	E [MPa]	ρ [kg/m^3]	ν [−]
Monopile	Polycarbonate	2,300	1,200	0.300
Top mass	Steel	210,000	7,850	0.305

24.4.1 Finite Element Model

A FE model was created in ANSYS representing the experimental monopile model as shown in Fig. 24.1c. The monopile has an outer diameter of 50 mm, a thickness of 2 mm and length of 800 mm. It was discretized with 16 beam elements. Shell elements were used to model the top mass, which is composed by a horizontal plate (220 × 150 × 20.7 mm) and two vertical plates (80 × 80 × 8 mm). The thickness of the horizontal plate was defined such that the top mass was 6.17 kg. The material properties of the experimental monopile model are provided by Table 24.3.

The FE model has 318 DOFs. The monopile is assumed to be fixed, i.e. no rotation or translation at the support. Furthermore the connection between the horizontal steel plate and the polycarbonate-monopile was modelled as a rigid one. The first two natural frequencies, derived from the eigenvalue analysis of the FE model, were found to be $f_{FE_1} = 2.075$ Hz and $f_{FE_2} = 2.083$ Hz.

24.4.2 Loads and System Response

A code was developed to simulate wave kinematics and the corresponding Morison forces based on the theory described in Sect. 24.2.2. Providing the measured wave elevation and water depth, the wave kinematics were calculated at specific points for every 2.5 cm along the height of the monopile model. The wave simulation code developed herein is based on linear wave theory; hence,the Wheeler modification outlined in Sect. 24.2.2 is further applied to distribute the forces to the monopile model according to the true sea surface. A study by Stansberg, Gudmestad, and Haver [12] reports that the Wheeler stretching method approximates well the fluid kinematics near the free surface when using a measured wave elevation with linear wave theory stretched to the instantaneous surface. On the other hand, it is stated that the Wheeler stretching method underssestimates the kinematics at lower depths [12]. However, the responses are primarily governed by the wave forces close to the free surface.

The Morison forces were calculated according to Eq. (24.5) considering the hydrodynamic coefficients set to the recommended values by [4], i.e. drag $C_D = 1.0$ and inertia coefficient $C_M = 2.0$. The Morison equation, used to estimate the hydrodynamic forced exerted to a submerged structural element, has been found to be applicable if the wavelength is more than 5 times larger than the cross-section diameter [4]. As the Morison formulation provides the wave forces per unit length, numerical integration is used to obtain the point loads at the FE model nodes. By determining the external nodal load vector, the system response is simulated according to Eq. (24.2)—the procedure is implemented by using the OMA toolbox [11].

24.5 Modal Identification and FE Model Updating

Nine different datasets were analyzed, each one representing a different sea state. For each dataset, the modal identification, undertaken by the TDPR technique, was performed. The sampled data from accelerometers were detrended and decimated. Detrending is performed to force the signal to have a zero mean while the decimation is normally applied to account only for modes of the structure that are of interest in each investigation [5]. The measured data were also processed with a sampling frequency of $f_{s,a} = 27.5$ Hz.

In Fig. 24.2, the singular values of the spectral density matrix of the structural response (accelerations) and the wave elevations are shown. Especially, from Fig. 24.2a, two closely spaced modes can be observed while the signal-to-noise ratio is seen to be increased in the low frequency range. A bandpass filter was, therefore, applied to remove noise in the high frequency range and noise due to quasi-static response from wave loads. The bandpass filter applied herein was adopted to be flat from 1.2 to 3.0 Hz and zero outside this period range. The experimental natural frequencies were eventually estimated to be between 2.06–2.07 Hz for the first mode, and 2.10–2.12 Hz for the second mode.

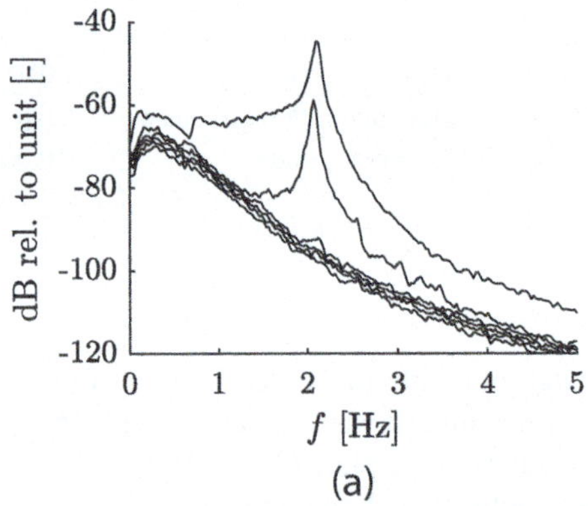

Fig. 24.2 Singular values of spectral density matrix for dataset 1. **(a)** Accelerometers and **(b)** Wave gauges

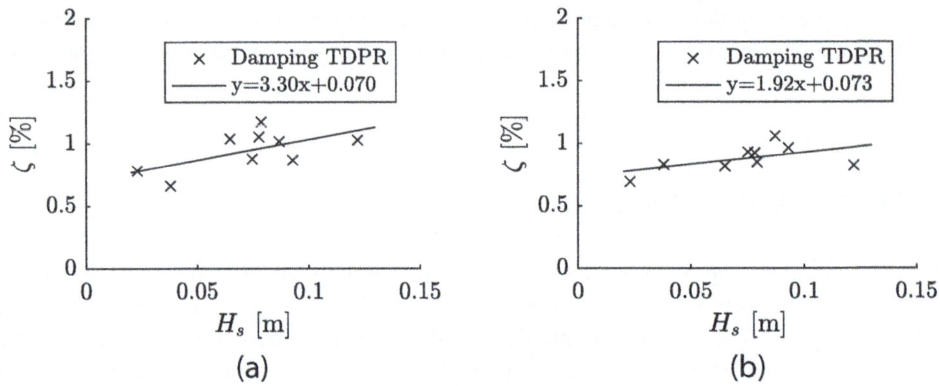

Fig. 24.3 Damping ratio ζ from TDPR as function of significant wave height. **(a)** Mode 1 and **(b)** Mode 2

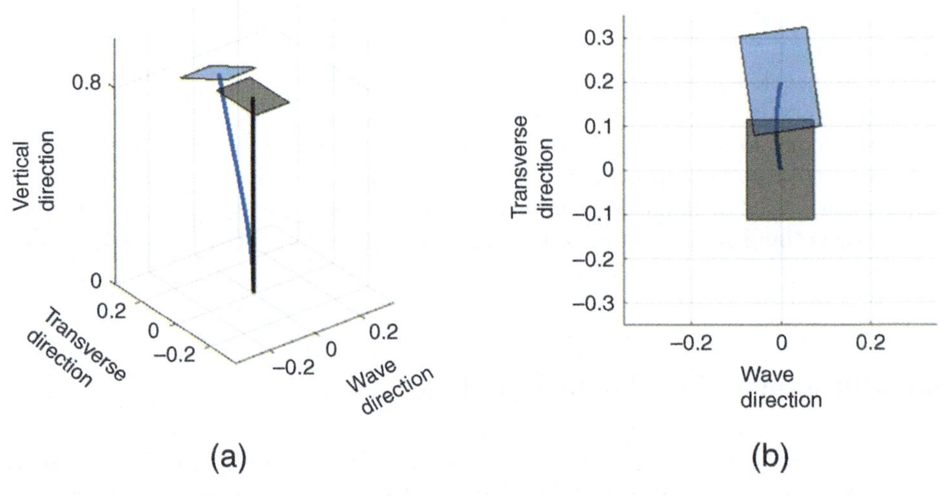

Fig. 24.4 1st mode shape identified by TDPR for dataset 1. **(a)** Isometric view and **(b)** top view

In Fig. 24.3 the damping ratio for the first and second mode are shown. It is noted that the damping ratio increases as the significant wave height, H_s, increases. This is expected since the modal identification based on the TDPR technique captures the combined structural and hydrodynamic damping [7]. It can be also noted that the estimated damping ratio for both first two vibration modes is, on average, 1% and the projection of the fitted line to $H_s = 0$ provides almost identical damping ratio for modes 1 and 2. The latter is expected due the symmetry of the structure and the absence of waves at this point. In Figs. 24.4 and 24.5 the first two mode shapes are shown. The first bending mode is transverse to the wave direction, while the second is in alignment with the wave direction.

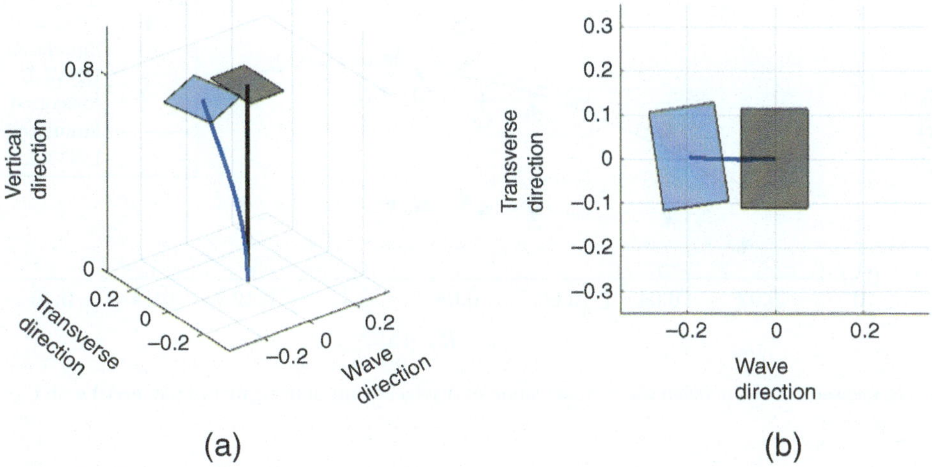

Fig. 24.5 2nd mode shape identified by TDPR for dataset 1. **(a)** Isometric view and **(b)** top view

The experimentally identified modal properties of the monopile enables its FE model to be updated and hence, capture with higher accuracy the actual response of the structure. To this end, the current study accounted for the FE model updating technique that was recently introduced by Amador et al. [13] and uses the Local Correspondence (LC) principle [14] to update the mass and stiffness matrix of the structure under study. Based on this principle, the FE model is updated in order its modal properties to match the OMA-identified modal properties of the experimental model. In terms of natural frequencies, the FE model has reached, after updating, the same natural frequencies as the ones identified by OMA.

The performance of FE model updating can be assessed in terms of the similarity between the experimentally identified mode shapes, **a**, and the ones calculated on the basis of the updated FE model, $\mathbf{b_a}$. To do so, the Modal Assurance Criterion (MAC) can be applied [5, 15]:

$$\mathrm{MAC}(\mathbf{a}, \mathbf{b_a}) = \frac{|\mathbf{a}^H \mathbf{b_a}|^2}{(\mathbf{a}^H \mathbf{a})(\mathbf{b_a}^H \mathbf{b_a})} \tag{24.8}$$

If the mode shapes are very similar, the MAC value is reaching the unity. On the other hand, higher deviation between the experimentally and numerically defined mode shapes leads to MAC values that are close to zero [15]. Considering the current study and the nine datasets of responses that were recorded and used, MAC values of 0.98 were calculated for the first two modes when comparing the experimental modes shapes with the numerical ones before the FE model updating. The MAC values were found to be slightly increased to 0.99 after updating the FE-model.

24.6 Response Simulation Results

The data obtained from wave gauge measurements were detrended, decimated and processed with a sampling frequency of $f_{s,w} = 13.3\,\text{Hz}$. The same bandpass filter applied to the accelerometers-captured data (Sect. 24.5) was also applied to the wave gauges data. After calculating the wave kinematics, the KC-number was found to be lower than 10 for all datasets. According to Journée and Massie [8], a value of KC-number being lower than 3 indicate that the drag force can be neglected. Contrarily, KC-number that is calculated between 3 and 15 indicates that the drag force can be linearized, however its contribution to the total hydrodynamic force is significantly low compared to the governing inertia-driven force. Such an approach (i.e., linearization of the drag force) was followed herein for the all the datasets, despite the fact that few of those were found to be associated with KC number being even lower than three.

By applying the estimated Morison forces to the updated FE model, the structural response, in terms of displacement at the deck of the monopile, was calculated by undertaking the relevant simulations. The standard deviation σ_c of the simulated and measured displacements of each dataset were compared for the DOF in the middle of the platform along the wave direction. The measured displacements at the center of the platform were derived by modal expansion [16]. Figure 24.6 shows that the simulated displacements are significantly larger than the measured displacements. It is also noted that by extending the trend

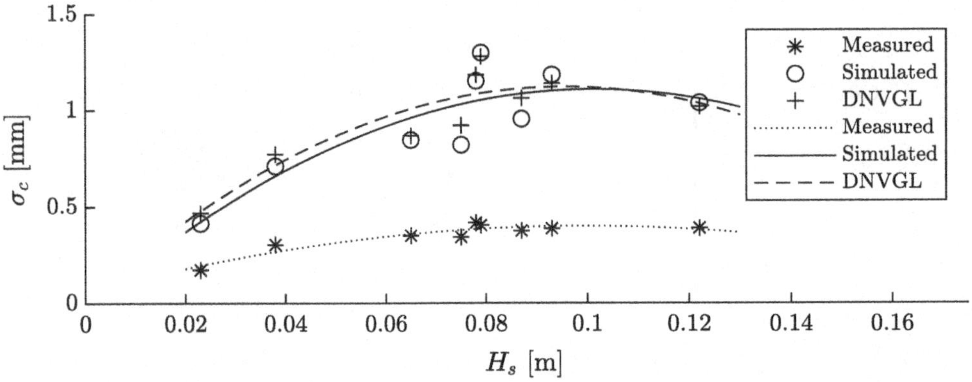

Fig. 24.6 Comparison of measured and simulated standard deviation of displacements at the center of the model with $C_M = 2.0$ and $C_D = 1.0$

lines, the response for both simulated and measurements is nearly equal to zero for $H_s = 0$. The latter is expected to be observed since the current simulation scheme (and the pertinent experimental one) accounted only for waves-induced loads disregarding any additional source load that would induce structural response. The ratio between simulated and measured displacements were found to be between 2.2 and 3.0 for the nine datasets considered herein. Such a result is an indication of conservatism that is introduced to the estimation of the structural response and hence, the hydrodynamic forces by using the code-recommended and widely used values for the drag and inertia coefficients.

To investigate further the validity of the aforementioned finding, simulations were performed by using the DNV GL Digital Solutions tools. This second simulation scheme, being independent on the first one, was undertaken to provide an additional basis of comparison between simulated and experimental response. To this end, an FE model has been created in GeniE [17], in which the monopile structure was modelled with identical geometry and material properties to those presented in 24.4.1 while the deck (top) structure was approximated by a nodal mass assigned to the top of the monopile. The wave loads were calculated and applied by using the loads simulation code Wajac [18] and adopting the linear wave theory and the assumptions described in Sect. 24.3. Structural dynamic analysis has been performed by using the Sestra analysis code [19] and the structural damping was considered by using the Rayleigh damping, in which the viscous damping matrix C is calculated as a linear combination of the stiffness and mass matrices, $C = \alpha_1 M + \alpha_1 K$. The mass and stiffness proportional coefficients, α_1 and α_2, were calculated accounting for the frequencies of modes 1 and 4.

As it is seen by Fig. 24.6, the standard deviation of the simulated responses by following the second simulation scheme was found to be highly similar with the response results obtained by the first simulation scheme adopted herein. Such a finding corroborates further the conservatism that was found to be related with the use of the recommended values for $C_D = 1.0$ and $C_M = 2.0$ in order to calculate the hydrodynamic forces and the hence, the response of the specific structural model.

24.7 Conclusion

An experimental investigation, being supported also by numerical analysis, was performed by exposing a bottom-fixed monopile structure to irregular wave loads generated within a wave flume. Nine datasets, including wave elevations and response from accelerometers at the deck of the monopile structure, were appropriately recorded and analyzed. Each dataset corresponds to different combinations of peak period and significant wave height, the latter being consistent with the JONSWAP wave spectrum. The recorded response was used to identify the modal properties of the monopile model by exploiting the robust OMA and especially, the TDPR identification technique. Apart from the experimental investigation, a numerical one was carried out and a FE model of the monopile structure was created and then successfully updated by using the LC principle and the experimentally identified modal properties. Such an approach led to almost identical modal properties between the experimental structure and the updated FE model.

The wave kinematics were also simulated on the basis of the measured wave elevation and the liner wave theory was applied herein. The Morison forces were then, calculated using the code-recommended and widely used values for hydrodynamic coefficients, i.e. $C_D = 1.0$ and $C_M = 2.0$. By considering the Wheeler modification, the already calculated Morison forces were applied to the updated FE model to simulate displacements. The obtained simulated displacements were found to be 2.2 up to 3.0 times higher than the measured displacements. This finding, which was further corroborated by undertaking a second, independent simulation scheme, is an indication of conservatism that can be introduced to the calculation of the hydrodynamic loads and the corresponding responses by following the conventional, Morison-based approach of estimating the hydrodynamic forced with the use of the commonly considered values for C_D and C_M. However, additional research should be dedicated to strengthen the validity of the results found herein. For example, different structural configurations of varying size, complexity and geometry need to be exposed to different wave loads under the framework of well controlled laboratory environment (i.e., wave flumes). Such experimental investigation needs also to be supported by refined numerical study enabling, in such a way, the comparison of the simulated and the measured responses. Additionally, the forces, subjected to the experimental model, can be directly measured and compared to the ones predicted by using the Morison equation and the relevant drag and inertia coefficients.

References

1. Morison, J.R., Obrien, M.P., Johnson, J.W., Schaaf, S.A.: The force exerted by surface waves on piles. Trans. Am. Inst. Min Metall. Eng. **189**, 149–154 (1950). https://doi.org/10.2118/950149-G
2. Keulegan, G.H., Carpenter, L.H.: Forces on cylinders and plates in an oscillating fluid. J. Res. Natl. Bur. Stand. **60**(5), 423–440 (1958). ISSN 0091-0635. https://doi.org/10.6028/jres.060.043
3. Wolfram, J., Naghipour, M.: On the estimation of Morison force coefficients and their predictive accuracy for very rough circular cylinders. Appl. Ocean Res. **21**(6), 311–328 (1999). ISSN 0141-1187. https://doi.org/10.1016/S0141-1187(99)00018-8
4. DNV GL. Dnvgl-rp-c205 environmental conditions and environmental loads (2017)
5. Brincker, R., Ventura, C.E.: Introduction to Operational Modal Analysis, 3rd edn. Wiley (2015). ISBN 9781119963158
6. Nabuco, B., Tarpø, M., Aïssani, A., Brincker, R.: Reliability analysis of offshore structures using OMA based fatigue stresses. In: Proceedings of the ASME 2017 36th International Conference on Ocean, Offshore and Arctic Engineering, Trondheim (2017). ISBN 978-0-7918-5765-6. https://doi.org/10.1115/OMAE2017-61730
7. Vigsø, M., Kabel, T., Tarpø, M., Brincker, R., Georgakis, C.: Operational modal analysis and fluid-structure interaction. In: Procedings of the International Conference on Noise and Vibration Engineering, ISMA (2018)
8. Journée, J.M.J., Massie, W.W.: Offshore Hydromechanics, 1st edn. Delft University of Technology, Delft (2001)
9. Wheeler, J.D.: Method for calculating forces produced by irregular waves. J. Pet. Technol. **22**(03), 359–367 (1970). https://doi.org/10.2118/2712-PA
10. Sumer, B.M., Fredsøe, J.: Hydrodynamics Around Cylindrical Structures. World Scientific, Singapore (2006)
11. Brincker, R.: Oma toolbox (2019). http://www.brinckerdynamics.com/oma-toolbox/
12. Stansberg, C., Gudmestad, O., Haver, S.: Kinematics under extreme waves. J. Offshore Mech. Arct. Eng. -transactions of The Asme **130**(05), (2008). https://doi.org/10.1115/1.2904585
13. Amador, S., Juul, M., Friis, T., Brincker, R.: Finite element model updating using the local correspondence principle. pp. 309–314, 01 (2019). https://doi.org/10.1007/978-3-319-74700-2-34
14. Brincker, R., Skafte, A., López-Aenlle, M., Sestieri, A., D'Ambrogio, W., Canteli, A.: A local correspondence principle for mode shapes in structural dynamics. Mech. Syst. Signal Process. **45**(1), 91–104 (2014). ISSN 0888-3270. https://doi.org/10.1016/j.ymssp.2013.10.025. http://www.sciencedirect.com/science/article/pii/S0888327013005669
15. Allemang, R.J.: The modal assurance criterion – twenty years of use and abuse. Sound Vib. **37**(08), 14–23 (2003)
16. Skafte, A., Tygesen, U., Brincker, R.: Expansion of mode shapes and responses on the offshore platform valdemar. In: Catbas, F.N. (ed.) Dynamics of Civil Structures, vol. 4, pp. 35–41. Springer International Publishing, Cham (2014). ISBN 978-3-319-04546-7
17. DNV GL. SESAM User Manual GeniE – Conceptual modelling of offshore and maritime structures (2019)
18. DNV GL. SESAM User Manual WAJAC – Wave and Current Loads on Fixed Rigid Frame Structures (2019)
19. DNV GL. SESAM User Manual SESTRA – Finite element analysis solver (2019)

Evangelos Katsanos is Assistant Professor at the Dept. of Civil Engineering of the Technical University of Denmark. His research interests are focused on structural dynamics, operational modal analyss and earthquake engineering.

Chapter 25
A Concept for the Estimation of Displacement Fields in Flexible Wind Turbine Structures

Johannes Luthe, Andreas Schulze, János Zierath, Sven-Erik Rosenow, and Christoph Woernle

Abstract The design of wind turbines is based on conservative load assumptions compared to the actual load conditions at the designated erection site. As a consequence, structural reserves are very likely at the end of the turbine's approved lifetime. Driven by this, a research collaboration between different institutions from wind industry and research facilities was initiated to exploit these structural reserves using continuous, model-based fatigue monitoring of individual wind turbines. This contribution focuses on the dynamic state estimator as a central part of the monitoring process chain. Arising challenges are the generally unknown system excitation, e.g. by wind loads, and the usually small set of available structural responses due to the high installation effort and maintenance costs of adequate sensors. On that account, the authors propose a Modal Decomposition and Expansion (MDE) approach for the displacement field reconstruction utilizing detailed Finite Element (FE) models of the structures under consideration in combination with inertial measurements. As a limiting factor, the MDE approach can only be applied to linear dynamic systems. To also account for the generally large nonlinear motion of wind turbines, the authors propose to subdivide the full system into a number of substructures mutually connected by interface definitions in the first stage, followed by a sequential implementation of single state estimators for each substructure. The quality of this implementation framework will be demonstrated in on a detailed flexible multibody model of a small-scale wind turbine test stand.

Keywords Fatigue monitoring · Wind turbines · State estimation · Displacement field · Inertial measurement unit

25.1 Introduction

The global challenge of limited fossil resources causes a growing significance of renewable energies. Due to their comparatively high energy efficiency, wind turbines play an important role to accomplish the transition to sustainable energy sources. Following the guidelines from DNVGL or IEC, the design process of modern wind turbines is governed by a rather rough classification of annual mean wind speed and turbulence intensity resulting in very conservative design loads compared to the actual load conditions at the specific erection site of the wind turbine. As a consequence, structural reserves are very likely at the end of the turbine's approved lifetime.

With the objective to exploit these reserves, reliable tracking of the actually endured fatigue loads at critical spots is fundamental. While the direct measurement of strains in operating wind turbines imposes high requirements on the measurement equipment and is in any case only possible at accessible measurement positions, an indirect, model-based derivation of fatigue indicators by rather simply obtainable motion quantities, e.g. accelerations, appears more reasonable in this context. Driven by this, a research collaboration between different institutions from wind industry and research facilities was initiated in 2019 with the aim to develop a continuous, model-based fatigue monitoring concept for an individual wind turbine. The underlying process chain consists of three main parts: 1. Estimation of the displacement fields of important structural components by a state estimator unit, 2. Derivation of stresses at critical fatigue spots within the components, 3. Online fatigue calculation based on custom damage models. This contribution introduces the dynamic state estimator as

J. Luthe (✉) · A. Schulze · C. Woernle
Chair of Technical Dynamics, University of Rostock, Rostock, Germany
e-mail: johannes.luthe@uni-rostock.de; andreas.schulze2@uni-rostock.de; woernle@uni-rostock.de

J. Zierath · S.-E. Rosenow
W2E Wind to Energy GmbH, Rostock, Germany
e-mail: JZierath@wind-to-energy.de; SERosenow@wind-to-energy.de

B. Dilworth (ed.), *Topics in Modal Analysis & Testing, Volume 8*, Conference Proceedings of the Society for Experimental Mechanics Series, https://doi.org/10.1007/978-3-030-47717-2_25

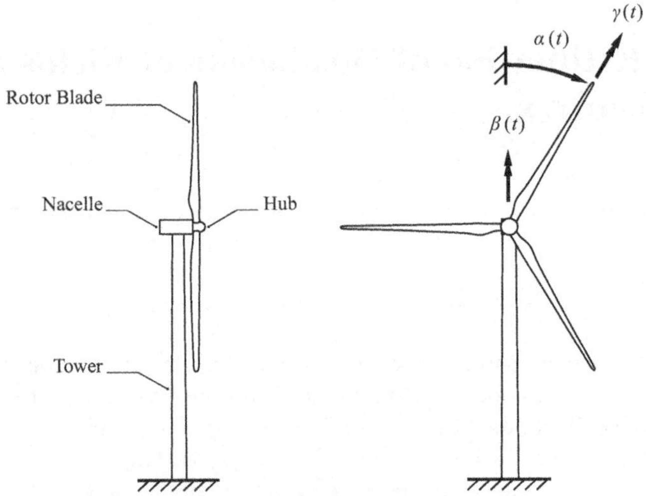

Fig. 25.1 Basic structural components and essential rigid body motion of wind turbine systems

a central part of the process chain. With a minimal number of measurements available, the main objective is to recover the displacement fields at critical spots in the support structure. While the term state estimation originates from control theory, it is being used consistently within the field of structural mechanics to describe the estimation of structural displacements. With the concepts of an Unknown Input Observer (UIO) and Modal Decomposition and Expansion (MDE) two approaches have already been investigated and applied to a multibody simulation model [1] and to a cantilever beam structure [2], respectively. Both investigations yield quite accurate results, however, the UIO approach revealed to be not directly applicable to structures under unknown distributed excitation. On that account, this contribution focuses on the MDE concept, which is inherently restricted to linear systems. To account for the generally large nonlinear motion of wind turbine structures, a sequential implementation framework for linear substructures is presented. The quality of this approach will be demonstrated by a detailed flexible multibody model of a small-scale wind turbine test stand presented in [3].

25.2 System Analysis

From a mechanical point of view, a wind turbine system is primarily composed of the support structure components of tower, nacelle, hub and rotor blades, see Fig. 25.1. During operation, the system's structural dynamics is typically characterized by a large nonlinear motion of the rotor (defined by rotor angle $\alpha(t)$, yaw angle $\beta(t)$ and pitch angle $\gamma(t)$) superposed by small elastic deformations in the structural components. Generally unknown, distributed forces excite the entire system. These comprise external forces such as wind excitation and gravity as well as internal forces like centrifugal and Coriolis effects. Classifying the excitation forces according to their frequency content shows quasi-static components due to the aerodynamic thrust on the rotor and the tower, periodic quantities that depend on the rotor speed (1p/3p) and stochastic components according to the wind turbulence spectrum.

The superior objective of the state estimator is to recover the displacement field around fatigue hot spots regardless of any unknown excitation loads and based only on a minimal number of measurements.

25.3 Basic Framework

Due to the large rigid body motion of the rotor, wind turbines need to be regarded as nonlinear dynamical systems. However, the rotor angle $\alpha(t)$, the azimuth angle $\beta(t)$ and the pitch angle $\gamma(t)$ are continuously monitored by the Supervisory Control And Data Acquisition (SCADA) system of the turbine and are thus available online, so that the nonlinear rotor motion can be regarded as rheonomic constraints prescribed by the turbine controller. Considering nacelle and hub as rigid bodies due to their high specific stiffnesses, the wind turbine can be subdivided into flexible substructures for tower and blades, see Fig. 25.2. Assuming small elastic deformations and a linear stress-strain relationship, the substructures can each be

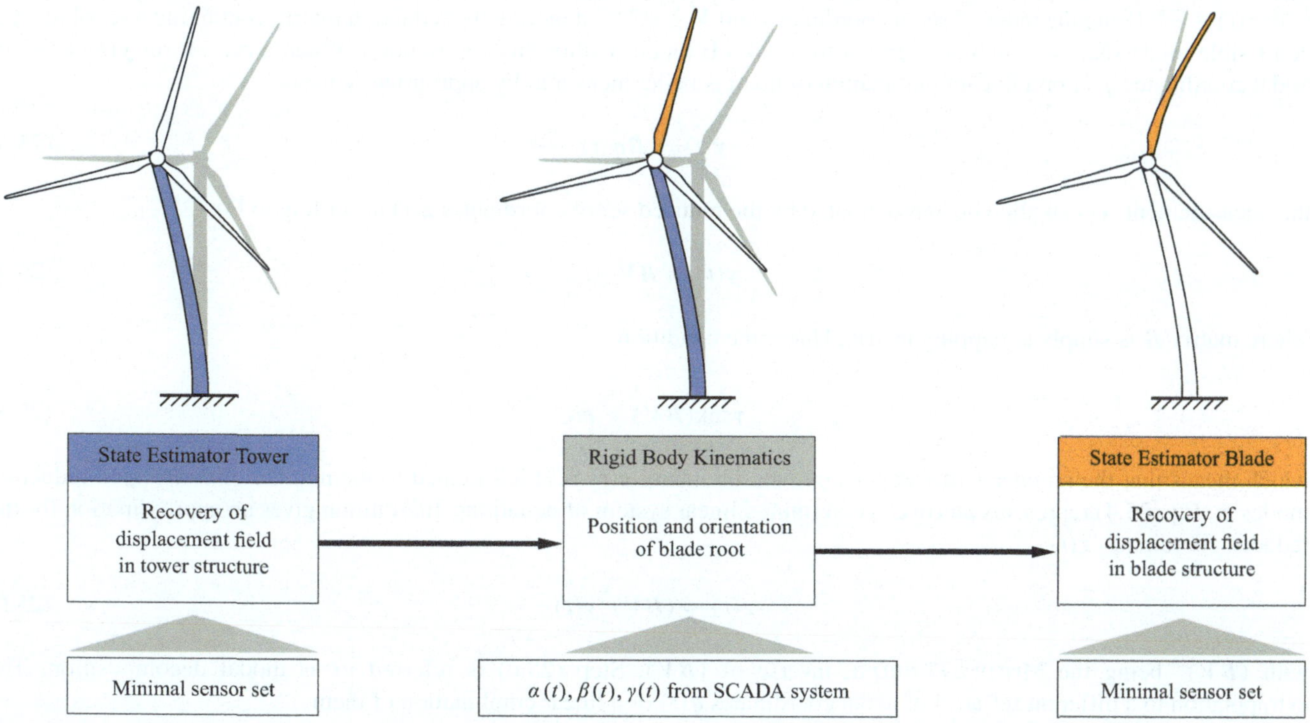

Fig. 25.2 Basic framework for the sequential state estimation

represented by linear Finite Element (FE) models with respect to their local reference frames. Since the blade FE model needs to be formulated with respect to a moving reference frame, additional rotor speed dependent damping and stiffness values would have to be considered, see e.g. [4]. Due to the comparably low rated speed limits of wind turbines, however, these terms are neglected in a first stage. Following these assumptions, a sequential state estimation framework can be set up with separate linear estimators for each substructure according to Fig. 25.2.

25.4 State Estimation

The objective of the state estimator is to provide an accurate displacement field approximation based on a limited number of measurements. Within this contribution the concept of Modal Decomposition and Expansion (MDE) is presented which is a widely accepted method for full-field stress and strain recovery and has been applied on a wide range of mechanical and civil structures [5–8]. A comparison with other state-of-the-art strain estimation methods like Kalman filtering and joint-input filtering techniques is given in [9]. The fundamental idea is that the dynamic response of linear structural systems can be represented by a linear combination of a limited number of appropriate mode shapes, which can be derived either by experimental modal analysis [10], finite element (FE) models or multibody systems [11, 12]. All these approaches assume that the experimental or numerical mode shapes sufficiently correspond to their physical counterparts. Within this contribution, FE-based mode shapes are applied.

Considering an FE model of a flexible structure, the displacements $\boldsymbol{u}(\boldsymbol{r}, t)$ at a particular point $\boldsymbol{r}$ are expressed by the nodal displacements $\boldsymbol{q}(t) \in \mathbb{R}^n$ in combination with global shape functions $\boldsymbol{\Phi}(\boldsymbol{r})$, which depend on the specific finite element formulation,

$$\boldsymbol{u}(\boldsymbol{r}, t) = \boldsymbol{\Phi}(\boldsymbol{r})\boldsymbol{q}(t)\,. \tag{25.1}$$

Following classical model order reduction concepts, the nodal displacements can be approximated by a lower dimensional approach

$$\boldsymbol{q}(t) \approx \boldsymbol{V}\boldsymbol{z}(t)\,, \tag{25.2}$$

with $z(t) \in \mathbb{R}^m$ being the reduced set of coordinates and $\boldsymbol{V} \in \mathbb{R}^{n \times m}$ denoting the reduction matrix containing a set of $m \ll n$ admissible mode shapes, which are derived from the FE model within this contribution. When a certain set $\boldsymbol{y}(t) \in \mathbb{R}^\ell$ of nodal coordinates $\boldsymbol{q}(t)$ or a linear combination of them is being measured by appropriate sensors,

$$\boldsymbol{y}(t) = \boldsymbol{B}\boldsymbol{q}(t) \tag{25.3}$$

this measurement set can alike be approximated by the reduced set of coordinates $z(t)$ according to Eq. (25.2),

$$\boldsymbol{y}(t) \approx \boldsymbol{B}\boldsymbol{V}z(t), \tag{25.4}$$

where matrix $\boldsymbol{B}$ is simply a mapping matrix. Under the condition

$$\operatorname{rank}(\boldsymbol{B}\boldsymbol{V}) \overset{!}{\geq} m\,, \tag{25.5}$$

which means that the number ℓ of independent measurements $\boldsymbol{y}(t)$ is at least equal to the number of considered reduction modes m, Eq. (25.4) represents an (over-) determined linear system of equations. Its solution gives an approximation for the reduced coordinates $z(t)$,

$$z(t) \approx (\boldsymbol{B}\boldsymbol{V})^\dagger \boldsymbol{y}(t) \tag{25.6}$$

with $(\boldsymbol{B}\boldsymbol{V})^\dagger$ being the MOORE-PENROSE inverse of $(\boldsymbol{B}\boldsymbol{V})$. Step (25.6) is referred to as modal decomposition. The extrapolation to a different set $\boldsymbol{w}(t)$ of nodal coordinates $\boldsymbol{q}(t)$ or a linear combination of them

$$\boldsymbol{w}(t) = \boldsymbol{C}\boldsymbol{q}(t) \tag{25.7}$$

can be accomplished by the identified reduced coordinates from Eq. (25.6) in combination with the m given mode shapes $\boldsymbol{V}$

$$\boldsymbol{w}(t) \approx \boldsymbol{C}\boldsymbol{V}z(t) \approx \boldsymbol{C}\boldsymbol{V}(\boldsymbol{B}\boldsymbol{V})^\dagger \boldsymbol{y}(t)\,. \tag{25.8}$$

25.5 Sensor Concept

To obtain reliable approximations for the displacement field in the flexible wind turbine substructures, the underlying FE models need to be supported by measurements according to Eq. (25.3). The required measurement of absolute displacements with respect to an undeformed reference configuration, however, is under practical conditions on wind turbines not realizable without a considerable technical and financial effort. Out of these considerations the projected sensor concept comprises Inertial Measurement Units (IMU) on a low-cost basis, which provide accelerations and angular velocities with respect to the three independent IMU axes. To obtain the required displacement information, the IMU raw signals need to be pre-processed by means of signal fusion, well known from integrated navigation technology, see e.g. [13, 14]. Since the pre-processing unit is currently under development, the required measurements in Sect. 25.6 are assumed to be readily available in a first stage.

25.6 Application to a Flexible Multibody System

As experimental investigations on operating wind turbines are inherently connected to high economical expense due to necessary stand-still periods on the one hand and the need for robust measurement equipment on the other hand, a scaled test rig of a wind energy turbine has been developed (tower height 3.68 m, blade length 1.81 m), which allows for fast analyses of different sensor concepts before their implementation on a real turbine. For validation purposes, also a flexible multibody model of the test rig structure has been built-up, which is utilized in this section as a reference system allowing for virtual measurements and reference results for the displacement field. The flexible bodies for tower and blades are included based on beam theory. Due to limited amount of space, the interested reader is referred to [3] for further details.

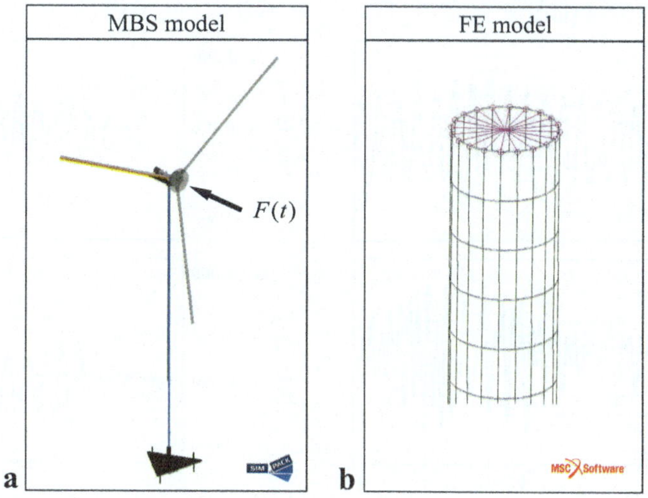

Fig. 25.3 Application to a flexible multibody system. (**a**) Multibody model of the wind turbine test rig according to [3], system is loaded with stochastic force $\boldsymbol{F}(t)$ at the rotor center (**b**) Detailed FE model of the tubular tower substructure with multi-point constraints at the tower top

25.6.1 Load Case

For the following test case it is assumed that the multibody system, depicted in Fig. 25.3a, is loaded by the external protected space: force $\boldsymbol{F}(t)$ acting on the rotor center. To represent operating conditions, the spectral density of $\boldsymbol{F}(t)$ is specified according to the KAIMAL wind turbulence spectrum [15]. Furthermore, the wind turbine model performs large nonlinear motions, i.e. the rotor angle $\alpha(t)$ and the azimuth angle $\beta(t)$ are rheonomically constrained by constant angular velocities $\dot{\alpha}$ and $\dot{\beta}$. The objective is to estimate the displacement fields in the tower and blade substructures. For this purpose, detailed FE models were built up of shell elements, see Fig. 25.3b.

For accurate estimation results, the mode shapes in matrix $\boldsymbol{V}$ of Eq. (25.2) need to be selected according to the anticipated operational deflection shapes (ODS) of the structure under consideration. Due to the dominant aerodynamic damping effects and the resulting rapidly decaying eigen behavior, the ODS of a real operating wind turbine are mainly affected by the unknown distributed loads according to Sect. 25.2. Taking the limited relevant spectral content of these loads of less then 1 Hz into consideration, the selection of purely static mode shapes appears reasonable in this context. Since the projected sensor concept aims at the installation of only one IMU per substructure (Sect. 25.5), i.e. only three independent measurements are available with the angular velocities about the IMU axes, the maximal number of admissible mode shapes is likewise restricted to $l = m = 3$, see Eq. (25.5). For the present test case, virtual IMU measurements are conducted at the tower top and the outer blade tips. Within this contribution we assume perfect knowledge of the required measurements in a first stage, independent of possible challenges connected with the practical sensor setup. To link the nodal coordinates of these measurement points to the surrounding structure, additional rigid elements were introduced in the form of multi-point constraints, see Fig. 25.3b.

25.6.2 Results

In order to compare reference and estimated states, the Time Response Assurance Criterion (TRAC) [7] can be applied as suitable means. With $\boldsymbol{q}_i = \left[q_i(\Delta t),\, q_i(2\,\Delta t),\, \ldots\right]^{\mathrm{T}}$ being the sample vector of the i-th reference state $q_i(t)$ sampled with the fixed time step Δt and $\hat{\boldsymbol{q}}_i$ being the corresponding estimated quantity, the TRAC is obtained by

$$\mathrm{TRAC}_i = \frac{\left(\boldsymbol{q}_i^{\mathrm{T}}\hat{\boldsymbol{q}}_i\right)^2}{\left(\boldsymbol{q}_i^{\mathrm{T}}\boldsymbol{q}_i\right)\left(\hat{\boldsymbol{q}}_i^{\mathrm{T}}\hat{\boldsymbol{q}}_i\right)}, \qquad i = 1, \ldots, n. \tag{25.9}$$

A TRAC value close to 1 indicates a strong correlation between both signals. However, the TRAC is only useful for a qualitative comparison. In order to also account for amplitude differences, the relative error $e(t)$ is introduced,

$$e_i(t) = \frac{q_i(t) - \hat{q}_i(t)}{\max(\hat{q}_i(t))}, \qquad i = 1, \ldots, n. \tag{25.10}$$

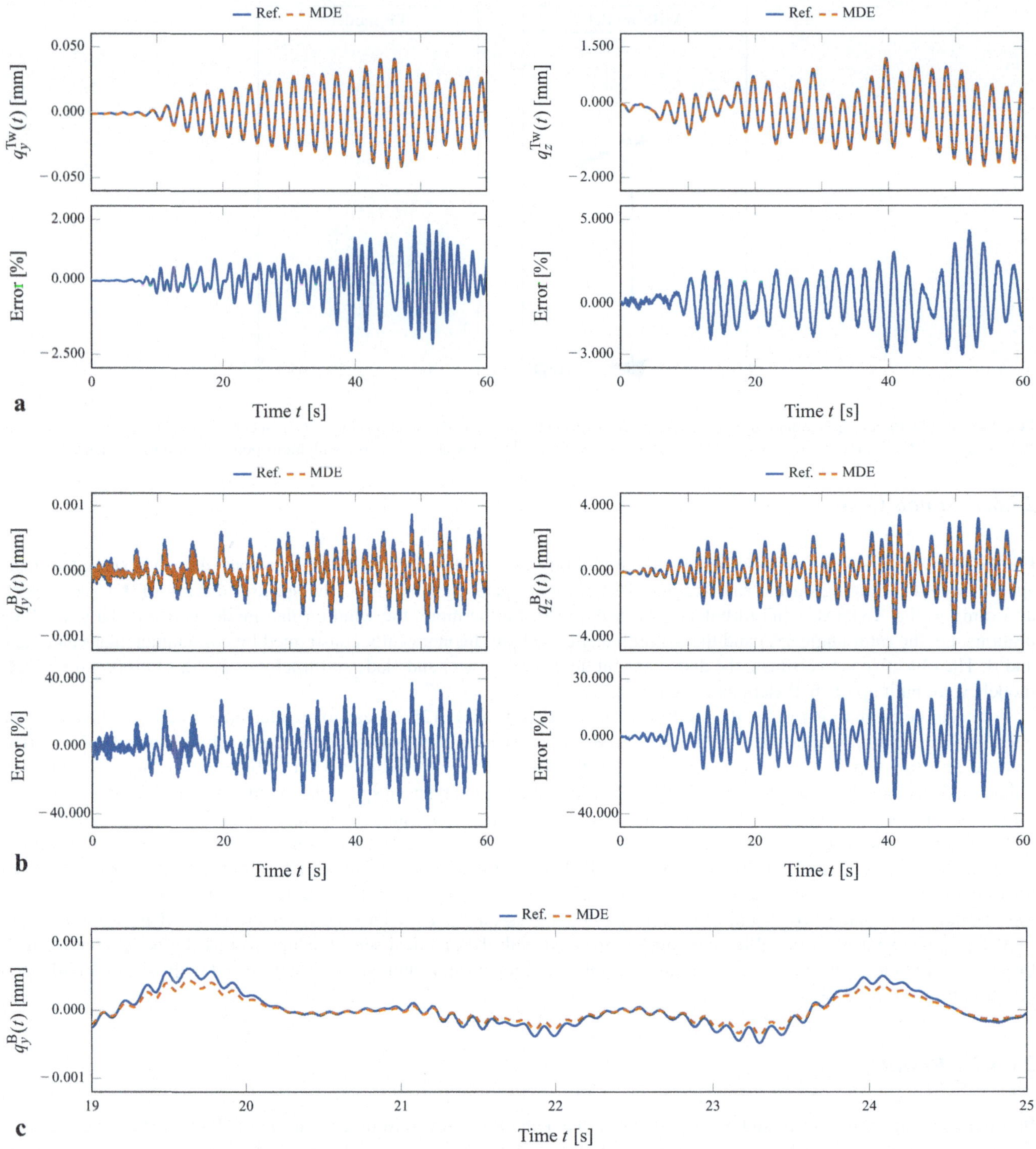

Fig. 25.4 Reference and estimated transversal displacements (**a**) within the tower substructure at 20% off the tower base in y- and z-direction, $q_y^{\mathrm{Tw}}(t)$ and $q_z^{\mathrm{Tw}}(t)$, (**b**) within the blade substructure at 20% off the blade root in y- and z-direction, $q_y^{\mathrm{B}}(t)$ and $q_z^{\mathrm{B}}(t)$, (**c**) Detailed view of $q_B^{\mathrm{B}}(t)$.

Results for the selected load case are shown in 25.4. Due to the limited amount of space, not the entire displacement field of the substructures is presented here. Instead, two specific locations 20% off the tower base and 20% off the blade root, respectively, are arbitrarily selected for comparing the estimated to the reference lateral displacements.

With a TRAC value of 0.9998, the two estimated displacements in the tower structure, $q_y^{\mathrm{Tw}}(t)$ and $q_z^{\mathrm{Tw}}(t)$, show a very good correlation with the reference curves, see Fig. 25.4a. In addition, relative errors of less than 3% confirm that even by applying only one single IMU, the displacement field can be accurately recovered within the tower structure.

Regarding the blade structure in Fig. 25.4b, a TRAC value of 0.9988 indicates a very good correlation as well, but the relative errors of up to 35% reveal larger magnitude differences between estimated and reference displacements. The latter effect can be ascribed to an insufficient representation of the actual ODS by the selected static modes only and needs to be investigated carefully. Nonetheless, the sole number of load cycles is retained correctly, compare the detailed view of Fig. 25.4c, which is important for a subsequent fatigue analysis.

25.7 Conclusion and Outlook

As a central part of model-based fatigue monitoring in structurally loaded mechanical systems, the displacement fields at critical spots need to be recovered accurately on the basis of a limited number of measurements. For that purpose, the present contribution proposes a sequential framework to apply the well-known concept of Modal Decomposition, inherently restricted to linear models, to nonlinear wind turbine systems. The high quality of this approach is demonstrated by a numerical multibody simulation of a small-scale wind turbine test rig based on virtual IMU measurements in combination with static mode shapes.

Within the next project phases, the proposed framework will be implemented on a detailed flexible multibody model of a full wind turbine to account for more realistic operating conditions. Furthermore, the appropriate definition of a modal basis for the MDE concept will be carefully analyzed to ensure accurate state estimations within the entire operating range. Up until now, perfect knowledge of the required state measurements, independent of a specific sensor concept, has been assumed. Future investigations will aim at utilizing Intertial Measurement Units for data acquisition on the real turbine.

Acknowledgments This research is part of the *DynAWind*^{2} project funded by the German Federal Ministry for Economic Affairs and Energy under grant number 0325228E/F/G.

References

1. Luthe, J., Schulze, A., Rachholz, R., Zierath, J., Woernle, C.: State observation in beam-like structures under unknown excitation. In: Kecskeméthy, A., Geu Flores, F. (eds.) Multibody Dynamics 2019, Computational Methods in Applied Sciences, vol. 53, pp. 283–291. Springer Nature (2019). https://doi.org/10.1007/978-3-030-23132-3_34
2. Luthe, J., Schulze, A., Rachholz, R., Zierath, J., Woernle, C.: Acceleration-based strain estimation in a beam-like structure. In: Proceedings of the 5th Joint International Conference on Multibody System Dynamics, Lisboa (2018)
3. Rachholz, R., Bartkowiak, R., Schulze, A., Luthe, J., Zierath, J., Woernle, C.: Development of a dynamically scaled wind turbine. J. Phys. Conf. Ser. **1037**, 052033 (2018). https://doi.org/10.1088/1742-6596/1037/5/052033
4. Schwertassek, R., Wallrapp, O.: Dynamik flexibler. Vieweg+Teubner Verlag, Wiesbaden (1999). https://doi.org/10.1007/978-3-322-93975-3
5. Iliopoulos, A., Weijtjens, W., Van Hemelrijck, D., Devriendt, C.: Fatigue assessment of offshore wind turbines on monopile foundations using multi-band modal expansion. Wind Energy **20**(8), 1463–1479 (2017). https://doi.org/10.1002/we.2104
6. Pelayo, F., Skafte, A., Aenlle, M.L., Brincker, R.: Modal analysis based stress estimation for structural elements subjected to operational dynamic loadings. Exp. Mech. **55**(9), 1791–1802 (2015). https://doi.org/10.1007/s11340-015-0073-6
7. Avitabile, P., Pingle, P.: Prediction of full field dynamic strain from limited sets of measured data. Shock Vib. **19**(5), 765–785 (2012)
8. Aenlle, M.L., Hermanns, L., Pelayo, F., Fraile, A.: Stress estimation in a scale model of a symmetric two story building. In: 5th International Operational Modal Analysis Conference (IOMAC 2013). Curran, Red Hook (2014)
9. Maes, K., Iliopoulos, A., Weijtjens, W., Devriendt, C., Lombaert, G.: Dynamic strain estimation for fatigue assessment of an offshore monopile wind turbine using filtering and modal expansion algorithms. Mech. Syst. Signal Process. **76-77**, 592–611 (2016). https://doi.org/10.1016/j.ymssp.2016.01.004
10. Chierichetti, M., Grappasonni, C., Coppotelli, G., McColl, C.: A modal approach for dynamic response monitoring from experimental data. Mech. Syst. Signal Process. **48**(1-2), 199–217 (2014). https://doi.org/10.1016/j.ymssp.2014.04.003
11. Zierath, J., Rachholz, R., Rosenow, S.E., Bockhahn, R., Schulze, A., Woernle, C.: Experimental identification of modal parameters of an industrial 2-MW wind turbine. Wind Energy **89**, 338–356 (2018). https://doi.org/10.1002/we.2165
12. Woernle, C.: Mehrkörpersysteme: Eine Einführung in die Kinematik und Dynamik von Systemen starrer Körper, 2. Auflage Springer Vieweg, Berlin/Heidelberg (2016). https://doi.org/10.1007/978-3-662-46687-2
13. Konrad, T., Gehrt, J.J., Lin, J., Zweigel, R., Abel, D.: Advanced state estimation for navigation of automated vehicles. Annu. Rev. Control. **46**, 181–195 (2018). https://doi.org/10.1016/j.arcontrol.2018.09.002

14. Wendel, J.: Integrierte Navigationssysteme: Sensordatenfusion, GPS und Inertiale Navigation. Oldenbourg, München (2007). http://www.oldenbourg-link.com/isbn/9783486581607
15. Kaimal, J.C., Wyngaard, J.C., Izumi, Y., Coté, O.R.: Spectral characteristics of surface-layer turbulence. Q. J. R. Meteorol. Soc. **98**(417), 563–589 (1972). https://doi.org/10.1002/qj.49709841707

Johannes Luthe 2013: Semester abroad at the University of Limerick, Ireland
2014: Master's degree in the field of Mechanical Engineering at the University of Rostock, Germany
since 2014: Research assistant at the Chair of Technical Dynamics, University of Rostock, Germany

Chapter 26
Towards Population-Based Structural Health Monitoring, Part III: Graphs, Networks and Communities

Julian Gosliga, Paul Gardner, Lawrence A. Bull, Nikolaos Dervilis, and Keith Worden

Abstract Population-based structural health monitoring opens up the possibility of using information from a population of structures to provide extra information for each individual structure. For example, population-based structural health monitoring could provide improved damage-detection within a homogeneous population of structures by defining a normal condition across a population of structures, which was robust to environmental variation. Furthermore, in cases where structures are sufficiently similar, damage location, assessment, and classification labels could be transferred, increasing the damage labels available for each structure. To determine whether two structures are sufficiently similar requires the comparison of some representation of the structure. In fields such as bioinformatics and computer science, attributed graphs are often used to determine structural similarity. This paper will describe methods for comparing the topology attributes of two such graphs. The algorithm described is suited to population-based structural health monitoring as it provides matches between two graphs which have physical significance. This paper will also describe the process of comparing hierarchical attributes to determine the level of knowledge transfer possible between two structures.

Keywords Population-based structural health monitoring · Irreducible element model · Attributed graph · Structural similarity

26.1 Introduction

The ideal case for structural health monitoring (SHM) is where both the normal condition and the various damage states for a structure are known, and corresponding data sets are available. If the normal condition (including any benign variations) is known, then damage detection is trivial, since any deviation must come from damage to the structure. Additionally, if data were available for all possible damage states, then for any signal measured from the structure the damage location and classification labels would be known. Also, if detailed data for each damage state were available, it would be possible to determine the extent of the damage. A situation where damage can be detected, classified, located, and the extent calculated would be extremely favourable, and with the aid of a good model could allow for damage prediction [1].

In reality, data sets are scarce, for both the normal condition and the damage states of an individual structure. A solution to this is to move away from individuals and perform structural health monitoring using populations of structures instead [2, 3]. Population-based structural health monitoring (PBSHM) seeks to overcome the limitation on available data for structural health monitoring, which to date has been a major obstacle in the industrial uptake of SHM. By examining a population of structures, rather than simply looking at each structure individually, it may be possible to share information about the normal condition and damage states, provided certain conditions are met. An example of this is the Lillgrund wind farm [4], where the nominally identical structures form a homogeneous population [5]. In the Lillgrund wind farm, it was possible to make the detection of performance anomalies robust to variations in the normal condition by sharing SCADA data between wind turbines. In homogeneous populations, it is also possible to describe the population using a single model, called a *form*, which effectively captures the variation within the population [5].

Homogeneous populations are a special case within PBSHM, as for a population to be homogeneous, all structures within the population must have identical topology, and nominally-identical geometry and materials. For a population to be considered strongly homogeneous, the boundary conditions also need to match. Otherwise, a population is considered to be

J. Gosliga (✉) · P. Gardner · L. A. Bull · N. Dervilis · K. Worden
Dynamics Research Group, Department of Mechanical Engineering, University of Sheffield, Sheffield, UK
e-mail: j.gosliga@sheffield.ac.uk; p.gardner@sheffield.ac.uk; l.a.bull@sheffield.ac.uk; n.dervilis@sheffield.ac.uk; k.worden@sheffield.ac.uk

B. Dilworth (ed.), *Topics in Modal Analysis & Testing, Volume 8*, Conference Proceedings of the Society for Experimental Mechanics Series, https://doi.org/10.1007/978-3-030-47717-2_26

heterogeneous, which has implications for the level of information transfer possible [6]. Within a heterogeneous population, certain members may share a greater or lesser degree of similarity with another. The degree of similarity between two structures will determine the information transfer possible. In order to calculate the degree of similarity between structures, it is first necessary to create an Irreducible Element (IE) model of the structure (as described in [7]) and then convert this IE model into an attributed graph (AG) [8]. This paper will describe how the AGs from two separate structures are compared in order to determine the degree of similarity.

This paper will also describe the *network* of structures, which is formed from the AGs of various structures. The idea of the network is that it will create a space where the distance between members of a heterogeneous population is determined by the degree of similarity [9]. Since the distance is determined by the degree of similarity between two structures, it is expected that similar structures will be closer together, and that this can be used to create *communities* of structures. Within communities, the structures should be similar enough that information transfer is possible. For example, within a community of aeroplanes, some information regarding damage in a wing should be transferable, as all aeroplanes have wings.

Since the distance between structures in the network is determined by how similar they are, it is necessary to have some method for quantifying structural similarity. The first step in quantifying the structural similarity is describing the objects as IE models. These IE models are then converted into an AG. It is then possible to use AGs for two structures to find common substructures or *common subgraphs*. For this application, the search is limited to *induced* and *connected* subgraphs. Thankfully the problem of finding common subgraphs to quantify structural similarity exists in other fields – most notably chemistry, bioinformatics and computer science – which means that tools are already available [10]. Methods for finding the induced common subgraph [11] and connected induced common subgraph [12] will be explored in this paper. A method for calculating a similarity score based on the size of the subgraph and attribute matching, relevant to this application, will be developed.

26.2 Matching Using the Attributed Graphs

An IE model of a structure is a way of abstracting and representing aspects of a structure that are important for transferring information about that structure. By comparing the IE representations of two structures, it is possible to determine the level of inference possible between them. To allow this comparison to be performed automatically, it is more efficient to restructure the information into the form of an AG. In the AG, the same information is used as for the IE representation, but whereas the properties for the IE representation are organised in a tabular format to improve readability by humans, the information in the attributed graph is organised so that it can be more efficiently processed by a graph-matching algorithm. The process for creating the AG for a structure is illustrated in Fig. 26.1 and described in [7, 8].

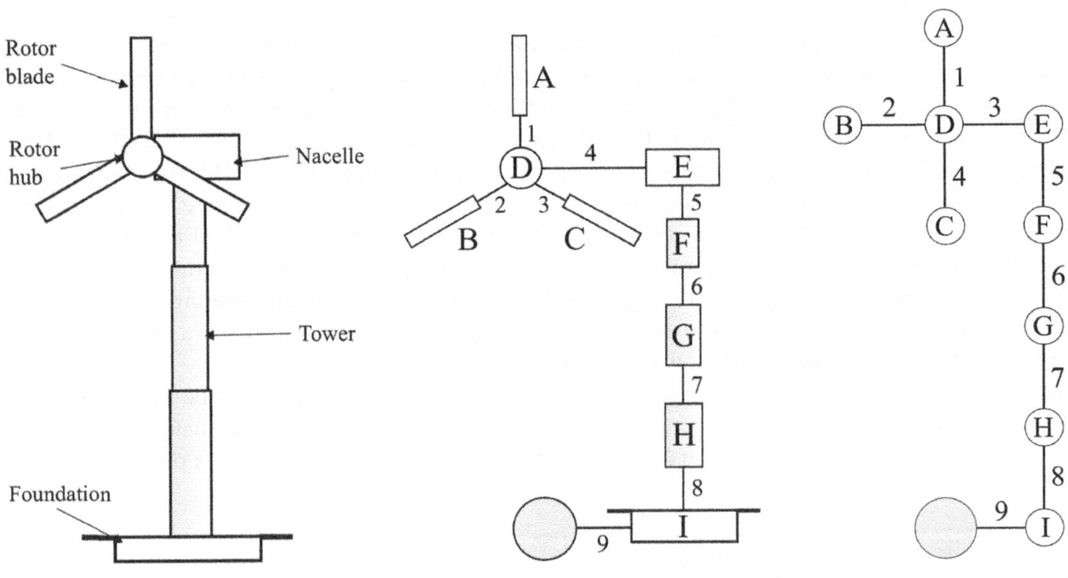

Fig. 26.1 The evolution of an attributed graph. The structure in this case is a wind turbine, and shown (from left to right) are the IE model; an expanded version of the IE model, showing the element and joint designations; and finally, the AG

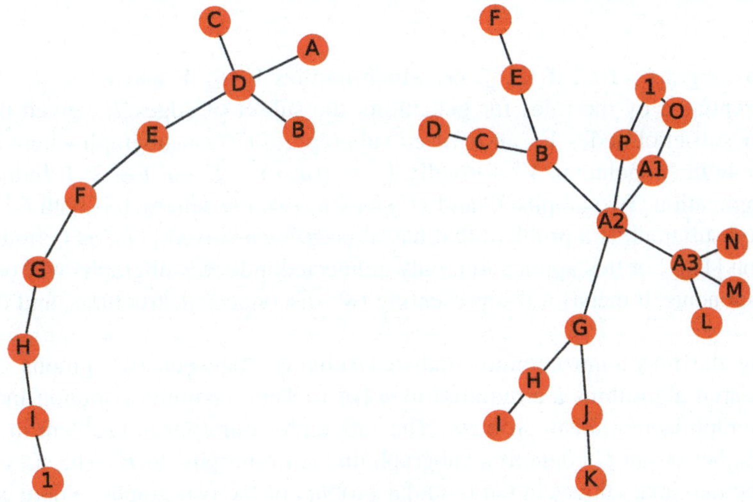

Fig. 26.2 The graphs prodcued from list of elements in Tables 26.2 and 26.1 and list of joints in Tables 26.4 and 26.3. The graph shown on the left is from the wind turbine (compare with Fig. 26.1) and the graph on the right is from the aeroplane

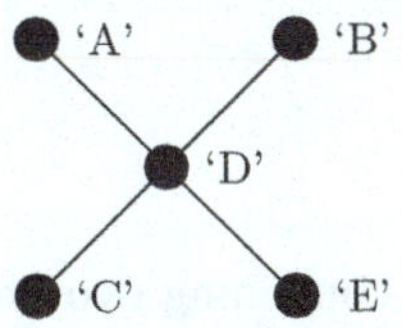


```
graph = { 'A' :  ['D']
          'B' :  ['D']
          'C' :  ['D']
          'D' :  ['A', 'B', 'C', 'E']
          'E' :  ['D'] }
```

Fig. 26.3 A graph and the corresponding Python dictionary. The dictionary defines the neighbourhood for each node

For this paper, it is assumed that the previous steps have been followed, generating the attributed graphs for an aeroplane and a wind turbine as shown in Fig. 26.2. The attributes are taken from the IE representations; where the elements and joints for the aeroplane come from Tables 26.2 and 26.4 respectively, and the elements and joints for the wind turbine come from Tables 26.1 and 26.3.

26.2.1 Introduction to Graph-Matching Algorithms

Definition

A graph G is defined by a set of nodes V and edges E, and so $G = (V, E)$ [13]. The edge and node set of a graph can alternatively be represented as $E(G)$ and $V(G)$, respectively. It is also possible to define a graph by its neighbourhood N, where the neighbourhood is defined as $N(v) = \{u \in V \mid (u, v) \in E\}$. This is a common approach when writing code which incorporates graphs, and an example of a graph defined using the dictionary datatype in Python is shown in Fig. 26.3. The neighbourhood N as defined does not contain the node itself; this is known as the *open neighbourhood* of v. The *closed neighbourhood* of a node v, contains the node itself and is defined as $N[u] = N(v) \cup v$. Neighbourhoods are assumed to be open unless otherwise specified. If two nodes from $V(G)$, say v_1 and v_2, are adjacent, then $(v_1, v_2) \in E(G)$. A graph is *connected* if there is a path from any node to any other node.

Isomorphism

Two graphs, G and H, are said to be isomorphic if there exists a one-to-one mapping from $V(G) \rightarrow V(H)$, and a one-to-one mapping from $E(G) \rightarrow E(H)$. Isomorphic graphs have the same topology, but have different node labels. Therefore, all structures within a homogeneous population will have isomorphic AGs.

Subgraphs

A graph G' is said to be a subgraph of G, if $G' \subseteq G$, which implies $V' \subseteq V$ and $E' \subseteq E$. There are different types of subgraph, which are differentiated by the rules for generating the subset of edges E', given the subset of nodes V'. For example, given an arbitrary subset of nodes V', an *induced* subgraph of G is a subgraph where the edge set E' contains all the edges from E that have both endpoints in V', formally $E' = \{(u, v) \in E \mid u, v \in V'\}$. Induced subgraphs are the most useful for this particular application. Two graphs, G and H, share a *common* subgraph if both $G' \subseteq G$ *and* $G' \subseteq H$. Finding the largest common induced subgraph is a problem that has already been solved [11], as is finding the maximum common connected induced subgraph [12]. For this application, only connected induced subgraphs will be considered. If an AG falls into two disconnected components, it means it is representing two disconnected structures, and the two structures should be considered separately.

The problem of finding the maximum common induced subgraph between two graphs G and H is NP-complete, meaning that exhaustive search algorithms are required to solve it. The maximum common induced subgraph problem is a generalisation of the subgraph isomorphism problem. The subgraph isomorphism problem involves taking two graphs G and H and determining whether or not G contains a subgraph that is isomorphic to H. The subgraph isomorphism problem can be solved by finding the maximal cliques in the modular product of the two graphs, where a clique is a subset of nodes of an undirected graph such that every two distinct nodes in the clique are adjacent. Since the problem of enumerating all maximal cliques is NP-hard, the computational time of any algorithm used to solve this problem will not scale for larger graphs. However, the graphs for structures are unlikely to become so large that this is an issue.

The Modular Product Graph

The *modular product* is a graph (represented by $G \diamond H$) produced by combining two graphs G and H. An example of a modular product graph is shown in Fig. 26.4. The modular product has the node set, generated by the cartesian product $V(G) \times V(H)$ in which two nodes (x, u) and (y, v) are adjacent if:

- $(x, y) \in E(G)$ and $(u, v) \in E(H)$, or
- $(x, y) \notin E(G)$ and $(u, v) \notin E(H)$

Examining Fig. 26.4 it can be seen that the cliques correspond to subgraphs which are common to both G and H. For example, the nodes A_1B_2 and A_2B_3 in the modular product correspond to the subgraphs G' and H', where $V(G') = \{A_1, A_2\} \subseteq V(G)$ and $V(H') = \{B_1, B_2\} \subseteq V(H)$, with corresponding edges $(A_1, A_2) \in E(G)$ and $(B_1, B_2) \in E(H)$. It is possible to verify that this is an induced subgraph.

However, induced subgraphs are not always connected. If the nodes A_1, B_1 and A_3, B_3 were chosen instead, then $V(G') = \{A_1, A_3\}$. There are no edges in $E(G)$ with both endpoints in $V'(G)$, since the edge (A_1, A_3) is not in $E(G)$. The subgraph G' is still induced, but since no edges exist in G', it is unconnected. Likewise for the subgraph H', where $V(H') = \{B_1, B_3\}$.

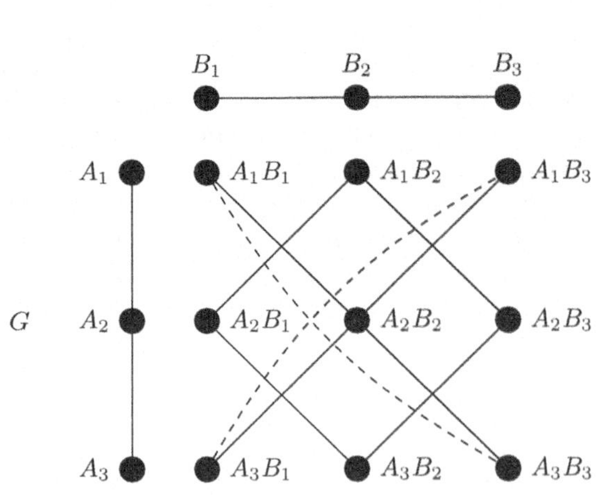

Fig. 26.4 The modular product graph for G and H. The solid lines represent c-edges and the dashed lines represent d-edges

26.2.2 Bron-Kerbosch Algorithm

As mentioned previously, finding cliques in the modular product of two graphs is equivalent to finding the common subgraphs. The Bron-Kerbosch algorithm [11], described in Algorithm 1, is a clique-finding algorithm that will report all maximal cliques, and therefore all subgraphs between two graphs. This algorithm is guaranteed to generate induced subgraphs but does not guarantee that they are connected. The Bron-Kerbosch algorithm is considered one the most efficient backtracking algorithms for maximal clique enumeration is widely used in chemical and life science applications [10]. The complexity of the Bron-Kerbosch algorithm is $\mathcal{O}(3^{n/3})$.

Algorithm 1 Bron-Kerbosch

R: set of nodes to be reported, initially $R = \emptyset$
P: set of nodes which can be added to R, initially $P = V$
X: set of nodes which cannot be added to R, initially $X = \emptyset$

1: **procedure** ENUMERATECLIQUES(R, P, X)
2: Let P be the set $\{u_1, \ldots, u_k\}$
3: **if** $P = \emptyset$ **and** $X = \emptyset$ **then**
4: report R
5: **else**
6: **for** $i \leftarrow 1$ **to** k **do**
7: $P \leftarrow P \backslash \{u_i\}$
8: $N \leftarrow \{v \in V \mid (u_i, v) \in E\}$
9: ENUMERATECLIQUES($R \cup \{u_i\}, P \cap N, X \cap N$)
10: $X \leftarrow X \cup \{u_i\}$
11: **end for**
12: **end if**
13: **end procedure**

A simple example using the graph shown in Fig. 26.5 will be used to illustrate how the algorithm works, although the recursive nature of the algorithm makes it hard to describe, so it is recommended to work through the example, writing out the sets at each step.

Recursive level 1: when the procedure ENUMERATEC-CLIQUES is first called $P \neq \emptyset$. The procedure then iterates through the nodes in P, which in this example makes $u_1 = A$. The node u_1 is then removed from the set nodes P which can be added to the clique.
The open neighbourhood $N(A)$ is generated; in this case, $N = \{B, C\}$. The algorithm is then called recursively with A now added to the current clique R. The set of nodes which can be added to the clique is now limited to nodes within the neighbourhood of A. The set of nodes which are excluded from being added is also now limited to nodes within the neighbourhood of A.

Recursive level 2: for the second recursive call, $P = \{B, C\} \neq \emptyset$, and so the next node B is chosen; this is now removed from P and the neighbourhood is generated. The next recursive call is made, where B is added to R, $P = \{C\}$ only, and $X = \emptyset$.

Recursive level 3: for the third recursive call, $P = C$ initially, so the algorithm continues. C is removed from P, the neighbourhood $N(C)$ is generated. In the next call, C is added to R, but $P \cap N$ and $X \cap N$ are now empty, so when Line 3 is reached, the clique $R = \{A, B, C\}$ is reported, and the procedure terminates. This returns the algorithm to the previous level of recursion as C is added to X. Since there are no more nodes in P, the procedure terminates.

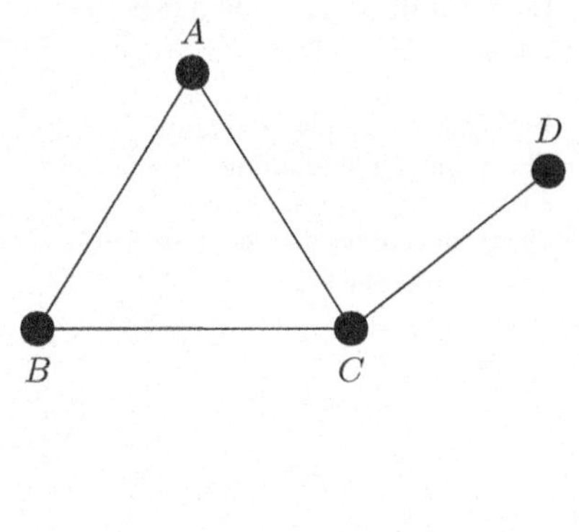

Fig. 26.5 A simple graph for illustrating clique-finding algorithms

Recursive level 2: the algorithm returns to the previous level, where $P = \{B, C\}$ originally and $R = \{A, B\}$. B is now added to X and $u_2 = C$.

Recursive level 3: the algorithm is called with $P \cap N(C) = \emptyset$, $X \cap N(C) = \{B\}$. Even though there are no more nodes in P to add, the clique $R = \{A, B, C\}$ is not reported again since X is not empty. Also, since P is empty, there are no nodes to loop over, and so the procedure is terminated and the algorithm returns to the previous recursive level.

Recursive level 2: C is now added to X and the procedure terminates and returns to the previous level.

Recursive level 1: A is added to X and now $u_2 = B$. Now $P = \{C, D\}$.

Recursive level 2: in this call $P \cap N(B) = \{C\}$, $X \cap N(B) = \{A\}$, $R = \{A, B\}$. No clique is reported, and $u_1 = C$. C is then removed from P and ENUMERATEC-CLIQUES is called again.

Recursive level 3: in this call $P \cap N(C) = \emptyset$ and $R = \{A, B, C\}$, but since $X \cap N(C) = \{A\}$, the clique R is not reported. Since P is empty, the loop cannot be performed and the procedure terminates.

Recursive level 2: C is now moved to X. Since C was the only member of P, the loop has now finished and the procedure terminates.

Recursive level 1: now $u_3 = C$, $P = \{D\}$ and $X = \{A, B\}$.

Recursive level 2: in this call $R = \{C\}$, $P \cap N(C) = \{D\}$ and $X \cap N(C) = \{A, B\}$. $u_1 = D$ and D is removed from P.

Recursive level 3: in this call $P \cap N(D) = \emptyset$ and $X \cap N(D) = \emptyset$ so the clique $R = \{C, D\}$ is reported. The procedure then terminates.

Recursive level 2: D is added to X and since P is empty, the loop ends and the procedure terminates.

Recursive level 1: now $X = \{A, B, C\}$, which prevents the clique $R = \{C, D\}$ being reported again when $u_4 = D$. From visual inspection of Fig. 26.5 it is possible to determine that no more maximal cliques are present.

The algorithm will report all of the cliques in the graph. To find the largest subgraph, the cliques are then ranked in order of size. Due to node re-labelling any clique-finding algorithm will report all graphs that are isomorphic to a given subgraph. The alignment of the subgraph which gives the best physical match is made possible through the use of attributes.

26.2.3 C-Clique-Finding Algorithm

This algorithm is a modification of the Bron-Kerbosch algorithm which ensures only connected induced subgraphs are found. This is not only useful for ensuring they are valid in a physical sense, but also significantly cuts down the search time. In order to limit the search to connected subgraphs, it is first necessary to define the concept of *c-edges*.

C-Edges

Two nodes (x, u) and (y, v) in the modular product $G \diamond H$ are adjacent via a c-edge (where the c stands for connected) if $(x, y) \in E(G)$ and $(u, v) \in E(H)$. This is the first condition for adjacency in the modular product graph. The rest of the edges are d-edges (where the d stands for disconnected) and are defined by the second adjacency condition in the modular product graph. The c-edges and d-edges are shown in Fig. 26.4.

Cliques which include c-edges are called c-cliques. Limiting the clique-finding algorithm to only find c-cliques ensures that any reported subgraphs are connected. This can also decrease the computation time [12] as the set of c-cliques is a subset of the set of all cliques in a graph. Therefore a smaller number of cliques needs to be considered, which reduces the size of the search tree. If the graph contains d-edges, as shown in Fig. 26.6, the only clique of interest is $\{A, B, C\}$ as the clique $\{C, D\}$ does not contain any c-edges.

To limit the search to only c-cliques, it is necessary to modify Algorithm 1. The major modification is the creation of a new set D which contains the set of nodes which cannot be directly added to the current clique because they are adjacent to the current node u via a d-edge.

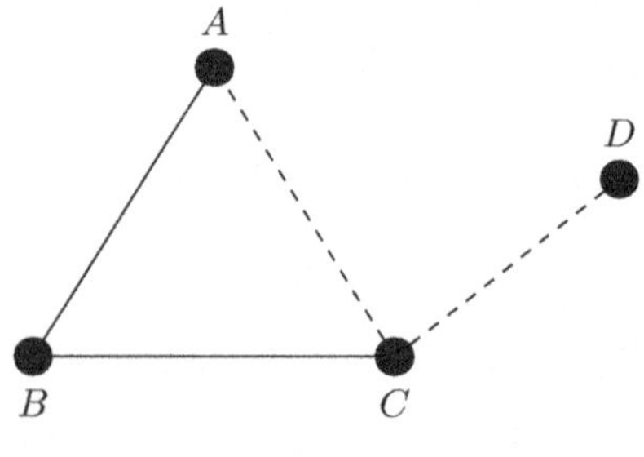

Fig. 26.6 A simple graph for illustrating clique-finding algorithms. The solid lines represent c-edges and the dashed lines represent d-edges

This problem now requires an initialisation algorithm, shown in Algorithm 2. This algorithm iterates over all nodes in the graph, gradually adding them to T, which is the set of nodes that cannot be added to a clique because they have already been used to intialise the algorithm. It is necessary to exclude nodes that have already been used to initialise the algorithm because otherwise every clique of size n would be reported n times.

Algorithm 2 Initialisation for Enumerate C-Cliques

R: set of nodes to be reported
P: set of nodes which can be added to R, because they are adjacent to node u via c-edges
D: set of nodes which cannot be directly added to R, because they are adjacent to u via d-edges
T: set of nodes which have already been used to initiate the ENUMERATEC-CLIQUES algorithm
E_c: set of c-edges for the graph G E_d: set of d-edges for the graph G

```
for u ∈ V do
    P ← ∅
    D ← ∅
    X ← ∅
    N ← {v ∈ V | (u, v) ∈ E}
    for v ∈ N do
        if (u, v) ∈ E_c then
            if v ∈ T then
                X ← X ∪ {v}
            else
                P ← P ∪ {v}
            end if
        else if (u, v) ∈ E_d then
            D ← D ∪ {v}
        end if
    end for
    R ← {u}
    ENUMERATEC-CLIQUES(R, P, D, X, T)
    T ← T ∪ {u}
end for
```

The initialisation algorithm iterates through all nodes in the neighbourhood $N(v)$ and determines whether a node u is adjacent to v via a c-edge, in which case – provided that node has not been previously used to initialise ENUMERATEC-CLIQUES – the node is added to P. Nodes in P may be directly added to R in the next procedure call. If the node u has been previously used to initialise ENUMERATEC-CLIQUES, it is added to X instead. If the node u is adjacent to v via a d-edge, it is added to D. The node u is then added to R and the procedure ENUMERATEC-CLIQUES, described in Algorithm 3 is called. After the procedure has finished, u is added to T so that it is excluded from being added to future cliques. This procedure excludes cliques which do not contain a simple path of c-edges that connect all vertices within that clique (for example, $\{C, D\}$ in Fig. 26.6).

A node in D cannot be added to R at the point where ENUMERATEC-CLIQUES is called. However, nodes in D may at some point be adjacent to a node in P via a c-edge. This check is performed on Lines 9 and 10 in Algorithm 3, where each node in D is checked for adjacency via a c-edge with the current node u_i. If the node v is adjacent to u_i via a c-edge, and has not previously been used to intialise the procedure ENUMERATEC-CLIQUES (i.e. $v \notin T$), then the node is added to P and can potentially be added to R in the next procedure call. If the node $v \in T$ then v is added to X. The node is then removed from D. Aside from this check to see whether or not nodes can be moved from D to P, Algorithm 3 is essentially the same as Algorithm 1.

For the structures shown in Fig. 26.2, the maximum common connected induced subgraph found by Algorithm 3 is shown in Fig. 26.7. For the remainder of the discussion all subgraphs are assumed to be connected and induced.

26.3 Similarity Scores

The similarity scores are calculated by examining the node attributes in each graph that form the subgraph in each question. This is achieved by extracting the nodes in the subgraph and querying the dictionary containing the node attributes. For the sample subgraph shown in Fig. 26.8, the node labels can be used to look up the attributes for the corresponding nodes in the original AGs, shown in Fig. 26.2.

Algorithm 3 Modified Bron-Kerbosch

R: set of nodes to be reported
P: set of nodes which can be added to R, because they are adjacent to node u via c-edges
D: set of nodes which cannot be directly added to R, because they are adjacent to u via d-edges
T: set of nodes which have already been used to initiate the ENUMERATEC-CLIQUES algorithm
E_c: set of c-edges for the graph G

procedure ENUMERATEC-CLIQUES(R, P, D, X, T)
  Let P be the set $\{u_1, \dots, u_k\}$
  if $P = \emptyset$ **and** $X = \emptyset$ **then**
    report R
  else
    for $i \leftarrow 1$ **to** k **do**
      $P \leftarrow P \backslash \{u_i\}$
      $N \leftarrow \{v \in V \mid (u_i, v) \in E\}$
      for all $v \in D$ **do**
        if $(v, u_i) \in E_c$ **then**
          if $v \in T$ **then**
            $X \leftarrow X \cup \{v\}$
          else
            $P \leftarrow P \cup \{v\}$
          end if
          $D \leftarrow D \backslash \{v\}$
        end if
      end for
      ENUMERATEC-CLIQUES($R \cup \{u_i\}, P \cap N, D \cap N, X \cap N, T$)
      $X \leftarrow X \cup \{u_i\}$
    end for
  end if
end procedure

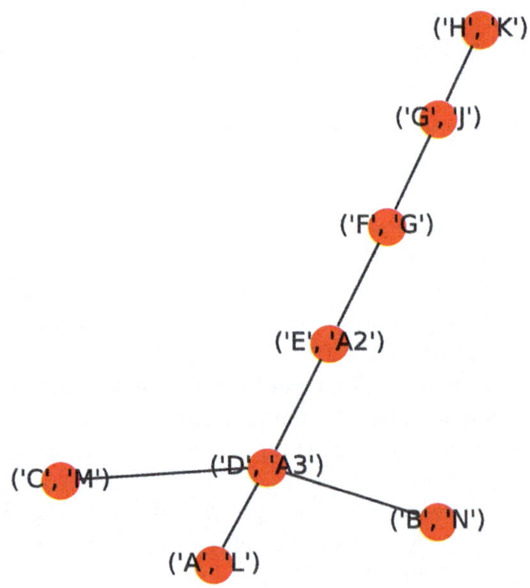

Fig. 26.7 The maximum common connected induced subgraph of the two graphs shown in Fig. 26.2

The left-hand entries in node labels (A, B, C, D, E) correspond to the wind turbine, and so node and edge attributes can be found in Tables 26.1 and 26.3, whereas the right-hand entries in the node labels (A2, A3, N, M, L) correspond to the aeroplane and can be used to find node attributes from Tables 26.2 and 26.4. For example, the geometry class and shape for node in the wind turbine is 'Beam, Aerofoil', and for node N in the aeroplane the geometry class and shape is also 'Beam, Aerofoil' so there is a match. However, node D in the wind turbine represents *complex* geometry, while node A3 in the aeroplane represents a *shell*, so these do not match.

The comparison is performed at the greatest possible level of resolution, if the geometry class matches, then the shape is compared. If the shape matches, then the dimensions for each are examined. If the elements match on all levels, then they are essentially identical. This is shown as a flowchart in Fig. 26.9. This is true for the case of materials as well. If the material class, specific material and material properties are all the same, then the two elements are made from the same material.

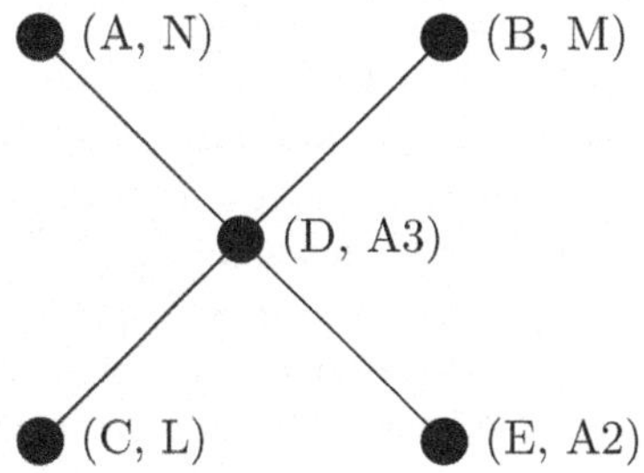

Fig. 26.8 A possible subgraph from the two graphs shown in Fig. 26.2

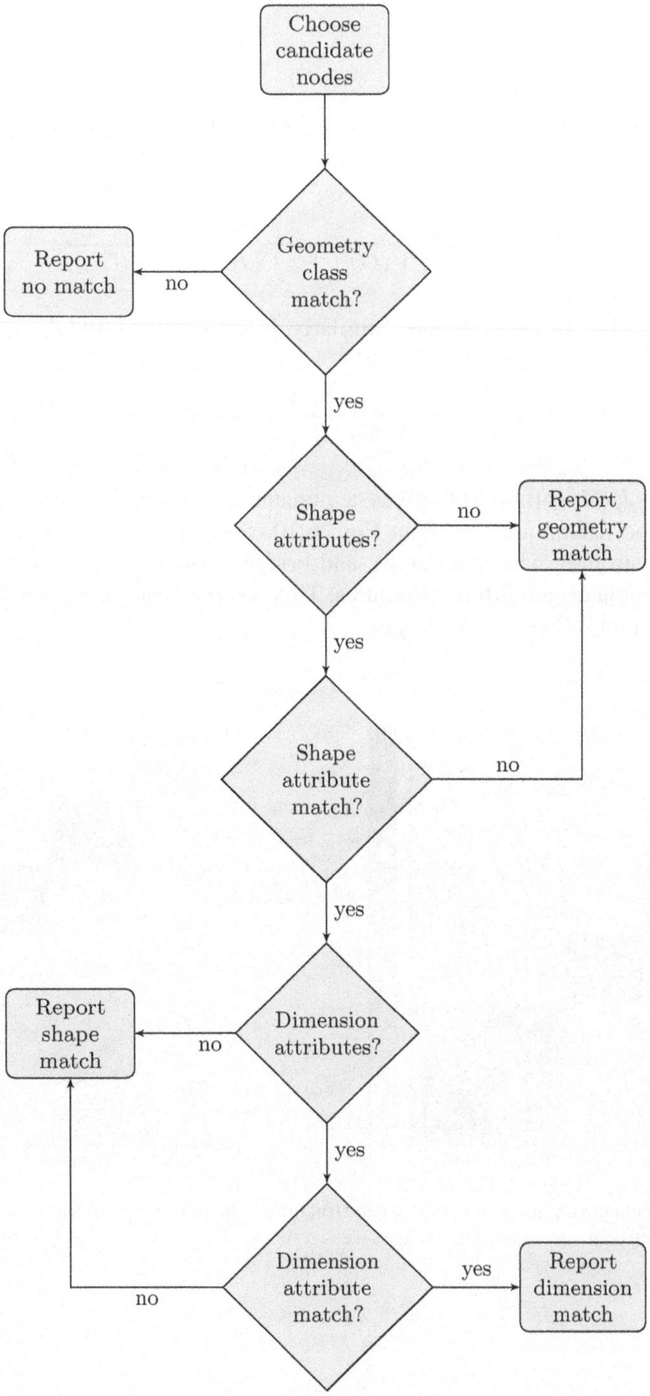

Fig. 26.9 A flowchart for comparison of the geometry attributes of a pair of nodes in the AG

The same matching can be performed for joints. In this case, the joint to examine in each graph is determined by examining the edges in the subgraph. which are adjacent in the subgraph. This gives a set of nodes which are connected by the edge in the original graphs. From this, the edge attributes can be extracted. The joint type is compared, and if the joint is kinematic, the restricted degrees of freedom can also be compared.

If all elements and joints in two graphs are identical (all of the hierarchical attributes are available and match) then these two graphs form a homogeneous population. If all elements and joint in the subgraph of two graphs are identical, then they share a subcomponent and transfer is possible within this subcomponent [6]. The more likely case is that only partial matches are seen, and in this case the level of match determines the appropriate transfer learning approach.

26.4 Communities

It is possible to determine a similarity score (Jaccard similarity coefficient) based purely on the size of this maximum common subgraph G'. This is calculated using the following equation,

$$J_v(G,H) = \frac{\mid V(G') \mid}{\mid V(G) \mid + \mid V(H) \mid - \mid V(G') \mid} \tag{26.1}$$

where $J_v(G, H)$ is the Jaccard index for the node sets. Similarly, it is possible to find the Jaccard index for the edge sets,

$$J_e(G,H) = \frac{\mid E(G') \mid}{\mid E(G) \mid + \mid E(H) \mid - \mid E(G') \mid} \tag{26.2}$$

Multiplying $J_v(G, H)$ and $J_e(G, H)$ by 100 gives a percentage similarity score. This was calculated for several representative structures and the results are shown in Fig. 26.10. Structures tend to match more with similar structures, for example aeroplanes match strongly with aeroplanes, and bridges match strongly with bridges. However, some bridge graphs match more with the aeroplane and turbine structures. This could be caused by there being a higher chance of finding large subgraphs when the two graphs G and H are larger.

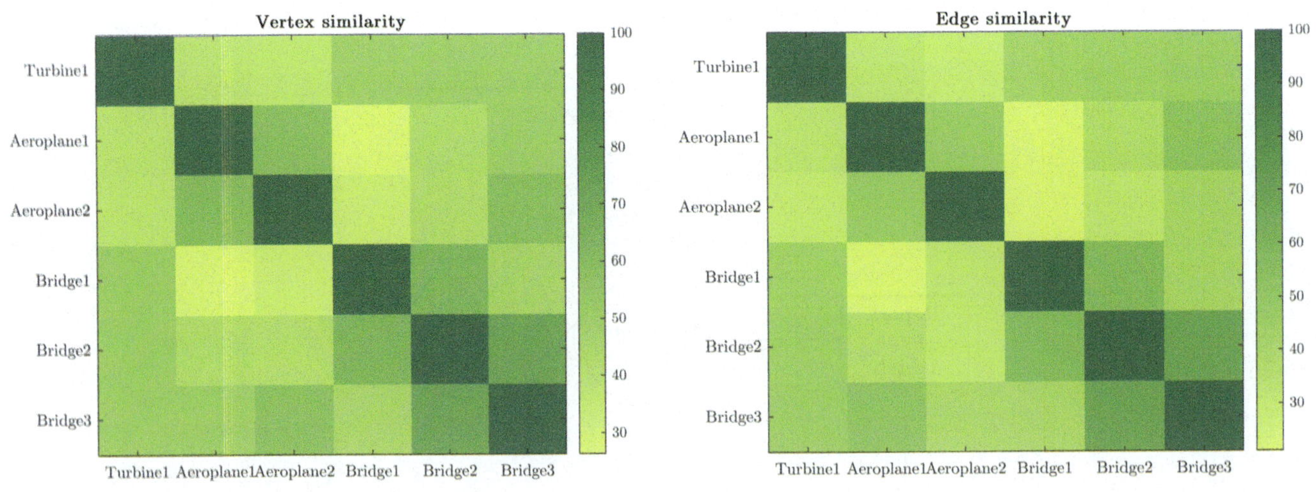

Fig. 26.10 Figures showing the percentage topology match between structures. The percentage score is based on the Jaccard index $J_v(G, H)$ and $J_e(G, H)$ for nodes and edges respectively

This highlights one of the main issues with matching on topology alone. Topology aids in finding similar structures with consistent location labels, but topology matching alone does not guarantee that nodes in the subgraph have similar geometry or material attributes. For example, looking at topology alone may lead one to match the deck of a bridge with the wing of an aeroplane. Equally, when matching an aeroplane with an aeroplane there are many orientations of the subgraph that do not produce a sensible match. For example, the fuselage may be aligned with a wing of the aeroplane. The key to producing sensible matches is using similarity scores which take the node attributes into account.

By grouping structures according to their similarity it should be possible to create *communities* of structures. Within these communities, it is expected that a certain level of information transfer is possible. The communities which are formed may change based on the particular SHM problem, depending on which attributes are determined to be the most relevant.

26.5 Conclusion

To compare the similarity of two structures, they are first converted into IE models. The IE models are then converted into AGs. These graphs are then compared to find a list of possible substructures. The method for doing this is to first find the modular product of the two graphs, and then use a clique finding algorithm to generate a list of subgraphs. The attributes of the corresponding nodes in each graph can then be compared. Comparing node attributes is simple as the nodes in each graph are uniquely identified (within that graph) and the information can be stored using a Python dictionary, or in a database. Transferring this methodology to a database will form the basis of future work.

Acknowledgments The authors would like to thank the UK EPSRC for funding through the Established Career Fellowship EP/R003645/1 and the Programme Grant EP/R006768/1.

Appendix

Tables listing the element and joint properties for the IE models.

Table 26.1 List of elements and their properties for Turbine 1

Element designations for Turbine 1				
Name	**Element ID**	**Material**	**Geometry**	**Shape**
Rotor blade	A	FRP	Beam	Aerofoil
Rotor blade	B	FRP	Beam	Aerofoil
Rotor blade	C	FRP	Beam	Aerofoil
Rotor hub	D	FRP	Complex	Rotor hub
Nacelle	E	FRP	Shell	Cuboid
Tower section 1	F	Metal	Beam	Cylindrical
Tower section 2	G	Metal	Beam	Cylindrical
Tower section 3	H	Metal	Beam	Cylindrical
Foundation	I	Concrete	Plate	Cylindrical
Name	**Element ID**	**Boundary**	–	–
Footing	1	Ground	–	–

Table 26.2 List of elements and their properties for Aeroplane 1

Element designations for Aeroplane 1				
Name	**Element ID**	**Material**	**Geometry**	**Shape**
Fuselage	A1	FRP	Shell	Truncated cone
Fuselage	A2	FRP	Beam	Cylindrical
Fuselage	A3	FRP	Shell	Cone
Wing 1	B	FRP	Beam	Aerofoil
Pylon 1	C	FRP	Complex	Pylon
Engine 1	D	Assembly	Shell	Cylinder
Pylon 2	E	FRP	Complex	Pylon
Engine 2	F	Assembly	Shell	Cylinder
Wing 2	G	FRP	Beam	Aerofoil
Pylon 3	H	FRP	Complex	Pylon
Engine 3	I	Assembly	Shell	Cylinder
Pylon 4	J	FRP	Complex	Pylon
Engine 4	K	Assembly	Shell	Cylinder
Vert stabiliser 1	L	FRP	Beam	Aerofoil
Vert stabiliser 2	M	FRP	Beam	Aerofoil
Horz stabiliser	N	FRP	Beam	Aerofoil
Front landing gear	O	Assembly	Complex	Assembly
Rear landing gear	P	Assembly	Complex	Assembly
Name	**Element ID**	**Boundary**	–	–
Tarmac	1	Ground	–	–

Table 26.3 List of joints and their properties for Turbine 1

Joint designations for Turbine 1					
Joint ID	Element set	Coordinate	Type	Disp. DoF	Rot. DoF
1	A, D	8, 15, 235.75	Bearing	[x, y, z]	[y, z]
2	B, D	8, 14, 254	Bearing	[x, y, z]	[y, z]
3	D, E	10, 15, 253	Bearing	[x, y, z]	[y, z]
4	D, C	8, 16, 254	Bearing	[x, y, z]	[x, y]
5	E, F	15, 15, 250	Bearing	[x, y, z]	[x, y]
6	F, G	15, 15, 183	Bolted	–	–
7	G, H	15, 15, 105	Bolted	–	–
8	H, I	15, 15, 5	Bolted	–	–
9	I, 1	15, 15, 0	Soil	–	–

Table 26.4 List of joints and their properties for Aeroplane 1

Joint designations for Aeroplane 1					
Joint ID	Element set	Coordinate	Type	Disp. DoF	Rot. DoF
1	A1, A2	34.2, 14.68, 5.165	Perfect	–	–
2	A2, A3	34.2, 60.96, 5.165	Perfect	–	–
3	A2, B	32.2, 29.79, 2.89	Lug	–	–
4	B, C	13.2, 42.67, 4.74	Complex	–	–
5	C, D	13.2, 40.17, 4.74	Complex	–	–
6	B, E	23.2, 30.79, 3.57	Complex	–	–
7	E, F	23.2, 28.29, 3.57	Complex	–	–
8	A2, G	36.2, 29.79, 2.89	Lug	–	–
9	G, H	45.2, 30.79, 3.57	Complex	–	–
10	H, I	45.2, 28.29, 3.57	Complex	–	–
11	G, J	55.2, 42.67, 4.74	Complex	–	–
12	J, K	55.2, 40.17, 4.74	Complex	–	–
13	A3, L	33.2, 68.58, 7.55	Lug	–	–
14	A3, M	35.2, 68.58, 7.55	Lug	–	–
15	A3, N	34.2, 64.58, 9.16	Lug	–	–
16	A1, O	34.2, 7.75, 1.75	Complex	–	–
17	A2, P	34.2, 29.67, 1.75	Complex	–	–
18	O, 1	34.2, 7.75, 0	Plane	[z]	[x, y]
19	P, 1	34.2, 29.67, 0	Plane	[z]	[x, y]

References

1. Rytter, A.: Vibrational based inspection of civil engineering structures. Ph.D. thesis, Deptartment of Building Technology and Structural Engineering, Aalborg University, Denmark (1993)
2. Antoniadou, I., Dervilis, N., Papatheou, E., Maguire, A.E., Worden, K.: Aspects of structural health and condition monitoring of offshore wind turbines. Philos. Trans. R. Soc. A Math. Phys. Eng. Sci. **373**, 1–14 (2015)
3. Worden, K., Cross, E.J., Dervilis, N., Papatheou, E., Antoniadou, I.: Structural health monitoring: from structures to systems-of-systems. IFAC-PapersOnLine **48**, 1–17 (2015)
4. Papatheou, E., Dervilis, N., Maguire, A.E., Antoniadou, I., Worden, K.: A performance monitoring approach for the novel Lillgrund offshore wind farm. IEEE Trans. Ind. Electron. **62**, 6636–6644 (2015)
5. Bull, L.A., Gardner, P.A., Gosliga, J., Rogers, T.J., Haywood-Alexander, M., Dervilis, N., Cross, E.J., Worden, K.: Towards population-based structural health monitoring, part I: homogeneous populations and forms. In: Proceedings of IMAC XXXVIII – the 38th International Modal Analysis Conference, Houston (2020)
6. Gardner, P.A., Worden, K.: Towards population-based structural health monitoring, Part IV: heterogeneous populations, matching and transfer. In: Proceedings of IMAC XXXVIII – the 38th International Modal Analysis Conference, Houston (2020)
7. Gosliga, J., Worden, K.: A general representation for assessing the similarity of structures. In: Proceedings of the 12th International Workshop on Structural Health Monitoring, Palo Alto (2019)
8. Gosliga, J., Gardner, P.A., Bull, L.A., Dervilis, N., Worden, K.: Towards population-based structural health monitoring, Part II: heterogeneous populations and structures as graphs. In: Proceedings of IMAC XXXVIII – the 38th International Modal Analysis Conference, Houston (2020)
9. Worden, K.: Towards population-based structural health monitoring, Part VI: structures as geometry. In: Proceedings of IMAC XXXVIII – the 38th International Modal Analysis Conference, Houston (2020)
10. Duesbury, E., Holliday, J.D., Willett, P.: Maximum common subgraph isomorphism algorithms. Match **77**, 213–232 (2017)
11. Bron, C., Kerbosch, J.: Algorithm 457: finding all cliques of an undirected graph. Commun. ACM **16**, 575–577 (1973)
12. Koch, I.: Enumerating all connected maximal common subgraphs in two graphs. Theor. Comput. Sci. **250**, 1–30 (2001)
13. Diestel, R.: Graph Theory, 3rd ed. Springer, New York (2006)

Julian Gosliga completed his Masters degree in Mechanical Engineering at the University of Sheffield. He then went on to complete a PhD on Energy Harvesting within the same department. Julian now works as a Research Associate for the Dynamics Research Group at the University of Sheffield, focusing on population-based structural health monitoring.

Chapter 27
Utilization of Experimental Data in Elastic Multibody Simulation: Case Study on the Ampair 600 Turbine Blade

Andreas Schulze, Johannes Luthe, János Zierath, and Christoph Woernle

Abstract The development of detailed elastic multibody models is a standard practice for the transient analysis of complex mechanical and mechatronic systems. With the constant increase in complexity of modern engineering problems, however, the development of representative multibody models may not be feasible. Consequently, the utilization of Experimental Dynamic Substructuring is necessary to integrate experimentally derived data into the modeling process. Within this contribution, Experimental Dynamic Substructuring is carried out on an Ampair 600 Wind Turbine blade. A particular interest of the work lies in the consideration of the strong nonlinearity of the inertia-variant multibody model. A first investigation of the derived multibody model exhibits an exact representation of the measured mode shapes, natural frequencies, and damping ratios.

Keywords Elastic multibody simulation · Experimental dynamic substructuring · Experimental modal analysis · Floating frame of reference formulation · Ampair 600 turbine blade

27.1 Introduction

Over the last two decades, elastic multibody dynamics has emerged as a standard practice for efficient and reliable numerical simulation of complex mechanical systems. Within its numerous application fields, Elastic Multibody Simulation (EMBS) particularly excels in transient analysis of mechanical systems that undergo large, non-linear rigid body motions accompanied by small elastic deformations, e.g., space structures, robotics, and wind turbines. However, the increasing complexity of modern simulation tasks leads to unavoidable deviations between the mass, stiffness, and damping distribution of the real structure and the assumptions made in the simulation process. Consequently, experimental identification of the system's dynamics on a prototype of the mechanical system becomes necessary to identify and to adjust simulation uncertainties.

The integration of experimentally derived data into the simulation process sets the focus of Experimental Dynamic Substructuring (EDS) [1]. The various methods that were developed in the context of EDS are commonly classified into frequency-based (FEDS) and modal-based (MEDS) methods. In FEDS, an experimentally derived substructure is coupled to an analytical model by directly coupling the frequency response functions of the two substructures. The coupling in MEDS is accomplished in the modal domain via Component Mode Synthesis utilizing modes shapes identified through measurements on the physical substructure. A substantial amount of research, e.g. [2, 3], concerning the application and further enhancement of EDS on an Ampair 600 Wind Turbine testbed has been carried out by the various researchers of the SEM Substructure Focus Group [4].

In elastic multibody simulation, the elastic body is generally integrated into the multibody model framework as a reduced-order dynamic substructure [5], which makes MEDS methods seem a desirable option to integrate experimental data into a multibody model. A direct application of MEDS methods in the elastic multibody domain, however, requires special consideration due to the strong non-linear system dynamics. Early investigations concerning the build-up of elastic multibody models based on experimental data from modal data were carried out in [6], where the importance of correct representation of

A. Schulze (✉) · J. Luthe · C. Woernle
Chair of Technical Dynamics, University of Rostock, Rostock, Germany
e-mail: andreas.schulze2@uni-rostock.de; johannes.luthe@uni-rostock.de; woernle@uni-rostock.de

J. Zierath
W2E Wind to Energy GmbH, Rostock, Germany
e-mail: jzierath@wind-to-energy.de

B. Dilworth (ed.), *Topics in Modal Analysis & Testing, Volume 8*, Conference Proceedings of the Society for Experimental Mechanics Series, https://doi.org/10.1007/978-3-030-47717-2_27

the inertia-variant system matrices was pointed out. In [7], the integration of modal data into the modeling process of elastic multibody system is partially resumed and tested on numerical examples. The challenges accompanied by the integration of spatial incomplete modal data are investigated in [8]. Recently, the experimental modeling approach was further refined and tested on a simple beam-like structure in [9] by the authors of this contribution.

The aim of this contribution is the application of the experimental modeling approach in [9] on an Ampair 600 Wind Turbine rotor blade to further investigate the approach on a complex mechanical structure. The first section provides a brief overview of the proposed method and outlines the prerequisites to synthesize a multibody model from experimental data. To simplify the nomenclature within this contribution, the experimental dynamic substructuring method for multibody systems is referred to as experimental synthesis Sect. 27.3 gives a detailed description of the blade measurements, arising challenges, and the data processing necessary for experimental synthesis. Finally, a multibody model of the turbine blade is built up by utilizing the experimental data, and first investigations of the model's dynamic behaviour are carried out and discussed.

27.2 Theoretical Background

The configuration of an elastic body in a multibody system is commonly described using the Floating Frame of Reference Formulation (FFRF). In this approach, the absolute position $\boldsymbol{r}^P \in \mathbb{R}^3$ of an arbitrary body point P is given by

$$\begin{aligned} \boldsymbol{r}^P(\boldsymbol{R},t) &= \boldsymbol{r}(t) + \boldsymbol{d}(\boldsymbol{R},t) \\ &= \boldsymbol{r}(t) + \boldsymbol{R} + \boldsymbol{u}(\boldsymbol{R},t) \end{aligned} \tag{27.1}$$

where $\boldsymbol{r} \in \mathbb{R}^3$ describes the absolute position of the body reference frame $\mathcal{K}^1$ relative to the inertia frame $\mathcal{K}^0$, $\boldsymbol{R} \in \mathbb{R}^3$ is the position vector of P relative to $\mathcal{K}^1$ in the undeformed reference configuration, and $\boldsymbol{u} \in \mathbb{R}^3$ is the vector of elastic deformation relative to $\mathcal{K}^1$, whereby all vectors are represented in $\mathcal{K}^1$, see Fig. 27.1. A central advantage of the kinematic description (27.1) is the decoupling of the body deformation $\boldsymbol{u}$ from the rigid body motion described by the reference frame $\mathcal{K}^1$. For small deformations, the vector $\boldsymbol{u}$ can then be approximated using a Ritz approach so that

$$\boldsymbol{u}(\boldsymbol{R},t) = \boldsymbol{\Phi}(\boldsymbol{R})\boldsymbol{q}(t) \tag{27.2}$$

with the shape function matrix $\boldsymbol{\Phi} \in \mathbb{R}^{3\times n}$ and the vector $\boldsymbol{q} \in \mathbb{R}^n$ of the n time-dependent deformation coordinates.

Further derivation of the kinematic equations (27.1) with respect to time and following Jourdain's principle, see, e.g. [10, 11], the equations of motion of a single, unconstrained body are formulated as

$$\underbrace{\begin{bmatrix} \boldsymbol{M}^{tt} & \boldsymbol{M}^{tr} & \boldsymbol{M}^{te} \\ & \boldsymbol{M}^{rr} & \boldsymbol{M}^{re} \\ sym. & & \boldsymbol{M}^{ee} \end{bmatrix}}_{\boldsymbol{M}} \underbrace{\begin{bmatrix} \ddot{\boldsymbol{r}} \\ \dot{\boldsymbol{\omega}} \\ \ddot{\boldsymbol{q}} \end{bmatrix}}_{\boldsymbol{b}} + \underbrace{\begin{bmatrix} \boldsymbol{0} \\ \boldsymbol{0} \\ \boldsymbol{K}^e\boldsymbol{q} + \boldsymbol{D}^e\dot{\boldsymbol{q}} \end{bmatrix}}_{\boldsymbol{k}^i} + \underbrace{\begin{bmatrix} \boldsymbol{k}^t \\ \boldsymbol{k}^r \\ \boldsymbol{k}^e \end{bmatrix}}_{\boldsymbol{k}^c} - \underbrace{\begin{bmatrix} \boldsymbol{h}^t \\ \boldsymbol{h}^r \\ \boldsymbol{h}^e \end{bmatrix}}_{\boldsymbol{h}} = \begin{bmatrix} \boldsymbol{0} \\ \boldsymbol{0} \\ \boldsymbol{0} \end{bmatrix} \tag{27.3}$$

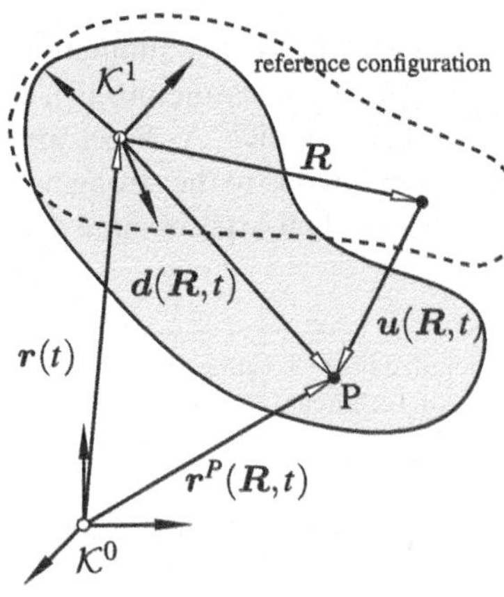

Fig. 27.1 Kinematics of a flexible body in floating frame of reference formulation

where the superscripts t, r, and e refer to rigid body translation, rotation, and elastic deformation, respectively. $\boldsymbol{M}$ is the total mass matrix of the body with the sub-matrices $\boldsymbol{M}^{tt}$, $\boldsymbol{M}^{tr}$, $\boldsymbol{M}^{rr} \in \mathbb{R}^{3\times 3}$ known from rigid body dynamics, the structural mass matrix $\boldsymbol{M}^{ee} \in \mathbb{R}^{n\times n}$ associated with the elastic coordinates and the inertia coupling matrices $\boldsymbol{M}^{te}$, $\boldsymbol{M}^{re} \in \mathbb{R}^{3\times n}$. The vectors $\ddot{\boldsymbol{r}}$ and $\dot{\boldsymbol{\omega}} \in \mathbb{R}^3$ in the body acceleration vector $\boldsymbol{b}$ describe the absolute accelerations of the reference frame. $\boldsymbol{k}^i \in \mathbb{R}^{6+n}$ is the vector of generalized internal forces, with the constant stiffness and damping matrices $\boldsymbol{K}^e$, $\boldsymbol{D}^e \in \mathbb{R}^{n\times n}$. Finally, $\boldsymbol{k}^c \in \mathbb{R}^{6+n}$ represents generalized gyroscopic forces and $\boldsymbol{h} \in \mathbb{R}^{6+n}$ is the vector of generalized external forces.

The final step in deriving the equations of motion of a single, unconstrained body in elastic multibody simulation constitutes a linearization of the total mass matrix $\boldsymbol{M}$ and the vectors $\boldsymbol{k}^c$, $\boldsymbol{h}$, which are highly nonlinear in the elastic coordinates $\boldsymbol{q}$. Linearization of equations (27.3) up to first order in $\boldsymbol{q}$, see [11], results in

$$\begin{aligned} \boldsymbol{M}^{tt} &= m\boldsymbol{E}, & \boldsymbol{M}^{ee} &\cong \boldsymbol{M}^{ef0}, \\ \boldsymbol{M}^{tr}(\boldsymbol{q}) &\cong \boldsymbol{M}^{tr0} + \boldsymbol{M}^{tr1}\boldsymbol{q}, & \boldsymbol{K}^{e} &\cong \boldsymbol{K}^{e0}, \\ \boldsymbol{M}^{te} &\cong \boldsymbol{M}^{te0}, & \boldsymbol{D}^{e} &\cong \boldsymbol{D}^{e0}, \\ \boldsymbol{M}^{rr}(\boldsymbol{q}) &\cong \boldsymbol{M}^{rr0} + \boldsymbol{M}^{rr1}\boldsymbol{q}, & \boldsymbol{k}^{c} &\cong \boldsymbol{k}^{c}(\boldsymbol{\omega}, \boldsymbol{q}, \dot{\boldsymbol{q}}), \\ \boldsymbol{M}^{re}(\boldsymbol{q}) &\cong \boldsymbol{M}^{re0} + \boldsymbol{M}^{re1}\boldsymbol{q}, & \boldsymbol{h} &\cong \boldsymbol{h}(\boldsymbol{R}, \boldsymbol{q}, \ldots). \end{aligned} \tag{27.4}$$

Experimental synthesis aims to calculate the unknown quantities in (27.4) solely from experimentally derived data of the real physical body. According to the investigations in [9], the integration of the experimental data into the multibody framework sets specific prerequisites regarding the extent and the processing of the experimental data as well as the choice of the body reference frame:

- The body mass m, the location of the center of gravity $\boldsymbol{r}^c \in \mathbb{R}^3$ and the inertia tensor $\boldsymbol{J} \in \mathbb{R}^{3\times 3}$ of the body in reference configuration are known from experiments or derived from, e.g., a CAD model.
- The body reference frame is attached to the center of gravity of the body in reference configuration so that $\boldsymbol{r}^c = 0$ in $\mathcal{K}^1$.
- A modal model of the unconstrained body

$$\boldsymbol{\Phi}^e(\boldsymbol{R}) = \left[\boldsymbol{\Phi}^e_1, \ldots, \boldsymbol{\Phi}^e_n\right], \quad \boldsymbol{\Lambda} = \operatorname{diag}\left(\bar{\omega}_1^2\sqrt{1+\zeta_1}, \ldots, \bar{\omega}_n^2\sqrt{1+\zeta_n}\right) \tag{27.5}$$

is derived from experimental modal analysis (EMA [12]), where $\boldsymbol{\Phi}^e_i \in \mathbb{R}^{6mp\times 1}$ is the mode shape scaled to unity modal mass of the i-th body mode, and $\bar{\omega}_i$, $\zeta_i \in \mathbb{R}^1$ are the corresponding undamped natural frequencies and damping ratios. Further, n denotes the number of modes derived by EMA, and mp is the number of measurement locations on the body. Note that the mode shapes in (27.5), are described in reference frame $\mathcal{K}^1$.

By fulfilling these conditions, the full set of system matrices in (27.4) can be derived from the measured data. Here, the key idea is to approximate the flexible coordinates $\boldsymbol{q}$ by utilizing the experimentally derived mode shapes so that

$$\boldsymbol{q} = \boldsymbol{\Phi}^e(\boldsymbol{R})\boldsymbol{q}^e \tag{27.6}$$

with the mode shape matrix $\boldsymbol{\Phi}^e \in \mathbb{R}^{6mp\times n}$ and the vector of modal coordinates $\boldsymbol{q}^e \in \mathbb{R}^n$. The relationship in (27.6) implies that the elastic coordinates $\boldsymbol{q} \in \mathbb{R}^{6mp}$ of the experimentally derived multibody model represent the deformations of material points located at the measurement locations described in $\boldsymbol{R} \in \mathbb{R}^{3mp}$.

The modal reduction of the body in (27.6) facilitates the calculation of the structural sub matrices $\boldsymbol{M}^{ee0}$, $\boldsymbol{K}^{e0}$, and $\boldsymbol{D}^{e0}$ directly from the modal model in (27.5)

$$\boldsymbol{M}^{ef0} = \boldsymbol{E}, \quad \boldsymbol{K}^{e0} = \operatorname{diag}\left(\bar{\omega}_i^2, \ldots, \bar{\omega}_n^2\right), \quad \boldsymbol{D}^{e0} = \operatorname{diag}\left(2\zeta_1\bar{\omega}_1, \ldots, 2\zeta_n\bar{\omega}_n\right). \tag{27.7}$$

The sub-matrices in (27.7) are associated with the elastic deformation of the body relative to the reference frame $\mathcal{K}^1$. In a next step, rigid body motion of the reference frame $\mathcal{K}^1$ is considered by calculating the rigid body modes of the body from

$$\boldsymbol{\Phi}^t = \left[\left[\boldsymbol{E}\ \boldsymbol{0}\right]_j^\mathsf{T}\right], \quad \boldsymbol{\Phi}^r = \left[\left[\tilde{\boldsymbol{R}}\ \boldsymbol{E}\right]_j^\mathsf{T}\right], \quad j = 1, \ldots, mp \tag{27.8}$$

with the matrix of rigid body translation $\boldsymbol{\Phi}^t \in \mathbb{R}^{6mp\times 3}$ and the matrix of rigid body rotation $\boldsymbol{\Phi}^r \in \mathbb{R}^{6mp\times 3}$. Further, $\boldsymbol{E} \in \mathbb{R}^{3\times 3}$ is the identity matrix, and $\tilde{\boldsymbol{R}}_j \in \mathbb{R}^{3\times 3}$ is the skew symmetric matrix of the j-th measurement location $\boldsymbol{R}_j$ relative to the reference frame.

Finally, the calculation of the remaining quantities in (27.4) requires information about the distribution of the body mass to the grid of measurement locations. This is achieved by the calculation of a pseudo mass matrix using a system-equivalent reduction expansion process (SEREP, see, e.g. [13]) of the form

$$\bar{\boldsymbol{M}} = \left[\boldsymbol{\Phi}^t\ \boldsymbol{\Phi}^r\ \boldsymbol{\Phi}^e\right]^{+\top} \begin{bmatrix} m\boldsymbol{E} & \boldsymbol{0} & \boldsymbol{0} \\ & \boldsymbol{J} & \boldsymbol{0} \\ sym. & & \boldsymbol{E} \end{bmatrix} \left[\boldsymbol{\Phi}^t\ \boldsymbol{\Phi}^r\ \boldsymbol{\Phi}^e\right]^{+} \tag{27.9}$$

where the superscript $+$ denotes the Moore-Penrose pseudo inverse. As pointed out in [7], the pseudo mass matrix $\bar{\boldsymbol{M}} \in \mathbb{R}^{6mp\times 6mp}$ is not a unique solution, and other valid mass distributions exist that fulfill (27.9). In the scope of experimental synthesis, however, $\bar{\boldsymbol{M}}$ holds two essential features necessary for the approach. First, $\bar{\boldsymbol{M}}$ is consistent with the system mass m and inertial properties $\boldsymbol{J}$, and second, the mode shape matrices $\boldsymbol{\Phi}^t$, $\boldsymbol{\Phi}^r$, and $\boldsymbol{\Phi}^e$ are orthogonal about $\bar{\boldsymbol{M}}$.

With the mass distribution $\bar{\boldsymbol{M}}$, the remaining unknown matrices in (27.4) are calculated following the standard framework of elastic multibody systems [11]. The zero-order matrices in (27.4) are then given by

$$\begin{aligned} \boldsymbol{M}^{tt} &= \boldsymbol{\Phi}^{t\top}\bar{\boldsymbol{M}}\boldsymbol{\Phi}^t, & \boldsymbol{M}^{tr0} &= \boldsymbol{\Phi}^{r\top}\bar{\boldsymbol{M}}\boldsymbol{\Phi}^t, & \boldsymbol{M}^{te0} &= \boldsymbol{\Phi}^{t\top}\bar{\boldsymbol{M}}\boldsymbol{\Phi}^e, \\ \boldsymbol{M}^{rr0} &= \boldsymbol{\Phi}^{r\top}\bar{\boldsymbol{M}}\boldsymbol{\Phi}^r, & \boldsymbol{M}^{re0} &= \boldsymbol{\Phi}^{r\top}\bar{\boldsymbol{M}}\boldsymbol{\Phi}^e. \end{aligned} \tag{27.10}$$

The calculation of the first-order terms in (27.4) as well as of the vectors $\boldsymbol{k}^c$, $\boldsymbol{h}$ of generalized centrifugal and external forces follows similarly to equations (27.10). For a detailed description of the calculation of theses terms, the authors refer to the investigations in [11].

27.3 Wind Turbine Blade Case Study

The aim of this contribution lies in the application of the experimental synthesis approach to an Ampair 600 wind turbine blade, see Fig. 27.2. The turbine blade is part of the Ampair 600 Wind Turbine testbed of the Substructuring Focus Group [4, 14] within the Society of Experimental Mechanics (SEM). This Focus Group initiated thorough investigations of the testbed by various researches intending to enhance experimental dynamic substructuring methodologies and theories, e.g., [2, 3]. Extensive experimental investigations of the turbine blade are found, e.g., in [2, 15–18].

Within this contribution, Experimental Modal Analysis (EMA) is performed on the rotor blade to capture the blade's dynamic behavior in flapwise direction. Overall, $mp = 27$ measurement locations are distributed over the blade surface, see Fig. 27.2 a. As shown in the previous section, the measurement grid will later define the multibody model's geometric representation and may provide coupling nodes to couple the blade model to other bodies in the multibody model. Therefore, special attention is paid to correct determination of the measurement locations on the blade by utilizing a MicroScribe-3D coordinate measurement machine, see Fig. 27.2b. Following the investigation in [16] the blade is fixed on both ends to the table, to minimize blade deformation during the MicroScribe-3D measurements. Furthermore, two additional sets of

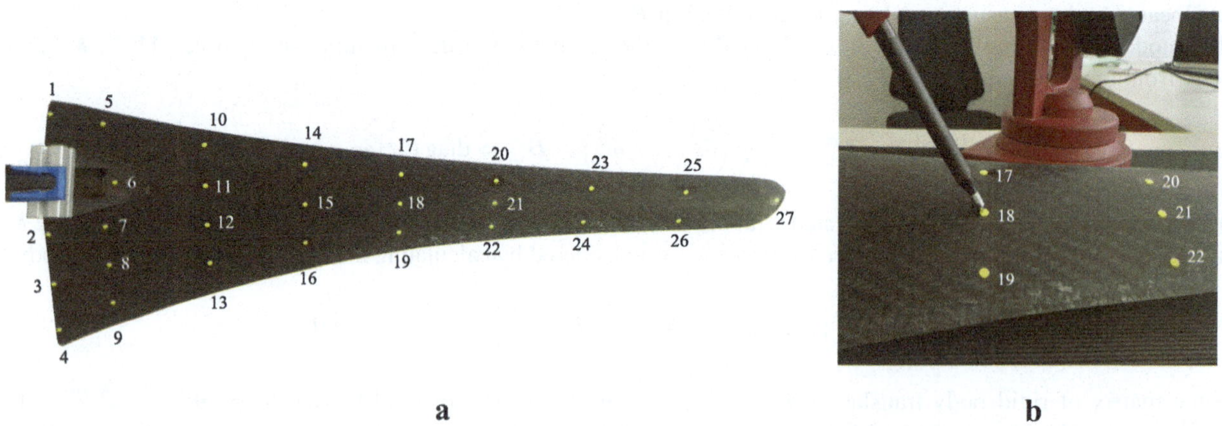

Fig. 27.2 Measurement locations on the turbine blade: (**a**) Overall view (**b**) MicroScribe-3D to determine measurement location

coordinates in the close vicinity of each measurement location were determined to approximate the surface normal vector at each measurement location.

27.3.1 Test Setup

Experimental synthesis requires mode shapes of the unconstrained structure as a prerequisite. Thus, the blade is suspended by a very flexible rubber band, which effectively decouples the near rigid body motion of the blade due to the suspension from the desired deformation modes, see Fig. 27.3a, b. Additionally, a driving point measurement has to be included in the measurement set to scale the mode shapes to unity modal mass.

The modal testing of the turbine blade is carried out by exciting the blade structure with an impact hammer equipped with PCB 208B02 force transducer. The blade's response is captured by a PCB 333B40 accelerometer (uni-axial, 7.5 g). As described in [2], the quality of the measurement results may suffer from a roving hammer measurement due to the high flexibility of the blade structure, and instead, the hammer excitation is kept stationary (at measurement location 6, see Figs. 27.2a and 27.3b) while the accelerometer is roved. Additionally, dummy masses equal to the accelerometer mass are attached to each measurement location to account for the mass deviation caused by the roving accelerometer, Fig. 27.3a, b. Naturally, the additional masses alter the dynamic behavior of the blade structure. Within the scope of this paper, however, the additional masses are treated as part of the blade structure since this is the cause of limitations given by the utilized measurement hardware and not of the experimental synthesis approach.

Figure 27.3 gives an overview of the measurement geometry. At every measurement point, a local coordinate system is defined and rotated so that the z-axis is parallel to the surface normal derived from the MicroScribe-3D measurements. The measurement direction of the uni-axial accelerometer is then set to be equal to the z-axis of the corresponding coordinate system to account for the complex shape of the blade structure. Finally, the frequency bandwidth of interest was set to 0–800 Hz with a frequency resolution 0.125 Hz. Overall $n = 12$ modes are identified in the frequency range. Table 27.1 lists the damped natural frequencies and damping ratios of each mode. Exemplary mode shapes are illustrated in Fig. 27.4. Mode 1 is characterized by a pure flapwise bending of the blade, and mode 3 is the first torsional mode. Similarly, Mode 4 is the third flapwise bending mode and mode 5 the second torsional mode. The higher modes are characterized by coupled torsional and bending motion.

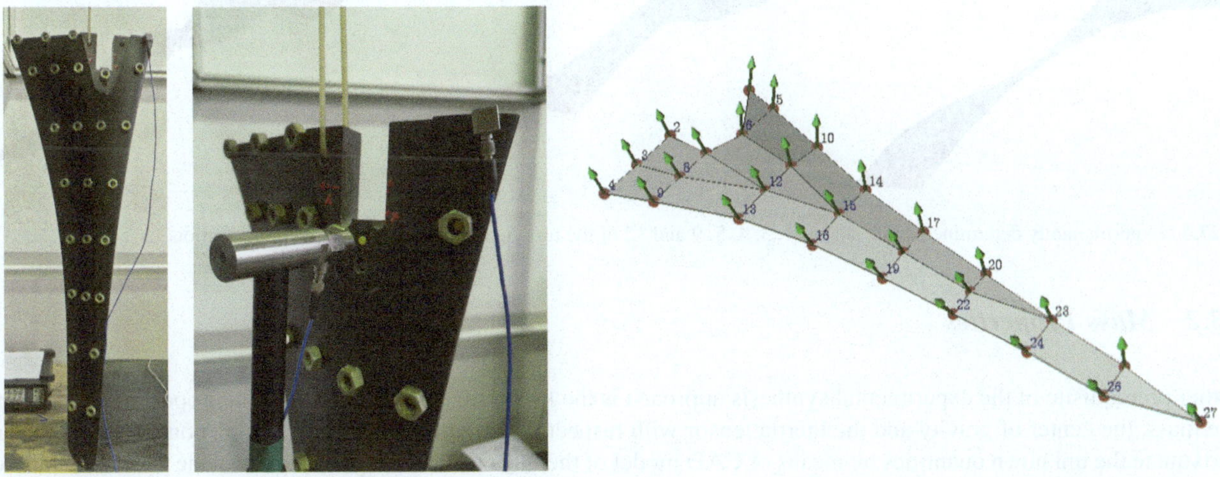

a b c

Fig. 27.3 Measurement setup of the turbine blade and added masses: **(a)** Overall view **(b)** Impulse hammer and roving sensor **(c)** Measurement geometry

Table 27.1 List of identified modes from the test setup

Mode nr. #	Damped frequencies [Hz]	Damping [%]
1	35.36	1.59
2	101.45	1.32
3	148.51	2.08
4	198.12	1.15
5	270.14	2.24
6	312.06	0.34
7	333.96	0.36
8	389.52	2.15
9	444.05	1.65
10	499.10	1.31
11	553.70	2.20
12	711.27	1.32

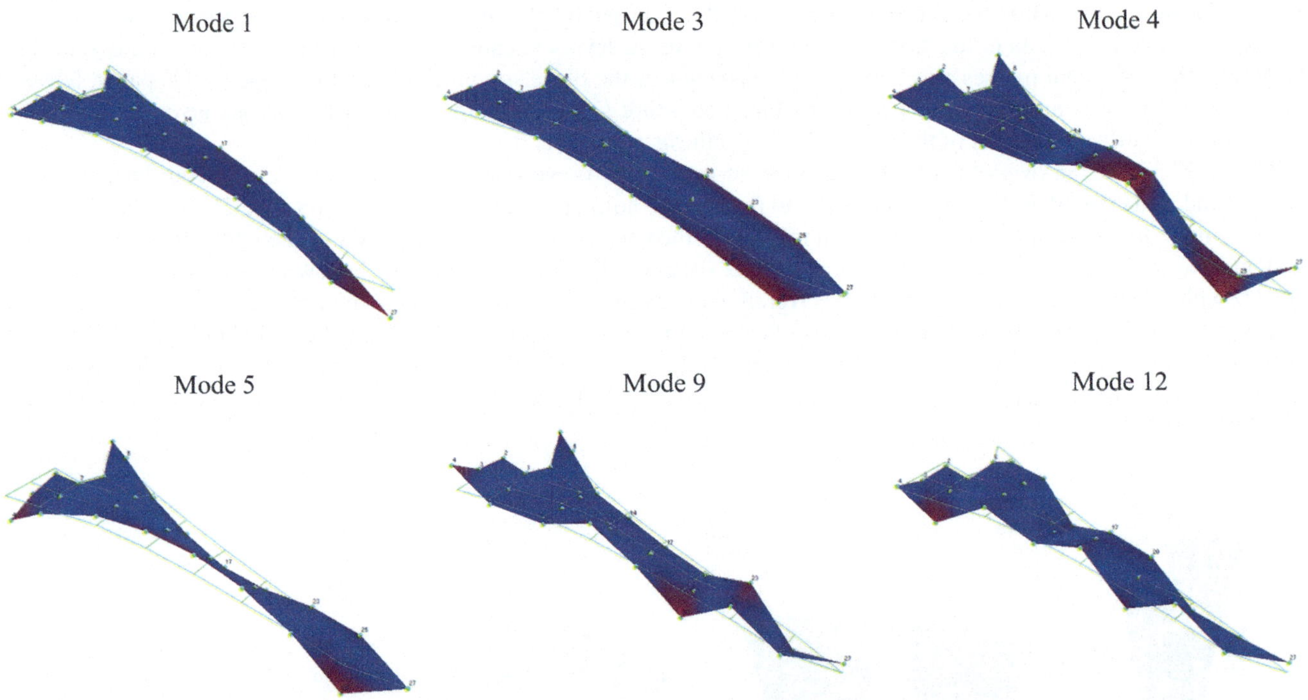

Fig. 27.4 Experimentally determined mode shapes 1, 3, 4, 5, 9 and 12 of the turbine blade equipped with compensation masses

27.3.2 *Mass Properties*

Another prerequisite of the experimental synthesis approach is the knowledge of the blade inertia properties, comprising the blade mass, the center of gravity and the inertia tensor with respect to the center of gravity. In this contribution, the authors approximate the unknown quantities by means of CAD model of the turbine blade, Fig. 27.5. The blade geometry data for the CAD model is kindly provided by the Substructuring Focus Group [4] and the efforts of M. Allen and The Atomic Weapons Establishment [19] which averaged the blade geometry over multiple measurements on different Ampair 600 turbine blades.

In a first step, the rotor blade is weighed, and the density of the CAD model is scaled to fit the mass m of the blade. Secondly, the dummy masses on the measurement positions are fitted to the CAD model. Under the assumption that the mass distribution in blade structure is homogeneous, the unknown location of the center of gravity and the inertia tensor can then be drawn from the model. This procedure is associated with unavoidable deviations between the inertia properties of the real blade structure and the CAD model. Within the scope of this contribution, these deviations are considered neglectable, and the authors refer, e.g., to [12, 20], for alternative, more powerful methods to estimate the unknown inertia properties from direct measurements.

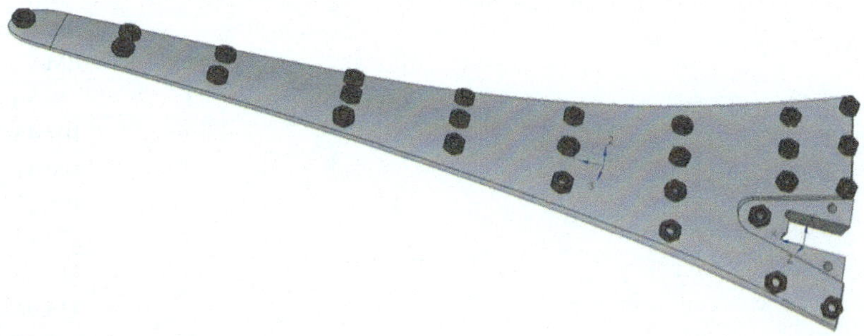

Fig. 27.5 CAD model of the measurement setup

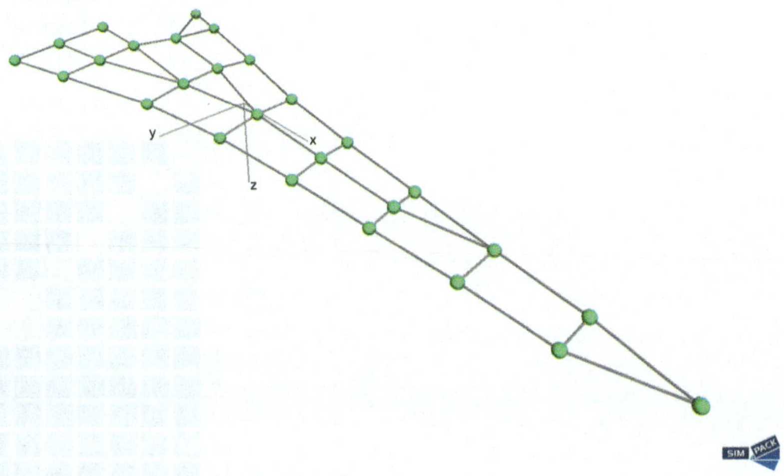

Fig. 27.6 Multibody model of the turbine blade

27.3.3 Multibody Model Build-Up

Based on the measurement results, an elastic multibody model of the turbine blade is build up by following the experimental synthesis framework. First, the reference frame $\mathcal{K}^1$ is positioned at the center of gravity of the body. Furthermore, the reference frame is rotated so that the axis of $\mathcal{K}^1$ coincide with the principal inertia axes of the blade. In the next step, the modal model, derived from EMA, is further processed to be integrated into the experimental synthesis framework. In general, experimentally derived mode shapes of a mechanical structure are complex. In experimental synthesis, however, these mode shapes are utilized to calculate a real matrix $\bar{\boldsymbol{M}}$ of the body's mass distribution, and orthogonalization of the complex mode shapes is necessary. It is important to note that this step implies the damping behavior of the blade to be of proportional type, see [12]. With the calculation of the rigid body modes in (27.8) and the pseudo mass matrix $\bar{\boldsymbol{M}}$ from (27.9), all necessary sub-matrices in equations (27.10) can be calculated to build up the equations of motion of the elastic multibody model.

A multibody model of the turbine blade is built up in the commercial multibody software SIMPACK using the Standard Input Data Format (SID) [11], Fig. 27.6. The green dots represent the $mp = 27$ measurement positions on the blade. These dots represent the elastic coordinates $\boldsymbol{q} \in \mathbb{R}^{6mp}$ of the blade. Constraining or coupling the blade to other bodies in a multibody system, as well as the application of forces to the blade, is only possible via these coordinates. Overall, the unconstrained model has 18 degrees of freedom (DOF), 6 DOF for the rigid body motion described by the reference frame $\mathcal{K}^1$ and 12 DOF corresponding to the 12 modal coordinates for the elastic deformation of the blade relative to $\mathcal{K}^1$.

A linear analysis is performed of the unconstrained multibody model as a first investigation of the dynamic behavior of the blade model. Table 27.2 lists the damped natural frequencies and the damping ratios of the multibody model and compares them to the results from experimental modal analysis. Experimental synthesis directly incorporates the experimental data into the multibody framework so that the results reveal no deviations in the natural frequencies and damping ratios between model and experiment. This also holds for the experimentally derived mode shapes of the blade and the mode shapes from multibody simulation. Figure 27.7 displays the MAC comparison [12] of the mode shapes from multibody simulation and

Table 27.2 Comparison of damped natural frequencies and damping ratios between multibody model (MBS) and modal testing (EMA)

	Damped freq. [Hz]		Damping [%]	
Mode no.	EMA	MBS	EMA	MBS
1	35.36	35.36	1.59	1.59
2	101.45	101.45	1.32	1.32
3	148.51	148.51	2.08	2.08
4	198.12	198.12	1.15	1.15
5	270.14	270.14	2.24	2.24
6	312.06	312.06	0.34	0.34
7	333.96	333.96	0.36	0.36
8	389.52	389.52	2.15	2.15
9	444.05	444.05	1.65	1.65
10	499.10	499.10	1.31	1.31
11	553.70	553.70	2.20	2.20
12	711.27	711.27	1.32	1.32

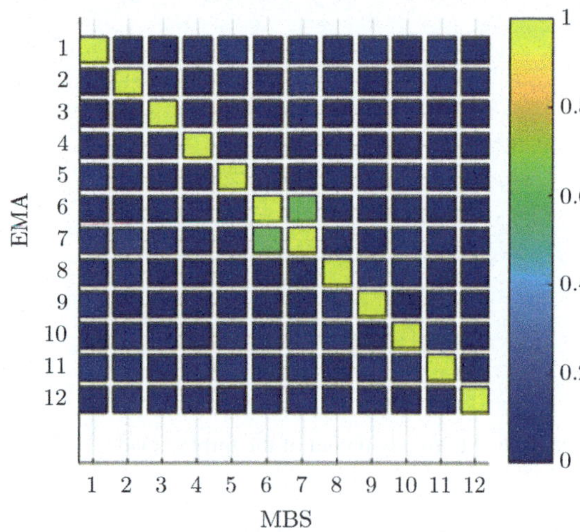

Fig. 27.7 MAC comparison between deformation mode shapes from multibody model (MBS) and modal testing (EMA)

modal testing. The diagonal MAC values are equal to 1 since the multibody model mode shapes are identical to measured mode shapes. Furthermore, Fig. 27.8 gives an overview of the mode shapes derived from multibody simulation.

The results from the multibody simulation reveal the substantial advantages of the experimental synthesis. The approach offers the possibility to built up a multibody model of a complex mechanical structure solely from experimental results without the need for an underlying numerical model, e.g., a finite element model. Due to the direct implementation of the measurement data into the built-up process, the modal characteristics of the model are identical to results from modal testing. This, however, also points out potential drawbacks of the method. Direct incorporation of measured data implies that every error made during the measurements is directly transferred to the multibody model. In the scope of this paper, these errors comprise, e.g., geometry errors due to deviations between the desired measurement points and the actual accelerometer positions, mass deviations due to the added dummy masses, errors in the inertia properties, measurement noise, and identification errors. Additionally, the damping behavior of the structure has to be assumed to be of proportional type. Experimental synthesis, thus, may put rigorous requirements concerning accuracy of the measurement hardware and execution to result in a multibody model of desired quality. Nevertheless, experimental synthesis is a compelling alternative to the conventional modeling approach, especially for complex mechanical structures where an underlying numerical model is prone to significant errors concerning mass, stiffness, and damping distribution.

27.4 Conclusion and Outlook

This contribution demonstrates the application of the experimental synthesis approach to an Ampair 600 turbine blade. The approach offers the possibility to derive an inertia-variant elastic multibody model of the blade entirely from measurement results. The prerequisites of experimental synthesis are a modal model of the unconstrained blade, the blades inertia

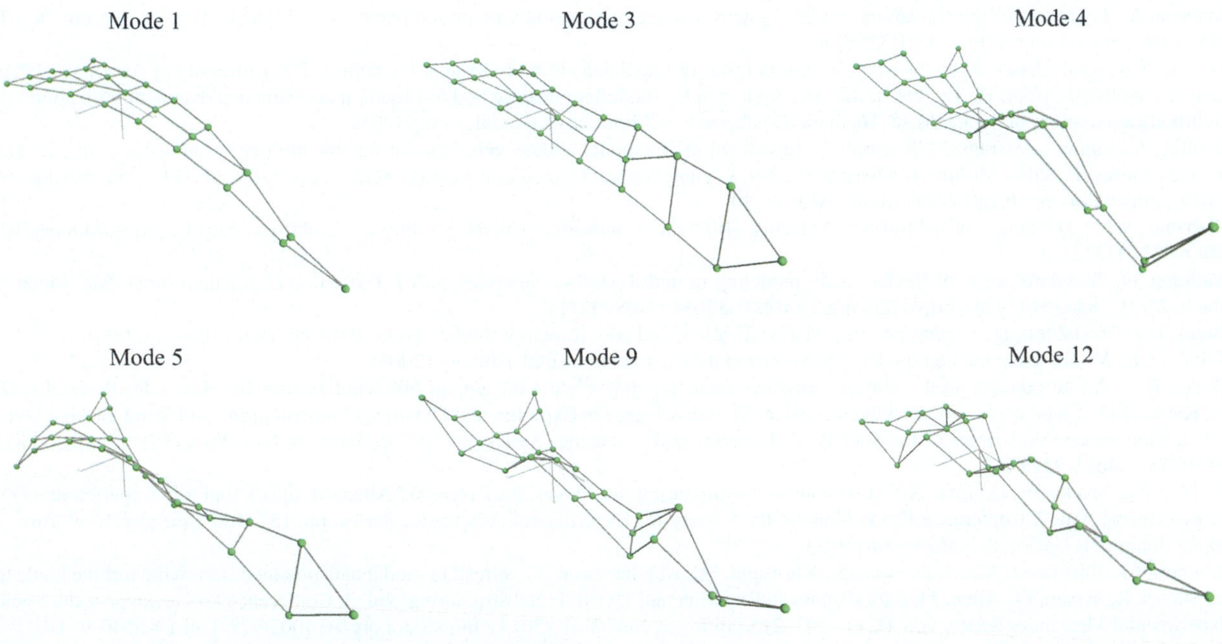

Fig. 27.8 Mode shapes 1, 3, 4, 5, 9 and 12 of the multibody model

properties, and the positioning of the reference frame in the center of mass. For the modal model, experimental modal analysis is carried out on the blade utilizing a total of 27 measurement locations. The inertia properties are estimated based on a CAD model of the blade. A linear analysis of the derived multibody model exhibits an exact representation of the measured mode shapes, natural frequencies, and damping ratios.

Future work will address the following topics:

- Although it was shown in [9] by the authors of this contribution that experimental synthesis results in a full set of system matrices to account for the strong nonlinear dynamics of the body caused by gross rigid body motion together with elastic deformations, transient analysis of the blades multibody model will be conducted to validate the approach further.
- So far, the investigations concentrated on the unconstrained multibody model. Future work will focus on the coupling of the blade model to other bodies in a multibody system. As pointed out in [8], the correct representation of rotation in the set of elastic coordinates has to be accounted for, which is not a result of conventional modal analysis.

Acknowledgments This research is part of the *DynAWind*[2] project funded by the German Federal Ministry for Economic Affairs and Energy under grant number 0325228E/F/G.

References

1. Klerk, D.D., Rixen, D.J., Voormeeren, S.N.: General framework for dynamic substructuring: History, review and classification of techniques. AIAA Journal **46**(5), 1169–1181 (2008). https://doi.org/10.2514/1.33274
2. Rahimi, S., de Klerk, D., Rixen, D.J.: The ampair 600 wind turbine benchmark: Results from the frequency based substructuring applied to the rotor assembly. In: Mayes, R., Rixen, D., Allen, M. (eds.) Topics in Experimental Dynamic Substructuring, vol. 2, Conference Proceedings of the Society for Experimental Mechanics Series, vol. 329, pp. 179–192. Springer, New York (2014). https://doi.org/10.1007/978-1-4614-6540-9_15
3. Macknelly, D., Nurbhai, M., Monk, N.: Imac XXXI: Additional modal testing of turbine blades and the application of transmission simulator substructuring methodology for coupling. In: Mayes, R., Rixen, D., Allen, M. (eds.) Topics in Experimental Dynamic Substructuring, vol. 2, Conference Proceedings of the Society for Experimental Mechanics Series, vol. 329, pp. 145–155. Springer, New York (2014). https://doi.org/10.1007/978-1-4614-6540-9_12
4. Dynamic substructuring wiki – dynamic substructuring focus group wiki (15 Mar 2018). http://substructure.engr.wisc.edu/substwiki/index.php/Dynamic_Substructuring_Wiki
5. Fehr, J., Eberhard, P.: Error-controlled model reduction in flexible multibody dynamics. J. Comput. Nonlinear Dyn. **5**(3), 031005 (2010). https://doi.org/10.1115/1.4001372

6. Shabana, A.: Dynamics of inertia-variant flexible systems using experimentally identified parameters. J. Mech. Transm. Autom. Des. **108**(3), 358 (1986). https://doi.org/10.1115/1.3258740
7. Yi, T.Y.: Structural Identification based on Vibration Data for Flexible Multibody System Dynamics. The University of Arizona (1996)
8. Lein, C., Woller, J., Hopf, H., Beitelschmidt, M.: Approach for modelling flexible bodies based on experimental data with utilization in elastic multibody simulation. In: ECCOMAS Thematic Conference on Multibody Dynamics Prag (2017)
9. Schulze, A., Luthe, J., Zierath, J., Woernle, C.: Investigation of a model update technique for flexible multibody simulation. In: Kecskeméthy, A., Geu Flores, F. (eds.) Multibody Dynamics 2019, Computational Methods in Applied Sciences, vol. 53, pp. 247–254. Springer Nature (2019). https://doi.org/10.1007/978-3-030-23132-3_30
10. Shabana, A.A.: Dynamics of Multibody Systems, 4th edn. Cambridge University Press, Cambridge (2014). https://doi.org/10.1017/CBO9781107337213
11. Wallrapp, O.: Standardization of flexible body modeling in multibody system codes, part I: Definition of standard input data. Mech. Struct. Mach. **22**(3), 283–304 (1994). https://doi.org/10.1080/08905459408905214
12. Ewins, D.J.: Modal testing: Theory, practice, and application. 2nd edn, Research Studies Press, Baldock, Hertfordshire (2000)
13. Avitabile, P.: Model reduction and model expansion and their applications part 1-theory (2004)
14. Mayes, R.L.: An introduction to the sem substructures focus group test bed – the ampair 600 wind turbine. In: Mayes, R., Rixen, D., Griffith, D., Klerk, D.D., Chauhan, S., Voormeeren, S., Allen, M. (eds.) Topics in Experimental Dynamics Substructuring and Wind Turbine Dynamics, vol. 2, Conference Proceedings of the Society for Experimental Mechanics Series, pp. 61–70. Springer, New York (2012). https://doi.org/10.1007/978-1-4614-2422-2_7
15. Nurbhai, M., Macknelly, D.: Imac XXXI: Dynamic substructuring. In: Mayes, R., Rixen, D., Allen, M. (eds.) Topics in Experimental Dynamic Substructuring, vol. 2, Conference Proceedings of the Society for Experimental Mechanics Series, pp. 157–166. Springer, New York (2014). https://doi.org/10.1007/978-1-4614-6540-9_13
16. Gibanica, M., Johansson, A.T., Rahrovani, S., Khorsand, M., Abrahamsson, T.: Spread in modal data obtained from wind turbine blade testing. In: Mayes, R., Rixen, D., Allen, M. (eds.) Topics in Experimental Dynamic Substructuring, vol. 2, Conference Proceedings of the Society for Experimental Mechanics Series, vol. 41, pp. 207–215. Springer, New York (2014). https://doi.org/10.1007/978-1-4614-6540-9_417
17. Gross, J., Oberhardt, T., Reuss, P., Gaul, L.: Model updating of the ampair wind turbine substructures. In: Allen, M., Mayes, R., Rixen, D. (eds.) Proceedings of the 32nd IMAC, a Conference and Exposition on Structural Dynamics, 2014, Conference Proceedings of the Society for Experimental Mechanics Series. Springer, Cham (2014)
18. Harvie, J., Avitabile, P.: Comparison of some wind turbine blade tests in various configurations. In: Mayes, R., Rixen, D., Griffith, D., Klerk, D.D., Chauhan, S., Voormeeren, S., Allen, M. (eds.) Topics in Experimental Dynamics Substructuring and Wind Turbine Dynamics, vol. 2, Conference Proceedings of the Society for Experimental Mechanics Series, pp. 73–79. Springer, New York (2012). https://doi.org/10.1007/978-1-4614-2422-2_9
19. Allen, M.: Scanned_averaged_blade_geometry.txt (2013). http://substructure.engr.wisc.edu/substwiki/index.php/Main_Page
20. Doniselli, C., Gobbi, M., Mastinu, G.: Measuring the inertia tensor of vehicles. Veh. Syst. Dyn. **37**, 301–313 (2002). https://doi.org/10.1080/00423114.2002.11666241

Andreas Schulze 2015 Master of Science in mechanical engineering, University of Rostock, Germany.
Since 2015 Scientific Assistant, Chair of Technical Dynamics, University of Rostock, Germany.
Research topics include wind turbine dynamics, flexible multibody dynamics.

Chapter 28
On the Use of PVDF Sensors for Experimental Modal Analysis

Tomaž Bregar, Blaž Starc, Gregor Čepon, and Miha Boltežar

Abstract Polyvinylidene fluoride (PVDF) piezoelectric thin film sensors are already well-known as low-cost vibrational sensors. One of the primary reasons, in addition to the overall cost, is also the PVDFs high chemical resistance, thermal stability and flexibility, which makes them a compelling alternative to piezoelectric ceramic. The PVDF sensors can also be used to measure strain response, similarly to the piezoelectric strain gauges. In this paper a performance evaluation of a PVDF sensor and a commercially available piezoelectric strain gauge is made for the mode shape identification on a simple-beam structure. The estimated mode shapes using the PVDF sensor are compared to the mode shapes estimated from a piezoelectric strain gauge. Mode shapes for both sensors are estimated using a roving hammer excitation. The two sensors are compared using a coherence criterion for the FRFs and the identified mode shapes are compared using a MAC criterion. Promising results can be observed with the PVDF sensor, which can be regarded as a replacement for the standard piezoelectric strain gauge for specific applications.

Keywords PVDF sensor · Piezoelectric strain sensor · Strain response · Experimental modal analysis · Mode shapes

28.1 Introduction

The experimental modal analysis (EMA) is usually performed using a standard piezoelectric accelerometer to obtain the natural frequencies and associated damping. A special approach in the field of EMA is based on the identification of the strain FRFs [1]. With stress EMA an experimental investigation of the stress-strain distribution can be obtained. This characteristic is especially useful in the field of vibration fatigue [2].

In this paper, an application of PVDF sensors on identification of modal parameters will be shown. The measured FRFs and identified parameters using the PVDF sensor will be compared to the well-established commercially available piezoelectric strain sensors.

28.2 Piezoelectric Polyvinylidene Fluoride (PVDF) Sensor

Polyvinylidene fluoride is a semi-crystalline polymer and is one of the most popular piezopolymer due to its large piezoelectric coefficient (20–28 pC/N) which is still relatively small compared to the most common piezoelectric ceramic. However, the overall low-cost, flexibility, wide frequency response, low acoustic resistance have made PVDF viable for multiple industrial applications [3, 4].

T. Bregar · M. Boltežar
Gorenje d.d., Velenje, Slovenia
e-mail: Tomaz.Bregar@gorenje.com; miha.boltezar@fs.uni-lj.si

B. Starc · G. Čepon (✉)
Faculty of Mechanical Engineering, University of Ljubljana, Ljubljana, Slovenia
e-mail: blaz.starc@fs.uni-lj.si; gregor.cepon@fs.uni-lj.si

B. Dilworth (ed.), *Topics in Modal Analysis & Testing, Volume 8*, Conference Proceedings of the Society for Experimental Mechanics Series, https://doi.org/10.1007/978-3-030-47717-2_28

28.3 Measurement Setup

Measurements were done on a simple beam with dimensions 12 × 40 × 300 mm in an approximate free-free boundary condition. A schematic representation of the experimental setup is depicted in Fig. 28.1. Altogether 24 different impact locations were used, depicted with small red circles.

The PVDF sensor used (LDT0–028K) has a 28 μm thick piezoelectric polymer film, laminated with 0.125 mm polyester substrate. As a reference a PCB 740B02 piezoelectric strain sensor was used. Both of the sensors were glued to the structure using LOCTITE 454. A close-up view of both sensor on the opposite side of the beam is shown in Fig. 28.2.

A Least-Squares Complex Frequency (LSCF) [5] method was used to identify the eigenvalues of the system and the Least-Squares Frequency-Domain (LSFD) [6] method was used to determine mode-shapes and lower and upper residuals for both sensors.

28.3.1 Results

Frequency response functions were measured by roving modal hammer with a vinyl tip at 24 impact locations. In Fig. 28.3a comparison of both sensors is shown. An increased levels of noise can be observed on the PVDF sensor especially in the low-frequency range where the 50 Hz harmonics are clearly seen. However, at the resonances the measurements of PVDF sensor are consistent with the PCB sensor.

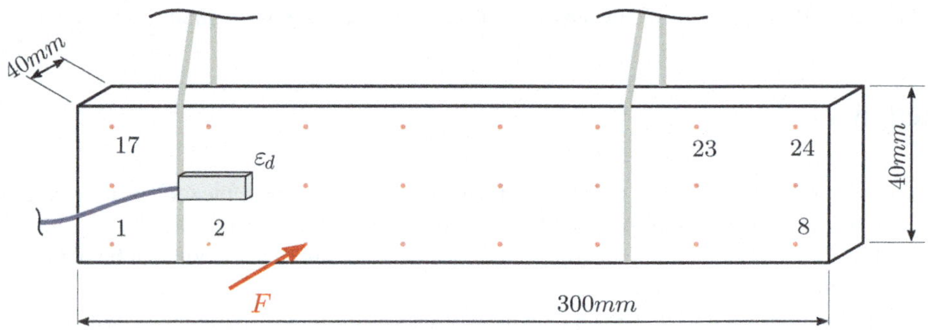

Fig. 28.1 Schematic representation of experimental setup with different impact locations depicted in red

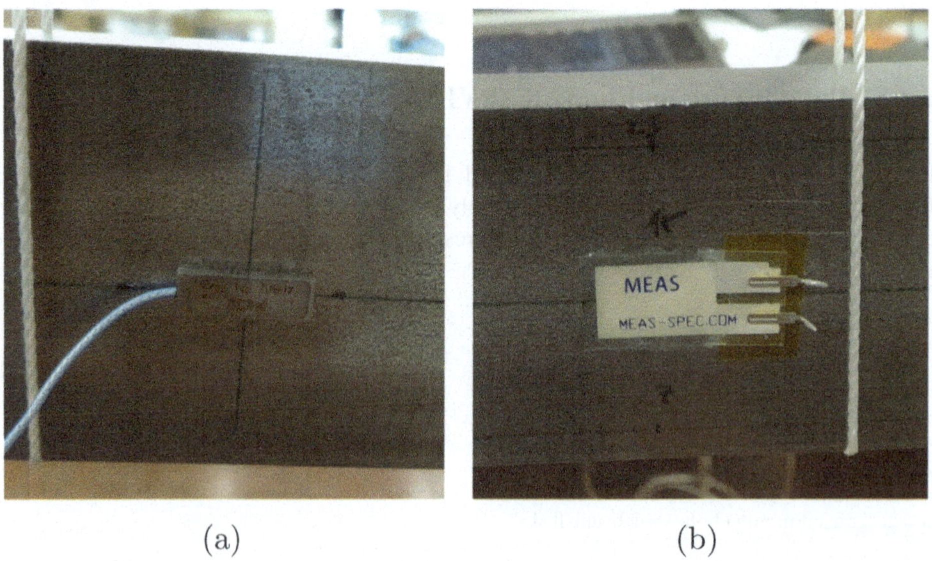

Fig. 28.2 Attachment of deformation sensors on the beam: (**a**) PCB 740B02; (**b**) PVDF LDT0-028K

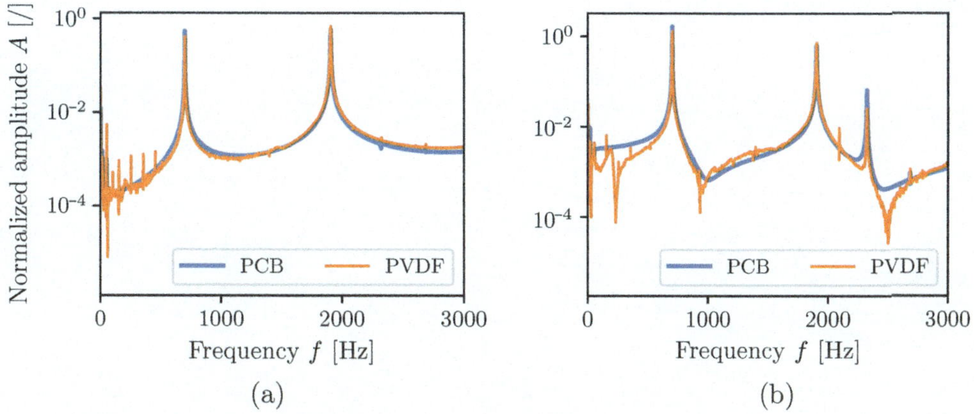

Fig. 28.3 A comparison of the FRFs between the PCB and PVDF sensors at different impact locations: (**a**) FRFs ε_{10}/F11; (**b**) FRFs ε_{10}/F24

Table 28.1 Comparison of identified modal parameters from both PVDF and PCB sensor up to 3 kHz

	PVDF		PCB	
Mode no.	Nat. freq. [Hz]	Damping [%]	Nat. freq. [Hz]	Damping [%]
1	696.2	0.03	696.6	0.02
2	1904.9	0.06	1904.9	0.06
3	2322.6	0.13	2321.4	0.04

The LSCF/LSFD method was used to identify modal parameters for both sensors. In Table 28.1 the identified modal parameters for first three mode shapes are shown. It can be observed that both sensors yield similar identification. Therefore, the PVDF sensor can be used as a low-cost alternative to high-performance piezoelectric strain sensor for modal analysis.

References

1. Yam, L., Leung, T., Li, D., Xue, K.: Theoretical and experimental study of modal strain analysis. J. Sound Vib. **191**(2), 251–260 (1996)
2. Mršnik, M., Slavič, J., Boltežar, M.: Frequency-domain methods for a vibration-fatigue-life estimation – application to real data. Int. J. Fatigue **47**, 8–17 (2013)
3. Hu, Y., Kang, W., Fang, Y., Xie, L., Qiu, L., Jin, T.: Piezoelectric poly(vinylidene fluoride) (PVDF) polymer-based sensor for wrist motion signal detection. Appl. Sci. **8**(5), 836 (2018)
4. Xin, Y., Sun, H., Tian, H., Guo, C., Li, X., Wang, S., Wang, C.: The use of polyvinylidene fluoride (PVDF) films as sensors for vibration measurement: A brief review. Ferroelectrics **502**(1), 28–42 (2016)
5. Guillaume, P., Verboven, P., Vanlanduit, S.: Frequency-domain maximum likelihood identification of modal parameters with confidence intervals. Proc. ISMA **23**, 16–18 (1998)
6. Auweraer, H., Leurs, W., Mas, P., Hermans, L.: Modal parameter estimation from inconsistent data sets, In: Proceeding of IMAC XVIII, pp. 763–771 (2000)

Blaz Starc is researcher at the Faculty of Mechanical Engineering, University of Ljubljana. His research focus is orientated towards model reduction, experimental techniques and numerical and experimental substructuring

Chapter 29
Predicting Tool Wear Using Linear Response Surface Methodology and Gaussian Process Regression

Chandula T. Wickramarachchi, Timothy J. Rogers, Wayne Leahy, and Elizabeth J. Cross

Abstract Predicting damage on cutting tools is extremely useful to machining operators, as tool wear is time consuming and expensive to directly measure. It is possible to use supervised methods to learn the state of damage on a tool by using features from outputs of the machining process that can be continually measured, such as acoustic emissions. Subsequently, operators are able to make vital decisions on whether to continue machining or to retire the tool. In this paper, the authors suggest a linear response surface methodology to predict tool wear using multiple linear features from acoustic emission signals collected during a turning operation. Here, the model fits a linear plane through the input features in order to make predictions about the output. The nonlinear case is also considered here where the authors apply Gaussian process regression to evaluate whether nonlinear features are able to predict tool wear more accurately than the linear model.

Keywords Tool wear · Linear response surface · Gaussian process regression · Regression · Supervised learning · PcBN

29.1 Introduction

Currently, there is an increased drive for automation in the manufacturing sector, embracing Industry 4.0. For a fully automated production facility, predicting wear of tools is key. Many authors have attempted to monitor and predict tool wear using supervised learning techniques such as neural networks [1–4], support vector machines [5–7], Bayesian methods [8] and, in particular, Gaussian processes [9–12] in the past. In general, although most of the literature has shown that theses models can capture trends in the training data well, they do not perform well in the testing phase where prediction of wear on previously unseen tools (i.e. tools that have not been used for training) takes place.

It is the aim of this work to predict wear on unseen tools from an outer diameter turning process by using a linear response surface (RSL) methodology and Gaussian process regression (GPR). These techniques are well suited for machining data as the datasets are usually small in size due to large costs associated with collecting damage labels (responses in the model). In the field of machining research, response surfaces are usually used in conjunction with design of experiments to quantify the effect of machining conditions and parameters on the outputs of the process, such as tool wear or surface roughness [13–16]. They have rarely been used to predict tool wear on unseen tools. In structural health monitoring research, however, RS models have been shown to obtain high accuracies when predicting variables such as the lowest modal frequency of a bridge when using operational and environmental conditions as inputs [17]. Equation 29.1 shows a generic linear response

C. T. Wickramarachchi (✉)
Dynamics Research Group, Department of Mechanical Engineering, University of Sheffield, Sheffield, UK
Industrial Doctorate Centre in Machining Science, Advanced Manufacturing Park, Catcliffe, Rotherham, UK
e-mail: ctwickramarachchi1@sheffield.ac.uk

T. J. Rogers · E. J. Cross
Dynamics Research Group, Department of Mechanical Engineering, University of Sheffield, Sheffield, UK
e-mail: tim.rogers@sheffield.ac.uk; e.j.cross@sheffield.ac.uk

W. Leahy
Element Six Global Innovation Centre, Harwell Campus, Didcot, UK
e-mail: wayne.leahy@e6.com

B. Dilworth (ed.), *Topics in Modal Analysis & Testing, Volume 8*, Conference Proceedings of the Society for Experimental Mechanics Series, https://doi.org/10.1007/978-3-030-47717-2_29

surface, with β_i the parameters of the plane that fits through the input features x_i for a given response y. When considering features which are linearly correlated, the simplicity of this regression model allows for fast computation of predicted values and lessens the chance of overfitting that may occur with a more flexible model.

$$y = \beta_0 + \beta_1 x_1 + \beta_2 x_2 + \ldots + \beta_n x_n \tag{29.1}$$

Where nonlinear interactions occur, this simple form may be adapted and extended by considering polynomial terms and cross products of the input features. However, in situations where there exists many challenges related to feature collection, preprocessing and selection, it may be difficult to ascertain the most suitable model form to represent the underlying trend in the data. On occasions such as these, GPR proves invaluable as the functional form of the model is not specified a-priori, instead a family of functions is considered through selection of a covariance function. When evaluating GPR models, Eq. 29.2 is used. This is the standard result when the joint distribution under the prior is conditioned on the data. Equation 29.2 shows target distribution for the mean function (top part) which gives an indication of the prediction and the covariance function (bottom part) which permits the calculation of the model confidence. For an in depth explanation of GPR, please refer to [18].

Here, K is the covariance matrix, $\boldsymbol{X}$ is the training input matrix (design matrix), testing data is $\boldsymbol{X}^*$, σ_n^2 is the variance from the noise, $\boldsymbol{y}$ is the training response data and $\boldsymbol{y}^*$ is the prediction.

$$\begin{aligned} \boldsymbol{y}^*|\boldsymbol{X}^*, \boldsymbol{X}, \boldsymbol{y} \sim \mathcal{N}\big(&K(\boldsymbol{X}^*, \boldsymbol{X})[K(\boldsymbol{X}, \boldsymbol{X}) + \sigma_n^2 I]^{-1}\boldsymbol{y}, \\ &K(\boldsymbol{X}^*, \boldsymbol{X}^*) - K(\boldsymbol{X}^*, \boldsymbol{X})[K(\boldsymbol{X}, \boldsymbol{X}) + \sigma_n^2 I]^{-1}K(\boldsymbol{X}, \boldsymbol{X}^*)\big) \end{aligned} \tag{29.2}$$

29.2 Experimental Set-Up

The experimental set up for this work used six Polycrystalline cubic Boron Nitride (PcBN) inserts manufactured with 50% cBN content. The workpieces used were case hardened steel bars. In one pass of the turning operation, the tool travels across the outer diameter the workpiece. Piezoelectric acoustic emissions (AE) transducers (Mistras Micro 30D) were used with 1MHz sampling rate. Figure 29.1 displays the machine and sensor set-up. Wear measurements were taken after every 4 passes. Wear images were taken using a 3D surface scanner. Standard cutting conditions were used.

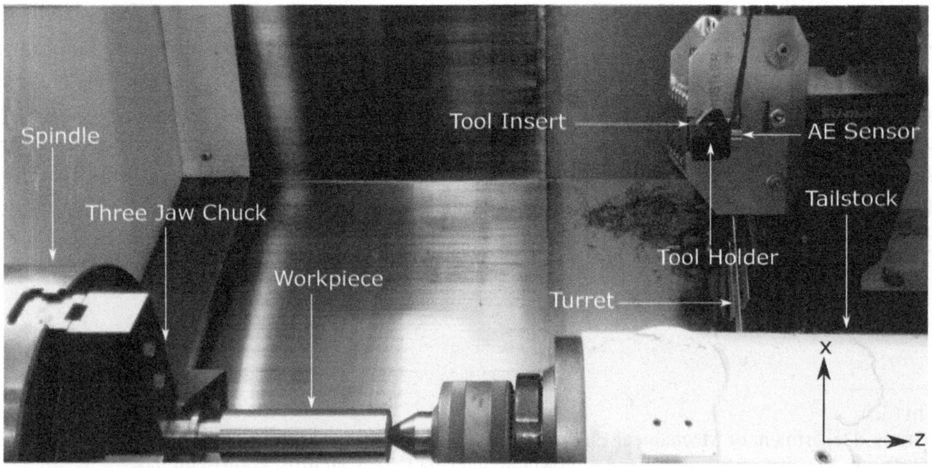

Fig. 29.1 The set up of the machine and AE sensor

29.3 Results

Data from three tools were used for training the algorithm, with the remaining three used for testing. As a pre-processing step, a discrete wavelet transform was applied to the AE signal to isolate the frequencies at which the chips were formed (62.5–125 kHz) and in another band representing transient behaviour of the machining process (250–500 kHz). Candidate input features for regression are statistical moments of the reconstructed signal in each of these frequency bands.

By studying the linear correlations between the statistical moments from the AE (inputs) and flank wear area (response), a (down-selected) feature set for the RSL was found. To explore the nonlinear correlations, mutual information between the AE and tool wear features was also calculated to aid with input selection for GPR. In both cases, the features with strongest relationships were used in the models and can be found in Table 29.1. Here the number of features that can be used in each model is limited by the number of observations in the data.

The model predictions for both RSL and GPR can be seen visually in Fig. 29.2 where the GPR confidence intervals have been included. In both cases, the model has generalised well from training to testing. The normalised mean square error (nMSE) is used as a measure to evaluate model performance. If instead of a prediction model, only the mean of the data was used to make predictions, the resulting nMSE would be 100%. Therefore, nMSE values below 100% signify the existence of correlation [17]. For the models here, the nMSEs were found to be $RSL_{train} = 22.19\%$, $RSL_{test} = 28.64\%$, $GPR_{train} = 14.3\%$, $GPR_{test} = 27.7\%$.

From these percentages and studying Fig. 29.2, one can see that both the linear response surface and GPR have been able to predict the tool wear behaviour reasonably well on the unseen data. The GPR has a much lower training nMSE compared with RSL likely due to increased flexibility (there are more input features in the model and the model is not bound by linearity), this has not been translated into an improved predictive capability for the unseen tool data.

Table 29.1 Features with highest MI chosen as inputs to the GPR model

RSL	GPR
σ of chip formation frequencies	σ of chip formation frequencies
σ of all frequencies	σ of all frequencies
σ of high frequencies	σ of high frequencies
–	RMS of high frequencies
–	Kurtosis of high frequencies

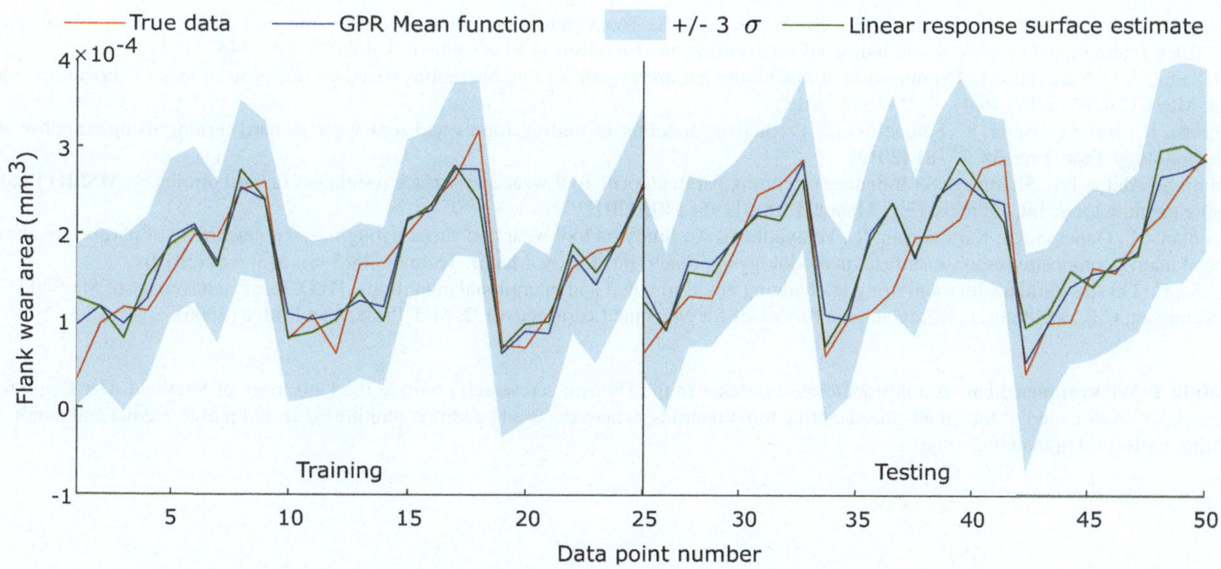

Fig. 29.2 RSL and GPR results predicting flank wear area using the AE features from Table 29.1

29.4 Conclusions

In order to predict tool wear on PcBN tools, this work applied a RSL methodology to model linear relationships and Gaussian process regression for nonlinear relationships between AE features and tool wear. With low training and testing nMSE scores, it was found that both of these models performed well and can generalise well between training and testing. In conclusion, these methods may be suitable for predicting tool wear in real time.

Acknowledgments Element Six Ltd. funded this work and also provided the machine, workpieces and tools used in the experiment. This work was also part funded by the EPSRC grant (EP/I01800X/1). The AMRC also extended help and guidance throughout the experiment.

References

1. Das, S., Roy, R., Chattopadhyay, A.B.: Evaluation of wear of turning carbide inserts using neural networks. Int. J. Mach. Tools Manuf. **36**, 789–797 (1996)
2. Das, S., Bandyopadhyay, P.P., Chattopadhyay, A.B.: Neural-networks-based tool wear monitoring in turning medium carbon steel using a coated carbide tool. J. Mater. Process. Technol. **63**(1–3), 187–192 (1997)
3. Dimla, D.E., Lister, P.M.: On-line metal cutting tool condition monitoring. II: tool-state classification using multi-layer perceptron neural networks. Int. J. Mach. Tools Manuf. **40**(5), 769–781 (2000)
4. Sun, J., Rahman, M., Wong, Y.S., Hong, G.S.: Multiclassification of tool wear with support vector machine by manufacturing loss consideration. Int. J. Mach. Tools Manuf. **44**(11), 1179–1187 (2004)
5. Binsaeid, S., Asfour, S., Cho, S., Onar, A.: Machine ensemble approach for simultaneous detection of transient and gradual abnormalities in end milling using multisensor fusion. J. Mater. Process. Technol. **209**, 4728–4738 (2009)
6. Benkedjouh, T., Medjaher, K., Zerhouni, N., Rechak, S.: Health assessment and life prediction of cutting tools based on support vector regression. J. Intell. Manuf. **26**(2), 213–223 (2015)
7. Dutta, S., Pal, S.K., Sen, R.: On-machine tool prediction of flank wear from machined surface images using texture analyses and support vector regression. Precis. Eng. **43**, 34–42 (2016)
8. Karandikar, J., McLeay, T., Turner, S., Schmitz, T.: Tool wear monitoring using naïve Bayes classifiers. Int. J. Adv. Manuf. Technol. **77**(9–12) 1613–1626 (2015)
9. Wang, G., Qian, L., Guo, Z.: Continuous tool wear prediction based on Gaussian mixture regression model. Int. J. Adv. Manuf. Technol. **66**(9–12), 1921–1929 (2013)
10. Kong, D., Chen, Y., Li, N.: Gaussian process regression for tool wear prediction. Mech. Syst. Signal Process. **104**, 556–574 (2018)
11. Wang, M., Wang, J.: CHMM for tool condition monitoring and remaining useful life prediction. Int. J. Adv. Manuf. Technol. **59**(5–8), 463–471 (2012)
12. Hong, J., Zhou, J.H., Chan, H.L., Zhang, C., Xu, H., Hong, G.S.: Tool condition monitoring in deep hole gun drilling: a data-driven approach. In: IEEE International Conference on Industrial Engineering and Engineering Management, vol. 2017, pp. 2148–2152 (2018)
13. Makadia, A.J., Nanavati, J.I.: Optimisation of machining parameters for turning operations based on response surface methodology. Meas. J. Int. Meas. Confederation **46**(4), 1521–1529 (2013)
14. Suresh, R., Basavarajappa, S., Samuel, G.L.: Predictive modeling of cutting forces and tool wear in hard turning using response surface methodology. Proc. Eng. **38**, 73–81 (2012)
15. Saini, S., Ahuja, I.S., Sharma, V.S.: Influence of cutting parameters on tool wear and surface roughness in hard turning of AISI H11 tool steel using ceramic tools. Int. J. Precis. Eng. Manuf. **13**(8), 1295–1302 (2012)
16. Seeman, M., Ganesan, G., Karthikeyan, R., Velayudham, A.: Study on tool wear and surface roughness in machining of particulate aluminum metal matrix composite-response surface methodology approach. Int. J. Adv. Manuf. Technol. **48**(5–8), 613–624 (2010)
17. Cross, E.: On structural health monitoring in changing environmental and operational conditions. Ph.D. thesis, University of Sheffield (2012)
18. Rasmussen, C.E., Williams, C.K.I.: Gaussian Processes for Machine Learning, vol. 2. MIT Press, Cambridge (2006)

Chandula T. Wickramarachchi is a postdoctoral researcher in the Dynamics research group at the University of Sheffield. Her EngD focused on the topic of 'Automated testing of advanced cutting tool materials' where she used condition monitoring techniques to predict and monitor wear on cutting tools used in manufacturing.

Chapter 30
Computer Aided Measurement Uncertainty Calculation by Modern DAQs for Raw Acceleration and Force Data in Modal Analysis

David Kuntz, Thomas Petzsche, Martin Stierli, and William Zwolinski

Abstract Reliable modal analysis testing results are closely related to precise acceleration and force raw data for magnitude and phase. For this reason, measurement uncertainty calculations are necessary based on a routine described in the ISO guideline "Expression of the Uncertainty of Measurement in Calibration" (ISO/BIPM ,Guide to the Expression of Uncertainty in Measurement" 2008; Technische Regel ISO/IEC Guide 98-3:2008-09 Messunsicherheit – Teil 3: Leitfaden zur Angabe der Unsicherheit beim Messen; Guide of the Joint Committee for Guides in Metrology: JCGM 100:2008 Evaluation of measurement data — Guide to the expression of uncertainty in measurement).

An uncertainty of measurement is not an exception. Every measurement result in research and industry is associated with a certain degree of inaccuracy. Even small fluctuations in the ambient temperature can lead to deviations that make the result unreliable, and unusable. Whenever decisions are based on a measurement result, it is important to have an indication of the quality of the data being used. Thus, the knowledge of the respective measurement uncertainty is indispensable in achieving meaningful results.

The more components the measurement chain consists of, the more complex the determination of uncertainty becomes, as every potential influencing factor needs to be taken into account. Up until recently, it has been necessary to review the respective data sheets for each possible source of uncertainty, check individual specifications, and factor in additional external influences, and include them in the calculation. This procedure requires a considerable effort. To keep the measurement processes economical, and manageable, in many cases generous assumptions are made, or the calculation of the uncertainty is even omitted completely. On the other hand, if an excessive allowance is made for the uncertainty, this has a negative influence on the cost-efficiency of the entire process. When measurement uncertainty can be reliably determined, tolerance limits can be defined more precisely, and the process is more efficient as a result.

Research results are only valid if they are based on precise, and above all, reliable measurement results. In this respect, Computer Aided Measurement Uncertainty Calculations by Modern DAQs can ease the burden on the research budget, and provide accurate data in a timely manner.

Keywords Accelerometer · Force sensor · Computed aided measurement · Uncertainty calculation · Modal analysis

---

D. Kuntz (✉)
Kistler Instrument Corp., Amherst, NY, USA
e-mail: david.kuntz@kistler.com

T. Petzsche
App.-Exp. Acceleration, Thermoakustik, Kistler Instrumente GmbH, Sindelfingen, Germany
e-mail: thomas.petzsche@kistler.com

M. Stierli
Team Lead DAQ Product Management, Kistler Instrumente AG, Winterthur, Switzerland
e-mail: martin.stierli@kistler.com

W. Zwolinski
T&M Manager, Kistler Instrument Corp., Novi, MI, USA
e-mail: bill.zwolinski@kistler.com

B. Dilworth (ed.), *Topics in Modal Analysis & Testing, Volume 8*, Conference Proceedings of the Society for Experimental Mechanics Series, https://doi.org/10.1007/978-3-030-47717-2_30

30.1 Consideration of Measurement Deviations in the Overall Result

Measurements are generally associated with imperfections which cause a measurement deviation in the measurement result. Traditionally, a measurement deviation has two components: a random component and a systematic component.

A random measurement deviation can occur due to unpredictable or random temporal and spatial changes of influencing variables. The influences of such changes, referred to as random influences, can cause variations when the measurand is observed several times.

Similar to a random deviation, a systematic deviation also cannot be eliminated but can often be reduced. A systematic deviation stems from the known effect of an influencing variable on the measurement result and is subsequently referred to as a systematic influence. If this influence can be quantified and if it is significant with respect to the required accuracy of the measurement, a correction or a correction factor can compensate for this influence.

The calculation of the uncertainty of measurement for a specific measurement task today is usually carried out by drawing up a budget for uncertainty of measurement according to the ISO Guide "Guide of the Expression of the Uncertainty of Measurement" (also known as GUM) [1]. It has been implemented in national guidelines so that a strong degree of standardization can be observed internationally. According to expediency, either method A (Calculation of the measurement uncertainty by statistical analysis of the measurements) or method B (Calculation of uncertainty of measurement by means other than statistical analysis) is used.

With method B it is assumed for the calculation of measurement uncertainties that the result of a measurement has been corrected with respect to all identified significant systematic influences and that efforts have been made to identify such influences. With the extensive elimination of systematic influences on the measurement uncertainty budget, the remaining influences are considered random. The mathematical formula for the calculation of the combined standard uncertainty follows in principle the Gaussian error propagation law, whereby correlations between the influencing variables can occur. However, this can often be excluded.

The uncertainty of a measurement reflects the insufficient knowledge of the value of the measurand. The measuring result is still an estimate of the measurand after the mentioned correction of the recognized systematic influences. These result from the uncertainty of coincidental influences and the imperfect correction of the result of the systematic influences.

Due to the traceable calibration to national standards of each component in the measurement system used, the influences of the metrological components on the measurement uncertainty of the overall result can be quantified. In addition, procedural measurement deviations may occur. With both influences, a measurement uncertainty can usually be specified for the overall result.

Modern measurement systems, such as the KiDAQ system from Kistler Instruments enable the computer-aided inclusion of measurement uncertainties that affect components known to the manufacturer, typically the sensor, the cable, and the amplifier or data acquisition stage.

This is done under the functional name "KiXact" by accessing a growing database with measurement uncertainties of various known system components. These values are on one hand based on the results of systematic calibration measurement series, but also on influence models elaborated over many years. For example, it is possible to minimize the measurement uncertainty for the specific measurement task by specifying narrower parameters (temperature, measuring range, etc.).

The user then only has to consider the influence of the measurement setup on the measurement uncertainty for the overall result.

As a representative of the measurement variables commonly used in structural analysis, a measurement uncertainty budget for acceleration measurement by magnitude and phase is to be a derived and automated process demonstrated by using KiXact and the KiDAQ measurement system from Kistler Instruments. The proof of the smallest measurement uncertainty is theoretically provided by the calculated measurement uncertainty budget and verified practically by suitable follow-up measurements.

30.2 Model for a Measurement Uncertainty Budget for Vibration Transducers

For acceleration measurements, measuring chains consisting of a sensor, a cable, a measuring amplifier and an A/D conversion are usually used. The acceleration is determined as a function of time at a defined point in the test structure. The measurement result is influenced by numerous factors, such as the coupling of the sensor (roughness, alignment, stiffness of the coupling, capacitance of the cable, measurement uncertainty of the gain and A/D conversion, etc.) (Fig. 30.1). The

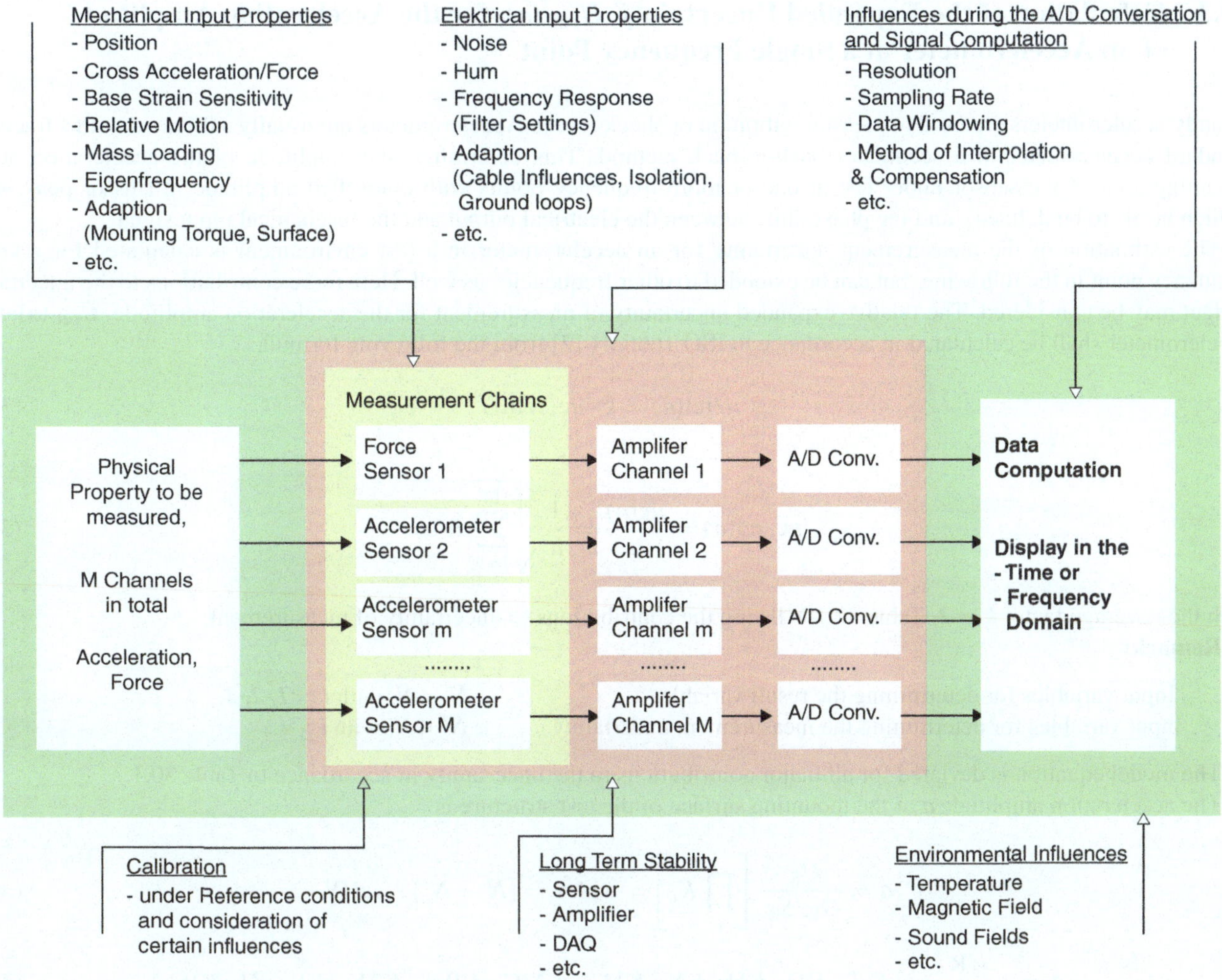

Fig. 30.1 Influences on the measurement uncertainty budget to an acceleration and force measurement chain

determinants and influencing variables considered in the following calculation are selected as examples. In general, the relationship between the result variable $\boldsymbol{Y}$ and the input variables is determined by the determination equation:

$$Y = f\left(X_1, X_2, \ldots X_N\right) \tag{30.1}$$

The estimated value $\boldsymbol{y}$ of the result variable $\boldsymbol{Y}$ is generally calculated by using the estimated values $\boldsymbol{x}_i$ of the input variables $\boldsymbol{X}_i$:

$$y = f\left(x_1, x_2, \ldots x_N\right) \tag{30.2}$$

It is assumed that the estimates are "best estimates". Systematic measurement deviations that manifest themselves in asymmetrical distribution functions have already been taken into account by corrections. For the measured variable acceleration, the estimated value $\boldsymbol{y}$ of the result variable is obtained by inserting $\boldsymbol{N'}<\boldsymbol{N}$ input variables. When determining the measurement uncertainty however, all $\boldsymbol{N}$ input quantities $\boldsymbol{X}_i$ are taken into account.

30.3 Calculation of the Expanded Uncertainty, U_{rel} (*a*), for the Acceleration Amplitude of an Accelerometer at a Single Frequency Point

Usually accelerometers are used to measure vibration or shock levels. Accelerometers are usually calibrated with a traceable standard accelerometer in a secondary back-to-back method. This determines the amplitude of the acceleration at the mounting area of the sensor under test at one or more frequency points with controlled amplitude in rms or peak value (which needs to be defined), and the phase shift between the electrical output and the mechanical input signal.

The estimation of the measurement uncertainty for an accelerometer in a test environment is calculated for a single frequency point in the following, but can be extended to other frequencies as well. Here more contributions to the uncertainty budget may be considered. The relative expanded uncertainty of measurement for the acceleration amplitude, $\boldsymbol{U_{rel}}$ **(*a*)** of an accelerometer shall be calculated in accordance to ISO 16063-1 [2] from the following formula:

$$U\text{rel}(a) = k \cdot uc, \text{rel}(a) \tag{30.3}$$

$$uc, \text{rel}(a) = \frac{uc(a)}{a} = \frac{1}{a}\sqrt{\sum_{\text{i}=1}^{14} ui(a)} \tag{30.4}$$

with the coverage factor $\boldsymbol{k} = \boldsymbol{2}$. Table 30.1 is listing the contributions to uncertainty of measurement.

Remark:

N' Input variables for determining the result variable. $\boldsymbol{N' < N}$, with $\boldsymbol{i = 1, 2, 3}$.

N Input variables for determining the measurement uncertainty. $\boldsymbol{N' < N}$, with $\boldsymbol{i \geq 4}$.

The model equation is deviated for all major contributions to the uncertainty in accordance to Table 30.1.

The acceleration amplitude $\boldsymbol{a}$ at the mounting surface of the test structure is

$$a = \frac{U_R}{G_R \cdot S_R} \cdot \left[\prod_i K_i\right] = \text{with } \boldsymbol{i} = (N - N'), \ldots, N \tag{30.5}$$

$$a = \frac{UR}{GR \cdot SR} \cdot [K\,\text{T} \cdot K\text{D} \cdot K\text{H} \cdot K\text{N} \cdot K\text{MT} \cdot K\text{MC} \cdot K\text{Rel} \cdot K\text{TK} \cdot K\text{L} \cdot K\text{I} \cdot K\text{Res}] \tag{30.6}$$

Table 30.1 List of the contributions to uncertainty of measurement for acceleration amplitude

N_i	Property X_i	Best estimate x_i	Standard uncertainty component $u(x_i)$	Source of uncertainty $w_i(y)$	Uncertainty contribution
1	U_R	U_R	$u(U_R)$	Voltage at the output of the measurement chain, consisting of transfer standard, amplifier, voltmeter	$u1(UR)$
2	S_R	S_R	$u\,(S_R)$	Sensitivity of the accelerometer in use	$u2(SR)$
3	G_R	G_R	$u\,(G_R)$	Gain of the amplifier	$u3(GR)$
4	K_T	1	$u\,(K_T)$	Effect of cross motion	$u4(KT)$
5	K_D	1	$u\,(K_D)$	Effect of higher harmonics	$u5(KD)$
6	K_H	1	$u\,(K_H)$	Effect of hum	$u6(KH)$
7	K_N	1	$u\,(K_N)$	Effect of noise	$u7(KN)$
8	K_{MT}	1	$u\,(K_{MT})$	Effect of sensor mounting	$u8(KMT)$
9	K_{MC}	1	$u\,(K_{MC})$	Effect of the cable	$u9(KMC)$
10	K_{Rel}	1	$u\,(K_{REL})$	Effect of relative motion	$u10(KRel)$
11	K_{TK}	1	$u\,(K_{TK})$	Effect of thermal sensitivity drift of the transfer standard	$u11(KTK)$
12	K_L	1	$u\,(K_L)$	Effect of non-linearity	$u12(KL)$
13	K_I	1	$u(K_I)$	Effect of long term stability of the transfer standard sensitivity	$u13(KI)$
14	K_{Res}	1	$u(K_{Res})$	Effect of residual influences (e.g. magnetic flux)	$u14(KRes)$

It is assumed that the input values are the best estimates and the suitible corrections has been made, if necessary. There are no relevant correlations of the input properties thus far.

The standard measurement uncertainty $u_c(y)$ for the best estimate of the result variable y can be determined to belong to ([1, 2]) an approximate Taylor series of Equ. 30.1 in general and Equ. 30.6 in detail interrupting after the linear term to:

$$uc(y) = \sqrt{\sum_{i=1}^{N} \left(\frac{\partial f}{\partial xi}\right)^2 \cdot u^2\,(xi) + 2\sum_{i=1}^{N-1} \sum_{J=i+1}^{N} \frac{\partial f}{\partial xi}\frac{\partial f}{\partial xj} \cdot u\,(xi, xj)} \tag{30.7}$$

As we do not have significant correlations between the influences all mixed terms can be neglected and Equ. 30.6 follows to:

$$uc(y) = \sqrt{\sum_{i=1}^{N} \left(\frac{\partial f}{\partial xi}\right)^2 \cdot u^2\,(xi)} \tag{30.8}$$

The following measurement uncertainty budget is based on the calculation of relative uncertainties. Due to the selected linear product model and the consideration of relative measurement uncertainties, the sensitivity coefficients are always equal to 1 ($c_i = 1$), i.e. they apply:

$$w^2(a) = \frac{u\,(a)^2}{a^2} = \sum_{i=1}^{N} \left(\frac{u_i * c_i}{|\,x_i|}\right)^2 \tag{30.9}$$

30.4 Determination of the Measurement Uncertainty

As an example the determination of the amplitude of an acceleration measurement chain is defined as it is frequently used in practice. The measurement uncertainty is determined using the following concrete components as examples. For the determination of the relative measurement uncertainty, the following contributions listed in the Table 30.2 shall be taken into account.

Table 30.2 General measurement uncertainty budget for a single frequency point

No.	Measurand	Best estimate	Uncertainty	Sensitivity-coefficient	Uncertainty contribution	Variance
	X_i	x_i	$w(x_i)$	c_i	$w_i(y)$	$w_i^2(y)$
1	U_R	U_R	$w(U_R)$	1	$w1(UR)$	$w^2 1(UR)$
2	S_R	S_R	$w(S_R)$	1	$w2(SR)$	$w^2 2(SR)$
3	G_R	G_R	$w(G_R)$	1	$w3(GR)$	$w^2 3(GR)$
4	K_T	1	$w(K_T)$	1	$w4(KT)$	$w^2 4(KT)$
5	K_D	1	$w(K_D)$	1	$w5(KD)$	$w^2 5(KD)$
6	K_H	1	$w(K_H)$	1	$w6(KH)$	$w^2 6(KH)$
7	K_N	1	$w(K_N)$	1	$w7(KN)$	$w^2 7(KN)$
8	K_{MT}	1	$w(K_{MT})$	1	$w8(KMT)$	$w^2 8(KMT)$
9	K_{MC}	1	$w(K_{MC})$	1	$w9(KMC)$	$w^2 9(KMC)$
10	K_{Rel}	1	$w(K_{REL})$	1	$w10(KRel)$	$w^2 10(KRel)$
11	K_{TK}	1	$w(K_{TK})$	1	$w11(KTK)$	$w^2 11(KTK)$
12	K_L	1	$w(K_L)$	1	$w12(KL)$	$w^2 12(KL)$
13	K_I	1	$w(K_I)$	1	$w13(KI)$	$w^2 13(KI)$
14	K_{Res}	1	$w(K_{Res})$	1	$w14(KRes)$	$w^2 14(KRes)$
–	a	a	–		$w(a) = \sqrt{\sum_{i=1}^{N} w_i^2(a)}$	
–	a	a	–		$W(a) = k \cdot w(a)$	

Table 30.3 Calculated best measurement capability for f = **159.2 Hz**,a = **10.0 m/s²**

No.	Measurand	Best estimate	Uncertainty	Sensitivity-coefficient	Uncertainty contribution	Variance
	X_i	x_i	$w(x_i)$	c_i	$w_i(y)$	$w_i^2(y)$
1	U_R	1000 mV	$1.00 \cdot 10^{-3}$	1	$1.00 \cdot 10^{-3}$	$1.00 \cdot 10^{-6}$
2	S_R	1000 pC/(m/s²)	$5.00 \cdot 10^{-3}$ (k = 2)	1	$2.50 \cdot 10^{-3}$	$6.30 \cdot 10^{-6}$
3	G_R	100 mV/pC	$2.00 \cdot 10^{-3}$ (k = 2)	1	$1.00 \cdot 10^{-3}$	$1.00 \cdot 10^{-6}$
4	K_T	1	$1.00 \cdot 10^{-4}$	1	$1.00 \cdot 10^{-4}$	$1.00 \cdot 10^{-8}$
5	K_D	1	0.00	1	0.00	0.00
6	K_H	1	$1.00 \cdot 10^{-4}$	1	$1.00 \cdot 10^{-4}$	$1.00 \cdot 10^{-8}$
7	K_N	1	$1.00 \cdot 10^{-5}$	1	$1.00 \cdot 10^{-5}$	$1.00 \cdot 10^{-10}$
8	K_{MT}	1	$1.20 \cdot 10^{-3}$	1	$1.20 \cdot 10^{-4}$	$1.40 \cdot 10^{-8}$
9	K_{MC}	1	$1.00 \cdot 10^{-3}$	1	$1.00 \cdot 10^{-3}$	$1.00 \cdot 10^{-6}$
10	K_{Rel}	1	0.00	1	0.00	0.00
11	K_{TK}	1	$3.00 \cdot 10^{-4}$	1	$3.00 \cdot 10^{-4}$	$9.00 \cdot 10^{-8}$
12	K_L	1	$2.90 \cdot 10^{-5}$	1	$2.90 \cdot 10^{-5}$	$8.41 \cdot 10^{-10}$
13	K_I	1	$3.50 \cdot 10^{-4}$	1	$3.50 \cdot 10^{-4}$	$1.20 \cdot 10^{-7}$
14	K_{Res}	1	$1.00 \cdot 10^{-3}$	1	$1.00 \cdot 10^{-4}$	$1.00 \cdot 10^{-8}$
–	a	10.0 m/s²	$w(a) = \sqrt{\sum_{i=1}^{N} w_i^2(a)}$		**$3.083 \cdot 10^{-3}$**	**$9.505 \cdot 10^{-6}$**
–	a	–	$W(a) = k \cdot w(a) k = 2$		**$6.166 \cdot 10^{-3}$**	–
Complete measurement result			a = **10.0 m/s² ± 1.0%**			

As an instance a measurement uncertainty budget has been calculated for a single frequency point at 160 Hz in Table 30.3. For an uncertainty budget over the full range of frequency, the sensitivity deviation of the sensor under test has to be considered. Contributions to the uncertainty calculation of the A/D conversion have not been considered.

Table 30.3 indicates a measurement uncertainty for the measured acceleration amplitude with an accelerometer for 160 Hz of $\boldsymbol{a} = \mathbf{(10.0 \pm 0.1)\ m/s^2}$ for a coverage factor of $\boldsymbol{k} = \mathbf{2}$. For test results on a wider frequency range the uncertainty calculation needs more contributions to considerate the frequency response deviation for the sensor sensitivity and the amplifier gain.

30.5 Calculation of the Expanded Uncertainty, U_{rel} (φ_X), for the Acceleration Phase Shift of an Accelerometer at a Single Frequency Point

The phase accuracy results from the following model calculation, and the treatment of these measurement deviations as random via Gaussian error propagation law without consideration of correlations. For the phase shift φ_X of an accelerometer or force sensor the following model equation can be deviated:

$$\begin{aligned} \varphi\,\mathrm{x} &= \varphi\,\mathrm{S} + \varphi\,\mathrm{I,SS} + \varphi\,R + (\varphi\,\mathrm{o} - \varphi\,\mathrm{i})\,\mathrm{S} + (\varphi\,\mathrm{o} - \varphi\,\mathrm{i})\,\mathrm{X} + \varphi\,\mathrm{H} + \varphi\,\mathrm{Res} = \\ \varphi\,\mathrm{x} &= \varphi\,\mathrm{S} + \varphi\,\mathrm{I,SS} + \varphi\,R + \Delta\varphi\,\mathrm{S} + \Delta\varphi\,\mathrm{X} + \varphi\,\mathrm{H} + \varphi\,\mathrm{Res} \end{aligned} \tag{30.10}$$

herby is:

φ_X total phase shift of the measurement chain
φ_S phase shift of the sensor
$\varphi_{I,Ss}$ long term stability of the sensors phase shift
$\Delta\varphi = \varphi_o - \varphi_i$ phase shift of the amplifier between φ_o phase at output and φ_i phase at input,
φ_H phase shift caused by noise and hum influences, and
φ_{Res} phase shift caused by residual influences, latency time, instabilities of the amplifier, etc.

The calculation of the total measurement uncertainty would be similarly derived with the previous magnitude calculation.

30.6 Example of an Automatic KiXact Measurement Uncertainty Calculation with the KiDAQ Data Acquisition System

30.6.1 Introduction To KiDAQ

KiDAQ from Kistler Instruments is a versatile data acquisition system that offers a wide selection of modules for various sensor technologies and more than 20 different measurands – from simple voltage signals across measurement bridges and IEPE sensors to charge signals from piezoelectric sensors. Thanks to the modular design and the different housing options, KiDAQ can be used in a wide range of applications. Depending on the scope of the application, measurement technicians and engineers can choose from versatile designs for laboratory applications, permanent installations, and mobile use.

Measurement devices are configured efficiently with the intuitive KiStudio Lab Software, which provides an overview of the whole measurement setup and enables quick navigation (Fig. 30.2).

30.6.2 Introduction To KiXact

KiXact is the name of the concept for automatic measurement uncertainty calculations in the KiStudio Lab software. This is based on detailed know-how about the measuring components on a type level in combination with conditions the user specifies for his application. Further integration of calibration data on instance level is planned as well as the continuous increase of the actual component data base.

As a first step, the measurement technician models his measuring chain by selecting the exact sensor, cable and the input channel of the data acquisition system. In the following example this is a Kistler PiezoStar accelerometer 8766A050A in combination with a Type 1734A cable for IEPE sensors. It is connected to an input of a 5501A KiDAQ measuring module (Fig. 30.3).

After configuring the actual data acquisition channel correctly, additional parameters for the KiXact measurement uncertainty calculation can be specified in the section "KiXact Ranges". In case of an expected signal value of 20 . . . 50 g, a signal frequency between 1 . . . 5000 Hz and an operating temperature of 10 . . . 50 °C, the measurement uncertainty of 6.88% is immediately displayed above the sketch of the measuring unit. Further narrowing of the KiXact parameters can further reduce the calculated uncertainty value (Fig. 30.4 and Table 30.4).

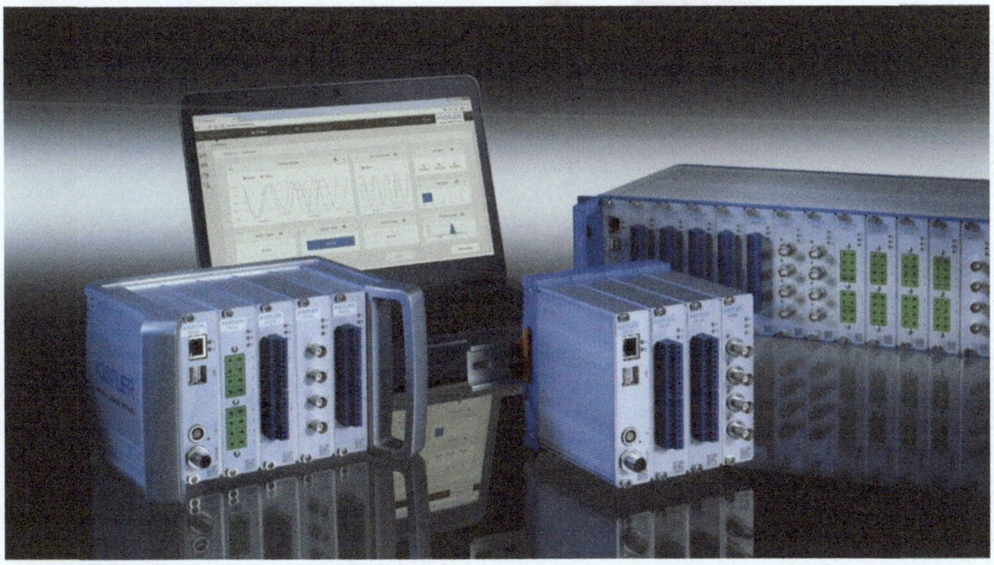

Fig. 30.2 The Kistler KiDAQ system

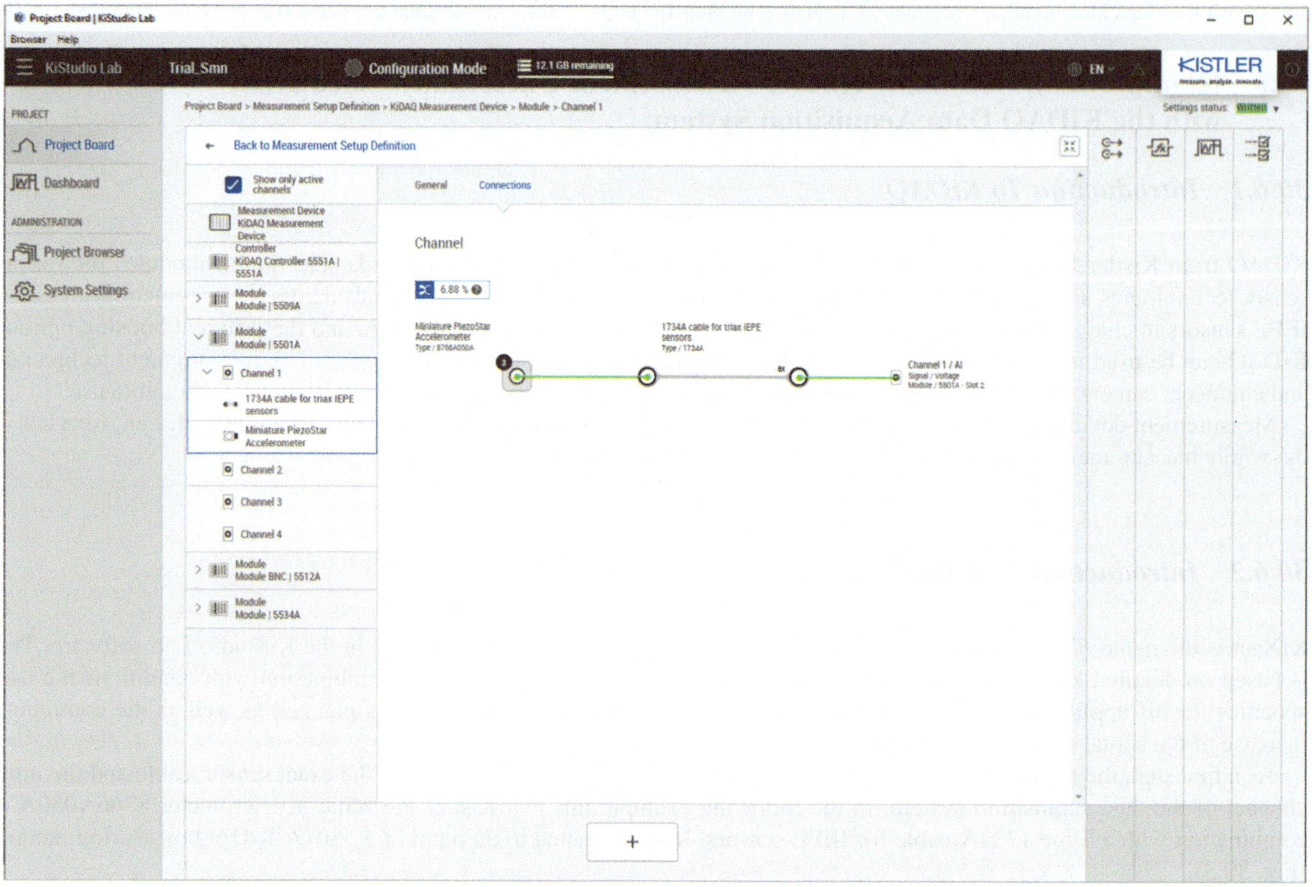

Fig. 30.3 Model of the measuring chain in the KiStudio Lab software

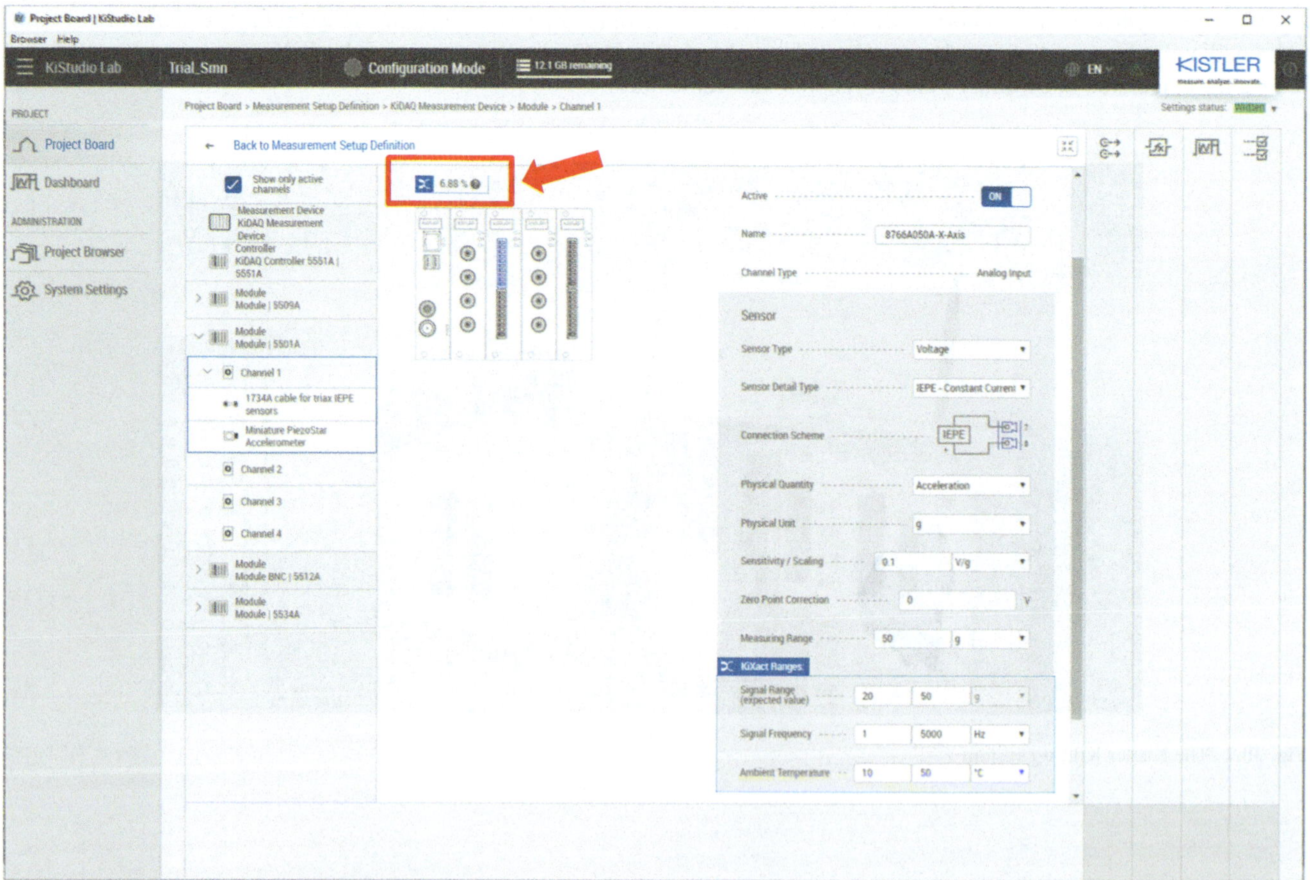

Fig. 30.4 Channel configuration in the KiStudio Lab software

Table 30.4 Influence factors included in the KiXact model for a piezoelectric force measuring chain

List of KiXact Influence Factors – PE Force	
Working standard	
1	Reference calibration system
Unit under test	
2	Repeatability
3	Reproducibility
4	Hysteresis
5	Zero deviation
6	Linearity
Charge amplifier	
7	Linearity
8	Drift tolerance
9	Long-term stability
10	Temperature
11	Humidity
Data acquisition	
12	Uncert.Meas. Charge calibrator
13	Calibration coefficient
14	Temperature A/D converter
15	Signal-noise-ratio A/D converter
16	Long-term stability A/D converter

30.7 Summary

Both approaches in terms of a fully theoretical and practical configured measurement chain have been shown. Measurement uncertainty calculations can be supported by computer aided software and data base tools, like KiXact. Complex calculations can easily be formed using the software which provides a better understanding of measurement deviations in practice as well as an increase of quality insurance improvements in daily measurements (e.g. modal analysis with multiple channel requirements).

Thanks to the impressive ease and time saving function for the user, an automated calculation concept like Kistler's KiXact can be a substantial help to increase the systematic consideration of the uncertainty in daily measurements. Nevertheless, it needs to be understood that such an automated calculation does not cover the influence of the individual technical setup. Other contributing influences considering the test set-up or measurement uncertainty budget may be implemented by the user to get a complete uncertainty calculation, although the most substantial parts are already included by the KiXact computed aided design tools and are easily accessible.

References

1. ISO/BIPM ‚Guide to the Expression of Uncertainty in Measurement“2008; Technische Regel ISO/IEC Guide 98-3:2008-09 Messunsicherheit – Teil 3: Leitfaden zur Angabe der Unsicherheit beim Messen; Guide of the Joint Committee for Guides in Metrology: JCGM 100:2008 Evaluation of measurement data – Guide to the expression of uncertainty in measurement
2. ISO 16063-1: Methods for the calibration of vibration and shock transducers – part 1: Basic Concepts
3. ISO 16063-21: Methods for the calibration of vibration and shock transducers – part 21: Secondary vibration calibration by comparison

Chapter 31
Towards Population-Based Structural Health Monitoring, Part VII: EOV Fields – Environmental Mapping

Weijiang Lin, Keith Worden, Andrew Eoghan Maguire, and Elizabeth J. Cross

Abstract In a population-based structural health monitoring setting, data from one structure in a population, where the health state is known, may be used to make inferences about the health state in any nominally-identical structure. Any deviation from the learned 'healthy response' potentially indicates damage. However, as in standard applications of structural health monitoring, the healthy response from different structures also varies with the changes in environmental conditions across the population. This paper investigates the modelling of the change in environment across a population of structures located in one geographical region, such as a wind turbine farm. A data-driven mapping method (based on Gaussian process regression) will be introduced that aims to quantify and normalise variation stemming from the environment, such that the remaining response is only sensitive to damage or performance anomalies. The way in which environmental maps are constructed and implemented is demonstrated via a case study from an offshore wind farm. The ideas introduced here will constitute a *field* in the framework of population-based structural health monitoring presented elsewhere in the conference.

Keywords Population-based structural health monitoring (PBSHM) · Environmental and operational variation (EOV) · Gaussian process (GP)

31.1 Introduction

Population-based structural health monitoring ((PB)SHM) extends the subject of SHM from an individual structure to a group (or population) of structures. One of the motivations behind this is to address the issue caused by the lack of damage state data. Many would agree that supervised learning is one of the most well-suited approaches to damage identification tasks at the higher levels of Rytter's hierarchy, i.e. damage classification, assessment and prediction [1]. To this end, labelled data for various damage cases are crucial in inferring detailed diagnostic information about a structure. However, it is often difficult to obtain these data, especially for high-value structures such as wind turbines and aircraft. A possible solution can be provided by the population-based approach as it allows the damage state data from one structure to aid inference on damage states for all structures in the population [2]. In this paper, the discussion is focussed on the knowledge transfer between nominally-identical structures, referred to as a homogeneous population [3]. An almost ideal example is given by an offshore wind farm, based on which, the analysis in this paper is carried out.

The data collected from operating structures, unlike that from highly-controlled laboratory experiments, are subject to complex environmental and operational variations (EOVs). The influence on measured responses due to EOVs can be comparable to, or sometimes more significant than, that due to damage [4]. Since it is almost impossible to find a damage-sensitive response that is not heavily influenced by EOVs [5], understanding and normalisation of the effect due to EOVs become necessary steps to implement SHM in reality.

Much effort has gone into studying how EOVs may affect the responses of aerospace and civil structures. Some EOVs directly affect how a structure vibrates. This is demonstrated by the flutter or buffeting in long-span bridges as a result of wind-induced vibrations [4, 6]. There are also factors that tend to alter the material properties and boundary conditions of a structure, which then lead to a different dynamic response. For example, [4] summarised that thermal-induced vibrations in

---

W. Lin (✉) · K. Worden · E. J. Cross
Dynamics Research Group, Department of Mechanical Engineering, University of Sheffield, Sheffield, UK
e-mail: wlin17@sheffield.ac.uk; k.worden@sheffield.ac.uk; e.j.cross@sheffield.ac.uk

A. E. Maguire
Vattenfall Research and Development, Edinburgh, UK

B. Dilworth (ed.), *Topics in Modal Analysis & Testing, Volume 8*, Conference Proceedings of the Society for Experimental Mechanics Series, https://doi.org/10.1007/978-3-030-47717-2_31

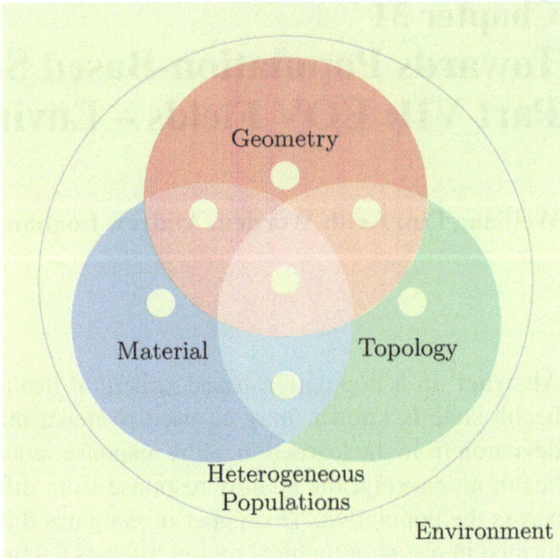

Fig. 31.1 Categories of heterogeneous population within population-based SHM [8], amended for EOV effects

bridges may stem mainly from changes in stiffness and boundary conditions (e.g. thermal contraction). In the case of wind turbines, various EOVs exhibit a combined effect on the dynamic responses of components. An example discussed in [7] showed how modal frequencies negatively correlated with temperature (given a rotor speed within a specific range), and how this negative correlation could also be moderated by the wind-induced aerodynamic damping.

When an individual structure is assessed, as in traditional SHM, the changes in EOVs are often mainly thought of as temporal. For example, in the monitoring campaigns of bridges, the EOVs can be considered as the daily or seasonal variations in temperature, wind loading, and traffic loading [6]. However, in PBSHM, the spatial dimension also needs to be considered. As well as the temporal trends, the structures in a population can also be subject to different EOVs based on where they are located.

The additional difficulties in the wind farm case arise from the fact that, the *homogeneous* population of nominally-identical structures, may have significant differences in their responses as a result of their positions within the EOV *field*. An example can be seen in a wind farm; although the turbines are all located in the same geographical region, the wind conditions at different turbines can still differ due to wake effects. This issue adds to the complexity of the problem, and new methods are called for. In particular, this issue means that a more sophisticated approach to the *transfer* of information between the structures may be needed [3, 8]. If a population is *heterogeneous* [9, 10], the problem of *transfer* is compounded. As observed in [8, 9], individuals in a heterogeneous population may differ in terms of *topology*, *geometry* or structure, and all of these characteristics are sensitive to EOVs. For example, geometrical dimensions may change with thermal expansion and material properties can be sensitive to temperature and humidity (as noted above). It is not so obvious how EOVs might induce topology change; however, this is possible, e.g. a bridge may change its topology if an expansion joint closes due to temperature increase, and a new ground node (boundary condition) appears in the representation [9]. In summary, the Venn diagram from [8], on sensitivity of transfer, is modified to that in Fig. 31.1.

The spatial distribution of environmental conditions across a wind farm has been the subject of numerous studies in the field of wind turbine aerodynamics. A large number of computer experiments have been carried out to model the process when turbine rotors extract kinetic energy from the wind, referred to as wake modelling. These physics-based models typically consist of two parts: the first part of the model characterises the aerodynamics of turbine rotors [11–13], and the second part simulates the entire turbulent wind field and provides the input blade loads to all rotors in the modelled farm [13, 14]. As a result, the models simulate how much power each turbine can generate, given a specific wind farm configuration, and in other words, how the wind is disturbed due to the existence of operating turbines. However, such models are primarily developed with the purpose of improving the rotor aerodynamic efficiency or optimising wind farm configuration. In this context, a model is usually considered as sufficient so long as it is able to reproduce most of the physical phenomena, i.e. the simulated environment can produce an effect similar to the real wind field over a period of time. In contrast, models with higher accuracies are required for the purpose of SHM, where ideally the model would be able to predict the exact time and location of a physical phenomenon. Such high level in accuracy is difficult to achieve through physics-based models alone (if at all) without a high computational cost. On this account, a data-driven approach to modelling the wind farm environment has been considered here.

This paper aims to understand, through data-based modelling, how EOVs vary temporally and spatially across a population. It is the first step towards understanding the effect of spatially changing EOVs on dynamic responses, after which a new method to remove the EOV-induced trends can be attempted. The work in this paper is based on data from an operating wind farm, Lillgrund, with a brief introduction given in the next section. Section Two will provide a detailed description of the field-mapping approach and the associated data-driven algorithm that are used to model the spatially changing EOVs. The section that follows will present and discuss the results from the environmental mapping model. The paper will finish with conclusions and future implications.

31.1.1 Lillgrund Wind Farm

The Lillgrund offshore wind farm is located in the middle of the Öresund region between Denmark and Sweden. It consists of 48 turbines rated at 2.3MW, with the farm arrangement seen in Fig. 31.2. The turbines are spaced slightly further away (4.3D) in the prevailing wind direction of 225°–255° compared to the spacing in the perpendicular direction (3.3D). Such spacings are generally considered dense, because the layout was originally designed for smaller turbines [15]. This compact layout was preserved so as to facilitate the investigation into how turbine wakes can affect power production.

The data used in this paper were collected during a year of operation from the Supervisory Control and Data Acquisition (SCADA) system in Lillgrund. The SCADA data are presented as a statistical summary, including the mean, maximum, minimum, and standard deviation, of the actual recorded data every 10 min. In the analysis that follows, the 10-min mean values are used as an indication of the actual measured quantities.

Fig. 31.2 Turbine arrangement in Lillgrund wind farm [15]

31.2 Environmental Mapping

31.2.1 Motivation

According to the unique mechanism in wind turbines, most operational factors are driven by the environmental conditions. In particular, turbine rotor speed is directly proportional to wind speed across a specified operational range, and nacelles should be facing the incoming wind direction by default. The strong correlation between environmental and operational factors indicates that, in the first instance, the environment can be viewed as the dominant driving force behind the variations in an EOV field. As such, the method presented here is called an environmental mapping as it creates a spatial map of the environment across a farm as an indication of the changing EOV field across a population.

As well as being an important step in understanding the effect of EOVs on vibration signals, the environmental map can be applied in two other ways: for SHM purposes, this map can be used as the environmental input to individual turbine models in order to predict the normal, yet different, responses across a farm; another possibility lies in adapting it to a model that optimises the overall power production of the farm by means of wake steering or induction control. Taking into account these potential applications, wind loading becomes the most important aspect in the environment as it has the most direct impact on the structures and the strongest correlation with power. Therefore, the environmental mapping method described here focusses on modelling the spatial variations of wind conditions in a farm.

31.2.2 Wake Pattern in Wind Farms

Wind conditions vary across an offshore wind farm mainly due to wake effects. The fact that turbines are designed to extract kinetic energy from the wind explains why the wind becomes slower after passing through a turbine rotor. The moving rotor also brings extra turbulence into the wind, resulting in a region behind the rotor with reduced wind speed and intensified turbulence. Further downstream, the wake tends to spread and eventually return to free stream condition, yielding a cone-shaped wake region [16]. In Fig. 31.3a, the different shades of red indicate the variation in characteristics of the wake, which means that the wake effect on any downstream turbine is highly dependent on the relative position of the turbine in shadow with respect to its upstream neighbour. When multiple wakes superimpose on one another, with an example given in Fig. 31.3b, the combined effect needs to be considered.

It is self-explanatory that the turbine layout and wind direction play important roles in the overall wake pattern, as they affect the position and direction of individual turbine wakes. Given a fixed wind farm layout in this analysis, Fig. 31.4 shows some examples of the wind speed patterns based on different wind directions, visualising the mean wind speed deficit. The turbulence variations will be accounted for later in the modelling process. It is to be noted that the wind direction at each turbine location is estimated by the nacelle position, given that a good match is found between the wind direction measured on the available weather mast and the nacelle position at the nearest turbine location [15].

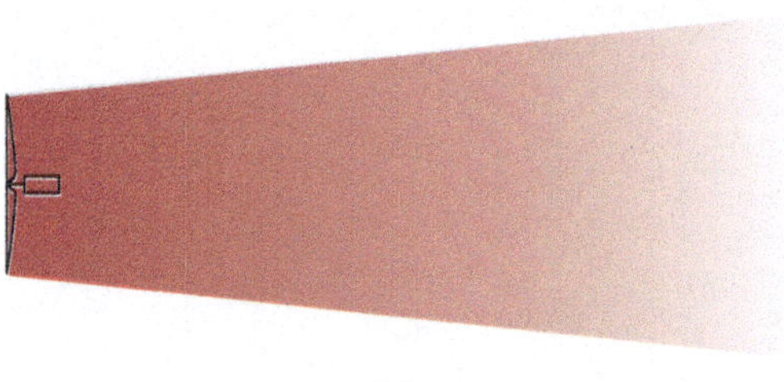

(a)

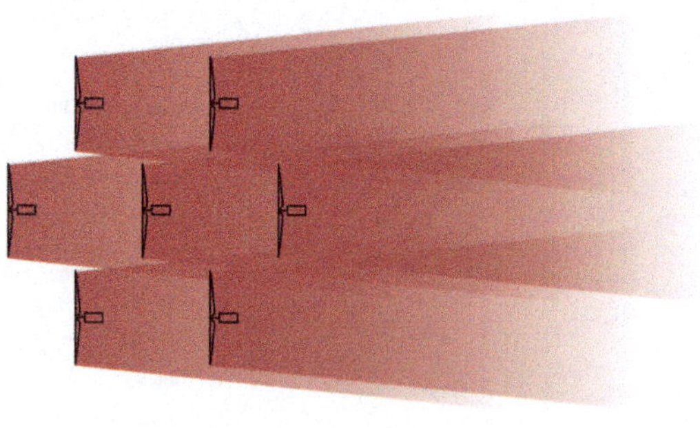

(b)

Fig. 31.3 Schematic illustration of turbine wakes. (**a**) Single turbine wake (top view). (**b**) Wake superposition (top view). (Adapted from [16])

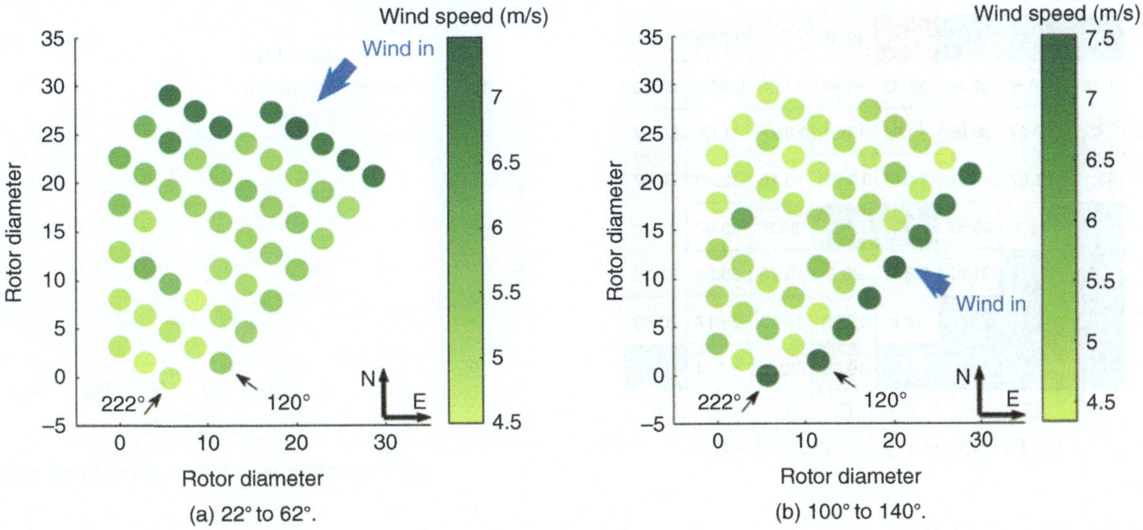

Fig. 31.4 Mean wind speed distribution across the Lillgrund farm for two incoming wind directions. (**a**) 22° to 62°. (**b**) 100° to 140°

31.2.3 Data-Based Model

To the best of the authors' knowledge, this paper demonstrates the first attempt to build a model that predicts the temporally and spatially-changing EOV fields for SHM purposes. In line with the recent trend in SHM [1], a data-driven approach is preferred for environmental mapping, since it allows for easier integration with other steps in the process of damage identification. Another reason is that, as the physics-based approach has been developed for decades, the current advancement in this field favours the inclusion of more data for better model calibration, namely data-enhanced physics-based models. Given the abundant availability of data in this analysis, an opportunity is given to explore the fully data-based approach in an attempt to achieve relatively high accuracy at a much lower computational cost compared to the physics-based alternatives. As the first step in constructing a reliable model, the model described in this paper is trained to predict only one spatial pattern in wind speed, with the wind coming from 100° to 140° (Fig. 31.4b) specifically.

The relation between wind speeds at various turbine locations is considered nonlinear due to the complex wake effects, especially with regard to turbulence. Gaussian process (GP) regression is used here as it provides an efficient way to learn the nonlinear dependence between continuous variables. GP regression is also a stochastic method that gives a prediction distribution based on which a mean prediction and its associated confidence interval can be computed. The simplest version of the GP is applied here, with a zero mean function and a squared-exponential covariance function. Detailed explanations for GP training and inference are provided by [17].

31.3 Wind Speed Predictions

The GP model seeks to predict a map of wind conditions by interpolating across the space. That is, it aims to predict the wind speed at all turbine locations, given the wind speed from a fixed subset of the locations as inputs. In the following, the training and testing data sets correspond to different time windows, ranging from 13 to 14 h' worth of data, that are subjected to similar wind directions (100° to 140°) and wind speeds (4 to 8 m/s). This choice is to make sure that the training and testing sets indicate similar spatial patterns on separate occasions. The training set is also selected such that the model prediction represents the wind speed variations under normal operational conditions.

It is a standard practice in SHM to use the error of predictive models as an indicator of structural performance. One of the most commonly-used error metrics is the normalised mean-square error (MSE) [6, 18], which is defined as,

$$MSE = \frac{100}{N\sigma_y^2}\sum_{i=1}^{N}(y_i - \hat{y}_i)^2 \tag{31.1}$$

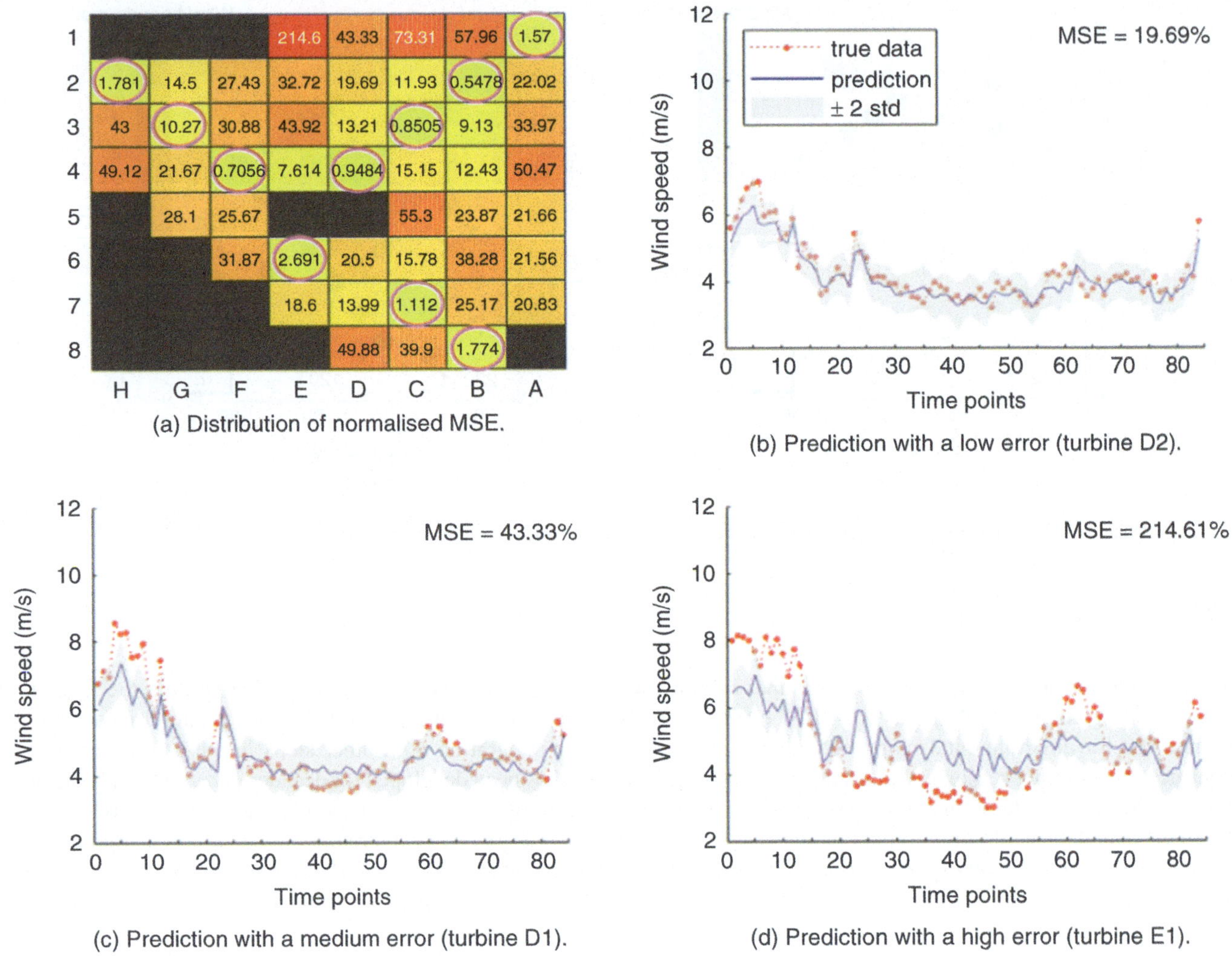

Fig. 31.5 Prediction error distribution, (**a**), and examples of time series predictions, (**b**)–(**d**), for a testing data set

where y and $\hat{y}$ represent target data and model prediction respectively, each with a size N. The error is normalised by the variance of target data, σ_y^2, in order to marginalise the effect due to discrepancy between training and testing data. The scaling factor of 100 ensures that the mean of the data will score 100% if used as the model. This serves as a threshold below which correlation is indicated between prediction and target. On a similar note, error values higher than 100% potentially indicate performance anomalies.

The error distribution for a testing data set can be seen in Fig. 31.5a. It is worth noting that the heat map is created to reflect on the relative positions between turbines rather than the exact wind farm layout, therefore the map appears to be a rotated and distorted version of the farm. Based on this view, the incoming wind of 100° to 140° appears to proceed from right to left in the heat map. The first thing to be observed was that the normalised MSEs at the ten reference turbine locations, as indicated by pink circles in Fig. 31.5a, were generally small. This gives an idea of how accurate the GP predictions can possibly be – a mean MSE of about 2.5% was obtained when the model attempted to predict what were given as inputs. For the region where the GP had to interpolate, the majority, except for the highlighted turbine with an error greater than 100%, demonstrated reasonably good predictability, with a mean error of 24%.

In respect of the testing set shown in Fig. 31.5a, examples of time-series predictions with low, medium and high errors are demonstrated in Fig. 31.5b–d. In the low-error case (Fig. 31.5b), the predicted line followed the trend of the target data well, with most of the data fluctuations captured by the confidence interval of ± twice the standard deviation. A short period of slight under-prediction could be seen from time point 0 to 10. The fact that most data points in this period are bounded by the confidence limits means that they are considered by the GP as fluctuations within the acceptable range.

When these fluctuations grew out of the confidence bounds, as seen in Fig. 31.5c, a higher error was assigned by the GP, showing an example of a medium-error prediction. In addition to the under-prediction at the beginning of the time window, the higher error can also be attributed to the over-prediction during time instances 35–50 and the under-prediction during 55–60. Since turbine D2 (Fig. 31.5b) and D1 (Fig. 31.5c) are the same distance away from the first row of turbines encountered by free-stream wind (i.e. D1 and D2 are both in Row D), it is reasonable that predictions with similar trends are made at these two turbine locations. However, such predictions provide a worse fit at D1 than at D2 because they do not take into account how the turbines on the edge of a wind farm may be influenced by the environment from the outside. In this instance, the influence from the outside manifests itself as larger-scale fluctuations in wind speed. This explains the relatively-higher errors at turbine locations along Row 1 (Fig. 31.5a), demonstrating an edge effect.

The most prominent edge effect was found at turbine E1, where the largest-scale fluctuations occurred (Fig. 31.5c). Once again the GP does not expect the fluctuations due to the outside environment, and tends to make a prediction that is similar to that in the neighbourhood (e.g. at turbine location E2). However, in most (if not all) testing data sets that correspond to the same spatial pattern, the highest normalised MSE along Row 1 is found at turbine E1, which suggests a change (in structure or control strategy) at this location. Although the reasons behind it require further investigation, given that very few reference turbines are placed on the edge of the farm and/or that only the information about wind speed is used to train the GP, it is understandable that the model gives a high error to address the unexpected change at turbine E1. To take a step back, putting aside the explanations, it is evident in Fig. 31.5c that the model has the ability to highlight a location, with a prediction error higher than 100%, where an unexpected trend in wind speed occurs.

31.4 Conclusions

This paper has introduced environmental mapping as a data-based method to understanding the temporally and spatially-changing EOVs for the implementation of PBSHM. The work is based on data from an offshore wind farm, Lillgrund, which exemplifies the concept of homogeneous population. At this stage, an environmental map can be produced by a data-driven model, based on GP regression, which is trained on data from one specific spatial pattern. The GP model has demonstrated its ability to interpolate across the space, obtaining a full map of wind speed on the basis of inputs from a few reference points. Furthermore, if the training data represents the normal wind/operating conditions, the model can also be used to indicate unexpected environmental conditions that might result from different control sequences not present in the training set.

The data-based environmental mapping introduced here can be applied in a manifold manner. Firstly, for the purpose of SHM, environmental mapping can be used as a cost-efficient wind field model for the specific wind farm(s) that the model is trained on. The wind speed predictions can be served as environmental inputs to individual turbine models, allowing them to predict normal structural responses subject to environmental variations. A second opportunity that the authors will be working on is to use a similar mapping approach on power production data, creating a map that aims to establish the relations between power and a number of environmental (e.g. wind speed and direction) and operational (e.g. nacelle position, pitch angle and rotor speed) factors. This new map is then applicable to the optimisation of wind farm power production through wake steering and/or induction control. Similarly, the methodology developed here can be adapted to damage sensitive features (e.g. derived from acceleration data), with the intention of building a more reliable indicator of structural conditions.

Taking a step further, it is also planned to modify the current model to account for a range of various spatial patterns in the environment, aiming for a more generalised model.

Acknowledgments The authors would like to acknowledge the support of the EPSRC, particularly through grant reference numbers EP/R004900/1, EP/S001565/1 and EP/R003645/1.

References

1. Farrar, C.R., Worden, K.: Structural Health Monitoring: A Machine Learning Perspective. John Wiley and Sons Ltd, Chichester (2013)
2. Worden, K., Cross, E.J., Dervilis, N., Papatheou, E., Antoniadou, I.: Structural health monitoring: from structures to systems-of-systems. IFAC **48**, 1–17 (2015)
3. Bull, L.A., Gardner, P.A., Gosliga, J., Maguire, A.E., Campos, C., Rogers, T.J., Haywood-Alexander, M., Dervilis, N., Cross, E.J., Worden, K.: Towards population-based structural health monitoring, Part I: homogeneous populations and forms. In: Proceedings of IMAC XXXVIII – the 38th International Modal Analysis Conference, Houston (2020)
4. Sohn, H.: Effects of environmental and operational variability on structural health monitoring. Philos. Trans. R. Soc. A Math. Phys. Eng. Sci. **365**, 539–560 (2007)
5. Worden, K., Farrar, C.R., Manson, G., Park, G.: The fundamental axioms of structural health monitoring. Proc. R. Soc. A Math. Phys. Eng. Sci. **463**, 1639–1664 (2007)

6. Cross, E.J.: On Structural Health Monitoring in Changing Environmental and Operational Conditions. Ph.D. thesis, Department of Mechanical Engineering, University of Sheffield (2012)
7. Hu, W., Thöns, S., Rohrmann, R., Said, S., Rücker, W.: Vibration-based structural health monitoring of a wind turbine system part II: environmental/operational effects on dynamic properties. Eng. Struct. **89**, 273–290 (2015)
8. Gardner, P.A., Worden, K.: Towards population-based structural health monitoring, Part IV: heterogeneous populations, matching and transfer. In: Proceedings of IMAC XXXVIII – the 38th International Modal Analysis Conference, Houston (2020)
9. Gosliga, J., Gardner, P.A., Bull, L.A., Dervilis, N., Worden, K.: Towards population-based structural health monitoring, Part II: heterogeneous populations and structures as graphs. In: Proceedings of IMAC XXXVIII – the 38th International Modal Analysis Conference, Houston (2020)
10. Gosliga, J., Gardner, P.A., Bull, L.A., Dervilis, N., Worden, K.: Towards population-based structural health monitoring, Part III: graphs, networks and communities. In: Proceedings of IMAC XXXVIII – the 38th International Modal Analysis Conference, Houston (2020)
11. Vermeer, L.J., Sørensen, J.N., Crespo, A.: Wind turbine wake aerodynamics. Prog. Aerosp. Sci. **39**, 467–510 (2003)
12. Brand, A.J., Peinke, J., Mann, J.: Turbulence and wind turbines. J. Phys. Conf. Ser. **318**(7), 072005 (2011). https://iopscience.iop.org/issue/1742-6596/318/7
13. Bianchi, F.D., de Battista, H., Mantz, R.J.: Wind Turbine Control Systems: Principles, Modelling and Gain Scheduling Design. Springer, London (2007)
14. Veers, P.S.: Three-dimensional wind simulation. Technical report, Sandia National Laboratories, Albuquerque ABQ (1988)
15. Dahlberg, J.-A.: Assessment of the Lillgrund windfarm: Power performance and wake effects. Technical report, Vattenfall Vindkraft AB (2009)
16. González-Longatt, F., Wall, P., Terzija, V.: Wake effect in wind farm performance: steady-state and dynamic behavior. Renew. Energy **39**, 329–338 (2012)
17. Rasmussen, C.E., Williams, C.K.I.: Gaussian Processes for Machine Learning. The MIT Press, Cambridge, MA (2006)
18. Worden, K.: Parametric and Nonparametric Identification of Nonlinearity in Structural Dynamics. Ph.D. thesis, Heriot-Watt University, Edinburgh (1989)

Weijiang Lin A first-year PhD student at the University of Sheffield with a focus on grey-box modelling and/or data-enhanced physics-based modelling, population modelling, multi-fidelity modelling, machine learning, renewable energy, and computational fluid dynamics.

Chapter 32
Investigation of Resistive Forces in Variable Recruitment Fluidic Artificial Muscle Bundles

Jeong Yong Kim, Nicholas Mazzoleni, and Matthew Bryant

Abstract This paper experimentally investigates the mechanical behavior of inactive and low-pressure fluidic artificial muscle (FAM) actuators under applied axial load. In most cases, the active characteristics of an actuator are of interest because they provide valuable information about its force-strain relationship. However, a system of actuators requires attention to the interaction between individual units. One such configuration is a bundle of McKibben artificial muscle actuators arranged in parallel and used for load-adaptive variable recruitment. This bio-inspired actuator bundle sequentially increases the number of actuators activated depending on the load required, which is analogous to how motor units are recruited in a mammalian muscle tissue. While using the minimum number of actuators allows the bundle to operate efficiently, the resistive force of inactive elements acts against total bundle contraction due to their inherent stiffness. In addition, when the bundle transitions between recruitment levels, motor units for a given recruitment level may be gradually pressurized; these low-pressure motor units can also cause resistive forces. Experiments were conducted to characterize the complex interaction between the bladder and braided mesh that cause the resistive force and deflection of inactive and low-pressure elements. Based on observations made from experiments, the paper proposes the initial criteria for developing a model of the resistive forces of a McKibben actuator, both individually, and within the context of a variable recruitment bundle.

Keywords McKibben actuator · Fluidic artificial muscle · Variable recruitment · Inactive actuator · Buckling

32.1 Introduction

In recent years, to address human-robot interface issues in robotic systems, the use of soft actuators have become a topic of great significance. The term "soft actuator" refers to a broad classification of actuators that consist of a compliant component. The inherent compliance of these actuators more closely mimics actuation systems found in nature than traditional rigid robotic linkages and piston-cylinders, which has led to a number of bio-inspired robots that mimic the motion of a caterpillar, snake, and fish, to name a few [1–4]. In addition, because the compliance of soft actuators is so similar to that of mammalian muscles, many of them have been integrated with human physiology and used for human and animal rehabilitation in the form of soft grippers for a hand and soft ankle-foot orthotics [1, 5, 6].

One of the limitations of many current soft bio-inspired actuators is that they generate low force outputs in comparison to their rigid counterparts. Fluidic artificial muscles (FAMs), or McKibben actuators, are a class of soft actuator that balances the tradeoffs between force and deflection. These actuators are characterized by a braided mesh helically wound around an elastic tube. When pressurized with fluid (either pressurized air or hydraulic fluid), it expands radially and contracts axially. Its high force-to-weight ratio, inherent compliance, and simple design has made it favorable to robotic applications.

Although McKibben actuators themselves have existed since the 1960s, they have recently received increased attention due to their soft, compliant nature and favorable force capabilities. Applications of McKibben actuators have also become more bio-inspired, as multiple actuators can be combined and actuated in a variable recruitment bundle. A variable recruitment bundle was first introduced by Bryant et al. [7] and consisted of McKibben actuators configured in parallel, as shown in Fig. 32.1. Just as mammalian muscle consists of motor units that are recruited sequentially based on the required load, in a variable recruitment bundle, individual actuators are divided into motor units, and these motor units are activated

J. Y. Kim (✉) · N. Mazzoleni · M. Bryant
Department of Mechanical & Aerospace Engineering, North Carolina State University, Raleigh, NC, USA
e-mail: jkim84@ncsu.edu; mbryant@ncsu.edu

B. Dilworth (ed.), *Topics in Modal Analysis & Testing, Volume 8*, Conference Proceedings of the Society for Experimental Mechanics Series, https://doi.org/10.1007/978-3-030-47717-2_32

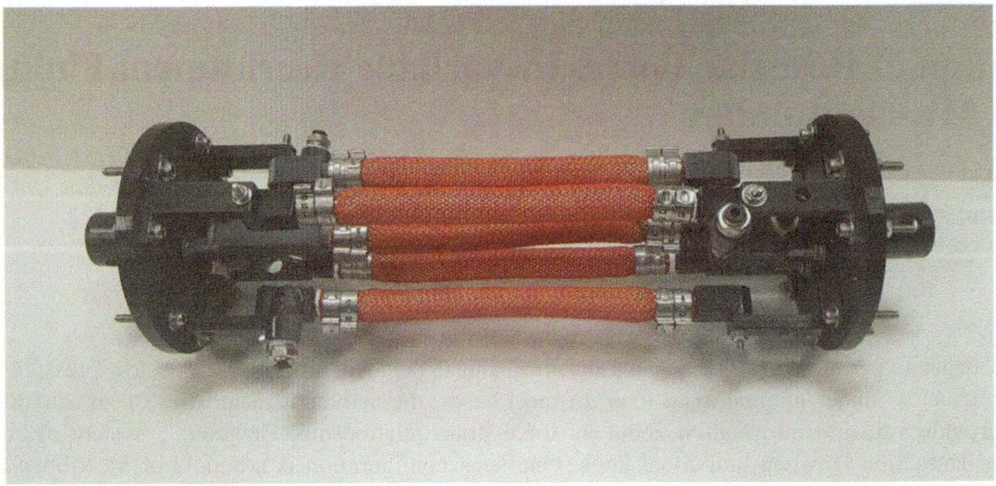

Fig. 32.1 One example of a system with multiple soft actuators: a variable recruitment fluidic artificial muscle bundle consisting of multiple McKibben actuators in parallel

in distinct recruitment levels based on the load requirements of the system. Compared to a single McKibben actuator, the variable recruitment bundle operates at a higher average efficiency and bandwidth [7, 8].

A set of studies have given insight into the design and control of variable recruitment bundles. Jenkins et al. designed variable recruitment controls and analytically demonstrated efficiency gains and chatter reduction [9], Meller et al. and De La Hunt et al. demonstrated these gains experimentally on both hydraulic and pneumatic artificial muscles, respectively [10, 11]. These studies, while promising, have made a number of simplifying assumptions that make it difficult to capture the complex interactions between individual recruitment levels and actuators within a bundle. For example, in variable recruitment, there will be scenarios in which not all actuators are activated simultaneously, allowing active elements to interact with inactive elements. These studies assumed that the resistive force due to these inactive elements was negligible. The purpose of this study is to characterize the forces due to (1) inactive elements at specific recruitment levels and (2) low-pressure elements at partial recruitment levels during recruitment level transitions.

32.2 Effect of Inactive Elements in a Variable Recruitment Bundle

To observe the effects of inactive elements within a variable recruitment bundle, an experiment was devised to measure the quasi-static force output of a single McKibben actuator from its blocked force condition to its free strain condition. A McKibben actuator with a bladder outer diameter of 12.7 mm, slenderness ratio of 24, and an initial braid angle of 26° was used. A universal material tester (Interactive Instruments Model 1 K-16) was used to constrain the actuator at its free length while pressure was applied thus creating its blocked force condition. A pneumatic pressure supply was used to apply a pressure of 275.8 kPa at the blocked force condition. The testing machine allowed the McKibben actuator to contract sufficiently slowly for the measurement to be assumed as quasi-static until the free strain state was achieved. To compare, another experiment was conducted with the same actuator but with the addition of four inactive actuators centered around and parallel to the active actuator. The inactive actuators are of the same dimension as the active element. The pressure ports of the inactive actuators were open to allow them to vent freely. The actuators are spaced sufficiently to eliminate any effects from the units contacting each other. The result shows that although the blocked force remains the same, as the bundle contracts, the force output was significantly lower than that of a single actuator as shown in Fig. 32.2. Although the center active unit is producing the same amount of force, the inactive elements apply a force resisting the contraction of the bundle. Consequently, the range of strain during which the bundle outputs positive force decreases. In addition to the reduced force output, a change in slope can be observed which suggests the forces caused by the inactive elements are non-linear.

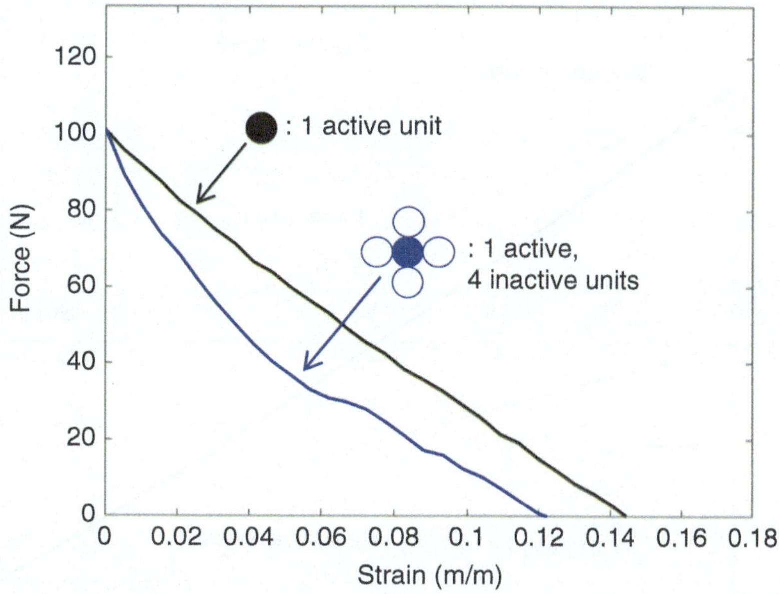

Fig. 32.2 Quasi-static force output comparison between a single McKibben actuator at 275.8 kPa pressure with and without inactive actuators

32.3 Resistive Force Past Free Strain

The force exerted by inactive elements can be attributed to the elastic forces of the inner bladder. An analytical model was proposed by Ball et al. that expresses the total force output of a McKibben actuator as the sum of the force output due to the kinematic constraints of the braided mesh and the force due to the elastic bladder [12]. The intent of this model was to predict the force-strain behavior of a McKibben actuator up until free strain. The Ball model expresses the total force output of a McKibben actuator unit as:

$$F_{unit} = F_{mesh} + F_{elastic} \tag{32.1}$$

where F_{mesh} is the force due to the kinematic constraints of the braided mesh and $F_{elastic}$ is the force due to the elastic bladder. Figure 32.3 illustrates how each force component in Ball's model add up to the total force of a McKibben actuator. The force due the braided mesh decreases as the unit contracts until the neutral braid angle of 54.7° is reached. This relationship is also dependent on pressure which decides the initial blocked force value. In many McKibben actuator literatures, this relationship is known as the ideal force-strain behavior and is expressed as

$$F_{mesh} = F_{ideal} = \left(\pi {r_0}^2\right) P \left[a(1-\varepsilon)^2 - b\right] \tag{32.2}$$

$$a = \frac{3}{\tan^2(\alpha_0)},\ b = \frac{1}{\sin^2(\alpha_0)} \tag{32.3}$$

where P is the applied pressure, r_0 is the initial outer diameter, and α_0 is the initial braid angle [13, 14]. The elastic force due to bladder starts at zero and is always negative. The sum of the two forces become the final output of the McKibben actuator. The force output of a McKibben actuator is associated with the positive portion of the graph, typically characterized by the blocked force and free strain conditions. However, the forces due to inactive elements, as seen in the previous section, are associated with the negative portion of the graph past free strain.

At this point, it is necessary to clearly define the condition for when the total force output of a unit is referred to as *resistive force*. It is possible to refer to the elastic force as resistive force as they are negative components contributing to the total force. However, it is more intuitive to define resistive force according to the direction of the total force of a unit. By doing this, the overall force output of a bundle can be expressed as the difference between mesh forces and elastic forces. The consideration of elastic force as a part of the total force output as proposed by Ball's model provides a base model from which the resistive forces can be understood. The condition for resistive force is defined as:

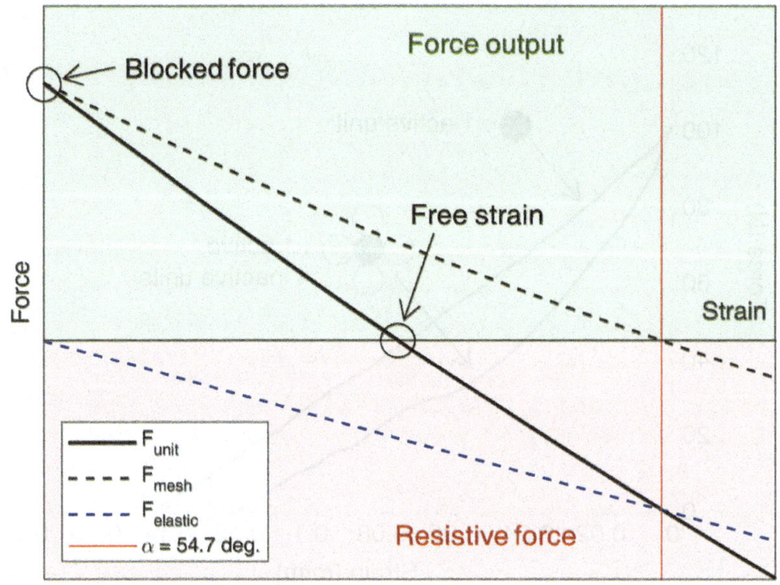

Fig. 32.3 Component-wise breakdown of the total force output of a McKibben actuator according to Ball's model

$$F_{mesh} + F_{elastic} = F_{unit} < 0 \tag{32.4}$$

According to this definition, resistive forces are not bound to the zero applied pressure condition and manifest whenever the force from the braided mesh is less than the opposing elastic force. The resistive force generated by an inactive unit is a special case in which the applied pressure is equal to zero (i.e., the force generated by the braided mesh is zero). For a pressurized unit, past free strain, the kinematic constraint imposed by the braided mesh produces positive force until the neutral angle is reached, but not enough to overcome the opposing force of the elastic bladder. This has significance in the characterization of the different components of resistive force, as this force cannot be considered separately from the positive output force of a unit but rather as a continuation of the same force as the unit passes through the free strain condition.

32.4 Experimental Observation of Resistive Forces with Varying Pressure

A typical study on McKibben actuators will measure forces until free strain. However, to understand the overall output force of a variable recruitment bundle, it is necessary to predict the strain at which an actuator unit starts to exert resistive force. For this purpose, a separate set of experiments was conducted measuring the force output of one unit in order to observe the resistive forces. The test setup is the same as the previous experiment, but the stroke of the testing machine was prescribed to intentionally measure the forces past the free strain condition. Additionally, the actuator unit was activated at varying pressures from 34.5 kPa (5 psi) to 103.4 kPa (15 psi) at 34.5 kPa (5 psi) intervals to observe the pressure dependency of resistive force.

The results are shown in Fig. 32.4. The positive forces show the typical force-strain profiles of a McKibben actuator at varying pressure. As defined in the previous section, the resistive forces are the forces measured past free strain. The resistive force when the muscle is inactive is the resistive force at 0 kPa. As observed in the previous experiment, the profiles are non-linear and exhibit a distinct 'knee' at which the magnitude of the resistive force reaches a maximum value, $|F_{resistive}|_{max}$. Past the strain, ε_{min} at which this 'knee' occurs, the force slightly increases to reach a steady value. It is also noteworthy that the $|F_{resistive}|_{max}$, ε_{min} and the final value are dependent on the applied pressure. The resistive force is divided into distinct regions based on these observations.

After free strain, the McKibben actuator behaves like a hollow, cylindrical column under axial load. However, the resistive force is the sum of elastic force from the bladder and force transmitted by braided mesh constraint. Therefore, the force from the braided mesh needs to be taken into consideration, as well as the dynamically changing dimensions of the bladder due to pressure. From the experimental observations, this paper proposes a criterion for future model development by dividing the resistive force into three regions as shown in Fig. 32.5.

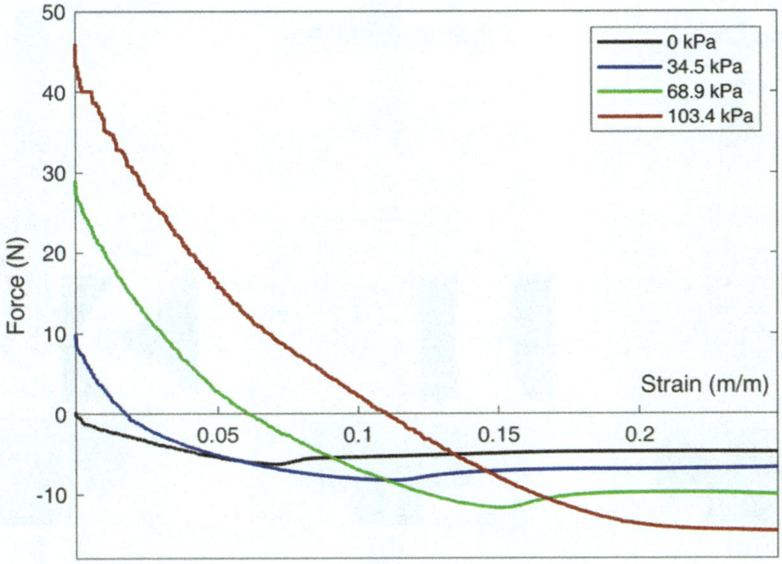

Fig. 32.4 The measurement of forces from blocked force to past free strain at varying pressure

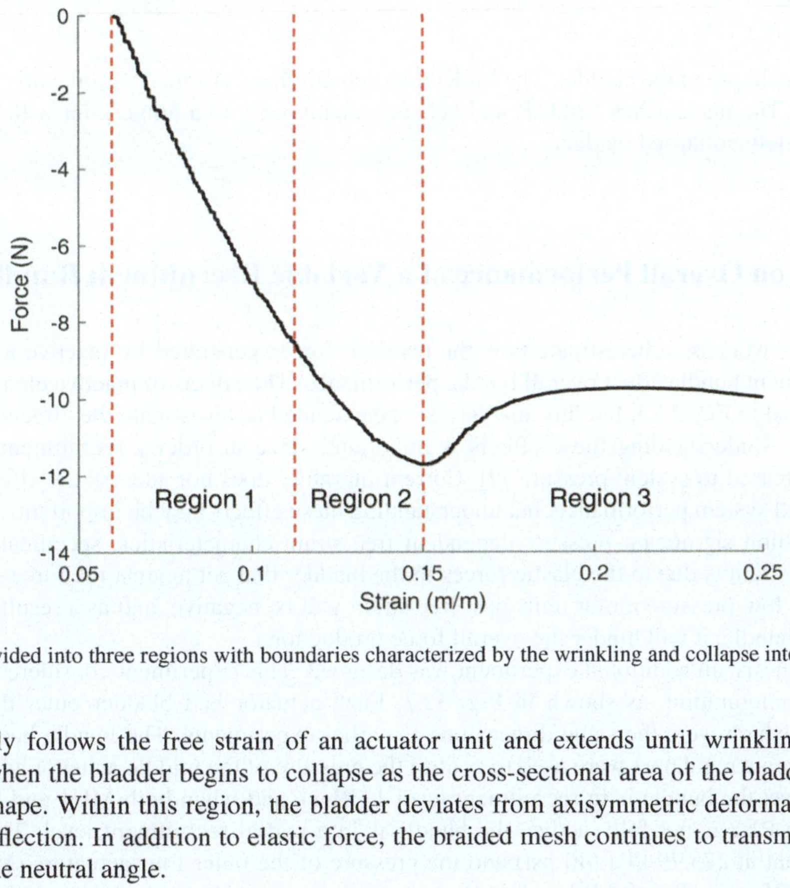

Fig. 32.5 Resistive force divided into three regions with boundaries characterized by the wrinkling and collapse internal moment

- **Region 1** immediately follows the free strain of an actuator unit and extends until wrinkling occurs in the bladder. Wrinkling refers to when the bladder begins to collapse as the cross-sectional area of the bladder starts to deform from its original circular shape. Within this region, the bladder deviates from axisymmetric deformation and starts to buckle, causing transverse deflection. In addition to elastic force, the braided mesh continues to transmit positive force until the braid angle reaches the neutral angle.
- **Region 2** begins as wrinkling occurs in the bladder and extends until its collapse. In this region, the bladder can be considered an inflatable column with thick walls. Collapse occurs as deformation of the bladder cross-section progresses significantly enough that it no longer behaves as a column. If the neutral braid angle has not been reached, the braided mesh continues to exert positive force opposing the force from the bladder. Figure 32.6b illustrates the bladder within the mesh as its cross-section deforms from its circular shape.

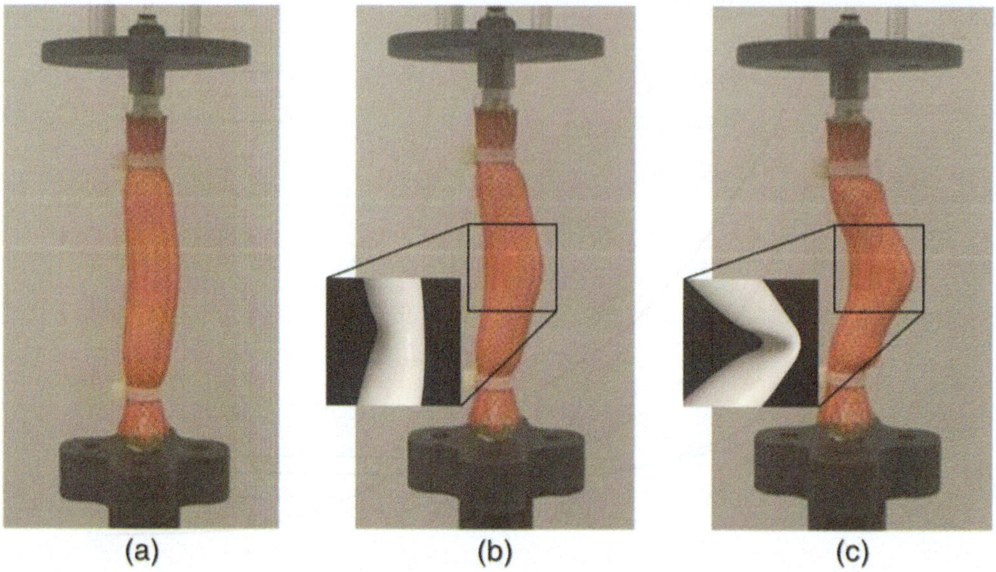

Fig. 32.6 Representative pictures of the McKibben actuator at different regions (**a**) region 1, (**b**) onset of region 2 as wrinkling occurs, and (**c**) onset of region 3 as bladder collapses

- **Region 3** follows the collapse of the bladder. The McKibben actuator has deformed significantly in the transverse direction from its linear shape. The bladder has 'folded' and behaves analogously to a hinge joint with a torsional spring. Figure 32.6c shows a completely collapsed bladder.

32.5 Implications on Overall Performance of a Variable Recruitment Bundle

The ultimate goal of this work is to investigate how the resistive forces generated by inactive and low-pressure elements within a variable recruitment bundle affect overall bundle performance. The effects of inactive elements within a bundle have already been demonstrated in Fig. 32.2, but this analysis can be extended to investigate the effects of low-pressure elements within a bundle as well. Understanding these effects is important, since in orderly recruitment, the pressure of a given motor unit is slowly increased to system pressure [9]. Current literature does not address the effects that these recruitment transitions have on overall system performance, but understanding these effects may be important, since McKibben actuators with elastic bladders exhibit significant pressure-dependent free strain characteristics, specifically at low pressures. This pressure-dependent free strain is due to the elastic forces in the bladder that act against the force exerted by the mesh. Any force generated by these low-pressure motor units past free strain will be negative, and as a result, it is hypothesized that if the actuator is used in a bundle, it will hinder the overall force production.

To validate this hypothesis, an additional experiment was designed. This experiment considered five identical McKibben actuators in a bundle configuration, as shown in Fig. 32.7. Each actuator had bladder outer diameter of 12.7 mm and slenderness ratio of 24, which were the same dimensions as earlier experiments. The bundle had two distinct motor units (MU1 and MU2), as separate regulators were used to control the pressure of the middle actuator and the outer four actuators. When MU1 is fully active, the bundle is in recruitment level 1 (RL1), and when both MU1 and MU2 are fully active, the bundle is in RL2. When RL2 is not fully active, the bundle is in a partial recruitment level. The pressure of the middle actuator was kept constant at 275.79 kPa (40 psi) and the pressure of the outer four actuators (MU2) was varied between 0 kPa and 241.32 kPa (35 psi). To establish a baseline, the force-strain behavior of the middle actuator at 275.79 kPa (40 psi) was characterized without the outer four actuators attached. Then, the MU2 actuators were added and force-strain characterizations were performed for MU2 pressures of 0 kPa to 241.32 kPa (35 psi) in increments of 34.5 kPa (5 psi), as the bundle transitioned from RL1 to RL2. These force-strain plots can be seen in Fig. 32.8. As expected, overall bundle free strain increases from 5 to 35 psi, where it is almost equal to the free strain of the middle actuator at 40 psi. More significantly, there seems to be a range of MU2 pressures during partial recruitment between RL1 and RL2 for which the overall bundle performs better with MU2 completely inactive than with MU2 at low pressures. This phenomenon seems to exist for the

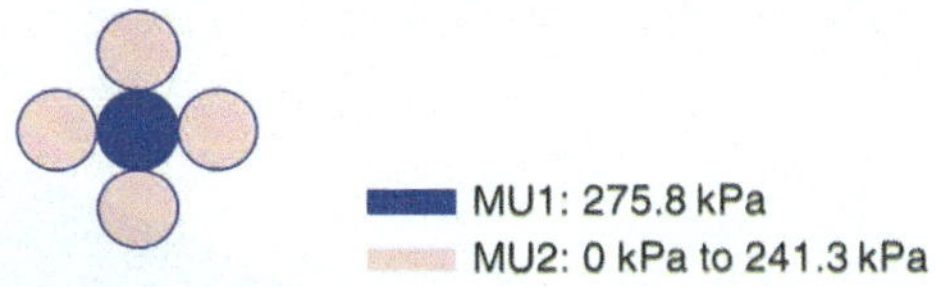

Fig. 32.7 Motor Units 1 and 2 and their testing pressure

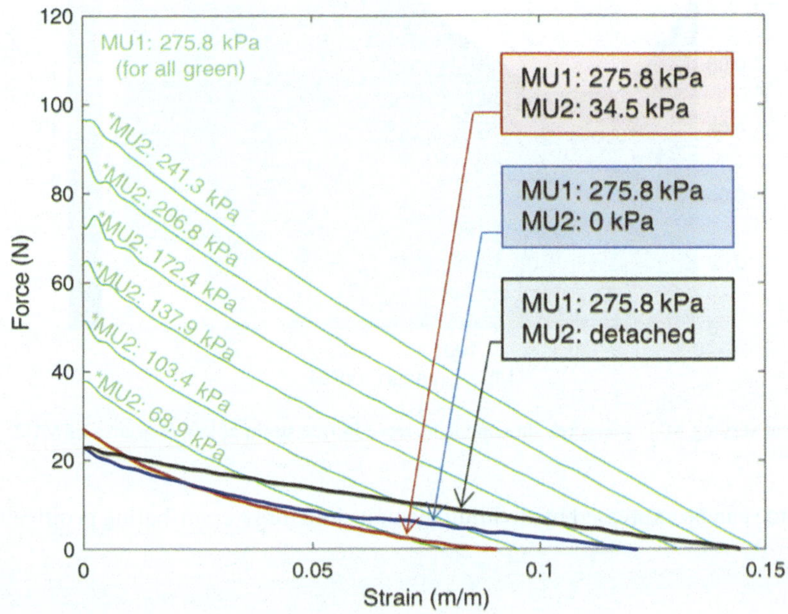

Fig. 32.8 Force-strain behavior of a variable recruitment bundle during recruitment level transition

5–15 psi cases, while at 20 psi, the bundle generates more force with an active MU2 than with an inactive MU2. This is visually illustrated in Fig. 32.9. It is hypothesized that this happens because at lower pressure, the magnitude of the elastic forces exceeds that of the forces generated by the mesh, resulting in resistive forces that decrease overall bundle force. In a bundle with an inactive MU2, there are no pressure-dependent elastic forces that act against the mesh force; there are only axial forces due to the compression of the bladder. Based on the data, these axial compressive forces must be lower in magnitude than the magnitude of the low-pressure elastic forces, a hypothesis is supported by Fig. 32.4, which shows that at higher strains, resistive forces of low-pressure actuators are greater than those of inactive actuators.

It is important to note that these results do not imply that variable recruitment would be infeasible or would hurt performance; rather, these results show that the contributions of individual actuators in partial recruitment levels is more complex than was previously believed. To address the resistive forces contributed by actuators in partial recruitment levels, it will be necessary to either design controllers that account for these effects or design hardware that prevents actuators from contributing any force to the overall bundle if they are not generating positive force. Using a more inelastic bladder would also help alleviate the elastic contributions at low pressures [15].

Previous literature on variable recruitment bundles have considered the overall force output of a bundle as the sum of the force output of individual elements. When resistive forces are considered, there are some interesting implications especially during transition between recruitment levels. First, let us express the force output of a bundle in the context of Ball's model as

$$F_{bundle} = \sum_{i=1}^{N} F_{unit,i} = \sum_{i=1}^{N} \left(F_{ideal,i} + F_{elastic,i}\right) \tag{32.5}$$

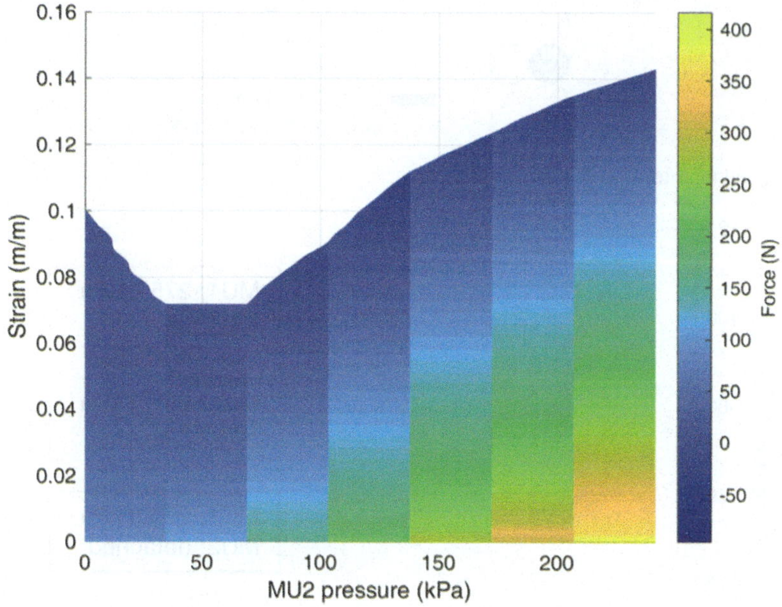

Fig. 32.9 Force-strain space at varying MU2 pressures showing a decrease in free strain at low pressures of MU2

Similarly, the same force can be expressed by dividing the units into those contributing positive force and the others acting against the overall force.

$$F_{bundle} = \sum_{j=1}^{n} F_{output,j} + \sum_{k=1}^{N-n} F_{resistive,k} \tag{32.6}$$

where n is the number of units in a bundle exerting a force, F_{output}, which is the positive force output of those units. In order to calculate the bundle force, it is necessary to accurately calculate the resistive forces generated by each actuator in the bundle past free strain, either by using an analytical formulation (incorporating elastic and compressive force expressions) or by empirically measuring the resistive force generated by each actuator in a bundle as a function of pressure. The latter method would provide a lookup table to calculate exact bundle force as a function of recruitment state, but it would require extensive campaigns of experiments. Future work will involve the use of both of these methods (analytical and empirical) to predict bundle force as a function of recruitment state.

32.6 Conclusion

This study experimentally investigated the contribution of the resistive force of inactive and low-pressure actuators to the overall force output of a variable recruitment bundle. It was found that these resistive forces can significantly affect the overall performance of a variable recruitment bundle at partial recruitment levels. An experimental study was carried out to characterize resistive force at varying pressures and some significant trends were identified to guide future attempts to model its behavior. As demonstrated in this paper, the resistive force within a variable recruitment bundle is a component of its overall performance which has the potential to utilize these resistive forces in optimizing hardware design and control methods. Although the scope of this study was focused on variable recruitment bundles, this investigation can potentially be extended to any system with multiple soft actuators that exhibit similar resistive behavior.

Acknowledgements This work was supported primarily by the Faculty Early Career Development Program (CAREER) of the National Science Foundation under NSF Award Number 1845203 and Program Manager Irina Dolinskaya. Additionally, this material is based upon work supported by the National Science Foundation Graduate Research Fellowship Program under Grant No. 1650114. Any Opinions, findings and conclusions or recommendations expressed in this material are those of the author(s) and do not necessarily reflect those of the National Science Foundation.

References

1. Rus, D., Tolley, M.T.: Design, fabrication and control of soft robots. Nature. **521**(7553), 467–475 (2015). http://dx.doi.org.prox.lib.ncsu.edu/10.1038/nature14543
2. Lin, H., Leisk, G., Trimmer, B.: Soft robots in space: a perspective for soft robotics. Acta Futura. **6**, 69–79 (2013)
3. Onal, C.D., Rus, D.: Autonomous undulatory serpentine locomotion utilizing body dynamics of a fluidic soft robot. Bioinspir. Biomim. **8**, 026003 (2013)
4. Marchese, A.D., Onal, C.D., Rus, D.: Autonomous soft robotic fish capable of escape maneuvers using fluidic elastomer actuators. Soft Robot. **1**, 75–87 (2014)
5. Polygerinos, P., Wang, Z., Galloway, K.C., Wood, R.J., Walsh, C.J.: Soft robotic glove for combined assistance and at-home rehabilitation. Robot. Auton. Syst. **73**, 135–143 (2015). https://doi.org/10.1016/j.robot.2014.08.014
6. Park, Y., Chen, B.: Design and control of a bio-inspired soft wearable robotic device for ankle-foot rehabilitation. Bioinspir. Biomim. **9**, 016007 (2014)
7. Bryant, M., Meller, M.A., Garcia, E.: Variable recruitment fluidic artificial muscles: modeling and experiments. Smart Mater. Struct. **23**, 074009 (2014). https://doi.org/10.1088/0964-1726/23/7/074009
8. Chapman, E.M., Bryant, M.: Bio-inspired passive variable recruitment of fluidic artificial muscles. In: Proceedings of SPIE 10593, Bioinspiration, Biomimetics, and Bioreplication VIII, 105930Y (2018)
9. Jenkins, T., Chapman, E.M., Bryant, M.: Bio-inspired online variable recruitment control of fluidic artificial muscles. Smart Mater. Struct. **25**, 125016 (2016). https://doi.org/10.1088/0964-1726/25/12/125016
10. Meller, M.A., Chipka, J., Volkov, A., Bryant, M., Garcia, E.: Improving actuation efficiency through variable recruitment hydraulic McKibben muscles: modeling, orderly recruitment control, and experiments. Bioinspir. Biomim. **11**, 065004 (2016). https://doi.org/10.1088/1748-3190/11/6/065004
11. De La Hunt, S.A., Pillsbury, T.E., Wereley, N.M.: Variable recruitment in bundles of miniature pneumatic artificial muscles. Bioinspir. Biomim. **11**, 056014 (2016). https://doi.org/10.1088/1748-3190/11/5/056014
12. Ball, E., Garcia, E.: Effects of bladder geometry in pneumatic artificial muscles. ASME J. Med. Devices. **10**(4), 041001 (2016). https://doi-org.prox.lib.ncsu.edu/10.1115/1.4033325
13. Chou, C., Hannaford, B.: Measurement and modeling of McKibben pneumatic artificial muscles. IEEE Trans. Robot. Autom. **12**(1), 90–103 (1996)
14. Tondu, B.: Modelling of the McKibben artificial muscle: a review. J. Intell. Mater. Syst. Struct. **23**(3), 225–253 (2012). https://doi.org/10.1177/1045389X11435435
15. Meller, M.A., Bryant, M., Garcia, E.: Reconsidering the McKibben muscle: energetics, operating fluid, and bladder material. J. Intell. Mater. Syst. Struct. **25**(18), 2276–2293 (2014). https://doi.org/10.1177/1045389X14549872

Jeong Yong Kim received the B.S (2012) in mechanical engineering from University of California, Berkeley, M.S. (2017) in mechanical engineering from University of Minnesota, Twin Cities. He is currently working on his Ph.D in mechanical engineering from North Carolina State University since 2019.

Chapter 33
Numerical and Experimental Study on the Modal Characteristics of a Rotor Test Rig

Verena Heuschneider, Florian Berghammer, and Manfred Hajek

Abstract A Mach scaled 85 kW rotor test rig for the investigation of Dynamic Stall on highly loaded helicopter rotor blades up to a diameter of 1.80 m and a tip speed up to 220 m/s is built at the Institute of Helicopter Technology, Technical University of Munich. The current state is the complete assembly of the drive train components including the engine cradle, wind tunnel adaptor, rotor shaft, and bearings as well as strain gauge load cells. To ensure a safe entry into service and operation at the nominal rotor speeds between 40 and 50 Hz, we develop a numerical finite element model of the test rig structure in order to study its dynamic characteristics. The numerical analysis results show critical eigenmodes at 44.21 and 54.98 Hz, modes 5 and 6. By means of a complementary experimental modal analysis on the steel frame structure natural frequencies, mode shapes, and damping characteristics of the critical modes are identified. The measurement results for three different excitation points and their synthetization prove the identification of ten eigenmodes between 0 and 70 Hz, which mainly emerge from the wind tunnel adaptor. Modes 6 and 7 have eigenfrequencies that are in and critically close to the nominal speed range, 46.97 and 54.88 Hz. They match with the numerically calculated eigenfrequencies but show different mode shapes, as the Modal Assurance Criterion (MAC) values for the comparison of numerical and experimental model convey. As closer investigations show, the numerical eigenfrequencies have a tendency to higher values from the fifth mode onwards.

Keywords Structural dynamics · Experimental modal analysis · Finite element model · Modal Assurance Criterion · Rotor test rig · Impact excitation

33.1 Introduction

Munich Experimental Rotor Investigation Testbed (MERIT) is a Mach scaled 85 kW rotor test rig for the investigation of Dynamic Stall on highly loaded helicopter rotor blades under centrifugal force influence. Designed and built at the Institute of Helicopter Technology, Technical University of Munich (TUM), its robust design allows for very high structural dynamic loads. Size and structure are due to the main design target, i.e. applicability in TUM's wind tunnel (Institute of Aerodynamics and Fluid Mechanics) for investigations in forward flight conditions. Divisibility and compactness simplify assembly as well as maintenance. The wind tunnel adaptor is necessary for bridging the wind tunnel attachment pads and is designed with the restriction of confined space conditions in height. Its structure as well as the four load cells mounted between bearing and middle frame imply eigenfrequencies within the rotational speed range of 0–50 Hz. Closer investigations of the system's modal parameters will ensure a safe entry into service and operation at the nominal rotor speeds between 40 and 50 Hz.

The objective of this paper is to identify and discuss the frequencies and shapes of critical eigenmodes by means of an experimental modal analysis on the steel frame structure with the complete drive train installed. A subsequent comparison of the modal analysis results with respective finite element model (FEM) analysis is used for verification and validation of the numerical model. Rotor and swashplate actuation components are not included for the investigations shown here.

V. Heuschneider (✉) · F. Berghammer · M. Hajek
Institute of Helicopter Technology, Department of Aerospace and Geodesy, Technical University of Munich, Munich, Germany
e-mail: v.heuschneider@tum.de; flo.berghammer@tum.de; hajek@tum.de

B. Dilworth (ed.), *Topics in Modal Analysis & Testing, Volume 8*, Conference Proceedings of the Society for Experimental Mechanics Series, https://doi.org/10.1007/978-3-030-47717-2_33

33.2 Experimental Setup

33.2.1 Mechanical Structure

The experimentally investigated structure consists of four different welded steel frames, Fig. 33.1:

- the wind tunnel adaptor welded on clamping pads
- the middle frame bolted to the wind tunnel adaptor
- the engine cradle bolted to the middle frame
- the bearing frame connected to the middle frame by four three component load cells

The bearing frame contains bearings and rotor shaft and serves as fundament for the attachment of actuators, swashplate, pitch links and the rotor itself. A torsion stiff, bending flexible double gimbal coupling transfers the engine torque to the rotor shaft. It allows for small radial, axial and angular displacements (3.7 mm, 2.0 mm, 1.0°) and contains a strain gauge torque measurement unit. The four three component strain gauge load cells measure rotor forces and moments. Each load cell allows for a deformation of 0.06 mm at a nominal force of +/− 2 kN. The 85 kW synchronous engine is side mounted to the engine cradle attached to the middle frame. The wind tunnel adaptor's clamping pads are pinched to the foundation's metal bars.

33.2.2 Data Acquisition

The investigated frequency scope ranges from 0 to 70 Hz, since the nominal rotational speed is located between 40 and 50 Hz and all eigenmodes up to and surrounding 50 Hz are considered. As a compromise between setup simplicity and time efficiency on the one hand, and excitability of the relevant modes on the other hand, we use impact excitation with a PCB impact hammer of 0.23 mV/N sensitivity, 0.32 kg weight and a vinyl tip. According to Berninger et al., the impact hammer with vinyl tip excites frequencies up to at least 120 Hz with almost constant amplitude, whereas a rubber tip shows a fast amplitude loss from 100 Hz onwards and the metal tip amplitude is half as high as the vinyl tip [1]. Own measurements show a constant hammer amplitude for up to 250 Hz, as presented in Fig. 33.2.

Four PCB triaxial accelerometers of 1.02 mV/g sensitivity and 4 g weight are used in a roving accelerometer approach [3] to measure the structural vibrations on 91 different discretization points as shown in Fig. 33.3. In relation to the overall structural weight of 1260 kg, the sensors' weight influence on eigenmodes is negligibly small. The impact hammer and accelerometer sensor signals were processed using the 24-channel Siemens LMS SCADAS A/D converter. The recording

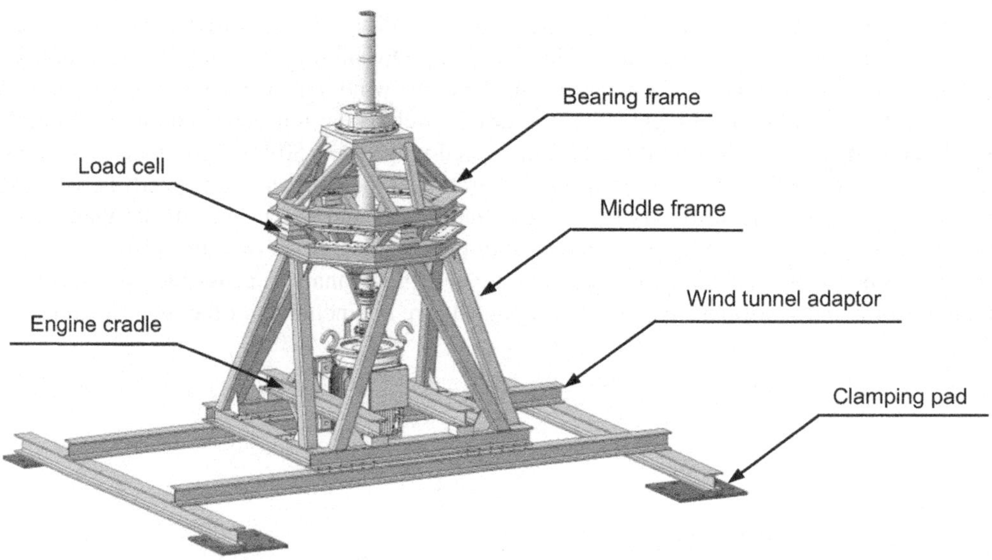

Fig. 33.1 Basic structural components of MERIT including drive train system

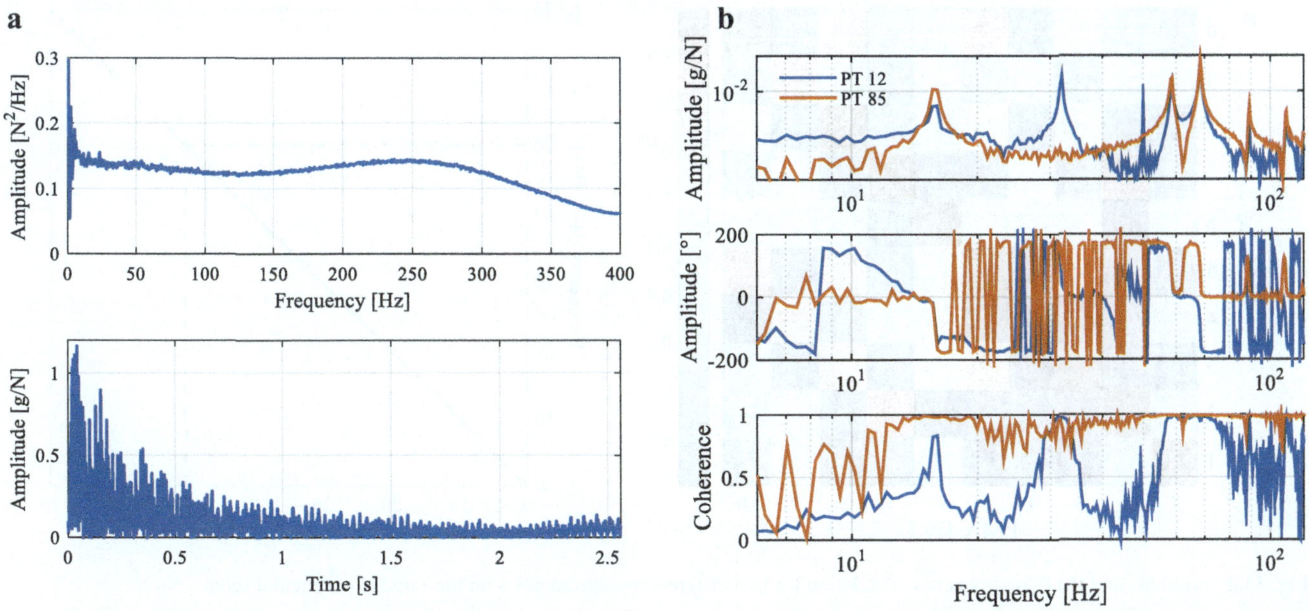

Fig. 33.2 (**a**) Frequency spectrum of exemplary hammer signal and time signal of acceleration sensor (**b**) Bode plot and coherence function for two different exemplary response paths in x-direction: from upper shaft end excitation point to sensor point PT12 on the wind tunnel adaptor (blue), and to sensor point PT85 on the lower shaft end (orange)

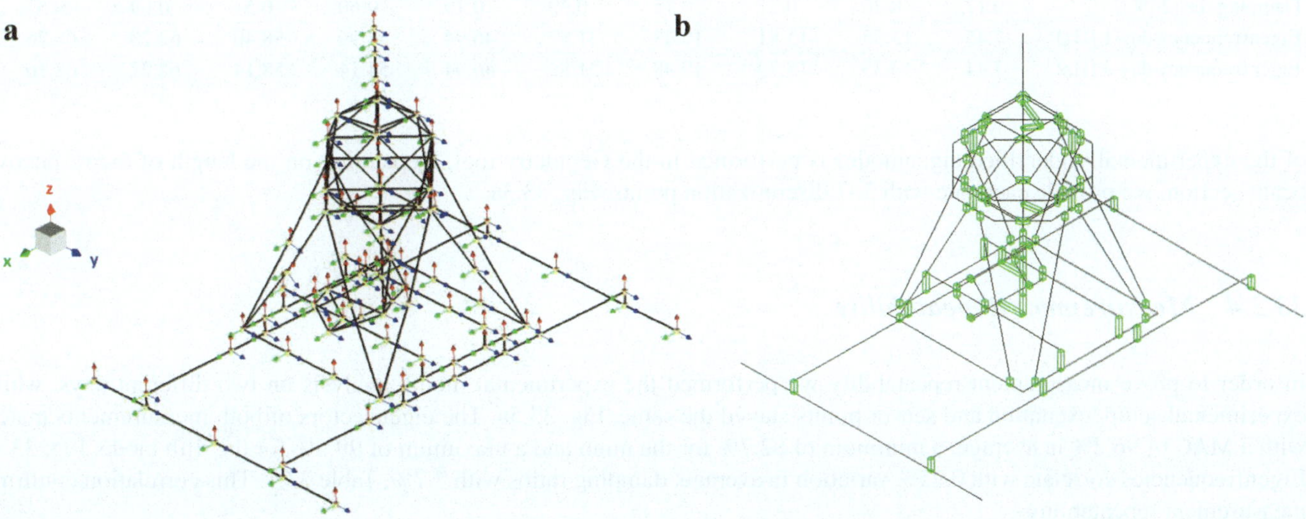

Fig. 33.3 Model geometry (**a**) Model geometry in Siemens Simcenter Testlab, 91 points (**b**) Finite element model geometry in Dymore with rigid body connections marked green

time for the sensor signals is 2.56 s and allows the structural vibrations to decay, as seen in Fig. 33.2a. Bode plot and coherence functions for two different exemplary response paths show completely different response quality, Fig. 33.2b. This confirms the need of more than one excitation point for eigenmodes identification.

33.2.3 *Experimental Modal Analysis Software*

The software used for the experimental modal analysis (EMA) is the Impact Testing Tool in the Siemens LMS Simcenter Testlab. We applied the PolyMax algorithm therein to get a sum function of all frequency response functions (FRF) at a resolution of 0.4 Hz and the deflection images for the identified eigenmodes [6]. The geometry generation for the extraction

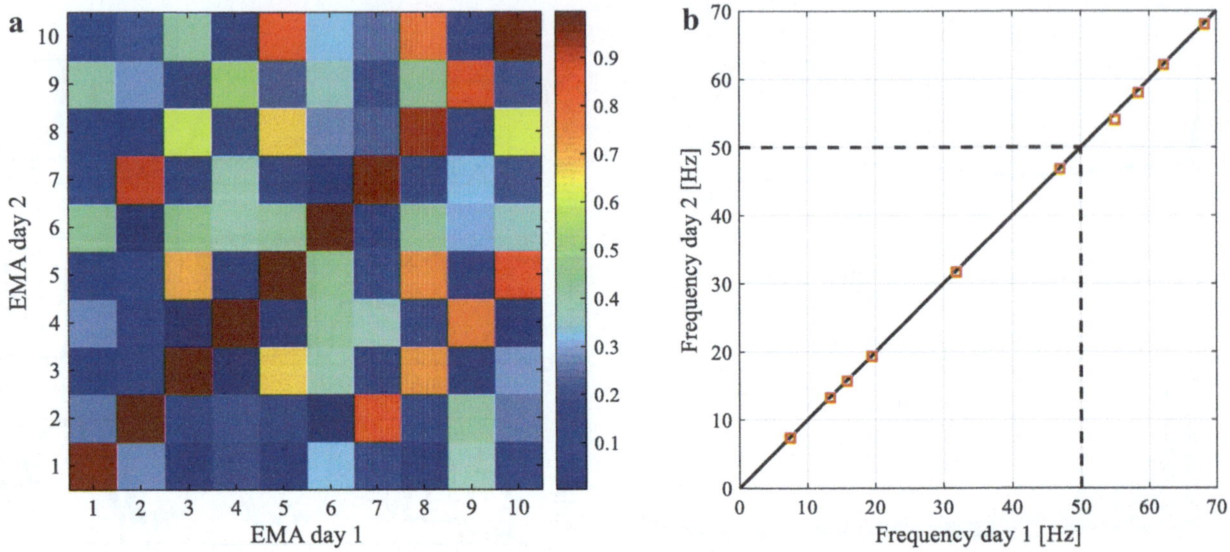

Fig. 33.4 (**a**) MAC and (**b**) Eigenfrequency correlation for two different measurements with the same experimental setup

Table 33.1 Damping and eigenfrequency correlation for two different measurements with the same experimental setup

Mode No.	1	2	3	4	5	6	7	8	9	10
Damping day 1 [%]	0.20	0.18	0.35	0.16	0.27	0.17	1.70	0.47	0.53	0.51
Damping day 2 [%]	0.17	0.26	0.31	0.15	0.30	0.19	0.60	0.53	0.69	0.57
Eigenfrequency day 1 [Hz]	7.45	13.35	15.81	19.48	31.87	46.94	54.99	58.40	62.20	68.26
Eigenfrequency day 2 [Hz]	7.44	13.35	15.79	19.48	31.82	46.94	54.14	58.14	62.22	68.16

of the experimentally identified eigenmodes is performed in the Geometry tool. Depending on the length of the respective beam section, we provide each one with 3–7 discretization points, Fig. 33.3a.

33.2.4 Measurement Repeatability

In order to prove measurement repeatability we performed the experimental modal analysis on two different days, while experimental setup, excitation and sensor points stayed the same, Fig. 33.5a. The eigenvectors of both measurements match with a MAC of 96.2% in average, a minimum of 82.7% for the ninth and a maximum of 99.5% for the fifth mode, Fig. 33.4. Eigenfrequencies correlate with 0.25% variation in average, damping ratios with 7.7%, Table 33.1. This correlation confirms measurement repeatability.

33.3 Experimental Modal Analysis

33.3.1 Excitation Variation

Preliminary measurements show that energy dissipation through material and structural damping, such as friction in bearings and screw connections, involve a combination of rotor shaft and wind tunnel adaptor excitation and successive modal synthesis in order to ensure the excitability of the first ten eigenmodes within 0 and 70 Hz. These are assumed by means of the finite element model calculated in advance. We chose three different impact positions for each measurement, as shown in Fig. 33.5. The excitation point for the first measurement is the upper shaft end, point A, where the highest amplitudes in all ten eigenmodes are expected. In order to ensure the excitation of the eigenmodes emanating from the wind tunnel adaptor, the second measurement involved point B1 on the middle frame for y and z excitation, and B2 for x excitation. Point C1 and C2 are symmetrical to point B1, B2 referring to the x-z-plane.

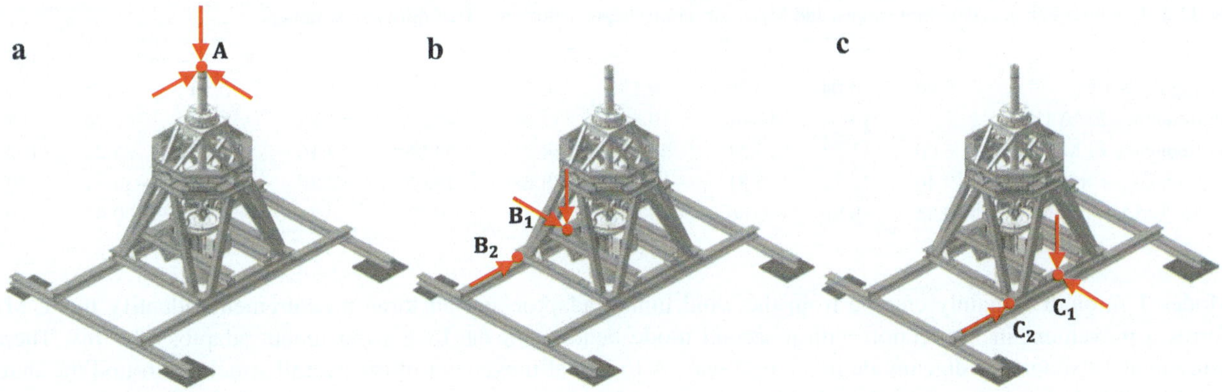

Fig. 33.5 Excitation variation for the three measurements **(a)** Excitation in point A **(b)** Excitation in point B **(c)** Excitation in point C

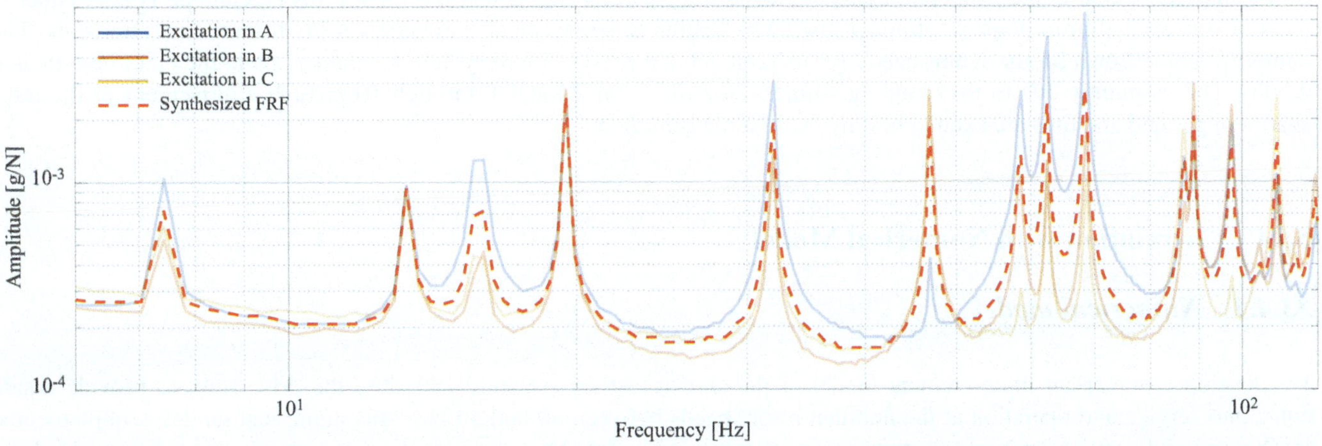

Fig. 33.6 FRF for the different measurements with excitation in points A, B, and C and synthesized FRF for all three measurements

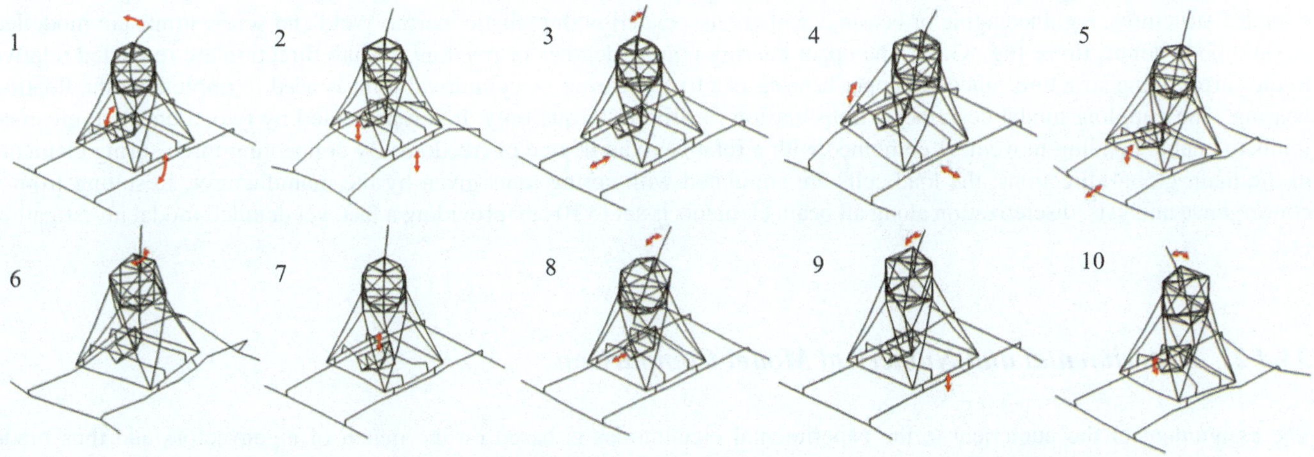

Fig. 33.7 Mode shapes 1 through 10 identified in the synthetization of all three measurements, red arrows indicate main movement

33.3.2 Experimental Modal Analysis Results

Figure 33.6 shows the frequency response functions of the three measurements with excitation in A, B, and C, as well as the synthesized FRF of all measurements, whereas Fig. 33.7 conveys the first ten mode shapes up to 70 Hz, resulting from the synthetization of all three measurements. Red arrows indicate the main movement origin.

Table 33.2 Eigenfrequencies, damping values, and MAC for mode shapes 1 through 10 of third measurement

Mode No.	1	2	3	4	5	6	7	8	9	10
Damping EMA [%]	0.06	0.04	0.20	0.13	0.24	0.16	1.36	0.50	0.35	0.41
Eigenfrequency EMA [Hz]	7.45	13.34	15.81	19.48	31.88	46.97	54.88	58.39	62.26	68.36
Eigenfrequency FEM [Hz]	6.00	11.58	13.31	23.20	44.21	54.98	59.16	72.32	85.24	102.76
Eigenfrequency Error [%]	19.46	13.19	15.81	−19.10	−38.68	−17.05	−7.80	−23.86	−36.91	−50.32
MAC for EMA vs. FEM	0.88	0.96	0.86	0.87	0.89	0.71	0.74	0.70	0.52	0.74

Modes 1 through 5 mainly emerge from the wind tunnel adaptor and all three measurements identify them. Mode 3 performs a movement in x direction with a second mode bending in the long wind tunnel adaptor's beams. Therefore, measurement 1 excites this eigenmode in the best way. A torsional movement of the overall structure around the shaft axis dominates mode 6, emanating from the wind tunnel adaptor. This mode is hard to be excited with the first measurement only as the small amplitude in the FRF confirms. The seventh Mode mainly results from the engine cradle oscillating in direction of the shaft axis. Due to the axial softness of the coupling between engine and rotor shaft this eigenmode is not visible in the FRF for excitation in point A, although the LMS Testlab Software clearly identifies it in all three measurements. The damping value, shown for the synthesized FRF in Table 33.2, for mode 7 is very high in comparison to the other investigated modes. This is mainly due to the coupling softness in z-direction. Modes 8 through 10 involve a movement of the rotor shaft's upper end and thus are excited best by the first measurement.

33.4 Correlation with Numerical Model

33.4.1 Numerical model

We develop a numerical finite element model of the shown test rig structure including the drive train to provide a safe entry into service and operation at the nominal rotor speeds between 40 and 50 Hz. This numerical model is implemented in Dymore and consists of beam sections, point masses and joints. Dymore uses the beam theory according to Hodges, which enables consideration of transverse shear effects and is an expansion of Timoschenko's beam theory [4]. The beam properties, such as mass distribution, stiffness and damping, are set for each beam segment. Point masses represent more complex structures, e.g. the engine or bearings and are associated with realistic inertia. Weld and screw joints are modelled as rigid body connections, Fig. 33.3b. The upper bearing points' degrees of freedom in shaft direction are restricted relative to the surrounding structure, since the upper bearing is a fixed bearing. A cylindrical joint is used to implement the floating bearing. The coupling model describes a transmission of driving torque only. It is represented by two connected universal joints to enable bending movement combined with a relative axial degree of freedom. By connecting three spring elements in the main global directions, the load cells are simulated with spring rates given by the manufacturer. Resulting from a convergence analysis, discretization along all beam elements is set to 10 cm, providing a fast, yet detailed modal investigation [1].

33.4.2 Experimental and Numerical Model Comparison

The assignment of the numerical to the experimental eigenmodes is based on the match of eigenvectors and thus mode shapes. The experimentally identified modal model is the synthesized model from all three measurements. The correlation by means of the MAC shows that the first five eigenmodes match unambiguously, Fig. 33.8a. Modes 6 to 8 and 10 show an average MAC of 72.3 %, which still needs improvement. The FE model does not describe the ninth eigenmode sufficiently.

Eigenfrequencies correspond very well for modes 1 through 4, Table 33.2. For higher modes, a clear shift to higher stiffness in the FEM is visible. We assume the reason for this shift is the use of rigid body connections between attachment points.

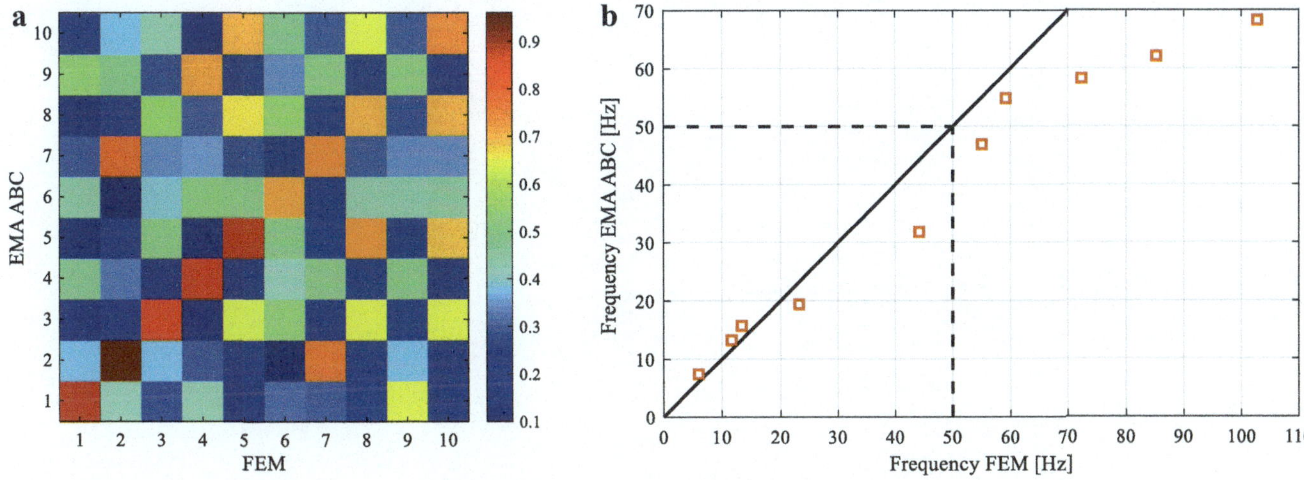

Fig. 33.8 (**a**) Modal Assurance Criterion and (**b**) Frequency comparison for FEM and EMA measurements 1, 2, and 3

33.5 Conclusion

The results of the Experimental Modal Analysis show that there are 10 eigenmodes between 0 and 70 Hz. Two of them lie critically close to and in the nominal rotor speed range of 40–50 Hz: mode 6 at 46.97 Hz and mode 7 at 54.88 Hz. Mode 7 is highly damped and therefore less critical than mode 6. Because of its oscillation around the shaft axis a passive damping construction is hard to apply. Therefore, a stiffening of the wind tunnel adaptor's long beams is worth considering. The damping value for the seventh eigenmodes, as is shown in Table 33.1, varies strongly and points to a nonlinear damping behavior resulting from the coupling, which allows for small axial displacements. Stiffness and damping adjustment or identification need to be performed in order to increase model accuracy for the eigenmodes higher than the first five.

Acknowledgements The Siemens LMS Simcenter software and data acquisition equipment were used with the kind support of the Chair of Applied Mechanics, Prof. Dr. Daniel Rixen, PD Dr.-Ing. habil. Thomas Thümmel and Ahmed El Mahmoudi, M.Sc. All experiments were performed at the Laboratory for Product Development and Lightweight Design, both at Technical University of Munich, Germany. Their support is gratefully acknowledged.

References

1. Berghammer, F.: Vibration behaviour modelling and analysis of a helicopter rotor test rig. Master's Thesis, Institute of Helicopter Technology, Technical University of Munich (2019)
2. Berninger, T. F. C., Fuderer, S., and Rixen D. J.: Modal analysis of a 7 DoF Sweet Pepper Harvesting Robot. In Society *for Experimental Mechanics, Inc. 2020, Topics in Modal Analysis & Testing, Volume 8,* Proceedings of the 37th IMAC, A conference and exposition on structural dynamics 2019, pp. 163–170. (2019). https://doi.org/10.1007/978-3-030-12684-1_16
3. Di Lorenzo, E., Manzato, S., Peeters B., Ruffini V., Berring, P., Haselbach, P. U., Branner, K., and Luczak, M.M.: Modal analysis of wind turbine blades with different test setup configurations. In *Society for Experimental Mechanics, Inc. 2020, Topics in Modal Analysis & Testing, Volume 8,* Proceedings of the 37th IMAC, A conference and exposition on structural dynamics 2019, pp. 143–152. (2019). https://doi.org/10.1007/978-3-030-12684-1_14
4. Hodges, D.H., Yu, W.: Generalized Timoschenko Theory of the Variational Asymptotic Beam Sectional Analysis. Journal of the American Helicopter Society. **50**, 46–55 (2005)
5. Larsen, G.C., Hansen, M.H., Baumgart, A., Carlén, I.: Modal analysis of wind turbine blades. In: Forskningcenter Risoe, Risoe-R, No. 1181 (EN). Forskningscenter Risoe, Roskilde (2002)
6. Peeters, B., Van Der Auweraer, H.: PolyMAX: a revolution in operational modal analysis. In: International operational Modal Analysis Conference (IOMAC), p. 13. (2005)

Verena Heuschneider (i) Aerospace Engineering Bachelor and Master at TUM. (ii) Master's Thesis on the topic of nonlinear stiffness and damping characteristics of a rotor blade attachment and their influence on helicopter ground resonance. (iii) Research Associate at Institute of Helicopter Technology since March 2016. (iv) Rotor test rig design and construction.

Chapter 34
A Comparison of Different Boundary Condition Correction Methods

Peter A. Kerrian

Abstract Fixed base correction methods have been increasingly used to transform flexible or dynamically active boundary conditions into fixed boundaries in modal tests. This paper examines the extraction of fixed base modes from a demonstration test article utilizing different frequency response function (FRF) and mode shape based methods. First, fixed base FRFs are directly calculated utilizing the base acceleration measurements or constraint shapes as the references. Second, constraints shapes are applied to the nominal mode shapes. Finally, a hybrid method is investigated that calculates FRFs from baseline mode shapes and then calculates fixed base modes using the FRF-based methods.

Keywords Modal testing · Vibrations · Base-shake · Fixed base · Constraint shapes · Boundary condition correction

Nomenclature

BCCM	Boundary condition correction method
CMIF	Complex mode indicator function
DOF	Degree of freedom
DP	Drive point
FB	Fixed base
FBCM	Fixed base correction method
FEM	Finite element model
FRF	Frequency response function
SMURF	Structural modification using frequency response functions
SVD	Singular value decomposition
TA	Test article
TDM	Test display model

34.1 Introduction

Recently, multiple boundary condition correction methods (BCCMs) [1–4] have been developed to remove the boundary dynamics of test articles (TAs). This ability is very desirable, as extracting fixed base (FB) modes shapes from the results of modal tests performed on TAs with dynamically active boundary conditions can save time and effort during any subsequent model correlation task. The fixed base correction method (FBCM) [5–13] uses acceleration data or constraint shapes as references when calculating frequency response functions (FRFs) to transform flexible or dynamically active boundary conditions into fixed boundaries. An alternative approach [6, 14, 15] calculates and applies constraint shapes directly to the mode shapes via the modal mass, stiffness, and damping matrices; an eigensolution is then computed with the transformed mass, stiffness, and damping matrices to calculate the FB modes directly. Finally, a hybrid approach utilizes the extracted mode shapes to synthesize the measured FRFs and subsequently applies the FCBM to the synthesized FRFs. The primary motivation of this paper is to apply different BCCMs to a common set of test data to compare the effectiveness of the

P. A. Kerrian (✉)
ATA Engineering, Inc., San Diego, CA, USA
e-mail: peter.kerrian@ata-e.com

B. Dilworth (ed.), *Topics in Modal Analysis & Testing, Volume 8*, Conference Proceedings of the Society for Experimental Mechanics Series, https://doi.org/10.1007/978-3-030-47717-2_34

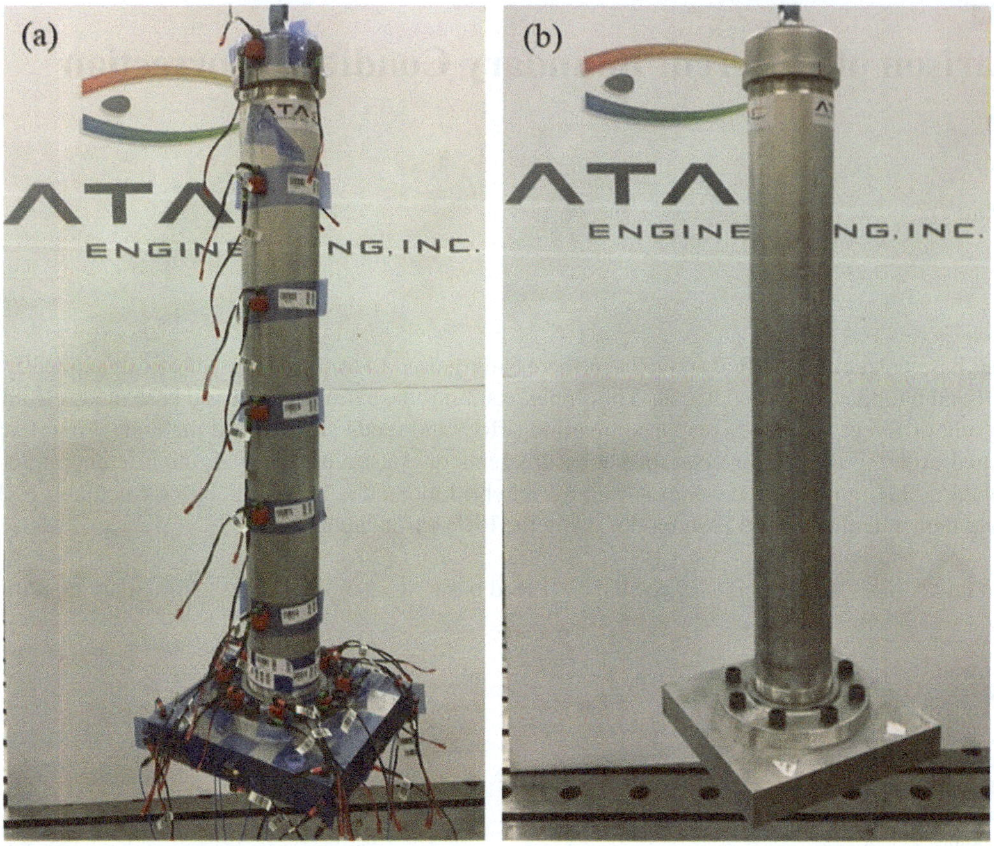

Fig. 34.1 Photographs of the TA with (**a**) and without (**b**) instrumentation

methods and identify any challenges or additional considerations that arise when FB mode shapes are calculated from test data collected in a free-free boundary condition.

34.2 Demonstration Test Article

Figure 34.1 shows a photograph of the demonstration TA with and without instrumentation. The TA consists of a 3 ft. aluminum pipe that is 0.5 in. thick, has a diameter of 4.5 in., and is threaded into a steel flange attached to a 1 ft^2, 2 in. thick aluminum plate. The top of the pipe is threated into an aluminum cap with approximately a 10 lb. weight firmly attached to the center of the cap inside the pipe. A bungee suspension system was used to attach the TA to an overhead gantry crane via an eyelet screwed into the cap. A shell finite element model (FEM) of the TA without the suspension system was developed to identify target modes in both a free-free and an FB boundary condition.

34.3 Summary of Test and Free-Free Modes

The FEM was utilized in a traditional pretest analysis to calculate the reduced mass and stiffness matrices as well as a constraint matrix for back-expanding the measured degrees of freedom (DOFs) to all DOFs in the test display model (TDM). The TA was instrumented with 95 channels of accelerometers, and a roving impact hammer test was completed. Figure 34.2 shows the TDM with the measurement DOFs and select node numbers. For the baseplate, redundant DOFs were placed at the corners and midpoints to allow for a reference and response set of DOFs at the same location. Impacts were made in all three directions at the corners of the baseplate, in the vertical direction at the midpoints and center of the baseplate, in the X direction at node 330, and in the X and Y directions at top of the cylinder at nodes 101 and 104, respectively. During

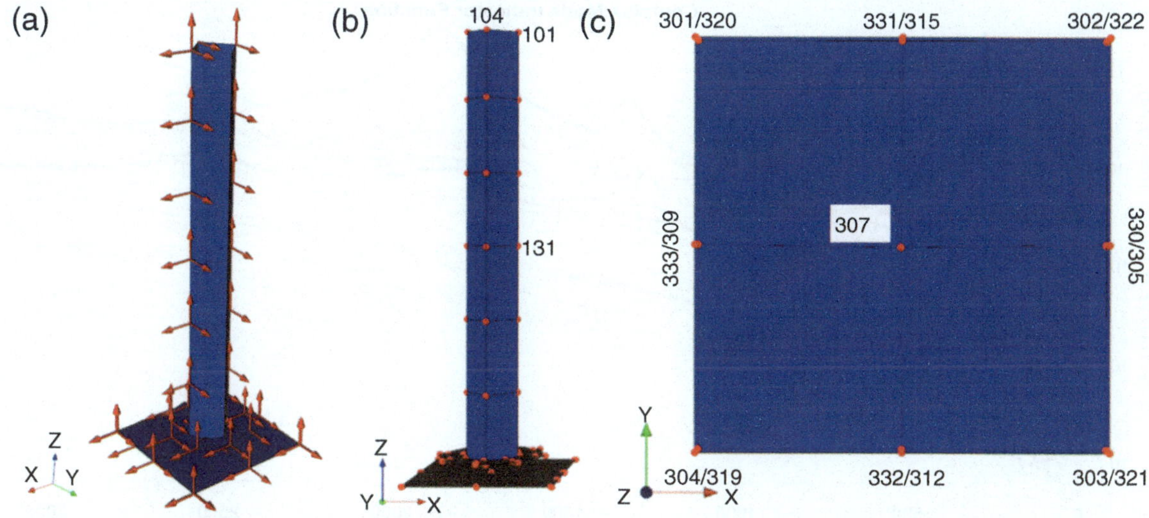

Fig. 34.2 TDMs showing (**a**) the measurement DOFs of the full model, and select node numbers for the (**b**) full model and (**c**) baseplate

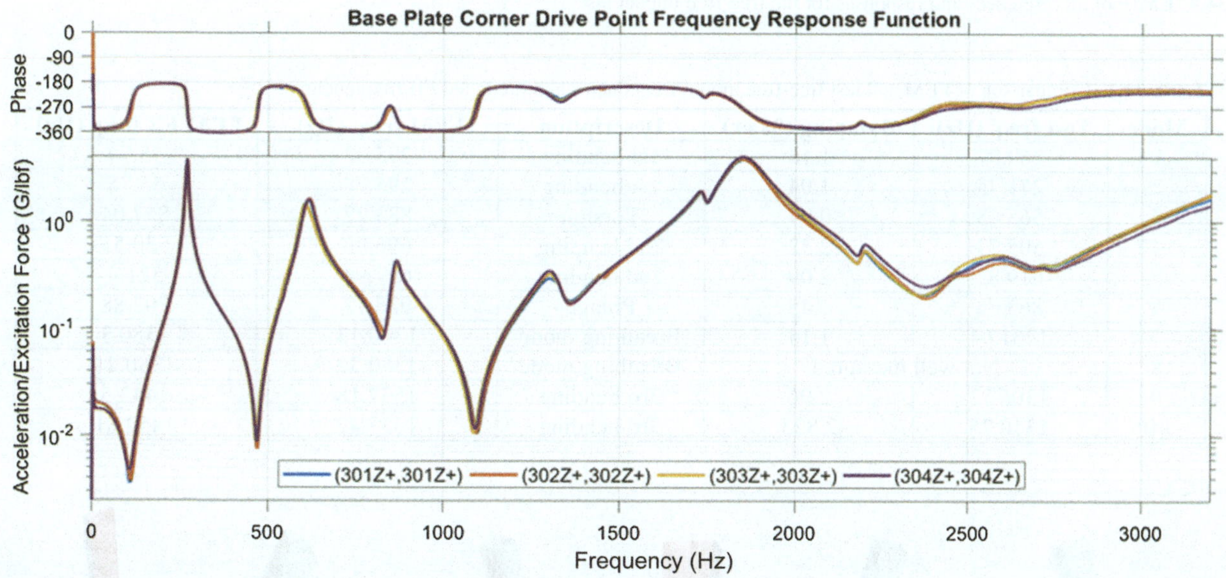

Fig. 34.3 DP FRFs for vertical impacts at the four corners of the baseplate

the data collection, special attention was paid to ensuring the quality of the antiresonances of the baseplate drive point (DP) FRFs. Figure 34.3 shows the consistency of the antiresonances of the DP FRFs for the four corners of the TA. Figure 34.4 shows the complex mode indicator function (CMIF) of all references and responses. The higher-order curves of the CMIF appearing to show multiple closely spaced modes near the primary bending modes of the structure are actually small shifts in frequency of the first two modes associated with the roving impact hammer.

Twenty-three free-free modes of the TA were extracted from three different curve fits over different frequency ranges to capture sufficient higher-frequency modes for subsequent use in different BCCMs. Table 34.1 summarizes the predicted FEM free-free, FEM FB, and measured free-free test modal frequencies for the target modes (green) and additional modes (orange). The FEM was not updated with the free-free mode shapes before the FB modes were predicted. Figure 34.5 shows plots of the measured free-free target mode shapes.

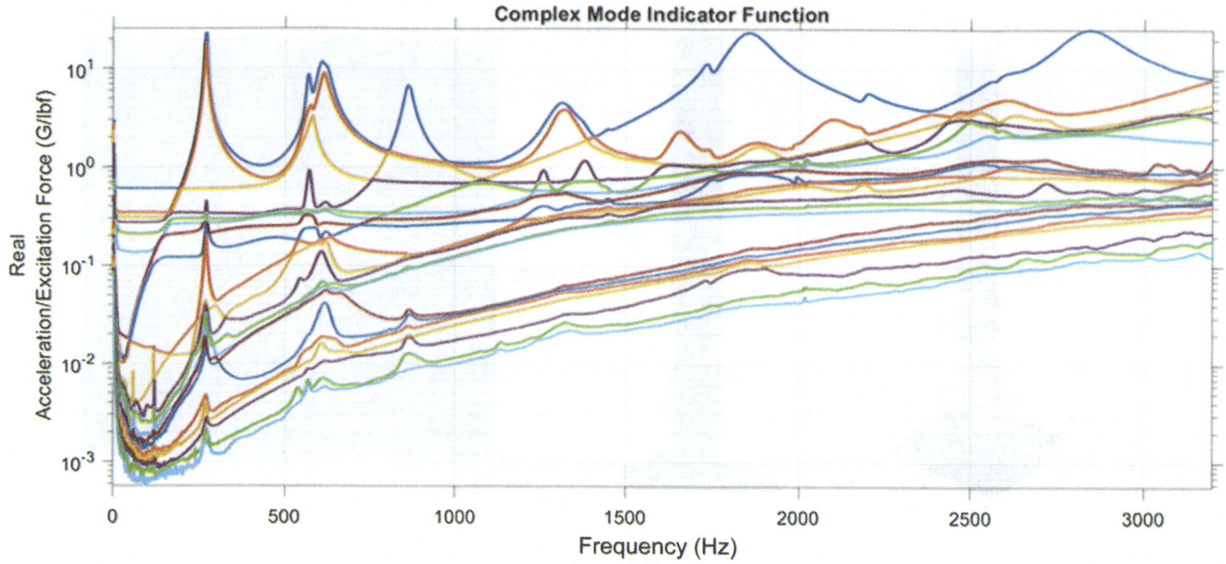

Fig. 34.4 CMIF of all references and responses for the free-free impact test

Table 34.1 Comparison of FEM and test free-free modal frequencies and predicted FB frequencies

Mode	Test freq. (Hz)	Damping (% cr)	Description	FEM freq. (Hz)	FEM FB freq. (Hz)
1	269.07	1.16	1st bending	295.32	55.74
2	271.75	1.04	1st bending	295.78	55.75
3	565.68	0.94	Torsion	883.22	857.04
4	605.73	2.32	2nd bending	696.96	520.55
5	619.82	2.09	2nd bending	697.94	521.3
6	863.76	1.45	Plunge	935.72	702.88
7	1261.64	1.14	Breathing mode	1380.13	1380.3
8	Not well measured		Breathing mode	1380.32	1380.21
9	1306.52	2.96	3rd bending	1537.18	1442.5
10	1310.75	2.823	3rd bending	1537.42	1452.81

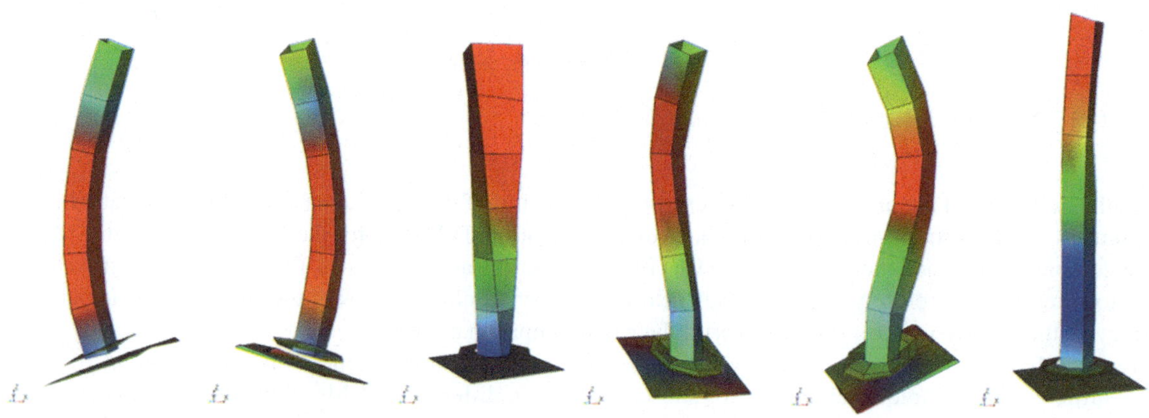

Fig. 34.5 Plots of the free-free target modes

34.4 Method 1: Fixed Base Correction Method

The fixed base correction method (FBCM) directly calculates FB FRFs from the test data by using base acceleration data or constraint shapes as references in either the initial calculation of the FRF matrix utilizing the H_d method [16], or through a partial matrix inversion of the FRF matrix with the structural modification using frequency response functions method (SMURF). For this work, constraint shapes $[\Psi]$ were calculated and used as the references. The constraint shapes relate the base DOFs $\{x_B\}$ to a set of DOFs associated with each constraint shape, ϵ:

$$\{x_B\} = [\Psi]\{\epsilon\} \tag{34.1}$$

The measured FRF matrix that contains the TA $\{x_I\}$ and base DOFs can be back-expanded to include the DOFs associated with each constraint shape:

$$\begin{Bmatrix} x_I \\ x_B \end{Bmatrix} = \begin{bmatrix} H_{II} & H_{IB} \\ H_{BI} & H_{BB} \end{bmatrix} \begin{Bmatrix} f_I \\ f_B \end{Bmatrix} \quad \rightarrow \quad \begin{Bmatrix} x_I \\ \epsilon \end{Bmatrix} = \begin{bmatrix} H_{II} & H_{IB} \\ \Psi^+ H_{BI} & \Psi^+ H_{BB} \end{bmatrix} \begin{Bmatrix} f_I \\ f_B \end{Bmatrix} \tag{34.2}$$

A partial matrix inversion of the FRF matrix can be performed to utilize the constraint shape DOFs as references to remove the base motion of the interface and obtain the FB FRFs:

$$f_B = -\left(\Psi^+ H_{BB}\right)^T \Psi^+ H_{BI} f_I + \left(\Psi^+ H_{BB}\right)^T \epsilon \tag{34.3}$$

The resulting FRF matrix relation is as follows:

$$\begin{Bmatrix} x_I \\ x_B \end{Bmatrix} = \begin{bmatrix} H_{II} - H_{IB}\left(\Psi^+ H_{BB}\right)^T H_{BI} & H_{IB}\left(\Psi^+ H_{BB}\right)^T \\ H_{BI} - H_{BB}\left(\Psi^+ H_{BB}\right)^T \Psi^+ H_{BI} & H_{BB}\left(\Psi^+ H_{BB}\right)^T \end{bmatrix} \begin{Bmatrix} f_I \\ \epsilon \end{Bmatrix} \tag{34.4}$$

The constraint shapes were calculated from a singular value decomposition (SVD) of the FRF matrix partitioned down to the 45 base DOFs. To calculate the FB FRFs, the SMURF method was used to invert all the impacts on the base, with eight constraint shapes encompassing the six rigid body modes and two flexible plate modes. The eight constraint shapes encompassed 93% of the total baseplate motion. Figure 34.6 contains a CMIF of the FB FRFs for only the response DOFs

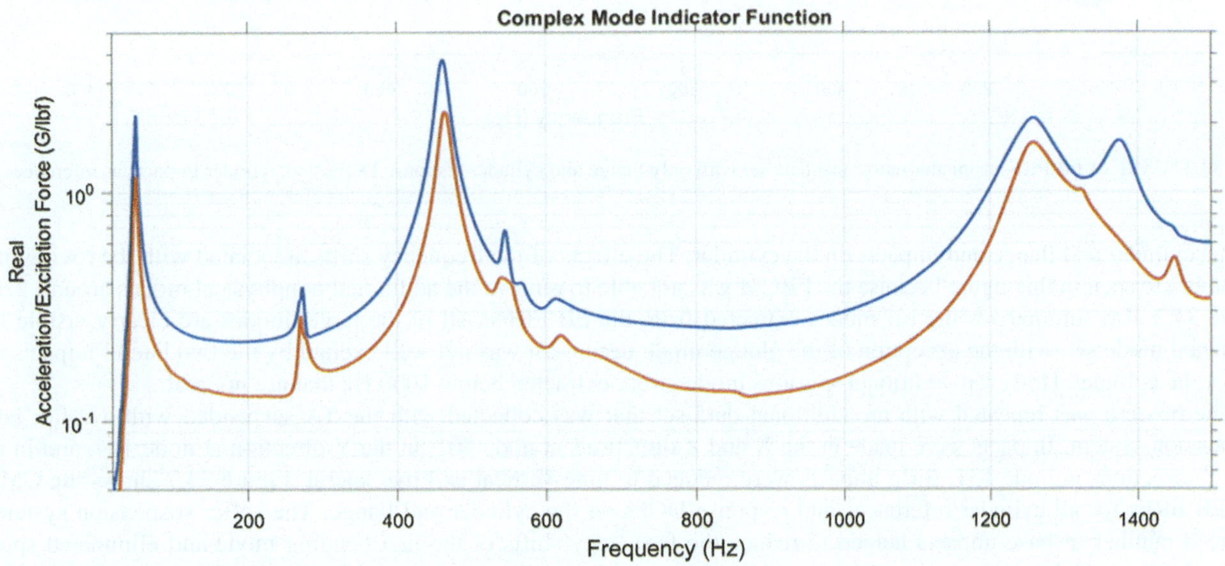

Fig. 34.6 CMIF of FB FRFs with only flange and cylinder response DOFs and cylinder impacts as references

Table 34.2 Extracted FB modal frequencies for method 1 from two sets of test data

			Method 1	
Mode	FEM FB freq. (Hz)	Description	FB freq. (Hz)	FB freq. (Hz)[a]
1	55.74	1st bending	49.60	49.55
2	55.75	1st bending	50.07	50.4
3	520.55	2nd bending	461.35	461.24
4	521.3	2nd bending	465.90	465.0
5	857.04	Torsion	544.9	543.78
6	702.88	Plunge	606.27[b]	660.8
7	1380.3	Breathing mode		
8	1380.21	Breathing mode		
9	1442.5	3rd bending	1250.92	1258.87
10	1452.81	3rd bending	1257.13	1258.94
A	Spurious 1st bending		271.57	
B	Spurious torsion		568.57	
C	Spurious plunge		617.96	

[a]Alternative set of test data
[b]Strongly coupled with torsion

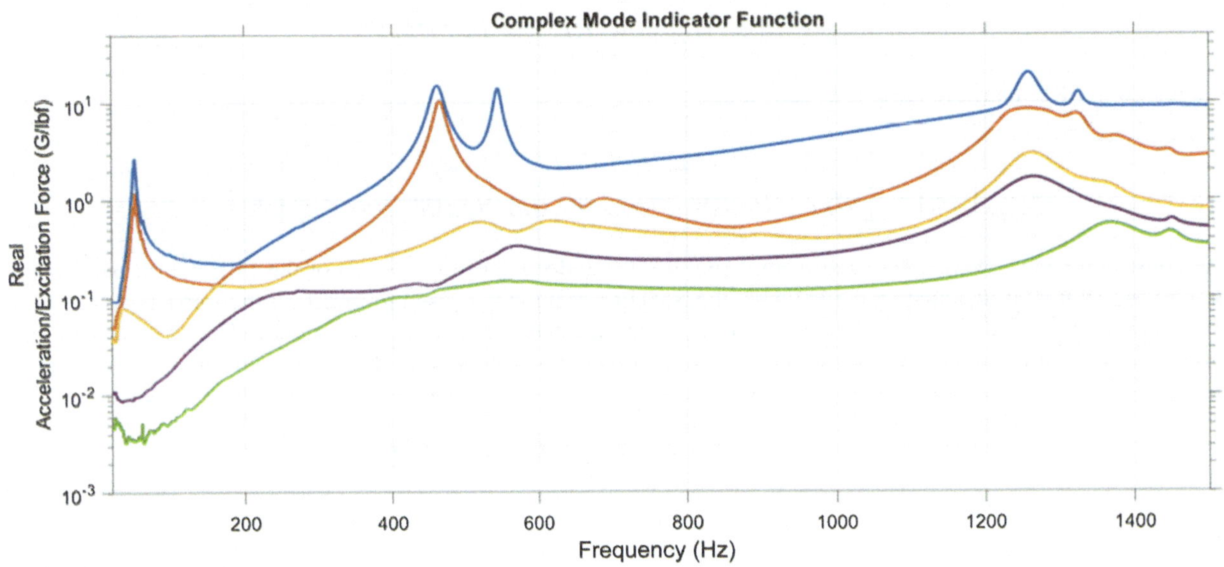

Fig. 34.7 CMIF of FB FRFs from alternative test data set with only flange and cylinder response DOFs and cylinder impacts as references

on the cylinder and flange and impacts on the cylinder. The effects of the frequency shifts associated with the roving impact hammer are seen in this figure because the FBCM was not able to remove the additional nonphysical modes around 270 Hz. Table 34.2 also summarizes the FB modes extracted from the FB FRFs. All of the target modes are clearly visible in the extracted mode set, with the exception of the plunge mode because it was not well excited by the two lateral impacts at the top of the cylinder. However, additional spurious modes were extracted below 1000 Hz that are not real.

The process was repeated with an additional data set that was collected with the TA suspended with a softer bungee suspension system. Impacts were made in the X and Z directions at node 101, in the Y direction at node 104, and in the X and Y directions at node 131. Base impacts were reduced to nine vertical and five lateral. Figure 34.7 shows the CMIF of the FB FRFs for all cylinder references and response DOFs on the cylinder and flange. The softer suspension system and reduced number of base impacts helped to reduce the frequency shifts of the first bending mode and eliminated spurious modes below 1000 Hz. The extracted FB modes from the alternative data set are contained in the last column of Table 34.2. Also, the vertical impact at node 101 enabled a clean extraction of the plunge mode. Figure 34.8 contains plots of the FB mode shapes extracted.

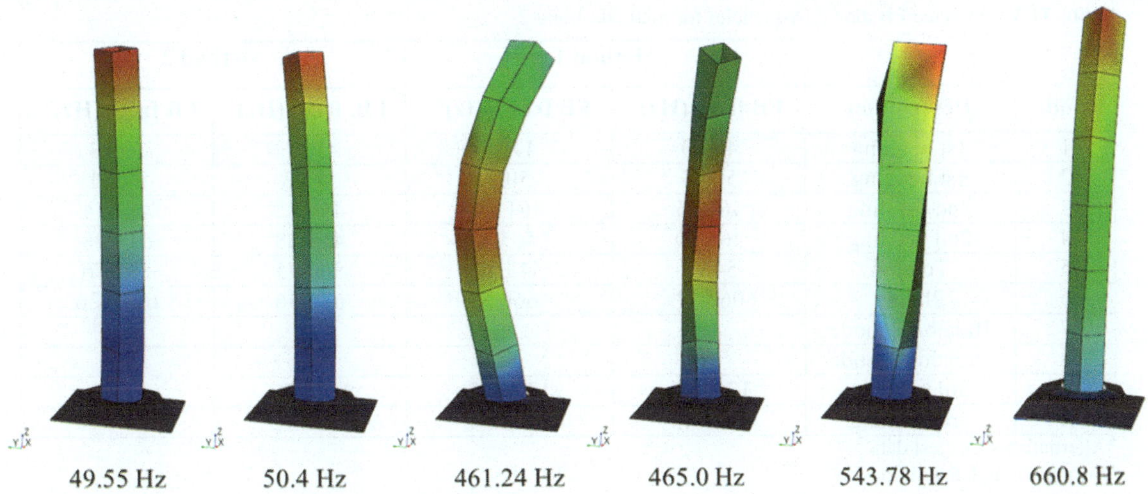

Fig. 34.8 FB mode shapes collected from the alternative data set

34.5 Method 2: Constraint Shape Applied Directly to Mode Shapes

The second method [15] applies rigid body constraint shapes directly to the measured mode shapes by modifying the test-measured mass, damping, and stiffness matrices with a transformation matrix and calculating an eigensolution of the modified dynamic system. The physical DOFs, $\{x\}$, are related to the modal DOFs, $\{q\}$, with the mode shapes, $[\Phi]$. For this method, the mode shapes need to include the rigid body modes of the system with accurate modal mass. In the modal space, the mass, damping, and stiffness matrices can be calculated from the modal parameters:

$$\begin{aligned} M_m &= \Phi^T \mathrm{M}\ \Phi = m_m \\ C_m &= \Phi^T \mathrm{C}\ \Phi = 2\zeta\omega m_m \\ K_m &= \Phi^T \mathrm{K}\ \Phi = \omega^2 m_m \end{aligned} \tag{34.5}$$

As in Method 1, constraint shapes can be calculated from an SVD of the modal coefficients. Only the constraint shapes associated with the six rigid body modes, $[\Psi_c]$, are used in the calculation of the transformation matrix, T. The transformation matrix is calculated from the null space of the product of the pseudoinverse of the six rigid body constraint shapes multiplied by the free-free mode shapes partitioned down to the base DOFs.

$$\mathrm{T} = \varnothing\left(\Psi_c^+ \Phi_c\right) \tag{34.6}$$

The transformation matrix is applied to the modal space mass, damping, and stiffness matrices, and an eigensolution of the modified system is solved via a traditional matrix coordinate transformation (i.e., $\mathrm{T}^T \mathrm{M}_m\ \mathrm{T}$ for the modal mass matrix). The resulting eigenvalues correspond to the FB modal frequencies, and the eigenvectors, $[\Phi_\epsilon]$, correspond to the FB modes. The FB mode shapes in physical domain can be recovered with the following coordinate transformation:

$$[\Phi_{FB}] = [\Phi]\,[\mathrm{T}]\,[\Phi_\epsilon] \tag{34.7}$$

The six rigid body mode shapes for the TA were synthesized based on the estimated center of gravity and scaled to have accurate modal mass based on the measured FRFs for the free-free condition. Constraint shapes were calculated from the SVD of baseplate DOFs of the combined rigid body and free-free mode set. Table 34.3 contains the frequencies associated with the eigensolution for both data sets, and Fig. 34.9 shows the FB modes back-expanded to the physical modal domain. Overall, there is good agreement between the two methods relative to both frequency and mode shape of the first six mode shapes. There is more of a separation between the bending pairs, perhaps because the uncertainty in the mass normalized rigid body modes calculated from the free-free test modes. Table 34.3 also shows that the FB modal frequencies calculated with the alternative data set were not as consistent with the other FB modal frequencies. However, Table 34.4 shows the

Table 34.3 Extracted FB modal frequencies for methods 1 and 2

		Method 1		Method 2	
Mode	Description	FB freq. (Hz)	FB freq. (Hz)[a]	FB. freq. (Hz)	FB freq. (Hz)[a]
1	1st bending	49.60	49.55	49.46	53.83
2	1st bending	50.07	50.4	58.83	61.31
3	2nd bending	461.35	461.24	456.82	489.14
4	2nd bending	465.90	465.0	479.58	511.28
5	Torsion	544.9	543.78	560.75	549.76
6	Plunge	606.27[b]	660.8	688.36	699.626
7	Breathing mode				
8	Breathing mode				
9	3rd bending	1250.92	1258.87	1210.57	
10	3rd bending	1257.13	1258.94	1268.92	

[a]Alternative set of test data
[b]Strongly coupled with torsion

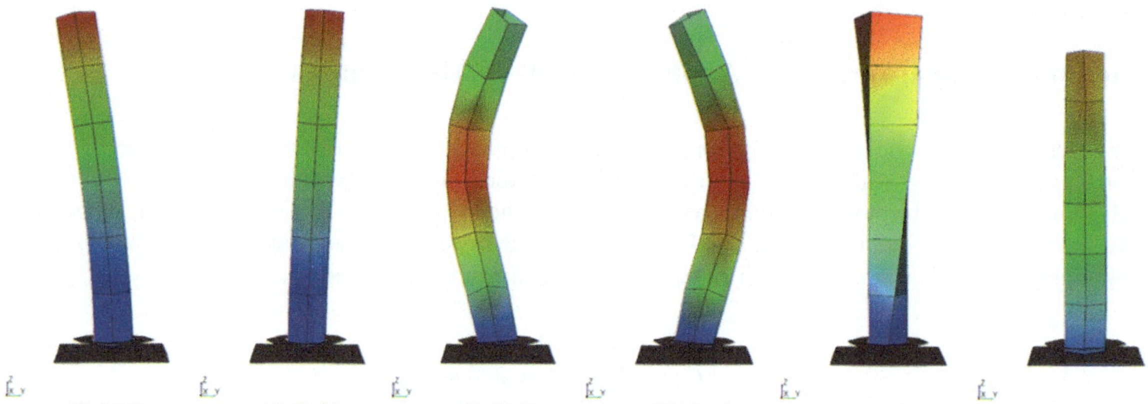

Fig. 34.9 FB mode shapes calculated from method 2, original data set

Table 34.4 Cross-orthogonality of FB modes calculated with methods 1 and 2

		Method 2						Test	Test	90–100
		1	2	3	4	5	6	CRSS	CRSS	50–89
Method 1[a]		53.83	61.31	489.14	511.28	549.76	699.63	3%	All	0–49
1	49.55	82	57					82	100	
2	50.40	22	98					98	100	
3	461.24			99				99	100	
4	464.99				99			99	100	
5	543.78					100		100	100	
6	660.81						100	100	100	
CRSS	3%	100	100	100	99	100	100			
CRSS	All	100	100	100	100	100	100			

[a]FB modes from alternative data set

cross-orthogonality matrix between the FB mode shapes extracted via methods 1 and 2 for the alternative data set, and there is excellent agreement between the mode shape sets. To investigate the discrepancy, the method was performed on simulated data from a free-free FEM, and the results matched a fixed base FEM. Therefore, the discrepancy is likely associated with the test-generated modal mass properties of the rigid body modes.

34.6 Method 3: Hybrid Approach

The final approach applies the FBCM, method 1, to FRFs synthesized from the free-free mode shapes. In theory, the synthesized FRFs should help improve the overall quality of the FRFs by removing noise and cleaning up the phase. The challenge, however, is obtaining quality free-free mode shapes and accurate residuals to match the test-measured FRFs. Deviations in the antiresonances will result in frequency shifts of the FB modes due to the partial matrix inversion. The set of 23 free-free mode shapes were used to synthesize the FRF matrix for all test-measured references and responses. Figures 34.10, 34.11, and 34.12 compare the measured and synthesized DP FRFs for impact locations on the cylinder and in the corners of the baseplate. The synthesized FRFs represent the measured FRFs accurately up to approximately 1000 Hz, with the presence of the plunge mode appearing in the 101X+ synthesized DP FRF as the obvious anomaly. The antiresonances line up well with respect to frequency, with the second antiresonance for the odd-corner baseplate as the largest source of error. Above 1000 Hz, the synthesized FRFs follow the trend of the measured FRFs but do not have as strong of a correlation in shape. As a note, the synthesized FRFs of the alternative data set did not represent the measured FRFs as well as the primary data set, so no FB modes were extracted for that set.

Method 1 was applied to the synthesized FRF matrix with constraint shapes calculated from all base DOFs and inverted with eight reference FRFs: four vertical impacts on the baseplate corners and four lateral impacts. Figure 34.13 shows CMIF of the synthesized FB FRF matrix with only reference impacts on the cylinder and response DOFs on the cylinder and flange. FB modes were extracted from the synthesized FB FRFs. The modal frequencies are tabulated in Table 34.5, and the mode shapes are plotted in Fig. 34.14.There is good agreement between the modal frequencies for the extracted modes. The cross-orthogonality between methods 1 and 3, shown in Table 34.6, indicates that the modes are fairly well correlated. Unfortunately, only one of the first bending modes was extracted from the synthesized FB FRFs.

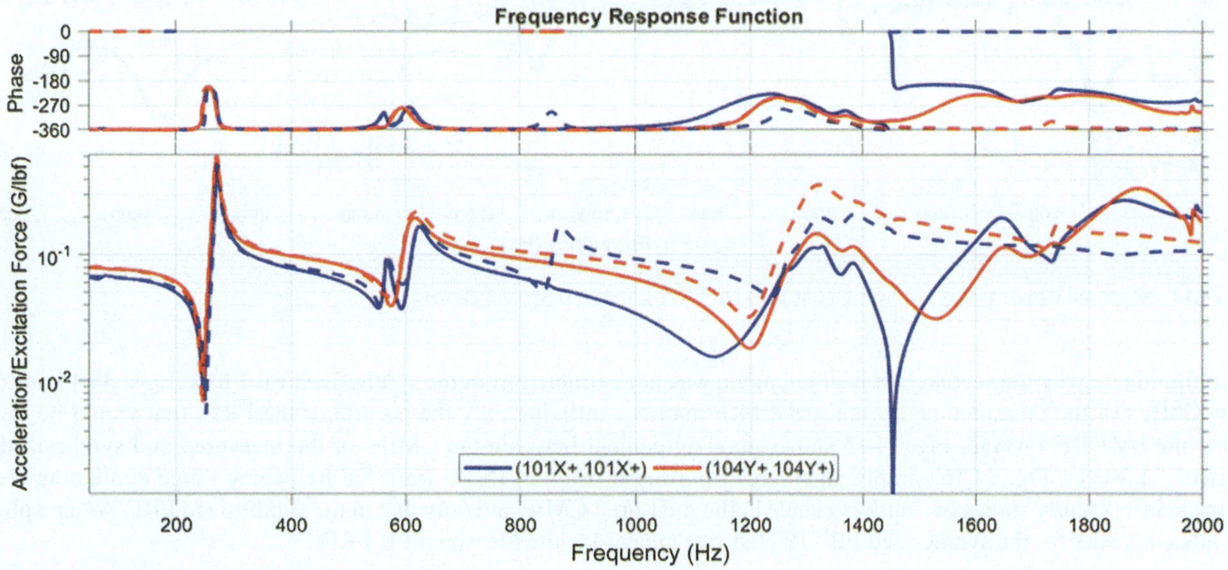

Fig. 34.10 Measured (solid) versus synthesized (dashed) DP FRFs for cylinder impact locations

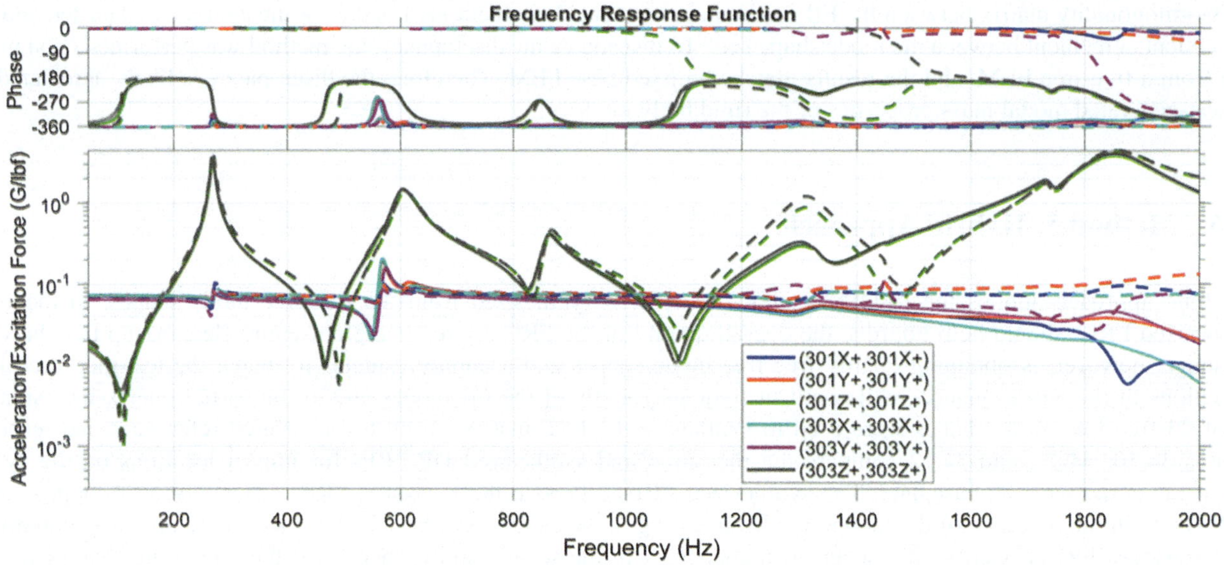

Fig. 34.11 Measured (solid) versus synthesized (dashed) DP FRFs for odd baseplate corners

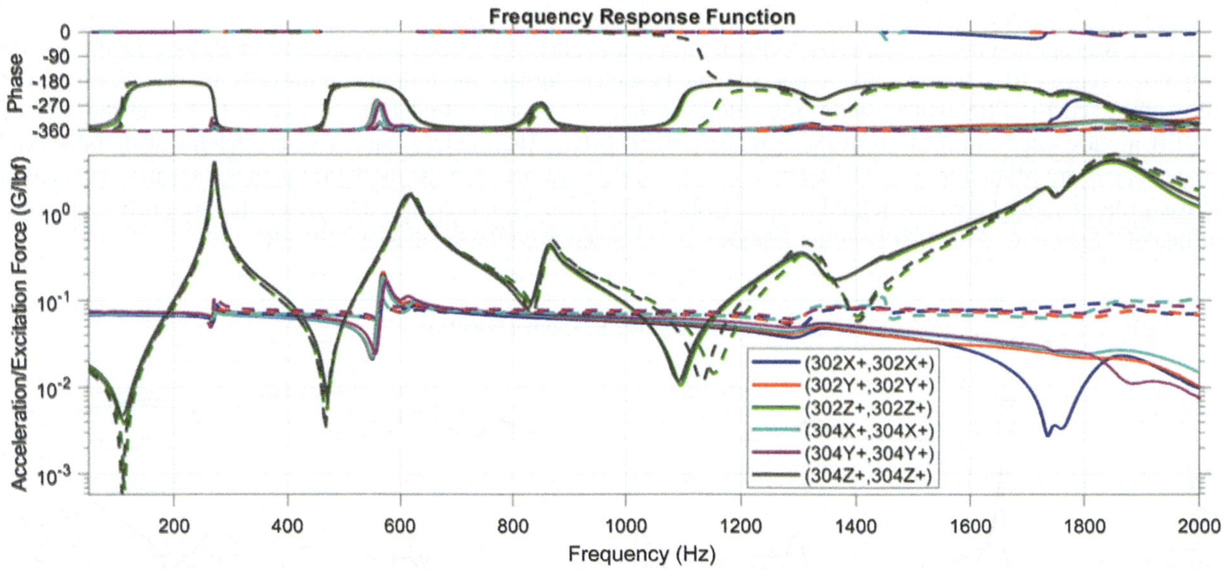

Fig. 34.12 Measured (solid) versus synthesized (dashed) DP FRFs for even baseplate corners

To illuminate why the second first bending mode was not extracted from the synthesized FB FRFs, Figs. 34.15 and 34.16 show CMIFs of the measured and synthesized FRF matrices utilizing only the portion of the FRFs that would be inverted during the SMURF. Overall, Fig. 34.15 shows good agreement between the CMIFs of the measured and synthesized FRF matrices. However, Fig. 34.16, simply an enlarged portion of Fig. 34.15 showing the frequency range containing the first antiresonance, clearly shows two antiresonance in the measured CMIF and only one in the synthesized FRF. As a result, only one mode appears for the synthesized FB FRF, but two appear for the measured FB FRFs.

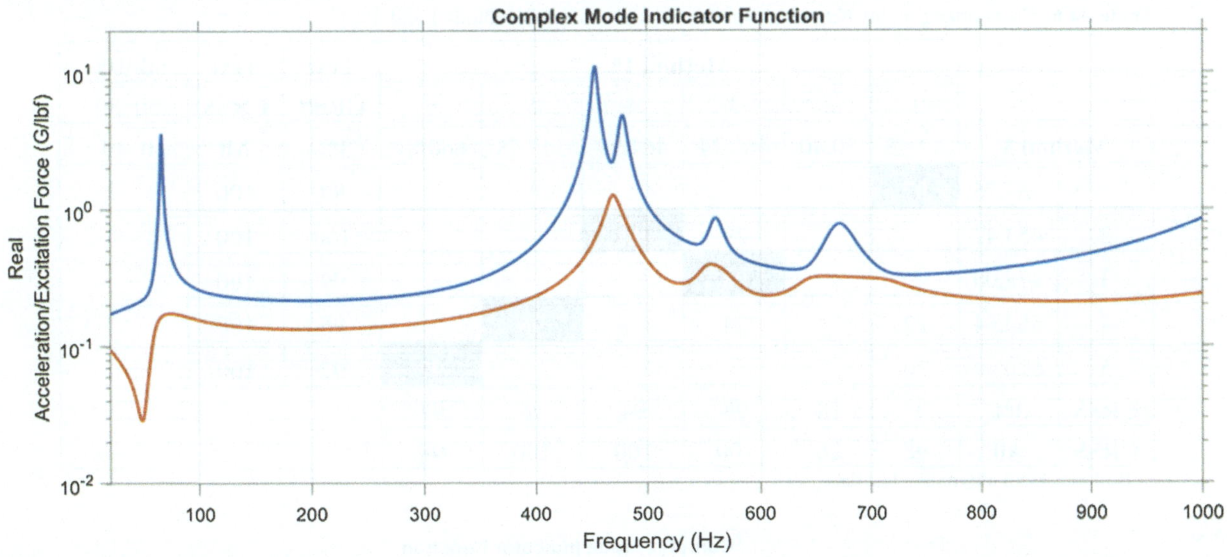

Fig. 34.13 CMIF of synthesized FB FRFs with only flange and cylinder response DOFs and cylinder impacts as references

Table 34.5 Extracted FB modal frequencies for methods 1–3

		Method 1		Method 2		Method 3
Mode	Description	FB freq. (Hz)	FB freq. (Hz)[a]	FB. freq. (Hz)	FB freq. (Hz)[a]	FB freq. (Hz)
1	1st bending	49.60	49.55	49.46	53.83	65.75
2	1st bending	50.07	50.4	58.83	61.31	N/A
3	2nd bending	461.35	461.24	456.82	489.14	453.42
4	2nd bending	465.90	465.0	479.58	511.28	487.0
5	Torsion	544.9	543.78	560.75	549.76	560.64
6	Plunge	606.27[b]	660.8	688.36	699.626	670.69

[a]Alternative set of test data
[b]Strongly coupled with torsion

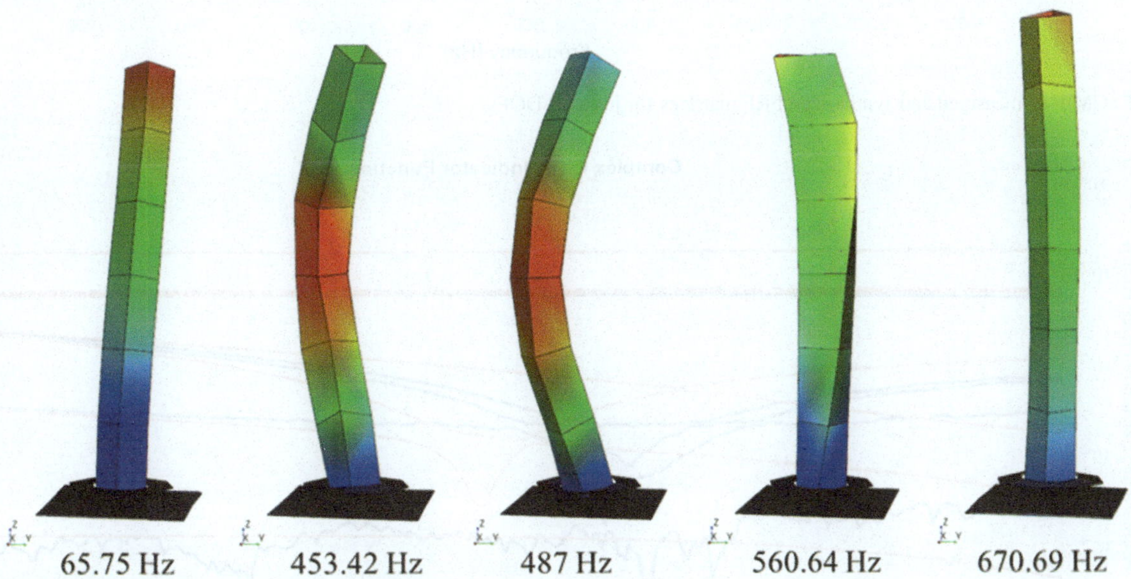

Fig. 34.14 FB mode shapes calculated from method 3, original data

Table 34.6 Cross-orthogonality between FB modes calculated from methods 1 and 3

		Method 1[a]						Test	Test	90–100
		1	2	3	4	5	6	CRSS	CRSS	50–89
Method 3		49.55	50.40	461.24	464.99	543.78	660.81	3%	All	0–49
1	65.75	97						99	100	
2	453.42			29	94			100	100	
3	478.00			98				99	100	
4	560.64	22		24		96		96	100	
5	670.69	20					97	97	100	
CRSS	3%	97	18	98	94	96	97			
CRSS	All	98	26	100	100	100	98			

[a]FB modes from alternative test data

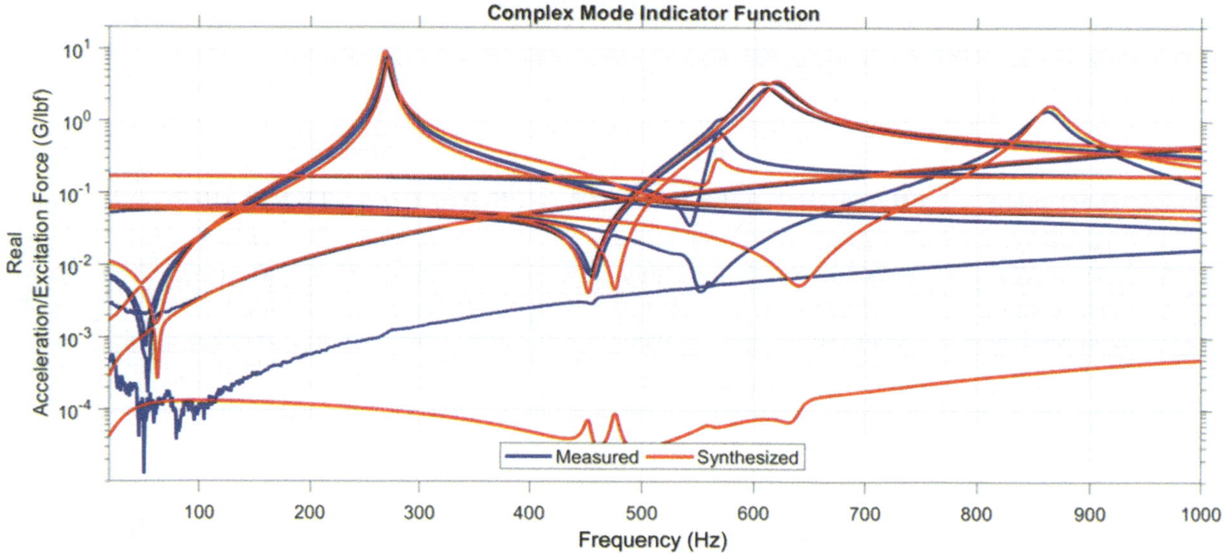

Fig. 34.15 CMIF of measured and synthesized FRF matrices for inverted DOFs

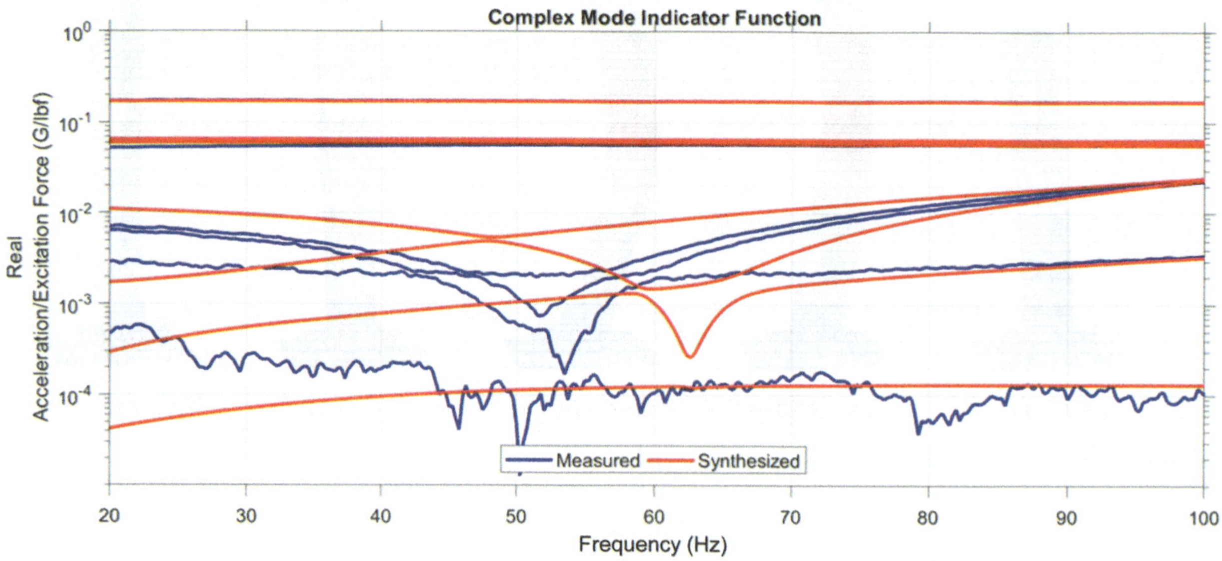

Fig. 34.16 CMIF of measured and synthesized FRF matrices for inverted DOFs, showing missing antiresonances in synthesized FRF matrix

34.7 Summary

This paper has presented a comparison of three different BCCMs applied to the same TA and data set. In general, all three methods were able to extract FB modes from the measured free-free data set. The FBCM, method 1, appeared to be the most robust and easiest to implement because all the required inputs to the methods were calculated from the test data. Additionally, the method was able to remove the rigid body motion and some of the motion of the baseplate when eight constraint shapes were used as references. Method 2, applying the constraint shapes directly to the mode shapes, was able to extract all mode shapes, but there was a larger separation in the modal frequencies of the bending pairs. The most likely explanation for the deviation was a discrepancy in the mass normalized rigid bodes that were estimated from the test measured free-free mode shapes. Finally, method 3, the hybrid approach, was able to produce FB modes but failed to extract the second first bending mode. Additionally, the method was more sensitive to poor fits in the free-free mode shapes because any errors propagated to the synthesized FRFs. In conclusion, all three methods were successful and are worthy of consideration for calculating FB modes of a TA on a dynamic boundary condition.

References

1. Carne, T.G., Martinez, D.R., Nord, A.R.: A comparison of fixed-base and driven base modal testing of an electronics package. Proceedings of the 7th International Modal Analysis Conference, 1989
2. Beliveau, J.G., Vigneron, F.R., Soucy, Y., Draisey, S.: Modal parameter estimation from base excitation. J. Sound Vib. **107**, 435–449 (1986)
3. Fullekrug, U.: Determination of effective masses and modal masses from base-driven tests. Proceedings of the 14th International Modal Analysis Conference, 1996
4. Sinapius, J.M.: Identification of fixed and free interface normal modes by base excitation. Proceedings of the 14th International Modal Analysis Conference, 1996
5. Imregun, M., Robb, D.A., Ewins, D.J.: Structural modification and coupling dynamic analysis using measured FRF data. Proceedings of the 5th International Modal Analysis Conference, 1987
6. Mayes, R.L., Bridgers, L.D.: Extracting fixed base modal models from vibration tests on flexible tables. Proceedings of the 27th International Modal Analysis Conference, 2009
7. Napolitano, K., Yoder, N.: Fixed base FRF using boundary measurements as references – analytical derivation. Proceedings of the 30th International Modal Analysis Conference, 2012
8. Mayes, R., Rohe, D., Blecke, J.: Extending the frequency band for fixed base modal analysis on a vibration slip table. Proceedings of the 31st International Modal Analysis Conference, 2013
9. Napolitano, K., Yoder, N.: Extraction of fixed-base modes of a structure mounted on a shake table. Proceedings of the 31st International Modal Analysis Conference, 2013
10. Staab, L., Winkel, J., Suárez, J., Jones, T., Napolitano, K.: Fixed base modal testing using the mechanical vibration facility 3-axis base shake system. Proceedings of the 34th International Modal Analysis Conference, 2016
11. Winkel, J., Akers, J., Suarez, V., Staab, L., Napolitano, K.: Modal survey of the MPCV Orion European service module structural test article using a multi-axis shake table. Proceedings of the 37th International Modal Analysis Conference, 2018
12. Napolitano, K.: Fixing degrees of freedom of an aluminum beam by using accelerometers as references. Proceedings of the 38th International Modal Analysis Conference, 2019
13. Kerrian, P., Napolitano, K.: Pretest analysis for modal survey tests using fixed base correction method. Proceedings of the 38th International Modal Analysis Conference, 2019
14. Allen, M.S., Gindlin, H., Mayes, R.: Experimental modal substructuring to extract fixed-base modes from a substructure attached to a flexible fixture. Proceedings of the 28th International Modal Analysis Conference, 2010
15. Mayes, R.L., Allen, M.S.: Converting a driven base vibration test to a fixed base modal analysis. Proceedings of the 29th International Modal Analysis Conference, 2011
16. Napolitano, K.: Using singular value decomposition to estimate frequency response functions. Proceedings of the 29th International Modal Analysis Conference, 2011

Chapter 35
Deflection Shape Balancing: An Alternative to Modal Balancing

W. Jason Morrell, B. Damiano, K. Hylton, and C. Jordan

Abstract Rotor balancing is the final step in the rotor manufacturing process and may be necessary during rotor refurbishment. The process can be time consuming and costly, especially for supercritical monolithic rotors, impacting schedule and increasing costs. Modal balancing is a commonly used practice and is effective but requires multiple runs and at least one high-speed run. Deflection shape balancing is introduced as an alternative that may offer a more efficient method of balancing making the process faster with potentially fewer balancing runs. This paper explores the performance of the deflection shape fit balancing technique by utilizing MATLAB code simulations.

Keywords Rotor dynamics · Modal balancing · Deflection shape balancing · MATLAB · Rotor balancing

35.1 Introduction

This paper compares Deflection Shape Fit (DSF) and modal balancing approaches through simulation using MATLAB. A 28-in. steel beam spinning at a speed above the third critical speed and measured by probes at five axial locations is used as a simulated test piece. The first three critical speeds of the simulated rotor are 113, 454, and 1003 Hz. Reduction of modal magnitudes at all three of these speeds is compared for DSF and Modal balancing when only the first two modes are balanced, i.e. the modal magnitude at 1003 Hz is compared when the two methods being examined are used to balance at only 113 and 454 Hz. The simulated results for a 28-in. steel beam are presented in this paper.

35.2 Background

The rotor balancing process can be time consuming and costly making it a good target for process improvement. The process typically involves accelerating a rotor until a predetermined limit is reached at which time the rotor is stopped and the source of unbalance is addressed so the rotor may be accelerated to a higher speed on the next balance run. As the rotor is accelerated it takes on the shape of its modes as its natural frequencies are excited by unbalance. The point where the natural frequency and spin speed of a rotor intersect are called critical speeds. The unbalance is corrected at these critical speeds

W. J. Morrell
Department of Electrical Engineering and Computer Science, College of Engineering, The University of Tennessee, Knoxville, TN, USA
e-mail: wmorrell@vols.utk.edu

B. Damiano · K. Hylton · C. Jordan (✉)
Dynamic Systems Analysis Group, Electrical and Electronics Systems Research Division, Oak Ridge National Laboratory, Oak Ridge, TN, USA
e-mail: damianob@ornl.gov; hyltonkw@ornl.gov; jordanca@ornl.gov

B. Dilworth (ed.), *Topics in Modal Analysis & Testing, Volume 8*, Conference Proceedings of the Society for Experimental Mechanics Series, https://doi.org/10.1007/978-3-030-47717-2_35

by applying balance weights at weight planes, which requires knowledge of influence coefficients to select the appropriate weight. Influence coefficients simply provide information on how a rotor will respond to weight applied at different locations. Modal balancing has proven very effective for balancing rotors when the rotor has been accelerated through the critical speed being balanced, but more efficient ways of balancing rotors are desired to reduce the cost and time needed. Thus, DSF balancing is introduced as alternative to modal balancing for investigation.

Modal and deflection shape balancing are governed by the same equations of motion but arrive at their solutions differently. Equation (35.1) shows the equations of motion in matrix form for a rotor at steady state with symmetric stiffness and damping.

$$\begin{bmatrix} M & 0 \\ 0 & M \end{bmatrix} \begin{Bmatrix} \ddot{x} \\ \ddot{y} \end{Bmatrix} + \begin{bmatrix} D_S & 0 \\ 0 & D_S \end{bmatrix} \begin{Bmatrix} \dot{x} \\ \dot{y} \end{Bmatrix} + \begin{bmatrix} K & 0 \\ 0 & K \end{bmatrix} \begin{Bmatrix} x \\ y \end{Bmatrix} = \begin{Bmatrix} F cos\,(\omega t + \delta) \\ F sin\,(\omega t + \delta) \end{Bmatrix} \tag{35.1}$$

In the above equation M is the rotor or modal mass, D_S is the lateral external modal damping, and K is lateral isotropic stiffness. Modal balancing takes deflection data gathered during a balance test run, uses that data to identify the mode being addressed, and individual modes are balanced. DSF balancing takes the same deflection data and treats the rotor's shape as a combination of deflections needed to create the measured rotor shape.

Correct placement and magnitude of the corrective weights for both modal and DSF balancing requires weight planes and influence coefficients, and the weight planes do not need to be the same for the two methods. Modal balancing weight planes are in close proximity to antinodes of the mode being balanced while DSF balancing simply uses some number of weight planes along the length of the rotor. Influence coefficients for the two approaches consist of one influence coefficient for each mode in modal balancing and would correspond to the number of weight planes in use for DSF balancing. For instance, in the case of the 28-in. steel beam spinning above the third critical speed the rotor has gone through three critical speeds, corresponding to three mode shapes, and would require three sets of influence coefficients for the modal balancing approach. The number of influence coefficients for the DSF balancing approach would depend on the number of weight planes being used.

35.3 Analysis

DSF and modal balancing approaches are compared through simulation in MATLAB. A 28-in. long 0.8 in. outside diameter steel beam with a 0.025-in. wall thickness is used as a simulated rotor. A randomly distributed imbalance is applied to the beam and modal magnitudes are determined at the first three critical speeds: 113, 454, and 1003 Hz. The modal magnitudes are then balanced using the modal and the DSF approaches. The two approaches are compared by

- The reduction in modal magnitude at 113 Hz when balancing at 113 Hz
- The reduction in modal magnitude at 454 Hz when balancing at 454 Hz
- The reduction in modal magnitude at 1003 Hz when balancing at 113 and 454 Hz together

These comparisons are performed for 5, 10, 15, 20, and 28 DSF weight planes equally spaced along the length of the rotor. The MSF approach always uses three weight planes corresponding to the antinodes of the first and second flexural modes.

A total of 20,000 random imbalances were simulated and the final modal magnitude, normalized against the initial modal magnitude from the applied imbalance, was logged to create a distribution of the balancing results. Each balancing approach, Modal or DSF, is applied only once to each imbalance. The probability of each result is also plotted. Additionally, the effect of the number of weight planes used for the DSF approach is shown.

35.3.1 5 Weight Planes

Plots comparing the reduction in modal magnitudes using 5 DSF weight planes are presented below. Results from balancing at 113 Hz seen in Fig. 35.1 show very slightly better performance from the DSF approach when compared to the modal approach and Fig. 35.2 shows very similar performance when balancing at 454 Hz, but in both cases the magnitudes are so small that both approaches might be considered functionally equivalent. The biggest difference between approaches is shown in Fig. 35.3, where the impact on the modal amplitude of the third mode when balancing occurs only at the first and second critical speeds. The DSF produces clearly lower modal amplitudes than the modal approach for this condition.

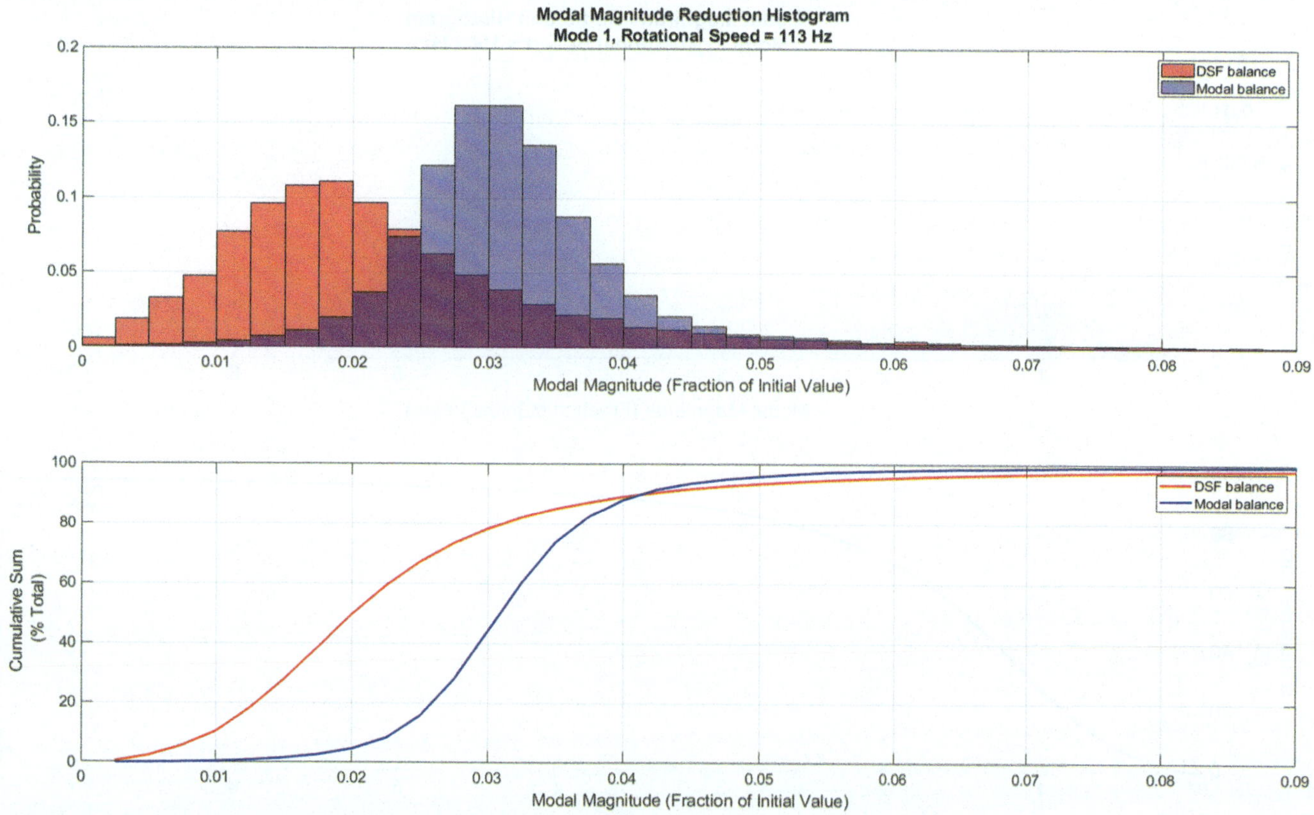

Fig. 35.1 Mode 1 modal magnitude after DSF and modal balancing approaches at 113 Hz using 5 weight planes

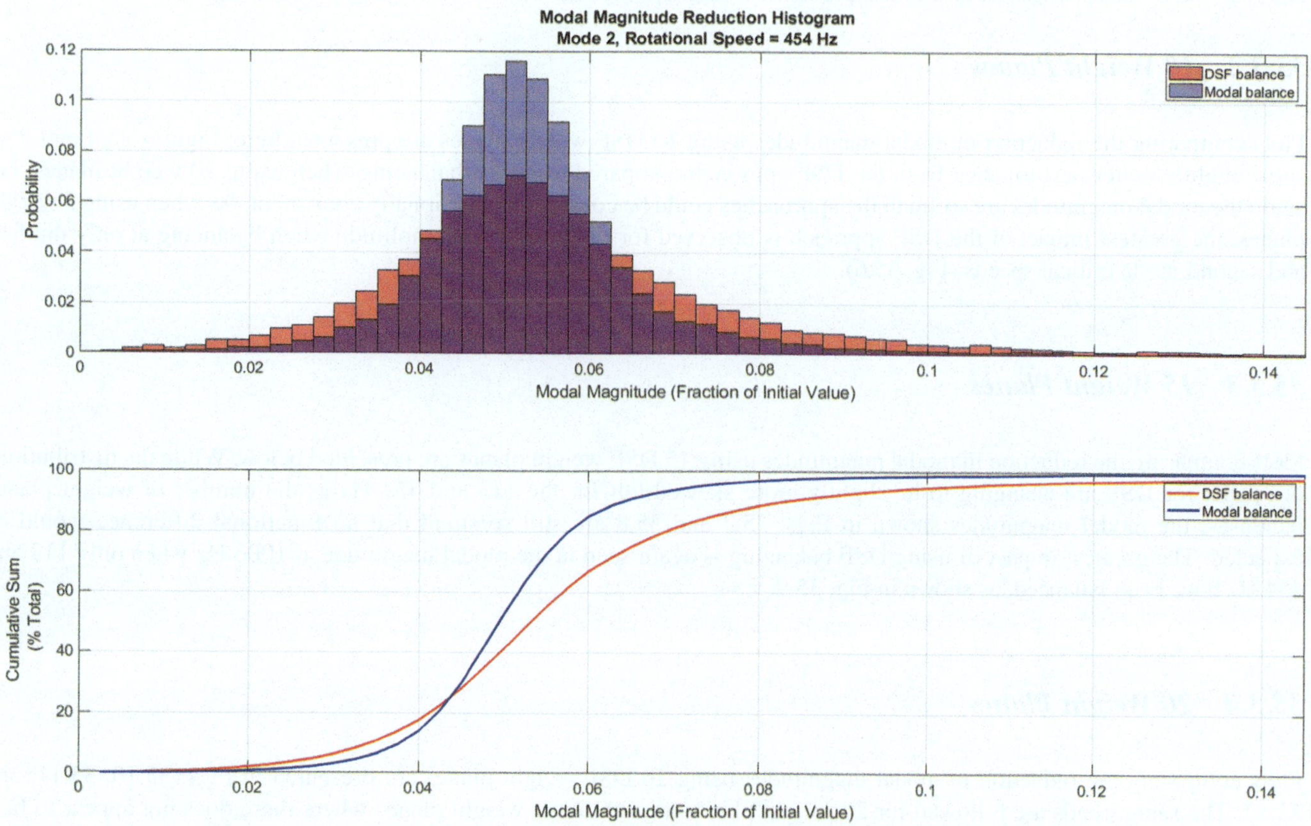

Fig. 35.2 Mode 2 modal magnitude after DSF and modal balancing approaches at 454 Hz using 5 weight planes

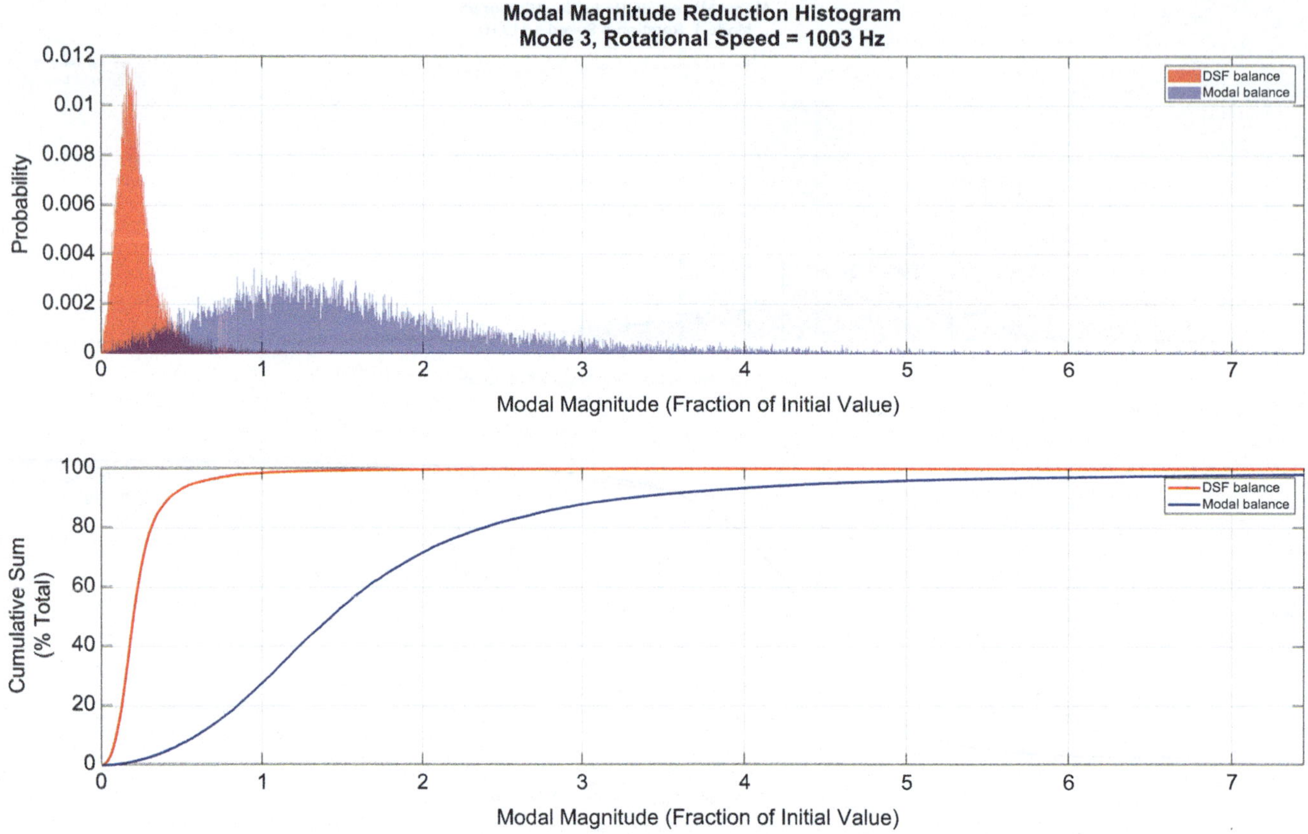

Fig. 35.3 Mode 3 modal magnitude after DSF and modal balancing approaches at 113 and 454 Hz using 5 weight planes

35.3.2 10 Weight Planes

Plots comparing the reduction in modal magnitudes using 10 DSF weight planes are presented here. Figures 35.4 and 35.5 show slightly better performance from the DSF approach compared to modal balancing when using 10 weight planes, but again the modal magnitudes are so small the approaches could be considered functionally equivalent. As when using 5 weight planes, the greatest impact of the DSF approach is observed for the third mode amplitude when balancing at only the first and second mode critical speeds (Fig. 35.6).

35.3.3 15 Weight Planes

Plots comparing the reduction in modal magnitudes using 15 DSF weight planes are presented below. While the distributions of results for DSF are changing to be slightly more skewed left for the 113 and 454 Hz as the number of weight planes increases, the modal magnitudes shown in Figs. 35.7 and 35.8 are still so small that no functional difference would be expected. The greatest impact of using DSF balancing is again seen in the modal magnitude at 1003 Hz when only 113 and 454 Hz have been balanced, as shown in Fig. 35.9.

35.3.4 20 Weight Planes

Plots comparing the reduction in modal magnitudes using 20 DSF weight planes are presented in Figs. 35.10, 35.11, and 35.12. The same trends are followed for 20 weight planes as with fewer weight planes where there does not appear to be a

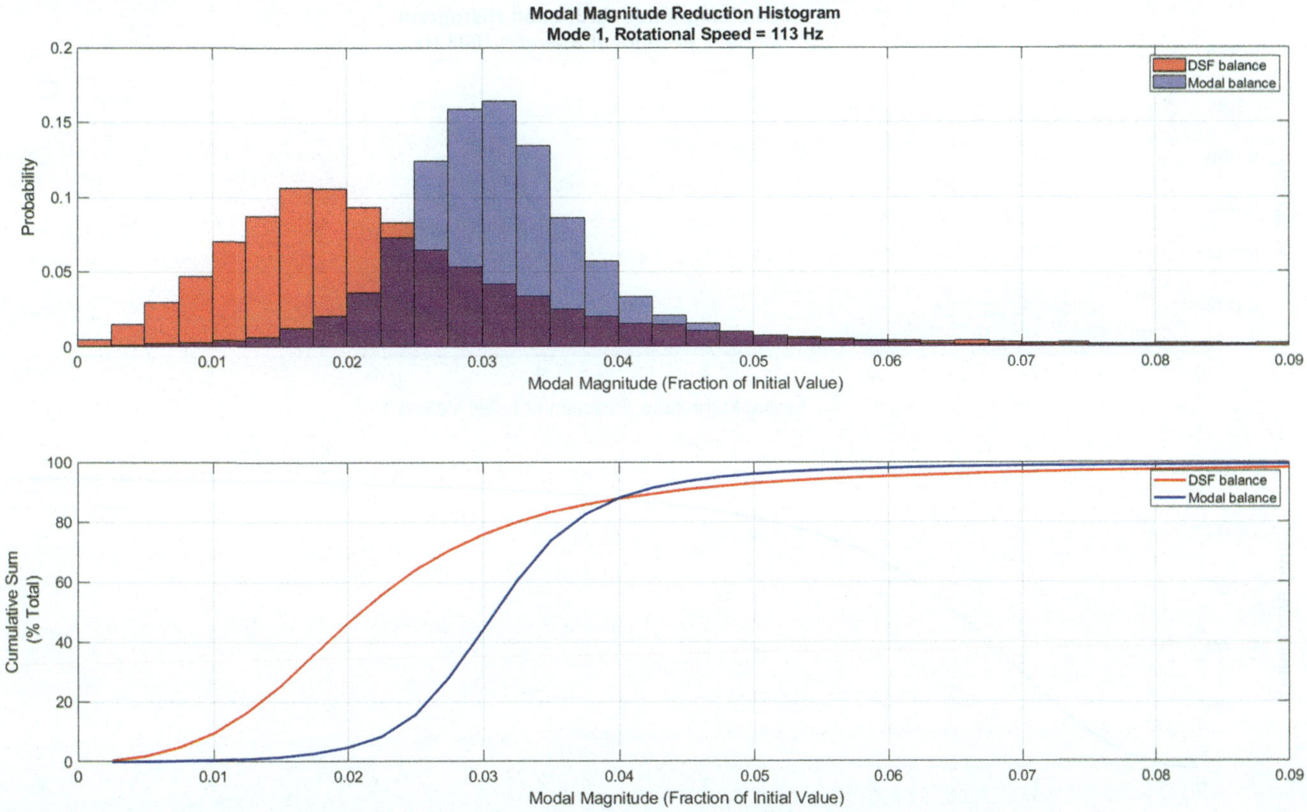

Fig. 35.4 Mode 1 modal magnitude after DSF and modal balancing approaches at 113 Hz using 10 weight planes

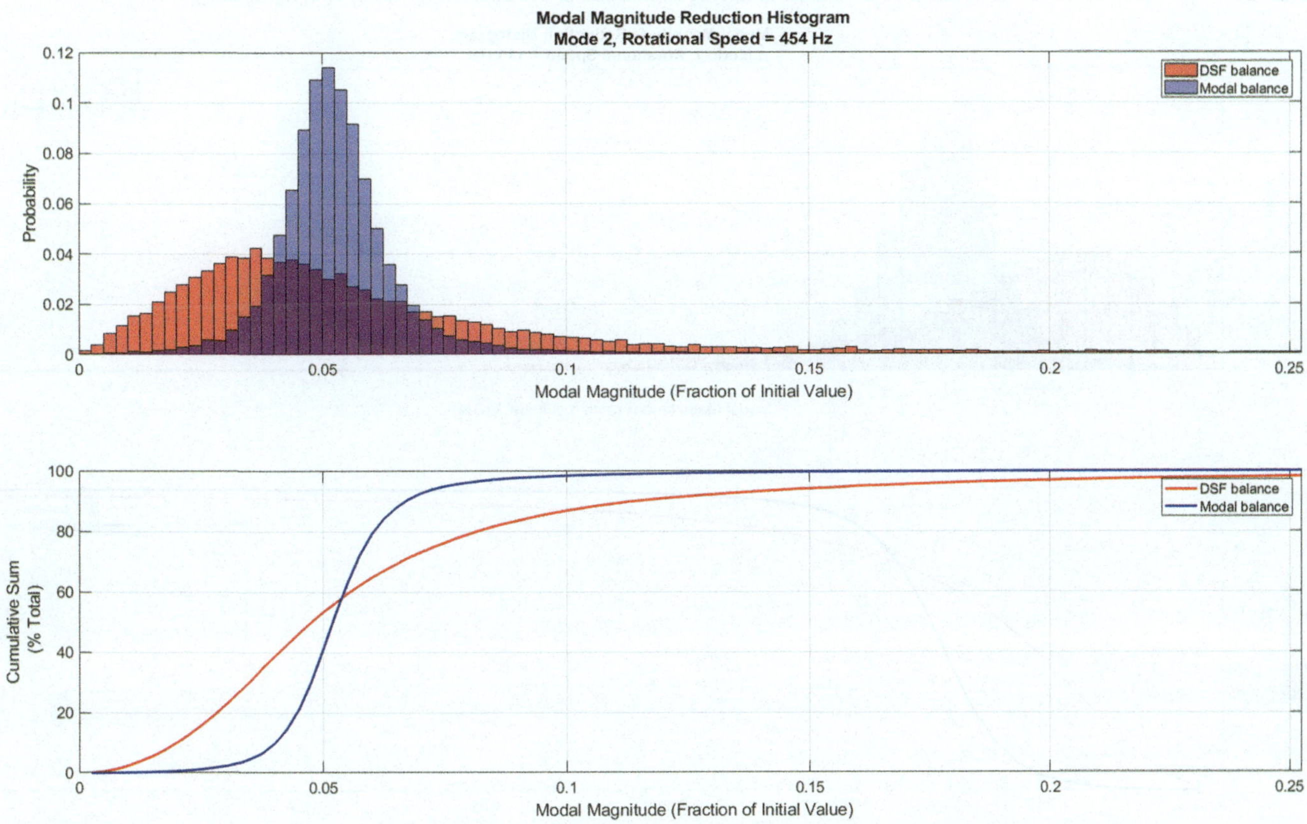

Fig. 35.5 Mode 2 modal magnitude after DSF and modal balancing approaches at 454 Hz using 10 weight planes

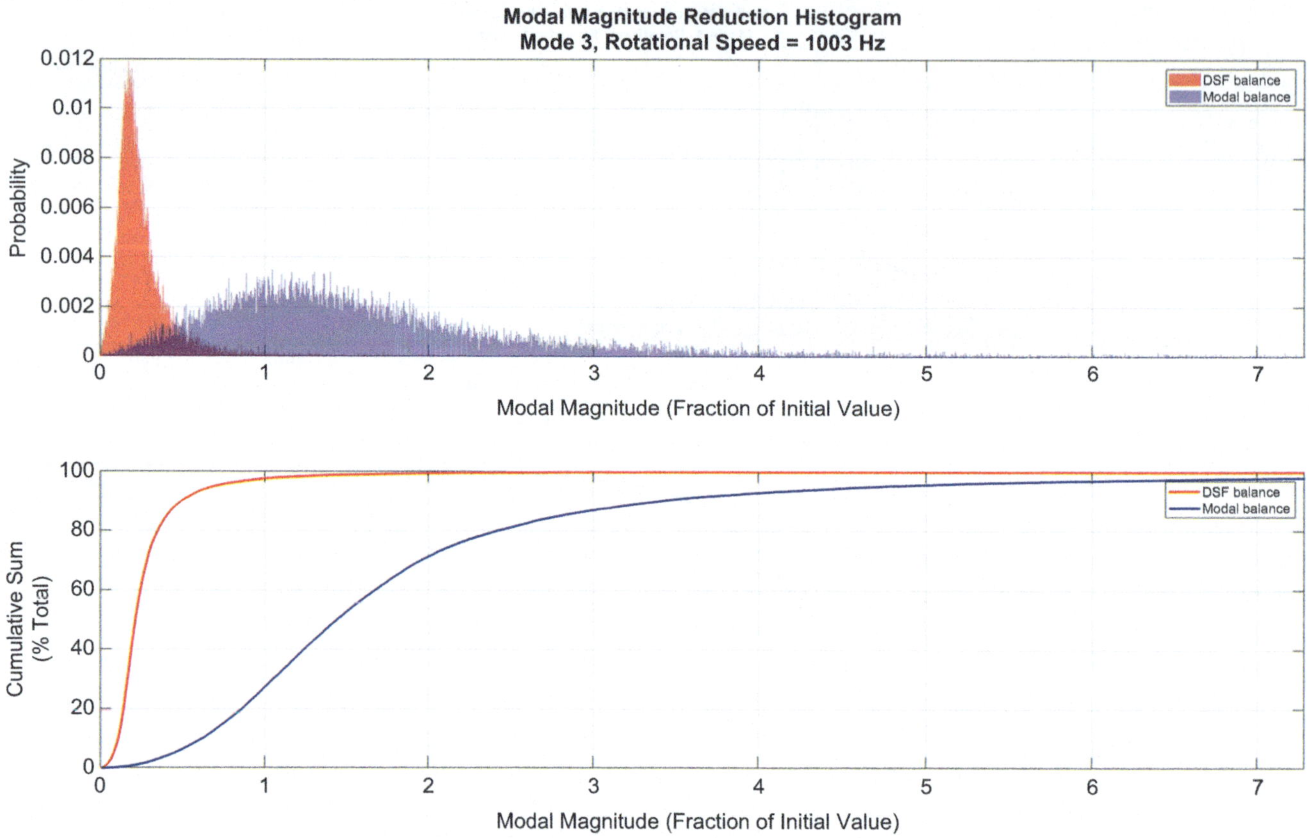

Fig. 35.6 Mode 3 modal magnitude after DSF and modal balancing approaches at 113 and 454 Hz using 10 weight planes

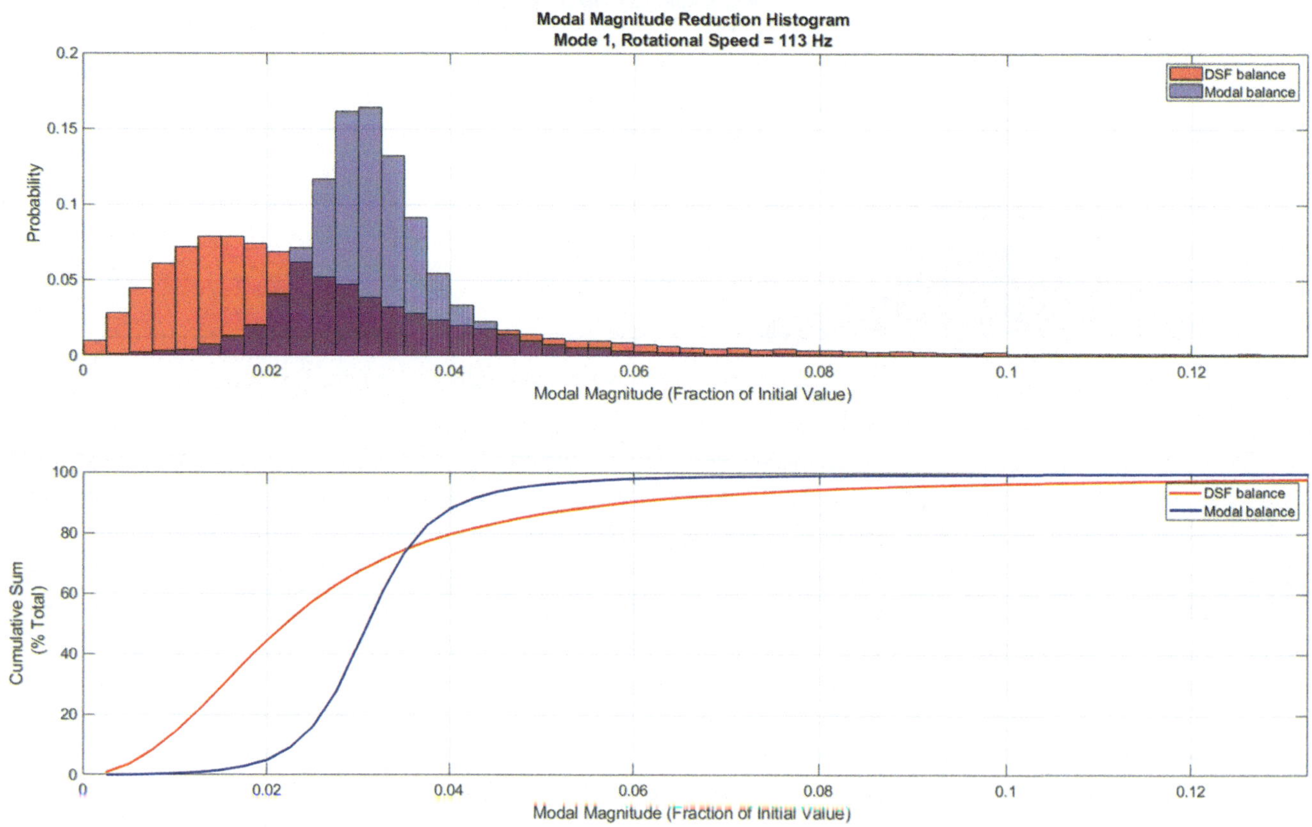

Fig. 35.7 Mode 1 modal magnitude after DSF and modal balancing approaches at 113 Hz using 15 weight planes

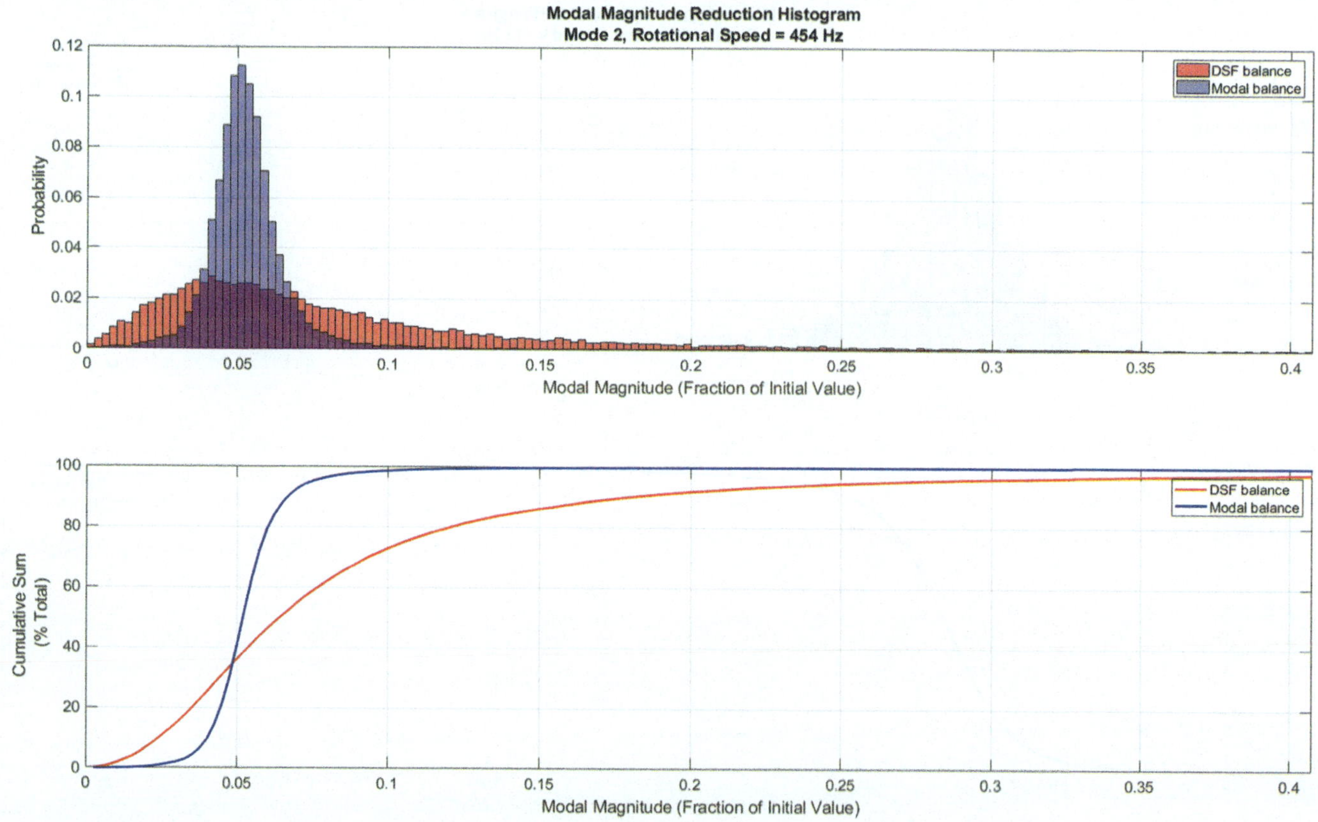

Fig. 35.8 Mode 2 modal magnitude after DSF and modal balancing approaches at 454 Hz using 15 weight planes

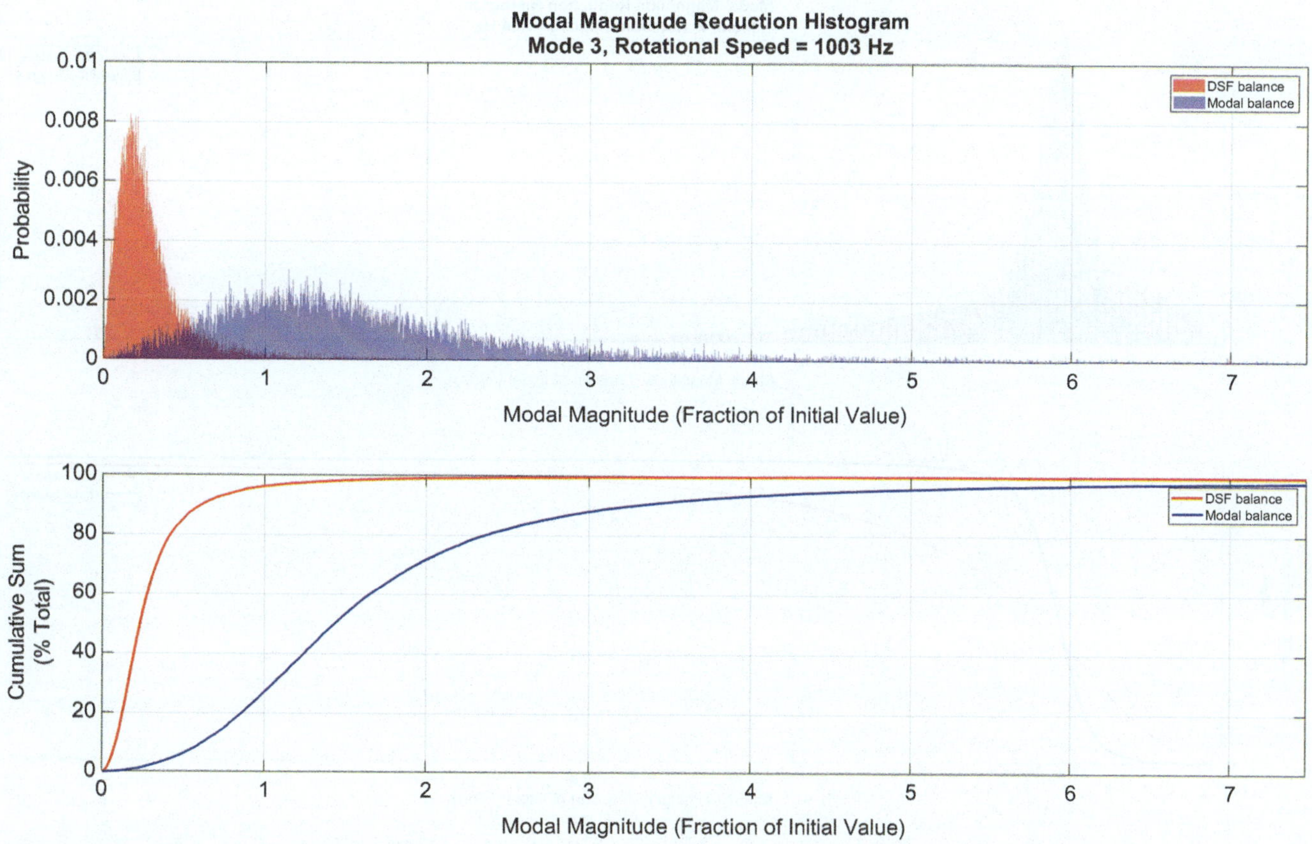

Fig. 35.9 Mode 3 modal magnitude after DSF and modal balancing approaches at 113 and 454 Hz using 15 weight planes

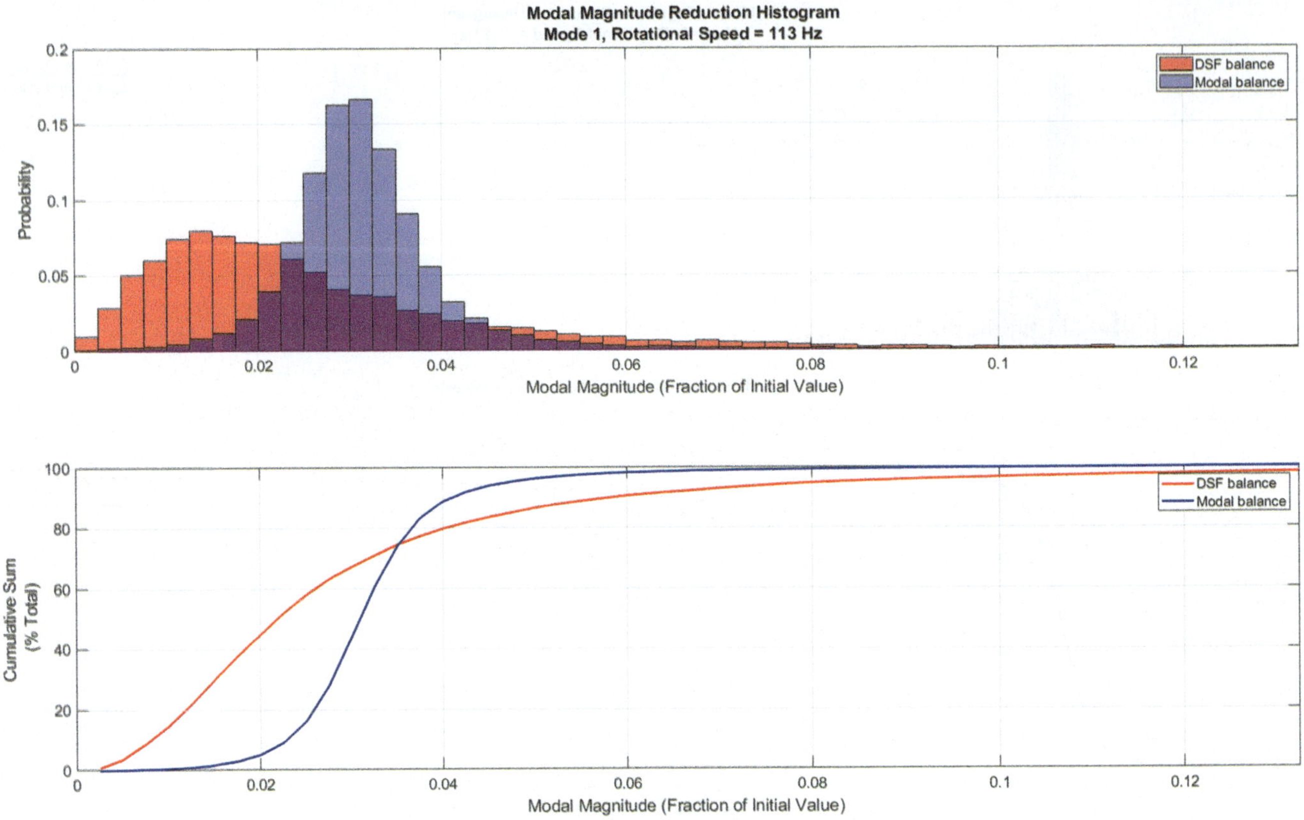

Fig. 35.10 Mode 1 modal magnitude after DSF and modal balancing approaches at 113 Hz using 20 weight planes

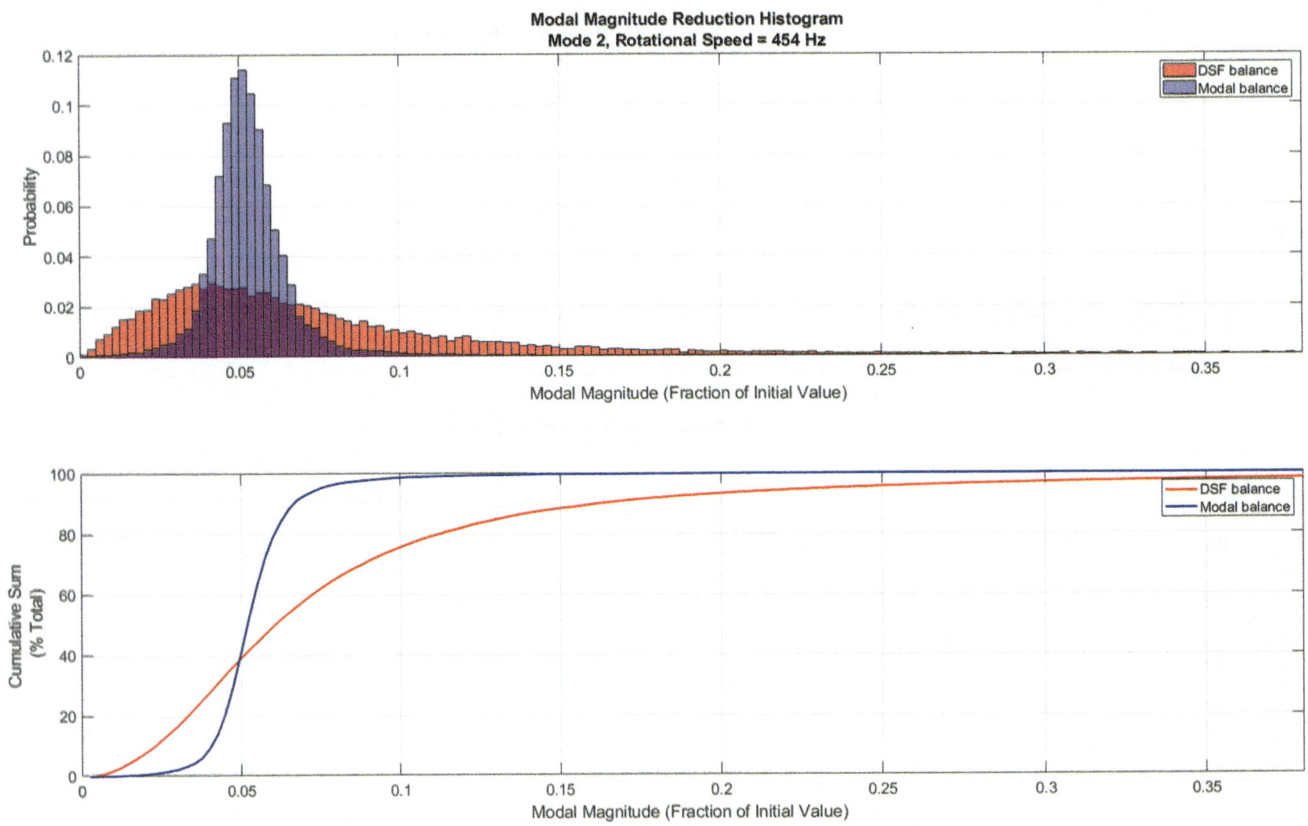

Fig. 35.11 Mode 2 modal magnitude after DSF and modal balancing approaches at 454 Hz using 20 weight planes

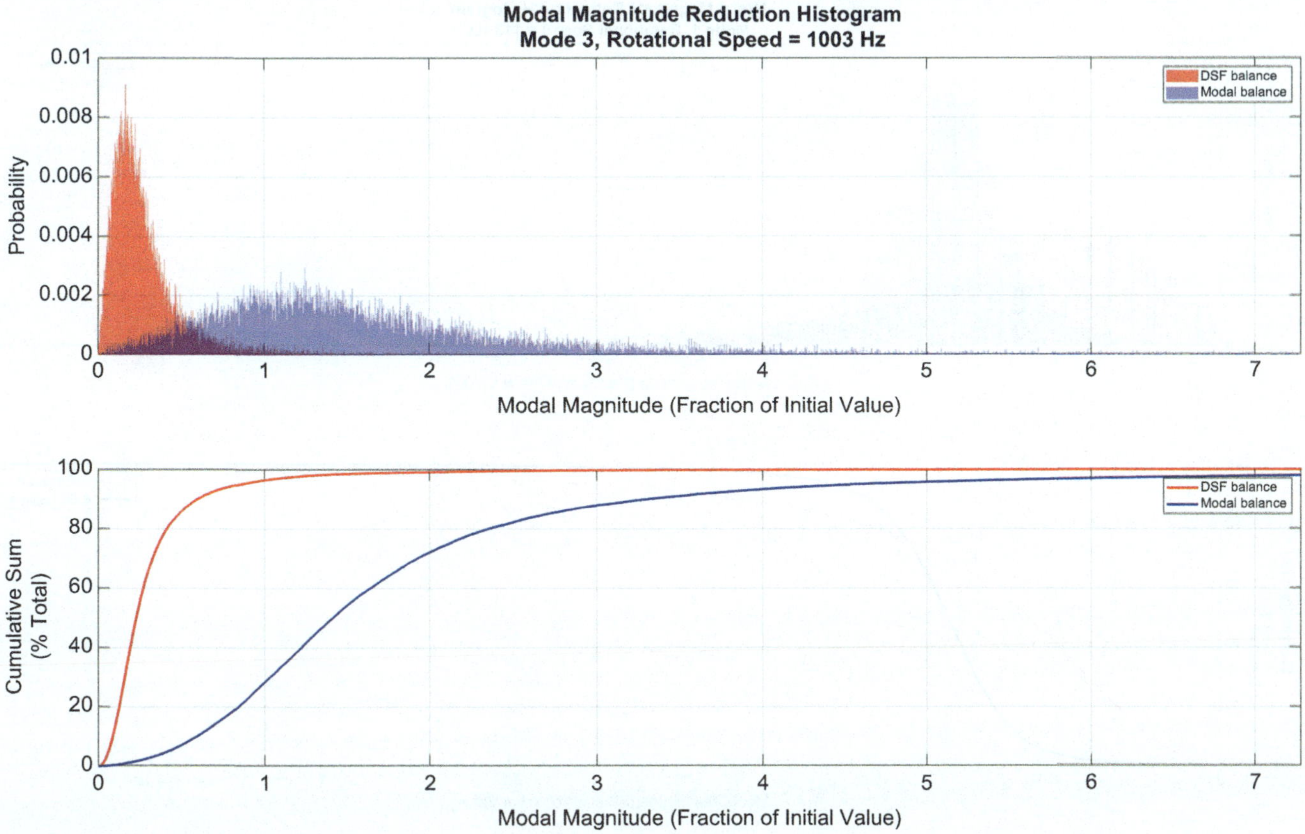

Fig. 35.12 Mode 3 modal magnitude after DSF and modal balancing approaches at 113 and 454 Hz using 20 weight planes

functional difference between the modal and DSF approaches when balancing at 113 and 454 Hz, but the two approaches perform very differently with regards to their impacts on the third mode at 1003 Hz when balancing at 113 and 454 Hz.

35.3.5 28 Weight Planes

Plots comparing the reduction in modal magnitudes using 28 DSF weight planes are presented in Figs. 35.13, 35.14, and 35.15. The same trends are observed for 28 weight planes as with fewer weight planes where there does not appear to be a functional difference between the modal and DSF approaches when balancing at 113 and 454 Hz, but the two approaches perform very differently with regards to their impacts on the third mode at 1003 Hz when balancing at 113 and 454 Hz.

In general, the DSF and modal approaches are expected to be functionally equivalent when balancing at the first two critical speeds. The DSF approach appears to be out-perform the modal approach with regard to the third mode when balancing at the first two critical speeds. The normalized modal magnitude of the third mode when balancing at the first two critical speeds is less than one for the DSF approach and more than one for the modal approach. This means the modal approach may increase the magnitude of the third mode when balancing at the first two critical speeds, while the DSF approach appears to reduce the magnitude of the third mode when balancing at the first two critical speeds. This indicates, depending upon a specific application's balancing requirements, it may be possible to sufficiently balance higher speed modes without actually spinning a rotor to those higher speeds.

35.3.6 Impact of the Number of Weight Planes

The number of weight planes used for the DSF approach was varied during the simulation and the results were unexpected. It was anticipated that more weight planes would generally produce a greater reduction in modal magnitude, but as observed in Fig. 35.16 that is not the case. For the first and second modes there does not appear to be a benefit to increasing the number of

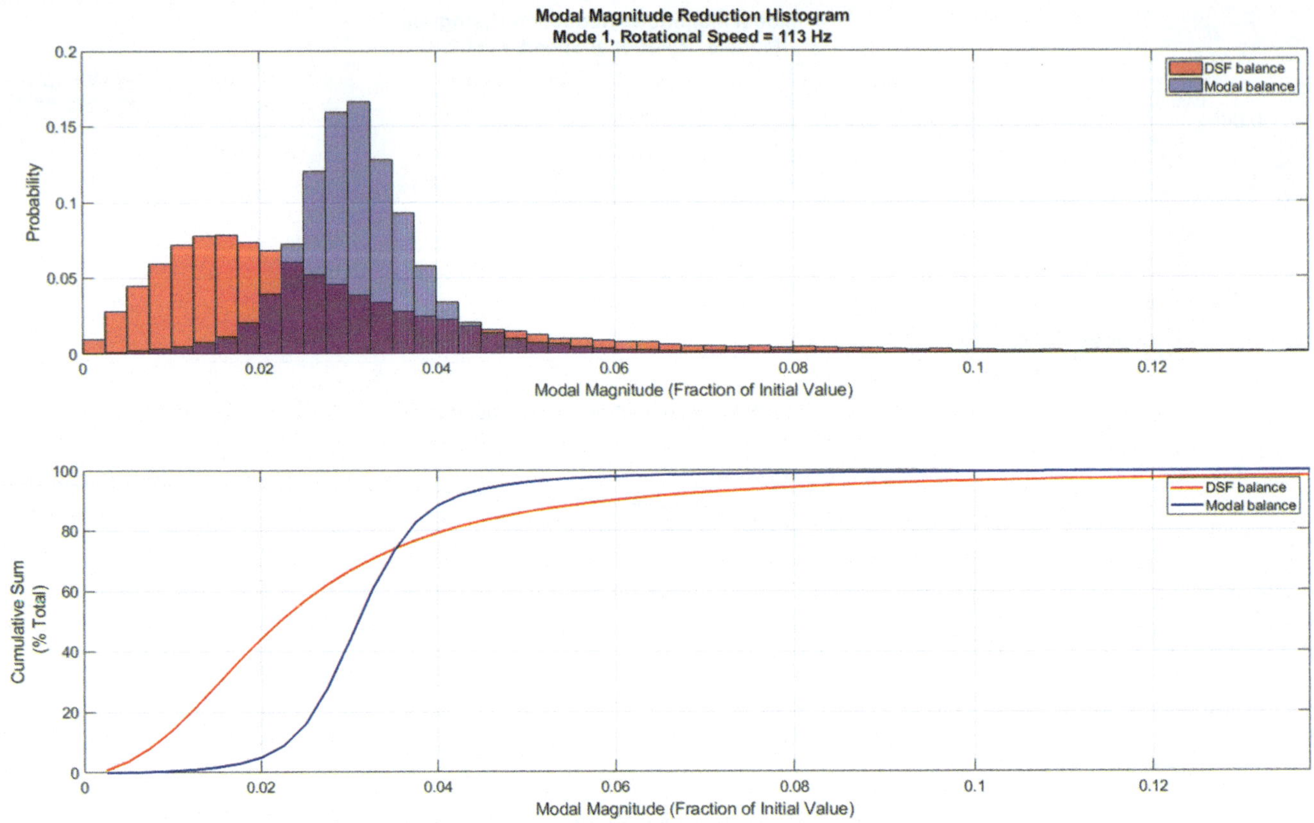

Fig. 35.13 Mode 1 modal magnitude after DSF and modal balancing approaches at 113 Hz using 28 weight planes

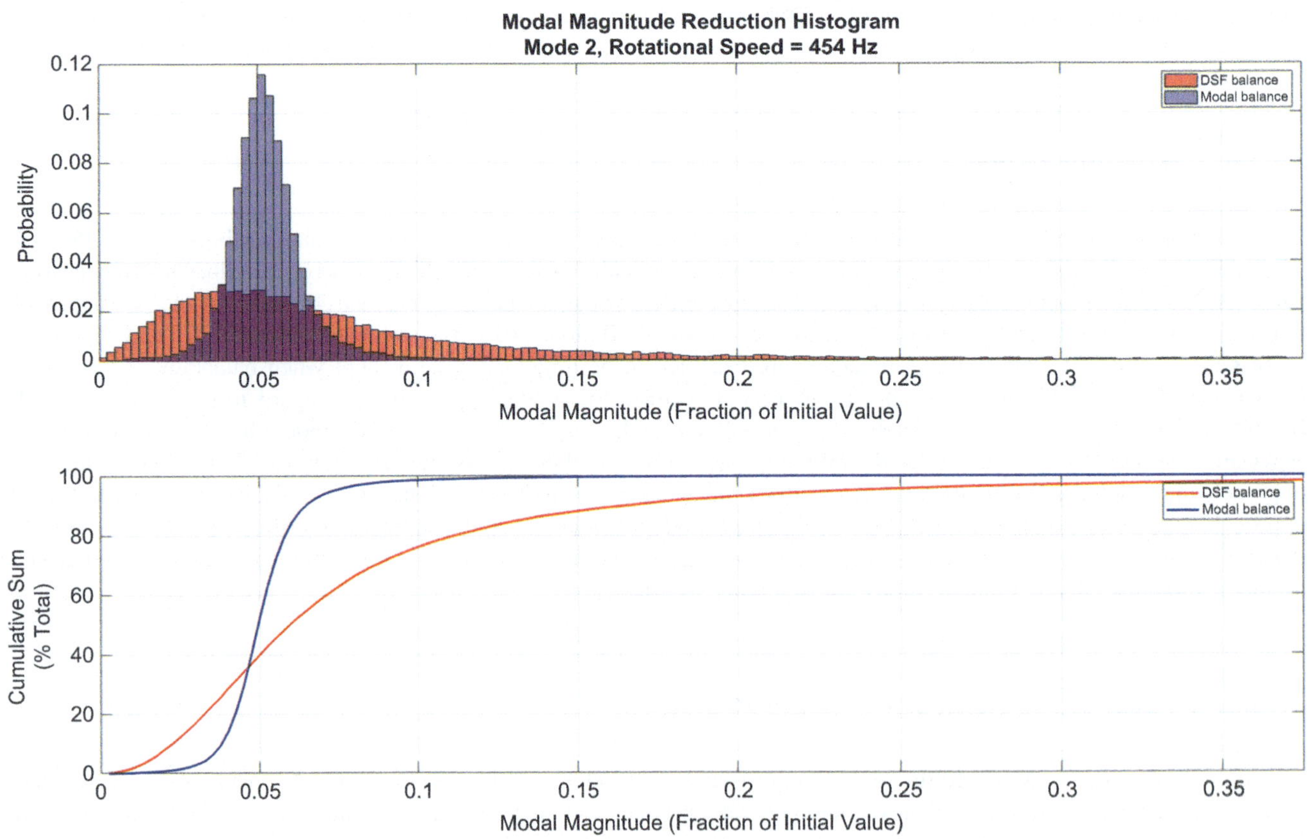

Fig. 35.14 Mode 2 modal magnitude after DSF and modal balancing approaches at 454 Hz using 28 weight planes

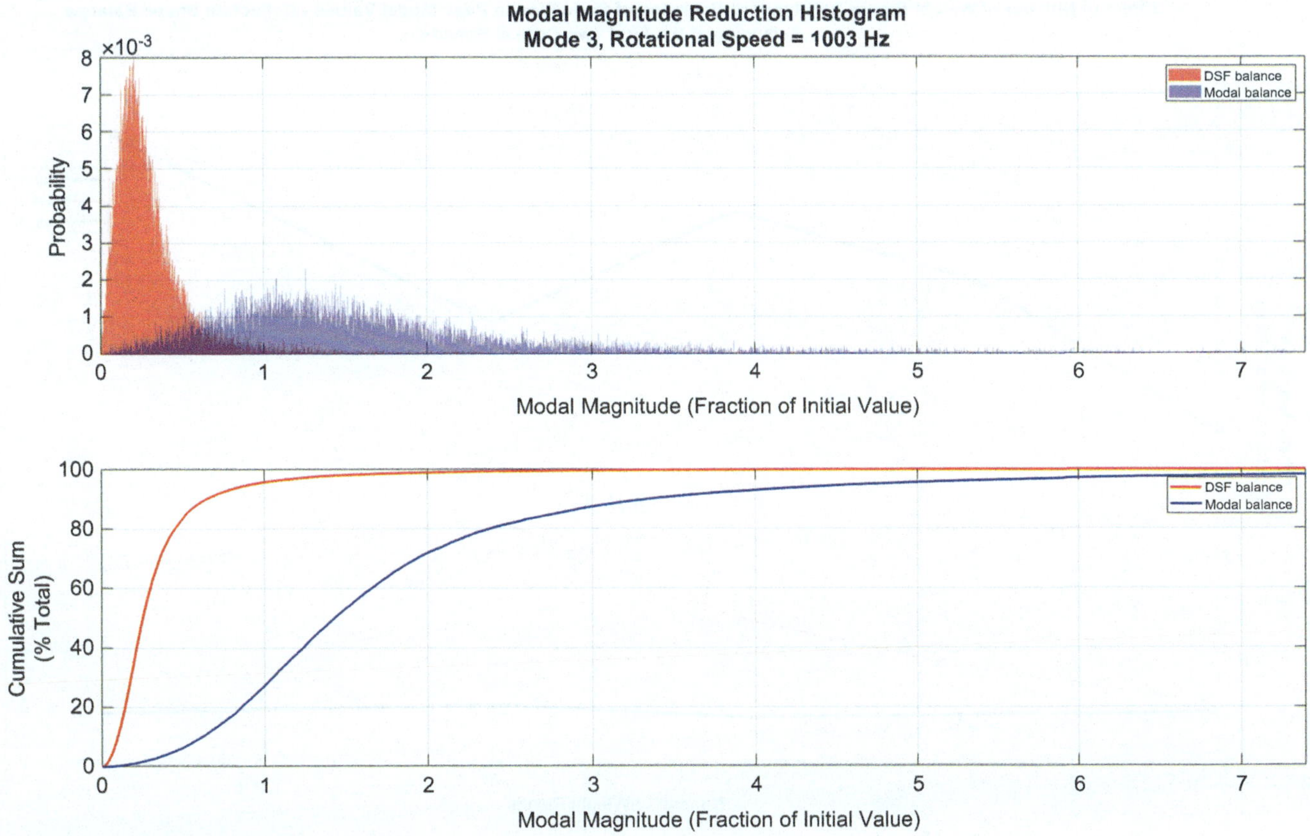

Fig. 35.15 Mode 3 modal magnitude after DSF and modal balancing approaches at 113 and 454 Hz using 28 weight planes

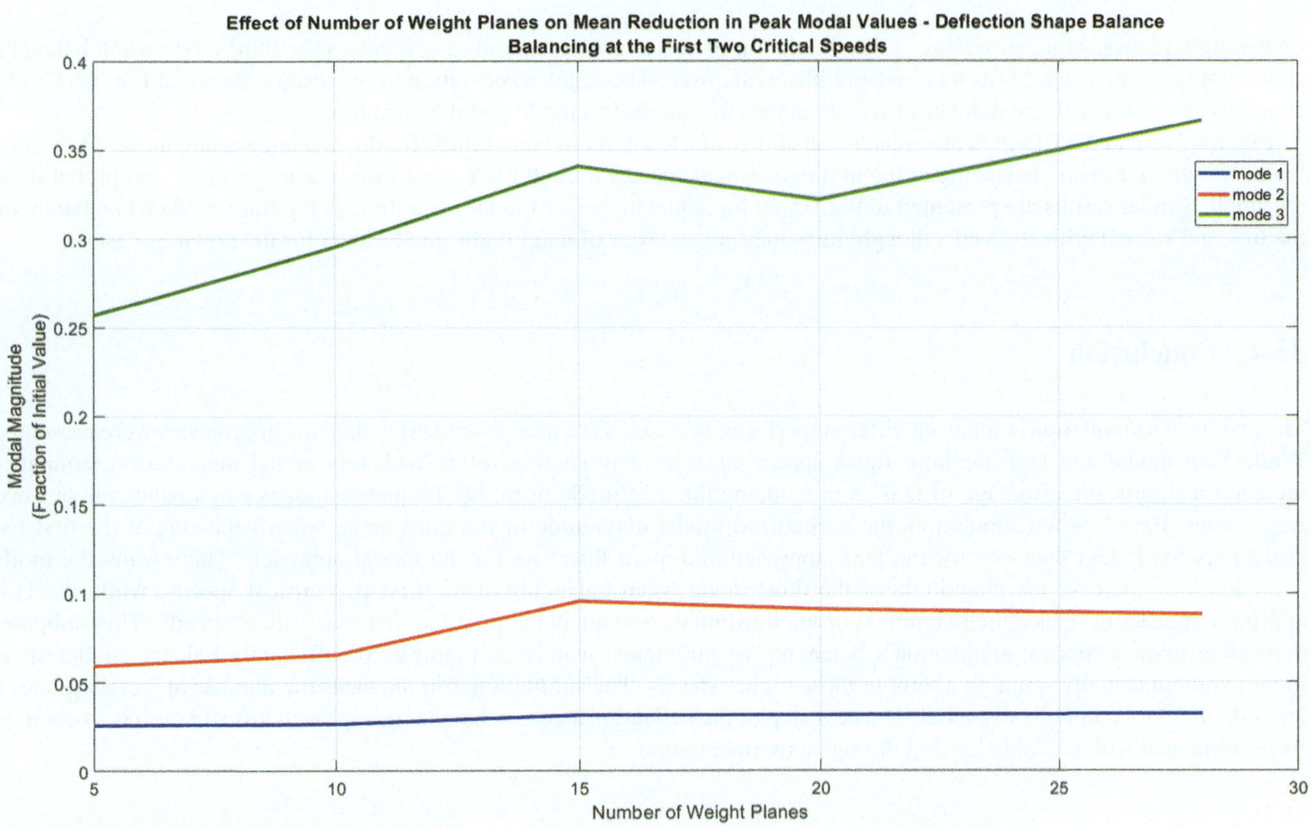

Fig. 35.16 Effect of the number of weight planes on the achieved mean modal magnitude using the DSF approach

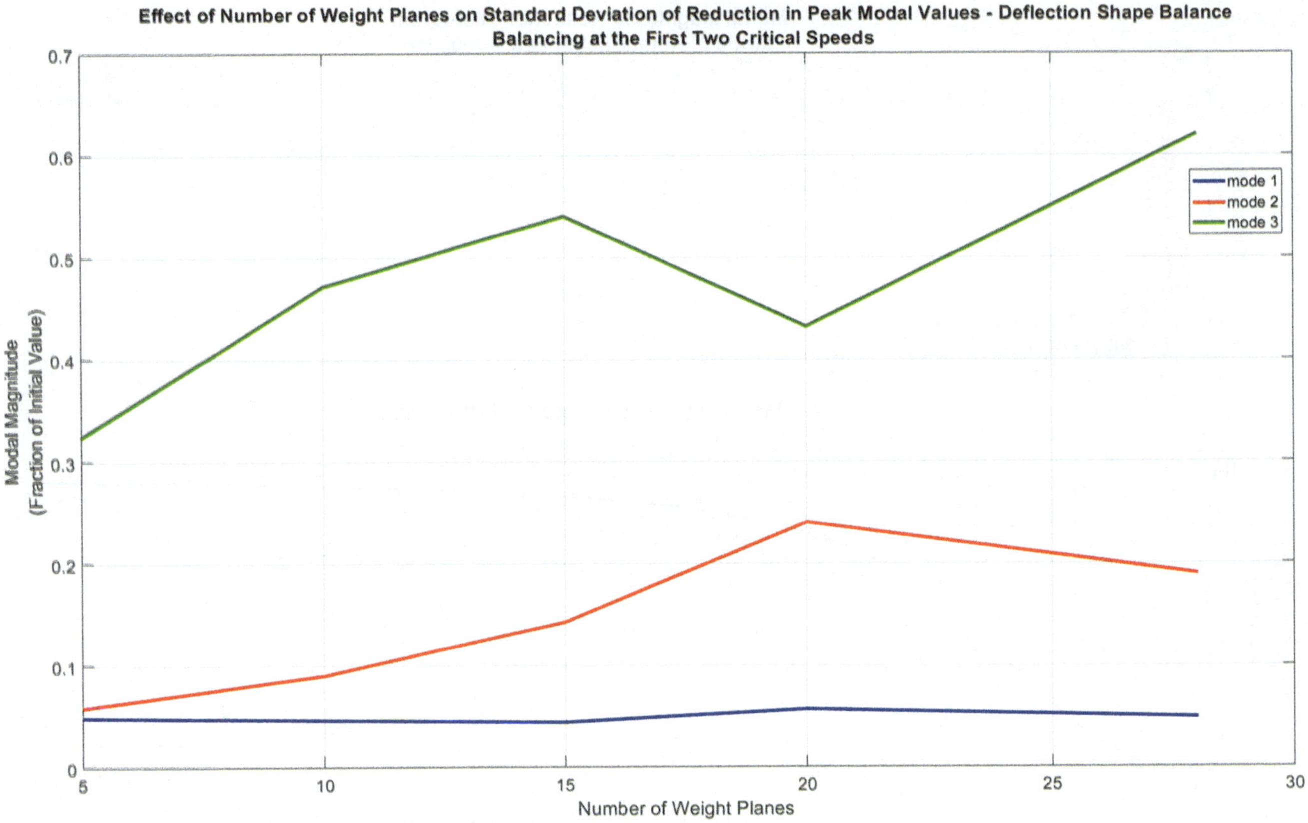

Fig. 35.17 Effect of the number of weight planes on the standard deviation of the achieved mean modal magnitude using the DSF balancing approach

DSF weight planes. More surprising is the simulation indicates greater modal magnitude in the third mode when balancing at the first two critical speeds as more weight planes are used. The standard deviation of the results, shown in Fig. 35.17, also generally increases with the number of weight planes for the second mode and the third mode.

Figures 35.18 and 35.19 indicate greater probability of a lower modal magnitude for the first and second mode when using fewer weight planes and balancing at the first two critical speeds, though both cases show the magnitudes and probabilities are small. Similar results are presented in Fig. 35.20 for achieving a low modal magnitude in the third mode when balancing the first and second critical speeds, though the values are an order of magnitude smaller than for the previous cases.

35.4 Conclusion

The results of a simulation comparing different performance aspects of modal and DSF balancing approaches were presented. While both modal and DSF methodologies appear capable of producing rotors with low modal magnitudes, simulation indicates a significant advantage of DSF is in reducing the magnitude of higher frequency modes when balancing at lower frequencies. Based on the simulation, the normalized modal magnitude of the third mode when balancing at the first two critical speeds is less than one for the DSF approach and more than one for the modal approach. This means the modal approach may increase the magnitude of the third mode when balancing at the first two critical speeds, while the DSF approach appears to reduce the magnitude of the third mode when balancing at the first two critical speeds. This indicates, depending upon a specific application's balancing requirements, it may be possible to sufficiently balance higher speed modes without actually spinning a rotor to those higher speeds. The simulation also indicates the number of weight planes is less important than initially expected. The next step of the deflection shape fit balancing approach investigation is to compare its performance with modal balancing through experimentation.

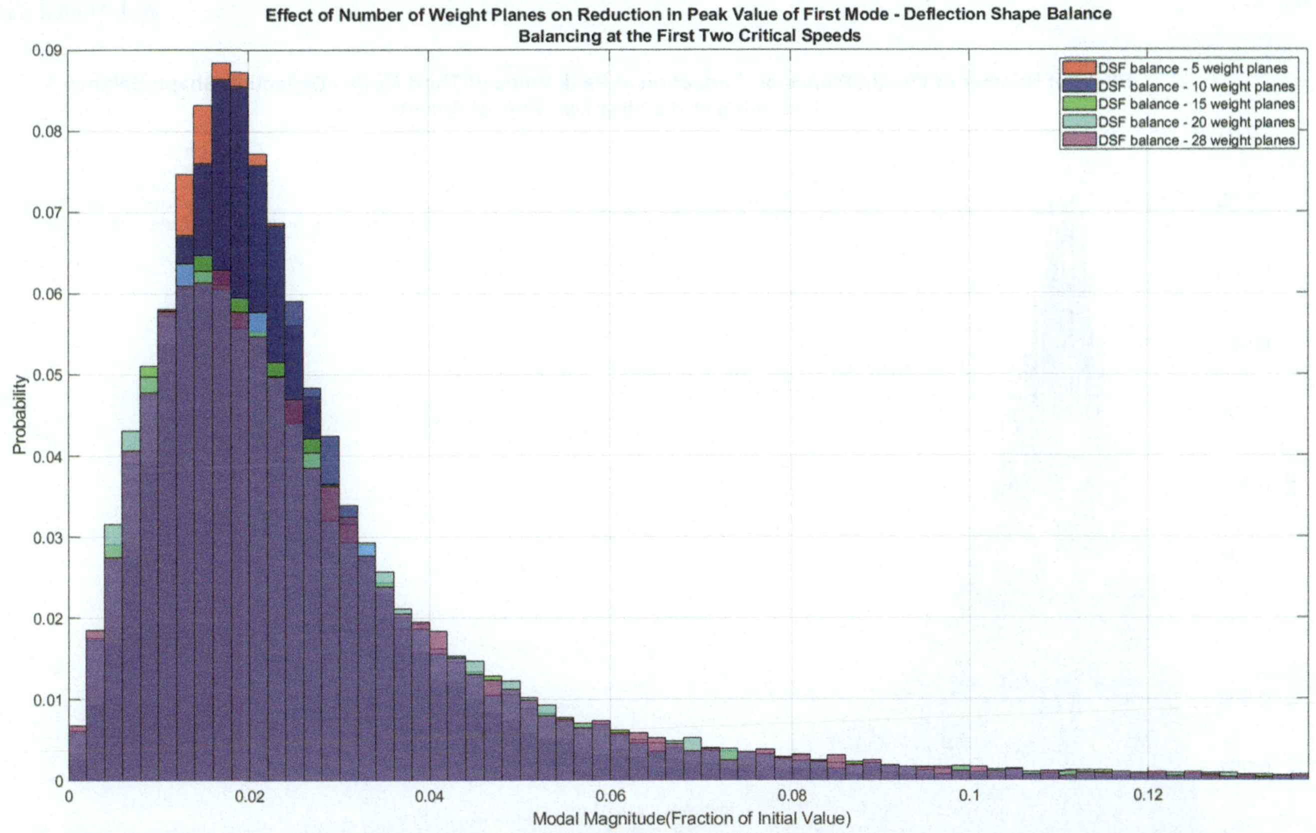

Fig. 35.18 Effect of the number of weight planes on the achieved modal magnitude for the first mode when balancing at the first two critical speeds

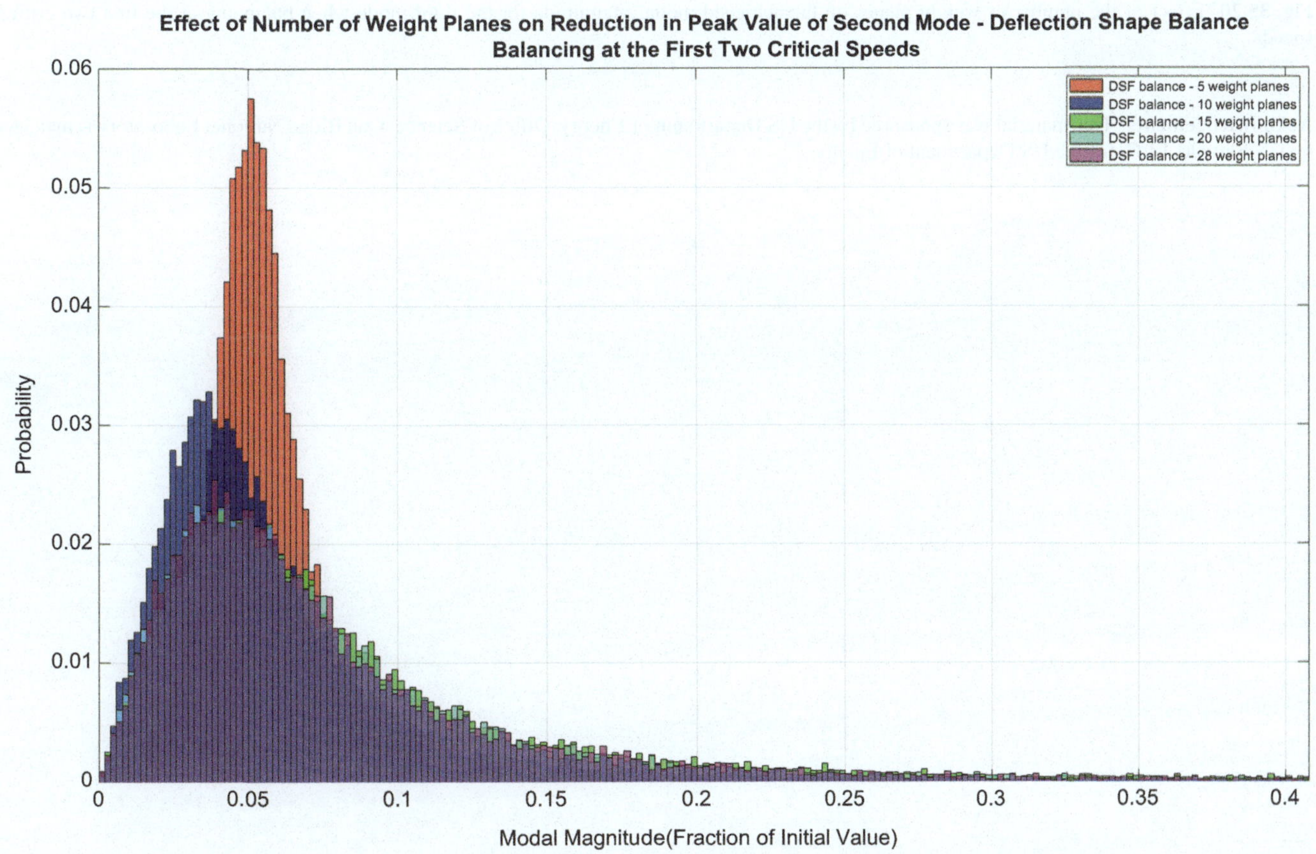

Fig. 35.19 Effect of the number of weight planes on the achieved modal magnitude for the second mode when balancing at the first two critical speeds

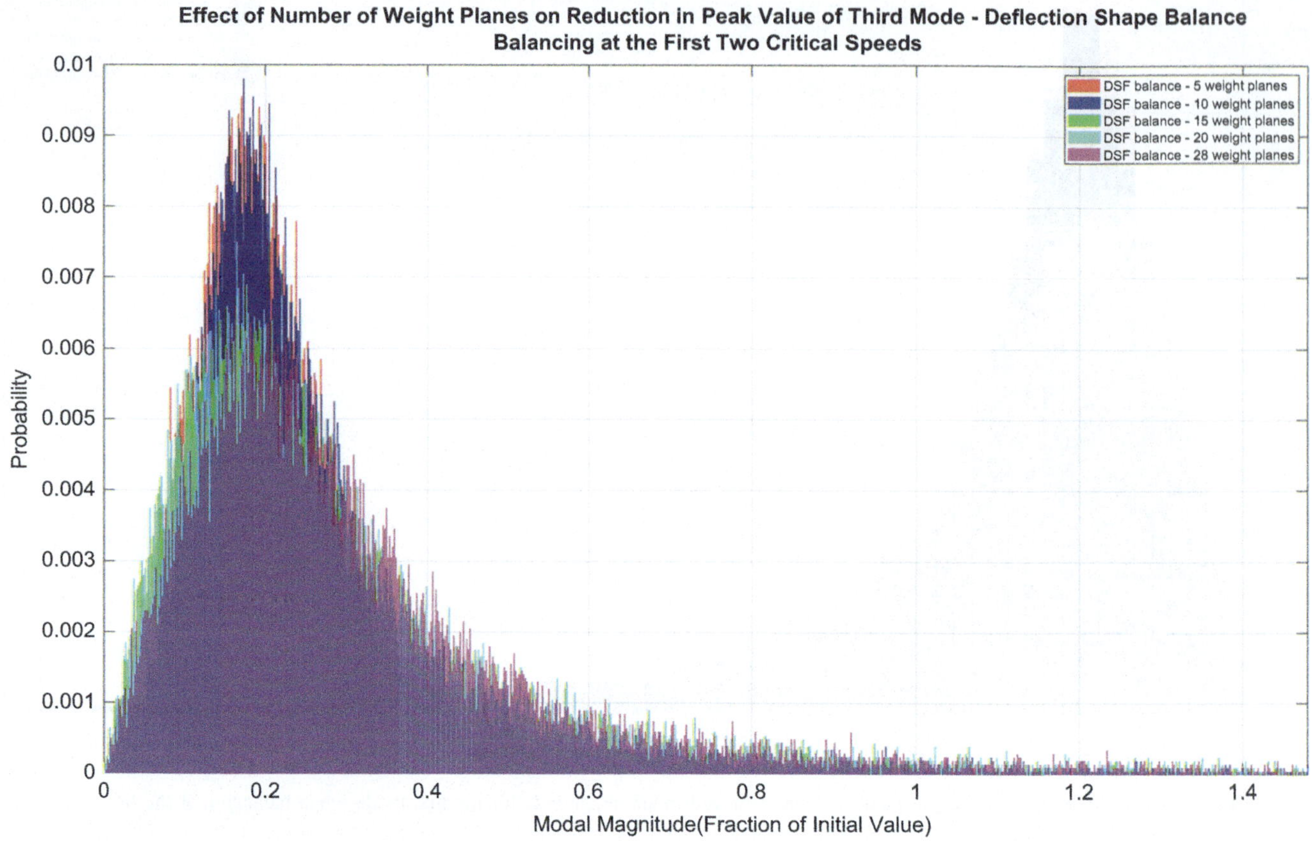

Fig. 35.20 Effect of the number of weight planes on the achieved modal magnitude for the third mode when balancing at the first two critical speeds

Acknowledgements This material was sponsored by the US Department of Energy, Office of Science. Oak Ridge National Laboratory is managed by UT-Battelle, LLC, for the US Department of Energy.

Chapter 36
Real-Time Theoretical and Experimental Dynamic Mode Shapes for Structural Analysis Using Augmented Reality

Maimuna Hossain, John-Wesley Hanson, and Fernando Moreu

Abstract Researchers conduct dynamic experiments to quantify changes in frequencies and mode shapes of interest for the safety and performance of structures. Even with sensor data and advanced computing techniques, valuable information of the experiment is always best observed with physical presence during the experiment. However, as of today, there exists no method for engineers to observe the dynamics experiments in the physical field and conduct dynamic analysis real-time. This paper describes a software application that conducts dynamic analysis in order to resolve this issue. The research consists of theoretical and experimental modal analyses of a cantilever beam using Augmented Reality (AR). The physical experiment involves a continuous aluminum beam attached to a shake table. First, a theoretical modal analysis is performed using Euler-Bernoulli equations in order to determine the modal parameters of this beam. A software application is built for the AR headset which contains a scaled hologram of the beam and depicts the first three theoretical mode shapes. This hologram is anchored to shake table to superimpose the physical beam. Using sinusoidal vibrations, the shake table can then excite the experimental mode shapes. The hologram and the actual beam are overlaid to compare real time structural data with theoretical mode shapes. By incorporating AR visualization of mode shapes, this prototype enables a new interface for structural analysis. Use of AR in the field of structural dynamics can expand human cognition by introducing a new interface for humans to interact with.

Keywords Augmented reality · Structural dynamics · Mode shapes · Euler-Bernoulli · Human-machine interface

36.1 Introduction

The goal of this project is to introduce a way to visualize dynamic properties of a structure using Augmented Reality (AR). Here, AR refers to the process of taking one's surroundings and enhancing it by overlaying digital content. A Generation 1 Microsoft HoloLens serves as the main tool for visualization. The project investigates the dynamic property called mode shape. Mode shapes refer to a pattern of vibration created by a structure associated with its natural frequency.

The use of AR as a form of engineering analysis and simulation has progressively increased in the past decade [1]. Previous works in the field include real-time finite element structural analysis in AR [2]. Although their paper demonstrates Finite Element Analysis using loads, it does not indicate mode shapes or natural frequencies of the model. Closest to mode shapes is demonstrated in an iOS application that visualized 3D frame deflection [3]. This application also enables mobile structural dynamics through the phone. When it comes to wearable technology, the AR headset allows developers to incorporate hand gestures, head motions, as well as a tangible user interface to determine location, data points and environment [4].

This paper describes a new first step towards visualizing mode shapes of a structure in real-time. The approach is innovative because of its contribution to the fields of AR and Structural Dynamics. The visualization of mode shapes has been

M. Hossain (✉)
Department of Mechanical Engineering, University of New Mexico, Albuquerque, NM, USA
e-mail: hossainm@unm.edu

J.-W. Hanson
Department of Business Information Technology, Central New Mexico Community College, Albuquerque, NM, USA

F. Moreu
Department of Civil, Construction, and Environmental Engineering, University of New Mexico, Albuquerque, NM, USA
e-mail: fmoreu@unm.edu

B. Dilworth (ed.), *Topics in Modal Analysis & Testing, Volume 8*, Conference Proceedings of the Society for Experimental Mechanics Series, https://doi.org/10.1007/978-3-030-47717-2_36

accomplished through computational fluid software as well as 3D modeling software [1]. Similar previous work also includes enabling a smart nuclear infrastructure using barcode scanners for AR headsets [5] as well as connecting sensors wirelessly to stream data for dynamic analysis [6]. However, since the AR headset is mostly a visualization tool, incorporating it with structural dynamics enables a stronger, newer method of seeing experimental data.

36.2 Experiment

Mode Shape refers to a pattern of vibration caused by a structure under dynamic load. Each mode shape is associated with a natural frequency. Exciting the natural frequencies of the beam creates the corresponding mode shape. For this experiment, a rectangular beam is chosen to represent a simple structure. The beam has uniform cross-section and is made of Aluminum Alloy 6061.

The software application for the AR headset shows both theoretical and experimental mode shapes. The theoretical mode shapes are generated using the Euler-Bernoulli equations for calculating dynamic properties of a beam. The experimental data is gathered using a shake table that induces sinusoidal input force on the beam.

This investigation considers the first three mode shapes of the structure only. More theoretical mode shapes can be visualized using the equations and also various structural configurations would enable the interface once programmed in the AR headset. For our example, since the mode participation factor decreases as the mode shape number increases, the first three mode shapes provide sufficient data for analyzing a structure.

36.2.1 Specimen

Figure 36.1 summarizes the material and geometric properties of an Aluminum beam. The Euler-Bernoulli theory considers a continuous cantilever beam and applies the equation of motion in order to obtain the natural frequencies and mode shapes.

36.2.2 Theoretical Analysis

In order to solve the Euler-Bernoulli equations, the beam must be modeled as a set of finite elements, each consisting of a node with mass and stiffness values. Figure 36.2 below shows an illustration of the beam reduced to 10 nodes.

The equation for natural frequency obtained using Euler-Bernoulli beam:

$$\omega_n = \frac{(\lambda_n L)^2}{L^2}\left(\frac{EI}{\rho A}\right)^{\frac{1}{2}} \tag{36.1}$$

Where n = 1,2,3,.., $I = \frac{bh^3}{12}$, A = bh, and $\lambda_n L$ is calculated numerically from the characteristic equation such that $\lambda_1 L = 1.875$, $\lambda_2 L = 4.6941$ and $\lambda_3 L = 7.8548$.

Similarly, the mode shape equation for a continuous cantilever beam:

$$V_n(x) = C\left[(\cosh \lambda_n x - \cos \lambda_n x) - k_n\,(\sinh \lambda_n x - \sin \lambda_n x)\right] \tag{36.2}$$

Material Properties	Geometric Properties
Aluminum Alloy 6061	L = .8414 m
E = 68.9 GPa	b = .0318 m
ρ = 2690 kg/m^3	h = .0031877 m

Fig. 36.1 Summary of beam properties

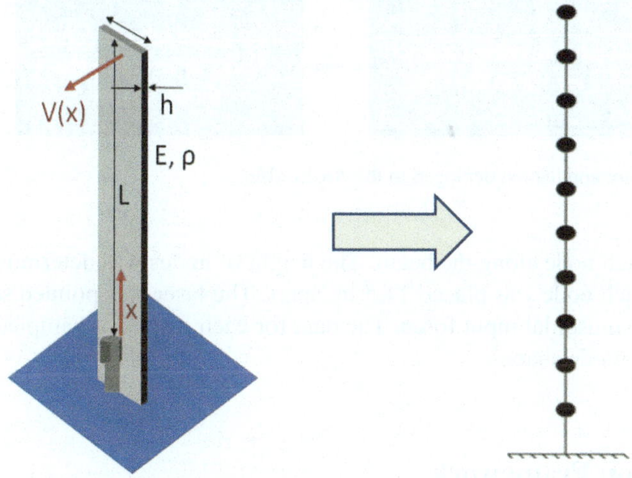

Fig. 36.2 Aluminum beam reduced to 10 nodes for theoretical calculations

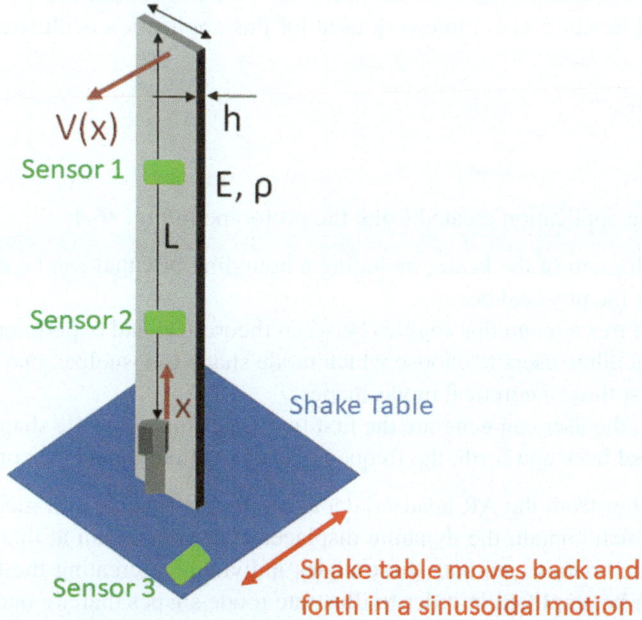

Fig. 36.3 Experimental setup

Where n = 1,2,3,... C is an arbitrary constant, chosen to be 1 in this calculation. λ_n values are calculated numerically from characteristic equation.

$$k_n = \frac{\cosh(\lambda_n L) - \cos(\lambda_n L)}{\sinh(\lambda_n L) - \sin(\lambda_n L)} \tag{36.3}$$

Equation 36.3 determines the stiffness factor associated with each mode shape. Using these equations, the mode shapes can be determined for the beam.

36.2.3 Experimental Analysis

The experimental data uses a displacement laser sensor. The setup involves a beam attached to a shake table, a laser for displacement, a data acquisition system, cables, a power source and a computer to collect recorded data (Fig. 36.3). Data was

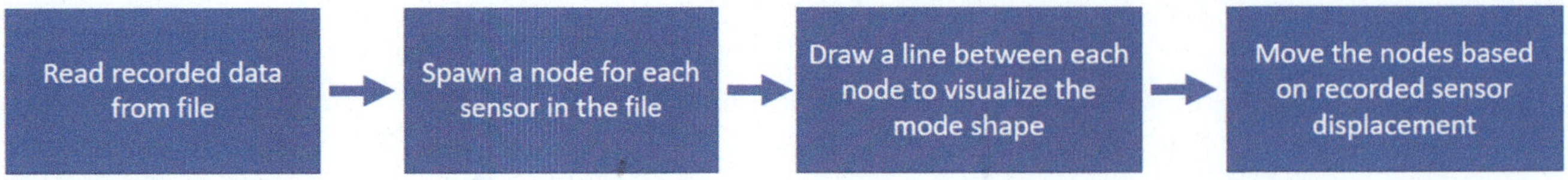

Fig. 36.4 Program flow for the software application deployed to the AR headset

collected three times, once for each node along the beam. The height of nodes was determined by dividing the total height of beam, 33.5 in., in three. Thus, each node was placed 11.5 in. apart. The laser was pointed straight at the beam using a level, and the shake table was given a sinusoidal input force. The data for each node was sampled for 3.5 s. This data was used to create the first two experimental mode shapes.

36.3 Software Application Framework

After collecting data using theoretical and experimental approaches, the project utilized a few software development tools to create an application for the AR headset. The framework used for this application is illustrated below (Fig. 36.4):

36.4 Results

Here are the main features of the application created using the prototype in Fig. 36.4:

1. The application creates a hologram of the beam, including a bounding box that can be used to intuitively to scale, adjust and rotate the beam to match the physical beam.
2. The software application features a menu that toggles between theoretical and experimental data.
3. There are two sub menus that allow users to choose which mode shape to visualize, one by one
4. The user can generate the first three theoretical mode shapes.
5. After selecting experimental, the user can generate the first two experimental mode shapes
6. Every mode shape is animated back and forth, the frequency of the animation can be controlled.

Figure 36.5 shows a screenshot from the AR headset, depicting the first theoretical mode shape of the beam. The mode shape has 10 nodes, each of which contain the dynamic displacement of the beam at that location. It is important to note that the application uses recorded data to animate these nodes individually, creating the hologram of each specific mode shape. These displacements can be amplified in order to illustrate mode shapes that are undetectable by the human eye. For example, the maximum displacement of the beam at the top node's displacement was exaggerated to illustrate the concept. This amplification does not compromise the accuracy of the mode shapes, only adopts it for easier viewing. The same is true for the second mode shape, which occurs at 18 Hz and the third mode shape which occurs at 54 Hz.

The third theoretical mode shape can be applied to determine what happens to the beam under a frequency of 54 Hz. This mode shape is specifically important because the limitations of the shake table used in this experiment do not allow researchers to excite the beam at that frequency. Therefore, the only way engineers can inspect the mode shapes higher than 20 Hz is through theoretical analysis. In this context, the development of a holographic interface with higher modes provides a useful tool for human cognition of structural dynamics not possible before. The future stages of this research will specifically outline a human-structure interface in the area of structural modal analysis using AR. In summary, the theoretical mode shapes generated by this AR application are a first step to provide humans with the ability to visualize structures mode shapes of a structure and superimpose those mode shapes on top of the physical structure. These preliminary results are being expanded to a human-structure interface that enables cognition of modal analysis in a human-centered framework. Allow meaningful interaction between humans and machine interface but it also expands the applications of two different fields: structural dynamics and augmented reality. This work is inherently innovative because these two fields have never been combined for the purpose of structural analysis before.

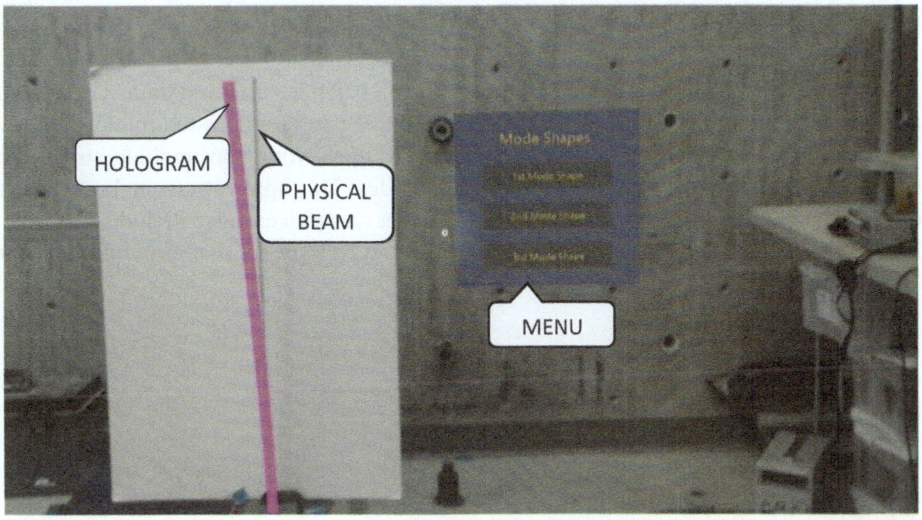

Fig. 36.5 First theoretical mode shape

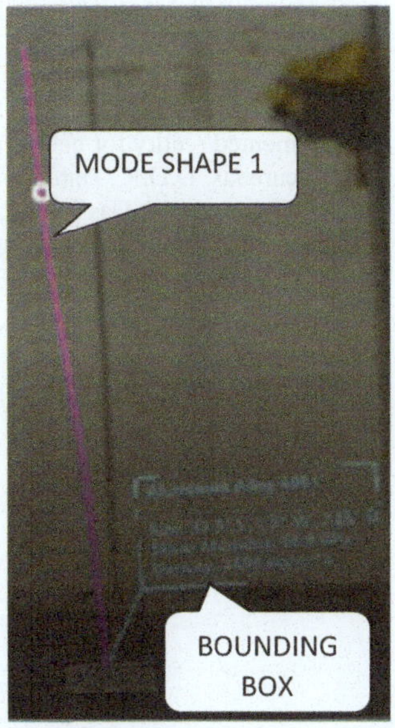

Fig. 36.6 Experimental mode shape 1

36.5 Ongoing Work

The first two experimental mode shapes are generated from running the experiments mentioned in the experimental analysis section. Although theoretical analysis and predictions are valuable, the experimental results show the actual behavior of the beam. The sensor data collected for the each of the three nodes capture the mode shapes associated with the first two natural frequencies. This data from which can be inserted into the AR application in order to animate the first two experimental mode shapes. Figure 36.6 depicts the two mode shapes, as well as the bounding box containing the properties of the beam.

The impact of the experimental mode shape is that it provides a proof of concept of linear dynamic analysis of a cantilever beam. This can be extended to more complex geometry and non-linear dynamic analysis as well. Currently the experimental mode shapes contain data for three nodes, but the number of nodes can be increased to produce a smoother shape. The AR application also enables engineers on the field to interact with the experimental mode shape in real time, adjusting scales, and reading properties. Future research will involve establishing a connection between the AR headset and the sensors used in the experiment, so that one can read the data and generate the mode shapes in real-time rather than looking at recorded data.

36.6 Conclusion

This project lays the groundwork for a software application that can visualize structural data. The AR application can take a flexible number of nodes and data points to create a noodle-like shape that emulates the mode shapes of the real structure. The flexibility of this application allows room for improvement. Future work may include streaming live data from the acquisition system as well as generating mode shapes with more nodes. This application can be successfully combined with structural inspections for maintenance workers and engineers. Further research can also include more complex and non-linear structures. Overall, the application is a success and calls for further development that combines structural engineering and AR in meaningful ways.

References

1. Li, W., Nee, A.Y.C., Ong, S.K.: A state-of-the-art review of augmented reality in engineering analysis and simulation. Recent Adv Augment Reality. (2017). https://doi.org/10.3390/mti1030017
2. Huang, J.M., Ong, S.K., Nee, A.Y.C.: Real-time finite element structural analysis in augmented reality. Adv. Eng. Softw. (2015). https://doi.org/10.1016/j.advengsoft.2015.04.014
3. Turkan, Y., Radkowski, R., Karabulut-Ilgu, A., Behzadan, A.H., Chen, A.: Mobile augmented reality for teaching structural analysis. Adv. Eng. Softw. (2017). https://doi.org/10.1016/j.aei.2017.09.005
4. Chai, L., Hoff, W., Vincent, T.: Three-dimensional motion and structure estimation using inertial sensors and computer vision for augmented reality. Presence Telep Virtual Environ. (2006). https://doi.org/10.1162/105474602320935829
5. Ballor, J.A.P., McClain, O.L., Mellor, M.A., Cattaneo, A., Harden, T.A., Shelton, P., Martinez, E., Narushof, B., Moreu, F., Mascarenas, D.D.L.: Augmented reality for next generation infrastructure inspections. (2018). https://doi.org/10.1007/978-3-319-74793-4_23
6. Mascarenas, D.D.L., Harden, T.A., Morales Garcia, J.E., Boardman, B.L., Sosebee, E.M., Blackhart, C., Cattaneo, A., Krebs, M.S., Tockstein, J.J., Green, A.W., Dasari, S.R., Bleck, B.M., Katko, B.J., Moreu, F., Maharjan, D., Aguero, M., Fernandez, R., Trujillo, J.B., Wysong, A.R.: Augmented reality for enabling smart nuclear infrastructure. Front Built Environ. (2019). https://doi.org/10.3389/fbuil.2019.00082

Chapter 37
A Bottom-Up Approach to FE Model Updating of Industrial Structures

Daniel J. Alarcón, Fabian Keilpflug, and Peter Blaschke

Abstract The main aim of FE model updating is to iteratively correct the inaccurate parameters in a structural model so that the agreement between numerical predictions and modal test results is improved. Modeling the boundary conditions and contact interfaces between the different components is a challenge that needs to be taken into account in regards to the computational cost of the model for the iterative stage of the study. For instance, an extremely detailed model is of no use if each iteration takes hours to solve.

A bottom-up approach can be followed to generate a highly accurate FE model by gradually increasing the complexity of the studied structure. This requires of a careful test planning phase to acquire experimental data from the structure in different boundary conditions and assembly stages. An overhead crane runway beam FE model is updated in this study case. Three FE models of increasing complexity in their boundary conditions and contact interfaces are optimized and updated with their three corresponding modal experimental datasets.

This bottom-up approach allows dividing a complex problem in smaller phases, which results in a smaller number of design variables present on each model updating cycle. This is essential for the successful application of optimization algorithms with a limited computational power. In a qualitative sense, it allows the analyst to have a deeper understanding of the model from its conception and to have a solid grasp of the different what-if scenarios.

Keywords Model updating · Optimization · Finite element analysis · Modal analysis · Structural health monitoring

37.1 Project Motivation and Scope

This project is focused on the study of heavy duty overhead crane runway beams, which are an essential part of the manufacturing and logistics of many industrial plants. Figure 37.1, left, describes the different parts of an overhead crane for clarity. Runway beam, runway rail and downshop conductors are of interest in this paper.

These overhead cranes typically transport heavy loads along an industrial plant for storage of to prepare deliveries. According to maintenance representatives of a specific German plant, the crane rails and runway beams holding the rails have been exposed to different loads, damages and fatigue processes during decades. In their experience, runway beams are increasingly prone to the generation of cracks, especially localized on the upper flanges and extending to the beam's web due to material fatigue, as shown in Fig. 37.1, right. Replacing these beams is a work-intensive procedure, which requires of a careful planning and coordination of several teams of technicians. Unplanned plant downtimes needed to replace the runway beams are, however, what truly drive up the replacement costs; especially because the uncertainty about the plant downtime duration is generally difficult to estimate. During the replacement time, the crane simply cannot operate with the obvious impacts in plant productivity and delivery delays. It is therefore interesting for production management to have a service life prognosis to plan these downtimes ahead of time and minimize the effects on downtime in plant productivity.

D. J. Alarcón (✉) · F. Keilpflug
Laboratory for Machine Dynamics and NVH, Technische Hochschule Wildau, Wildau, Germany
e-mail: daniel.alarcon@th-wildau.de; fabian.keilpflug@th-wildau.de

P. Blaschke
Laboratory for Machine Dynamics and NVH, Technische Hochschule Wildau, Wildau, Germany
NV Tech Design GmbH, Steinheim, Germany
e-mail: peter-g.blaschke@th-wildau.de

B. Dilworth (ed.), *Topics in Modal Analysis & Testing, Volume 8*, Conference Proceedings of the Society for Experimental Mechanics Series, https://doi.org/10.1007/978-3-030-47717-2_37

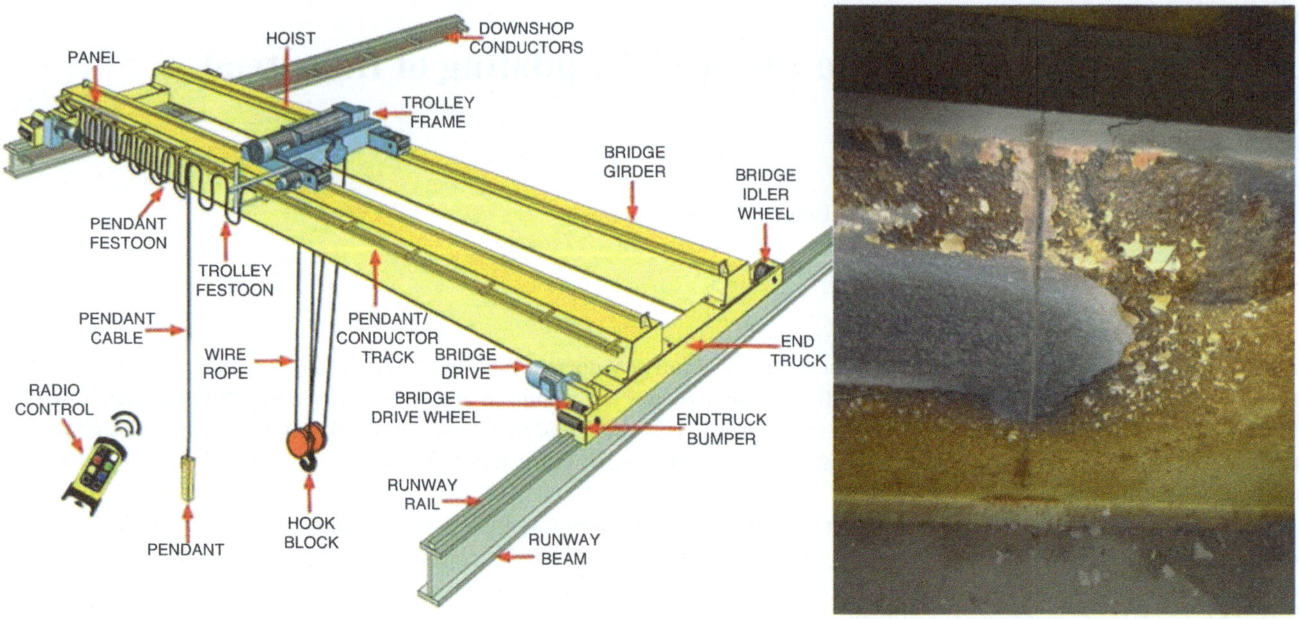

Fig. 37.1 Left – parts of an overhead crane for illustrative purposes [1]. Right – example of a crack on the upper flange of a runway beam

The methods in use at the project partner's facility rely on visual inspections. This procedure suffers of the same drawbacks as ultrasonic methods (static or rolling), magnetic field methods, radiography, eddy-current methods or thermal field methods: these techniques require a priori knowledge of the damaged area. Structural damage detection based in changes in modal properties is still a debated topic in the structural dynamics community, and research on the effectiveness of these methods is still ongoing, as shown in the extensive literature review by Almeida et al. [2]. Doebling et al. [3] discuss that resonance frequency shifts alone have practical limitations in larger applications. The low sensitivity of resonance frequency shifts to damage requires either very precise measurements or larger levels of damage. Methods based in changes in mode shape changes show more promising results [4], with some ideas explored in the work of Mayes [5]. Some previous investigations based on the evolution of changes in mode shapes [6] have shown promising results in the past in proof-of-concept, smaller laboratory-scale setups and models when acquiring data with precise equipment.

The work described in this paper is a small part of a larger long-term research and development project. The goal of this project is the creation of a predictive digital model for the continuous structural health monitoring of these runway beams. The precise realization of these ideas is still a matter under discussion among all stakeholders in the project and funding agencies. One of the ideas under discussion is based on the continuous SHM of each single runway beam in a given facility. Vibrational data from each beam is to be continuously logged by transducers and fed into a digital twin [7] to qualitatively and quantitatively evaluate changes in the vibrational behavior of each beam in real time. In the future, artificial intelligence algorithms are expected to generate predictive models from these digital twins, in order to generate maintenance models about the runway beam's life expectancy. The creation of an accurate parametric beam model which can represent the structure accurately at a low computational cost is hence a crucial step in this process.

The main aim of this paper is to propose a bottom-up process to model, optimize and update a finite elements (FE) model of a runway beam under installed conditions by correlating FE analytical modal data with experimental modal data. This approach has been successfully applied to a variety of investigations during the last decades [8, 9]. The updated FE model is to be later applied to further stages of the project as needed and, possibly to the future generation of a digital twin of the studied beam.

37.2 Theoretical Background

The Finite Elements Method (FEM) is a tool to approximate solutions for mathematical constructs that resemble real phenomena under real loads and working conditions. The formulation of a suitable mathematical model of the physical problem at hand, considering all factors and assumptions that will affect the response of the model, is the analyst's main task. Having a solid understanding of the limitations of the FEM and striking the right balance between model accuracy, feasibility and computational efficiency is of paramount importance for a successful application of the FEM. This is an inherent optimization problem by itself, which underlies in any FE model optimization, as described in this paper.

Establishing controls to test the validity of FE models is essential. Mottershead and Friswell [11] define this process in a simple way: model updating seeks to correct the inaccurate parameters in the model so that the agreement between

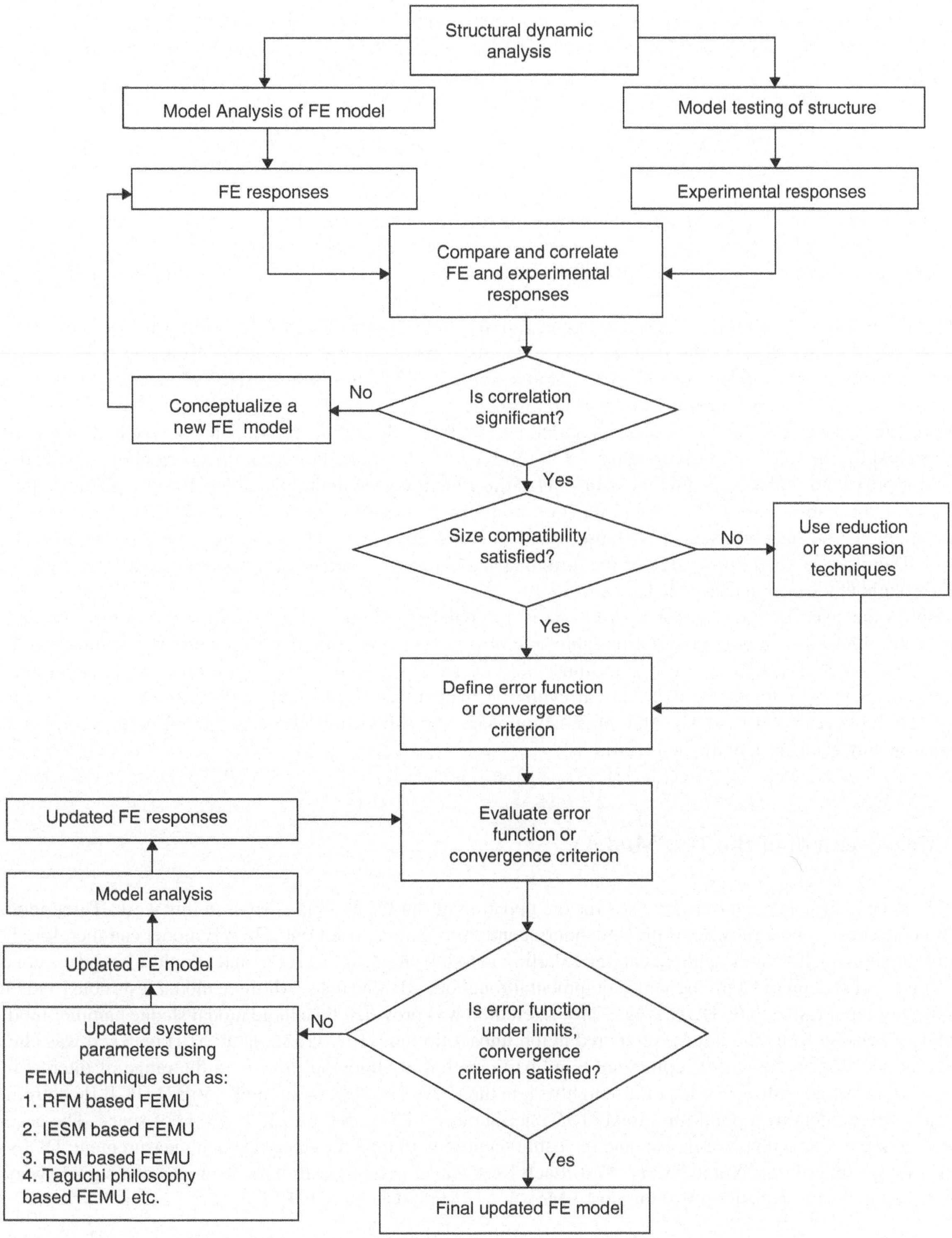

Fig. 37.2 Algorithm for FE model updating [10]

predictions and test results is improved. Dynamic experimental results are used to check that the numerical simulation correlates to the experimental data iteratively, until the model is said to be updated. Sehgal and Kumar [10] propose the algorithm in Fig. 37.2, chosen for this paper because it describes in a visually simple but rigorous manner the complete model updating process followed in this study, and emphasizes in the iterative aspect of the process, which starts with the formulation of an error function or convergence criterion.

In FE model updating, the deviation between numerical and experimental dynamic models is calculated with the aim of reducing the differences between models. When these differences are lower than a certain threshold (an error function), the model is said to be updated. Marwala [12] proposes an objective function to be minimized for this purpose. It represents the distance between test modal data and FE modal data. This iterative process describes mathematically the loop in the flow chart in Fig. 37.2.

$$E = \sum_{i=1}^{N} \gamma_i \frac{\omega_i^m - \omega_i^{calc}}{\omega_i^m} + \beta \sum_{i=1}^{N} \left(1 - \mathrm{diag}\left(MAC\left(\phi_i^{calc}, \phi_i^m\right)\right)\right)$$

This equation is, however, not formally implemented in this paper due to the complexity of implementing the calculation of a cross-correlation MAC between an experimental dataset and a FE model in ANSYS® Mechanical APDL and then implementing this calculation in ANSYS® DesignXplorer, but the rationale behind the methods used in this paper follows this objective function equation.

Minimizing this objective function poses an optimization problem. The objective function equation seeks a FE solution that fits the experimental data in the best way possible. As a consequence, knowledge from the field of mathematical optimization needs to be applied, specifically, iterative optimization algorithms. ANSYS® DesignXplorer has several algorithms built in, they can be generally categorized as gradient-like and evolutionary algorithms.

The Non-Linear Programming by Quadratic Lagrangian (NLPQL) algorithm is a gradient-like method based on Newton's iterative method for the solution of linear equations. It relies on iterative calculations of the gradient of an objective function using local approximations based in Taylor expansions of this function. Due to these mathematical foundations, the algorithm requires initial conditions close to the global function minimizer and convergence to a single minimizer, which may be a local one. Complex design spaces, such as those found in real industrial problems, typically present more than one local minimizer. Therefore, a good knowledge of the design space is required beforehand. Mathematical smoothness is as well required throughout the design space [13].

The Multi-objective Genetic Algorithm (MOGA) is a population-based, probabilistic algorithm that operates to find a solution to a problem from a population of possible solutions, where concepts such as inheritance, mutation, selection and crossover are applied. The idea of these algorithms is to iteratively evolve a population of possible candidate values of the design vector using rules that mimic natural evolution, with fitness defined according to the values of the objective function. The drawbacks of the MOGA and other evolutionary algorithms are related to their very high computational cost in comparison with gradient-like methods [14].

37.3 Measurement of the Test Model

Three different experimental datasets are used for the updating of the FE models created in this work. Three modal tests of increasing complexity in the runway beam's boundary conditions were carried out. The FE model can therefore be updated in a bottom-up approach with these three datasets, starting with the simplest test setup and simplest boundary conditions.

The datasets were acquired with the same equipment throughout all three tests. All three modal tests consisted on a single-input multiple-output test with 65 DOFs (Fig. 37.3). Excitation was provided by a large modal sledge hammer mod. Endevco 2305 (Meggit Sensing Systems) fitted with its red nylon impact tip mod. EHM1658. DOF 51 (Fig. 37.3) was chosen as the driving point due to its accessibility, compliance and participation in almost all modes in the range of interest in all tests. Response was measured with a general purpose lightweight three-axis accelerometer mod. 356B21 (PCB Piezotronics), roved throughout each measurement point for a total of 195 accelerance FRFs, 65 for each direction of space. The accelerometer was in all cases attached with paraffin wax and the cable secured with tape for safety at each measurement. DOFs are about 1 m apart of each other in the X axis in Fig. 37.4. Each DOF was measured with three averages and signals were acquired with a four-channel data acquisition system mod. vMeas4-2.1 (Maul-Theet GmbH).

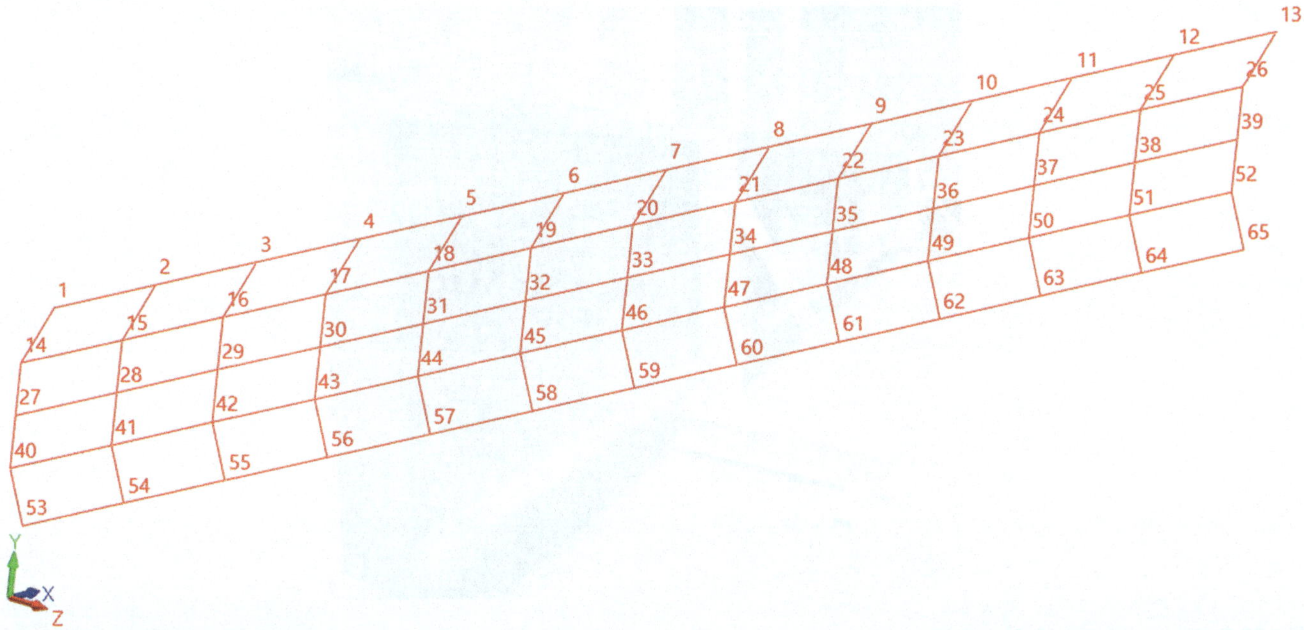

Fig. 37.3 Test mesh used for all three modal tests

Fig. 37.4 Left – runway beam under study. Engineers and equipment depicted for scale. Right – detail of the beam supported on L-shaped profiles around DOF 55

37.3.1 Test I

The first modal test was carried out on the runway beam simply double-supported at positions L = 2000 mm and L = 9500 mm on two robust support rigs on the plant's floor (Fig. 37.4). This setup added the least artificial frictional damping to the beam as possible and represented the simplest study case. Data was acquired with a bandwith of 0.5 kHz and 25,600 FFT lines, resulting in a time block of 51.6 s and a resolution of 0.02 Hz. This was required to measure the entire unwindowed vibration decay of the beam, which is very lightly damped under these boundary conditions.

37.3.2 Test II

This test was conducted analogously to Test I due to the similarities in boundary conditions. The only changes were the addition of the downshop conductor fittings onto the runway beam's web, as Fig. 37.5 shows. The change in eigenfrequencies and damping due to the increase in mass and stiffness of the web was audible when impacting the beam. Measurement

Fig. 37.5 Runway beam with downshop conductor fittings screwed on its both sides

Fig. 37.6 Left – tested runway beam installed on its final position. Right – detailed view of the bolts between runway beams and downshop conductor fittings around DOFs 40 and 53

parameters were changed to a bandwidth of 1 kHz and 12,800 FFT lines for a resulting time block of 25.6 s and a resolution of 0.02 Hz, as the resonance time shortened to about 10–15 s. The fittings were not measured as they are irrelevant in the model correlation tasks.

37.3.3 Test III

This test differed heavily from Tests I and II as the boundary conditions in this setup were different. In this case, the runway beam was installed on its final place, bolted to its steel supports and to the other neighboring beams (Fig. 37.7). Rigid downshop conductors and runway rails were installed on their fittings, further increasing the system stiffness. Measurements were carried out from a scaffold placed right next to the runway beam, as shown in Fig. 37.6, left. Due to the radical change in the dynamic properties of the structure, parameters of 1 kHz and 1600 FFT lines for a resulting time block of 3.2 s and a resolution of 0.3 Hz were used. Again, the fittings remained unmeasured.

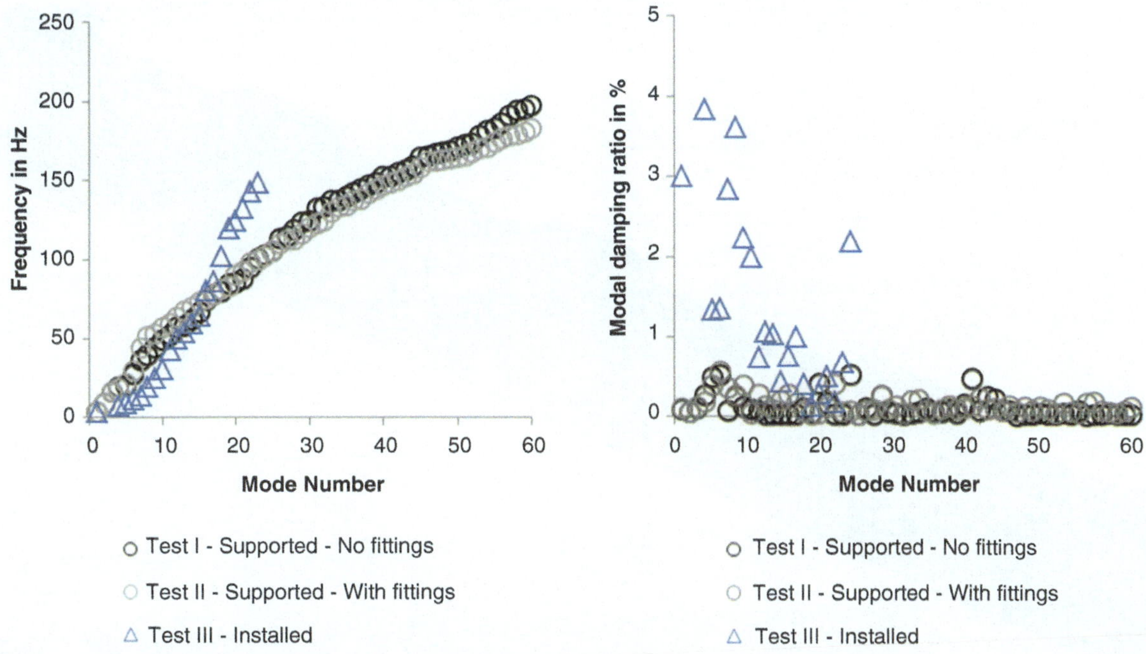

Fig. 37.7 Left – eigenfrequencies plot for each test. Right – modal damping ratios plot for each test

Modal parameters, namely eigenfrequencies, modal damping ratios and mode shapes, were estimated from the 165 measured FRFs. For Tests I and II, this task was carried out with FEMtools™ Modal Parameter Extractor (DDS NV) up to mode 65 on the Complex Mode Indicating Function (CMIF) with the poly-reference Least Squares Complex Frequency (pLSCF) algorithm built in the software. For Test III, the pLSCF yielded inadequate results and therefore modal parameters were estimated manually with a generally damped MDOF curve-fit method in vModal (Maul-Theet GmbH) on FRF 32Z/51Z. Most of the modes of interest for this work are found in a range between 0 and 50 Hz, and the excitation was clearly too broadband for this case, of up to 1 kHz. The range of interest was therefore not determined properly before the test. Using the softest hammer tip available would have been more appropriate to excite the lower and more interesting frequency ranges.

Figure 37.7 shows that differences between Test III and the other two tests are very remarkable.

37.4 FE Modeling and Updating

The bottom-up approach proposed in this paper is proven in this chapter by creating three models of increasing complexity. This is needed to reduce the number of design variables needed to update. Solving the very complex Model III all at once would prove very difficult and very computationally expensive without having the information first gathered in Models I and II.

37.4.1 Model I

This first FE model is the most basic of all three in this paper. Figure 37.8 shows Model I, which represents the runway beam as tested in Test I. This shell model yields a very computationally efficient mesh made of only 3690 nodes and 3510 SHELL181 4-node quadrilateral elements in their vast majority. The 82% of these elements have a quality metric above 0.75. Body sizings of 100 mm for the beam's web and flanges, and 50 mm for the gussets are applied.

Drillings in the real structure are not modeled. Meshing the drillings implies using a much more computationally expensive FE mesh just to represent tiny amounts of missing material in the structure that can be considered negligible at this stage. The negative impacts of defining a very computationally expensive FE model increase when iterative solving is required for model optimization.

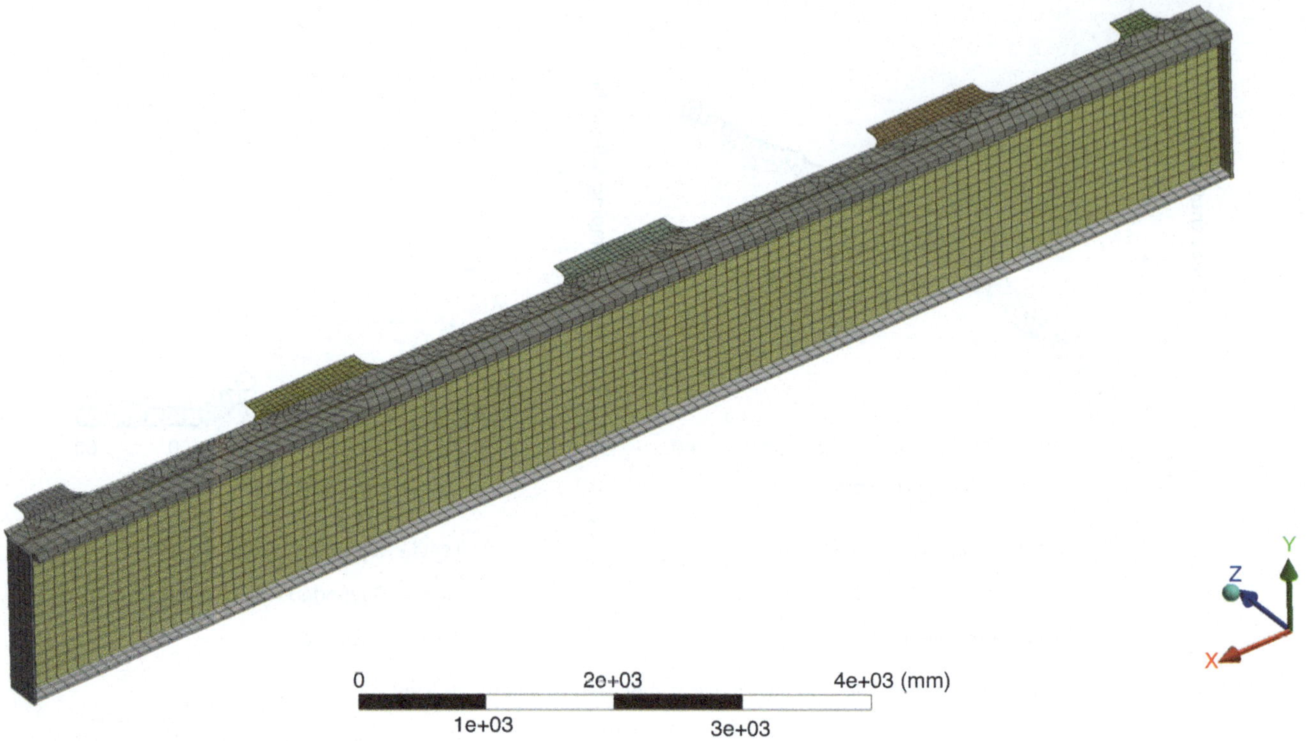

Fig. 37.8 FE model of the supported runway beam without conductor fittings attached

The boundary conditions of Model I consist in restricted nodal displacements at the rows of nodes corresponding to the X length dimensions of $L = 2000$ mm and $L = 9500$ mm, where the supports are found. Only 10 nodes, 5 per each support, are restricted of displacement in the Y direction in this way. Displacements along X and Z axes are left unrestricted to allow simulating the friction interface taking place in this contact. It is necessary to model this contact interface as well, because the beam is not bonded to its supports. This is done by means of springs, one spring directly attached to each contact node, as Fig. 37.9 shows. This setup is analogous for the contact interface at length L = 9500 mm. The only force applied in this case in the static structural model is the standard Earth gravity (9806.6 mm/s^2).

All springs are given an arbitrary baseline longitudinal stiffness value of 10 kN/mm and no longitudinal damping. These baseline stiffness values are later updated by means of a model optimization.

A baseline solve under these boundary conditions yields the eigenfrequencies detailed in Table 37.1. Note how modes 1 and 6 are missing from the experimental data dataset. Mode 1 could not be acquired reliably in the tests due to the low accelerometer sensitivity at such low frequency. Mode 6 seems to present rigid body-like motion in the numerical simulation due to the lack of stiffness on the X-axis springs. Table 37.1 shows some agreement between test and numerical models, but the FE model still requires of substantial updating.

Before updating the model, it is necessary to know the sensitivity of each design variable in the results. This allows simplifying the design space under study by removing insensitive variables, saving in turn computational resources and solving time. Two Spearman parameter correlations with 100 samples were carried out in ANSYS® DesignXplorer. Decisions in regards to the interval of the design space must be done here. All springs are given a stiffness constant between 100 and 10,000 kN/mm; this is decided after exploring previously these values with ANSYS® Mechanical. This represents a continuous and mathematically smooth design space. The material is assumed to be isotropic and homogeneous.

Figure 37.10 shows how parameters *X1_Left*, *X5_Left*, *X1_Right* and *X5_Right* are the most sensitive in this design space. It is reasonable to think that when the runway beam vibrates, it bends a few micrometers in the process and this generates a small friction in the outermost nodes of the supports (Fig. 37.10).

This FE model is therefore updated within the design space constrained by the six design variables: spring stiffness values *X1_Left*, *X5_Left*, *X1_Right* and *X5_Right* (all constrained between 100 and 10,000 N/mm) together with the material density (constrained between 7000 and 9000 kg/m^3) and the Young's modulus (constrained between 1.8 and 2.2 GPa). Eigenfrequencies 2 through 10 are all set as individual objective functions in ANSYS® DesignXplorer. These objective functions are all constrained by a $\pm$ 2% confidence interval, which is arbitrarily chosen as a reasonable error function. The

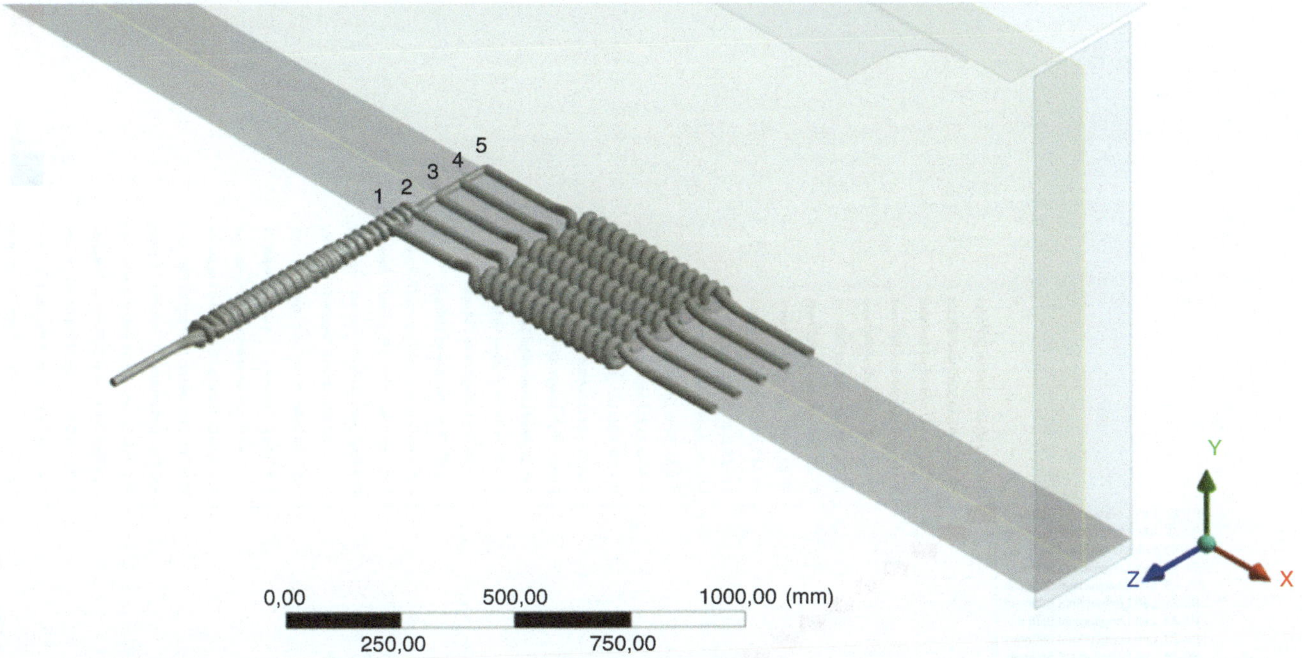

Fig. 37.9 The ten spring connections attached on the X and Z axes directions at the five contact nodes at length L = 2000 mm

Table 37.1 Comparison between the baseline numerical and objective (experimental) eigenfrequencies obtained for Model I and Test I

	Mode number									
	1	2	3	4	5	6	7	8	9	10
Objective (exp. EF) in Hz	N/A	3.58	9.28	16.26	19.51	N/A	20.69	27.40	36.04	39.46
Numerical EF in Hz	1.70	3.78	9.71	16.59	19.21	20.22	20.77	29.08	36.87	40.23
Relative difference in %	N/A	−5.3	−4.4	−2.0	1.6	N/A	−0.3	−5.8	−2.2	−1.9

MOGA is chosen for this optimization for a simple reason: this design space is very large and there are no hints of plausible starting values for the design variables that the NLPQL would require. After several preliminary attempts, an initial number of 30 initial samples and 20 samples per iteration are chosen. With a maximum number of 20 iterations, this results in a maximum of 410 design points, which results in turn in an iterative calculation about 4–5 h long. The maximum allowable Pareto percentage is increased to 80% to avoid premature convergence, to compensate for the allowable ±2% error in the objective eigenfrequencies.

The final of several attempts does not converge to a solution due to the unconverged eigenfrequencies 4 and 10, but the algorithm stops at iteration 16 because at this point the convergence stability percentage became lower than 2%. Some history charts of this optimization are shown in Fig. 37.11. All objective functions in these two figures except mode 4 (and modes 6 and 10, not shown for brevity), show a good degree of convergence, albeit some modes converge very closely to the set ±2% error boundaries. The lack of full convergence does not allow ANSYS® DesignXplorer to show a collection of candidate points, but there is enough raw data available to nominate a candidate point that allows carrying on with the model updating. The design point number 320 is chosen (Table 37.2), as it is the last data point evaluated by the MOGA.

Both test and the updated FE model with the design variable values of design point 320 are subsequently paired in FEMtools to check their degree of cross-correlation, as shown in Fig. 37.12, left. The cross-correlation MAC matrix plot between Model I and Test I is plotted in Fig. 37.13, right. A total of 24 well-correlated mode pairs (presenting MAC values higher than 60%, a threshold established by default in the software) are identified. This is a very high degree of correlation between models, even up to higher order modes, which describe much localized deflection. Note that there are many more FE modes than experimental modes in this XMAC matrix. This is due to the fact that more DOFs are used in the numerical model; many of the calculated modes describe local deflection in the gussets. The gussets are the plates welded to the top flange of the beam, which were not measured in the modal tests and therefore, these modes are missing from the experimental datasets. Therefore, the gusset modes are irrelevant in this work and can be disregarded. The measured global vibrational

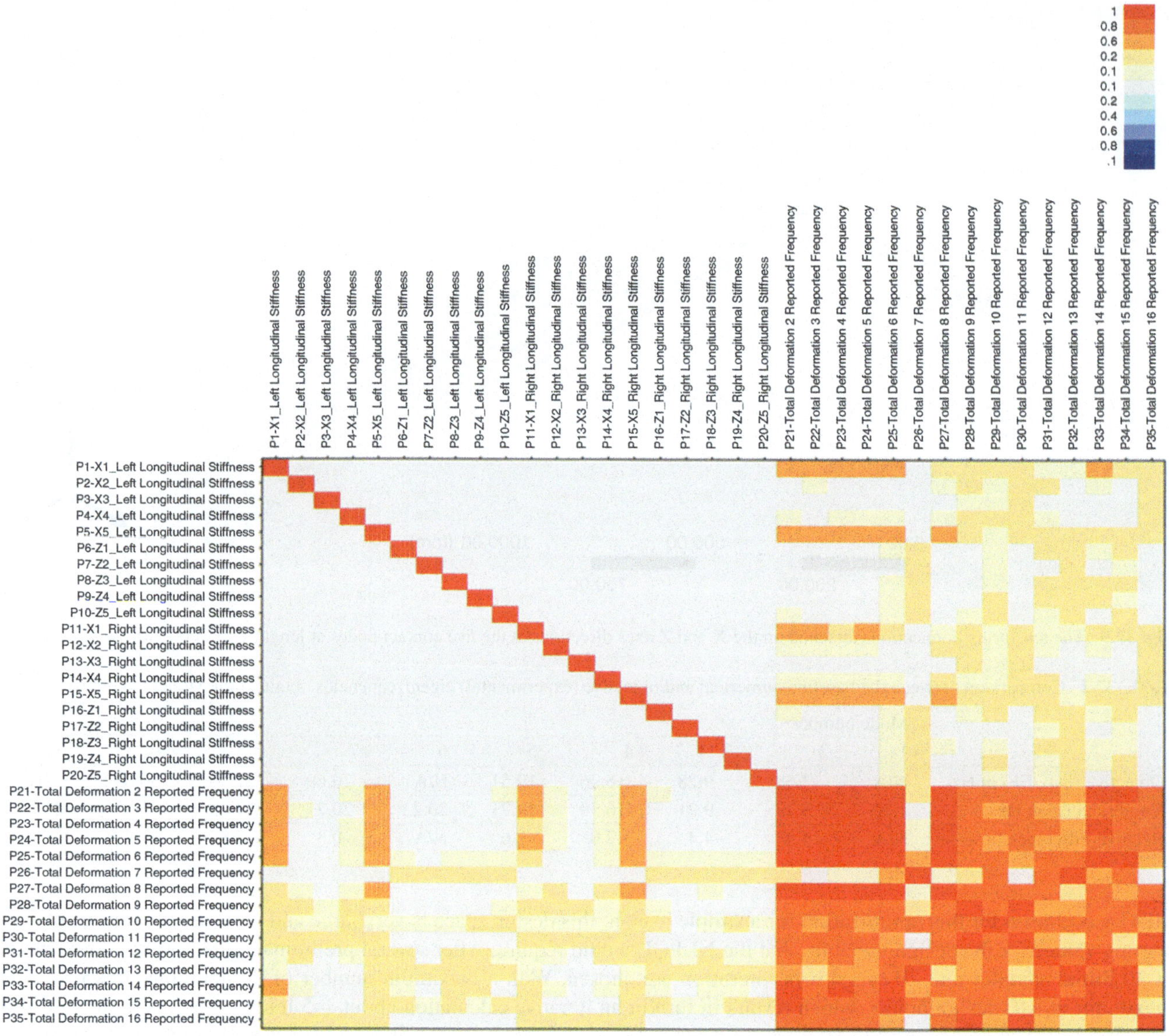

Fig. 37.10 Parameters correlation matrix between design variables and eigenfrequencies. Only the 20 stiffness values (K) appear as design variables, as Young's modulus and density are known to be very sensitive parameters

modes of the runway beam (modes where the web and/or the upper and lower flanges vibrate) are those that need to correlate with their numerical counterparts.

Figure 37.13 provides an overview of how Model I has been optimized in this section, with attention to the ±2% error tolerance between objective and optimized model. Figure 37.13 shows how differences between the objective (experimental) eigenfrequencies and numerical eigenfrequencies were drastically reduced after the optimization.

Modes 5 and 7 present higher differences with their objective counterparts even after being optimized. This is related to the fact that modes 5 and 7 are very closely spaced (about ±1 Hz) to mode 6, which presents this apparent rigid body behavior in the baseline Model I. Changes in the boundary conditions taking place during the iterative optimization make mode 6 to gain or lose stiffness. When the model becomes stiffer, mode 6 eigenfrequency increases and overtakes mode 7 in frequency. When positions are swapped between both modes, mode 6 becomes 7 and *vice versa*. The same phenomenon takes place when the model becomes more compliant and mode 6 swaps with mode 5. This continuous swapping of positions for these very closely spaced modes leads to non-convergence and explains the lack of overall improvement in the range of modes 5, 6 and 7.

As a conclusion for this section, Model I can be considered optimized after the MOGA cycle if mode 5 is disregarded. This serves as a "lesson learnt" to optimize Models II and III. Cross-correlation is high throughout a large number of modes.

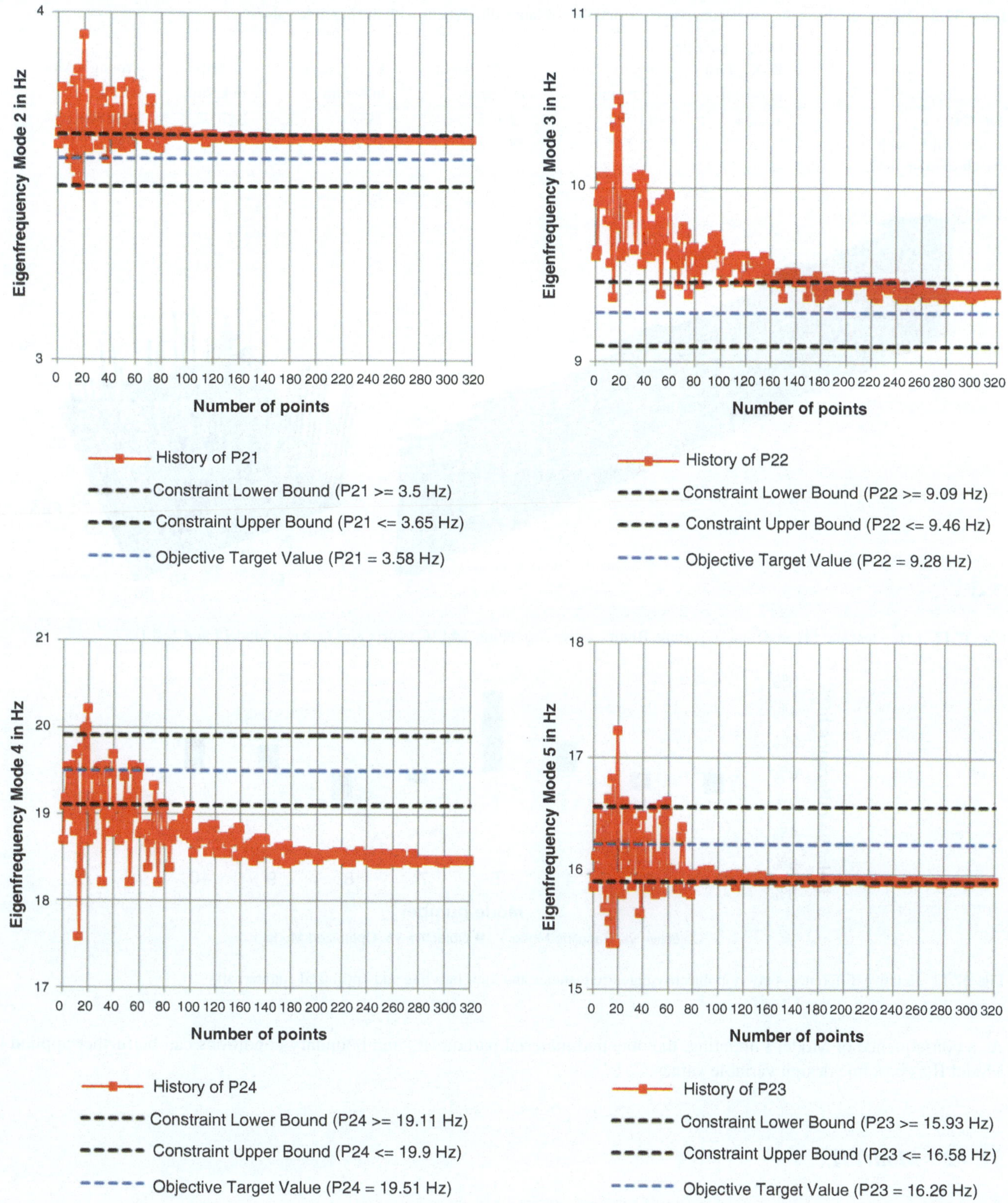

Fig. 37.11 Convergence charts for the objective modes 2 and 3 (top row) and 4 and 6 (bottom row)

Table 37.2 Design variable values in design point 320 and their relative differences with the baseline values

	Design variables					
	K_X1_Left in N/mm	K_X5_Left in N/mm	K_X1_Right in N/mm	K_X5_Right in N/mm	Density[3] in kg/m^3	Young's Mod. in Pa
DP 320	10,824	11,782	24,112	7432	8470.7	1.988e+11
Relative diff. with baseline value in %	7.6	15.1	58.5	25.7	7.9	0.6

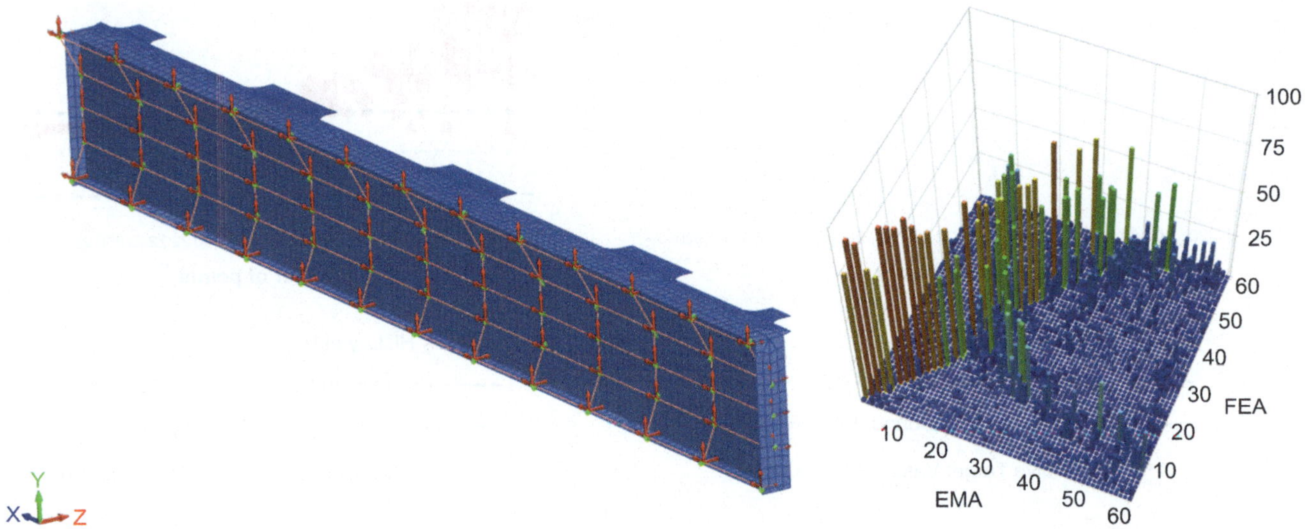

Fig. 37.12 Left – test and FE model mesh pairing. Right – cross-correlation XMAC matrix plot between Model I and Test I

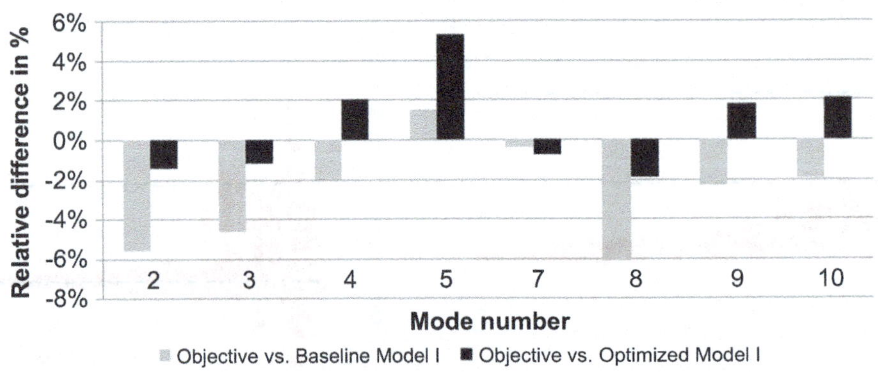

Fig. 37.13 Relative differences between objective eigenfrequencies and their baseline and optimized counterparts

As a consequence of Model I updating, the obtained material parameters and boundary conditions can be further applied in Model II as baseline design variable values.

37.4.2 Model II

Once Model I is considered to be updated, progress can be done to Model II. Density, Young's modulus and springs linear stiffness values are inherited from the updated Model I. The main difference between Models I and II are the downshop conductor fittings screwed to both sides of the beam's web. Optimizing Model II implies finding beforehand an optimal modeling approach for the screwed connections; in essence, an approach that minimizes the model's computational cost while being as close as possible to the objective eigenfrequencies. This represents an optimization problem inside another optimization problem. Different approaches to the modeling can be followed, each with their own effects, advantages and drawbacks. The first task towards establishing a baseline Model II is identifying which of the following modeling approaches is the most interesting to further carry on this work:

Approach 1: Screws, nuts and washers are simply left unmodeled. Fittings are connected to the beam's web by means of bonded contact interfaces. This results in a high computational efficiency but *a priori* low accurateness.

Approach 2: Screwed connections are modeled with beam elements. A Ø 16 mm beam bolt is individually modeled in ANSYS® Mechanical for each bolted connection. Beam-like bolts are extremely computationally efficient, as they are meshed with just a few BEAM188 nodes each, and their deformation behavior can be adapted in a case-by-case basis. Contact interfaces between fittings and web are bonded too, as any other type of interface contact leads to an unconstrained model. This is meant to be a compromise between approaches 1 and 3.

Approach 3: Normalized, unthreaded M16 screws, nuts and washers solid models are included in the assembly with and are meshed with SOLID187 10-node tetrahedral elements in ANSYS® Mechanical. This approach makes the model closer to reality in terms of mass, but modeling each individual component sharply increases the computational cost of the model. Modifying element sizings and meshing methods is necessary to keep reasonable meshing and solving times (Fig. 37.14).

Figure 37.15 shows a graphical plot of the relative differences between objective and numerical eigenfrequencies with each modeling approach.

It is evident that Model II gains an excessive amount of stiffness when modeled with the Approach 2 regardless what behavior is assigned to the beam connections. Deviations between objective eigenfrequencies and their counterparts obtained with Approaches 1 and 3 are still higher than ±2% in most cases, but they offer far superior results than Approach 2. The relative differences in the lower order modes between Approach 2 and the objective eigenfrequencies are very large, but in absolute values they are in the range of ±3 Hz for mode 1. Approaches 1 and 3 yield similar results, but Approach 3 is by far more computationally expensive and complex than Approach 1. Due to its nature, element quality metrics are also worse in Approach 3 and interpenetration between normalized elements and beam results an issue as well. This leaves Approach 1 as the most adequate to carry on with the optimization of Model II in this thesis, and is considered as the baseline Model II for the remainder of this section. Using Approach 1 has as well the advantage that there is no need to model the drillings in the beam's web or in the attachments in a simulation without bolts, as explained also in the previous section. Removing

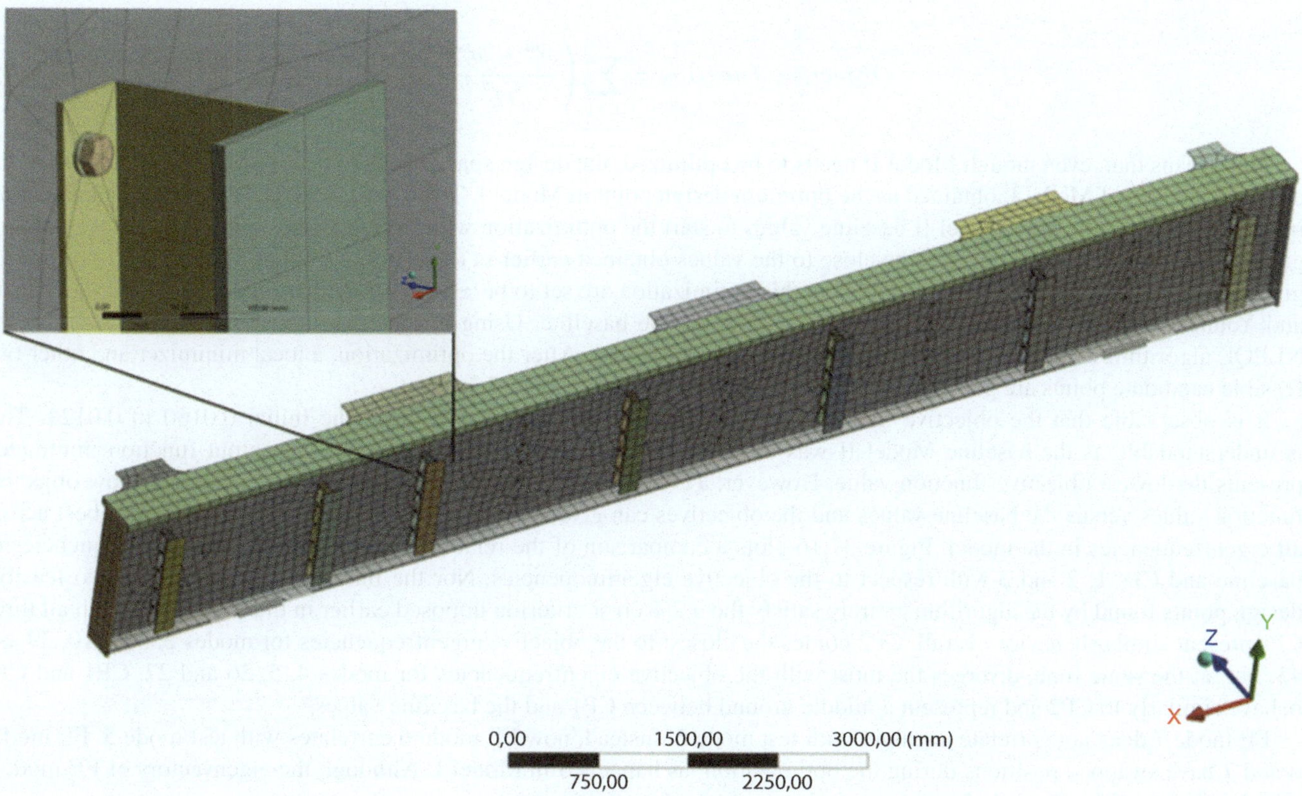

Fig. 37.14 Detailed view of the bolted connection and coarse meshing around the drillings as described in Approach 3

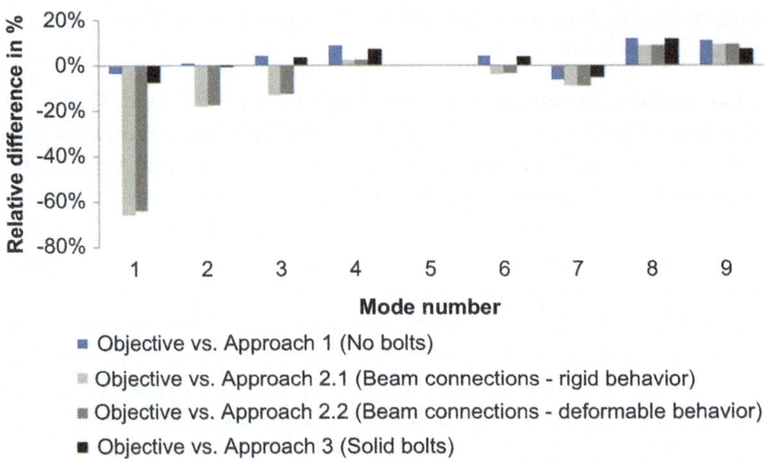

Fig. 37.15 Graphical plot of the relative differences between objective (experimental) eigenfrequencies and their numerical counterparts obtained with the three different modeling approaches proposed

the drillings from Model II drops the model's size by a factor of two thirds, to a total of 4073 nodes and 3620 elements. Increasing the model efficiency is of crucial importance to carry out this study.

A parameters correlation check for Model II confirms the conclusions obtained from Model I. FE modes are sensitive to variations in the stiffness of *X1* and *X5* springs for both supports and density and Young's modulus. It is initially assumed that Model II is simply a modified version of the already optimized Model I. This case allows the use of the gradient-like method NLPQL for the optimization. The initial conditions of the design variables are known, as they inherit the values obtained for DP 320 in Model I. The default parameters shown by ANSYS® DesignXplorer are used for this optimization: forward finite difference approximations, 0.1% allowable convergence percentage, a maximum of 20 iterations and 3 resulting candidate points.

$$Objective\ function = \sum_{i=1}^{N} \left(\frac{\omega_i^m - \omega_i^{calc}}{\omega_i^m} \right)^2$$

This means that, even though Model II needs to be optimized, the design space needs to be around a neighborhood of the design point 320 of Model I, obtained as the optimum design point in Model I. These design variable values (detailed earlier in Table 37.2) become now Model II baseline values to start the optimization with. In practice, these means that constraints given to the design variables need to be close to the values obtained earlier in design point 320 of Model I. The constraints given to the spring stiffness design variables in this optimization are set to be ±50% off the baseline values, and the density and Young's modulus constraints are set to be ±20% off the baseline. Using a tighter design space allows us to use the NLPQL algorithm without risk of falling in local minimizer values. After the optimization, a local minimizer and other two feasible candidate points are generated. These are CPs 1, 2 and 3 in Table 37.3.

It is observable that the objective function value has only improved slightly from the initial 0.0160 to 0.0124. This is understandable as the baseline Model II was already quite optimized. As it is logical, the output function minimizer, presents the lowest objective function value. However, a comparison between other two feasible points with low objective function values versus the baseline values and the objectives can give some insights in assessing what CP fares best across all eigenfrequencies in the model. Figure 37.16 plots a comparison of the relative differences between eigenfrequencies of baseline and CPs 1, 2 and 3 with respect to the objective eigenfrequencies. Nor the minimizer nor the other two feasible design points found by the algorithm entirely satisfy the ±2% error criterion imposed earlier in this thesis, although all three CPs present similar behavior overall. CP2 comes the closest to the objective eigenfrequencies for modes 2, 3, 7, 10, 39 and 43; but, at the same time, diverges the most with the objective eigenfrequencies for modes 4, 5, 26 and 27. CP1 and CP3 behave similarly to CP2 and represent a middle ground between CP1 and the baseline values.

FE mode 7 does not correlate anymore with test mode 5. Instead, now FE mode 6 correlates with test mode 5. FE modes 6 and 7 have swapped positions during the optimization, as happened in Model I. Although the eigenvectors of FE mode 6 correlate better with test mode 5, the frequencies are far one another.

It is clear that adjusting all individual numerical eigenfrequencies to their objectives can be a very difficult task. To continue with the model updating, it is thought necessary to evaluate how relevant is the each mode for this analysis, to

Table 37.3 Objective eigenfrequencies and starting point (DP 320), output function minimizer (CP 1) and two other feasible design points (CP2 and CP3) with their design variable values and their resulting eigenfrequencies for comparison

	Objective	DP 320 (baseline)	Minimizer (CP 1)	Feasible point (CP 2)	Feasible point (CP 3)
Density in kg/m^3		8470.7	7789.6	8559.4	8203.4
Young's Modulus in Pa		$1.988{\cdot}10^{11}$	$1.7414{\cdot}10^{11}$	$1.8661{\cdot}10^{11}$	$1.7924{\cdot}10^{11}$
X1_Left in N/mm		10,824	13,873	11,712	10,551
X5_Left in N/mm		11,782	18,331	22,666	20,651
X1_Right in N/mm		24,112	18,424	27,580	24,832
X5_Right in N/mm		7432	16,493	16,447	16,406
EF Mode 2 in Hz	3.58	3.75	3.67	3.62	3.62
EF Mode 3 in Hz	9.04	9.20	9.12	9.00	9.01
EF Mode 4 in Hz	16.21	15.92	15.59	15.38	15.41
EF Mode 5 in Hz	19.18	18.25	18.09	17.87	17.89
EF Mode 7 in Hz	22.13	21.59	22.23	22.15	22.13
EF Mode 8 in Hz	30.25	32.30	31.95	31.34	31.49
EF Mode 10 in Hz	44.69	46.88	45.99	45.31	45.42
EF Mode 26 in Hz	73.31	73.18	71.59	70.62	70.75
EF Mode 27 in Hz	74.94	73.61	71.89	70.96	71.05
EF Mode 39 in Hz	78.02	82.25	80.32	79.29	79.39
EF Mode 43 in Hz	81.27	84.42	82.42	81.37	81.47
Objective function output parameter value (dimensionless)	0 (theoretical)	0.0160	0.0124	0.0133	0.0133

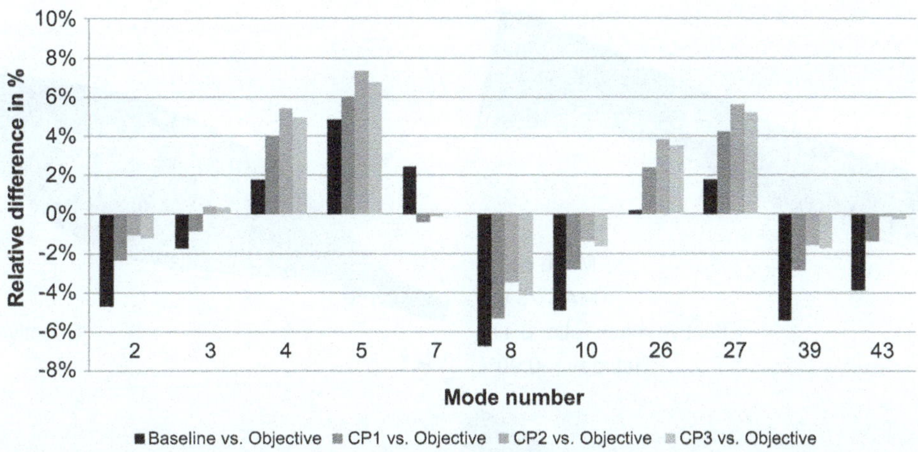

Fig. 37.16 Relative differences between the eigenfrequencies of the baseline Model II and the minimizer (CP 1) and the other two feasible design points (CP2 and CP3) with respect to the objective eigenfrequencies

focus the optimization on the most significant mode shapes. Importance has been given during this section to low order, global mode shapes due to a simple fact: these are the most relevant mode shapes for the study due to the "amount of mass" that's mobilized by each mode shape. Knowledge about the relevance of each mode shape and thus, their relevance in the correlation can be gathered if the mass participation metrics for each mode are observed (Fig. 37.17). In the case at hand, only participation in the Z axis is relevant for this analysis, as this is the direction where most of the vibration takes place.

Among the well-correlated mode shapes, mode pairs 5, 3, 2 and 7 in Fig. 37.18 are the ones with the highest mass participation metrics, and thus, the most important mode shapes in this analysis. Mode pair 3 (FE mode 4, test mode 3) presents a deviation between their eigenfrequencies higher than −5%, but this deviation is small in absolute terms, less than 1 Hz. Mode pair 9, although well correlated, is of a higher frequency and in consequence, small relative deviations (−3.8%) are related to larger absolute deviations in eigenfrequencies (−2.7 Hz).

From a starting array of 12 mode pairs, half of them present differences of less than ±2% in their eigenfrequencies and 9 of the pairs deviate less than ±2 Hz in absolute terms. In average, there is an eigenfrequency deviation between FE and test models of 1.08 Hz in absolute terms and 2.80% in relative values. This is above the 2% previously fixed as deviation limit, but still a reasonable value taking into account the increased complexity of Model II. Out of the top four mode shapes

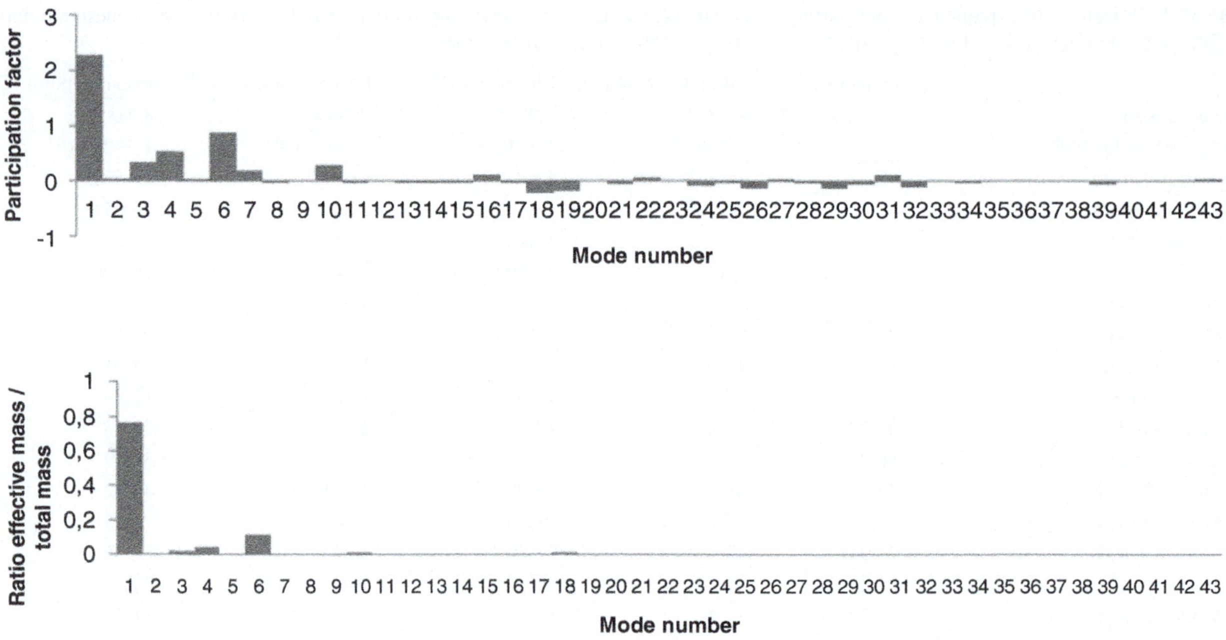

Fig. 37.17 Top – mass participation factor for Model II in direction Z. Bottom – ratio of effective mass vs. total mass for Model II in direction Z

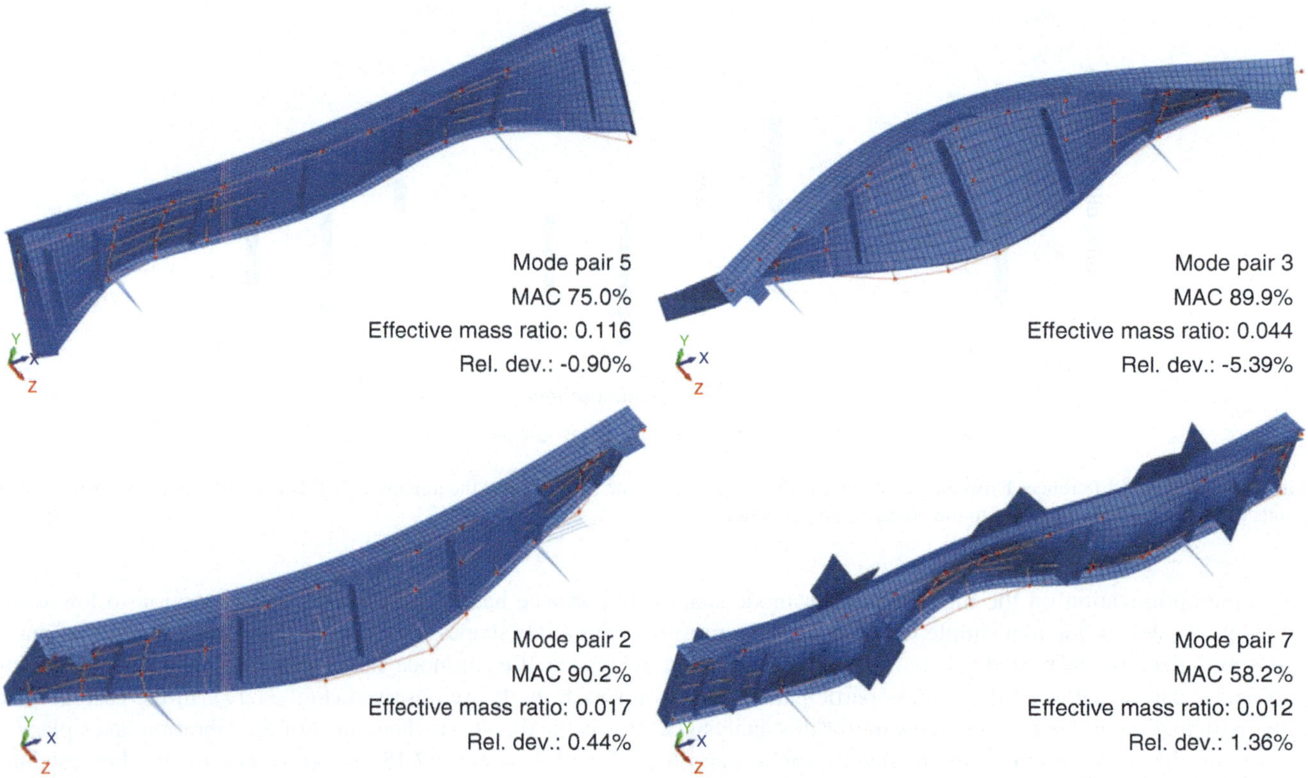

Fig. 37.18 Overview of the four well-correlated mode shape pairs with the highest effective mass ratios. The FE model is represented in solid blue, the test model is represented as a red mesh

with the highest mass participation (except mode 1), three of them present MAC values of 75% or above, which point to a generally good degree of correlation.

On the light of the optimization and correlation results in Model II, although not fully satisfactory under a strict standpoint, can be used to progress on to Model III.

37.4.3 Model III

Model III is the most complex FE model created in this work. In essence, it is Model II modified to be constrained with BCs simulating those found on the runway beam installed at the industrial plant, which are different of those used for Models I and II. The modeling of the BCs for this section is based on this drawing and in the observations made during the data acquisition phase for Test III. Several components' drawings are unavailable or incomplete, or the components require of severe simplifications to minimize their computational cost. Some of the components added in Model III have approximate geometries, mostly aiming to recreate the extra mass the structure is loaded with and to assign the appropriate restrictions in the model while keeping the computational cost of the model as low as possible. Figure 37.19 shows the baseline Model III with the runway rail, its rubber underlayer, the downshop conductors and their holders attached to their fittings. This model is made of a total of 90 bodies, all of them meshed with SHELL181 or BEAM188 elements. All contacts between bodies are set as bonded to avoid having an unconstrained model. A variety of boundary conditions are needed to recreate the real conditions in the installed beam. An exhaustive study of the different boundary conditions is carried out, too extensive to be summarized in this paper.

The most relevant results are derived from the parameters correlation study in the contact stiffness at the gussets. About 4 h of continuous iterative simulations are needed to calculate the 100 samples in such a computationally demanding model. Project documentation shows how each gusset is welded to two to three stiffening L-profiles, which are in turn welded onto the plant structure. It is necessary to simulate the contact between the gussets with the plant's structure. Three or four body-to-ground springs are scoped to each gusset depending on how each gusset is welded to the plant structure. Each spring is deliberately modeled to measure 1 m long to keep the model simple. Figure 37.20 illustrates this in an intuitive way. Five extra springs in the orthogonal Y direction are not visible in the figure due to the perspective. Each spring is assigned a baseline longitudinal stiffness value of 1 kN/mm.

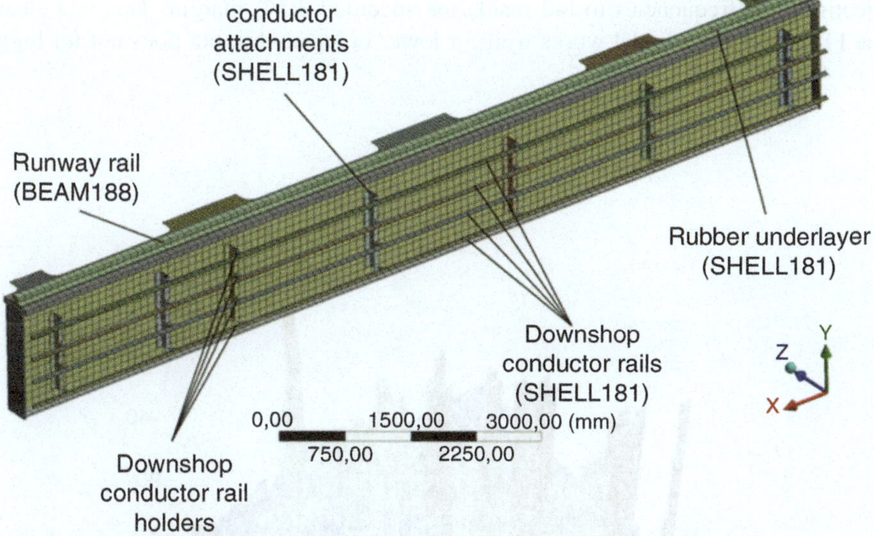

Fig. 37.19 Overview of all components added for Model III and their element types

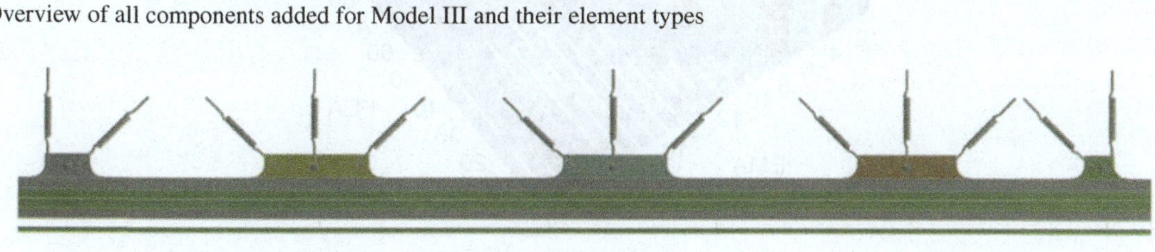

Fig. 37.20 Detailed view of the 13 springs scoped to the gussets in the XZ plane

The model is solved with these boundary conditions as a baseline, including certain constrains at both edges of the beam and at each edge of the downshop conductors. The baseline static structural solve requires of about 25 s be completed and the modal analysis solve requires an extra 1 min and 20 s because the modal solution is firstly expanded to 100 modes. This is done for the same reasons as in Model II – the vast majority of the modes calculated in Model III do not belong to the runway beam as a whole, but to its extra attached components (i.e., the downshop conductors, their holders, their fittings, etc.). The sharp increase in stiffness in the model, which is now essentially constrained in three of its sides, implies that very few global modes are found.

Both FE and experimental modal datasets are loaded in FEMtools and cross-correlated in the same manner as previously explained. The degree of cross-correlation between Test III and Model III is, as expected, quite low. This can be observed in the XMAC matrix plot in Fig. 37.21, where the matrix diagonal is difficult to identify due to the differences in the number of extracted modes, caused by the local deflection of the attached components in Model III. Only FE modes 1, 4, 21 and 37 take part in the model optimization, as they correlate well with their experimental counterparts. Although there is reasonable agreement between these mode shape pairs, their differences in terms of eigenfrequencies are unacceptably large. All numerical eigenfrequencies from mode pairs 1 to 6 are observed to be higher than their experimental counterparts, and deviate in similar percentages. This indicates that the FE model has a higher stiffness than the measured runway beam does. This higher stiffness could be caused by too restrictive BCs.

The previous investigations have served their purpose of gathering knowledge about the design space. It is now known that this is a design space with 15 design variables and mathematically smooth. This represents still a very large design space, probably with several existing local minima; which is therefore more efficiently explored using a genetic search with the MOGA, in a similar manner to the case in Model I.

Fifteen design variables (the 13 spring stiffness values in Table 37.4 detected in the parameters correlation plus the material density and the Young's modulus) take part in the optimization of the four objectives, one for each objective eigenfrequency. The stiffness values are constrained as in the parameters correlation, between 0 N/mm and 10 kN/mm, while the density and Young's modulus are constrained between ±10% of the values obtained for Model II. All eigenfrequencies are now bound at values ±5% off the objectives. After considering the results of the correlation with the baseline Model III and the simplifications done in the model, convergence in a range ± 2% on the lower order modes is thought very difficult to achieve.

This optimization leads to convergence by a very narrow margin for modes 4 and 37 (Fig. 37.22), which deviate just enough from the objective eigenfrequencies to fall inside the specified ±5% margin. This is a clear effect of the lack of modal damping in the FE model. The model works well for lower order modes, but does not for higher order modes, from

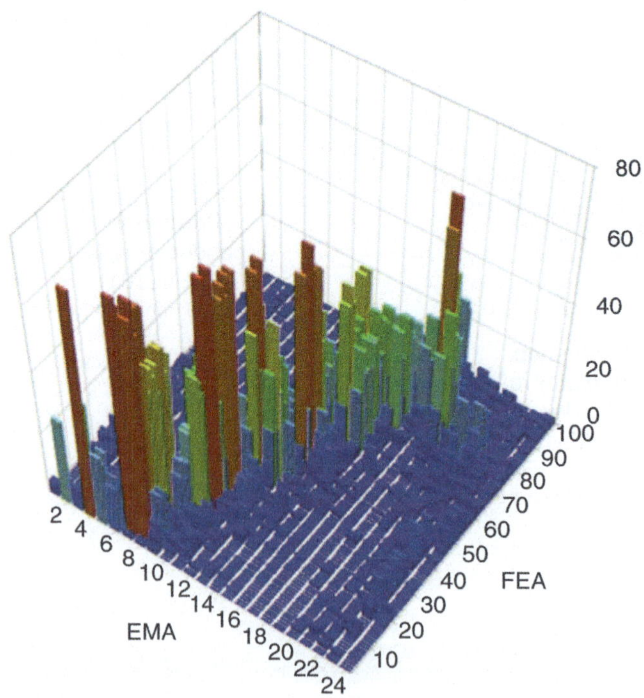

Fig. 37.21 Cross-correlation MAC plot between Test III and the baseline Model III

Table 37.4 List of stiffness variables used in this parameters correlation and their associated sensitivity

Gusset number	Spring	Directional component	Sensitive?	Springs diagram
1	1	+Z	Yes	
	2	+X/+Z	No	
	3	+Y	Yes	
2	1	+X/+Z	No	
	2	+Z	Yes	
	3	+X/+Z	No	
	4	+Y	Yes	
3	1	+X/+Z	Yes	
	2	+Z	Yes	
	3	+X/+Z	Yes	
	4	+Y	Yes	
4	1	+X/+Z	Yes	
	2	+Z	Yes	
	3	+X/+Z	No	
	4	+Y	Yes	
5	1	+X/+Z	No	
	2	+Z	Yes	
	3	+Y	Yes	

FE mode 86 onwards (mode pair 6). These order modes are however uninteresting for this investigation given their very low modal participation factor.

The optimization results meet therefore the requirements set at the beginning of the process for the first five modes as Fig. 37.23 shows and thus, the optimized Model III can be considered updated.

37.5 Conclusions and Further Work

This last chapter is devoted to summarize the conclusions extracted from the work carried out in the previous chapters, and to provide pointers for further work in this topic.

The quality of the experimental datasets has been discussed. Tests I and II were measured with the adequate means and the obtained experimental modal datasets are regarded as appropriate for the updating of Model I and II. Experimental data was, however, not measured adequately in Test III. The author claims full responsibility for this as the lead test engineer in this measurement. The eigenfrequencies and mode shapes estimated in Test III were curve-fitted from FRF datasets with very low signal-to-noise ratios and a relatively coarse measurement resolution given by the acquisition parameters. These parameters were originally chosen to be consistent with those of Tests I and II. Closely-spaced modes are unavoidable in the Test III FRFs for this reason. These issues have impact on the reliability of the experimental data and, hence, the conclusions extracted from it. A repetition of Test III with the appropriate measurement equipment and parameters would increase the confidence in the obtained experimental modal parameters and therefore, increase the overall confidence in Model III.

With the aforementioned assumptions, the FE model developed in this thesis can be considered updated for the moment with all the necessary caution. The imposed confidence intervals respond to the author's experience in the topic, not to guidelines or regulations. All procedures detailed have been carried out to the best of the author's knowledge following a logical training of thought, taking into account computational cost limitations and other factors such as the lack of damping in the model, which exists in the experimental modal analysis. Model I is found to deviate roughly 2% off the corresponding test model I, taking into account caveats like the swapping closely-spaced modes 6 and 7. Model II presents more mixed results with good overall correlation in terms of absolute eigenfrequency deviations (average: 1.08 Hz) and MAC values, but out of the $\pm 2\%$ allowed deviation in relative terms (average: 2.8%). Model III is the most complex model in this thesis, where the simplifications, assumptions and limitations of FEM have a more relevant role. The confidence interval in this correlation is raised from $\pm 2\%$ to $\pm 5\%$ for that reason, and the model does not deviate further from this $\pm 5\%$ for the well-correlated modes. The missing damping in the model influences the results in a way that only a wider confidence interval of $\pm 5\%$ can be ensured in the results.

The updated model developed from this work is fully parametric and should allow in the future its adaptation to represent each of the single runway beams installed at the industrial plant with the aim of following the studies this project is devoted to. These results shall not, however, be interpreted as a validation of the optimized FE model derived from this thesis. Processes of verification and validation (V&V) evaluate how evidence is collected, and documented, and help establish confidence in the results of complex numerical simulations [15]. Model III can only be validated once operational response data (i.e.

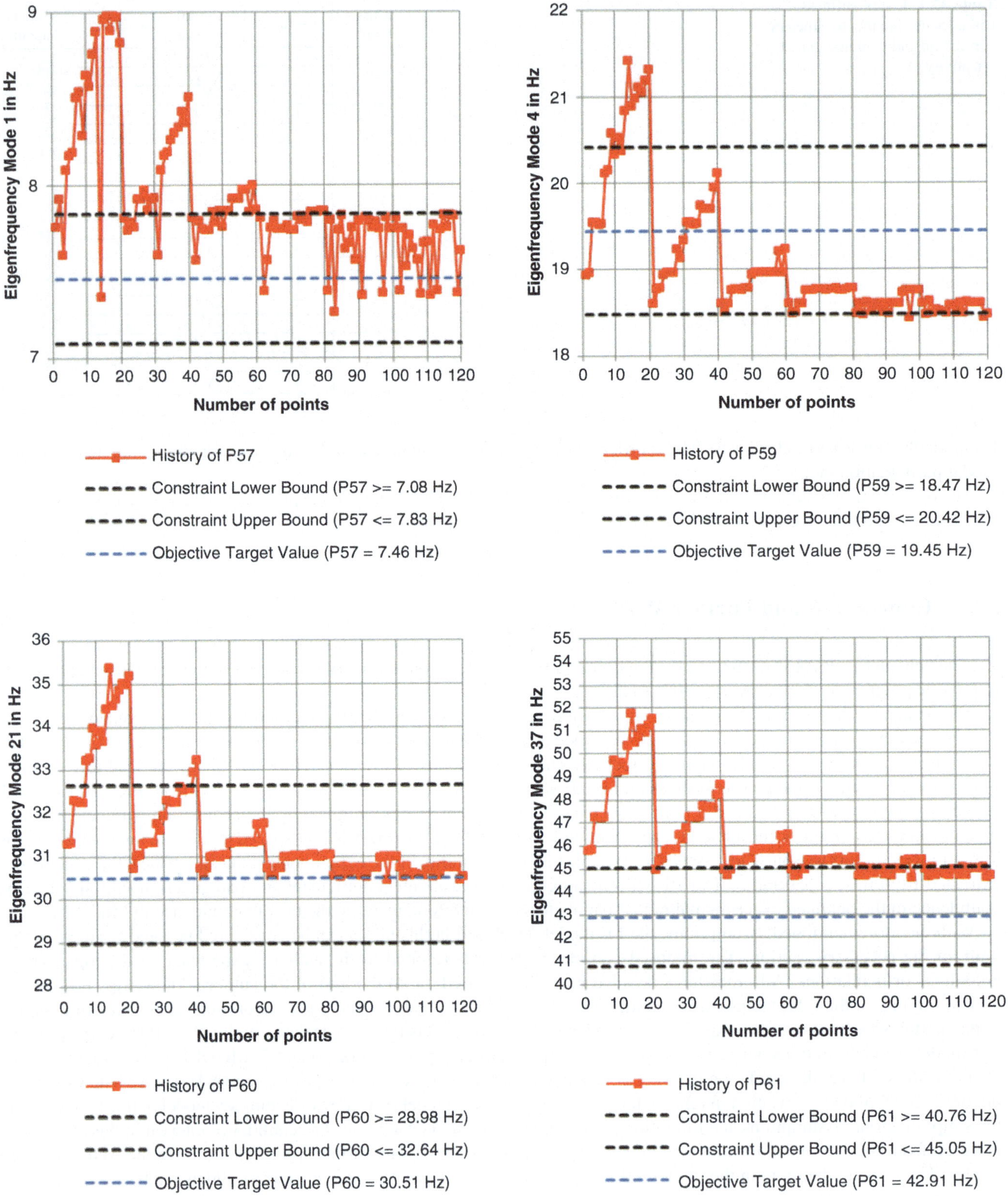

Fig. 37.22 History charts for modes 1 and 4 (top row), 21 and 37 (bottom row) for this model optimization

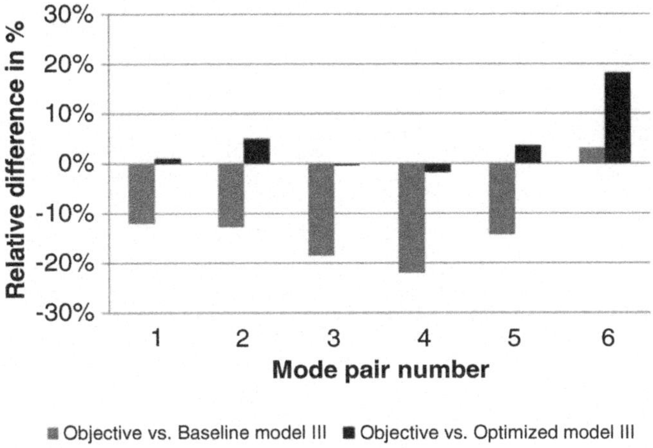

Fig. 37.23 Relative differences between the objective eigenfrequencies and their baseline and optimized counterparts

deflection by means of strain gauges) has been acquired from the real structure and compared to the model's numerical response [16]. V&V has not taken place in this work as it was outside of the scope of the current project, but may take place in further stages depending on the evolution of the project.

Acknowledgements The authors would like to express their gratitude to Prof. Ricardo Perera of the Polytechnical University of Madrid for the tutoring of the first author's Masters Thesis, where this research is originally published on its full length.

References

1. Munck Cranes, Inc.: Overhead crane components [Online]. Available: https://www.munckcranes.com/overheadcranecomponents.asp. Accessed 27 Feb 2019
2. Almeida, R., Urgueira, A., Morais, J., Maia, N.M.: An assessment of frequency response curvature methods for damage localization. Proceedings of the ISMA 2014, Leuven, Belgium
3. Doebling, S.W., Farrar, C., Prime, M.B.: A summary review of vibration-based damage identification methods. Shock Vib. Dig. **30**(2), 91–105 (March 1998)
4. Perera, R., Ruiz, A.: A multistage FE updating procedure for damage identification in large scale structures based on multiobjective evolutionary optimization. Mech. Syst. Signal Process. **22**(4), 970–991 (May 2008)
5. Mayes, R.L.: An experimental algorithm for detecting damage applied to the I-40 bridge over the Rio Grande. In: Proceedings of the XIII International Modal Analysis Conference, 1995
6. Keilpflug, F., Kamenzky, R., Alarcón, D.J., Mallareddy, T.T., Blaschke, P.: Structural health monitoring on industrial structures using a combined numerical and experimental approach. In: Topics in Modal Analysis & Testing, Vol. 8 – Proceedings of the 37th IMAC, A Conference and Exposition in Structural Dynamics, Orlando, FL, USA, 2019
7. Dascotte, E.: Modal analysis in the age of digital twins. Oral presentation at the XXXVII International Modal Analysis Conference, Orlando, FL, USA, 2019
8. Martínez Luengo, M., Shafiee, M.: Guidelines and cost-benefit analysis of the structural health monitoring implementation in offshore wind turbine support structures. Energies. **12**(1176), 1–26 (2019)
9. Dascotte, E., Guggenberger, J.: Anwendung von model updating mit FEMtools. In: 23rd CADFEM Users' Meeting, Bonn, Germany, 2005
10. Sehgal, S., Kumar, H.: Structural dynamic model updating techniques: a state of the art review. Arch. Comput. Methods Eng. **3**, 525 (September 2016)
11. Friswell, M., Mottershead, J.E.: Model updating in structural dynamics: a survey. J. Sound Vib. **167**(2), 347–375 (1993)
12. Marwala, T.: Finite Element Model Updating Using Computation Intelligence Techniques. Springer, London (2010)
13. Vega, J.M.: Chapter 2: Gradient-like optimization methods. Fundamentals and Applications of Finite Element Method in Mechanical Analysis. Documentation of the Master's Degree in Numerical Simulation with ANSYS, UPM, p. 22, Madrid, Spain
14. Vega, J.M.: Chapter 3: Additional optimization methods and multi-objective. Fundamentals and Applications of Finite Element Method in Mechanical Analysis. Documentation of the Master's Degree in Numerical Simulation with ANSYS, UPM, p. 82–85, Madrid, Spain
15. The American Society of Mechanical Engineers: Guide for Verification and Validation in Computational Solid Mechanics (PTC 60/V&V 10), New York, NY, USA, 2007
16. Avitabile, P.: What is a good MAC value so I know my model is right? SEM Experimental Techniques, August 2006, pp. 1–2

Chapter 38
Shaft Bending to Zero Nodal Diameter Disc Coupling Effects in Rotating Structures Due to Asymmetric Bearing Supports

G. Tuzzi, C. W. Schwingshackl, and J. S. Green

Abstract In a flexible shaft-disc assembly, coupled shaft-disc vibration modes are likely to occur, provided that the natural frequencies of the two components are close. It is well known that the shaft axial and bending modes can couple with the zero and one Nodal Diameter (ND) modes of the disc, respectively. In a previous work, it has been shown that in presence of asymmetric axial-radial bearing supports, combined axial-bending shaft modes can occur, which are further impacted by gyroscopic forces when the system is rotating.

Extending the previous findings, the impact of disc flexibility on this new coupling family has been investigated in more detail. The obtained results show the emergence of shaft whirling modes with an axial component, that can couple with 0ND or 1ND disc modes. As a result, a 0ND disc mode can possibly be excited by an out of balance mass on the shaft, leading to a previously unobserved vibration behaviour.

Keywords Rotor dynamics · Shaft-disc assembly · Modes coupling · Whirl · Axial vibration

38.1 Introduction

Rotor systems combining shafts and flexible bladed discs are extensively used in industry, with applications ranging from aircraft propulsion and power gas turbines to vacuum cleaners. Due to new requirements such as higher rotating speeds and lower weight, these components are subjected to higher dynamic loads, challenging today's vibration analysis. The dynamic analysis of shaft and discs has historically been carried out independently, first analysing the shaft with rigid discs, and the discs dynamics was then studied separately. This uncoupled analysis is considered valid as long as the disc resonances are well above the shaft ones. However, modern design trends, particularly in the aero engine sector, led to more flexible discs and blades [1], and consequently the dynamic interaction between the two components cannot be neglected. Detailed studies of the shaft and disc coupled vibration [2–5] have shown that axial and bending modes of the shaft can couple with the zero and one Nodal Diameter modes of the disc respectively. These investigations suggest more generally that shaft-disc mode coupling is the result of force/moment transmission at the interface between components. The effect of rotation on these coupled modes has also been investigated, showing that shaft whirling modes involving 1ND disc modes can appear. More recently Ma, Li et al. [6] studied the full coupled dynamics of shaft and bladed discs, showing that disc transverse and blade bending can interact with shaft bending and torsional vibration. It is important to point out that, throughout all these works, shaft lateral, torsional and axial modes have been considered as independent from each other. In [7], Berger showed that the nonlinear behaviour of a thrust hydrodynamic bearing can lead to a cross coupling between shaft axial and lateral displacements but its impact on the overall shaft dynamics has not been investigated further.

In a previous work [8], the authors followed up the observations from [7] and investigated the coupling behaviour of a shaft carrying a rigid disc in the presence of general asymmetric bearing supporting structures, which leads to a coupling of axial and lateral shaft displacements. It was found that under those circumstances shaft bending and axial modes can combine

G. Tuzzi (✉) · C. W. Schwingshackl
Department of Mechanical Engineering, Imperial College London, London, UK
e-mail: g.tuzzi17@ic.ac.uk; c.schwingshackl@imperial.ac.uk

J. S. Green
Rolls Royce PLC, Derby, UK
e-mail: Jeff.Green@Rolls-Royce.com

B. Dilworth (ed.), *Topics in Modal Analysis & Testing, Volume 8*, Conference Proceedings of the Society for Experimental Mechanics Series, https://doi.org/10.1007/978-3-030-47717-2_38

together, leading to the emergence of a new class of axial-bending mixed modes. When the system is rotating, the combined effect of axial-bending coupling and gyroscopic forces leads to shaft whirling modes involving axial motion as well.

To understand the impact of these new mode class on a general rotor-dynamic system, this work will extend these findings to the case of a shaft carrying a flexible disc, showing that those whirling-axial modes can potentially couple with disc 0ND and 1ND modes and that an out-of-balance load on the shaft is able to excite a disc 0ND mode.

38.2 Model Description

In order to capture the coupling effects described above, a simple shaft-disc assembly has been chosen (Fig. 38.1a). The shaft is solid and circular, with a length of 0.8 m and a diameter of 0.02 m. The disc has a 0.2 m diameter and a 3 mm thickness and is rigidly connected to the shaft at 0.3 m from the left end. The whole system is made of Aluminium, which is considered elastic and homogeneous. The shaft is supported by identical plain bearings at each end.

Each bearing is then supported by four flexible rods connected to ground. In the Y-Z plane, the two rods are perpendicular to the shaft, whilst in the X-Y plane they are inclined towards the axial direction (see Fig. 38.1b) to introduce a coupling with the axial displacement. The combined stiffness of the bearing and the supporting rod is described by a lumped spring of stiffness k_i, oriented by an angle α_i with respect to the shaft midline. This applies to the rods in both planes, even if in the Y-Z plane the angle α_i is fixed to 90°. The stiffness of the two springs in the X-Y plane varies with the following law $k_{1,2} = k_0(1 \pm \delta)$.

This choice has been made to introduce an asymmetry in the supports without changing the total axial and lateral stiffness of the system, which is proportional to the sum $k_1 + k_2$. In order to connect the mono-dimensional springs to the three dimensional shaft cross section, a Multi Point Beam constraint has been adopted.

In order to study in detail the system dynamics, 40,128 3D second order brick finite elements have been selected. The reason behind this choice can be explained with the three-dimensional nature of mode coupling; since it is the result of modal forces between the subsystems, displacements and stresses have to be computed and then integrated across the three dimensional interface surface to capture it correctly. All the analyses (modal analysis and forced response) were carried out in ABAQUS.

38.3 Numerical Results

In previous work [8] it was shown that, when the shaft supports are asymmetric ($\delta = 0.5$), the shaft second bending mode (occurring at 218.9 Hz) and the axial rigid mode (occurring at 279.67 Hz) combine themselves into two Mixed Modes at 205 Hz and 255 Hz, which exhibit both axial and bending vibration (Fig. 38.2a, b). When the system is rotating, the Mixed Modes combine with the corresponding orthogonal bending mode, giving rise to traveling waves (or whirling modes) which also involve axial vibration (Fig. 38.2c). It must be pointed out that these analyses were carried out with a rigid disc.

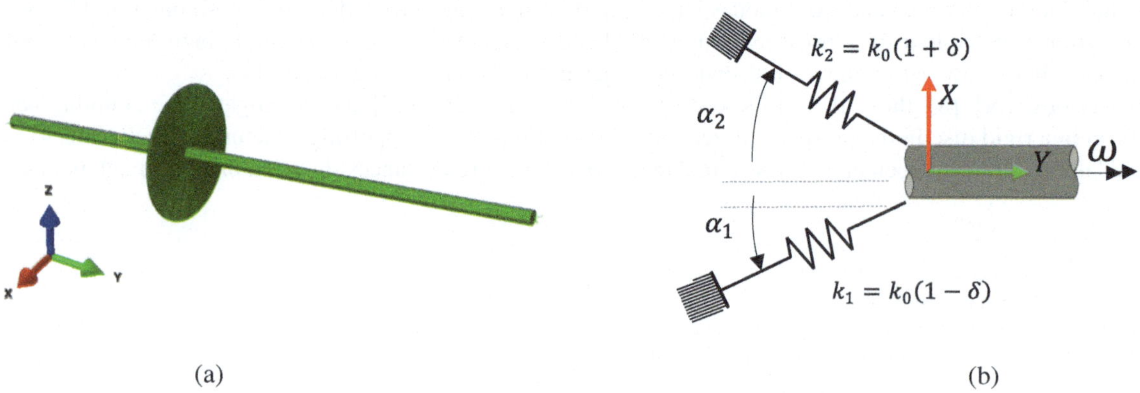

Fig. 38.1 Assembly under study – (**a**) solid geometry and mesh and (**b**) lumped parameters approximation of the bearing supporting structure

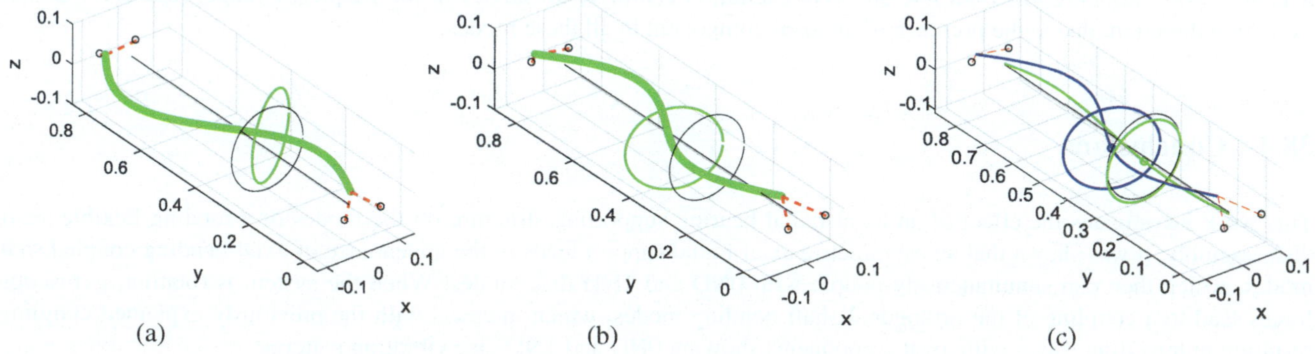

Fig. 38.2 Rigid disc model results (**a**) shaft axial-bending mode at 205.62 Hz, (**b**) shaft axial-bending mode at 255.02 Hz and (**c**) shaft forward travelling wave with axial component (real part in blue and imaginary part in green)

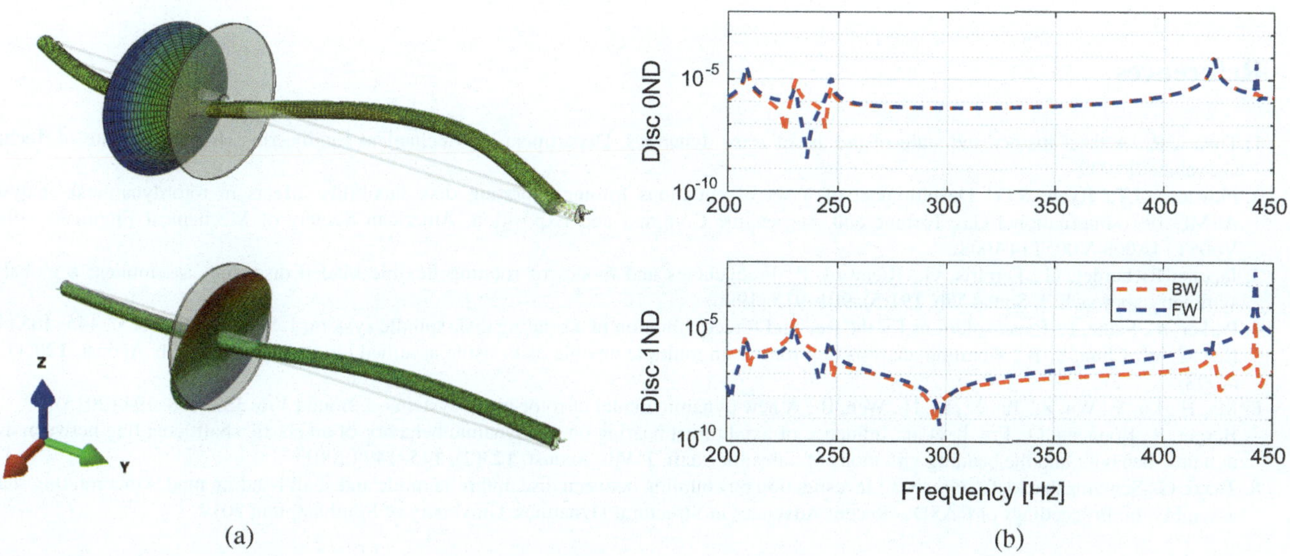

Fig. 38.3 (**a**) Real part (top) and imaginary part (below) of a XW whirling mode occurring at 246 Hz involving shaft axial and disc 0ND and 1ND vibration patterns and (**b**) Disc forced response with forward and backward rotating unitary force at left end bearing

When the disc flexibility is introduced in a rotordynamic system, shaft and disc modes couple. It is well known from the literature that shaft bending modes are prone to couple with disc 1ND modes, due to the transmission of an inertia moment between the two subsystems. On the other hand, shaft axial modes can couple with disc 0ND modes, as an axial force is transmitted. When both axial and bending shaft vibration are present at the same time, it can be seen that the disc vibrates on a superposed 0ND-1ND pattern.

A modal analysis of the rotating system is then performed considering the disc flexibility, with $\boldsymbol{\delta} = \mathbf{0.5}$ and $\boldsymbol{\Omega} = \mathbf{1000}\, \boldsymbol{rpm}$. Due to gyroscopic forces, eigenvectors are complex and whirling modes with axial components appear. Due to the disc flexibility, those shaft modes couple with disc 0ND and 1ND simultaneously, leading to travelling waves involving disc 0ND and 1ND components. An example of such a mode, occurring at 246 Hz, is shown in Fig. 38.3a. The real part (top) is dominated by a shaft first mixed mode and involves disc 0ND mode. The imaginary part (below) has a strong shaft bending component in the YZ plane and consequently disc 1ND is excited. Since the imaginary part is ahead of the real one ($j\ e^{j\omega t} = e^{j\left(\omega t + \frac{\pi}{2}\right)}$), this mode leads to a backward travelling wave.

In order to better understand these features, a forced response analysis has also been performed. The assembly is excited with a unitary forward and backward rotating force applied to the left end bearing (asynchronous excitation). Disc 0ND and 1ND components are then extracted at each frequency and shown in Fig. 38.3b. The most significant finding is that a rotating force in the XZ plane (which is the asynchronous equivalent of an out-of-balance load) is able to drive a 0ND disc mode, via the combined effect of bearing supports asymmetry, gyroscopic effects and shaft-disc coupling. This coupling effect has not been described previously but is suspected to be the source of some recently observed behaviour in an industrial application.

It is also worth noticing that both FW and BW excitation can drive all modes in the frequency range, regardless of their excitation direction, due to the presence of an axial component in all these modes.

38.4 Conclusions

This work investigates the effect of an axial-radial bearing supporting structure on the modes of a rotating flexible shaft-disc assembly. It was shown that an asymmetric axial-radial support leads to the appearance of axial-bending coupled shaft modes, which then can simultaneously couple with 0ND and 1ND disc modes. When the system is rotating, gyroscopic forces lead to a coupling of the orthogonal shaft bending modes, which interacts with the previously explained coupling, resulting in travelling waves with axial components showing 0ND and 1ND disc vibration patterns.

Acknowledgements The authors are grateful to Rolls-Royce plc for providing financial support for this project and for giving permission to publish this work.

References

1. Tiwari, R.: A brief history and state of the art of rotor dynamics. Department of Mechanical Engineering, Indian Institute of Technology Guwahati 781039
2. Flowers, G.T., Ryan, S.G.: Development of a set of equations for incorporating disk flexibility effects in rotordynamical analyses. In: ASME 1991 International Gas Turbine and Aeroengine Congress and Exposition, American Society of Mechanical Engineers, 1991, pp. V005T14A008-V005T14A008
3. Jacquet-Richardet, G., Ferraris, G., Rieutord, P.: Frequencies and modes of rotating flexible bladed disc-shaft assemblies: a global cyclic symmetry approach. J. Sound Vib. **191**(5), 901–915 (1996)
4. Parker, R., Sathe, P.: Exact solutions for the free and forced vibration of a rotating disk-spindle system. J. Sound Vib. **223**(3), 445–465 (1999)
5. Lee, C.-W., Chun, S.-B.: Vibration analysis of a rotor with multiple flexible disks using assumed modes method. J. Vib. Acoust. **120**(1), 87–94 (1998)
6. Ma, H., Lu, Y., Wu, Z., Tai, X., Li, H., Wen, B.: A new dynamic model of rotor-blade systems. J. Sound Vib. **357**, 168–194 (2015)
7. Berger, S., Bonneau, O., Fre, J., et al.: Influence of axial thrust bearing on the dynamic behavior of an elastic shaft: coupling between the axial dynamic behavior and the bending vibrations of a flexible shaft. J. Vib. Acoust. **123**(2), 145–149 (2001)
8. Tuzzi, G., Schwingshackl, C., Green, J.: Investigation on coupling between disc umbrella mode and shaft bending modes in a rotating shaft-disc assembly. In: Proceedings of RASD – Recent Advances in Structural Dynamics, University of Southampton, 2019

Chapter 39
Dynamic Characterization of a Pop-Up Folding Flat Explorer Robot (PUFFER) for Planetary Exploration

John Bell, Laura Redmond, Kalind Carpenter, and Jean-Pierre de la Croix

Abstract NASA's Jet Propulsion Laboratory is currently in the design phases of the new Pop-Up Flat Folding Explorer Robot (PUFFER) for planetary exploration. Inspired by origami, PUFFER is constructed of a printed circuit board (PCB) body which utilizes Nomex fabric as foldable joints. The structure can fold flat to stow for launch and to change shape according to the obstacles encountered. Of primary interest is how PUFFER will perform under impacts caused by drops into large pits on a planet's surface. In order to simulate PUFFER's dynamic performance, Finite Element Models (FEM) need to be constructed in an informed manner using the material properties and damping characteristics of the joints. This paper presents the first step in this process, the dynamic characterization of PUFFER via modal testing. The frequencies, associated mode shapes, and modal damping extracted by modal analysis will be correlated with the FEM to ensure that the dynamic behavior of the structure is properly captured.

Keywords Robotics · Flexible · Modal · Nomex · Aerospace

39.1 Introduction

As our understanding of our solar system and the universe increases, so does the capability and effectiveness of robotic technology. The implementation of robotics in the aerospace field has yielded a wealth of data and visuals of otherwise unseen and unanalyzed planetary bodies. For example, robotics has revealed many anomalies on the surface of Mars; such as recurring slope lineae, caves, and lava tubes. To name a few more examples, further studies on the characteristics of Europa, Enceladus, and other smaller celestial bodies have been made possible and are being pursued through similar technological advancements.

As science seeks to understand our universe deeper, the challenges only get larger. A primary consideration when preparing for interplanetary exploration is the terrain of the surface under analysis. Often, the high-interest areas found on various bodies are of unknown terrain patterns or prove to be difficult to maneuver using standard to-date autonomous methods. This motivates the creation of Pop-Up Flat Folding Explorer Robot (PUFFER), an origami-inspired, lightweight, and palm-sized rover designed to assist with larger spacecraft missions.

PUFFER is unique in that it is optimized to be both inexpensive and adaptable. Due to its design origins rooted in origami, PUFFER allows for folding and compaction not available with other robotic system designs. PUFFER's structure is constructed with printed circuit-board (PCB) composed with Nomex fabric, which serves as the folding joints of the system. The chassis is mounted on two wheels with a carbon fiber tail designed for rough terrain. The compact volume that PUFFER can attain allows for multiple units to be stored on a parent mission with little effect on the overall volume and mass of the system.

J. Bell (✉) · L. Redmond
Glenn Department of Civil Engineering, Clemson University, Clemson, NC, USA
e-mail: jsb6@clemson.edu; lmredmo@clemson.edu

K. Carpenter
Robotic Vehicles and Manipulators Group, Jet Propulsion Laboratory, California Institute of Technology, Pasadena, CA, USA
e-mail: Kalind.C.Carpenter@jpl.nasa.gov

J.-P. de la Croix
Robotic Systems Estimation, Decision, and Control Group, Jet Propulsion Laboratory, California Institute of Technology, Pasadena, CA, USA
e-mail: jean-pierre.de.la.croix@jpl.nasa.gov

B. Dilworth (ed.), *Topics in Modal Analysis & Testing, Volume 8*, Conference Proceedings of the Society for Experimental Mechanics Series, https://doi.org/10.1007/978-3-030-47717-2_39

The novelty of the conjunction of structural integrity and electrical components is what makes PUFFER as versatile and effective as it is. However, the flexible structure has unknown dynamic properties; indicating that performance under environments such as launch vibration, pitfalls, and other sources of shock are difficult to predict analytically. Modeling PUFFER's behavior under such loadings is a key step in refining the design and reaching a TRL (Technology Readiness Level) which is appropriate for flight. In order to develop reliable models of PUFFER subjected to dynamic events, the present research consists of conducting modal correlation to adequately tailor the properties of PUFFER based on experimental data.

39.2 Background

The primary unknown element needed for dynamic characterization of PUFFER is an understanding of the behavior of the composite PCB and Nomex structural system. In order to accurately represent PUFFER in a finite element model, the effective density, elastic modulus, and material damping is needed.

Literature on testing and correlation of layered circuit board is limited and still proves to have several different approaches. Arabi et al. [1], outlines a rigorous process of calculating effective density and orthotropic elastic properties for a circuit board layup, a method of modal testing, and the subsequent model correlation. This approach varies on a case-by-case basis, as small changes in the layup drastically effect the overall parameters and behaviors of the system. Krasnoveikin et al. [2] have shown that the components on a populated circuit board govern the frequency response. In order to most accurately characterize the effective density, elastic modulus and material damping of the PCB to be used on PUFFER, the authors have first conducted panel modal testing of an unpopulated PCB without the presence of the Nomex hinges.

Studies on modeling methods of flexible hinges and fabric take various approaches. A study [3] published in the Modelling and Simulation in Materials Science and Engineering journal highlights some challenges when performing advanced material modeling of fabric, involving the nonlinearities present when subjecting the fabric to compressive forces. Similar studies that perform FEA of fabrics find difficulties in the material model formulation and establishing realistic boundary conditions. The application of Nomex modeling for PUFFER is unique in that the limiting component in the system is not the fabric, but the PCB. Because of this, the primary goal is to capture the dynamic properties and damping effects of the Nomex when integrated with the PCB structure.

The Nomex hinges the authors have considered also has several approaches; and compared advantages and disadvantages with respect to solver method, solution convergence, and total runtime. The current approach utilizes plate materials with altered properties to effectively lower the bending stiffness, while preserving realistic tensile stiffness. The authors approach to modeling the Nomex hinges was refined based on a second panel modal test consisting of a PCB panel with a single Nomex hinge.

The overall research outline is as follows: first, modal correlation process and equipment checks were conducted via modal testing of an aluminum frame specimen that could be compared against a closed-form solution. Second, modal testing, analysis, and model correlation of a solid PCB sample is conducted. Third, modal testing is completed on a panel consisting of PCB with a single Nomex hinge, Finally, the updated parameters from the correlated FEMs of the PCB and PCB with Nomex panels are applied to the PUFFER FEM and modal results are presented (Fig. 39.1).

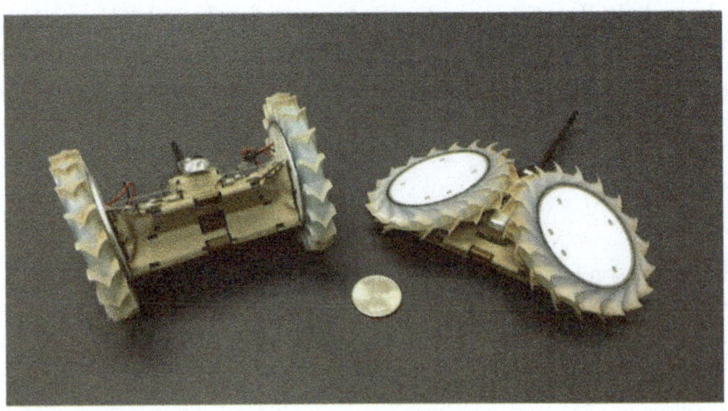

Fig. 39.1 PUFFER in standard and folded configurations

39.3 Initial Puffer Finite Element Model

All models were created in Hypermesh and executed in ABAQUS. Element types have been selected to allow for execution within both the implicit and explicit solvers for multiple types of simulation. PUFFER is initially modeled in a notional launch stow configuration. The model consists of four major components; Nomex hinges, PCB, an aluminum base, and an aluminum top cover. Each of these will be discussed individually (Fig. 39.2).

The initial Nomex joints are modeled as plate elements with modified moduli, density, and thickness to create negligible bending stiffness values. A primary drawback with the current formulation is the ability to carry small amounts of compression; however, this anomaly occurs only when perfectly flat, and is a transient state that unlikely to significantly effect model results. The advantages of this methodology are it allows for time-efficient modeling, has no convergence issues in an explicit solver, and allows for efficient tailoring of material properties to subsequently alter bending stiffness values based on test data.

Membrane elements for the Nomex joints were initially considered, but the elements must be pre-curved for the solver to initiate. Additionally, due to the zero-compression capability of membrane elements, joints often encounter numerical difficulties if they pass through a compressive state, resulting in solver divergence and excess low-energy modes (Fig. 39.3).

The material properties for the Nomex fabric are shown below (Table 39.1) [1];

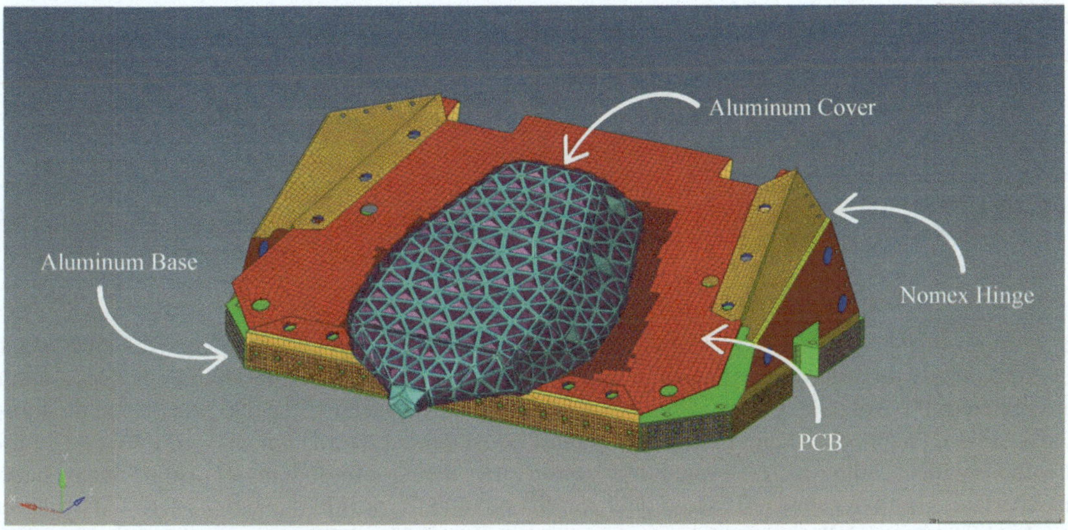

Fig. 39.2 Primary components of PUFFER FEM

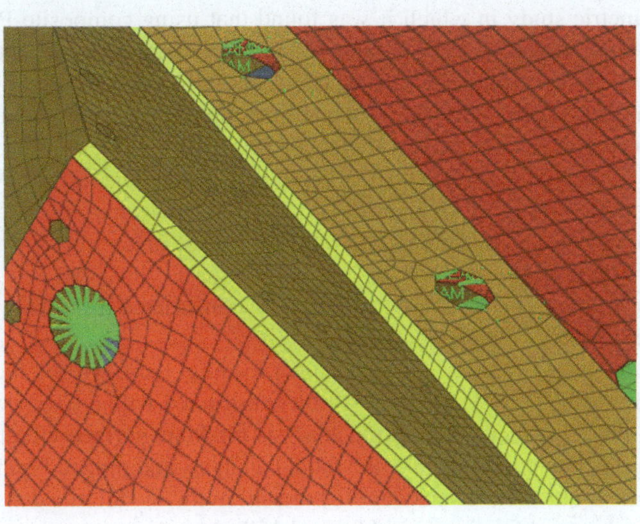

Fig. 39.3 Plate element Nomex joints

Table 39.1 Nomex joint material properties

Property	Value
Density	1.38 gm/cm^3
Elastic modulus	17 GPa
Poisson's ratio	0.36
Thickness	0.0003556 m

Table 39.2 Initial values for PCB material properties

Property	Value
Composite density	1.992 gm/cm^3
Elastic modulus	24 GPa
Poisson's ratio	0.2
Thickness	0.000856996 m

Table 39.3 Aluminum 6061 material properties

Property	Value
Density	2.70 gm/cm^3
Elastic modulus	68.25 GPa
Poisson's ratio	0.33

The dominating equations used when determining stiffness properties of the Nomex plate elements are given by Eqs. (39.1) and (39.2).

$$B_r = \frac{Et^3}{12\left(1 - \nu\right)} \tag{39.1}$$

$$A_r = \frac{Et}{\left(1 - \nu^2\right)} \tag{39.2}$$

Where B_r is the bending rigidity of a plate element, and A_r is the axial rigidity of a plate element. To achieve normal axial stiffness and mass with reduced bending stiffness, modifications to the elastic modulus, thickness, and density must be consistently altered. For example, to reduce the bending stiffness by a factor of 10,000, the elastic modulus and density must be increased by a factor of 100 and the thickness must be decreased by a factor of 100. Initial PUFFER models are constructed with very low bending stiffness values as a conservative solution for maximum displacement checks.

Based on experiences with dynamic tests and model correlation of PCBs at JPL, the authors selected to use an isotopically elastic material, which has typically captured the predominant modes of other PCB systems without necessitating a complex material model. This both allows for a time-efficient model correlation process and simpler material adjustments. The authors also conducted a separate parametric study in which it was found that using composite PCB properties with isotropic vs. orthotropic behavior had very little impact on the stresses within the system and only small differences in the frequency response. Any small differences between orthotropic and isotropic assumptions are typically negligible, as a larger source of modeling error is manifest in the mass and position variations of the distributed components on a fully populated board.

As a baseline, a typical value used for model correlation of PCB is 24 GPa, with the intention of further tuning this parameter to determine an effective modulus for PUFFER's layup based on modal testing. A sensitivity study was performed on the effects of changes to Poisson's ratio to the overall behavior of the system; the effects were minute. The value for Poisson's ratio was taken as an average of the orthotropic ratio properties given in [1]. The initial model implemented the following material properties for PCB (Table 39.2):

The last major components, the aluminum base and top cover, are constructed from Aluminum 6061. Using [4], we obtain the following material properties (Table 39.3):

PUFFER's chassis is connected to two wheels. During launch loading, these will be locked; to represent this, they have been left out of the model. Additional PUFFER models will be constructed with a fixed condition to represent this loading scenario. Further work can include the wheels in dynamic analysis for Mars-surface vibration and shock analysis; for now, PUFFER is modeled with free boundary conditions for the modal analysis and pinned boundary conditions for launch.

In Fig. 39.1, a tail on PUFFER is attached. To account for the tail in the model, the density of the aluminum top cover is increased by 20%. In addition to this modification, an additional 150 g mass is smeared over the part for an instrument load. The overall mass of the system is 256 g (not including wheel masses). Initial models have neglected the dynamic effects of the PUFFER tail because it is presumed to be folded or constrained during launch. Future modeling efforts may incorporate the tail depending on the intended launch configuration.

The PCB to PCB connections are modeled using rigid connector elements, and the PCB to aluminum connections are modeled using tie constraints. An initial assumption of 0.5% modal damping is assumed for conservative displacement predictions.

39.4 Modal Test Process Validation

As validation for a correct modal testing process, testing and model development of a single-story frame was first performed. An initial FEM of the frame was constructed assuming a fixed boundary condition that could effectively vary from the absolute base of the structure, to the top of the black brackets shown in Fig. 39.4. Further preliminary checks performed involved analyzing the stiffness of the steel brackets at the base; checking both the adequacy of the rotational stiffness and the effects of the bending length of the side members. The effective fixed location of the frame is most likely to be within the center of the bolt pattern shown in Fig. 39.4.

For modal testing validation, only the in-plane modes were considered. Uniaxial accelerometers were used, which required the addition of support blocks to rove accelerometers from X to Y orientations. These slight mass changes were updated in the model accordingly. Prior to modal testing, the FEM was run, and modal displacement data was extracted. Considering modes that contained up to 90% of the system's overall effective mass in both the X and Y directions, the most prominent modes were chosen by creating a scaling system. This scaling system exhibited the modes with the largest degree of in-plane motion. As outlined in [5], functions for the optimum driving point (ODP), the average driving point residue (ADPR), and the non-optimum driving point (NODP) were computed. The results suggested ideal stinger location, accelerometer placement, and areas of overall high and low response in the system over the modes chosen for model correlation.

The frame was mounted to a sand-filled box, weighing approximately 200 lb. Frequency based and time history random response models utilizing a random signal which met the testing specifications were run to ensure that the displacement of the system at the stinger attachment point was well under stroke length of the shaker. During setup, the alignment of the stinger was performed carefully using a laser level to prevent buckling and sloppy data. Prior to testing, the accurate calibration of the accelerometers used was tested and verified.

As tests were run, it was found that any local modes due to the flexure of the support boards had very low, insignificant response, and the motion of the base was less than 5% the prescribed acceleration of the stinger. The testing procedure involved three different roving accelerometer locations with a random excitation, exciting modes from 20–4500 Hz. A Hanning window was applied with no additional filtering.

Fig. 39.4 Constructed aluminum frame

Fig. 39.5 Shaker and frame setup

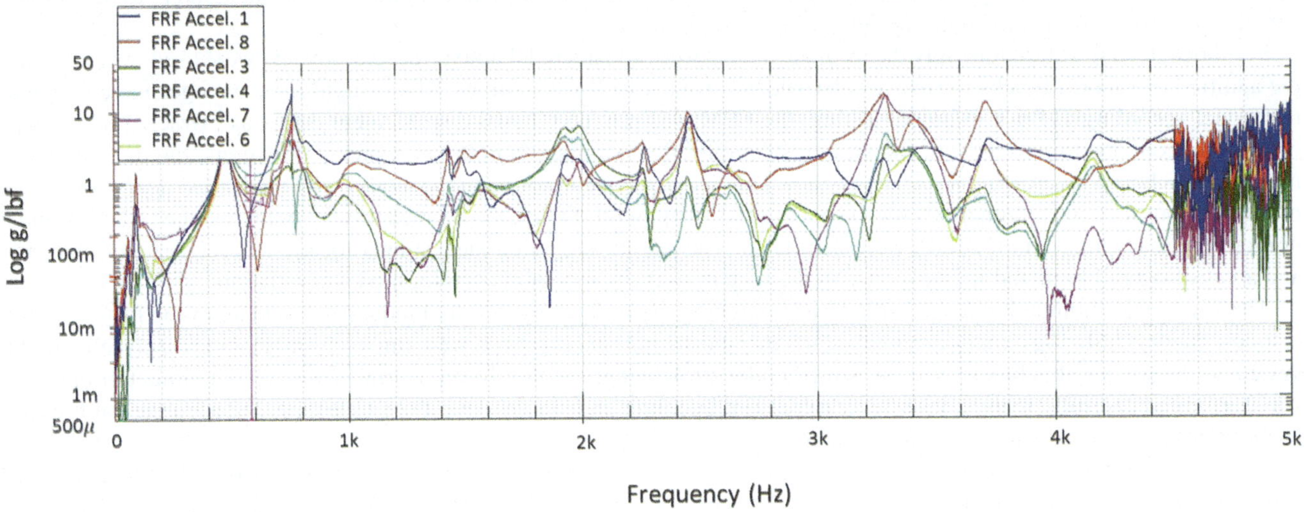

Fig. 39.6 Result data from a frame test run

The only necessary adjustments to the frame model to achieve correlation with the test data was a slight change in boundary condition. The effective fixed-end was roughly 1/3 up the length of the side bracket which falls between the bolt pattern shown in Fig. 39.4. After making this adjustment, the frequencies of the prominent modes (modes 1, 3, 4, 5, 16, and 17), and the mode shapes for both the test results and the FEM results were compared. The largest error between the measured frequency and the FEM result frequency was 2.43%. A Modal Assurance Correlation calculation, as demonstrated in [6], reinforces the conclusion that the model was correlated accurately (Figs. 39.5 and 39.6).

39.5 Test Procedure

Multiple test coupons of the printed circuit board, with two different configurations, were constructed and tested. The initial test specimen layup was PUFFER's solid PCB without any Nomex hinges, with an additional layer of Nomex attached to the underside of the board. The second coupon layup was PUFFER's PCB with Nomex backing, plus a single Nomex hinge. To best perform model correlation, the thickness of all boards of the same type were measured to the nearest 0.0127 mm and weighed to 0.05 g to ensure consistency and accuracy with the designated layup.

To determine the locations for accelerometers and the frequency range of each test, an initial FEM of the PCB layup was modeled. The density was calculated assuming the Nomex fabric contributes very little stiffness to the system. Thus, the effective modeled thickness of the layup constituted only the PCB thickness and only the Nomex mass was accounted for by dividing the measured mass by the modeled volume. The calculated densities for every sample are shown in Fig. 39.8, and the initial FEM used the average calculated thickness and density from all the boards.

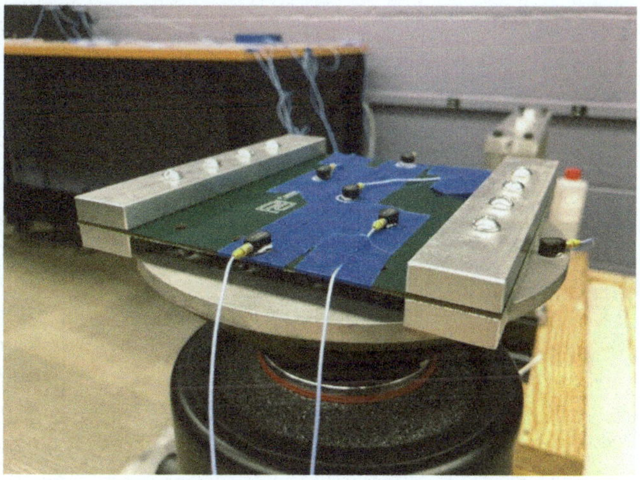

Fig. 39.7 Mounted solid PCB

The boundary conditions for both testing and FEM implementation was a fixed-fixed case. This was achieved through machining aluminum bars with a bolt pattern identical to the holes in the PCB and on the shaker expander head (see Fig. 39.7). The PCB was torqued directly to the expander head and was excited vertically. FEM models of the solid PCB utilized constraints in all translations and rotations over the length and width of the clamped area and included the mass of the accelerometers. The position of the accelerometers shown in Fig. 39.7 and the excitation range was selected based on the mode shapes and frequencies predicted by the initial FEM.

Testing of the solid PCB was performed both as a sine sweep and as a random 0.5 g excitation from 20–2000 Hz with a Hanning window. The FRF reference was an additional accelerometer mounted directly to the expander head, as shown in Fig. 39.7. Testing was performed on two solid boards, and the results were perfectly consistent (all modes within 2%). All the fabricated PCB samples had consistent thickness and density, giving confidence in frequency and mode shape uniformity between all samples.

Following the testing of the solid PCB samples, the Nomex-hinge samples were tested using the same setup and boundary conditions. The windowing, frequency range, and excitation of testing remained the same. The only changes made was a slight shift in accelerometer location. The FEM implementation utilized the same Nomex modeling process as the PUFFER model and accounted for the added weight of the accelerometers.

39.6 Test Results

For model correlation of the solid PCB sample, the first three modes and mode shapes were examined. Due to the sample's small size, higher-frequency modes yielded complex curvature patterns, which are difficult to capture because accelerometers must be placed very precisely, and the positioning of the accelerometers changes the overall frequency response of the board.

The PCB model used the average thickness of all the solid PCB samples, the average calculated density from Fig. 39.8, and allowed for the tuning of the elastic modulus. No adjustment to the boundary conditions was needed. The elastic modulus had to be increased only slightly from 24 to 25.5 GPa to correlate the model. The correlated PCB material properties are shown in Table 39.4. Using these properties, the maximum error between the FEM frequencies and the test results was 3.9% as shown in Table 39.5.

The two types of PCB samples being tested have a slight difference in average thicknesses. For the Nomex hinge PCB FEM, the thickness was set to the physical thickness of the samples without the underlying Nomex layer. The calibrated modulus value for the PCB was taken from the previous model correlation, leaving only the bending properties of the Nomex hinge to be adjusted. As discussed, an exercise in modifying the bending stiffness by consistently adjusting thickness, density, and elastic modulus was performed, as well as a sensitivity study on each parameter individually. The final model was correlated to test data by reducing the bending rigidity of the plate element (calculated from the physical measurements and properties of the Nomex fabric with a traditional plate element formulation) by 44%. This reduction in bending stiffness allowed for the accurate response of the fabric in a dynamic setting. The final correlated values for material properties of the Nomex fabric are shown in Table 39.6.

Fig. 39.8 Calculated densities of PCB

Type	Serial No.	Density (kg/m³)
Solid	016	2175.110723
Solid	011	2217.293522
Solid	012	2152.501138
Solid	013	2165.022238
Solid	015	2158.962939
Solid	014	2180.540824
Nomex	007	2239.371107
Nomex	008	2252.761961
Nomex	009	2220.366446
Nomex	010	2230.590502
Nomex	005	2237.875732
Nomex	006	2233.540183

Table 39.4 Final updated values for PCB material properties

Parameter	Value
E	25.5 GPa
ν	0.2
Thickness	0.000856996 m
ρ	2156.20 kg/m^3

Table 39.5 Frequencies from FEM and test for solid PCB samples

Mode no.	FEM frequency (Hz)	Sample 1 testing freq. (Hz)	Error 1 (%)	Sample 2 testing freq. (Hz)	Error 2 (%)
1	250.36	247.00	1.3	248.00	0.94
2	262.06	265.00	1.1	260.00	0.786
3	331	340	2.7	344.00	3.92

Table 39.6 Final updated Nomex parameters

Parameter	Value
E	25.5 GPa
ν	0.36
Thickness	0.00023706666 m
ρ	2070.0 kg/m^3

The first mode of the PCB with a Nomex hinge model; the flexing of the Nomex hinge, dominates. During testing, two different samples yielded a first mode frequency of 185.0 Hz. Upon making the adjustment to the Nomex stiffness properties in the FEM, the solution determined the first bending mode to be 185.

39.7 Updated Puffer Finite Element Model

The PUFFER model was refined with the newly determined PCB material properties and Nomex bending stiffness properties. The results of the PUFFER model's modes in a free-suspended condition are shown in Fig. 39.9. Upon obtaining an accurate model, PUFFER response can now be examined under various dynamic simulations. Future design adjustments can be guided by the results from launch, landing, and pitfall simulation. Future work includes the development of effective material damping of the PCB and PCB Nomex components. From the test results, modal damping was extracted, and the exercise of determining an equivalent material damping is currently being performed.

Fig. 39.9 Updated PUFFER FEM modes and frequencies

Mode No.	Freq. (Hz)
1	288.29
2	1003.6
3	1109.5
4	1575.1
5	1582.8
6	2442.4
7	2473.4
8	2708.9
9	3083.8
10	3115.9
11	3557.2
12	3601
13	3897
14	4228.5
15	4426.5
16	4640.6
17	4897.2
18	5040
19	5052.3
20	5396.9

Acknowledgements The funding sources for the project is the South Carolina Space Consortium Research and Education Award Program. The donor of the PCB layup panels for testing is Pioneer Circuits, based in Santa Ana, California, who works closely with Jet Propulsion Laboratory on many aerospace applications. Dale McKeeby, from Pioneer Circuits, provided extensive technical advice and information regarding the PCB layup panels. Technical information and guidance for modal testing was provided by Dale Schick with M + P International.

The PUFFER robot research was carried out at the Jet Propulsion Laboratory, California Institute of Technology, under a contract with the National Aeronautics and Space Administration (80NM0018D0004).

The opinions in this paper are those of the authors and do not represent the opinions of the sponsors.

References

1. Arabi, F., Gracia, A., Deletage, J.-Y., Fremont, H.: Vibration test and simulation of printed circuit board. 2018 19th International Conference on Thermal, Mechanical and Multi-Physics Simulation and Experiments in Microelectronics and Microsystems (EuroSimE), 2018. https://doi.org/10.1109/eurosime.2018.8369909
2. Krasnoveikin, V.A., Smolin, I.Y., Druzhinin, N.V., Kolubaev, E.A., Derusova, D.A.: Modal testing circuit board assembly of an electronic apparatus by laser vibrometry. IOP Conf. Ser. Mater. Sci. Eng. **156**, 012005 (2016). https://doi.org/10.1088/1757-899x/156/1/012005
3. Lin, H., Sherburn, M., Crookston, J., Long, A.C., Clifford, M.J., Jones, I.A.: Finite element modelling of fabric compression. Model. Simul. Mater. Sci. Eng. **16**(3), 035010 (2008). https://doi.org/10.1088/0965-0393/16/3/035010
4. MMPDS-05: Metallic Materials Properties Development and Standardization (MMPDS). [Washington, D.C.]: [Columbus, Ohio]: Federal Aviation Administration; Battelle Memorial Institute [distributor], 2010
5. Ewins, D.J.: Modal Testing: Theory and Practice. Research Studies Press, Baldock (2000)
6. Pastor, M., Binda, M., Harčarik, T.: Modal assurance criterion. Procedia Eng. **48**, 543–548 (2012). https://doi.org/10.1016/j.proeng.2012.09.551

Chapter 40
A Framework for the Design of Rotating Multiple Tuned Mass Damper

Kévin Jaboviste, Emeline Sadoulet-Reboul, Olivier Sauvage, and Gaël Chevallier

Abstract The acoustic and vibratory analysis represent an essential research axis in the automotive industry because these phenomena directly affect the appreciation of the customer when using a vehicle. Indeed, the combustion engine represents the main source of mechanical energy but it generates an acyclic torque because of the explosions. This acyclism is responsible for noise and vibration fatigue. To reduce NVH issues, a possible mean is to use a Multiple Tuned Mass Damper (MTMD) adapted to rotating machine.

The purpose of this paper is to propose a framework for the design of this type of MTMD. To achieve this goal, an optimization strategy is implemented in the non-rotating and rotating case to determine the optimal distribution of the MTMD frequencies. It is based on the minimization of the elastic strain energy of the transmission chain near the torsion mode. In addition, a dedicated reduced order method is proposed to reduce the calculation costs link to the modeling of the system by the finite element method and the optimization process. However, conventional reduction order methods are not suitable for this type of problem where the shape of the modes of the main structure and the MTMD are varying according to the rotation speed. To overcome this problem, a multi-model approach is employed. Finally, the influence of the number of absorbers composing the MTMD as well as the structural damping of the absorbers and their mass on the performance of the optimal solution is presented in the non-rotating and rotating case.

Keywords Rotating frame · Multiple tuned mass damper · Optimization · Model order reduction · Passive damping

40.1 Introduction

Nowadays, the noise generated by the vehicle operation is one of the important criteria taken into account by car manufacturers. Thus, many efforts have been made for several years to reduce the noise pollution produced by the various components of the vehicle. The overall goal is to meet environmental standards that become more severe, but also to improve the comfort of passengers within the vehicle. In most of the vehicle, the motor is the main source of mechanical energy and noises. These noises are generated by the presence of motor acyclism. More specifically, this acyclism generates an oscillation of the speed of rotation of the crankshaft around a given speed which causes the occurrence of shocks and noise. Hence, to reduce NVH issues, it is possible to use Multiple Tuned Mass Damper (MTMD) adapted to rotating machines.

The use of MTMD consisting in a network of TMDs with natural frequencies distributed around the frequency to control is a strategy to absorb vibration on a frequency range around this frequency of interest. The different possible distributions for the eigenfrequencies of the MTMD have been the subject of numerous studies. Many studies consider multiple absorbers with equally spaced eigenfrequencies [1, 7, 8, 20, 21]. The performance of MTMD with uniformly and linearly distributed masses are compared in [9, 16]. Optimization methodologies have been used to increase the efficiency of MTMDs and do not impose a specific profile distribution. Closed-form approximations of the TMD parameters are proposed using an integral form of the impedance in [7], optimal design theories are investigated as the minimax optimization in [22], a specific optimal design theory is proposed in [11], and a gradient-based method with unconstrained variables is used in [12]. It appears

K. Jaboviste (✉) · E. Sadoulet-Reboul · G. Chevallier
Department of Applied Mechanics, Univ. Bourgogne Franche-Comté FEMTO-ST Institute, Besançon, France
e-mail: kevin.jaboviste@univ-fcomte.fr; kevin.jaboviste@femto-st.fr; sadoulet-reboul@univ-fcomte.fr; chevallier.gael@univ-fcomte.fr

O. Sauvage
Groupe PSA Scientific and Future Technologies Department, Vélizy-Villacoublay, France
e-mail: olivier.sauvage@mpsa.com

B. Dilworth (ed.), *Topics in Modal Analysis & Testing, Volume 8*, Conference Proceedings of the Society for Experimental Mechanics Series, https://doi.org/10.1007/978-3-030-47717-2_40

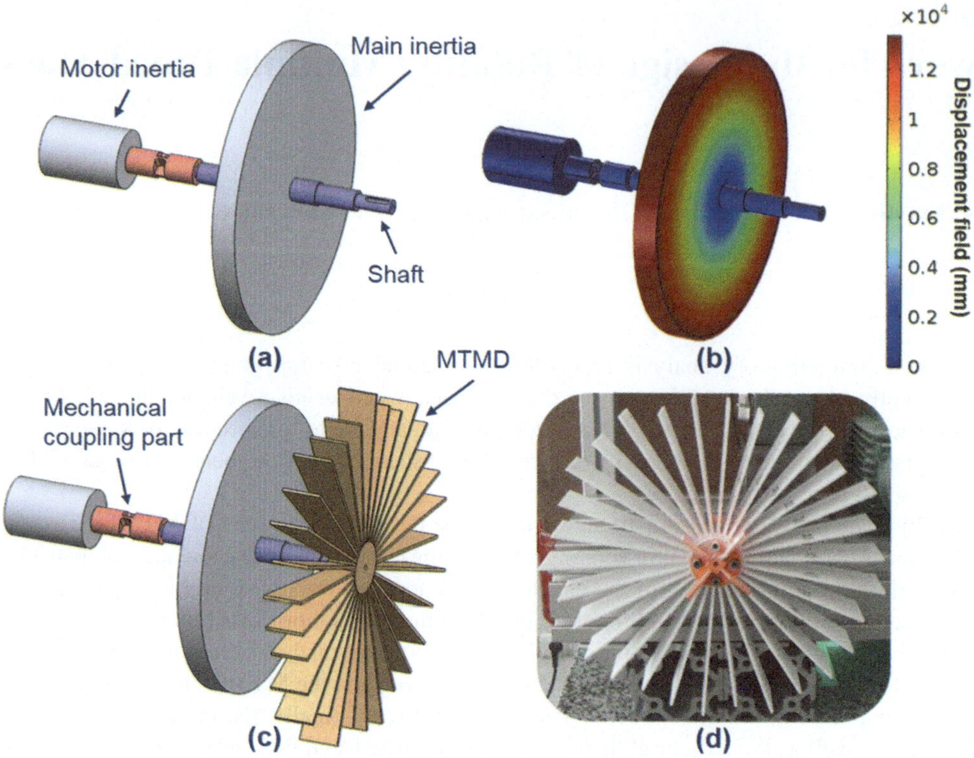

Fig. 40.1 Studied case: CAD of the main structure (**a**), first torsion mode of the main structure: displacement field from FE computation (**b**), CAD of the main structure with the MTMD (**c**), and prototype of the MTMD [14] (**d**)

through all these studies that optimal frequency distribution can be found to ensure the vibration control. Moreover, one can note that a particular frequency distribution characterized by a high modal density around the natural frequency of the master structure allows to absorb nearly irreversibly the energy of the master structure [3, 17].

On the other hand, the concept of absorber generally dedicated to rotating structures is commonly called order-tuned absorbers in the literature and allows, rather than being fixed to one and only frequency, to follow a particular motor rotation order and to be effective for a whole speed range. This particular tuning is achieved by exploiting the centrifugal force field [5] and this type of absorber has had some success in the field of rotating machines with applications on aircraft engines [10], helicopter rotors [19], camshafts [4] and car engines [13]. In 2005, Shaw and Pierre propose, in [18], an analytic study of the dynamic behavior of a flexible structure with 1 degree of freedom to which is fixed an order-tuned impact absorber. In the same year, Olson, Shaw and Pierre [15] improve this work by generalizing them in the case of cyclic symmetric structures such as bladed wheels.

This paper propose an optimal design methodology for a MTMD adapted to rotating machines instead of an order-tuned absorber [18]. The studied case is shown Fig. 40.1a and represents a simplified mechanical transmission chain. It is composed of a main inertia fixed on a shaft link to the main frame by a pivoting link and an adapted mechanical coupling part connecting this subsystem to a smaller inertia. The mechanical coupling part is designed in such a way that the first mode of vibration of the whole system is a torsion mode (Fig. 40.1b). A MTMD is used to reduce the vibration induced by the first torsion mode. The retained geometry [14] is a set of blades as shown Fig. 40.1d and the MTMD is attached to the master structure at the end of the shaft (Fig. 40.1c).

40.2 Optimization of a Multiple Tuned Mass Damper Adapted to Rotating Frame

Modeling The Multiple Tuned Mass Damper (MTMD) attached to the main structure is composed of 30 blades and each blade behaves like a dynamic vibration absorber. The Finite Element Method is used to model the whole structure and accurately predict its physical behavior. Damping is introduced using a structural damping model for both the main structure (η_s) and the MTMD (η_M). In our case, the rotation speed of interest is fixed at $1500 RPM$ and the only influencing effect is the centrifugal force which is comparable to a prechage modifying the stiffness matrices. In this context, the dynamic equation of the whole system in the frequency domain is written as,

$$-\omega^2 M\hat{U} + \left(K(\Omega) + K_e^*\right)\hat{U} = \hat{F} \tag{40.1}$$

where K_e^* and M are respectively the complex stiffness and the real constant mass matrices of the whole structure including the MTMD, $K(\Omega)$ is a matrix that represents the influence of the centrifugal force, $\hat{U}$ is the complex displacement vector and $\hat{F}$ is the excitation vector. The complex stiffness is written as,

$$K_e^* = (1 + j\eta_s)K_s + (1 + j\eta_M)K_M \tag{40.2}$$

where K_s is the stiffness matrix of the main structure and K_M is the stiffness matrix of the MTMD. Then, to reduce computational costs and perform the optimization of the MTMD it is possible to use a projection basis. Here, the stiffness matrices vary according to the rotation speed and generate changes in MTMD eigenfrequencies and can also impact the coupling between the blades. A way to take into account these parametric variations in a model order reduction is to use a multi-model basis [2]. The multi-model approach allows to build a projection basis representative of the whole system stiffening due to the rise of rotation speed. To create this multi-model basis, three different speeds are chosen and for each of them a modal basis is extracted. It is important to choose the first speed equal to the speed of interest, here 1500RPM and the two other basis are extracted at 0 and 3000RPM. A modified Gram-Schmidt ortho-normalization [6] is used to concatenate these three basis in one multi-model basis P where,

$$P = \left[\Phi^{1500} \quad \Phi^{0}_{\perp\Phi^{1500}} \quad \Phi^{3000}_{\perp\Phi^{1500}\cup\Phi^{0}_{\perp\Phi^{1500}}}\right] \tag{40.3}$$

The projected operators will be noted: m for the mass, k_s for the master structure stiffness, k_M for the MTMD stiffness and k_b^i for the reduced stiffness matrix of the ith blade.

MTMD Optimization at 1500RPM The design variables chosen for the MTMD optimization problem are the stiffnesses of the N blades. The structural damping of the main structure and the MTMD will be considered constants. The mass and inertia of the MTMD are ten times lower than those of the master system. A coefficient $\alpha_i(1 < i < N)$ is introduced for each blade of the MTMD such that the reduced stiffness matrix is modified as follow,

$$k_M = \sum_{i=1}^{N} \alpha_i k_b^i \tag{40.4}$$

Different optimization objective functions can be defined, it is chosen here to minimize the elastic strain energy on the frequency band around the resonant frequency of the torsion mode. It is defined as,

$$\Pi(\omega) = \frac{1}{2}\hat{q}^H(\omega)k_s\hat{q}(\omega) \tag{40.5}$$

where k_s is the reduced elastic real stiffness matrix of the main structure at $\Omega = 1500RPM$ and $\hat{q}$ the complex generalized coordinate.$\hat{q}^H$ is the Hermitian transpose of $\hat{q}$. The optimization problem can thus be written as,

$$\begin{cases} \min\limits_{\alpha_i} \bar{\Pi}(\alpha_i) = \frac{1}{\Delta\omega}\int_{\omega_{min}}^{\omega_{max}} \Pi(\omega)d\omega \\ -\omega^2 m\hat{q} + \left(k(\Omega) + k_e^*(\alpha_i)\right)\hat{q} = \hat{f} \\ \alpha_{inf} \leqslant \alpha_i \leqslant \alpha_{sup}, i = 1, \ldots, N \end{cases} \tag{40.6}$$

where α_{inf} and α_{sup} represent the lower and upper bounds for the parameters, ω_{min} and ω_{max} define the lower and upper bounds of the frequency band and $\bar{\Pi}$ is the mean value of the elastic strain energy on this frequency band.

Figure 40.2 presents the angular displacement Θ of the main inertia when a varying harmonic torque T is applied on its lateral surface with and without MTMD. In this simulation the MTMD is composed of 30 blades and it can be observed that the MTMD significantly reduces the amplitude of the response on a wide range of frequencies around the frequency to control. One can note the existence of an optimal loss factor for the MTMD ($\eta_M = 10\%$) that leads to a better attenuation efficiency. In the case of rotating machines, the rigid body mode is no longer at 0 Hz but higher in frequency because of the

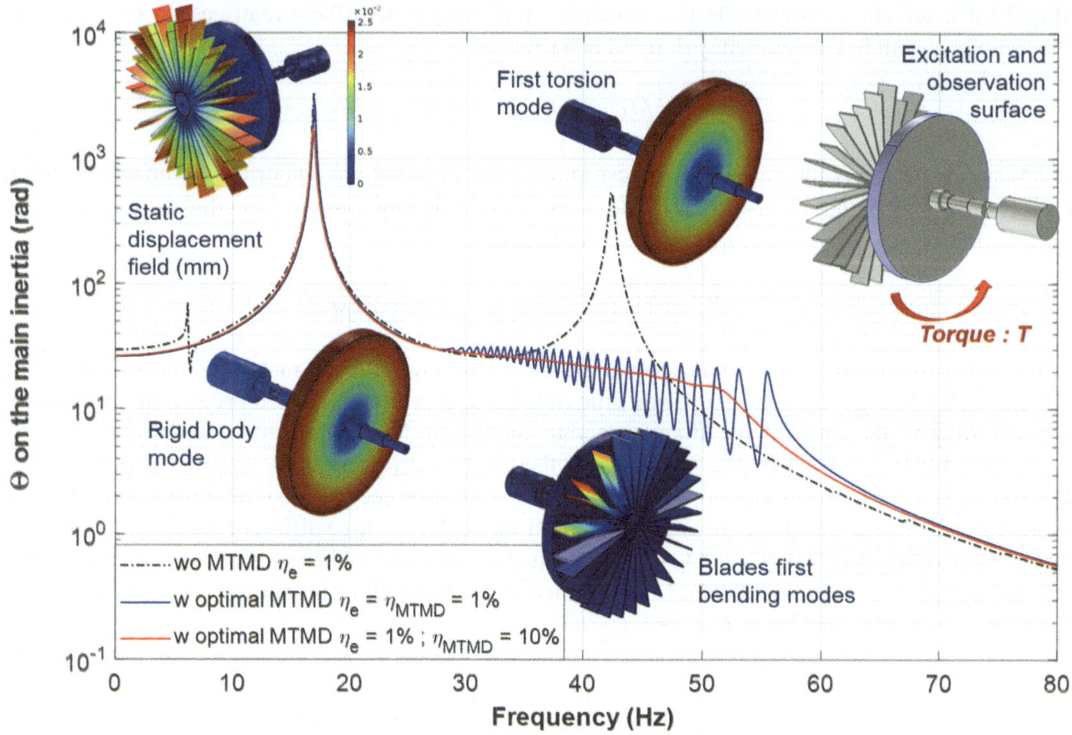

Fig. 40.2 Rotation of the main inertia according to the frequency without and with MTMD at $1500RPM$

matrix $K(\Omega)$ which becomes a non-zero matrix. Moreover, as previously mentioned in the paper, the centrifugal force acts as a precharge and thus modifies the shape of the static deformation.

40.3 Conclusion

This paper proposes a methodology to determine the optimal frequency distribution for a set of blades composing a Multiple Tuned Mass Damper adapted to rotating frame in order to reduce the vibration induced by the first torsion mode of the main structure. Each blade acts as a dynamic vibration absorber and its natural frequencies are controlled by a set of parameters α_i. Theses parameters are determined thanks to an optimization method based on the minimization of the mean elastic strain energy on a frequency band around the resonant frequency to control. The proposed methodology can be applied to the case of rotating machines as shown in this paper. Rotating effects such as centrifugal force are taken into account in the finite element model and the use of multi-model basis representative of the stiffness variation around the speed of interest drastically reduce the computation cost of the optimization process. Hence, it is possible to design an optimal MTMD for a given structure and speed as shown in Fig. 40.2.

Acknowledgments Authors are grateful to the french company Groupe PSA for their financial support.

References

1. Abé, M., Fujino, Y.: Dynamic characterization of multiple tuned mass dampers and some design formulas. Earthq. Eng. Struct. Dyn. **23**(8), 813–835 (1994)
2. Balmès, E.: Model reduction for systems with frequency dependent damping properties. Office National d'Etudes et de Recherches Aerospatiales ONERA-Publications-TP (1997)
3. Carcaterra, A., Akay, A., Bernardini, C.: Trapping of vibration energy into a set of resonators: theory and application to aerospace structures. Mech. Syst. Signal Process. **26**, 1–14 (2012)
4. Cook, M.: Absolute absorbtion. Car Craft Mag. **42**, 75 (1994)

5. Den Hartog, J.P.: Tuned pendulums as torsional vibration eliminators. In: Stephen Timoshenko 60th Anniversary Volume, pp. 17–26 (1938)
6. Festjens, H., Gaël, C., Franck, R., Jean-Luc, D., Remy, L.: Effectiveness of multilayer viscoelastic insulators to prevent occurrences of brake squeal: a numerical study. Appl. Acoust. **73**(11), 1121–1128 (2012)
7. Igusa, T., Xu, K.: Vibration control using multiple tuned mass dampers. J. Sound Vib. **175**(4), 491–503 (1994)
8. Joshi, A.S., Jangid, R.S.: Optimum parameters of multiple tuned mass dampers for base-excited damped systems. J. Sound Vib. **202**(5), 657–667 (1997)
9. Kareem, A., Kline, S.: Performance of multiple mass dampers under random loading. J. Struct. Eng. **121**(2), 348–361 (1995)
10. Ker Wilson, W.: Practical solution of torsional vibration problems. Chapman and Hall, London (1968)
11. Lee, C.-L., Chen, Y.-T., Chung, L.-L., Wang, Y.-P.: Optimal design theories and applications of tuned mass dampers. Eng. Struct. **28**(1), 43–53 (2006)
12. Li, H.-N., Ni, X.-L.: Optimization of non-uniformly distributed multiple tuned mass damper. J. Sound Vib. **308**(1–2), 80–97 (2007)
13. Nester, T.M., Haddow, A.G., Shaw, S.W., Brevick, J.E., Borowski, V.J.: Vibration reduction in a variable displacement engine using pendulum absorbers. Technical report, SAE Technical Paper (2003)
14. Olivier, S.: Dispositif de piege vibratoire a resonateurs distribues. FR Patent 3035939 (2017)
15. Olson, B.J., Shaw, S.W., Pierre, C.: Order-tuned vibration absorbers for a rotating flexible structure with cyclic symmetry. In: Proceedings of the ASME International Design Engineering Technical Conferences and Computers and Information in Engineering Conference 2005, vol. 1, pp. 2475–2484 (2005)
16. Park, J., Reed, D.: Analysis of uniformly and linearly distributed mass dampers under harmonic and earthquake excitation. Eng. Struct. **23**(7):802–814 (2001)
17. Roveri, N., Carcaterra, A., Akay, A.: Energy equipartition and frequency distribution in complex attachments. J. Acoust. Soc. Am. **126**(1), 122–128 (2009)
18. Shaw, S.W., Pierre, C.: The dynamic response of tuned impact absorbers for rotating flexible structures. J. Comput. Nonlinear Dyn. **1**(1), 13–24 (2006)
19. Wachs, M.A.: The main rotor bifilar absorber and its effect on helicopter reliability/maintainability. Technical report, SAE Technical Paper (1973)
20. Xu, K., Igusa, T.: Dynamic characteristics of multiple substructures with closely spaced frequencies. Earthq. Eng. Struct. Dyn. **21**(12), 1059–1070 (1992)
21. Yamaguchi, H., Harnpornchai, N.: Fundamental characteristics of multiple tuned mass dampers for suppressing harmonically forced oscillations. Earthq. Eng. Struct. Dyn. **22**(1), 51–62 (1993)
22. Zuo, L., Nayfeh, S.A.: Minimax optimization of multi-degree-of-freedom tuned-mass dampers. J. Sound Vib. **272**(3–5), 893–908 (2004)

Kévin Jaboviste is a new doctor who has defended his thesis in December 2018 on the experimental characterization and modeling of damping devices for the reduction of vibration at the FEMTO-ST institute. He is collaborating with PSA in order to design new Multiple Tuned Mass Damper adapted to rotating frame.

Chapter 41
Operational Modal Analysis of Wind Turbine Drivetrain with Focus on Damping Extraction

Pieter-Jan Daems, Cédric Peeters, Patrick Guillaume, and Jan Helsen

Abstract Due to changes in generator topology, wind turbines are operating in much wider speed ranges and thus at more varying operating conditions. This has a positive influence on the energy production, but results in much wider gear mesh excitation ranges that can lead to tonalities. As such, a high-quality characterization of the modal model of the gearbox housing and its interaction with the turbine nacelle is essential to pinpoint the specific resonances responsible for these tonalities. Not only the knowledge of the resonance frequencies is essential, also the damping should be well characterized. This allows to rank the modes with regard to importance for design modifications. This paper investigates the performance of different modal estimation methods to assess these damping values. An SSI, p-LSCF and Polymax Plus estimator are benchmarked with one another. A sensitivity study is done to optimize the configuration parameters of the different modal estimators specifically for damping extraction. A simulation example is first used to perform this optimization. Acceleration signals embedding the modal content of a rotating machine are simulated to serve as inputs for the modal estimators. Once detailed insights are gained, data of a wind turbine is used. Acceleration measurements along the gearbox housing together with encoder measurements serve as inputs. Damping estimates from rotating data sets are compared. The impact on the accuracy of the damping estimates is illustrated for the different modal estimators.

Keywords Automatic operational modal analysis · Wind turbine drivetrain · Harmonics · Noise vibration and harshness · Cepstrum

41.1 Introduction

In recent years, the installed capacity of wind energy has increased significantly. The worldwide capacity of wind energy has surpassed 600 GW, of which more than 50 GW was installed in 2018 [1]. Especially the construction of offshore farms has grown substantially in the last decade in Europe; in the last 10 years, the installed offshore capacity has grown from 3 to 40 GW. The main advantage of these wind farms is that the wind is stronger and more laminar offshore compared to onshore. Significant challenges however remain within this market, such as reducing the Operating and Maintenance (O&M) costs of the farms. A trend seen within wind industry is to increase the size of the wind turbines, as it is assumed by economics of scale that by decreasing the machines needed per MW, the overall O&M costs will also decrease [2]. Due to this up-scaling, the loads acting on the different subcomponents will however also increase, making the design more difficult [3]. For these new generations of wind turbines, there is moreover still a relatively high uncertainty present in the validity of the different design assumptions [4]. This is due to limited field data concerning failure cases, dynamic loading, etc. being available. These factors substantially increase the risk connected to these machines, warranting the need of methods that can validate the different design assumptions as soon as possible. Within the context of Industry 4.0 and the Internet of Things (IoT), manufacturers have been pushing towards equipping their machines with an increasing amount of sensors. Processing this information on a fleet-wide level would allow to gain insights in the validity of the design assumptions more quickly. A challenge that however remains within this context is to have algorithms available that can process such large data streams completely autonomous.

This paper focuses on the characterization of the dynamic behaviour of rotating machinery during normal operation. This is an issue relevant for both the onshore and offshore market, it is among other essential to be able to solve Noise,

---

P.-J. Daems · C. Peeters · P. Guillaume · J. Helsen (✉)
Vrije Universiteit Brussel (VUB), Brussel, Belgium
e-mail: pieter-jan.daems@vub.be; cedric.peeter@vub.be; patrick.guillaume@vub.be; jan.helsen@vub.be

B. Dilworth (ed.), *Topics in Modal Analysis & Testing, Volume 8*, Conference Proceedings of the Society for Experimental Mechanics Series, https://doi.org/10.1007/978-3-030-47717-2_41

Vibration and Harshness (NVH) issues. This is stringent for onshore turbines as noise regulations concerning these machines are becoming increasingly stringent [5]. The role of the gearbox is becoming ever more important in this context, as advancements in blade design have resulted into the fact that tonalities might become audible in certain operating regimes [6]. Being able to characterize the internal and external transfer paths is essential to efficiently make the needed design modifications. A big challenge in this context is the fact that a wind turbine is a complex system, meaning that is not adequate to do the modal testing in a single operating regime. To this end, manufacturers often do measurement campaigns over the span of different days in several conditions (wind speeds, wind directions, ...). It would however be of great benefit to test more continuously than at present day, since current measurement campaigns do not allow to capture all the different operating conditions.

Due to desiring to move towards the continuous extraction of a machine's modal behaviour, there is opted to use Operational Modal Analysis (OMA). The use of OMA is however not trivial within the context of wind turbine gearboxes, as OMA is based on the assumption that the input force is white noise within the frequency band of interest. The multi-stage gearbox of the turbine however introduces deterministic content at a multitude of different orders. To this end, it is needed to pre-process the data in order to reduce the influence of these harmonics. In literature, multiple methods have been proposed; for example cepstral editing [7], setting the amplitude levels of the harmonics to the neighbouring amplitude levels [8] and separation of the signal in a deterministic and stochastic part [9]. The advantage of OMA compared to more classical techniques such as Operational Deflection Shape (ODS) analysis is the fact that it allows to extract the damping ratios of the different modes. The accurate extraction of the damping is however not trivial. This is due to the fact that going from the measured vibration signals to the modal model involves several pre-processing steps. These steps require to impose several input parameters, of which several can strongly affect the estimated damping ratio [10]. To this end, it is vital to investigate how to set these parameters if these results are to be used to solve NVH issues.

41.2 Methodology and Theoretical Background

41.2.1 Reduction of the Influence of Harmonic Content

As mentioned, several algorithms are available in literature to reduce the influence of the harmonic content. An additional challenge within the context of rotating machinery is the fact that the harmonic content cannot be considered to be stationary (constant in amplitude, frequency and phase), as the harmonic content is often smeared over the spectrum due to speed fluctuations. To this end, it is often proposed to do an additional pre-processing algorithm, i.e. angular resampling. In this step, the vibration data is transformed from equally spaced samples in the time domain to equally spaced samples in the angular domain. Here, the harmonic orders will be identifiable as discrete peaks, even if speed fluctuation is present. Therefore, the editing often takes place on the angular signal. Afterwards, the signal is transformed back into the time domain in order to perform the modal parameter estimation [11].

In this work, a cepstrum-based time-domain signal editing procedure is adopted [7]. Cepstrum analysis is a procedure that brings the signal from the time domain to the quefrency domain by means of a double application of the Fourier algorithm.

$$C_c(\tau) = \mathscr{F}^{-1} log(\mathscr{F}(X(t))) = \mathscr{F}^{-1} ln(A(f)) + j\phi(f) \tag{41.1}$$

In this equation, $X(t)$ is the original signal in the time domain, and $A(f)$ and $\phi(f)$ are respectively the amplitude and the phase of the frequency domain signal. By setting the phase to zero in Eq. 41.1, the formulation of the real cepstrum can be obtained (41.2):

$$C_r(\tau) = \mathscr{F}^{-1} ln(A(f)) \tag{41.2}$$

In the cepstrum, the modal content is concentrated at low quefrencies, whereas the rahmonic content will be present as discrete peaks. Under the condition that these two regions do not overlap for the frequency band of interest, a short-pass lifter can thus be applied. This will result in reducing the influence of the deterministic content, while preserving the modal content. Multiple lifter types can be used in this domain, such as a rectangular, notch and exponential lifters. With the latter, the damping of the identified modes will be altered with a known amount:

$$\xi_r = \xi_m - \frac{1}{2\pi f_r \tau} \tag{41.3}$$

In this equation, ξ_r is the real damping [−], ξ_m is the measured damping [−], f_r is the real frequency [Hz] and τ is the time constant of the exponential window [s].

41.2.2 Modal Parameter Estimation

In this paper, three different modal parameter estimation algorithms will be benchmarked against one another: the poly-reference least-square complex frequency-domain (p-LSCF, also known as PolyMax), PolyMax Plus and the Covariance-driven Stochastic Subspace Identification (SSI-COV) method. The mathematical formulations of these three estimators are now briefly discussed.

PolyMax (p-LSCF)

PolyMax is a frequency domain estimator that fits the measured FRFs using a right matrix model (Eq. 41.4). Due to its computational speed and the generation of clean stabilisation diagrams, it has become one of the more popular modal estimators.

$$[H(\omega)] = [B(\omega)][A(\omega)]^{-1} \tag{41.4}$$

In the paper, an OMA formulation of the algorithm is used. In this case, an additional assumption in the mathematical formulation in the algorithm is the fact that the input force spectrum is constant: $[S_{UU}(\omega)] = [S_{UU}]$. The relationship between the input spectra and the output spectra is then reduced to:

$$[S_{yy}] = [H(\omega)][S_{UU}][H(\omega)]^H \tag{41.5}$$

In this equation, $H(\omega) \in \mathbb{C}^{l \times m}$ is the FRF matrix containing the FRFs between the m inputs and the l outputs, $B(\omega) \in \mathbb{C}^{l \times m}$ is the numerator matrix polynomial and $A(\omega) \in \mathbb{C}^{m \times m}$ is the denominator matrix polynomial. Under the mentioned assumption, the output spectra are modelled in the following way [12]:

$$S_{YY}(\omega) = \sum_{i=1}^{n} \frac{\{v_i\}\langle g_i\rangle}{j\omega - \lambda_i} + \frac{\{v_i^*\}\langle g_i^*\rangle}{j\omega - \lambda_i^*} + \frac{\{g_i\}\langle v_i\rangle}{-j\omega - \lambda_i} + \frac{\{g_i^*\}\langle v_i^*\rangle}{j\omega - \lambda_i^*} \tag{41.6}$$

Where λ_i are the poles of the system, $\langle g_i\rangle$ are the operational reference factors and $\langle v_i\rangle$ are the mode shapes.

PolyMax Plus

PolyMax Plus can be seen as an extension of the p-LSCF algorithm discussed in the previous paragraph. PolyMax was originally developed for modal testing taking place in laboratory conditions (e.g. hammer and shaker testing), where clean FRFs are obtained. When moving towards more noisy field data, it was however observed that the estimated damping ratios were being underestimated. To this end, the algorithm was extended to PolyMax Plus [13]. Among other, an additional pre-processing step was added in order to cope with the noisy data.

This pre-processing algorithm consists of a Maximum Likelihood Estimator with as main goal to 'smooth' or de-noise the measured FRFs or power spectra. Multiple formulations are available in literature. In this paper, this is done by minimizing the following cost function:

$$l^{ML}(\theta) = \sum_{o=1}^{N_o}\sum_{i=1}^{N_i}\sum_{k=1}^{N_f} \frac{|H_{oi}^{ML}(\omega_k,\theta) - H_{oi}(\omega_k)|^2}{\sigma_{H_{oi}}^2(\omega_k)} \tag{41.7}$$

In this equation, $H_{oi}(\omega_k)$ and $\sigma_{H_{oi}}^2$ are respectively the measured set of FRFs or power spectra and their standard deviation. A Levenberg-Marquardt algorithm is used in order to have a monotonically decreasing cost function. A common denominator model is used by the ML-estimator to synthesize the denoised set of FRFs $H_{oi}^{ML}(\omega_k, \theta)$.

In the next stage, these ML-generated FRFs are given as input to the classical p-LSCF algorithm. In this way, the advantages of the MLE and the p-LSCF approach are combined with one another: a clean stabilisation is obtained with punctual values of the modal parameters. In a final step, confidence bounds can also be calculated for the estimated poles.

Stochastic Subspace Identification

The Covariance-driven Stochastic Subspace Identification method (SSI-COV) is addressing the so-called stochastic realization problem, i.e. the problem of identifying a stochastic state-space model from output-only data [14]. The technique relies on the singular value decomposition of the output covariances, which are assembled in a Toeplitz matrix $T_{1|i}$, on which a Singular Value Decomposition (SVD) is performed.

$$T_{1|i} = O_i \Gamma_i^{ref} = U_1 S_1 V_1^T \tag{41.8}$$

In the equation above, i is the Hankel coefficient and O_i and Γ_i^{ref} are respectively the extended observability and the extended controllability matrix of the system. Based on Eq. 41.8, it can be seen that these can be calculated as follows:

$$\begin{aligned} O_i &= U_1 S^{1/2} T \\ \Gamma_i^{ref} &= T^{-1} S^{1/2} V_1^T \end{aligned} \tag{41.9}$$

Based on the definition of the observability and controllability matrix, the system matrix A and the output matrix C can readily be extracted. Afterwards, the modal parameters can be extracted using an eigenvalue decomposition on the system matrix:

$$A = \Psi \Lambda_d \Psi^{-1} \tag{41.10}$$

Where Ψ is the matrix containing the eigenvectors of the identified modes and Λ_d is a diagonal matrix containing the discrete time poles.

41.3 Simulations

In first instance, the methods are benchmarked on a simulation case of a rotating machine in order to be able to compare the retrieved set of modal parameters with their simulated value. A double supported beam is therefore simulated. A fixed amount of damping (3%) is imposed for the first five modes of the beam. Twenty different responses are simulated, representing accelerometers spread equidistantly along the length of the beam. Harmonic content is added in a later stage to the response signals, as particular focus has to be paid to the performance of the cepstral liftering to remove deterministic content. The time signals $x(t)$ are then given by:

$$x(t) = x_{beam}(t) + h(t) \circledast d_{harm}(t) + n(t) \tag{41.11}$$

Where $h(t)$ is the transfer function of the simulated system, $d_{harm}(t)$ are the deterministic signals, and $n(t)$ is the measurement noise. This paper wants to focus in particular on the extraction of the modal parameters in the presence of noise and on the reduction of the influence of the harmonic content. To this end, the simulation is done in two steps. First, no harmonic content is added in order to benchmark the performance of the different estimators in the presence of noise. Second, two common lifter types are compared with one another.

41.3.1 Comparison Between the Different Modal Parameter Estimators

In this section, particular focus wants to be paid to the accuracy and robustness of the modal parameter estimators in the presence of noise. For this simulation, the structure is excited using an impulse. The cross power spectra are then estimated

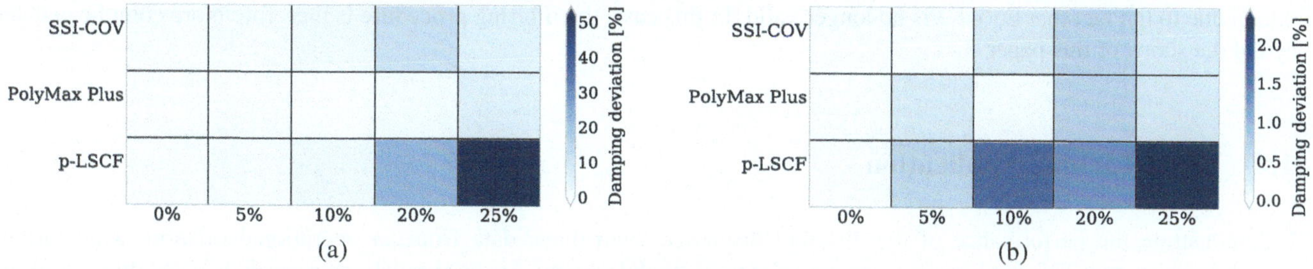

Fig. 41.1 Performance of the different modal parameter estimators as a function of the added noise power. (**a**) Absolute damping deviation for the first mode of the beam. (**b**) Absolute damping deviation for the fourth mode of the beam

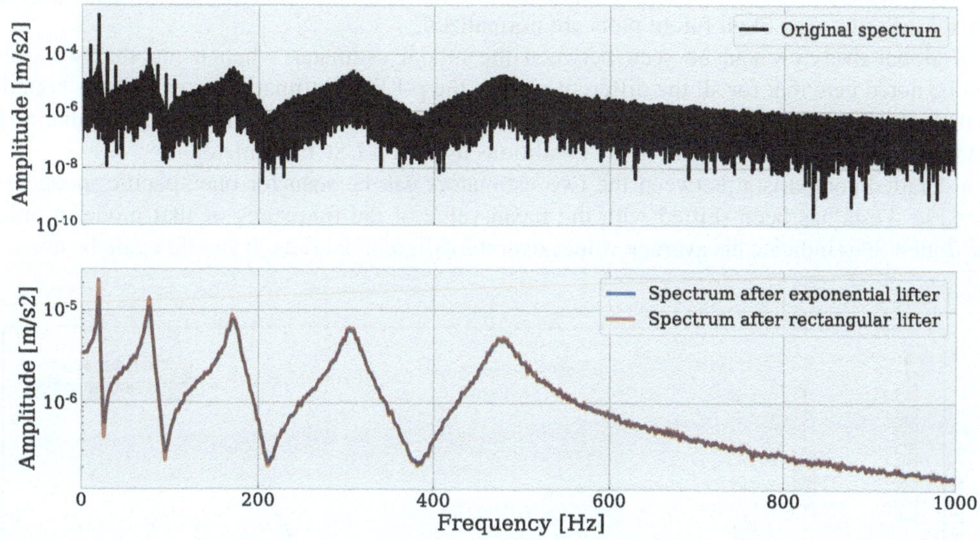

Fig. 41.2 Comparison between an exponential and rectangular lifter in case of constant speed

using a correlogram approach. This methodology is repeated for different noise levels. The results of this can be seen in Fig. 41.1. In the case of 25% added noise, the p-LSCF estimator underestimated the damping of all the modes. The error margin on the first mode was however substantially higher than the one on the others, which is due to the fact that this mode was the least well excited when looking at estimated power spectra. Even in case of noisy measurement data, accurate damping values were obtained with both the SSI-COV and the PolyMax Plus estimator. The big advantage of the latter is however that the obtained stabilisation diagram is much cleaner, resulting in easier automatization of the pole selection in the diagram.

41.3.2 Comparison Between the Different Lifter Types

As mentioned, the most common lifter types in literature are rectangular, exponential and notch lifters. When comparing the spectra before and after the cepstral liftering in Fig. 41.2, it can be seen that both the exponential and the rectangular lifter are successful in removing the harmonic content present between 0–150Hz. A big difference between the two different lifter types can be seen by comparing the amplitude levels of the first two modes. With the exponential lifter, the resonance peak levels are lower when compared to the rectangular lifter. This is because the exponential lifter increases the damping of the modes. This is especially noticeable for the lower frequency modes as the added damping is inversely proportional to the frequency of that mode (Eq. 41.3).

In this simulation case with little (<2%) speed fluctuation, the performance of the two lifter types is very similar. It is difficult to generalize which type of lifter to use, as this is a function of the modal content present in the signal and the rahmonic families present in the signal. It should also be noted that the exponential lifter should not be applied in the angular domain due to the fact that Eq. 41.3 is no longer valid. In this case, the liftering procedure is therefore more complicated and beyond the scope of this paper.

41.4 Wind Turbine Application

To demonstrate the performance of the different discussed algorithms, data from an operational offshore wind turbine gearbox is processed. This is a complex structure due to its modal density. As the simulations were done using quasi-constant speed, datasets were chosen where the turbine is at full power and where the speed fluctuation is less than 2%. Ten datasets have been investigated in this operating condition. Using this field data, no stable results were obtained with the SSI-COV method. Therefore, only the p-LSCF and the PolyMax Plus estimator are benchmarked in the discussion that follows. Due to confidentiality reasons, the axes of all future plots are normalized.

In Fig. 41.3, a global overview can be seen between the modal estimates when using the two different mentioned estimators. It can be noted here that for all the different modes, the p-LSCF estimator forms a lower bound for the damping ratio. The estimated resonance frequencies are however similar with the two estimators, indicating that no bias is present in the estimation of the resonance frequencies in noisy conditions using a p-LSCF estimator.

In Fig. 41.4, a detailed comparison between the two estimators can be seen for one specific mode of the system. The frequency axis of Fig. 41.4a has been shifted with the mean value of the frequency of that mode in order to be centered around zero. The dotted lines indicate the average values over the different datasets. It should again be noted that the retrieved

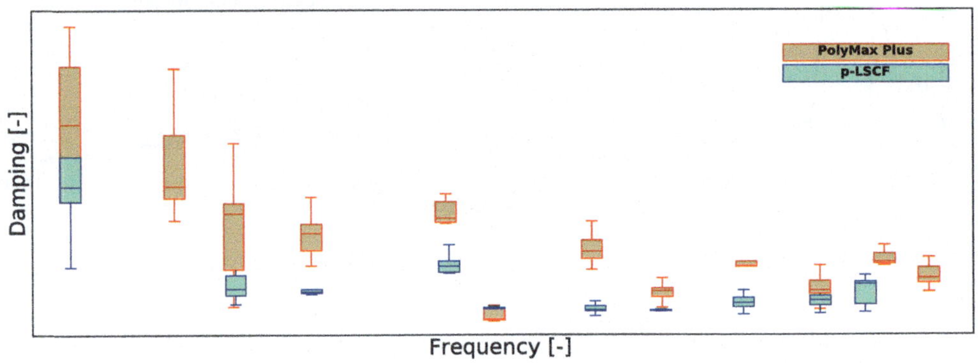

Fig. 41.3 Frequency-damping plane for the analysed frequency band. (Axes hidden for confidentiality reasons)

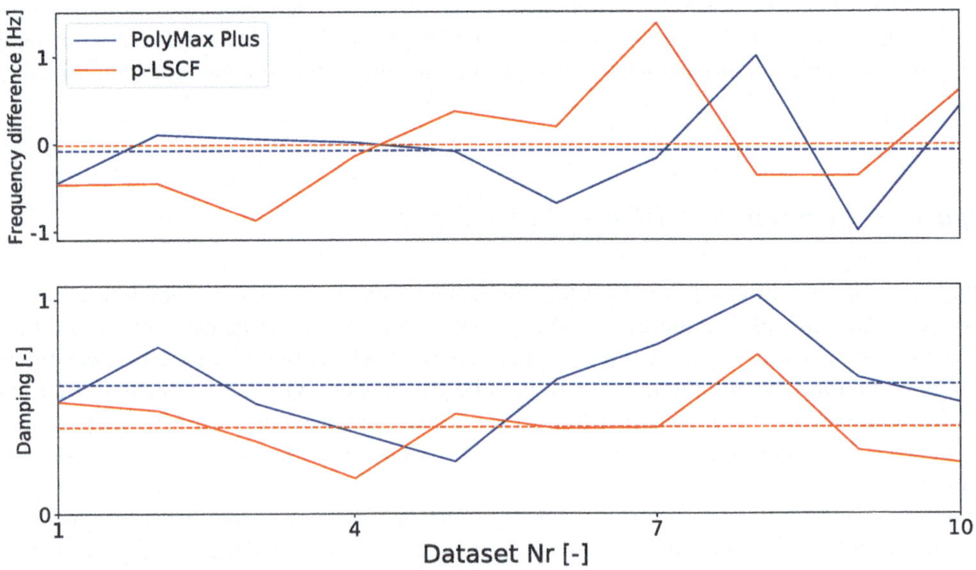

Fig. 41.4 Comparison between the estimated frequencies and damping ratios as a function of the dataset number for one specific mode

resonance frequencies match accurately. The damping of the PolyMax Plus is also consistently higher apart from one dataset. Moreover, the damping difference between the two estimators is relatively constant.

41.5 Conclusions

This paper has investigated the estimation of modal parameters of rotating machinery during their normal operating conditions. Three different modal parameter estimators were benchmarked between one another: the p-LSCF, PolyMax Plus and SSI-COV estimator. Particular attention was paid to the extraction of punctual damping values. In order to be able to compare the estimates with their correct value, simulation data was first used. This was done by simulating response data of a double supported beam. In a first step, the performance of the estimators were benchmarked between one another. Both the resonance frequencies and the mode shapes were correctly extracted with the investigated estimators. It was however observed that the damping ratios estimated using the p-LSCF estimator dropped significantly in the presence of noise.

In a second step, harmonic content was added to the simulated data in order to validate the performance of the cepstrum. The performance of the rectangular and the exponential lifter were compared to one another. No noticeable difference was however noticed in this case where the speed was nearly constant.

Finally, vibration data from an offshore wind turbine was used to extract the modal parameters in a frequency band that is typical for solving NVH problems. During this analysis, similar trends were seen compared to the simulation case.

The future work consists of better understanding the performance of the described pre-processing in case of (strong) speed fluctuations. Moreover, the results of the modal parameter estimation could be coupled with SCADA data of the turbine in order to better understand links between the operating regime of the machine and its NVH behaviour.

Acknowledgments The authors would like to acknowledge the financial support of VLAIO (Flemish Agency for Innovation & Entrepreneurship) through the SIM project MaSiWEC.

The authors also thank their partners for delivering the monitoring data.

References

1. Murdock, H., et al.: Renewables 2019 Global Status Report (2019)
2. Helsen, J., et al.: Some trends and challenges in wind turbine upscaling. In: Proceedings of ISMA International Conference on Noise and Vibration, vol. 6 (2012)
3. Sieros, G., et al.: Upscaling wind turbines: theoretical and practical aspects and their impact on the cost of energy. Wind Energy **15**(1), 3–17 (2012)
4. Witcher, D.: Uncertainty quantification techniques in wind turbine design. Presentation, DNV GL
5. Raman, G., et al.: A review of wind turbine noise measurements and regulations. Wind Eng. **40**(4), 319–342 (2016)
6. Rogers, A., et al.: Wind turbine acoustic noise. Renewable Energy Research Laboratory, University of Massachusetts at Amherst (2006)
7. Randall, R.B., et al.: A survey of the application of the cepstrum to structural modal analysis. Mech. Syst. Signal Process. **118**, 716–741 (2019)
8. Brandt, A.: A signal processing framework for operational modal analysis in time and frequency domain. Mech. Syst. Signal Process. **115**, 380–393 (2019)
9. Pintelon, R., et al.: Continuous-time operational modal analysis in the presence of harmonic disturbances. Mech. Syst. Signal Process. **22**(5), 1017–1035 (2008)
10. Gioia, N., et al.: Identification of noise, vibration and harshness behavior of wind turbine drivetrain under different operating conditions. Energies **12**(17), 3401 (2019)
11. Randall, R.B., et al.: A comparison of methods for separation of deterministic and random signals. Int. J. Cond. Monit. **1**(1), 11–19 (2011)
12. Peeters, B., et al.: PolyMAX: a revolution in operational modal analysis. In: Proceedings of the 1st International Operational Modal Analysis Conference, Copenhagen (2005)
13. Peeters, B., et al.: The new PolyMAX Plus method: confident modal parameter estimation even in very noisy cases. In: Proceedings of the ISMA (2012)
14. Peeters, B., et al.: Stochastic system identification for operational modal analysis: a review. J. Dyn. Syst. Meas. Control **123**(4), 659–667 (2001)

Jan Helsen Research team specialized in operational modal analysis and condition monitoring
Main focus on wind turbines
Prof at Vrije Universiteit Brussel
PhD from KU Leuven Department of Mechanical Engineering

Chapter 42
Investigation of Viscous and Friction Damping Mechanisms in a Cantilever Beam and Hanger System

Akhil Sharma, Aimee Frame, and Allyn W. Phillips

Abstract This paper presents an interesting mechanical system which can be used to experimentally demonstrate both viscous (Rayleigh) and friction (Coulomb) damping. The system consists of a cantilever beam, hanger and some weights. Although such a simple mechanical system is used in this experiment, two distinctly different damping mechanisms can be observed through minor changes in the structural configuration. The two structural configurations are the different orientations of hanger on the cantilever, since these orientations affect the primary mechanism of damping. During this investigation, the free response of the beam was measured in both the configurations, and damping mechanism and ratio were estimated from the time series data. For ensuring repeatability of results, the experiment was conducted on multiple combinations of three beams and hangers. The same type of phenomenon (two different damping mechanisms) was consistently observed on all three beams.

Keywords Damping mechanism · Viscous damping · Friction damping · Damping ratio · Cantilever beam

42.1 Introduction

Damping is the system property which is responsible for energy dissipation in a vibrating system. In most of the applications (automobiles, structures, machines, etc.) vibrations are not desirable; therefore, in order to minimize them, characterization and estimation of damping present in the system is important. With the ever-increasing need of lighter structures which produce minimal noise and vibration the research in this area has become more prominent. Even in the cases where vibrations are desirable, information about the damping mechanism of the system is the first step towards reducing the dissipation of kinetic energy. Although decades of research have been done in characterization of damping forces, the understanding of damping mechanisms is still far from complete. A major reason for this, unlike stiffness and inertia forces, is the state variables on which damping forces depend are generally not clear [1].

Due to the complex nature of damping phenomenon, it is very helpful if damping mechanisms can be experimentally demonstrated in an academic setting to give more insight to students. The work presented in this paper is an offshoot of an experiment conducted in the Mechatronics and Instrumentation Lab (MECH 4071C), University of Cincinnati. In this experiment, the time response (first bending mode) of a cantilever beam is measured using both an accelerometer and a strain gage by tapping the beam softly at the tip. The time response is taken for three conditions: beam only, beam with hanger and beam with known (generally 5 lbf) weight. The undamped natural frequency and damping ratio are estimated from the time response and are subsequently used for calculating the specific modulus (Modulus of Elasticity ÷Density) of the beam material, effective mass of the beam, and mass of the hanger. The damping ratio and the damping mechanism when the hanger is oriented towards the side of the beam are distinctly different when the hanger is oriented towards the front of the beam. This aspect of the experiment is the focus of this paper.

A. Sharma · A. Frame · A. W. Phillips (✉)
Structural Dynamics Research Lab (UC-SDRL), Department of Mechanical & Materials Engineering, College of Engineering and Applied Science, University of Cincinnati, Cincinnati, OH, USA
e-mail: sharmai@mail.uc.edu; aimee.frame@uc.edu; Allyn.Phillips@UC.Edu

B. Dilworth (ed.), *Topics in Modal Analysis & Testing, Volume 8*, Conference Proceedings of the Society for Experimental Mechanics Series, https://doi.org/10.1007/978-3-030-47717-2_42

42.2 Background

Dissipation of energy from a vibrating system is just the transformation of kinetic energy into some other form of energy. This transformation and the way energy is dissipated depends upon the physical mechanisms active in the structure which are complicated physical processes yet to be fully understood [1]. Moreover, there can be multiple damping mechanisms at play in a system which makes it even more difficult to understand the true nature of damping forces present in it.

The most widespread approach to model damping is to assume viscous damping, which was first introduced by Rayleigh (1877). Moreover, he proposed the damping matrix to be a linear combination of the mass and stiffness matrices. This damping model has proven to be a very good surrogate (mathematical approximation) for structural damping, and therefore, used in most commercial modal parameter estimation algorithms. However, it should be noted that structural damping is caused by internal material friction and physically, the mechanism is not similar to viscous damping mechanism. Viscous damping causes the vibration to decay exponentially as shown in Fig. 42.1a. Logarithmic decrement is used for finding damping ratio in a viscously damped system. This method postulates that the difference between the logarithms of successive peaks (or valleys) is constant. In this model the damping force is directly proportional to velocity

$$\mathrm{F_d} = \mathrm{c}\dot{x} \tag{42.1}$$

where $\mathrm{F_d}$ is damping force, c is damping coefficient and $\dot{x}$ is velocity.

This viscous damping model is not applicable in every condition, especially, when dry friction is present. Instead, the friction damping model is applicable in situations where two dry surfaces are rubbing against each other. Vibration decays in a linear manner when friction damping is predominantly present in the system as shown in Fig. 42.1b. Logarithmic decrement is not applicable to friction damping as the difference between the successive peaks, not their logarithms, is constant. In this model, the damping force is assumed to be independent of the velocity and proportional to the normal force between the sliding surfaces. The direction of force is opposite to the direction of velocity

$$\mathrm{F_d} = \mathrm{sgn}\,(\dot{x})\,\mu\mathrm{F_n} \tag{42.2}$$

where sgn is the signum (sign) function, $\mathrm{F_n}$ is the normal force, and μ is the coefficient of kinetic friction.

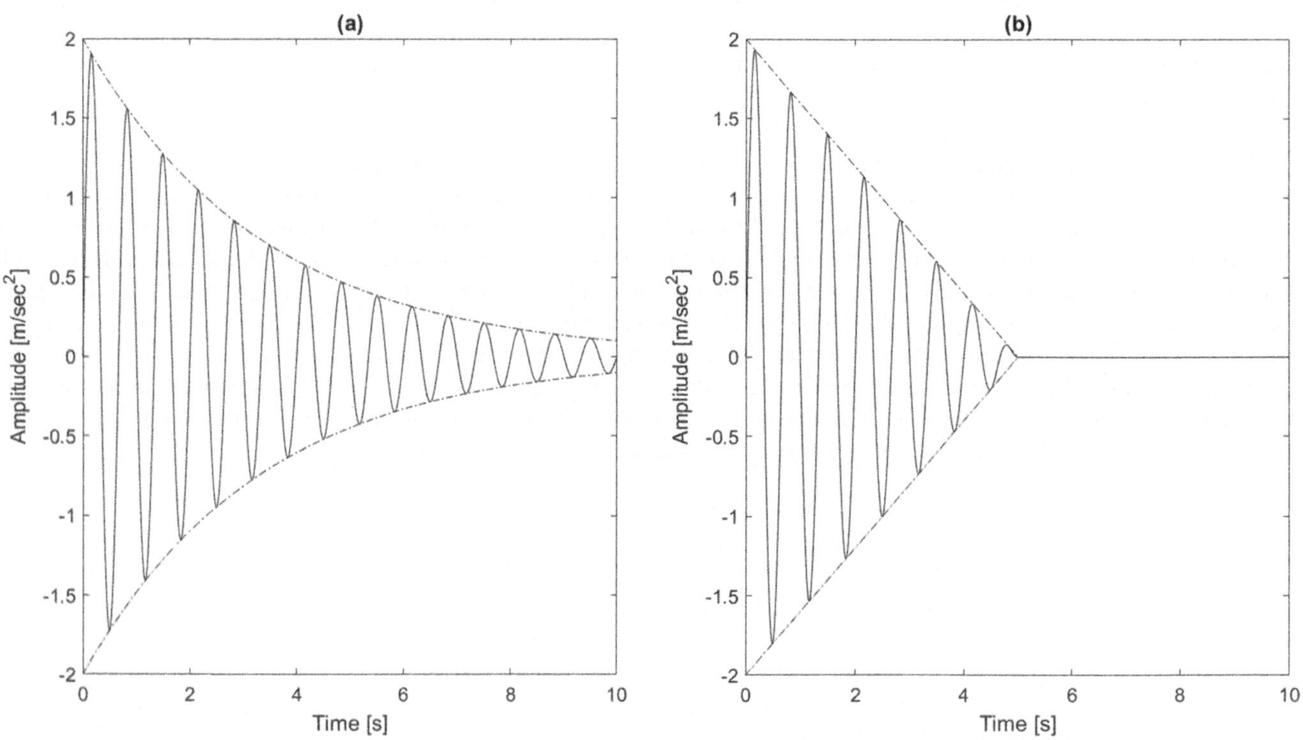

Fig. 42.1 (**a**) Exponential decay in a system with viscous damping and (**b**) Linear decay in a system with friction damping

42.3 Procedure

An acceleration measurement is taken at the tip of the cantilever beam using a small accelerometer (sensitivity 100 mv/g). A static load of approximately 1 kg force is applied using a string which is then cut, and the resulting free response of the beam is measured. The beam is tested in three types of conditions: beam only (Figs. 42.2, 42.3), beam with hanger oriented to the side (Figs. 42.4, 42.5), beam with hanger oriented to the front (Figs. 42.6, 42.7). As these two hanger configurations will be referred to often in this paper, from now onwards they will concisely be called as 'hanger side' and 'hanger front' In order to prevent rattling between the hanger and beam, a 2.83 lb. weight is also placed on the hanger. For ensuring consistency of the results, combinations of three different cantilever beams and hangers of the same nominal dimensions are measured.

Fig. 42.2 Beam only

Fig. 42.3 Detail of the hole through which hanger is suspended

Fig. 42.4 Beam with hanger oriented to the side

Fig. 42.5 Detail of hanger-beam contact for 'hanger side' case

Fig. 42.6 Beam with hanger oriented to the front

Fig. 42.7 Detail of hanger-beam contact for 'hanger front' case

42.4 Results

A representative measurement of each beam is given in the following figures (Fig. 42.8, 42.9 and 42.10). As can be seen in Fig. 42.8, the response without hanger (beam only) has a higher frequency (47.51 Hz) as compared to the cases with a hanger. This was expected because the effective mass of beam is increased due to addition of hanger, and as the mass increases the frequency will decrease. The frequency of the 'hanger side' case is 11.03 Hz and the frequency of 'hanger front' case is 10.65 Hz which are very close. However, the type of damping and the value of damping ratio in these two cases are markedly different. The damping ratio of the 'hanger front' case is significantly higher than the 'hanger side' case. Another important observation is the mechanism of damping present in these two cases: the 'hanger side' case shows exponential decay, however, the 'hanger front' case seems to have linear decay. These observations will be further explored in the analysis section. Although the values of damping ratio differs between the beams in Figs. 42.8, 42.9 and 42.10**,** it can be definitely said that the beams follow a similar trend.

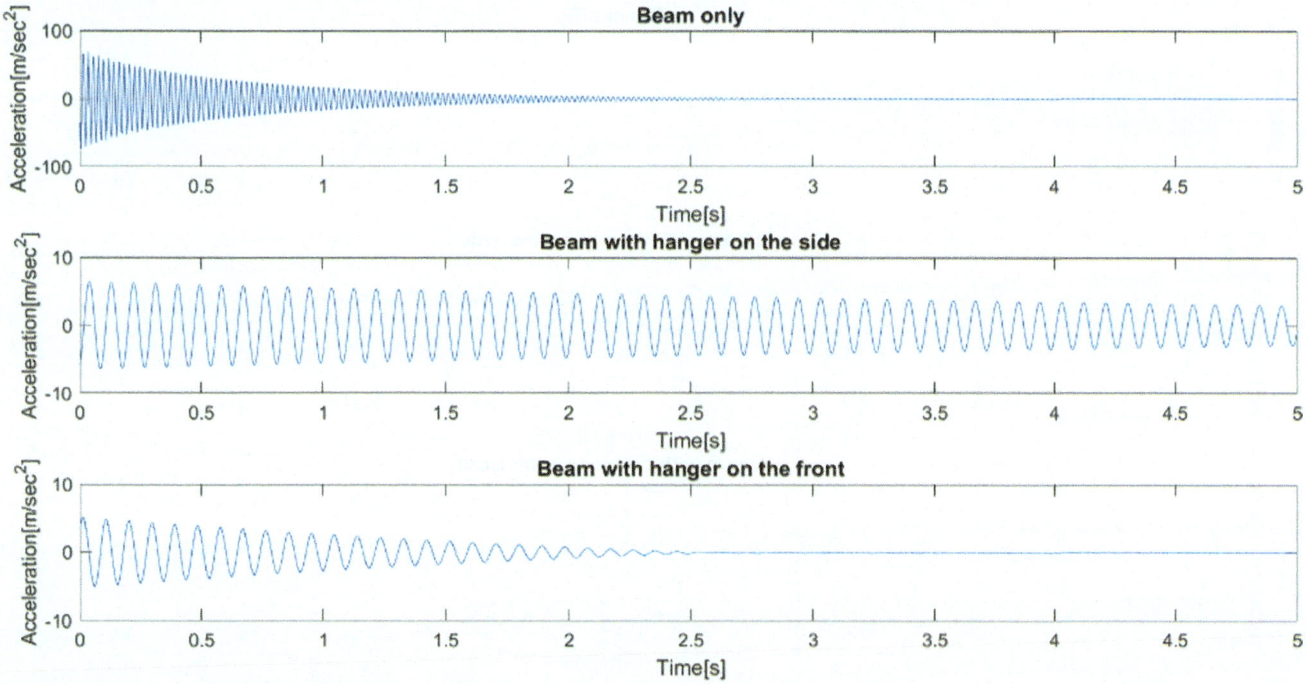

Fig. 42.8 Cantilever beam 1 free response

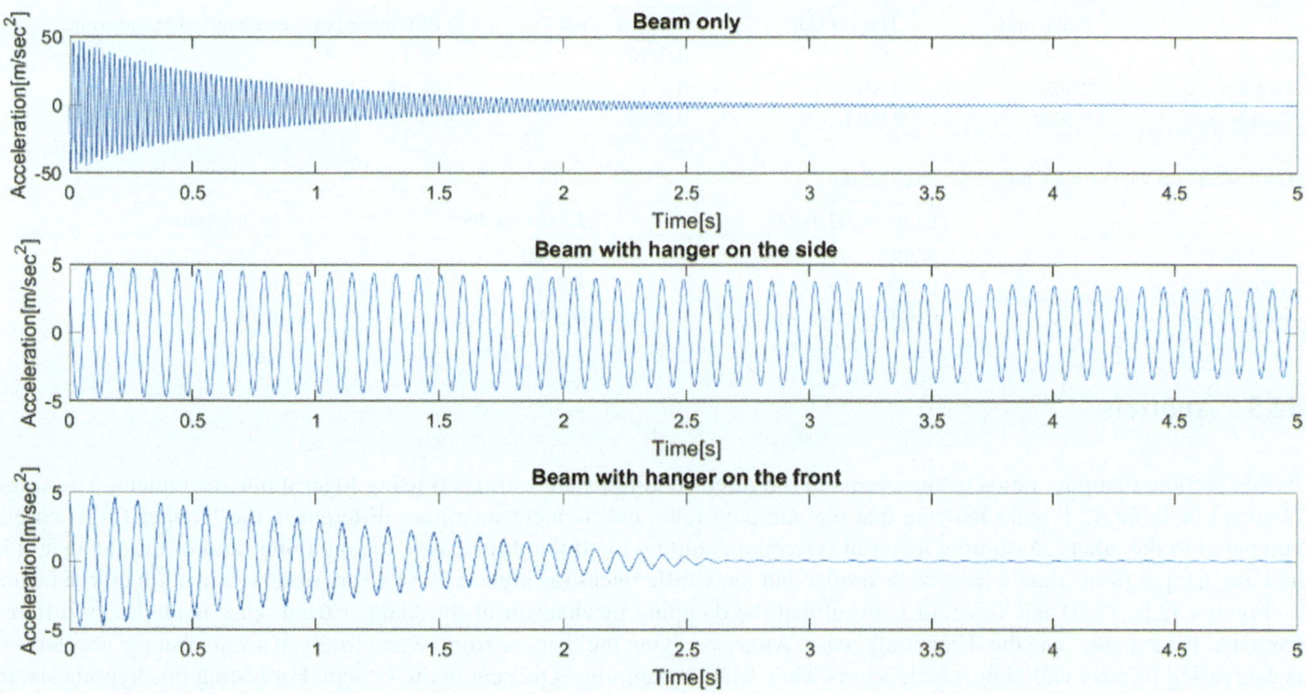

Fig. 42.9 Cantilever beam 2 free response

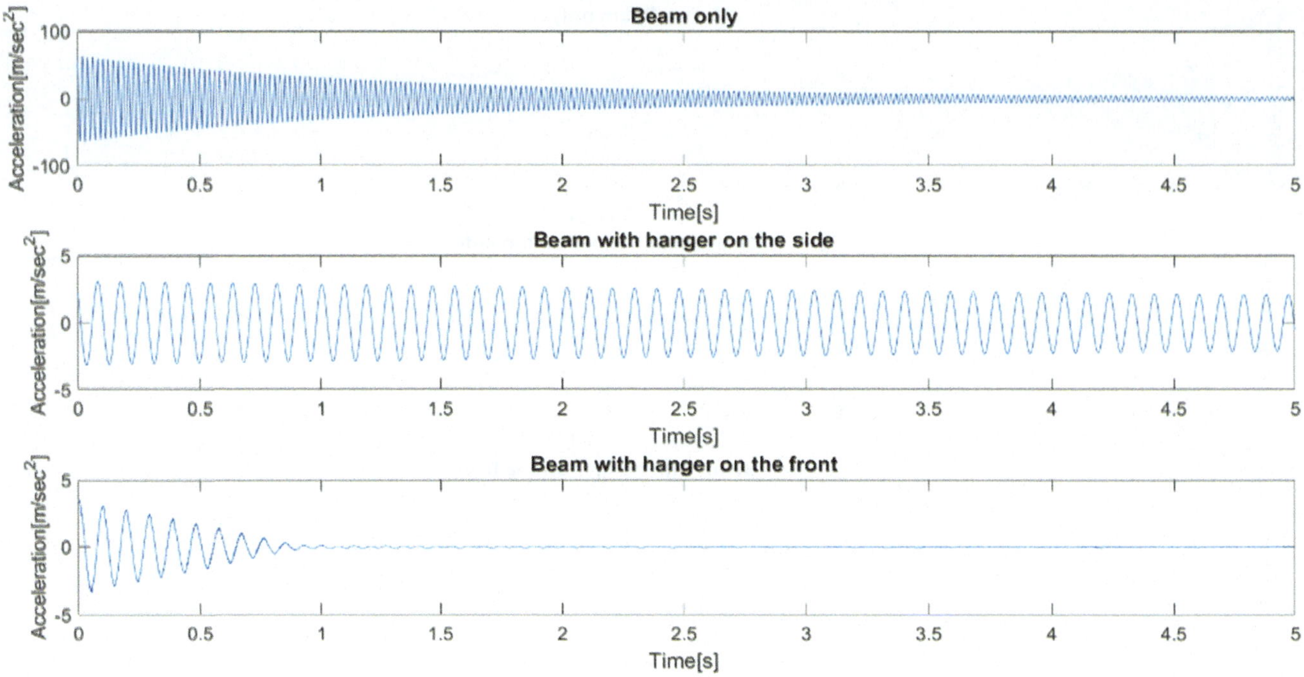

Fig. 42.10 Cantilever beam 3 free response

Table 42.1 Damping ratios of beams under different conditions

	Beam only	'Hanger side'	'Hanger front'	% difference between hanger side and front
Cantilever 1	0.0044	0.0024	0.0110	358.33
Cantilever 2	0.0038	0.0011	0.0100	809.09
Cantilever 3	0.0021	0.0011	0.0226	1954.55

Table 42.2 R2 value for the three cantilever beams

	Exponential model	Linear model	% difference
Cantilever 1	0.9918	0.9950	0.32
Cantilever 2	0.9904	0.9958	0.54
Cantilever 3	0.9416	0.9491	0.79

42.5 Analysis

In this section, damping ratios of the beams in the three conditions are estimated using logarithmic decrement. The values obtained in Table 42.1 again indicate that the damping ratio and its mechanism are different in the 'hanger front' case as compared to the others. Assuming material (viscous) damping in all the three cases, the damping ratio of the 'hanger side' and the 'hanger front' should have been similar, but large differences (as high as 1954%) are observed in all the three beams.

Figures 42.8, 42.10 and Table 42.1 reveal that the damping mechanism of the 'hanger front' case is markedly different from the 'hanger side' and the 'beam only' case. After observing the 'hanger front' case closely, it seems that the acceleration is decreasing linearly with time which occurs when friction damping is present in the system. For testing this hypothesis, the curve fitting tool in MATLAB (command: 'cftool') was used on the data and both types of model (exponential and linear) were fitted. This tool uses a non-linear least squares method to minimize noise in the data and produce the best possible fit. Custom equations (shown in the 'Results' pane of Figs. 42.11 and 42.12) have been provided as input to the tool for both type of models.

A cursory look at Figs. 42.11 and 42.12 gives the impression that both models identically fit the noisy data. However, after carefully noting down the goodness of fit parameters, it becomes obvious that the linear model is more suitable compared to the exponential model. For a better fit, the R^2 value (coefficient of determination) should be as high as possible and the SSE (Sum of Squares of the Errors) should be as low as possible. These values for the three cantilever beams are compared in Tables 42.2 and 42.3 respectively. In all the three cases the R^2 value is higher and the SSE value is lower for a linear model confirming that the linear model best fits the data.

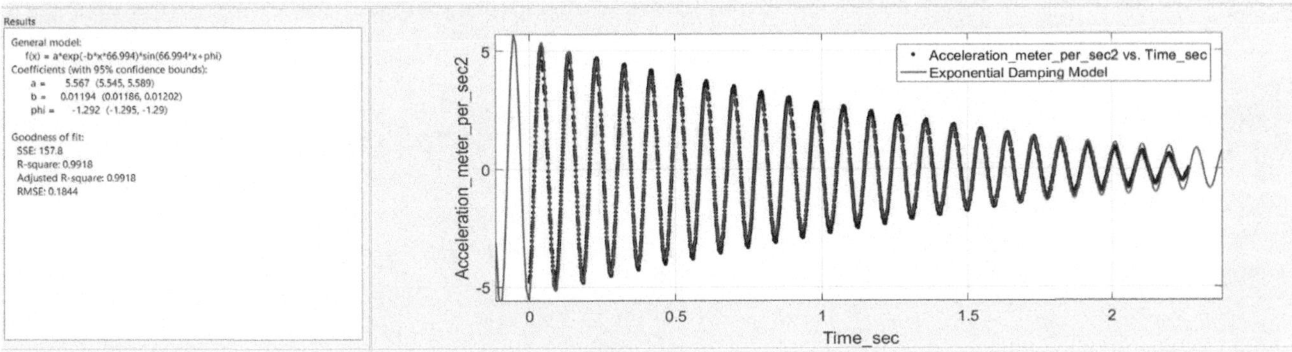

Fig. 42.11 Exponential damping model applied on 'hanger front' case

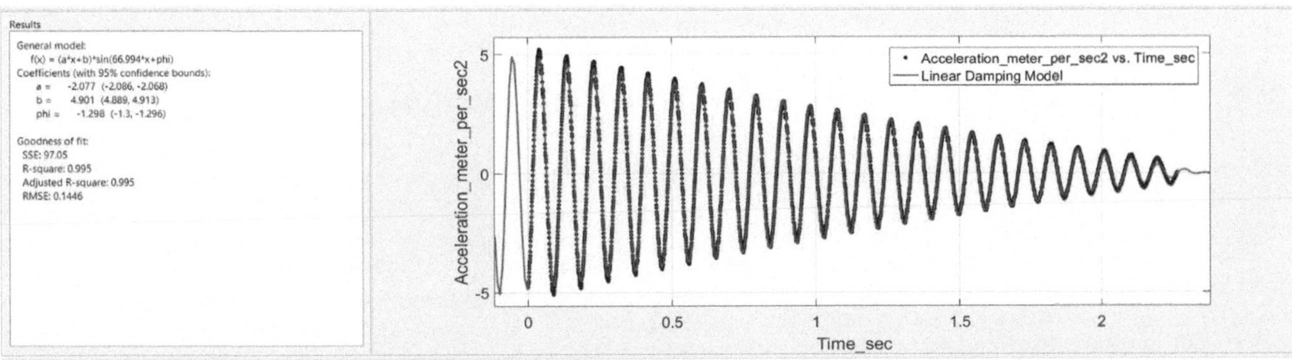

Fig. 42.12 Linear damping model applied on 'hanger front' case

Table 42.3 SSE value for the three cantilever beams

	Exponential model	Linear model	% difference
Cantilever 1	157.8	97.05	38.50
Cantilever 2	187.6	82.83	55.85
Cantilever 3	354.4	308.5	12.95

42.6 Conclusion and Future Work

The cantilever beam and hanger system presented in this paper seems to be a very good way of demonstrating both viscous (Rayleigh) and friction (Coulomb) damping mechanisms. It can also be effectively used as a demo in instructional courses on vibrations. It should be noted, however, that a combination of these mechanisms may be present in the system but only the predominant one is observable. Although using logarithmic decrement in a system with friction damping is not a perfect approach, nevertheless, it gave an idea about the approximate value of damping ratio of the system. The next goal is to investigate the reasons behind the dramatic change of damping mechanisms observed in this paper. It is postulated that the difference in damping mechanism for the two orientations is due to the type of two-point contact between the beam and the hanger. For the 'hanger side' case, the motion of the beam in oscillation results in a rocking motion against the hanger since the two points of contact (see detail of the contact in Fig. 42.5) have ideally the same motion. However, for the 'hanger front' case, the small differential motion between the two points of contact (see detail of the contact in Fig. 42.7) result in a friction contact. For testing this hypothesis, an analytical and experimental study of the contact details of the beam and the hanger is in process.

Reference

1. Adhikari, S.: Damping models for structural vibration. PhD dissertation, University of Cambridge, pp. 1–5, 2000

Chapter 43
Detecting Nonsynchronous Heart Cells from Video – An Unsupervised Machine Learning Approach

William Anderson, Lauren Schneider, Li-Ming Richard Yeong, Kent Coombs, Pulak Nath, Jennifer Harris, David Mascareñas, and Bridget Martinez

Abstract Cardiac myocytes possess the property of automaticity; however, irregularities in the automaticity can lead to cardiac arrhythmias and heart blocks, which can cause life-threatening health concerns. Currently, full-field, high-resolution, anomaly detection methods for non-contact, electro-mechanical dynamics of cardiac myocytes do not exist. This research uses emerging structural dynamic techniques coupled with cutting edge signal processing methods to examine the dynamics of beating heart cells. Applications of this research could offer novel advances in medical diagnostics for heart disorders. A properly functioning heart will have cells that contract together in one global motion. However, in certain pathologic conditions, one cell –or a collection of cells – will beat out of phase with the rest of the heart. Using a combination of blind source separation techniques (non-negative matrix factorization, sparse and low-rank matrix decompositions) and mutual information on videos, these nonsynchronous cardiac myocytes can be identified.

Keywords Signal processing · Machine learning · Non-negative matrix factorization · Sparse and low-rank matrix decompositions

43.1 Background

Cardiac muscle is composed of cardiac myocytes and proper function of these myocytes is critical for heartbeat regulation. Two characteristics that attribute to this regulation are the cells' auto-rhythmicity and connectivity. The former enables cardiac myocytes to excite themselves spontaneously and spark action potentials – electric signals; the latter propagates action potentials from one cell to another across gap junctions [1]. This is the essence of cardiac automaticity, which results in a natural, heart-beating rhythm.

In addition to the automaticity and ability of cardiac cells to beat alone, the heart has a conduction system that can regulate regional contraction. The sino-atrial (SA) node – the heart's pacemaker – triggers cardiac contraction and depolarizes in response to spontaneous calcium ion release from the sarcoplasmic reticulum (SR) within the cardiac myocytes [2]. Since depolarization occurs at a much faster rate than other parts of the conduction system, action potentials originate at the SA node and follow a pathway to the atrio-ventricular (AV) node, bundle branches, apex, and Purkinje fibers [1] [3]. These conduction myofibers directly stimulate the cardiac myocytes in the ventricles to contract, causing blood to pump and flow from the heart to the rest of the body [3].

W. Anderson
School of Biological and Health Systems Engineering, Arizona State University, Tempe, AZ, USA
e-mail: wanderson@lanl.gov

L. Schneider
Department of Computer Science, Maseeh College of Engineering & Computer Science, Portland State University, Portland, OR, USA
e-mail: lschneider@lanl.gov

L.-M. R. Yeong
Department of Mechanical Engineering, Bourns College of Engineering, University of California, Riverside, CA, USA
e-mail: lyeong@lanl.gov

K. Coombs · P. Nath · J. Harris · D. Mascareñas · B. Martinez (✉)
Los Alamos National Laboratory, Los Alamos, NM, USA
e-mail: coombske@lanl.gov; pulakn@lanl.gov; jfharris@lanl.gov; dmascarenas@lanl.gov; bmartinez26@ucmerced.edu

B. Dilworth (ed.), *Topics in Modal Analysis & Testing, Volume 8*, Conference Proceedings of the Society for Experimental Mechanics Series, https://doi.org/10.1007/978-3-030-47717-2_43

An electrocardiogram (ECG) trace displays the sequence of depolarizations and repolarizations of the heartbeat as a function of voltage versus time [4]. The three stages of a single heartbeat are as follows: atrial depolarization, ventricular depolarization, and atrial and ventricular repolarization [3]. On an ECG trace, they are represented as P waves, QRS complexes, and T waves, respectively [4]. Normal ECG traces have a fixed amplitude and period. Irregularities in the heart can be detected by observing an abnormal ECG trace. For example, in the case of a complete heart block, the P wave is independent from the QRS complex [5]. Another example is the absence of distinct P waves in ECG traces, which is indicative of atrial fibrillation [6]. While ECG traces are often successful in tracking the sequences, minute anomalies in the large-scale cardiac muscle pose a challenge in being seen in ECG traces. These anomalies may indicate minor cases of cardiac arrhythmias depending on their frequencies relative to the global beating of the heart.

43.2 Motivation

Currently, electrocardiography (ECG) is the main method of capturing electrical activity of the heart. For such cases of cardiac arrhythmias, it is more difficult for ECGs to obtain viable cell data before medical professionals can diagnose patients with a specific arrhythmic disorder. Several video processing methods exist that are designed to detect heart cell motion from videos to improve upon the conventional standards of measuring heartbeats with an ECG machine. One method employs a movement detection algorithm on video data of zebrafish (*Danio rerio*) to analyze their heartbeats [7]. This study was successful in extracting pixel intensity changes as a function of time from the video and using that data to quantify consecutive heartbeats of the zebrafish. However, several drawbacks arose from this method. If the zebrafish moved during the video, the data would be corrupted. Moreover, high image resolution (>640 × 480 pixels) slowed the detection procedure and reduced the number of images that could be analyzed.

Laboratory-based organs or organs on a chip have been developed for advanced pharmaceutical assessment and can be utilized to screen for potential cardiotoxicity. Microscopic assessment of these devices is utilized in assessment of beat frequency and motion. A motion detection algorithm known as MUSCLEMOTION was successful in measuring movement of cardiac myocytes from videos of one-dimensional *in vitro* models, two-dimensional *in vitro* models, and three-dimensional *in vitro* and *in vivo* engineering heart tissues and zebrafish [8]. The algorithm quantified a pixel's displacement in the current frame to a reference frame. From this data, the algorithm could calculate the velocity of the cardiac myocytes as well as their contractile force. However, MUSCLEMOTION could not detect vibrational or translational motion of cardiac myocytes, and the frame rate of the video was recommended to be greater than 70 frames per second to ensure accurate calculations. Another research group was able to track motion of sarcomeres treated with a fluorescent tracker in human induced pluripotent stem cell-derived cardiomyocytes (hiPSC-CMs) by means of a new MATLAB-based software called SarcTrack [9]. The current technology for detecting heartbeats through video has potential for improvement. Our proposed solution is to develop a more accurate and sensitive approach to detect unnoticeable anomalies of individual heart cell contractions via signal processing and unsupervised machine learning techniques in videos.

43.3 Methodology

43.3.1 Video/Data Collection

Videos of Cor4U cardiomyocyte cells (N-Cardia) in both a heart bioreactor and tissue culture polystyrene (TCPS) were obtained from Los Alamos National Laboratory's artificial organ program – the Advanced Tissue-engineered Human External Network Analyzer (ATHENA) program. The setup of capturing these videos included a ZEISS Axiovert 40C inverted microscope with an Edmund Optics (EO) color USB camera attached to one of the eyepieces (Fig. 43.1).

Each system comprised of a bioreactor or TCPS seeded with cardiac myocytes, which were then imaged using optical techniques. Pre-treatment videos showed heart cells beating spontaneously. The cells were then treated with different drugs to alter cardiac arrhythmic effects. Isoproterenol is a clinical drug given to patients who have either low blood pressure or a slow heart rate [10]. A dose of isoproterenol will increase the overall heart rate and treat cardiac arrhythmias such as bradycardia. Epinephrine is another drug that produces similar effects to isoproterenol. In addition, quinidine and propranolol serve to perform the opposite function, decreasing the overall heart rate. Post-treatment videos conveyed how the synthetic heart cells would react after a 30-min exposure to one of the drugs. All possible combinations of incubation and clinical drugs were considered. The difference in color from pre-treatment (purple) to post-treatment (red) was not due to the presence of a

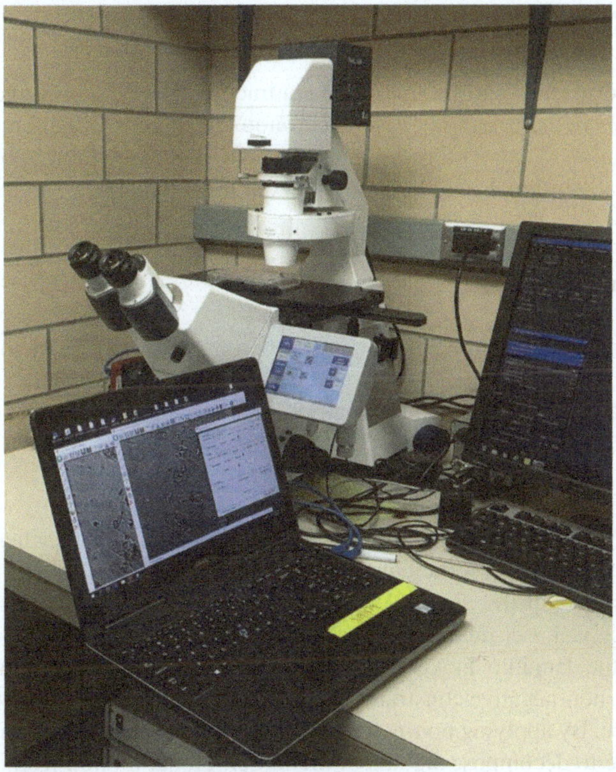

Fig. 43.1 Microscope setup to record videos of Cor4U cardiomyocyte cells

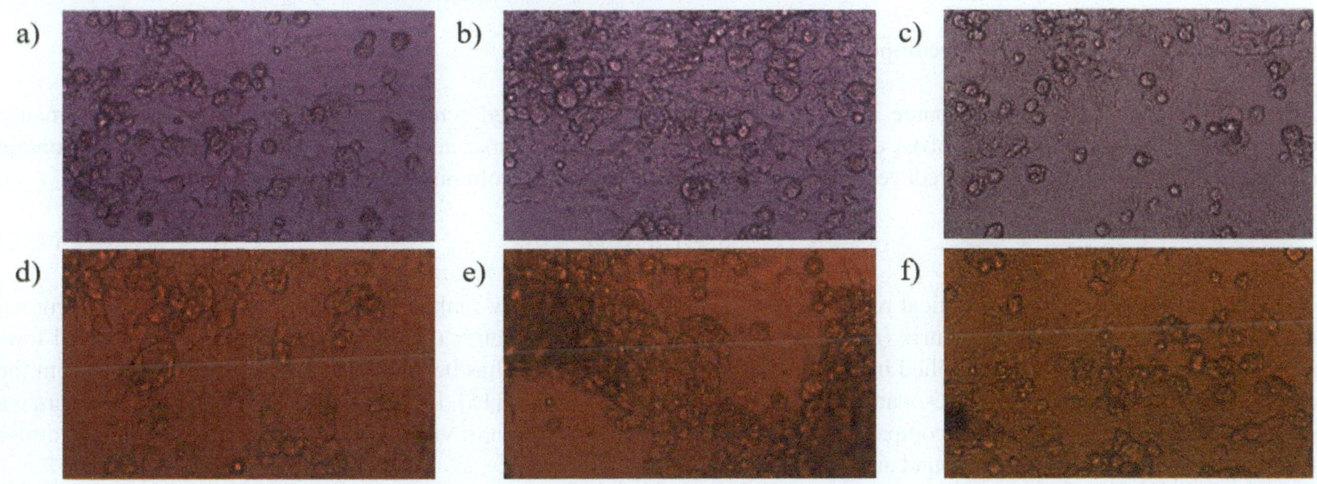

Fig. 43.2 Top row: screenshots of synthetic heart cells that were incubated in a (**a**, **b**) heart bioreactor and (**c**) TCPS; Bottom row: screenshots of synthetic heart cells injected with (**d**) isoproterenol, (**e**) epinephrine, and (**f**) quinidine

drug but was a result of an automatic color exposure from the uEye computer software (Fig. 43.2). The color would change depending on how much light was illuminated on the sample. Analysis was done on these videos to study pixel characteristics with the intention of isolating the local motion of the nonsynchronous heart cells. Results in this paper will primarily focus on (a) Video A from Fig. 43.2.

43.3.2 Signal Processing

Videos of the heart at a cellular level show cardiac myocytes contracting as a calcium signal is produced. To better study and quantify these small perturbations in the beating of nonsynchronous heart cells, we attempt to separate and process various signals from the videos. The solution to extract these signals requires implementation of blind source separation techniques, which are the decoupling of unknown signals that have been mixed in an unknown manner [11].

Non-negative Matrix Factorization

One common method to decompose videos into components is non-negative matrix factorization, which is a linear algebra algorithm that breaks a matrix into two matrices that only contain positive elements [12]. As a type of dimensionality reduction, this method is advantageous in both creating matrices that are much simpler to analyze and interpret.

$$V \approx WH \tag{43.1}$$

The governing equation for non-negative matrix factorization is listed as Eq. (43.1) where V is the original matrix ($n \times m$), W is a features matrix ($n \times r$), and H is a coefficients matrix ($r \times m$) [12]. In this case n is the number of frames in the video and m is the number of pixels in the video. The r value represents the rank or number of components to extract from the original V matrix, which is a user-defined input. Non-negative matrix factorization has been used for facial image reconstruction by decomposing the data into an $n \times r$ matrix where the n column contains non-negative pixel values of one of the r feature images [13]. The non-negative constraints on the V matrix ultimately allow for the combination of localized features to represent a human face. By applying non-negative matrix factorization on the intensity time series associated with each pixel in the videos, we hope to decompose signals of the larger, global motion from those of the smaller, local motion of the nonsynchronous heart cells.

Sparse and Low-Rank Matrix Decompositions

Another method used for blind source separation is a form of compressed sensing called sparse and low-rank matrix decompositions. This method involves decomposing a matrix into its sparse and low-rank components, which contain elements of mostly zeros and with reduced number of linearly independent columns, respectively [14].

$$S + LR \approx D \tag{43.2}$$

Equation (43.2) is the mathematical representation for the sparse and low-rank decompositions where D is the original matrix ($n \times m$), S is the sparse matrix ($n \times m$), and LR is the low-rank matrix ($n \times m$) [14]. Currently, sparse and low-rank decompositions have been applied in many disciplines. In radiology, it has been used to separate background from the dynamic components in magnetic resonance imaging (MRI) scans of organs [15]. In acoustics, it has been used to separate a singing voice from instrumental accompaniment in songs [16]. In application to video processing, this method has promise for decomposing video into foreground and background.

43.4 Results

To be able to successfully detect nonsynchronous heart cells in videos and isolate local motion, we have developed a new approach as shown in Fig. 43.3. This approach utilizes dimensionality reduction and decomposition techniques to manipulate the original videos into refined videos that carry more sensitive information about the nonsynchronous heart cells. After running the two signal processing techniques, mutual information is used to compare their differences.

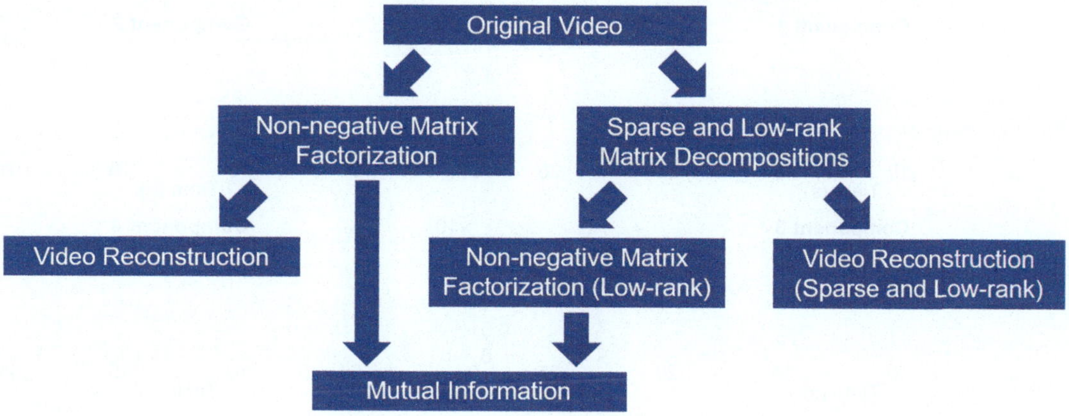

Fig. 43.3 Methodology

43.4.1 Non-negative Matrix Factorization on Videos

A V matrix of dimensions $n \times m$ was first constructed from the data for each video with n rows containing pixel intensities for m number of pixels. Matrix V was then decomposed into two approximated matrices, W and H, using a least squares approximation. Matrix W contains r observations that can be reconstructed into a new video using the weights of matrix H. Several values for rank r ranging between 2 and 60 were tested. The greater the rank, the more likely that local motion would be captured in some of the components. Signals from the weight matrix, H, were plotted to identify potential anomalies. Comparing Fig. 43.4 with Fig. 43.5, it is difficult to say whether a specific component corresponds to a nonsynchronous heart cell. However, observing and comparing peak patterns can narrow down the components to focus on. All ten components in Fig. 43.4 share similar patterns, which allude to the global motion of heart cells. Multiplying the extracted features, W, by its corresponding coefficients matrix, H, gives a video reconstruction containing only the signal from that individual component. When attempting to reconstruct Video A with the same ten components, the local motion was completely removed. On the other hand, Fig. 43.5 has several distinct components (i.e. components 20, 22, and 29), of which one may correspond to a nonsynchronous heart cell. Reconstructing all 30 components resulted in a new video that was similar to the original.

Since it is often challenging to isolate the local motion by reconstructing a new video with hand-selected components, a different approach to visualize cell motion for each component from non-negative matrix factorization is proposed. Instead of analyzing the weight matrix, H, to output signals for each component, we can focus on the spatial matrix, W, to produce images with scaled colors that are a measure of pixel contribution. As expected from Fig. 43.6, approximately the first half of the components correspond to global motion since most of the cardiac myocytes are present. Additionally, nonsynchronous heart cells are found in component 22 (Fig. 43.7). In Video A, since the nonsynchronous heart cell on the right is beating more profoundly than the one of the left, non-negative matrix factorization generates a more defined structure for that cell.

For videos with multiple cells beating at various frequencies, they are more likely to be scattered across multiple components. If a nonsynchronous heart cell is larger in size relative to the other cells in the videos, then its discrete motion can be perceived more clearly in the image with scaled colors. On the contrary, if the nonsynchronous heart cell is faint within the image, then it is suggested for visualization purposes to crop the video and rerun the program. Doing so magnifies the local motion at a greater scale, which in turn, enhances the features of the anomaly.

A consequence to using non-negative matrix factorization is that the algorithm is stochastic, meaning that the components – signals and images – created are always random each time the program is run. Depending on the number of components, the outputs will not consistently capture the anomalies. Moreover, there are tradeoffs when choosing the rank. Performing non-negative matrix factorization with a large rank yields more sensitive components that may likely capture the nonsynchronous heart cells; however, running the program for higher rank will require more computational resources. While this method has some success in pinpointing the exact location of the nonsynchronous heart cells, the next step is to reconstruct the video to isolate local motion.

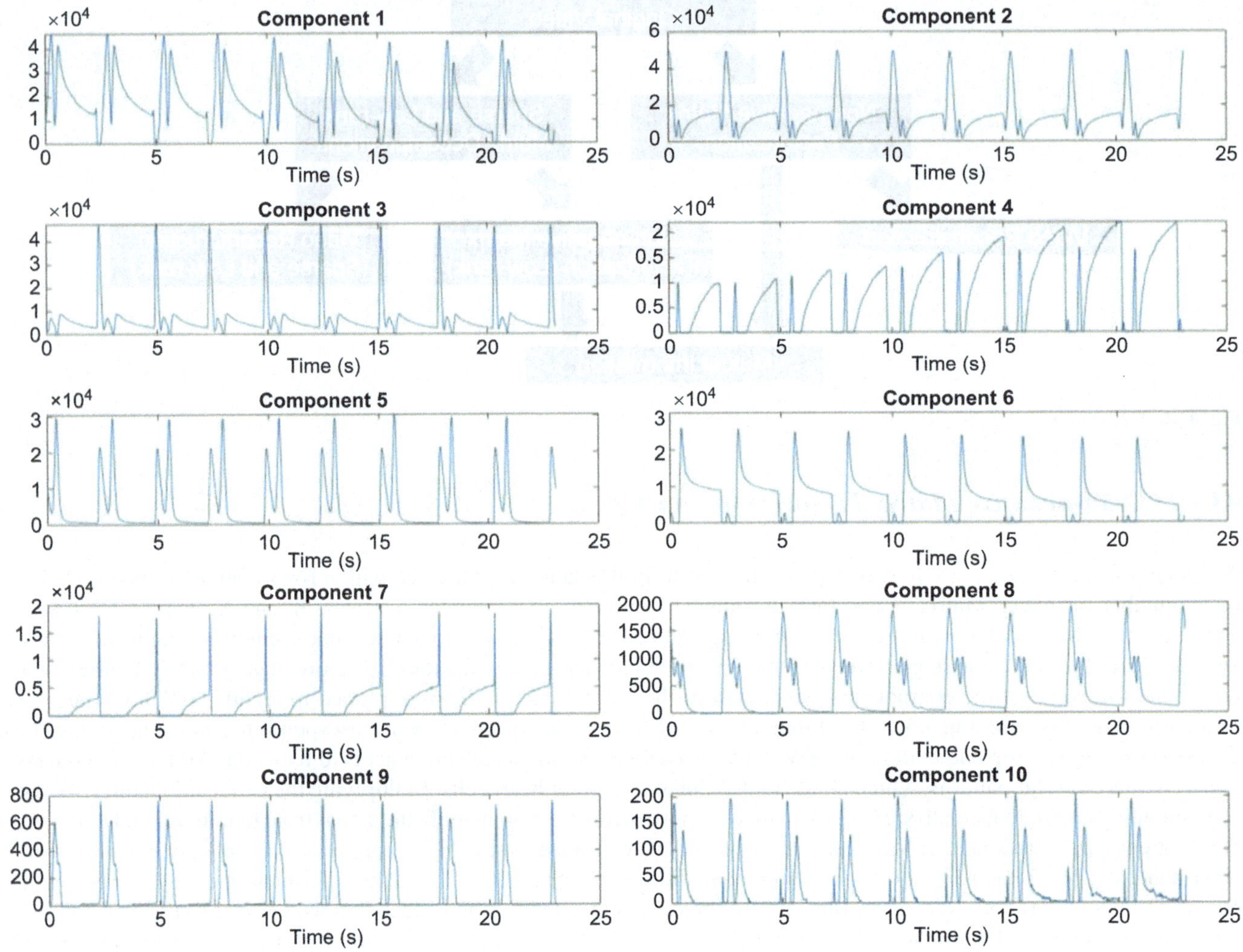

Fig. 43.4 The first ten components of the weight matrix, *H*, from non-negative matrix factorization for Video A

43.4.2 Sparse and Low-Rank Matrix Decompositions on Videos

In video processing, sparse and low-rank matrix decompositions are good for distinguishing between background (frame) and foreground (motion); however, they are unable to differentiate between local and global cell motion. Nonetheless, this method is an alternative to non-negative matrix factorization in extracting various signals. The mean was subtracted from each pixel in the video data before the sparse and low-rank matrices were obtained. When the low-rank data was reconstructed back into a video, only global motion was present. In contrast, the sparse data reconstruction proved to be better in conveying both local and global motions (Fig. 43.8).

43.4.3 Non-negative Matrix Factorization on Low-Rank Components

Since the low-rank components produced a video with a few shifting frames containing global motion, it is expected that performing non-negative matrix factorization on the low-rank components and plotting the new components would result in Fig. 43.9. Components 1 through 4 evidently show the sharp peak patterns relating to the global motion; components 5 through 30 have zero value. This is because the non-negative matrix factorization algorithm could not converge to a solution, most likely because the data only has rank 4 as opposed to 30.

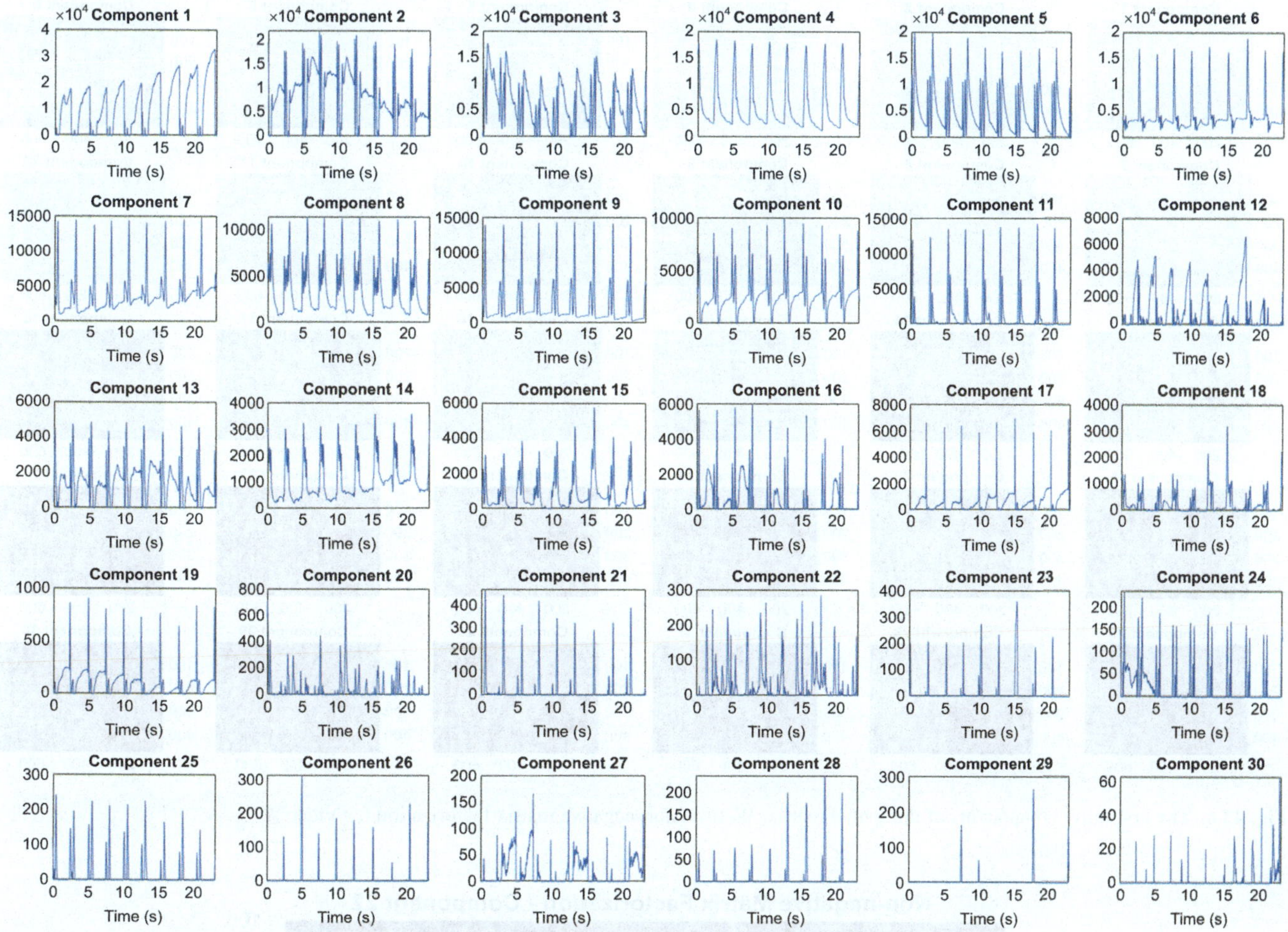

Fig. 43.5 The first thirty components of the weight matrix, H, from non-negative matrix factorization for Video A

43.4.4 Mutual Information

In order to compare the two blind source separation techniques, mutual information between the non-negative video components and the low-rank non-negative components was studied. Mutual information was calculated by finding the joint probabilities of each component of the original video to each component of the low-rank video in hopes of isolating components with distinct behaviors such as movement related to the nonsynchronous cell motion. In Fig. 43.10, the portions to highlight are the first four columns since they align with the first four low-rank components from Fig. 43.9. By observing the mutual information values corresponding to the color spectrum from yellow to blue, it was conjectured that high mutual information (between 0.5 and 1) translates to global motion whereas low mutual information (between 0 and 0.5) translates to either local or no motion.

By reconstructing the components in areas of high mutual information, we were able to view all cardiac myocytes beating together in a large, uniform motion. Areas of low mutual information were found to contain aspects of both local and global motion. For Video A, only component 23 revealed the motion of the nonsynchronous heart cell in both reconstructed videos. Although this method successfully narrowed down the data that contained the nonsynchronous motion, it was not ideal in isolating the local motion because many other components with low mutual information (i.e. components 12, 26, 27, 28, 29, and 30) still contained some aspects of global motion.

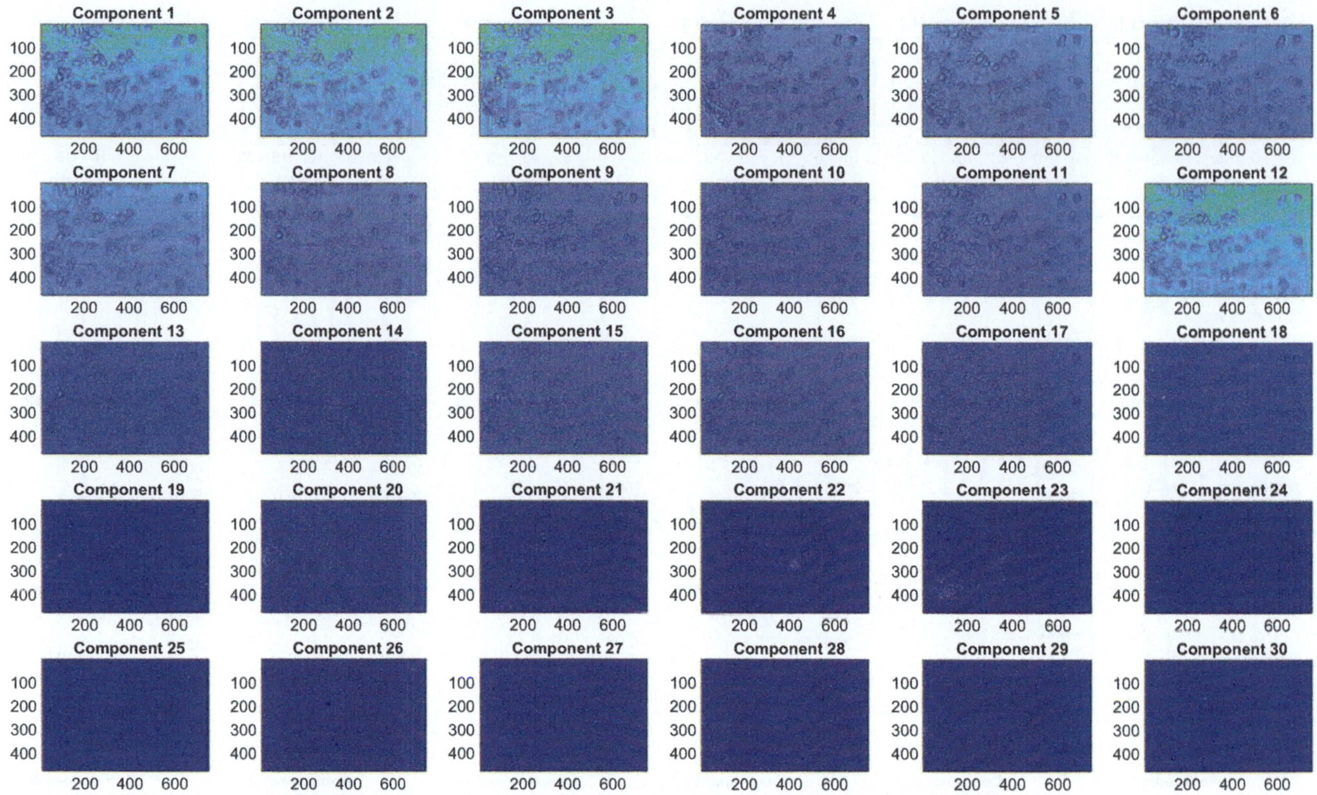

Fig. 43.6 The first thirty components of the spatial matrix, *W*, from non-negative matrix factorization for Video A

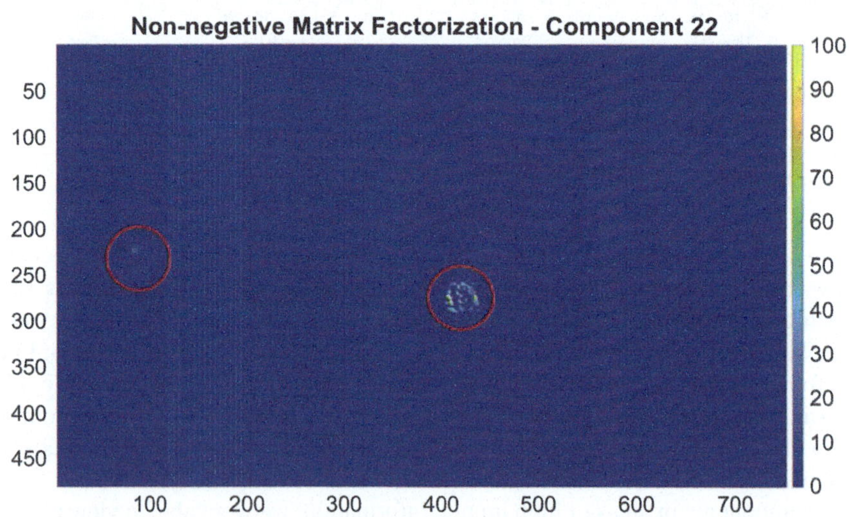

Fig. 43.7 Image with scaled colors displaying nonsynchronous heart cells

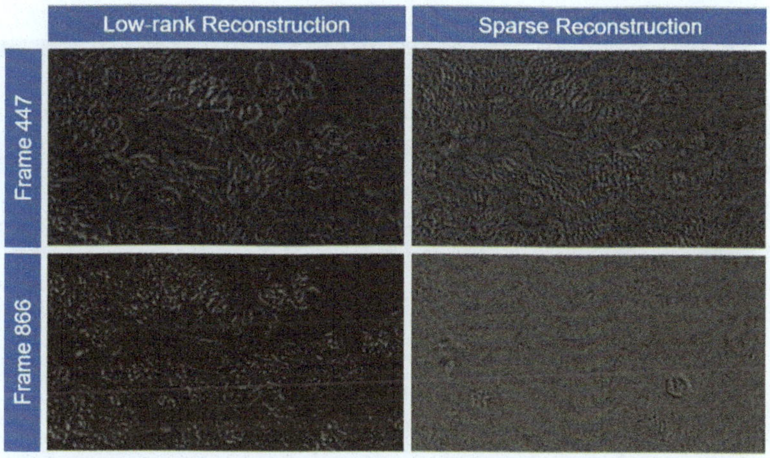

Fig. 43.8 Left column: two frames where only global motion is depicted after low-rank reconstruction for Video A; Right column: (top) global motion and (bottom) local motion are depicted after sparse reconstruction for Video A

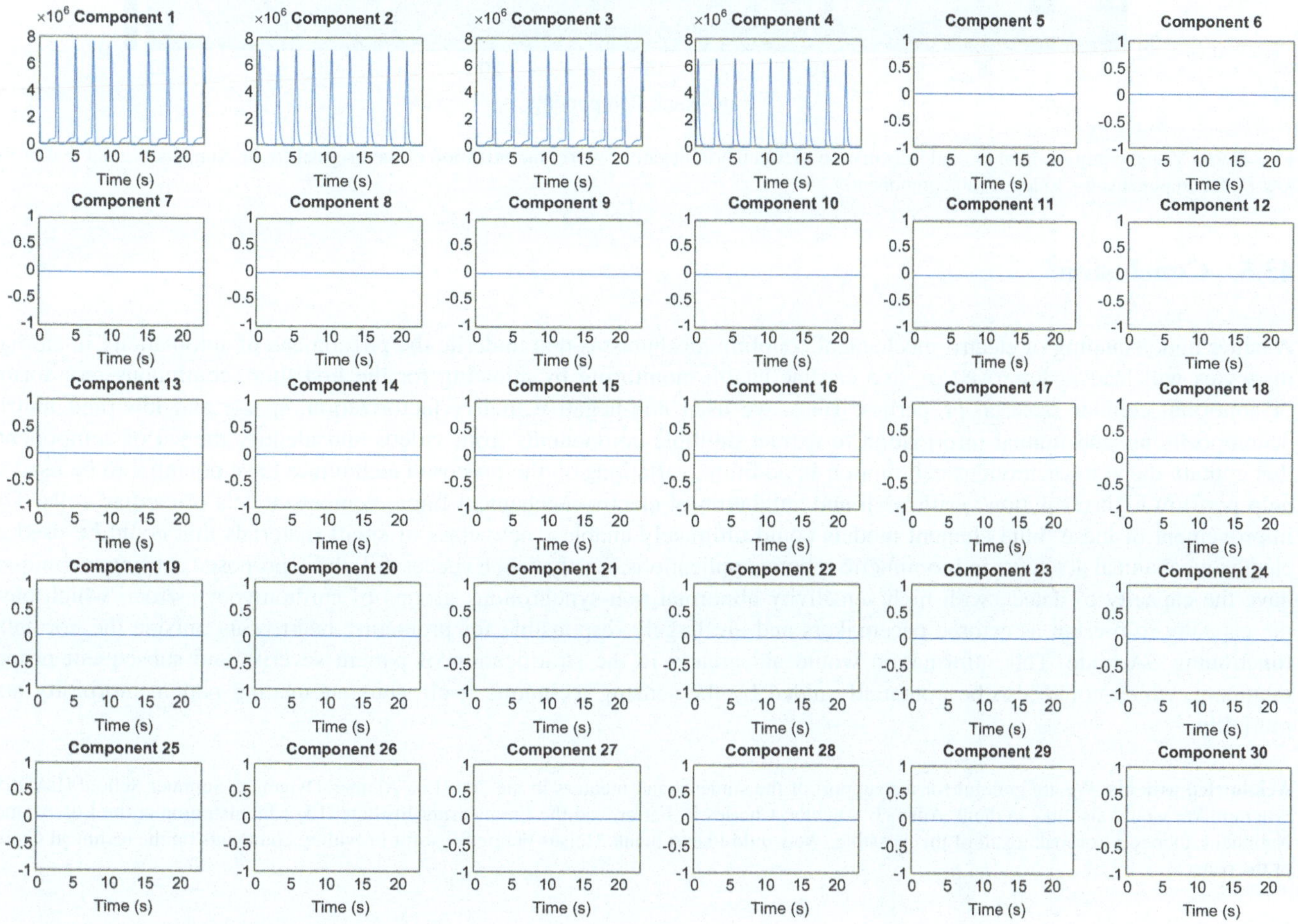

Fig. 43.9 The first thirty components of low-rank matrix, *LR*, from non-negative matrix factorization for Video A

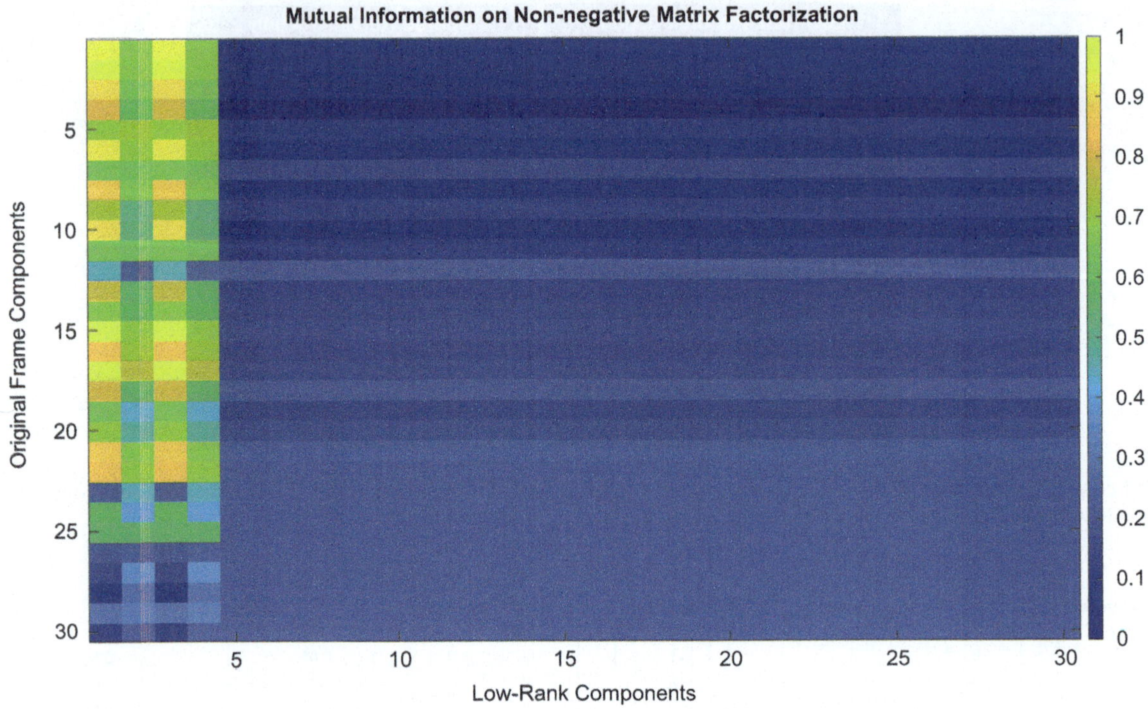

Fig. 43.10 Visual comparison of mutual information between non-negative matrix factorization of the original frame components and that of the low-rank components for Video A (30 components)

43.5 Conclusion

A better understanding of electro-mechanical coupling mechanisms that underlie the phenomena of automaticity in cardiac myocytes will lead to improved in-vivo cardiac health monitoring by allowing for the first time, continuous monitoring of abnormal cellular material properties. Here, we used non-negative matrix factorization, sparse and low-rank matrix decompositions, and mutual information to extract multiple components from videos and identify the set of components that contain the nonsynchronous cell motion In addition, variations on the proposed techniques have potential to be used to help perform high-resolution verification and validation of electro-mechanical finite element models of cardiac cells. The improvement of these finite element models could ultimately enable a new class of smart materials that could be used as electro-mechanical actuators in biomimetic robotic applications. Lastly, when successful, these proposed sensing techniques have the capacity to detect, with high sensitivity abnormal non-synchronous groups of cardiomyocytes/foci, which have the capacity to present as ectopic pacemakers and are largely responsible for premature heart beats outside the normally functioning SA node. This information would aid greatly in the stratification of patient severity and subsequent proper treatment, which could also be potentially aided by the sensing technique itself, thus attenuating patient morbidity and mortality.

Acknowledgements We are grateful for the support of the students and mentors in the 20th Los Alamos Dynamics Summer School (LADSS) program. We would also like to thank Adam J. Wachtor, Charles R. Farrar, and the Engineering Institute (EI) administration at the Los Alamos National Laboratory for making all of this possible. We would like to thank Moises Felipe Silva for providing comments on the technical details of the paper.

References

1. The Cardiovascular System: Cardiac Conduction. Wiley. https://www.johnwiley.net.au/highered/interactions/media/Distribution/content/Distribution/cardca1a/frameset.htm. Accessed 21 June 2019
2. Rossi, A., Dirksen, R.: Sarcoplasmic reticulum: the dynamic calcium governor of muscle. Muscle Nerve. **33**(6), 715–731 (2006)

3. Cardiac Conduction System. School of Health Sciences, The University of Nottingham. https://www.nottingham.ac.uk/nursing/practice/resources/cardiology/function/conduction.php. Accessed 23 June 2019
4. Ashley, E., Niebauer, J.: Chapter 3: conquering the ECG. In: Cardiology Explained. Remedica (2004)
5. Levis, J.: ECG diagnosis: complete heart block. Perm. J. **15**(2), 90 (2011)
6. German, D., et al.: Atrial fibrillation predictors: importance of the electrocardiogram. Ann. Noninvasive Electrocardiol. **21**(1), 20–29 (2016)
7. Pylatiuk, C., et al.: Automatic zebrafish heartbeat detection and analysis for zebrafish embryos. Zebrafish. **11**(4), 379–383 (2014)
8. Sala, L., et al.: MUSCLEMOTION: a versatile open software tool to quantify Cardiomyocyte and cardiac muscle contraction in vitro and in vivo. Circ. Res. **122**(3), e5–e16 (2018)
9. Toepfer, C., et al.: SarcTrack: an adaptable software tool for efficient large-scale analysis of sarcomere function in hiPSC-Cardiomyocytes. Circ. Res. **124**(8), 1172–1183 (2019)
10. Kieffer, C., Abel, P.: Isoproterenol. In: Reference Module in Biomedical Sciences (2016)
11. Clifford, G.: Chapter 15 – BLIND SOURCE SEPARATION: principal & independent component analysis. In: Biomedical Signal and Image Processing, pp. 1–47 (2008)
12. Lee, D., Seung, H.: Algorithms for non-negative matrix factorization. In: 13th International Conference on Neural Information Processing Systems, pp. 535–541 (2000)
13. Guillamet, D., Vitrià, J.: Non-negative matrix factorization for face recognition. Computer & Communications Industry Association, pp. 336–344 (2002)
14. Chandrasekaran, V., et al.: Sparse and low-rank matrix decompositions. In: 15th IFAC Symposium on System Identification, vol. 42, no. 10, pp. 1493–1498 (2009)
15. Otazo, R., Candès, E., Sodickson, D.: Low-rank plus sparse matrix decomposition for accelerated dynamic MRI with separation of background and dynamic components. Magn. Reson. Med. **73**(3), 1125–1136 (2015)
16. Yang, Y.: On sparse and low-rank matrix decomposition for singing voice separation. In: 20th ACM International Conference on Multimedia, pp. 757–760 (2012)

The manufacturer's authorised representative in the EU is Springer Nature Customer Service Centre GmbH, Europaplatz 3, 69115 Heidelberg, Germany. If you have any concerns regarding our products, please contact ProductSafety@springernature.com

Printed and bound by CPI Group (UK) Ltd, Croydon, CR0 4YY
30/09/2026
02225516-0011